Geographic Information Systems and Cartography for a Sustainable World

Geographic Information Systems and Cartography for a Sustainable World

Guest Editors

Andriani Skopeliti
Anastasia Stratigea
Vassilios Krassanakis
Apostolos Lagarias

MDPI

Basel • Beijing • Wuhan • Barcelona • Belgrade • Novi Sad • Cluj • Manchester

Guest Editors
Andriani Skopeliti
School of Rural, Surveying and Geoinformatics Engineering
National Technical University of Athens
Zographos
Greece

Anastasia Stratigea
Department of Geography and Regional Planning
National Technical University of Athens
Zographos
Greece

Vassilios Krassanakis
Department of Surveying and Geoinformatics Engineering
University of West Attica
Egaleo
Greece

Apostolos Lagarias
Department of Regional Planning and Development
University of Thessaly
Volos
Greece

Editorial Office
MDPI AG
Grosspeteranlage 5
4052 Basel, Switzerland

This is a reprint of the Special Issue, published open access by the journal *ISPRS International Journal of Geo-Information* (ISSN 2220-9964), freely accessible at: https://www.mdpi.com/journal/ijgi/special_issues/7G361IM802.

For citation purposes, cite each article independently as indicated on the article page online and as indicated below:

Lastname, A.A.; Lastname, B.B. Article Title. *Journal Name* **Year**, *Volume Number*, Page Range.

ISBN 978-3-7258-4753-2 (Hbk)
ISBN 978-3-7258-4754-9 (PDF)
https://doi.org/10.3390/books978-3-7258-4754-9

Contents

About the Editors

Andriani Skopelti

Andriani Skopeliti [Dipl. Ing. Surveyor Engineer NTUA and PhD NTUA] is an Assistant Professor at the School of Rural, Surveying and Geoinformatics Engineering in the National Technical University of Athens in Greece. She specializes in Cartography and the analytical processing and visualization of geospatial data. She teaches courses in the areas of Cartography, Geographic Information Science, and Web Mapping at both the undergraduate and postgraduate levels. She has authored numerous research papers and book chapters. She is also a reviewer in scientific journals and conferences. She is a member of the Reviewer Board of *ISPRS International Journal of Geo-Information* and Guest Editor of the Special Issues "Geovisualization: Current Trends, Challenges, and Applications" (in "Geographies"), "Geographic Information Systems and Cartography for a Sustainable World" & "Cartography and Geovisual Analytics" (in "*ISPRS International Journal of Geo-Information*"). She has participated in numerous research projects as a scientific manager and researcher, with funding from international and national sources. She is a member of the board of directors of the Hellenic Cartographic Society. Her main research interests focus on web cartography, AI and generalization, ethics in Cartography, VGI use in mapping, and VGI quality.

Anastasia Stratigea

Anastasia Stratigea is Professor in "Spatial Planning and Policy" at the Dept. of Geography and Regional Planning, School of Rural, Surveying and Geoinformatics Engineering, National Technical University of Athens (NTUA). She holds a Diploma in Rural and Surveying Engineering (Aristotle University of Thessaloniki, 1984) and a PhD (NTUA, 1996). She has participated in a number of National and EU funded Projects. She is member of scientific, national and international, associations/networks; and member of scientific committees of international/national conferences. She is Guest Editor in the following Special Issues: "Potential of Participatory Urban Planning and Governance", *International Journal of Electronic Governance* (IJEG); "Linking Land and Underwater Cultural Heritage Management to Technology in Smart Cities and Communities", *Heritage*; "Indulging in Preservation and Sustainable Management of Land and Maritime Cultural Resources in Current Times", Heritage; "Coping with Coastal and Maritime Sustainability Concerns in the Mediterranean", *Ocean and Coastal Management*. She has published her work in books, edited volumes, International Journals and Conference Proceedings. Her main research interests are: Sustainable Development, Spatial Planning and Policy, Integrated Urban and Regional Planning, Participatory Planning, Tourism and Urban/Regional Development, Tourism Planning, Culture and Urban/Regional Development, Cultural Planning, Culture and Tourism in island regions, Smart cities and communities, Foresight methodologies.

Vassilios Krassanakis

Vassilios Krassanakis, Diploma of Rural and Surveying Engineering, Doctor of Engineering, National Technical University of Athens (NTUA), is an Associate Professor at the Department of Surveying and Geoinformatics Engineering, University of West Attica, specialized in Cartography and Visual Perception of Maps. Currently, he teaches courses mainly in the areas of cartography, visual perception of maps, web mapping, and programming in geoinformatics. He has also taught as a Visiting Professor, Invited Speaker or Instructor, at the National and Kapodistrian University of Athens, the Harokopio University, the Hellenic Military Geographical Service, and the Technical Chamber of Greece. Previously, he worked as a Postdoctoral Researcher at the Université de Nantes and the NTUA, as a Visiting Researcher at the Ghent University, as an External Research Collaborator at the National Observatory of Athens and the Hellenic Centre for Marine, and as a Technical Researcher at the NTUA. He is a member of the Editorial Board of the "*Journal of Eye Movement Research*" and Guest Editor of Special Issues in "*Geographies*", "*ISPRS International Journal of Geo-Information*", and "*Applied Sciences*". He has participated in several research projects funded by European, French, or Greek grants. He has authored numerous scientific publications, including journal papers, conference proceedings, book chapters, and an academic book. His current research activities are mainly focus on the areas of Map and Landscape Perception, Eye Movement Analysis and Gaze Behavior Modeling, Geovisualization and Spatial Data Infrastructures (SDIs).

Apostolos Lagarias

Apostolos Lagarias is Assistant Professor of Spatial Planning & Climate Change in the Department of Planning & Regional Development, University of Thessaly (UTh). He holds a strong publication record on quantitative methods, spatial analysis and sustainable urban development, focusing on climate change adaptation and the particular case of coastal zones. He has long-term academic and research collaboration with four or the Universities of Greece (Aristotle University of Thessaloniki, National Technical University of Athens, University of Crete, University of Thessaly). He holds a Ph.D. (2013) from the Division of Urban and Regional Planning, Faculty of Architecture AUTH, a M.Sc. (2011) in Methods and techniques in analysis, Planning and Management of Space, School of Rural and Surveying Engineering AUTH and has a 6-year Postdoctoral experience at the Foundation for Research and Technology (FORTH), and the National Technical University of Athens (NTUA). He has participated in several national research projects, and he has been awarded for his scientific work. He has successfully claimed two research grants from the States Scholarship Foundation actions and conducted research as Principal Investigator (autonomous research of post-doctoral level). He is substituting representative of the Department of Planning & Regional Development into AESOP (Association of European Schools of Planning), and Guest Editor of the Special Issue "Integrating Spatial Analysis and Regional Science to Guide Urban Planning" (*Land*, MDPI, 2024–2026).

International Journal of
Geo-Information

MDPI

Editorial

Geographic Information Systems and Cartography for a Sustainable World

Andriani Skopeliti [1,*], Anastasia Stratigea [2], Vassilios Krassanakis [3] and Apostolos Lagarias [4]

1 Cartography Laboratory, School of Rural, Surveying and Geoinformatics Engineering, National Technical University of Athens, 15780 Athens, Greece
2 Department of Geography and Regional Planning, School of Rural, Surveying and Geoinformatics Engineering, National Technical University of Athens, 15780 Athens, Greece; stratige@central.ntua.gr
3 Department of Surveying & Geoinformatics Engineering, University of West Attica, Egaleo Park Campus, Ag. Spyridonos Str., Egaleo, 12243 Athens, Greece; krasvas@uniwa.gr
4 Department of Planning & Regional Development, University of Thessaly, 38334 Volos, Greece; lagarias@uth.gr
* Correspondence: askop@survey.ntua.gr

Abstract: This article summarizes the scope and content of the Special Issue (SI) entitled "Geographic Information Systems (GIS) and Cartography for a Sustainable World" and its contribution to the global discourse regarding sustainability concerns. At the heart of the discussion in this SI lies: (i) GIS, a valuable tool and a means for modeling, designing, and analyzing (spatial) data and processes related to the pursuance of sustainability objectives at both local and global scales; and (ii) Cartography as a discipline, which through maps and visualizations can convey the present state. The latter can play a vital role in educating, empowering, and raising public awareness with regard to sustainability concerns on the one hand, and can form a basis for policy-makers, scientists, and citizens for articulating effective sustainability strategies on the other. The fulfillment of the SI goals is attained through a collection of 26 papers that delve into and attempt to visualize sustainability achievements or concerns on a variety of themes in different parts of the world. More specifically, the content of this collection of papers can be categorized into the following sustainability-related themes: Urbanization, Transportation, Carbon Emissions Management, Infrastructure, Rural Development, and Climate Change. The main conclusion is that planning and implementing sustainability policies is a challenging and multi-level task, and must be carried out within a fully dynamic decision environment. Although some progress has already been made, more intensive and collective efforts from scientists, governments, the entrepreneurial community, and citizens are needed in order for the ambitious goals of Agenda 2030 to be reached.

Keywords: sustainability; Sustainable Development Goals (SDGs); climate change; carbon emissions; urbanization; infrastructure; rural development; transportation

1. Introduction

Humanity is currently confronted with the challenge of sustainable development, i.e., meeting current societal needs without compromising the ability of future generations to meet their own. The achievement of the Sustainable Development Goals is a complex and interdisciplinary task that aims to balance economic, environmental, and societal/cultural objectives. The United Nations' Sustainable Development Goals (SDGs), outlined in the 2030 Agenda, call for action from all the world's states. The 17 SDGs include the following policy fields: SDG 1—No Poverty, SDG 2—Zero Hunger, SDG 3—Good Health and Well-being, SDG 4—Quality Education, SDG 5—Gender Equality, SDG 6—Clean Water and Sanitation, SDG 7—Affordable and Clean Energy, SDG 8—Decent Work and Economic Growth, SDG 9—Industry, Innovation and Infrastructure, SDG 10—Reduced Inequality, SDG 11—Sustainable Cities and Communities, SDG 12—Responsible Consumption and Production, SDG 13—Climate Action, SDG 14—Life Below Water, SDG 15—Life on Land, SDG 16—Peace and Justice Strong Institutions, and SDG 17—Partnerships to achieve the Goals. The consequences of humanity's unsustainable development patterns are becoming evident and are therefore necessitating the fulfillment of the SDGS; consequently, the international community's interest in achieving these goals is piquing as the deadline of 2030 draws near.

In seeking to achieve the SDGs, various kinds of datasets and relevant tools for assessing, managing, and monitoring progress are nowadays a crucial topic. Towards this end, scientists and spatial data stakeholders worldwide are currently utilizing spatial data, Geographic Information Systems (GIS), and Cartography to analyze, assess, visualize, and highlight spatial and developmental inefficiencies in order to support planning efforts and related policy pathways, aimed at achieving sustainability objectives. Geographic Information Systems (GIS) are the ultimate tools for modeling, designing, and analyzing large datasets that can demonstrate progress or barriers to sustainability achievements at scales ranging from local to global. They offer insights into complex spatial relationships and facilitate more informed policy-making. At the same time, Cartography provides an opportunity for visualization and mapping of achievements in various topics, thereby enabling the communication of sustainability progress to scientists, citizens, and the policy community. In this respect, it plays a key role in educating, empowering, and raising awareness about sustainability concerns in society as a whole.

Within such a context, this Special Issue aims to present research works on the role of Geographic Information Systems (GIS) and Cartography as key pillars in the effort of progressing towards a sustainable world.

The Special Issue (SI) collected twenty-six distinct contributions. Based on their specific focus, these papers can be classified into six topics related to sustainability, which are as follows: Rural Development, Carbon Emissions Management, Climate Change, Infrastructure, Transportation, and Urbanization. In addition to the above thematic classification, the papers address the following Sustainable Development Goals (SDGs) as outlined in the 2030 Agenda by the United Nations:

- Rural Development: SDG 1 (No Poverty), SDG 4 (Quality Education), SDG 5 (Gender Equality), SDG 7 (Affordable and Clean energy), SDG 8 (Decent Work and Economic Growth), SDG 12 (Responsible Consumption and Production), SDG 13 (Climate Action), and SDG 15 (Life on Land).
- Carbon Emissions Management: SDG 7 (Affordable and Clean Energy), SDG 12 (Responsible Consumption and Production), and SDG 13 (Climate Action).
- Addressing Climate Change Implications: SDG 13 (Climate Action).
- Infrastructure: Industry, Innovation and Infrastructure (SDG 9).
- Transportation: SDG 3 (Good Health and Well-being), SDG 9 (Industry, Innovation and Infrastructure), SDG 11 (Sustainable Cities and Communities).
- Urbanization: SDG 11 (Sustainable Cities and Communities), SDG 12 (Responsible Consumption and Production), SDG 15 (Life on Land) [1].

Several case studies are presented in the SI articles, covering an area that spans all continents except North America and Oceania. Specifically, 13 case studies (50%) are dealing with topics situated in Asia, 9 in Europe (38%), 1 in South America (4%), and 1 in Africa (4%), while 1 article (4%) does not have a particular case study reference.

This diversity and global coverage highlights the efforts of the international scientific community to achieve the SDGs, through gaining insight and developing tools that can tackle the global challenges and overcome obstacles to SDGs, as an important facet of their research. Regarding the nationalities of the first authors, 14 (54%) are from Asia, 10 (38%) are from Europe, and 2 (8%) are from Africa.

The research articles integrate several software and tools for data acquisition, processing, (spatial) analysis, statistical analysis, visualization, and web dissemination. These include the following:

- Open-source and commercial GIS software, such as QGIS and plugins, ArcMap and toolboxes, as well as Supermap (e.g., [2–14]);
- Statistical software, such as SAS, SPSS, and R (e.g., [13,15,16]);
- Remote sensing software, e.g., ERDAS Imagine, platforms such as Google Earth Engine (GEE), and toolboxes such as Guidos Toolbox (e.g., [6,17,18])
- Web mapping technologies such as Open Layers, Leaflet, and Geoserver (e.g., [8,10,12,16]);
- Python scripting (e.g., [3,5,7,10,12]);
- Spatial database management, e.g., Postgres (e.g., [8]);
- Other specialized spatial data analysis tools, libraries, and custom tools (e.g., [5,7,8,10,11,13,16]).

AI and machine learning techniques are rarely employed across the case studies of the research. One study uses Support Vector Machine (SVM), a machine learning technique, for land use classification from satellite imagery [19]. Furthermore, in one study, cellular automata–artificial neural networks (CA-ANN) are utilized to monitor LULC changes over three decades (1990–2020) [11]. However, several papers suggest the future integration of advanced machine learning algorithms to improve research efforts [5,12,17,18,20].

The studies include a rich variety of data, such as the following:

- Satellite data (e.g., Sentinel-2, Landsat) (e.g., [5,13,19–21]).
- Volunteered Geographic Information (e.g., OSM) (e.g., [9–11,16]).
- Social and economic data (e.g., [9,11,14,17–19,21–26]).
- Climate and meteorological data (e.g., [5,7,11,17,20]).
- Land use–land cover (e.g., [2,5,10,11,14,17,23]).
- Topographic data such as road networks, hydrological networks, etc., (e.g., [2–4,6,8,9,11,18,19,25]) and Digital Elevation Models (e.g., [3,6,7,10,23]).
- Environmental data (e.g., [5,14,22,24,26]).
- Geotechnical and seismic data (e.g., [13]).
- Soil data (e.g., [4,13]).
- Protected Natural Areas and Biodiversity data (e.g., [2]).
- Cultural heritage data (e.g., [4,26]).
- Tourism data (e.g., [26]).
- Data collected especially for the case study, e.g., GNSS measurements, LIDAR (e.g., [8,27]).
- Expert and citizen survey data (e.g., [3]).

Cartography and geovisualization play an important role in supporting the analysis and presenting the results of the case studies. Most papers use traditional 2D paper maps and some use 3D maps [7,8,10], while web maps are also utilized [10,12,15,16].

In Section 2, the published papers are briefly presented according to their classification into the sustainability topics. As the major points and outcomes of each contribution are highlighted and discussed, a general discussion for each sustainability topic is formed. Finally, in Section 3, future research directions in sustainability assessments and visualization are drawn.

2. An Overview of the Published Papers

In this section, the main findings of each research article of the SI are summarized. Articles are grouped by class based on the main topic they address.

2.1. Rural Development

Hrustek et al. [15] analyze the transparency of government tenders for leasing state-owned agricultural land in Croatia. To improve transparency, the authors propose a national digital tendering platform and repository, hybrid public bid openings, randomized expert selection for commissions, full public disclosure of all bids, and an evaluation of the results to ensure the responsible and transparent management of agricultural land leases. Ge et al. [23] analyze the coordinated development of rural industry and employment (RIE) in Chongqing, China, from 2013 to 2023. This is achieved by examining the spatiotemporal evolution and driving mechanisms of the coordinated development of RIE. The study suggests employment-centered strategies to address industrial lag and spatial imbalances, providing insights for rural revitalization policies in developing regions. Cao et al. [19] evaluate the new quality productivity of agriculture and rural areas across 31 provinces in mainland China from 2010 to 2022 by examining spatial changes and evolutionary trends. The study emphasizes the importance of regional coordination, technological innovation, infrastructure investment, and ecological protection to promote balanced rural revitalization and quality productivity growth. Angelelli et al. [2] present a replicable methodology for assessing land suitability for agrivoltaic (agriPV) projects, focusing on the province of Viterbo, Italy. The case study reveals significant spatial variability in suitability and sensitivity within the province and emphasizes the need for nuanced, flexible approaches to planning agriPV installations that ensure the minimization of conflicts and enhance sustainability and public acceptance.

In summarizing key issues raised in the rural development context, all papers highlight the importance of spatially-aware, integrated, and multi-dimensional approaches to rural and agricultural development, highlighting regional disparities, the critical role of infrastructure and investment, the environmental and social sustainability, and transparency in governance. These insights collectively inform strategies for balanced rural revitalization and sustainable land use.

2.2. Carbon Emissions Management

Xie et al. [14] focus on the geographical pattern, evolution model, and driving mechanisms of carbon emission density from urban industrial land in the Yangtze River Economic Belt (YREB) of China. The research aims to provide key information and a basis for incorporating carbon emission density indicators into urban and territorial planning to achieve "carbon peaking and carbon neutrality" goals. Chen et al. [5] explore the factors influencing vegetation carbon sequestration across different regions of Beijing. They investigate the spatial and temporal variations in Annual Gross Primary Productivity (AGPP) in Beijing from 2000 to 2020, focusing on the impacts of urbanization and environmental factors. The study also discusses future urbanization and environmental management policies with the intention of promoting the carbon peak and the carbon neutrality process of ecological space management. Zhao et al. [22] analyze CO_2 emissions across different sectors (total, power, industry, transport) in 31 major Chinese cities from 2019 to 2022. Towards this end, they use a city-scale CO_2 emission correlation model and social network analysis. The study emphasizes the need for collaborative governance, targeted sector-specific policies, and technological innovation to achieve emission reduction and sustainable development goals. In the Cartography context, Touya et al. [12] elaborate on energy consumption related to the transfer of map tiles for web mapping. The article suggests focusing on the most frequent zoom levels and applying generalization to create lighter tiles, as well as promoting eco-friendly browsing behaviors to reduce energy use.

To summarize this section, all the papers conclude that spatial heterogeneity, multifactor influences, and dynamic interactions are key features of urban environmental phenomena, such as carbon emissions.

2.3. Handling Climate Change

Kotovs et al. [16] draw attention to beekeeping as a means for raising the productivity and profitability of agricultural crops and agricultural systems. In this respect, they present an experimental prototype that can be replicated at a universal scale; additionally, they demonstrate the feasibility of using crowdsourced spatial data in combination with web and GIS technologies to conduct a pre-exploration of proper and environmentally-responsible placement areas for apiaries, supporting the flourishing of precise beekeeping. Gaspari et al. [8] deal with glacier melting, which is deeply connected to mountainous local communities. The authors elaborate on web-based geospatial solutions that integrate qualitative and quantitative aspects of the transformation of the natural environment into a multi-level digital geospatial framework to enable the realization of the long-term changes of the Belvedere Glacier in the Italian Alps. The scope of the study is to provide an interactive visualization tool for displaying glacier retreat processes overtime, raising awareness of the tangible impacts of global warming. Gaia et al. [7] focus on the visualization of climate change (CC) impacts in Switzerland through an interactive digital map as part of the national Atlas of Switzerland (AoS). In this work, they elaborate on three specific types of CC impacts, namely the rise of the zero degree line, the evolution of glacial lakes, and the change in the flowering dates of plants. The ultimate goal of this effort is to use simple, clear-cut, and visual communication means—maps—in order to raise the awareness of laypeople to the implications of CC on the planet's stability in a comprehensible, visible, and easy to grasp manner. Jiang et al. [20] elaborate on an emerging threat to regional water quality safety caused by high concentrations of chlorophyll-a (Chl-a) as a result of hydrological extremes in Jiujiang City, Jiangxi Province, China. A quantitative inversion model for estimating Chl-a concentration is developed for assessing water quality in various inland water types for each quarter from 2020 to 2022. The results demonstrate seasonal changes in Chl-a concentrations in inland waters due to extreme hydrological events, a result that is valuable for the sustainable management of water quality for the population.

In summary, these papers use contemporary means for handling and visualizing geospatial data in order to identify the distinct implications of climate change and use them for both raising public awareness and making more informed policy decisions.

2.4. Infrastructure

Yang et al. [17] attempt to assess the impact of Coal Mining Intensity (CMI) on Ecosystem Service Value (ESV). This is carried out by obtaining deep insight into ecosystem services and studying their interaction with coal resource development, using the Yellow River Basin as a case study area. The results demonstrate that the overlap between the coal mining area and basin has expanded, resulting in an increasing negative impact on the ESV. These results highlight the need to address the adverse effects of resource extraction on ecosystem services and the risk of ESV loss, demonstrating the value of spatial overlapping analysis in identifying 'hotspots' and the articulation of relevant policies for the protection of ecologically fragile basins. Wahba et al. [13] elaborate on a currently critical issue, caused by land scarcity: the optimization of site selection for serving urban, industrial, and tourist development objectives. In seeking to fulfill this goal, Kharga Oasis, Egypt, is used as a case study; and a GIS and a multicriteria decision-making approach are combined with multiple datasets, integrating in situ and laboratory-based data. As a result, a detailed site suitability map is created, using the analytic hierarchy process to develop a weighted GIS model that accounts for numerous elements and ultimately influences civil project design and construction. Wei et al. [18] explore the spatial pattern of basic educational services in the Yellow River Basin of China. Basic educational services are perceived as a crucial component of public services, as well as a means for safeguarding individuals' fundamental rights to survival and development and facilitating comprehensive human development. The study is based on a big data analysis model for sketching the layout of basic public educational facilities. The results display certain disparities, mostly influenced by the permanent population and the large proportion of tertiary industry.

Outcomes of this part stress the use of a variety of means and geospatial data for informing decisions as to infrastructure development in a sustainable, socially-fair, and resource-respectful way.

2.5. Transportation

Astrat & Cho [27] explore recent advancements in the generation and updating methods and techniques of high-definition (HD) maps within the rapidly growing field of autonomous driving. In addition, they discuss the existing challenges and future prospects of the technologies associated with autonomous vehicles and driving systems. Bi et al. [24] examine the decoupling state of traffic carbon emissions in Guangdong Province from 1999 to 2019, utilizing the Tapio decoupling index model. Their results indicate that traffic carbon emissions increased within the study area during the examined period. The study shows that the decoupling effect of traffic carbon emissions is mainly "weakly decoupled", while the overall decoupling effect is not significant in the study area. The study also reveals that the production value of traffic and turnover volume are the most significant factors contributing to traffic carbon emissions. Miao et al. [25] describe the development of a comprehensive connectivity index system towards evaluating the inter-country connectivity of Latin America. The system is based on sea, road, air, and railroad transport modes, and uses actual trade volume as a comparison. They analyze the spatial pattern of connectivity in the study area by combining network analysis and visualization techniques. The results indicate the main cross-border transport types and the existing patterns of overall connectivity, as well as the roles of different countries in regional connectivity. Aprigliano et al. [3] present a method which combines a multicriteria GIS-based analysis with an experimental study in the municipality of Valparaíso (Chile) for the evaluation of potential routes and the possibility of increasing the power limitations for non-motorized mobility. Their results suggest that the variables of slope, traffic accidents, intersections, and street directionality impact the viability and safety of E-bike users. The study emphasizes the necessity of adapting urban mobility policies to each city's topographical characteristics. Li et al. [9] examine the characteristics of the bus network in Beijing during the period 2006–2014 by combining multi-source spatiotemporal data. Their results show that the bus network has extended from the central urban areas to peripheral regions within the study area. The study indicates that administrative boundaries and the layouts of ring roads greatly affect bus network characteristics. Additionally, the study demonstrates that bus stops are characterized by high levels of clustering and irregular development patterns. Misthos & Krassanakis [10] present an integrated method and a geoprocessing tool in Python for highway route characterization, based on landscape composition and visual significance. In particular, this method leads to the characterization of segments of highway routes according to the dominant element of the visible landscape by employing spatial aggregation approaches. The introduced method is applied in a peri-urban area in the Attica region (Greece), while the produced results are visualized through an interactive web map, which serves as an exploratory tool for summarizing the landscape character of the selected highway.

2.6. Urbanization

Chen et al. [6] put forward a methodology of geospatially analyzing a multi-level and multi-objective composite ecological network, based on red space (tourism resources for slow walking), green space (public/urban green space), and blue space (water-related natural and man-made spaces). This study constructs a composite ecological network under urban–rural integration in Dali City (China). A major contribution is in proposing the "point, line, and surface" combination of the optimization scheme to coordinate ecological protection. Kang et al. [21] employed satellite-derived indexes to investigate socio-economic development and eco-environmental changes across a system of cities in the Loess Plateau, China. Geospatial methods are used to assess the relationship between urbanization and ecological factors. The results are particularly interesting and geographically diverse, as there is a very uneven development of different types of cities, with developed cities exhibiting signs of improved alignment between urbanization and ecological conditions. Mumtaz et al. [11] focus on the integrated impacts of significant land use/land cover (LULC) attributed to urbanization processes in the riverine coastal megacities of semi-arid Southern Asian regions that are particularly affected by climate change. The megacity of Karachi in Pakistan is used as a case study. LULC changes are monitored over the past three decades and simulated until 2025, with the incorporation of various sets of geospatial data, while flood inundation scenarios are also created. The study's innovation lies in quantifying SDGs achievements and integrating LULC analysis with SDGs metrics through a quantitative methodology. Chatzi et al. [4] set out to examine spatial relations and potential conflicts, underscoring the insufficient protection of cultural heritage due to ineffective spatial planning and mass tourism development. Geospatial methods are used to quantify the impact of uncontrolled built-up area dispersion on the cultural heritage of islands, with a focus on the Southern Aegean region. The results lead to suggestions of regulating urban sprawl and tourism development, while taking into consideration the spatial impacts at various spatio-cultural levels. Leka et al. [26] introduce an innovative approach to evaluating overtourism by integrating geospatial technologies and the well-established PSR—Pressure/State/Response—model. The study applies this framework to assess overtourism impacts in insular regions, using Santorini as a case study. Geographic Information Systems (GIS), remote sensing techniques, and spatial indicators are used to map tourism pressures on several aspects of socio-spatial structure and the environment of the island. By integrating quantitative spatial data with qualitative insights, the framework offers a comprehensive tool for serving sustainable tourism planning purposes.

3. Future Research Directions

With humanity facing complex sustainability challenges, future research empowered by GIS and Cartography in the pursuit of sustainable development goals should leverage new technological capabilities, such as GeoAI, Digital Twins, and Smart Cities, while also investing in the ethical dimension of Geospatial Science. Several future research directions can be proposed:

- GeoAI and Machine Learning Monitoring: Future efforts should focus on developing more sophisticated methods for monitoring the SDGs using machine learning. The application of GeoAI in satellite image analysis significantly enhances the results [28] and improves the assessment of the SDGs' indicators. Scientists should develop explainable AI models [29], tailored to diverse geographic and socio-economic contexts.

- Digital Twins and Smart Cities: In promoting the Sustainable Development Goals (SDGs), digital twins, i.e., dynamic, real-time 3D models of urban or environmental systems, can be highly beneficial [30]. By monitoring and predicting urban development trajectories and their associated costs, they can contribute to reducing key resource consumption, such as land, water, and energy. Smart cities offer tremendous potential for advancing the SDGs, since they provide strong and positive support for sustainable development achievements [31].
- Participatory and Ethical Cartography: Citizen Science and VGI can be utilized to enhance data collection in areas facing severe SDGs challenges, as seen in the case of Humanitarian OSM [32]. Accessible, inclusive, and effective visualizations should communicate complex sustainability issues for every part of the world within the framework of Ethical Cartography [33,34].

Future research in GIS and Cartography is expected to have a significant impact on achieving the Sustainable Development Goals (SDGs). Furthermore, since developing countries are confronted with significant obstacles in achieving the SDGs [35], their support through access to the evolving scientific knowledge and financial resources is also a critical dimension in the course of the universal effort towards the SDGs achievement.

Author Contributions: All authors contributed equally to the conceptualization and writing of this editorial. All authors have read and agreed to the published version of the manuscript.

Funding: This research received no external funding.

Conflicts of Interest: The authors declare no conflicts of interest.

References

1. Chen, M.; Chen, L.; Cheng, J.; Yu, J. Identifying interlinkages between urbanization and Sustainable Development Goals. *Geogr. Sustain.* **2022**, *3*, 339–346. [CrossRef]
2. Angelelli, I.; Codato, D.; Pappalardo, S.E.; De Marchi, M. Developing a Methodology for Assessing Visual and Environmental Sensitivity for Agrivoltaics Land Suitability Projects: The Case Study of Viterbo Province (Italy). *ISPRS Int. J. Geo-Inf.* **2025**, *14*, 115. [CrossRef]
3. Aprigliano, V.; Toro, C.; Rojas, G.; Bastías, I.; Cardoso, M.; Santos, T.; da Silva, M.A.V.; Bustos, E.; de Oliveira, U.R.; Seriani, S. Mapping the Dream: Designing Optimal E-Bike Routes in Valparaíso, Chile, Using a Multicriteria Analysis and an Experimental Study. *ISPRS Int. J. Geo-Inf.* **2025**, *14*, 38. [CrossRef]
4. Chatzi, E.; Derdemezi, E.-T.; Tsilimigkas, G. The Impact of Built-Up Area Dispersion on the Cultural Heritage of the Region of the South Aegean, Greece. *ISPRS Int. J. Geo-Inf.* **2025**, *14*, 97. [CrossRef]
5. Chen, L.; Yu, S.; Shen, S.; Wan, Y.; Song, C. Spatial and Temporal Variation of GPP and Its Response to Urban Environmental Changes in Beijing. *ISPRS Int. J. Geo-Inf.* **2024**, *13*, 396. [CrossRef]
6. Chen, R.; Zhang, S.; Huang, X.; Li, X.; Peng, J. Ecological Network Construction Based on Red, Green and Blue Space: A Case Study of Dali City, China. *ISPRS Int. J. Geo-Inf.* **2024**, *13*, 279. [CrossRef]
7. Gaia, L.; Neumann, A.; Hurni, L. Climate Change Maps for the Atlas of Switzerland. *ISPRS Int. J. Geo-Inf.* **2025**, *14*, 99. [CrossRef]
8. Gaspari, F.; Barbieri, F.; Fascia, R.; Ioli, F.; Pinto, L.; Migliaccio, F. Strategies for Glacier Retreat Communication with 3D Geovisualization and Open Data Sharing. *ISPRS Int. J. Geo-Inf.* **2025**, *14*, 75. [CrossRef]
9. Li, X.; Wang, S.; Zhou, L.; Sun, Y.; Zheng, J.; Liu, C.; Zhou, J.; Su, C.; Xu, D. Geospatial Analytics of Urban Bus Network Evolution Based on Multi-Source Spatiotemporal Data Fusion: A Case Study of Beijing, China. *ISPRS Int. J. Geo-Inf.* **2025**, *14*, 112. [CrossRef]
10. Misthos, L.-M.; Krassanakis, V. RouteLAND: An Integrated Method and a Geoprocessing Tool for Characterizing the Dynamic Visual Landscape Along Highways. *ISPRS Int. J. Geo-Inf.* **2025**, *14*, 187. [CrossRef]
11. Mumtaz, M.; Jahanzaib, S.H.; Hussain, W.; Khan, S.; Youssef, Y.M.; Qaysi, S.; Abdelnabi, A.; Alarifi, N.; Abd-Elmaboud, M.E. Synergy of Remote Sensing and Geospatial Technologies to Advance Sustainable Development Goals for Future Coastal Urbanization and Environmental Challenges in a Riverine Megacity. *ISPRS Int. J. Geo-Inf.* **2025**, *14*, 30. [CrossRef]
12. Touya, G.; Courtial, A.; Kalsron, J.; Berli, J.; Le Mao, B.; Wenclik, L. Understanding the Carbon Footprint of Tile Transfer for Web Maps. *ISPRS Int. J. Geo-Inf.* **2025**, *14*, 107. [CrossRef]

13. Wahba, D.; Omran, A.A.; Adly, A.; Gad, A.; Arman, H.; El-Bagoury, H. Optimizing Site Selection for Construction: Integrating GIS Modeling, Geophysical, Geotechnical, and Geomorphological Data Using the Analytic Hierarchy Process. *ISPRS Int. J. Geo-Inf.* **2025**, *14*, 3. [CrossRef]
14. Xie, F.; Zhang, S.; Zhang, Q.; Zhao, S.; Lai, M. Research on the Geographical Pattern, Evolution Model, and Driving Mechanism of Carbon Emission Density from Urban Industrial Land in the Yangtze River Economic Belt of China. *ISPRS Int. J. Geo-Inf.* **2024**, *13*, 192. [CrossRef]
15. Hrustek, L.; Kević, K.; Varga, F. Open Data for Transparency of Government Tenders: A State Analysis in Croatian Agriculture Land Lease. *ISPRS Int. J. Geo-Inf.* **2024**, *13*, 401. [CrossRef]
16. Kotovs, D.; Krievina, A.; Zacepins, A. Enhancing Precision Beekeeping by the Macro-Level Environmental Analysis of Crowd-sourced Spatial Data. *ISPRS Int. J. Geo-Inf.* **2025**, *14*, 47. [CrossRef]
17. Yang, Y.; Gong, R.; Wu, Q.; Chen, F. Spatial-Temporal Changes in Ecosystem Service Value and Its Overlap with Coal Mining Intensity in the Yellow River Basin, China, During 2000–2030. *ISPRS Int. J. Geo-Inf.* **2024**, *13*, 412. [CrossRef]
18. Wei, H.; Ji, W.; Li, L.; Yang, Y.; Liu, M. Exploring the Equality and Determinants of Basic Educational Public Services from a Spatial Variation Perspective Using POI Data. *ISPRS Int. J. Geo-Inf.* **2025**, *14*, 66. [CrossRef]
19. Cao, X.; Lei, J.; Shi, D.; Yu, W.; Tao, T.; Zhang, X.; Wang, A. New Quality Productivity of Agriculture and Rural Areas at the Provincial Scale in China: Indicator Construction and Spatiotemporal Evolution. *ISPRS Int. J. Geo-Inf.* **2025**, *14*, 104. [CrossRef]
20. Jiang, W.; Ding, X.; Kong, F.; Luo, G.; Long, T.; Pang, Z.; Cui, S.; Liu, J.; Adam, E. How Hydrological Extremes Affect the Chlorophyll-a Concentration in Inland Water in Jiujiang City, China: Evidence from Satellite Remote Sensing. *ISPRS Int. J. Geo-Inf.* **2025**, *14*, 85. [CrossRef]
21. Kang, S.; Jia, X.; Zhao, Y.; Luo, M.; Wang, H.; Zhao, M. The Coupling Coordination Relationship Between Urbanization and the Eco-Environment in Resource-Based Cities, Loess Plateau, China. *ISPRS Int. J. Geo-Inf.* **2024**, *13*, 437. [CrossRef]
22. Zhao, Y.; Feng, Y.; Du, M.; Fraedrich, K.; Shen, Z. Spatial Network Analysis of CO_2 Emissions in Major Cities in China: Regional Structures and Influencing Factors. *ISPRS Int. J. Geo-Inf.* **2025**, *14*, 82. [CrossRef]
23. Ge, G.; Huang, Y.; Chen, Q. The Coordinated Development Characteristics of Rural Industry and Employment: A Case Study of Chongqing, China. *ISPRS Int. J. Geo-Inf.* **2025**, *14*, 48. [CrossRef]
24. Bi, H.; Zhang, S.; Lu, F. Analysis of Decoupling Effects and Influence Factors in Transportation: Evidence from Guangdong Province, China. *ISPRS Int. J. Geo-Inf.* **2024**, *13*, 404. [CrossRef]
25. Miao, C.; Zhang, Y.; Zhang, X.; Liu, J.; Zhang, S. Evaluation of Cross-Border Transport Connectivity and Analysis of Spatial Patterns in Latin America. *ISPRS Int. J. Geo-Inf.* **2025**, *14*, 22. [CrossRef]
26. Leka, A.; Lagarias, A.; Stratigea, A.; Prekas, P. A Methodological Framework for Assessing Overtourism in Insular Territories—Case Study of Santorini Island, Greece. *ISPRS Int. J. Geo-Inf.* **2025**, *14*, 106. [CrossRef]
27. Asrat, K.T.; Cho, H.-J. A Comprehensive Survey on High-Definition Map Generation and Maintenance. *ISPRS Int. J. Geo-Inf.* **2024**, *13*, 232. [CrossRef]
28. Zhang, L.; Zhang, L. Artificial intelligence for remote sensing data analysis: A review of challenges and opportunities. *IEEE Geosci. Remote Sens. Mag.* **2022**, *10*, 270–294. [CrossRef]
29. Xing, J.; Sieber, R. The challenges of integrating explainable artificial intelligence into GeoAI. *Trans. GIS* **2023**, *27*, 626–645. [CrossRef]
30. Tzachor, A.; Sabri, S.; Richards, C.E.; Rajabifard, A.; Acuto, M. Potential and limitations of digital twins to achieve the sustainable development goals. *Nat. Sustain.* **2022**, *5*, 822–829. [CrossRef]
31. Blasi, S.; Ganzaroli, A.; De Noni, I. Smartening sustainable development in cities: Strengthening the theoretical linkage between smart cities and SDGs. *Sustain. Cities Soc.* **2022**, *80*, 103793. [CrossRef]
32. Štampach, R.; Herman, L.; Trojan, J.; Tajovská, K.; Řezník, T. Humanitarian mapping as a contribution to achieving sustainable development goals: Research into the motivation of volunteers and the ideal setting of mapathons. *Sustainability* **2021**, *13*, 13991. [CrossRef]
33. Caudill, C.M.; Pulsifer, P.L.; Thumbadoo, R.V.; Taylor, D.F. Meeting the challenges of the UN sustainable development goals through holistic systems thinking and applied geospatial ethics. *ISPRS Int. J. Geo-Inf.* **2024**, *13*, 110. [CrossRef]
34. Monmonier, M. Ethics and map design: Six strategies for confronting the traditional one-map solution. *Cartogr. Perspect.* **1991**, *10*, 3–8. [CrossRef]
35. Barkat, K.; Mimouni, K.; Alsamara, M.; Mrabet, Z. Achieving the Sustainable Development Goals in developing countries: The role of remittances and the mediating effect of financial inclusion. *Int. Rev. Econ. Financ.* **2024**, *95*, 103460. [CrossRef]

International Journal of
Geo-Information

MDPI

Article

Research on the Geographical Pattern, Evolution Model, and Driving Mechanism of Carbon Emission Density from Urban Industrial Land in the Yangtze River Economic Belt of China

Fei Xie [1], Shuaibing Zhang [2], Qipeng Zhang [3], Sidong Zhao [4,5] and Min Lai [6,*]

1 School of Urban Construction and Environment, Dongguan City University, Dongguan 523419, China; xiefei@dgcu.edu.cn
2 School of Architecture and Urban Planning, Shenzhen University, Shenzhen 518060, China; zhangshuaibing2022@email.szu.edu.cn
3 School of Geography and Tourism, Zhengzhou Normal University, Zhengzhou 450044, China; z2221329@stu.zznu.edu.cn
4 School of Architecture, Southeast University, Nanjing 210096, China; 230189013@seu.edu.cn
5 School of Natural Resources and Surveying, Nanning Normal University, Nanning 530001, China
6 Guangxi-ASEAN Finance and Economics Research Center, Guangxi University of Finance and Economics, Nanning 530003, China
* Correspondence: lm65389@gxufe.edu.cn

Abstract: To achieve the goals of "carbon peaking and carbon neutrality", this paper puts forward the connotation and measurement method for the carbon emission intensity of urban industrial land and conducts an empirical study with the Yangtze River Economic Belt (YREB) as an example. We defined the carbon intensity of urban industrial land as the industrial carbon emissions per unit area of land, which is a spatial mapping of urban industrial economic development and carbon spillover and a key indicator for urban and territorial spatial planning oriented towards the "dual carbon" goal. Findings: The carbon emission density of industrial land in the YREB varied greatly between cities and exhibited significant positive spatial autocorrelation. In addition, the geographical pattern and spatio-temporal evolution model of the urban industrial land carbon emission density had a very complex driving mechanism, and different factors had significant synergistic effects. Therefore, it is suggested that while striving towards the goal of "dual carbon", the government should incorporate the carbon emission density indicator of urban industrial land into the urban and territorial spatial planning system, and based on the threshold of the medium suitable density, they should design differentiated management policies according to concrete urban policies and encourage cooperation among cities to jointly promote carbon emission management of urban industrial land. In policy design, emphasis should also be placed on highlighting the interactive effects of foreign direct investment, fiscal expenditure, and the number of patent authorizations as well as constructing a combination of policies centered around them to better leverage the impacts of globalization, government intervention, and innovation.

Keywords: carbon emission; density; industrial land; urban planning; Yangtze River Economic Belt

1. Introduction

1.1. Research Background

More and more governments, scholars, and businesses are now beginning to take notice of global warming caused by greenhouse gas emissions. Land-use change, as a principal element influencing global climate change and the carbon cycle, alters natural carbon processes at the surface and also affects the rate of the carbon cycle by changing the intensity of human energy consumption. The industrial sector is one of the most important sources of carbon emissions, and there is a close link between the spatial layout of industrial land and the types of industries and carbon emission density [1]. Therefore, it is a growing concern that strict control and the adjustment of land-use structures, especially strengthening the management of carbon emissions from industrial land use, can effectively mitigate global climate change. The World Cities Report 2022: Envisaging the Future of Cities, published by UN Habitat, stresses managing urban density in the context of multiple urban challenges as a key challenge for sustainable urban construction in the future. From the history of urban construction in China and abroad, it has been found that density has not only been applied to the measurement of urban form, but has also gradually evolved into an important starting point for dealing with urban problems, controlling urban form, and regulating the relationship between man and land in different periods [2,3]. Therefore, density research has outstanding theoretical and practical significance for urban construction management, and it is also the core proposition related to the sustainable construction of cities in the future.

Urban planning and land spatial planning, as the government's public intervention on the built environment, aim to ensure space supply, protect public interests, and overcome market failure. For urban and land spatial planning, density is an important indicator for measuring the coordination of land resource allocation, urbanization level, and spatial development quality [4,5]. Carbon emission density refers to the carbon emissions per unit land. As a new key indicator in urban and land spatial planning, the carbon intensity reflects the spatial evolution of cities and their regions in terms of land-use intensity, the level of economic development, and urban construction, serving as an index that represents how much of the city's "dual carbon" goal is achieved. The change in the carbon emission density of urban industrial land, in essence, is the result of the continuous inflow and outflow of urban industrial resources and enterprise clusters, employed population and human capital, industry and innovation capital, and other factors. It is accompanied by changes in the population density, employment density, economic density, and building density in urban spaces. Therefore, an empirical study on the geographical pattern, evolutionary pattern, and driving mechanism of carbon density of industrial land in urban agglomerations based on the spatial measurement model will provide key evidence and information for urban planning and territorial spatial planning to serve the goal of "carbon peaking and carbon neutrality", with a great theoretical and practical value.

1.2. Literature Review

The carbon emission density of industrial land is an emerging tool to measure the relationship between the scale of industrial carbon emissions and the area of industrial land occupied. Therefore, the literature review needs to take into account the research results on carbon emissions of land use and industrial development and use the ideas and methodological innovations from urban and territorial spatial densities, such as population density and building density.

1.2.1. Land Use and Carbon Emissions

Carbon emissions come from a variety of sources, with carbon emissions from land use ranking second after fossil fuel consumption. Economic development leads to changes in land use and has a significant impact on the carbon cycle in nature. Therefore, an in-depth analysis of the relationship between land-use carbon emissions and economic growth is of great significance to the scientific development of energy conservation and emission reduction programs, and it is also the key to promoting regional carbon neutrality as well as the sustainable development of the society [6]. Since the 1970s, an increasing number of scholars have been involved in research on land-use carbon emissions, summarized into the following areas and features: research related to land-use carbon sources and sinks from the perspective of the carbon cycle in terrestrial ecosystems, mainly dealing with the accounting of land-use carbon stocks and fluxes and their influencing factors [7,8]; correlation analysis of the relationship between carbon emissions and economic growth or factors of production based on land carbon emissions using econometric methods such as the gray correlation model [9], the coupled coordination degree model [10], and the decoupling model [11]; and research related to low-carbon development policies for land use under the "dual carbon" goal, that is, to portray the evolution models of land-use carbon emissions through methods such as spatial autocorrelation and cold–hot spot analysis and to further propose their policy implications. For example, Sidorczuk-Pietraszko [12] studied the spatial differences in household carbon intensities between different regions in Poland by means of exponential decomposition analysis (IDA); Rong [13] also found that there is an obvious spatial correlation between land-use carbon emissions in Chinese provinces, and thus suggested that China should strengthen the inter-provincial cooperation in order to achieve an overall emission reduction across the country.

1.2.2. Carbon Emissions from Industries

The industrial sector, as an important source of carbon emissions from human activities, shoulders the important responsibility of energy conservation and emission reduction, and it faces the important and urgent task of how to control and reduce carbon emissions while pursuing the high-quality development of industry and promote the mutual harmonization of economic development and resources and the environment. As a result, the study of industrial carbon emissions has become a hot topic. At present, the research on industry-related carbon emissions mainly focuses on the following three areas: The first is the research on carbon emission measurement and energy conservation and emission reduction in specific industrial sectors or industries, including the communications and transportation industry [14,15], tourism [16], agriculture [17], services [18], and construction [19], with topics involving carbon emission measurement and prediction methodology, influencing factors, spatio-temporal analysis, and the link between carbon emissions and economic development. The second is the research on the influencing factors of industrial carbon emissions, including the energy structure, industrial structure, economic output, population size, energy intensity, urbanization rate, technological progress, and foreign trade and is mostly based on index decomposition analysis, structural decomposition analysis, IPAT or STIRPAT models, and other econometrics and regression analysis methods [20,21]. The third is the research on industrial carbon transfer, carbon leakage, implied carbon emissions, and the differential responsibility for carbon reduction. With the evolution of regional economic development patterns, inter-regional industrial transfers have become more frequent, producing different environmental effects on the transferring places. In this context, some scholars have carried out research on the scale calculation, influencing factors, and mechanism of industrial carbon transfer and analyzed the relevance of industrial carbon emissions and its transmission network characteristics along the industrial chain [22]. Research on the differential responsibility for carbon emission reductions has mainly centered on the "polluter pays" principle, from which the principles of "responsibility within national territories" and "producer burden of pollution" have been derived [23]. Depending on the principles of "producer burden of pollution" and "consumer burden of pollution", scholars have put forward new principles such as "common sharing between importing and exporting countries" [12], "common sharing between producers and consumers" [13], and "common sharing between upstream and downstream of the industrial chain" according to different research objects [24,25].

1.2.3. Urban and Territorial Spatial Density

In human civilizations, cities are spatial carriers where factors such as population, resources, technology, and information are highly concentrated, and density, as a visual representation of the degree of spatial concentration of these factors, has become a key indicator for understanding the efficiency of the distribution of the various factors in cities and their spatial formations in a complex mega-system. Since the emergence of urban planning and territorial spatial planning, attention has been paid to the study of urban density, mainly on the following areas: The first is the research on the urban density measurement indicators and evaluation mechanism. Urban density research originated from population density analysis [26] and then expanded to building density [27], economic density [28], environmental density [29], employment density [30], land density [31], and road network density [32]. The second is the research on the spatio-temporal differentiation of urban density. Leveraging the exploratory spatial data analysis, Theil coefficient, coefficient of variation, Markov chain, kernel density estimation, and other models, it explores and empirically analyzes the laws of spatial differentiation, dynamic evolution, and spatio-temporal evolution of urban density at different scales [33,34]. The third is research on the interaction between urban density and other factors. It analyzes the impact of urban density on the economic efficiency, technological innovation, and land-use patterns and intensities. For example, Hong [35] found that when the urban population size is less than 1 million, increasing the urban density promotes carbon emission reduction; on the contrary, it will lead to more carbon emissions. Lee [36] and Li [37] analyzed the impact of urban morphology—particularly population density and building density—on household carbon emissions and household electricity consumption. It also explores the driving factors and mechanisms of the dynamic evolution and spatio-temporal evolution of urban density, revealing the key influencing factors and their pathways of action. For example, using geographically weighted regression models and Baidu heat maps, Zhang [38] found, in a case study of Wuhan, that residential workplaces, education levels, and the quality of urban landscapes are key indicators of real-time urban human activity density. Of note, the carbon emission density shares a similar concept with the population density, characterized by spatial clustering and mobility. Therefore, we can define the carbon intensity of industrial land as the ratio of the carbon emissions from the production of industrial enterprises, as well as the carbon emissions from the liquidity of the business and product transportation, to the area of industrial land. The measurement of carbon emissions and industrial land use is not limited to narrowly defined production processes and carriers but includes supporting links such as transportation and productive services [39].

1.3. Research Gap

In general, scholars have conducted in-depth analyses on land use and carbon emissions, industrial carbon emissions, and urban density, providing valuable references for this study. However, there are still some vulnerable spots, which are mainly as follows:

First, industrial development, land use, and carbon emissions are closely related, but existing studies generally separate the three to specifically study carbon emissions from industrial development or land use without incorporating them into a unified research framework. Simply calculating the carbon emissions of the industry without taking other factors into account often fails to provide a comprehensive understanding of the industry's carbon emissions, leading to the unsatisfactory implementation of carbon emission management policies for industrial land. With the factor of "land area occupied by industrial activity space" taken, the implementation of industrial carbon emissions to the land, focusing on the industrial carbon emission scale to the industrial carbon emission density, so as to organically integrate land use, industrial development and carbon emission management will be of great value to the policy design of energy conservation and emission reduction, industrial development, and land use under the "dual carbon" goal.

Second, most of the existing studies regard the study area as a separate entity and rarely discuss the spatial effect of carbon emission density from the perspective of geography and spatial interaction. No spatial econometric model has been selected for the regression analysis, which affects the accuracy of the results and conclusions. Cities and regions constitute an open system where information and resources flow and are exchanged constantly between cities or regions, and spatial elements tend to overflow. The first law of geography states that things that are closer in space are more dependent on each other; that is, the spatial autocorrelation of the attribute values is more significant. Wu [40] and Dong [41] have conducted exploratory research on carbon emission density, but it should be noted there are weaknesses in their studies. The former focused on the carbon intensity of agricultural land rather than industrial land, and its analytical results and conclusions are more applicable to rural planning than to urban and land spatial planning. The latter used the extended Kaya constant equation and the log-mean Divisia index model rather than a spatial econometric model and failed to capture the impact of spatial effects.

1.4. Research Question

The objective of this paper is to discuss the following issues: First, in the dimensions of urban planning and territorial spatial planning disciplines, what is the connotation and value of the concept of carbon emission density of industrial land, and how can it be measured quantitatively? Second, what are the regular characteristics of the geographical pattern and spatio-temporal evolution model of carbon emission density of industrial land? Third, what are the factors and how do they affect the carbon emission intensity of industrial land? This article takes the YREB as an example to empirically analyze the above issues through spatial econometric models, aiming to provide key information and a basis for dual-carbon-goal-oriented urban industrial planning and national spatial planning.

2. Materials and Methods

2.1. Study Area

The YREB is home to 11 provincial administrative regions, including Jiangsu, Zhejiang, Anhui, Jiangxi, Hubei, Hunan, Sichuan, Yunnan, Guizhou, Shanghai, and Chongqing, and the administrative units of this study area include 110 cities. In the study area, 16 autonomous prefectural regions and 3 provincial counties were excluded due to a significant amount of missing data or poor data quality, the former being Enshi in Hubei, Xiangxi in Hunan, Aba in Sichuan, Ganzi in Sichuan, Liangshan in Sichuan, Qianxinan in Guizhou, Qiandongnan in Guizhou, Qiannan in Guizhou, Chuxiong in Yunnan, Honghe in Yunnan, Wenshan in Yunnan, Xishuangbanna in Yunnan, Dali in Yunnan, Dehong in Yunnan, Nujiang in Yunnan, and Diqing in Yunnan and the latter being Xiantao, Qianjiang, and Tianmen in Hubei (Figure 1).

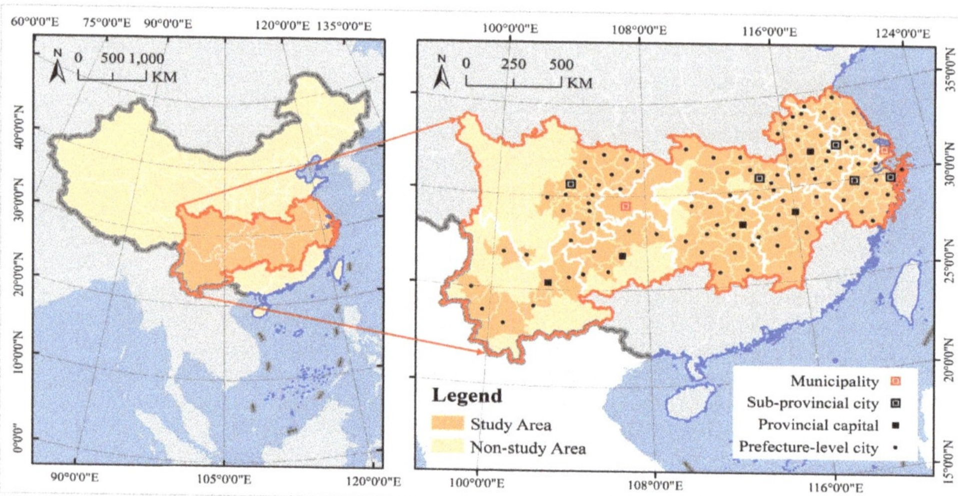

Figure 1. Study area.

The YREB is a major strategic space for regional development at the national level, alongside the Beijing–Tianjin–Hebei region, the Guangdong–Hong Kong–Macao Greater Bay Area, and the Yellow River Basin, with a good industrial base and urban construction. The YREB plays a crucial role in the construction of China's new plan and is highly representative in both global industrial clusters and urban agglomerations. The Chinese central government put forward the Outline of the YREB Development Plan in 2016, as well as the Implementation Plan for the Development of the YREB under the 14th

Five-Year Plan and the Opinions on Several Policy Measures to Further Promote the High-Quality Development of the YREB in 2021 and 2023. All these documents call for industry transformation and green development in the YREB, the synergistic promotion of ecological environmental protection, and sustainable economic development, so as to build a beautiful China model, where human beings and nature coexist in harmony. The YREB accounts for up to 40% of the country's population size and economic output and more than one-third of its carbon dioxide emissions. Therefore, strengthening the carbon emission management of industrial land for the high-quality green development of the YREB has become a key issue that needs to be solved urgently.

2.2. Research Methods

2.2.1. Carbon Emission Density of Industrial Land (CEDIL)

The concept of density has its origins in physics and is a measure of mass in a given volume, measured by dividing the mass of an object by its volume. Sociology, economics, geography, ecology, and urban and territorial spatial planning have drawn on the physical concept of density to introduce the concepts of population density, economic density, species density, and urban density, all using area instead of volume in the measurement of these emerging concepts. Drawing on the measurements of emerging concepts such as population density and economic density, this study proposes the concept of carbon intensity, which is measured by dividing the total amount of carbon emissions in a given area by its area. In particular, the carbon intensity of industrial land use is equal to the total industrial carbon emissions of each city divided by the total area of industrial land use, in tons per square kilometer. With ILA_i to represent the total industrial land area (square kilometer) of the ith city in the study area and CEI_i to represent the total industrial carbon emissions of the ith city, the carbon intensity (CEDIL_i) of the industrial land is calculated as follows:

$$\mathrm{CEDIL}_i = \frac{CEI_i}{ILA_i} \tag{1}$$

CEDIL_i stands for carbon emissions per unit of industrial land area, which is a relative concept compared to the absolute concept of total carbon emissions and represents the average carbon emissions from industrial land within a city. Carbon intensity and carbon emissions per capita are also common relative concepts, with the former referring to the total carbon emissions of a region divided by its GDP and the latter referring to the total carbon emissions of a region divided by its population. Total carbon emissions, carbon intensity, and per capita carbon emissions are commonly used indicators in urban and territorial spatial planning; however, they have a major drawback: weak correlation with land. Urban and territorial spatial planning manages the land and the elements it carries (population, buildings, businesses, etc.), and density control, especially density zoning, is a key tool for planning. Notably, the carbon intensity is further mapped by integrating both land and carbon emissions into a spatial zoning scheme for carbon intensity. Therefore, CEDIL_i is more advantageous than the indicators carbon emission scale, intensity, and per capita emissions and is more applicable to the preparation and management of urban and territorial spatial planning.

2.2.2. Boston Consulting Group (BCG) Matrix

The Boston Matrix originated in the field of corporate development strategy analysis and is a good method for quantitatively measuring the development trends of a company's business or product market. By integrating two indicators, such as the relative share and growth rate of a business or products, it is possible to select the most appropriate company plan from among four types of alternative development strategies (star, gazelle, cow, and dog) [42]. With the average growth rate (growth rate, GR_i) from 2010 to 2020 to represent the trend of change in the time dimension, and the relative share (relative share, RS_i) in 2020 to quantitatively measure the regional ecological niche of each city from a spatial perspective, the Boston Matrix enables the quantitative portrayal of the spatio-temporal evolutionary pattern of CEDIL in the YREB. CEDIL_{i-end} and CEDIL_{i-star} represent the carbon emission density of industrial land in the ith city in the end and base periods, respectively, and $\text{CEDIL}_{max-end}$ represents the maximum value in the study area at the end. GR_i and RS_i are calculated using the following equation [43]:

$$RS_i = RS_i \tag{2}$$

$$GR_i = GR_i \tag{3}$$

The median values of GR_i and RS_i were used as thresholds in this study, because they can eliminate artificial interference in the analyzed results and provide a more balanced classification scheme. According to the classification results based on the spatiotemporal evolution model, the 110 cities in the YREB are classified into four types, with high-carbon and high-growth cities (HCHG cities) representing the GR_i and RS_i of the YREB CEDIL_i, which are greater than the median, indicating a very bad state. Low-carbon and low-growth cities (LCLG cities) represent the GR_i and RS_i of the YREB CEDIL_i and are less than the median, indicating a desirable state. High-carbon and low-growth cities (HCLG cities) show that the GR_i of the YREB CEDIL_i is less than the median, while the RS_i is greater than the median, with a positive trend from bad to desirable conditions. Low-carbon and high-growth cities (LCHG cities) present a GR_i for the YREB CEDIL_i that is greater than the median, while the RS_i is less than the median, with a tendency to change from desirable to bad conditions.

2.2.3. Global Moran's Index and Cold–Hot Spot Analysis

Spatial effect detection is a prerequisite for the scientific selection of regression analysis models, and if the dependent variable has strong spatial autocorrelation and heterogeneity, the spatial measurement model is selected to have a higher accuracy than the traditional statistical regression model. In this study, we use Moran's I to quantify the global autocorrelation of the total industrial land area, total industrial carbon emissions, industrial land carbon emission density, and their spatiotemporal evolution models. Zero is a key threshold for measuring spatial autocorrelation properties, where values greater than zero represent positive autocorrelation. The maximum and minimum values of the Moran index are 1 and -1, respectively, with a larger absolute value indicating stronger spatial correlation. To further visualize the local spatial pattern of spatial autocorrelation, this study uses the cold–hot spot analysis tool and introduces the $Getis-OrdG_i^*$ index to identify the areas where high and low values are clustered. With W_{ij} as the spatial weight matrix, 1 representing spatial adjacency, and 0 representing spatial disjacency, Moran's I and $G_i^*(d)$ are calculated as follows [44,45]:

$$\text{Moran's I} = \frac{n}{S_0} \times \frac{\sum_{i=1}^{n}\sum_{j=1}^{n} W_{ij}(\text{CEDIL}_i - \overline{\text{CEDIL}})(\text{CEDIL}_j - \overline{\text{CEDIL}})}{\sum_{i=1}^{n}(\text{CEDIL}_i - \overline{\text{CEDIL}})^2},\ S_0 = \sum_{i=1}^{n}\sum_{j=1}^{n} = W_{ij} \tag{4}$$

$$G_i^*(d) = \frac{\sum_{i=1}^{n} W_{ij}(d)\text{CEDIL}_i}{\sum_{i=1}^{n}\text{CEDIL}_i} \tag{5}$$

2.2.4. Geodetector

Geodetector is the most commonly used spatial econometric model. Its main advantage is that it can measure the driving effect of a single factor and analyze the correlations between different factors. We employed Geodetector to analyze the driving mechanism of the carbon emission intensity of industrial land, quantifying the influences of measurement factors, pathways, and interactions [46]. Firstly, we discretized the source data of the independent variable (X_i). During the research process, the quantile method using Python was employed to divide the 110 cities in the study area into multiple zones and form multiple zoning schemes. Secondly, we overlayed the geographical distribution pattern of the dependent variable with the independent variable zoning scheme and analyzed the similarity between the two as measured using Geodetector and characterized by the q-index. Thirdly, we took the scenario with the largest q-index as the final calculation for each factor and conducted the impact analysis of the different factors. It should be noted that the maximum value of the q-index is 1, with larger values representing a stronger single-factor influence or dual-factor interaction. For example, for the driving factors X_i and X_j, $q(X_i)$ and $q(X_i \cap X_j)$ represent their single-factor driving force and dual-factor interaction, while $Min(q(X_i), q(X_j))$, $Max(q(X_i), q(X_j))$, $q(X_i) + q(X_j))$ represent the minimum, maximum, and sum of their single-factor driving forces. In the analysis of the interaction, the interactions were classified into five types by analyzing the dual-factor interaction in comparison with the maximum value of the one-factor interaction effect, the minimum value, and the sum of the two [47]. When $q(X_i \cap X_j) < Min(q(X_i), q(X_j))$ and $Min(q(X_i), q(X_j)) < q(X_i \cap X_j) < Max(q(X_i)), q(X_j))$, it is determined that the interaction between the driving factors X_i and X_j is nonlinear weakened and single nonlinear weakened, respectively. It indicates that there is antagonism between the two, and it is necessary to avoid the combination of the two in the future policy design. When $q\ (X_i \cap X_j) > Max(q(X_i), q(X_j))$ and $q(X_i \cap X_j) > q(X_i) + q(X_j)$, it is determined that the interaction is of bifactor enhancement and non-linear enhancement, respectively. It indicates that the two are synergistic, and it is necessary to encourage the combination of the two in the future policy design. When $q(X_i \cap X_j) = q(X_i) + q(X_j)$, it signifies that two factors are independent of each other and that factors can be freely combined with other factors during the policy design [48,49]. The following is the calculation formula for the relevant parameters [50]:

$$q = 1 - \frac{\sum_{h=1}^{l} N_h \sigma_h^2}{N\sigma^2} = q = 1 - \frac{\sum_{h=1}^{l} N_h \sigma_h^2}{N\sigma^2} \tag{6}$$

$$SSW = \sum_{h=1}^{l} N_h \sigma_h^2 \tag{7}$$

$$SST = N\sigma^2 \tag{8}$$

where h represents the number of partitions in the discretization result of the independent variable data (the value range is an integer from 2 to 10), N_h represents the number of cities in partition h, N = 110, σ_h^2 represents the variance of the dependent variable in partition h, σ^2 represents the variance of the dependent variable in the YREB study area, and SSW and SST represent partition variance and total variance, respectively.

2.3. Indicator Selection and Data Source

The geographical pattern of carbon emission density of urban industrial land and its change is a complex process, affected by the combined effect of economic scale, industrial structure and industrialization level, government intervention, degree of opening up, scientific and technological levels, and other factors. Based on the research experiences of related scholars, we chose the gross domestic product (GDP) to represent the impact of the scale effect of the urban economy in this paper [51,52], the proportion of the tertiary industry (PTI) and per capita GDP (PCGDP) to represent the industrial structure and the industrialization process [53,54], the fiscal expenditure (FE) to represent the impact of government intervention [55], the international trade goods export (ITGE) and foreign direct investment (FDI) to represent the influence of openness to the outside world and the degree of globalization [56], and the number of patent authorizations (NPA) to represent the impact of innovation [57]. It is notable that the distribution and evolution of carbon density come close to the natural environment, especially the geomorphology [58], but they are not included in the analytical framework of this paper due to the lack of the necessary complete data.

Urban industrial carbon emission data came from the China City CO_2 Emissions Dataset, released by the China City Greenhouse Gas Working Group (CCG) on its website. CCG is an urban greenhouse gas research and cooperation platform initiated and organized by the Center for Climate Change and Environmental Policy Research of the Academy of Environmental Planning of the Ministry of Ecology and Environment (MEP), established in June 2017. It regularly releases long time-series, full-caliber, and full-coverage data on greenhouse gas emissions from Chinese cities. The industrial carbon emission list covers carbon emissions from the ore, chemical, metal, and electronics industries, as well as carbon emissions related to the production and use of other products, including industrial energy consumption and industrial process emissions. The data on urban industrial land use came from the China Urban Construction Statistical Yearbook, released by the Ministry of Housing and Urban-Rural Development, which provides statistical data on urban construction land use over a long time series and across all land types. Urban industrial land is defined as land used for independently established factories, workshops, handicraft workshops, production sites for construction and installation, and slag (ash) discharge sites. There are three classes of industrial land according to the degree of disturbance to the residential and public environment, the degree of pollution, and the degree of safety hazards. And some cities, to meet the demand for the transformation and upgrading of traditional industries into a high-tech, synergistic production space, combined production space, and headquarters economy, have allocated land for new industries integrating research and development, creativity, design, pilot testing, non-polluting production, and other innovative industrial functions and their ancillary services to industrial land as well. The impact factor data came from the China Urban Statistical Yearbook and were released by the Department of Urban Socio-economic Survey of the National Bureau of Statistics (Appendix A).

3. Results

3.1. Geographical Pattern Analysis

3.1.1. Industrial Carbon Emissions

From 2010 to 2020, the city with the highest industrial carbon emissions in the YREB shifted from Shanghai to Suzhou-JS, with a decrease from 190.77 million to 143.67 million tons. Huangshan has been the city with the smallest industrial carbon emissions, with a decrease from 620,000 to 260,000 tons. Different from the decreasing trend of the maximum and minimum values, the average and median industrial carbon emissions in the YREB both increased to a certain extent, from 24.6636 and 17.2250 tons to 26.5923 and 20.3150 million tons, respectively. It should be noted that the coefficient of variation of industrial carbon emissions in the YREB decreased from 1.20 to 0.99, much higher than 0.36, showing high spatial heterogeneity despite the decline [59] (Table 1).

Table 1. Descriptive statistics on ICE, UIL, and CEDIL in the YREB.

Parameter	Industrial Carbon Emissions		Urban Industrial Land		Carbon Emission Density of Industrial Land	
	2010	2020	2010	2020	2010	2020
MAX	19,077	14,367	739	538	5971	1655
MIN	62	26	0.14	0.49	10.44	2.47
AVG	2466.36	2659.23	31.61	37.06	201.65	157.06
Median	1722.50	2031.50	15.26	22.50	91.87	97.08
CV	1.20	0.99	2.41	1.74	2.86	1.48

The geographic distribution pattern of industrial carbon emissions in the YREB was visualized through quartile spatial clustering analysis using GIS software(ArcGIS 10.4 produced by Esri company). It was found that in 2010, high-carbon cities formed four agglomerations in the Yangtze River Delta (YRD) city cluster, Chengdu–Chongqing city cluster, and southern Hunan, including Shanghai, Suzhou-JS, Chongqing, Ningbo, Wuxi, Wuhan, Nanjing, Xuzhou, Qujing, Bijie, and Liupanshui. However, low-carbon cities were agglomerated in western Yunnan, Anhui, and northern Sichuan, including Lu'an, Bozhou, Ya'an, Ziyang, Lincang, Baoshan, Nanchong, Fuzhou, Suining, Suizhou, Bazhong, Zhangjiajie, and Huangshan. The geographical distribution patterns of high-carbon cities showed that in 2020, the YRD urban cluster rapidly expanded to northern Jiangsu, and the carbon emissions of cities in central Zhejiang decreased rapidly; the Chengdu–Chongqing urban cluster was connected to the northern Hubei, with cities such as Shiyan and Xiangyang having transformed from low-carbon cities to high-carbon cities. Most of the high-carbon cities in Hunan, Yunnan, and Guizhou were transformed into low-carbon cities. In 2020, low-carbon cities were mainly found in the peripheral areas of provinces, especially in their border areas, including southwestern Yunnan, the northeast of Sichuan, the west and south of Hunan, the Yunnan–Sichuan–Guizhou border area, and the Jiangsu–Zhejiang–Anhui border area (Figure 2).

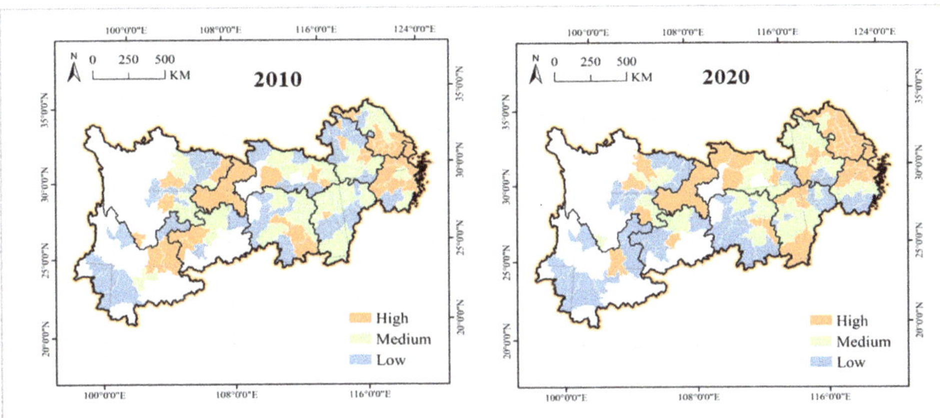

Figure 2. The geographical pattern of ICE in the YREB.

3.1.2. Urban Industrial Land

From 2010 to 2020, Shanghai remained the largest user of industrial land in the YREB, despite a decrease from 739 to 538 km^2. The city with the smallest industrial land use changed from Tongren to Pu'er, increasing from 0.14 to 0.49. The average and median industrial land in the YREB showed some growth, from 31.61 and 15.26 to 37.06 and 22.50 km^2, respectively. The coefficient of variation of industrial land in cities in the YREB decreased from 2.41 to 1.74 from 2010 to 2020, showing a spatial heterogeneity trend consistent with industrial carbon emissions, but with a higher degree of spatial differentiation (Table 1).

Most of the cities with a high value of industrial land in 2010 were clustered in the YRD, while Chongqing and the Yunnan–Guizhou border area were also the centers of industrial land clustering in the region, including Shanghai, Chongqing, Nanjing, Suzhou-JS, Wuhan, Ningbo, Chengdu, Kunming, Hefei, Hangzhou, and Wuxi. Cities with low industrial land value were mostly in Yunnan, southwestern Hunan, and northern Sichuan, including Huaihua, Lishui, Yibin, Shaoyang, Ya'an, Yingtan, Pu'er, Yuxi, Zhaotong, Baoshan, Zhangjiajie, Lijiang, Bazhong, Lincang, and Tongren. During the 2010–2020 period, the YREB pushed hard for industrial structure adjustment and upgrading, and Shanghai even proposed a development strategy based on resource reduction, prompting major changes in the distribution pattern of urban industrial land. Most of the cities with a high value of industrial land in 2020 were concentrated in the border areas of Sichuan, Yunnan, and Guizhou, including Shanghai, Chongqing, Wuhan, Chengdu, Ningbo, Nanjing, Suzhou-JS, Hangzhou, Changzhou, Hefei, and Nantong. Most of the low-value cities with industrial land were clustered in the northern Jiangsu–Anhui border area and the Chengdu–Chongqing urban cluster, including Dazhou, Zhaotong, Shaoyang, Yuxi, Lishui, Liupanshui, Ezhou, Ziyang, Zhangjiajie, Huaihua, Lijiang, Baoshan, Bazhong, Lincang, and Pu'er (Figure 2). Of note, the comparison between Figures 2 and 3 shows that the distribution pattern of industrial carbon emissions is inconsistent with the industrial land in the YREB, and there is a large spatial mismatch between the two. The results of the independent analysis separating the two do not well reveal the real condition of carbon emissions from industrial land.

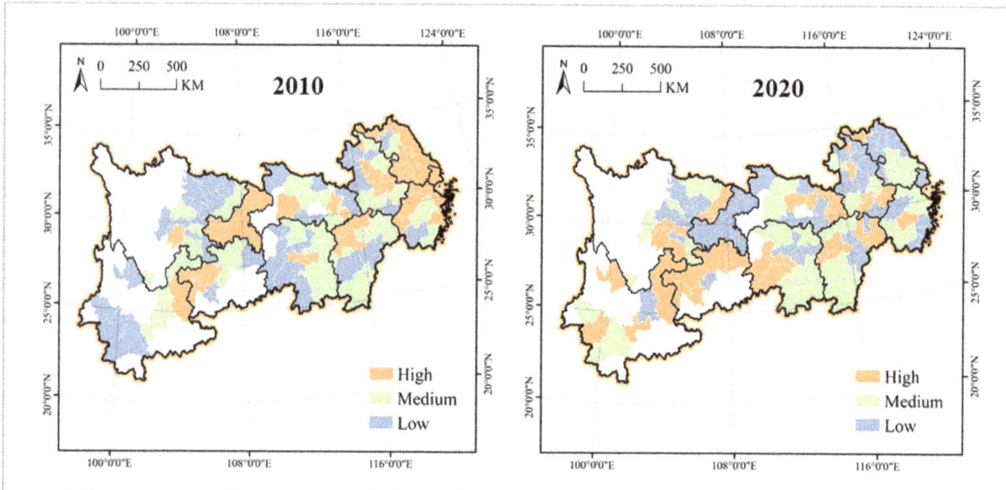

Figure 3. The geographical pattern of UIL in the YREB.

3.1.3. Carbon Emission Density of Industrial Land

From 2010 to 2020, the city with the largest industrial carbon emission density in the YREB changed from Tongren to Pu'er, with the value decreasing from 59.71 to 16.55 million tons/km^2. Huangshan remained the city with the lowest industrial carbon emissions, dropping from 104,400 to 24,700 tons/km^2. The average industrial carbon emission density in the YREB decreased from 2.0165 to 1.5706 million tons/km^2, while the median increased from 918,700 to 970,800 tons/km^2. It should be noted that the variation coefficient of industrial carbon emission density in the YREB decreased from 2.86 to 1.48, and the spatial heterogeneity level was much higher than that of industrial carbon emission and industrial land use (Table 1).

Most of the cities with a high carbon emission intensity of industrial land use in 2010 were concentrated in the YRD urban cluster, including Tongren, Bijie, Qujing, Yuxi, Lijiang, Lincang, Yibin, Liupanshui, Loudi, Shaoyang, Yichun, Huanggang, Neijiang, and Huainan. High-value cities in 2020 shifted from the east to the central and western regions, mostly concentrated in the border areas of Yunnan and southwestern Hunan, as well as Yunnan and Guizhou, including Pu'er, Liupanshui, Yuxi, Ezhou, Qujing, Loudi, Bijie, Dazhou, Leshan, Baoshan, Shangrao, Chizhou, Wenzhou, Lincang, and Huainan. In 2010, the low-carbon intensity cities of industrial land use were mainly in west–central Yunnan and Hunan, the border areas of Yunnan, Guizhou, Sichuan, and Chongqing, and central Jiangxi (and expanding to Anhui and Zhejiang), including Shiyan, Zigong, Bengbu, Suqian, Yancheng, Lianyungang, Bozhou, Nanchang, Chengdu, Shanghai, Ziyang, Fuzhou, Hefei, Lu'an, Suining, Nanchong, Suizhou, and Huangshan. Most of the low-value cities in 2020 were concentrated in the Chengdu–Chongqing urban area, with a relative concentration in the fringes of western Anhui and northern Hubei, including Xiangyang, Luzhou, Lu'an, Guiyang, Kunming, Deyang, Wuhan, Hangzhou, Shanghai, Shiyan, Nanchang, Suqian, Chengdu, Zigong, Suining, Nanchong, Suizhou, and Huangshan (Figure 4).

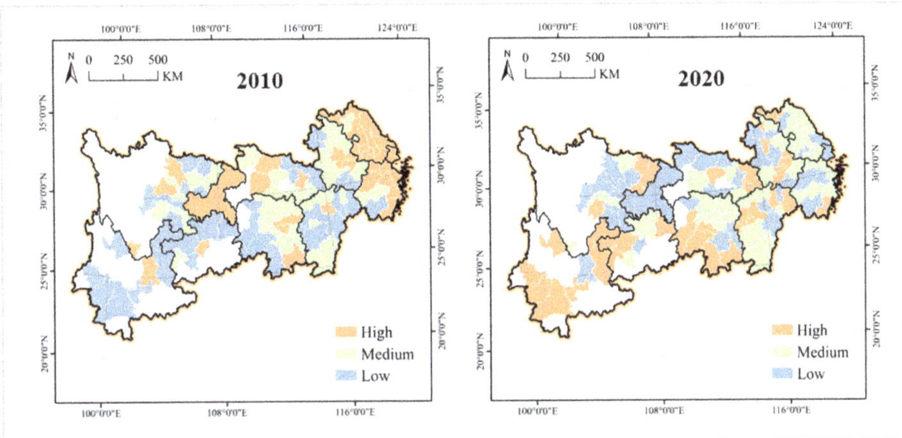

Figure 4. The geographical pattern of CEDIL in the YREB.

3.2. Spatiotemporal Evolution Model Analysis

From the perspective of industrial carbon emissions, the median relative share and growth rate were 0.14 and 12.46%, respectively. The classification based on these as thresholds showed that high-carbon and high-growth cities (HCHG cities) were mostly in the lower and middle reaches of the YREB, including Nanjing, Changzhou, Lianyungang, Yancheng, Yangzhou, Zhenjiang, Wenzhou, Jiaxing, Huzhou, Shaoxing, Hefei, Wuhu, Xuancheng, Jiujiang, Ganzhou, Yichun, and Shangrao. Low-carbon and low-growth cities (LCLG cities) developed into three agglomerations in eastern Sichuan, northwestern Hubei, and the border area of Hunan and Jiangxi provinces, including Suqian, Lishui, Huangshan, Fuyang, Pingxiang, Ji'an, Shiyan, Xiangyang, Xiaogan, Huanggang, Suizhou, Changsha, and Zhuzhou. High-carbon and low-growth cities (HCLG cities) were concentrated in the central part of Zhejiang, southern Jiangsu, the border areas of Yunnan and Guizhou, Chongqing, and its neighboring areas, including Wuxi, Xuzhou, Suzhou, Nantong, Huai'an, Ningbo, Quzhou, Huaibei, and Anqing. Low-carbon and high-growth cities (LCHG cities) were scattered in distribution except for some that were concentrated only in the southwest of Yunnan, including Zhoushan, Bengbu, Suzhou, Lu'an, Bozhou, Nanchang, Jingdezhen, Yingtan, Fuzhou, Jingzhou, Xianning, Shaoyang, Zhangjiajie, Yiyang, and Yongzhou (Figure 5).

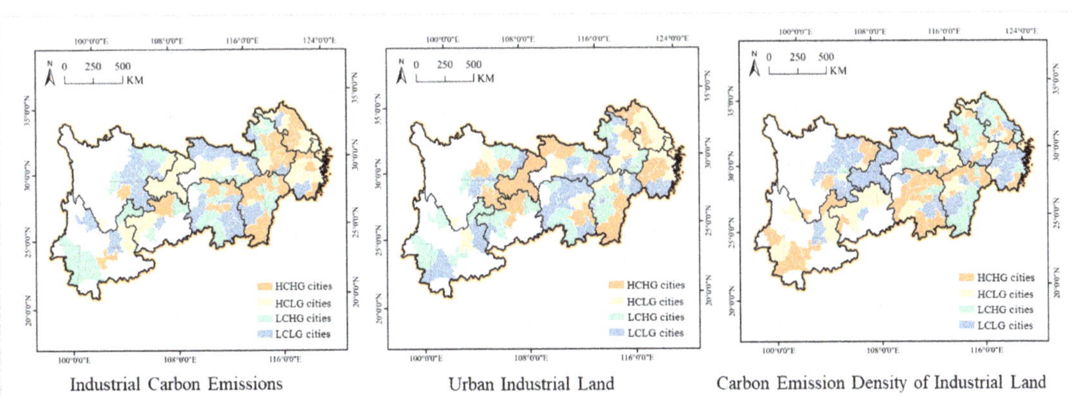

Figure 5. The spatio-temporal evolution model of ICE, UIL, and CEDIL in the YREB.

From the perspective of urban industrial land, the median relative share and growth rate were 0.04% and 35.82%. The classification based on these as thresholds showed that most of the high-carbon and high-growth cities (HCHG cities) were concentrated in the Chengdu–Chongqing urban cluster, including Xuzhou, Changzhou, Lianyungang, Yangzhou, Zhenjiang, Suqian, Hangzhou, Shaoxing, Jinhua, Quzhou, Tongling, Chuzhou, Nanchang, Ganzhou, Fuzhou, Wuhan, Shiyan, and Xiangyang. Low-carbon and low-growth cities (LCLG cities) were in the middle reaches of the YREB, especially in the central region of Jiangxi, including Wenzhou, Zhoushan, Lishui, Huainan, Huaibei, Suzhou, Lu'an, Chizhou, Pingxiang, Xinyu, Shangrao, Ezhou, and Jingmen. Most of the high-carbon and low-growth cities (HCLG cities) were clustered in the YRD, especially in the Jiangsu and Anhui provinces, including Shanghai, Huai'an, Nanjing, Ningbo, Wuxi, Nantong, Jiaxing, Huzhou, Yancheng, Hefei, and Wuhu. Low-carbon and high-growth cities (LCHG cities) were scattered in the provincial fringe areas, including Huangshan, Fuyang, Bozhou, Xuancheng, Yingtan, Ji'an, Yichun, Xiaogan, Suizhou, Shaoyang, Zhangjiajie, Yongzhou, Guangyuan, Suining, and Neijiang (Figure 5).

From the perspective of carbon emission density of industrial land, the median relative share and growth rate were 0.06 and −15.78%, respectively. The classification based on these as the thresholds showed that high-carbon and high-growth cities (HCHG cities) were mostly concentrated in the middle and upper reaches of the YREB, including Jingmen, Xianning, Xiangtan, Yueyang, Changde, Yiyang, Chenzhou, Yongzhou, Huaihua, Loudi, Panzhihua, Leshan, Dazhou, Bazhong, Liupanshui, Yuxi, Baoshan, and Zhaotong. Low-carbon and low-growth cities (LCLG cities) were mostly concentrated in the Chengdu–Chongqing urban cluster and the central part of Zhejiang, including Changzhou, Nantong, Suqian, Hangzhou, Ningbo, Shaoxing, Jinhua, Lishui, Anqing, Huangshan, and Fuyang. High-carbon and low-growth cities (HCLG cities) were in the north of Guizhou, including Yichun, Huangshi, Xiaogan, Huanggang, Shaoyang, Neijiang, Guang'an, Zunyi, Anshun, Bijie, Tongren, and Qujing. Low-carbon and high-growth cities (LCHG cities) were mostly clustered in the southeast of Anhui, Jiangsu, and Jiangxi, including Nanjing, Lianyungang, Huai'an, Yancheng, Yangzhou, Huzhou, Hefei, Bengbu, Chuzhou, Lu'an, Bozhou, and Jingdezhen (Figure 5).

3.3. *Driving Mechanism Analysis*

3.3.1. Spatial Effects Detection

Table 2 shows that Moran's I is greater than zero for all indicators, indicating that the geographic patterns and spatial and temporal evolution patterns of industrial carbon emissions, industrial land use, and industrial carbon intensity in the YREB are in significant positive spatial autocorrelation. The spatio-temporal evolution patterns of industrial carbon emissions from 2010 to 2020 and the geographic patterns of industrial carbon emissions in 2010 and 2020 showed a high autocorrelation index. From the industrial carbon perspective, the hot spots in 2010 and 2020 were concentrated in the Shanghai metropolitan area, with a shrinking spatial scope. In 2010, most of the cold spot cities were in eastern Sichuan and southwestern Yunnan, and in 2020, the cold spot cities expanded significantly, covering most of the YREB. They formed a large agglomeration in the middle reaches of the Yangtze River, centered on Hunan and Jiangxi, and extended into the lower reaches in western Anhui and southern Zhejiang. In addition, they formed a second agglomeration in the upper reaches of the Yangtze River, centered in Yunnan, Sichuan, and Guizhou and extending into southwestern Yunnan. The hot spots of the spatio-temporal evolution pattern from 2010 to 2020 were concentrated in the border area of Jiangsu province, Anhui province, and Zhejiang province (around Nanjing metropolitan area), while the cold spots formed three small clusters in eastern Sichuan, northern Hubei, and southern Hunan. The spatial pattern of "center-periphery" is formed by sub-cold and cold spots and sub-hot and hot spots in combination. It is worth noting that the sub-hot spots expanded in a beaded pattern by leaps to the middle and upper reaches of the Yangtze River region, forming a spatial structure of core-edge circles (Figure 6).

Table 2. Spatial autocorrelation analysis on ICE, UIL, and CEDIL in the YREB.

Dimension	Indicator	Moran's I	P	Z
Geographical patterns	Industrial carbon emissions in 2010	0.15	0.003	3.91
	Industrial carbon emissions in 2020	0.21	0.001	5.17
	Urban industrial land in 2010	0.08	0.006	3.44
	Urban industrial land in 2020	0.07	0.036	2.29
	Carbon emission density of industrial land in 2010	0.02	0.07	1.56
	Carbon emission density of industrial land in 2020	0.17	0.001	4.96
Spatiotemporal evolution models	Industrial carbon emissions in 2010–2020	0.26	0.001	6.56
	Urban industrial land in 2010–2020	0.09	0.026	2.16
	Carbon emission density of industrial land in 2010–2020	0.06	0.069	1.52

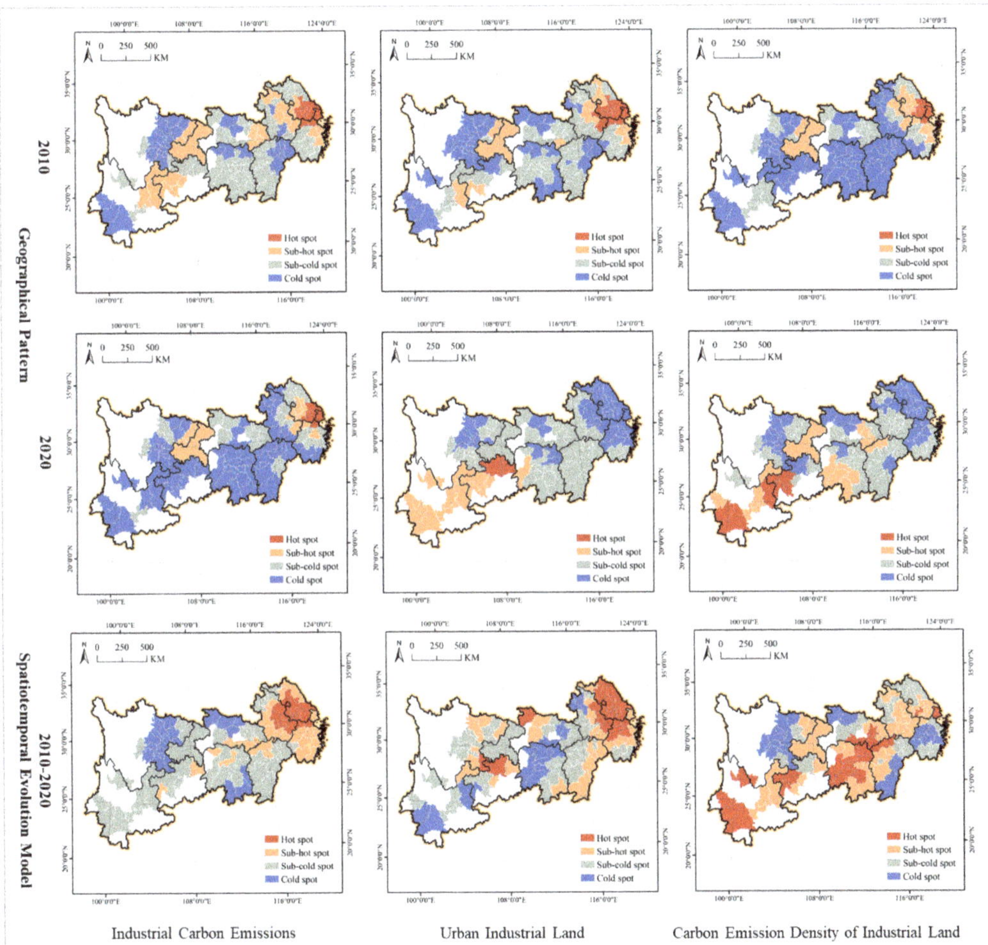

Figure 6. The spatial effect of ICE, UIL, and CEDIL in the YREB.

According to the industrial land use, hot and sub-hot cities in 2010 were relatively concentrated in the lower reaches of the YREB, especially in the border areas of Jiangsu, Zhejiang, and Anhui. There were hot spot clusters formed in 2020, and most of the sub-hot spot cities were distributed in Yunnan, with hot spots and sub-hot spots concentrated in the upper reaches of the YREB. Cold spot cities in 2010 were scattered, forming small clusters only in eastern Sichuan. Cold spot cities were mostly concentrated in the YRD urban cluster in 2020, while sub-cold spots were mostly concentrated in the middle reaches of the YREB and extended upward to the Chengdu–Chongqing urban cluster. The spatial pattern of hot and cold spots was characterized by an upstream and downstream gradient rather than a circle-edge structure. Most of the hot spot cities in the evolution model of urban industrial land use from 2010 to 2020 were concentrated in the YRD urban cluster, especially in Jiangsu province, while the cold spots were relatively concentrated in Jiangxi and Yunnan provinces (Figure 6).

In terms of the carbon intensity of industrial land use, the hot spots in 2010 were fewer and located in and around the Shanghai metropolitan area, while the sub-hot spots were spread out in their neighboring areas. The hot spots and sub-hot spots shifted from the lower to the upper YREB in 2020, mostly concentrated in Yunnan and its border area with Guizhou, with a slight increase in the number of hot cities. Cold spots in 2010 covered most areas and were mostly concentrated in the middle reaches of the Yangtze River, centered in Jiangxi and Hunan, extending upstream to northern Guizhou and eastern Sichuan and downstream to western Anhui and the southwest corner of Zhejiang. Cold spot cities in 2020 shrank significantly, mostly concentrated in the YRD urban cluster, with a small number concentrated in eastern Sichuan. Most of the hot spots in the spatio-temporal evolution model from 2010 to 2020 were concentrated in Hunan in the middle reaches of the Yangtze River, with a small part in the upper reaches in Yunnan and the sub-hot cities distributed in their periphery. There were few cold spots, mostly in eastern Sichuan, with a few in the central part of Zhejiang and southeastern Jiangxi.

3.3.2. Single Factor Driving Force

In terms of the geographic pattern, the FDI and FE showed the strongest influence on the carbon emission intensity of industrial land as key factors. The NPA, ITGE, PCGDP, and GDP were not less influential, and they were important factors. The PTI had a weak and not statistically significant influence as an auxiliary factor. In terms of spatio-temporal evolution patterns, the FE and NPA had the strongest influence on the carbon emission density of industrial land as key factors. The ITGE, PCGDP, FDI, and GDP had comparable influences as important factors. The PTI had the lowest influence as an auxiliary factor. Of note is that the FDI was not statistically significant, and both the PCGDP and GDP were not significant (Table 3). To sum up, there were significant differences in the influences of different factors on the carbon emission density of industrial land. From a similarity perspective, the FE was a key factor common to both geographic and spatio-temporal evolution patterns, while the NPA, FE, ITGE, and GDP were important factors, and the PTI was an auxiliary factor common to them. From the perspective of differences, the PCGDP and FDI had more significant influence on the geographical pattern than spatio-temporal evolution, while the PTI was just the opposite. In addition, most factors, including the PCGDP, FE, ITGE, and FDI, showed greater influence on the geographical pattern than the spatio-temporal evolution pattern. The PTI was the opposite of them, while the influence of the GDP and NPA on both remained stable overall.

Table 3. The impact of single factors on the geographical pattern and spatiotemporal evolution pattern of CEDIL in the YREB.

Indicator	Geographical Patterns		Spatio-Temporal Evolution Models	
	q	p	q	p
GDP	0.10	0.02	0.10	0.07
PTI	0.03	0.33	0.08	0.04
PCGDP	0.13	0.04	0.11	0.10
FE	0.21	0.02	0.16	0.04
ITGE	0.16	0.00	0.12	0.03
FDI	0.24	0.01	0.10	0.28
NPA	0.16	0.04	0.15	0.01

3.3.3. Multifactor Interaction Effect

Regarding the interaction type, most of the factor pairs were of nonlinear enhancement, with a few of bifactor enhancement, and there was a significant synergy between different factors. This suggests that the current pattern and change in the carbon emission density of industrial land is the result of a combination of factors. That is, different factors reinforce each other, and each factor has a significant direct impact on the carbon emission density. The additive effects produced by the interaction of different factors cannot be ignored. For geographical patterns, there was a large number of factor pairs of bifactor enhancement, including GDP and ITGE, GDP and FDI, GDP and NPA, PCGDP and ITGE, ITGE and FDI, ITGE and NPA, and NPA and FDI. However, for spatiotemporal evolution patterns, there was a very small number of factor pairs, only NPA and GDP and NPA and ITGE (Tables 4 and 5). In terms of interaction intensity, FE and FDI showed the strongest interaction on geographic pattern, followed by FE and ITGE and FE and GDP showing significant interactions, with a comprehensive force exceeding 0.8 or even 0.9, far higher than other factor pairs, considered as super-factor pairs. FE and NPA, PCGDP and FE, PCGDP and FDI, PCGDP and NPA, PTI and FE, and PTI and FDI also showed high levels of interaction and should therefore be taken into account together in future policy designs. Multifactor pairs showed weaker interactions on spatio-temporal evolutionary patterns than geographic patterns, with FE and FDI, FDI and PCGDP, PCGDP and FE, and PTI and FE leading the way in the interacting force, which can be considered as super-factor pairs. It is important to note that NPA and PCGDP, NPA and FE, NPA and FDI, and GDP and PCGDP also showed non-weak interaction forces, and they are factor combination options that should not be ignored in the policy design. Overall, multifactor interactions are dominated by nonlinear enhancements with significant synergistic effects, and more and more super-factor pairs are emerging. In particular, the interaction of FDI, FE, and NPA with other factors showed a leading advantage, and they are the factors that should be emphasized and prioritized during the portfolio policy design.

Table 4. The multifactor interaction effect on the geographical pattern of CEDIL in the YREB.

Indicator	GDP	PTI	PCGDP	FE	ITGE	FDI	NPA
GDP	0.10						
PTI	0.15	0.03					
PCGDP	0.31	0.25	0.13				
FE	0.81	0.39	0.59	0.21			
ITGE	0.19	0.22	0.29	0.86	0.16		
FDI	0.30	0.37	0.44	0.91	0.33	0.24	
NPA	0.20	0.30	0.40	0.62	0.27	0.38	0.16

Table 5. The multifactor interaction effect on the spatio-temporal evolution pattern of CEDIL in the YREB.

Indicator	GDP	PTI	PCGDP	FE	ITGE	FDI	NPA
GDP	0.10						
PTI	0.29	0.08					
PCGDP	0.42	0.33	0.11				
FE	0.33	0.47	0.49	0.16			
ITGE	0.30	0.28	0.32	0.38	0.12		
FDI	0.38	0.36	0.51	0.58	0.30	0.10	
NPA	0.26	0.30	0.44	0.43	0.28	0.40	0.15

4. Discussion

The analysis shows that there is not a synchronous linear relationship between industrial land and its carbon emissions. Introducing the concept of carbon emission intensity can well reveal the coupling relationship between the two, while better meeting the needs of enabling carbon peaking and carbon neutrality through spatial planning. The comparison of Figures 2–4 shows a large difference between industrial carbon emissions, industrial land use, and industrial land carbon emission density. For example, in 2020, Shanghai was the city with the largest industrial land use in the Yangtze River Economic Belt, but it was not the city with the highest carbon emissions. Instead, Suzhou had the highest carbon emissions, and Pu'er had the largest carbon emission density. The difference is universal, not an exception. For example, in the YRD city cluster, except Shanghai, all other cities have a high level of industrial land use, generally a medium or low level of industrial carbon emissions, and low or medium level of carbon emission densities. These three are not only different in their current patterns, but also increasingly in their evolution. For example, the evolution of industrial carbon emissions, industrial land use, and industrial land carbon intensity in Lincang present LCHG, LCLG, and HCHG patterns, respectively, completely differently. This result suggests that for the goal of carbon peaking and carbon neutrality, it is not enough to limit spatial planning to industrial land and the management of carbon emission scale, and a new concept of carbon emission density needs is needed.

The analysis shows that there is a significant spatial difference in the carbon emission density of industrial land in the YREB, with strong correlations between neighboring cities, and the evolution of density emerges in different patterns, including HCHG, HCLG, LCHG, and LCLG. Additionally, the spatial pattern and evolutionary pattern of carbon emission density of industrial land is the result of the joint action of factors. The FDI and FE play a key role in shaping spatial patterns, while the FE and NPA are key factors influencing evolution patterns. Different factors interact with each other to amplify their effects, so the final determination of the force of the factors should take into account rather than the direct force of a single factor, as well as the influence of the association between other factors, especially the strong interaction of the FE and FDI with other factors. It provides a scientific basis for combining strategies in the policy design.

Urban density gained attention as a major issue for cities after the Industrial Revolution [60]. To alleviate the problems of dirtiness and overcrowding in Western industrial cities, scholars from the early 20th century to the 1960s developed two schools of thought in the study urban density: decentralization and centralization [61]. The former, represented by Howard's Garden City, created the idea of low density, while the latter, represented by Corbusier's Radiant City, created the idea of high density by advocating higher building heights, more green space, and increased road widths. In the face of challenges such as suburban sprawl, inner-city decay, and urban disease, the 1970s and 1990s saw the rise of "new urbanism", "smart growth", and "compact city" concepts, with higher densities dominating urban development [62]. City governments hope to address urban disease by promoting the rebirth and revitalization of urban centers through high-density development, mixed-use land use, and transit-oriented development to conserve land and reduce energy consumption [63]. Since the beginning of the 21st century, scholars have begun to pay attention to and gradually attach importance to the negative effects of high-density urban development, such as traffic congestion caused by excessive high-density urban development, the declining living environment quality, insufficient green space, and serious air pollution [64]. Based on the value trade-off between "economy, environment and people" and the historical reflections on the development of cities with low and high densities, the inverted U-shaped relationship between urban density and urban development benefits has gradually become the consensus of academia and industry [65].

Carbon emission density is an important dimension of urban density, and based on the results of the above analysis, we believe that it is also consistent with parabolic theory. That is, the carbon emission density increases gradually in the early stage and begins to decrease after reaching a peak (critical value). Both too high and too low carbon emission densities pose problems for the sustainability of urban industrial land use, with a medium density being superior to high and low densities [66,67]. Medium-suitable density can balance the weaknesses of both high density and low density to maximize the economic, environmental, and human benefits of urban industrial land management. There are two possible reasons for the low carbon emission density of industrial land. First, the small scale of industrial carbon emissions as a molecule indicates a high level of development of greening and low carbonization of the industrial system. Significantly, for Chinese cities at the current stage, the excessive pursuit of low carbonization or even zero carbonization in industrial development implies high development costs and prices, which goes against the healthy development of industries and cities. Second, the large size of industrial land use as the denominator indicates extensive industrial land development. The non-intensive use of industrial land will lead to a waste of land resources, as well as increased transportation costs, high construction costs of supporting facilities and low interest rates. On the contrary, a high carbon emission density in industrial land may also lead to two unhealthy situations. On the one hand, the large scale of industrial carbon emissions as a numerator indicates that high-emission, high-energy-consumption, and high-pollution industrial sectors account for a high proportion of the industrial system, hindering industrial upgrading and high-quality urban development. On the other hand, the small size of industrial land use as the denominator indicates that the supply of industrial land resources is insufficient. In this context, enterprises can only seek to meet the demand by increasing the land plot ratio, and therefore, they put up high-rise buildings in large quantities, boosting the cost of development.

It is important to note that the medium density is not a specific and fixed number but a dynamic range, and the upper and lower limits may not be exactly the same across regions and cities (Figure 7a). The values at the two ends of the medium density can be determined in single-city sample studies using a time-series analysis; in large multi-city sample studies, they need to be obtained using regional comparative analysis. In addition to observing the pattern of change in the carbon intensity, the determination of values should also take into account the local development characteristics and stage needs of the study area. In addition, zoning can be traced back to the theory of locality and is a common tool for spatial planning of cities and territories [68]. It is of great value to quantitatively draw the carbon emission density planning zoning of urban industrial land and design and implement differentiated management policies under the goal of "carbon neutral carbon peak" for controlling the carbon emissions of urban industrial land. For specific study areas, their internal members can be divided into several planning zonings using quantile spatial clustering analysis, and the government needs to focus on ultra-high or ultra-low density urban governance. It should be noted that the number of members in different zonings should be in an olive-shaped structure of "small at both ends and large in the middle", and if, on the contrary, there is a dumbbell-shaped structure of being "large at both ends and small in the middle", it indicates unreasonable planning zoning of the carbon emission density of industrial land (Figure 7a). In the process of introducing the concept and algorithm of carbon density into spatial planning, planners need to pay attention to the differences between planning systems of different countries and regions, as well as the regional characteristics of different study areas, especially the industrial supporting features (transportation, logistics, and other productive services) [69].

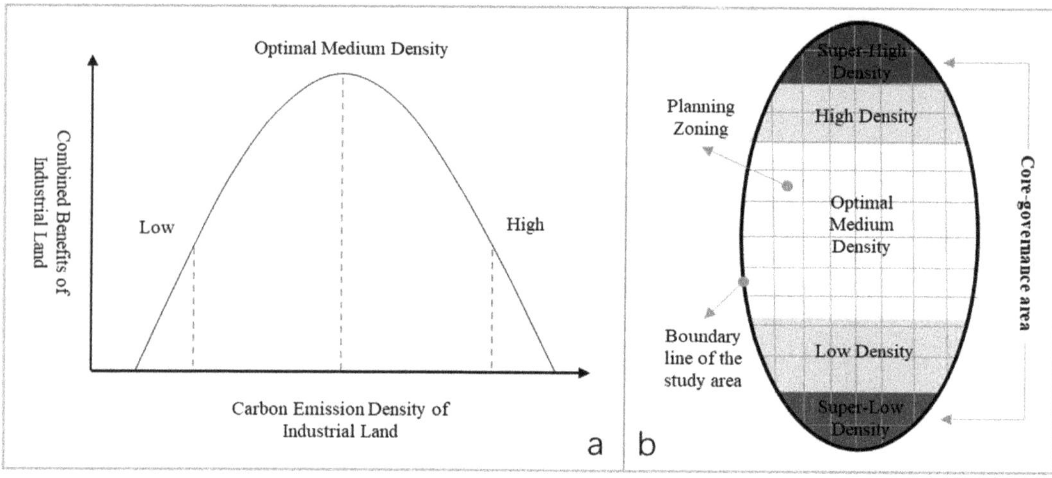

Figure 7. The functional relationship between carbon emission density and combined benefits of urban industrial land. (**a**) Carbon emission density benefit curve; (**b**) Carbon emission density zoning map.

The greatest contribution of this paper is to put forward the concept and measurement method of urban industrial land carbon emission density and discuss its connotation and application value from the theoretical level. Many studies in the past have discussed the relationship between urban density and carbon emissions [70,71]. For example, Norman and Ishii argued that the medium density is most conducive to reducing carbon emissions [72,73]. However, the variable used in these studies was total carbon emissions rather than density. From the previous analysis, it can be seen that total carbon emissions and density are two completely different concepts, and their geographical patterns and spatio-temporal evolution also have significant differences. In traditional studies, we pay more attention to the scale and intensity of carbon emissions (the relationship with economy) while neglecting the connection between carbon emissions and space (i.e., carbon emission density). Spatial planning requires us to find and solve problems from the perspective of space. The carbon emission scale and intensity indicators do not match the needs of spatial planning, while the carbon emission density solves the problem. To a certain extent, proposing the concept of carbon emission density and its measurement method and establishing the direct relationship between carbon emissions and space is of great academic innovation for spatial planning research. Therefore, the design of policies based on total carbon emissions alone is incomplete, as the insufficient consideration of carbon intensity is likely to cause failure in urban land-use planning under the "dual-carbon" goal. The introduction of spatial econometric modeling to improve the accuracy of the analytical results and conclusions is another marginal contribution of this paper. The industrial carbon emissions, urban industrial land, and carbon emission density of industrial land all have significant spatial effects, while most of the past studies chose linear regression and other traditional statistical models, ignoring the influence of spatial heterogeneity and autocorrelation on variables and their relationships.

As mentioned earlier, medium densities are more suitable for sustainable urban development than high and low densities, and therefore, carbon emission control on the YREB industrial sites should focus on both the low- and high-density ends. The identification of interval thresholds for medium density is a prerequisite for policy design. According to the numerical distribution structure of carbon emission density of industrial land in the YREB in 2020 and the balance of the number of members of different zoning schemes (small at both ends, large in the middle), the suggestion is to set the upper threshold to 1.5 million tons/km^2 and the lower threshold to 500,000 tons/km^2. Therefore, the proportion of 28.18%, 42.73%, and 29.09% of high-, medium-, and low-density cities in the YREB suggests that the distribution structure conforms to the "olive shape", which is reasonable (Figure 8).

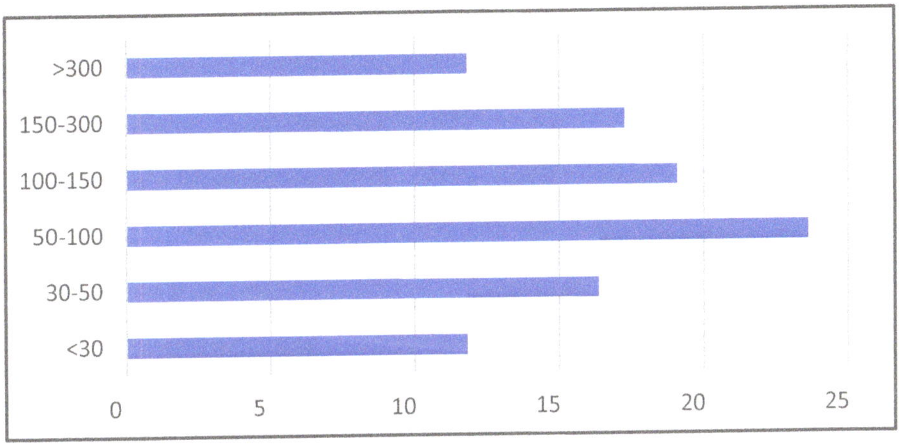

Figure 8. Threshold selection of CEDIL in the YREB.

For low-density cities including Nantong, Bozhou, Lishui, Bengbu, Jingzhou, Chongqing, Ya'an, Ziyang, Fuyang, Zhuzhou, Mianyang, Hefei, Shaoxing, and Xiangyang, they are faced up with a low-density spread of industrial land (Table 6). They need to be guided by the concept of compact cities and smart growth in the future to moderately control the spread and expansion of industrial land, coupled with industrial upgrading and transformation, to gradually increase the carbon emission density of industrial land. In particular, Chengdu, Zigong, Suining, Nanchong, Suizhou, and Huangshan have excessively low carbon densities, and they urgently need to control the spread of low densities on industrial land. In addition, low-density cities should be active in participating in the regional carbon emission trading for industrial land and enhance their capacity for sustainable development through inter-city ecological compensation transfers.

Table 6. The planning zoning of CEDIL in the YREB.

Zoning	Cities
High Density	Pu'er, Liupanshui, Yuxi, Ezhou, Qujing, Loudi, Bijie, Dazhou, Leshan, Baoshan, Shangrao, Chizhou, Wenzhou, Lincang, Huainan, Xinyu, Shaoyang, Zhaotong, Wuhu, Yichun, Chenzhou, Pingxiang, Huaihua, Zhoushan, Tongling, Bazhong, Huanggang, Jingmen, Xuzhou, Maanshan, Lijiang, Tongren
Medium Density	Huaibei, Huangshi, Yongzhou, Guang'an, Jiujiang, Xianning, Yiyang, Xuancheng, Neijiang, Xiaogan, Yueyang, Changde, Xiangtan, Suzhou-JS, Panzhihua, Zunyi, Jiaxing, Wuxi, Quzhou, Zhenjiang, Anshun, Suzhou-AH, Taizhou-JS, Huzhou, Taizhou-ZJ, Meishan, Yangzhou, Jinhua, Yingtan, Yancheng, Nanjing, Ji'an, Guangyuan, Yichang, Chuzhou, Anqing, Ganzhou, Lianyungang, Huai'an, Changsha, Ningbo, Jingdezhen, Zhangjiajie, Yibin, Hengyang, Changzhou, Fuzhou
Low Density	Nantong, Bozhou, Lishui, Bengbu, Jingzhou, Chongqing, Ya'an, Ziyang, Fuyang, Zhuzhou, Mianyang, Hefei, Shaoxing, Xiangyang, Luzhou, Lu'an, Guiyang, Kunming, Deyang, Wuhan, Hangzhou, Shanghai, Shiyan, Nanchang, Suqian, Chengdu, Zigong, Suining, Nanchong, Suizhou, Huangshan

For high-density cities including Zhoushan, Tongling, Bazhong, Huanggang, Jingmen, Xuzhou, Maanshan, Lijiang, and Tongren, they are characterized by high energy consumption and high emissions in industry types (Table 6). Therefore, the focus in the future is to increase the transformation of industrial structure and build a safe and clean energy system. Notably, Pu'er, Liupanshui, Yuxi, Ezhou, and Qujing have excessively high carbon intensities, especially Pu'er, which, as a tourist city, should push the low-carbon transformation of existing industries. For high-density cities, priority should be given to adjusting the land-use structure, promoting low-carbon and zero carbonization transformation of industrial space, and quickly building a circular economic system and spatial pattern. The influence of human economic behaviors on regional carbon emissions is to a large extent based on the changes to the way industries are spatially planned, with land use as the carrier. Meanwhile, they should also vigorously develop eco-industries and green services and promote the gradual utilization of energy between industries while reducing fossil energy consumption.

In addition, we suggest that all cities in the YREB include the carbon emission density of industrial land in their land transfer standards to strengthen their direct control over carbon emissions. The current criteria for the granting of industrial land in the YREB cities focuses mainly on economic efficiency and is not sufficiently responsive to "dual-carbon" objectives. For example, the policy provided in the "Standard Land" Control Indicators for Industrial Land in Mianyang City takes the intensity of fixed assets investment (10^4 yuan/mu), the average output value per mu (10^4 yuan/mu), the average tax per mu (10^4 yuan/mu), and the plot ratio of industrial land as mandatory indicators, and it has no environmental constraints related to carbon emissions. Therefore, it is recommended that all cities incorporate the density and total amount of carbon emissions from industrial land into land transfer standards and management policies as soon as possible, as a binding indicator for the selection of urban industries and structural optimization, so as to promote the low carbonization of the industrial system and economic development.

It is worth to note that close cooperation between different cities in the YREB has been established, and as a world-class urban agglomeration, the YREB is closely linked to other cities around the world [74]. And the analysis above shows that the carbon emission density of industrial land in the YREB shows significant spatial heterogeneity and autocorrelation. Therefore, it is necessary to develop policies to promote carbon peaking and carbon neutrality in an orderly manner based on the differences in the carbon emission densities of industrial land in different cities and to emphasize collaboration between different cities in achieving the goals, including carbon emission trading between internal cities and exchange and cooperation between external cities. Regional cooperation to promote the low-carbon transformation of industrial structure is to fully leverage the comparative advantages of each region under the constraint of carbon emission density, promote interregional industrial transfer and integrated development, jointly cultivate innovative and green industries, and optimize the combination of production and innovation factors in a larger spatial range.

In the end, we suggest that all cities in the YREB should highlight the synergy among multiple factors in designing policy systems and organically combine the policies to strengthen the indirect management of carbon emissions. For example, the FDI, FE, and NPA, as super-interacting factors, have a leading advantage in interacting with other factors, and significantly amplify their influence on the carbon emission intensity of industrial land. Therefore, for globalization, government intervention, and technological innovation represented by the FDI, FE, and NPA [75,76], it is necessary to design dedicated supporting policies, and more importantly, to highlight their combined relationship with related policies such as investment density, urban planning, and plot ratio control to optimize management performance [77,78].

5. Conclusions

Industry is one of the largest sources of carbon emissions, and the management of carbon emissions from industrial land is a key part to achieve the goal of "double carbon". This study conducted an empirical research based on the 2010–2020 carbon emission dataset of urban industrial land use in the YREB and proposed the concept of carbon emission density of urban industrial land use and its measurement method from the perspective of urban and territorial spatial planning. The findings are as follows:

First, the geographic patterns of industrial carbon emissions, urban industrial land, and carbon emission density of industrial land did not agree with their spatio-temporal evolution. The management of industrial land under the goal of "carbon neutrality and carbon peaking" is incomplete if it is only concerned with the total amount of carbon emissions, and it is necessary to incorporate the carbon emission intensity of industrial land into the control system as soon as possible. Second, the carbon emission density of urban industrial land showed significant spatial heterogeneity, with large differences between different cities. This suggests that the design of carbon emission intensity control systems for industrial land use needs to be city specific, with differentiated management policies for different zonings. Third, the carbon emission intensity of urban industrial land showed significant positive spatial autocorrelation. This demonstrates the need to strengthen intercity cooperation in carbon intensity management, with a focus on pushing spatial synergy in the upper, middle, and lower reaches of the YREB, especially in urban clusters, metropolitan areas, and interprovincial border areas. Fourth, the geographical pattern and spatio-temporal evolution model of urban industrial land carbon emission density had a very complex driving mechanism, with prominent features of diversification and ranking of influencing factors, and significant interaction effects among multiple factors. Therefore, in the scheme design of the control system, the influence of key factors should be emphasized, and attention should be paid to the coordination of multiple factors to realize the "organic combination of policies".

It should be noted that due to the difficulty of data quantification and the limitation of data availability, we neglected the empirical analysis of single-city cases at the micro-scale in this study when analyzing the geographic pattern, spatial, and temporal evolutionary patterns of carbon emission intensity of the YREB's multi-city industrial land at the macro-scale, as well as its spatial effects and driving mechanisms. Additionally, the dynamics of carbon reduction capacity are closely related to national, local, and urban strategies, but they have not been empirically analyzed due to the lack of suitable quantitative indicators.

Author Contributions: Conceptualization, Min Lai and Fei Xie; methodology, Min Lai, Sidong Zhao and Qipeng Zhang; software, Shuaibing Zhang and Qipeng Zhang; validation, Min Lai, Shuaibing Zhang and Qipeng Zhang; formal analysis, Min Lai, Sidong Zhao and Fei Xie; investigation, Fei Xie, Shuaibing Zhang and Qipeng Zhang; resources, Min Lai and Fei Xie; data curation, Shuaibing Zhang and Qipeng Zhang; writing—original draft preparation, Fei Xie, Shuaibing Zhang and Qipeng Zhang; writing—review and editing, Min Lai and Sidong Zhao; visualization, Shuaibing Zhang and Qipeng Zhang; supervision, Min Lai; project administration, Fei Xie; funding acquisition, Min Lai and Fei Xie. All authors have read and agreed to the published version of the manuscript.

Funding: This research was funded by the Guangdong Provincial Philosophy and Social Science Planning Project (GD24CYJ51) and the Open Research Project of the Research Center for University-City Integration Development (2024KF006). This work was funded by a new round of first-class disciplines (statistics) of Guangxi.

Data Availability Statement: The land data is sourced from the Statistical Yearbook of Urban Construction published by the Ministry of Housing and Urban Rural Development of the People's Republic of China, while the carbon emission data is sourced from the China City Greenhouse Gas Working Group's dataset on carbon dioxide emissions from Chinese cities (http://www.cityghg.com/toCauses?id=4, 2 November 2023), the remaining data is sourced from the China Urban Statistical Yearbook published by the National Bureau of Statistics.

Acknowledgments: Thank you to Kaixu Zhao, Weiwei Li and all the reviewers for their contributions and assistance in the process of improving the paper.

Conflicts of Interest: The authors declare no conflicts of interest.

Appendix A. Normalized Data on Industrial Land, Carbon Emissions, and Its Density

	Urban Industrial Land		Industrial Carbon Emissions		Carbon Emission Density	
	2010	2020	2010	2020	2010	2020
Shanghai	0.0703	0.0461	0.2126	0.1319	0.0012	0.0015
Nanjing	0.0295	0.0314	0.0465	0.0310	0.0022	0.0042
Wuxi	0.0298	0.0256	0.0163	0.0176	0.0064	0.0061
Xuzhou	0.0290	0.0288	0.0101	0.0123	0.0101	0.0097
Changzhou	0.0134	0.0162	0.0118	0.0221	0.0040	0.0030
Suzhou-JS	0.0514	0.0491	0.0451	0.0305	0.0040	0.0067
Nantong	0.0145	0.0124	0.0162	0.0179	0.0031	0.0029
Lianyungang	0.0044	0.0109	0.0098	0.0128	0.0016	0.0036
Huai'an	0.0073	0.0073	0.0133	0.0086	0.0019	0.0035
Yancheng	0.0039	0.0096	0.0083	0.0093	0.0016	0.0043
Yangzhou	0.0098	0.0124	0.0080	0.0097	0.0043	0.0053
Zhenjiang	0.0125	0.0188	0.0103	0.0131	0.0043	0.0060
Taizhou-JS	0.0080	0.0114	0.0082	0.0084	0.0034	0.0056
Suqian	0.0023	0.0021	0.0047	0.0075	0.0017	0.0012
Hangzhou	0.0152	0.0108	0.0183	0.0303	0.0029	0.0015
Ningbo	0.0360	0.0265	0.0358	0.0326	0.0035	0.0034
Wenzhou	0.0080	0.0086	0.0100	0.0019	0.0028	0.0191
Jiaxing	0.0115	0.0127	0.0079	0.0085	0.0051	0.0062
Huzhou	0.0091	0.0105	0.0101	0.0078	0.0032	0.0056
Shaoxing	0.0075	0.0087	0.0080	0.0164	0.0033	0.0022
Jinhua	0.0104	0.0074	0.0046	0.0062	0.0079	0.0050
Quzhou	0.0099	0.0102	0.0053	0.0071	0.0065	0.0060
Zhoushan	0.0015	0.0046	0.0021	0.0015	0.0024	0.0123
Taizhou-ZJ	0.0094	0.0093	0.0133	0.0072	0.0025	0.0053
Lishui	0.0012	0.0007	0.0010	0.0011	0.0043	0.0027
Hefei	0.0055	0.0116	0.0189	0.0214	0.0010	0.0023
Wuhu	0.0102	0.0196	0.0080	0.0056	0.0045	0.0146
Bengbu	0.0033	0.0042	0.0061	0.0065	0.0019	0.0027
Huainan	0.0168	0.0176	0.0045	0.0046	0.0132	0.0159
Maanshan	0.0134	0.0204	0.0092	0.0088	0.0051	0.0096
Huaibei	0.0179	0.0098	0.0055	0.0048	0.0115	0.0085
Tongling	0.0086	0.0162	0.0027	0.0059	0.0113	0.0114
Anqing	0.0077	0.0070	0.0069	0.0076	0.0039	0.0038
Huangshan	0.0002	0.0001	0.0017	0.0026	0.0005	0.0001
Chuzhou	0.0031	0.0070	0.0056	0.0074	0.0019	0.0039
Fuyang	0.0036	0.0029	0.0026	0.0051	0.0049	0.0024
Suzhou-AH	0.0030	0.0054	0.0038	0.0040	0.0028	0.0057
Lu'an	0.0010	0.0015	0.0035	0.0033	0.0010	0.0019
Bozhou	0.0010	0.0024	0.0023	0.0036	0.0015	0.0027
Chizhou	0.0045	0.0073	0.0015	0.0015	0.0104	0.0198
Xuancheng	0.0069	0.0074	0.0030	0.0044	0.0081	0.0070
Nanchang	0.0038	0.0048	0.0108	0.0150	0.0012	0.0013
Jingdezhen	0.0033	0.0051	0.0054	0.0063	0.0021	0.0034
Pingxiang	0.0063	0.0066	0.0022	0.0021	0.0101	0.0128
Jiujiang	0.0081	0.0152	0.0067	0.0076	0.0042	0.0083
Xinyu	0.0096	0.0084	0.0032	0.0022	0.0106	0.0158
Yingtan	0.0018	0.0026	0.0008	0.0024	0.0078	0.0046
Ganzhou	0.0039	0.0076	0.0043	0.0084	0.0032	0.0038
Ji'an	0.0038	0.0038	0.0021	0.0038	0.0062	0.0042
Yichun	0.0080	0.0107	0.0020	0.0031	0.0138	0.0145
Fuzhou	0.0006	0.0040	0.0019	0.0057	0.0010	0.0030
Shangrao	0.0059	0.0073	0.0016	0.0015	0.0127	0.0199
Wuhan	0.0295	0.0192	0.0445	0.0535	0.0023	0.0015
Huangshi	0.0136	0.0121	0.0056	0.0060	0.0086	0.0084
Shiyan	0.0028	0.0029	0.0052	0.0083	0.0019	0.0014
Yichang	0.0092	0.0093	0.0091	0.0097	0.0035	0.0040
Xiangyang	0.0071	0.0068	0.0077	0.0128	0.0032	0.0022
Ezhou	0.0081	0.0073	0.0037	0.0009	0.0077	0.0320
Jingmen	0.0076	0.0077	0.0036	0.0031	0.0076	0.0103
Xiaogan	0.0056	0.0057	0.0018	0.0035	0.0112	0.0068
Jingzhou	0.0029	0.0035	0.0047	0.0054	0.0022	0.0027
Huanggang	0.0061	0.0039	0.0016	0.0015	0.0137	0.0105
Xianning	0.0032	0.0065	0.0037	0.0036	0.0030	0.0076

Suizhou	0.0004	0.0004	0.0026	0.0054	0.0005	0.0003
Changsha	0.0072	0.0047	0.0087	0.0057	0.0029	0.0034
Zhuzhou	0.0069	0.0050	0.0068	0.0088	0.0036	0.0024
Xiangtan	0.0066	0.0126	0.0089	0.0078	0.0026	0.0067
Hengyang	0.0122	0.0054	0.0062	0.0073	0.0069	0.0031
Shaoyang	0.0041	0.0045	0.0010	0.0012	0.0148	0.0157
Yueyang	0.0064	0.0081	0.0049	0.0049	0.0045	0.0068
Changde	0.0052	0.0073	0.0054	0.0045	0.0034	0.0068
Zhangjiajie	0.0003	0.0006	0.0003	0.0007	0.0040	0.0033
Yiyang	0.0029	0.0041	0.0037	0.0023	0.0027	0.0073
Chenzhou	0.0101	0.0055	0.0073	0.0017	0.0049	0.0134
Yongzhou	0.0019	0.0040	0.0017	0.0020	0.0038	0.0084
Huaihua	0.0030	0.0016	0.0014	0.0005	0.0073	0.0125
Loudi	0.0133	0.0175	0.0029	0.0029	0.0164	0.0246
Chongqing	0.0504	0.0462	0.0578	0.0734	0.0031	0.0026
Chengdu	0.0088	0.0069	0.0259	0.0347	0.0012	0.0008
Zigong	0.0032	0.0014	0.0059	0.0072	0.0019	0.0008
Panzhihua	0.0095	0.0088	0.0070	0.0058	0.0048	0.0063
Luzhou	0.0056	0.0034	0.0041	0.0068	0.0049	0.0021
Deyang	0.0044	0.0024	0.0046	0.0061	0.0034	0.0016
Mianyang	0.0058	0.0043	0.0078	0.0079	0.0026	0.0023
Guangyuan	0.0020	0.0028	0.0017	0.0029	0.0041	0.0041
Suining	0.0005	0.0006	0.0022	0.0033	0.0008	0.0007
Neijiang	0.0076	0.0069	0.0020	0.0042	0.0134	0.0069
Leshan	0.0095	0.0107	0.0031	0.0022	0.0109	0.0201
Nanchong	0.0007	0.0006	0.0033	0.0058	0.0007	0.0005
Meishan	0.0034	0.0023	0.0016	0.0018	0.0075	0.0053
Yibin	0.0060	0.0060	0.0010	0.0080	0.0215	0.0031
Guang'an	0.0061	0.0060	0.0016	0.0030	0.0130	0.0084
Dazhou	0.0098	0.0071	0.0033	0.0014	0.0105	0.0206
Ya'an	0.0009	0.0014	0.0008	0.0023	0.0039	0.0026
Bazhong	0.0004	0.0012	0.0002	0.0004	0.0072	0.0113
Ziyang	0.0008	0.0005	0.0027	0.0008	0.0011	0.0026
Guiyang	0.0073	0.0066	0.0093	0.0147	0.0028	0.0018
Liupanshui	0.0216	0.0210	0.0043	0.0010	0.0176	0.0911
Zunyi	0.0074	0.0092	0.0023	0.0061	0.0113	0.0063
Anshun	0.0033	0.0040	0.0014	0.0028	0.0079	0.0059
Bijie	0.0250	0.0169	0.0020	0.0030	0.0450	0.0237
Tongren	0.0031	0.0039	0.0000	0.0019	0.2692	0.0087
Kunming	0.0156	0.0060	0.0211	0.0140	0.0026	0.0018
Qujing	0.0272	0.0169	0.0026	0.0024	0.0362	0.0295
Yuxi	0.0049	0.0090	0.0006	0.0011	0.0301	0.0332
Baoshan	0.0007	0.0022	0.0004	0.0005	0.0073	0.0199
Zhaotong	0.0013	0.0046	0.0006	0.0013	0.0084	0.0153
Lijiang	0.0018	0.0011	0.0003	0.0005	0.0226	0.0088
Pu'er	0.0011	0.0028	0.0007	0.0001	0.0059	0.0958
Lincang	0.0007	0.0018	0.0001	0.0004	0.0225	0.0171

References

1. Zhou, S.Y.; Li, W.; Lu, Z.G.; Lu, Z.Y. A technical framework for integrating carbon emission peaking factors into the industrial green transformation planning of a city cluster in China. *J. Clean. Prod.* **2022**, *344*, 131091. [CrossRef]
2. Coluzzi, R.; Bianchini, L.; Egidi, G.; Cudlin, P.; Imbrenda, V.; Salvati, L.; Lanfredi, M. Density matters? Settlement expansion and land degradation in Peri-urban and rural districts of Italy. *Environ. Impact Assess. Rev.* **2022**, *92*, 106703. [CrossRef]
3. Grosvenor, M.; O'Neill, P. The Density Debate in Urban Research: An Alternative Approach to Representing Urban Structure and Form. *Geogr. Res.* **2014**, *52*, 442–458. [CrossRef]
4. Claassens, J.; Koomen, E.; Rouwendal, J. Urban density and spatial planning: The unforeseen impacts of Dutch devolution. *PLoS ONE* **2020**, *15*, e0240738. [CrossRef] [PubMed]
5. Yeh, A.G.O.; Li, X. A cellular automata model to simulate development density for urban planning. *Environ. Plan. B-Plan. Des.* **2002**, *29*, 431–450. [CrossRef]
6. Shen, X.Z.; Huang, L.; Zhu, J.; Gao, J.J. Relationship between land use carbon emission and economic growth based on GIS. *Arab. J. Geosci.* **2021**, *14*, 457. [CrossRef]
7. Robinson, D.T.; Zhang, J.X.; MacDonald, D.; Samson, C. Estimating Settlement carbon stock and density using an inventory approach and quantifying their variation by land use and parcel size. *Urban For. Urban Green.* **2023**, *82*, 127878. [CrossRef]
8. Grassi, G.; Conchedda, G.; Federici, S.; Vinas, R.A.; Korosuo, A.; Melo, J.; Rossi, S.; Sandker, M.; Somogyi, Z.; Vizzarri, M.; et al. Carbon fluxes from land 2000–2020: Bringing clarity to countries' reporting. *Earth Syst. Sci. Data* **2022**, *14*, 4643–4666. [CrossRef]

9. Rehman, E.; Rehman, S. Modeling the nexus between carbon emissions, urbanization, population growth, energy consumption, and economic development in Asia: Evidence from grey relational analysis. *Energy Rep.* **2022**, *8*, 5430–5442. [CrossRef]
10. Huang, W.; Li, J.X. The Coupling Relationship Between Urbanization and Carbon Emissions from Land Use in Ningxia. *Front. Environ. Sci.* **2022**, *10*, 927798. [CrossRef]
11. Li, M.C.; Liu, H.M.; Yu, S.K.; Wang, J.S.; Miao, Y.; Wang, C.X. Estimating the Decoupling between Net Carbon Emissions and Construction Land and Its Driving Factors: Evidence from Shandong Province, China. *Int. J. Environ. Res. Public Health* **2022**, *19*, 8910. [CrossRef]
12. Sidorczuk-Pietraszko, E. Spatial Differences in Carbon Intensity in Polish Households. *Energies* **2020**, *13*, 3108. [CrossRef]
13. Rong, T.Q.; Zhang, P.Y.; Zhu, H.R.; Jiang, L.; Li, Y.Y.; Liu, Z.Y. Spatial correlation evolution and prediction scenario of land use carbon emissions in China. *Ecol. Inform.* **2022**, *71*, 101802. [CrossRef]
14. Sun, Y.P.; Kamran, H.W.; Razzaq, A.; Qadri, F.S.; Suksatan, W. Dynamic and causality linkages from transportation services and tourism development to economic growth and carbon emissions: New insights from Quantile ARDL approach. *Integr. Environ. Assess. Manag.* **2022**, *18*, 1272–1287. [CrossRef]
15. Russo, F.; Rindone, C.; Panuccio, P. External interactions for a third-generation port: From urban sustainable planning to research developments. *Int. J. Transp. Dev. Integr.* **2022**, *3*, 253–270. Available online: https://www.iieta.org/journals/ijtdi/paper/10.2495/EI-V6-N3-253-270 (accessed on 15 May 2024).
16. Xiong, G.B.; Deng, J.H.; Ding, B.G. Characteristics, decoupling effect, and driving factors of regional tourism's carbon emissions in China. *Environ. Sci. Pollut. Res.* **2022**, *29*, 47082–47093. [CrossRef]
17. Tang, Y.; Chen, M.H. Impact Mechanism and Effect of Agricultural Land Transfer on Agricultural Carbon Emissions in China: Evidence from Mediating Effect Test and Panel Threshold Regression Model. *Sustainability* **2022**, *14*, 13014. [CrossRef]
18. Wang, X.X.; He, A.Z.; Zhao, J. Regional disparity and dynamic evolution of carbon emission reduction maturity in China's service industry. *J. Clean. Prod.* **2020**, *244*, 118926. [CrossRef]
19. Zhang, P.; Hu, J.; Zhao, K.; Chen, H.; Zhao, S.; Li, W. Dynamics and Decoupling Analysis of Carbon Emissions from Construction Industry in China. *Buildings* **2022**, *12*, 257. [CrossRef]
20. Kara, F.; Bas, T.; Talu, N.H.T.; Alola, A.A. Investigating the carbon emission aspects of agricultural land utilization in Turkey. *Integr. Environ. Assess. Manag.* **2021**, *18*, 988–996. [CrossRef] [PubMed]
21. Rasheed, R.; Tahir, F.; Afzaal, M.; Ahmad, S.R. Decomposition analytics of carbon emissions by cement manufacturing—A way forward towards carbon neutrality in a developing country. *Environ. Sci. Pollut. Res.* **2022**, *29*, 49429–49438. [CrossRef]
22. Li, M.; Li, Q.; Wang, Y.A.; Chen, W. Spatial path and determinants of carbon transfer in the process of inter industrial transfer in China. *Environ. Impact Assess. Rev.* **2022**, *95*, 106810. [CrossRef]
23. Malik, A.; Lan, J. The role of outsourcing in driving global carbon emissions. *Econ. Syst. Res.* **2016**, *28*, 168–182. [CrossRef]
24. Liu, X.; Zhang, Y.C.; Wang, Q.W. Drivers of global carbon emission changes: A heterogeneity perspective of decomposition and attribution analysis. *Front. Environ. Sci.* **2023**, *10*, 1062500. [CrossRef]
25. Xu, D.X.; Zhang, Y.; Li, Y.G.; Wang, X.J.; Yang, Z.F. Path analysis for carbon transfers embodied in China's international trade and policy implications for mitigation targets. *J. Clean. Prod.* **2022**, *334*, 130207. [CrossRef]
26. Tikoudis, I.; Farrow, K.; Mebiame, R.M.; Oueslati, W. Beyond average population density: Measuring sprawl with density-allocation indicators. *Land Use Policy* **2022**, *112*, 105832. [CrossRef]
27. Cetin, M.; Adiguzel, F.; Gungor, S.; Kaya, E.; Sancar, M.C. Evaluation of thermal climatic region areas in terms of building density in urban management and planning for Burdur, Turkey. *Air Qual. Atmos. Health* **2019**, *12*, 1103–1112. [CrossRef]
28. Henderson, J.V.; Nigmatulina, D.; Kriticos, S. Measuring urban economic density. *J. Urban Econ.* **2021**, *125*, 103188. [CrossRef]
29. Perumal, A.; Timmons, D. Contextual Density and US Automotive CO_2 Emissions across the Rural-Urban Continuum. *Int. Reg. Sci. Rev.* **2017**, *40*, 590–615. [CrossRef]
30. Moghadam, A.S.; Soltani, A.; Parolin, B.; Alidadi, M. Analysing the space-time dynamics of urban structure change using employment density and distribution data. *Cities* **2018**, *81*, 203–213. [CrossRef]
31. Keeratikasikorn, C. A comparative study on four major cities in Northeastern Thailand using urban land density function. *Geo-Spat. Inf. Sci.* **2018**, *21*, 93–101. [CrossRef]
32. Cowie, C.T.; Ding, D.; Rolfe, M.I.; Mayne, D.J.; Jalaludin, B.; Bauman, A.; Morgan, G.G. Neighbourhood walkability, road density and socio-economic status in Sydney, Australia. *Environ. Health* **2016**, *15*, 58. [CrossRef]
33. Zhang, P.; Kohli, D.; Sun, Q.Q.; Zhang, Y.X.; Liu, S.X.; Sun, D.F. Remote sensing modeling of urban density dynamics across 36 major cities in China: Fresh insights from hierarchical urbanized space. *Landsc. Urban Plan.* **2020**, *203*, 103896. [CrossRef]
34. Li, J.G.; Li, J.W.; Yuan, Y.Z.; Li, G.F. Spatiotemporal distribution characteristics and mechanism analysis of urban population density: A case of Xi'an, Shaanxi, China. *Cities* **2019**, *86*, 62–70. [CrossRef]
35. Hong, S.F.; Hui, E.C.M.; Lin, Y.Y. Relationships between carbon emissions and urban population size and density, based on geo-urban scaling analysis: A multi-carbon source empirical study. *Urban Clim.* **2022**, *46*, 101337. [CrossRef]
36. Lee, S.; Lee, B. The influence of urban form on GHG emissions in the US household sector. *Energy Policy* **2014**, *68*, 534–549. [CrossRef]
37. Li, C.S.; Song, Y.; Kaza, N. Urban form and household electricity consumption: A multilevel study. *Energy Build.* **2018**, *158*, 181–193. [CrossRef]

38. Zhang, Z.; Xiao, Y.X.; Luo, X.; Zhou, M. Urban human activity density spatiotemporal variations and the relationship with geographical factors: An exploratory Baidu heatmaps-based analysis of Wuhan, China. *Growth Chang.* **2020**, *51*, 505–529. [CrossRef]
39. Russo, F.; Rindone, C. Smart City for Sustainable Development: Applied Processes from SUMP to MaaS at European Level. *Appl. Sci.* **2023**, *13*, 1773. [CrossRef]
40. Wu, H.Y.; Yan, W.Q.; Zheng, X.J.; Zhou, L.; Ma, J.S.; Liu, L.; Meng, Y. Carbon-Emission Density of Crop Production in China: Spatiotemporal Characteristics, Regional Disparities, and Convergence. *Agriculture* **2023**, *13*, 951. [CrossRef]
41. Dong, F.; Li, J.Y.; Zhang, Y.J.; Wang, Y. Drivers Analysis of CO_2 Emissions from the Perspective of Carbon Density: The Case of Shandong Province, China. *Int. J. Environ. Res. Public Health* **2018**, *15*, 1762. [CrossRef]
42. Kaya, M.C.; Persson, L. A Theory of Gazelle Growth: Competition, Venture Capital Finance and Policy. *N. Am. J. Econ. Financ.* **2020**, *50*, 101019. [CrossRef]
43. Zhang, P.; Chen, H.; Zhao, K.; Zhao, S.; Li, W. Dynamics, Risk and Management Performance of Urban Real Estate Inventory in Yangtze River Delta. *Buildings* **2022**, *12*, 2140. [CrossRef]
44. Zhao, S.; Zhao, K.; Yan, Y.; Zhu, K.; Guan, C. Spatio-Temporal Evolution Characteristics and Influencing Factors of Urban Service-Industry Land in China. *Land* **2022**, *11*, 13. [CrossRef]
45. Li, W.; Zhang, P.; Zhao, K.; Chen, H.; Zhao, S. The Evolution Model of and Factors Influencing Digital Villages: Evidence from Guangxi, China. *Agriculture* **2023**, *13*, 659. [CrossRef]
46. Ren, J.M.; Zheng, C.R.Y.; Guo, F.Y.; Zhao, H.B.; Ma, S.; Cheng, Y. Spatial Differentiation of Digital Rural Development and Influencing Factors in the Yellow River Basin, China. *Int. J. Environ. Res. Public Health* **2022**, *19*, 16111. [CrossRef]
47. Ma, Y.; Zhang, P.; Zhao, K.; Zhou, Y.; Zhao, S. A Dynamic Performance and Differentiation Management Policy for Urban Construction Land Use Change in Gansu, China. *Land* **2022**, *11*, 942. [CrossRef]
48. Zhao, S.; Zhang, C.; Qi, J. The Key Factors Driving the Development of New Towns by Mother Cities and Regions: Evidence from China. *ISPRS Int. J. Geo-Inf.* **2021**, *10*, 223. [CrossRef]
49. Zhao, S.; Li, W.; Zhao, K.; Zhang, P. Change Characteristics and Multilevel Influencing Factors of Real Estate Inventory—Case Studies from 35 Key Cities in China. *Land* **2021**, *10*, 928. [CrossRef]
50. Wang, J.F.; Hu, Y. Environmental health risk detection with GeogDetector. *Environ. Model. Softw.* **2012**, *33*, 114–115. [CrossRef]
51. Li, L.; Zhao, K.; Wang, X.; Zhao, S.; Liu, X.; Li, W. Spatio-Temporal Evolution and Driving Mechanism of Urbanization in Small Cities: Case Study from Guangxi. *Land* **2022**, *11*, 415. [CrossRef]
52. Zhao, S.; Yan, Y.; Han, J. Industrial Land Change in Chinese Silk Road Cities and Its Influence on Environments. *Land* **2021**, *10*, 806. [CrossRef]
53. Chen, H.; Zhao, S.; Zhang, P.; Zhou, Y.; Li, K. Dynamics and Driving Mechanism of Real Estate in China's Small Cities: A Case Study of Gansu Province. *Buildings* **2022**, *12*, 1512. [CrossRef]
54. Yang, X.; Li, W.; Zhang, P.; Chen, H.; Lai, M.; Zhao, S. The Dynamics and Driving Mechanisms of Rural Revitalization in Western China. *Agriculture* **2023**, *13*, 1448. [CrossRef]
55. Dehdar, F.; Silva, N.; Fuinhas, J.A.; Koengkan, M.; Nazeer, N. The Impact of Technology and Government Policies on OECD Carbon Dioxide Emissions. *Energies* **2022**, *15*, 8486. [CrossRef]
56. Li, W.; Wang, M.; Zhao, S. The Spatio-Temporal Dynamics, Driving Mechanism, and Management Strategies for International Students in China under the Background of the Belt and Road Initiatives. *ISPRS Int. J. Geo-Inf.* **2023**, *12*, 405. [CrossRef]
57. Liang, S.; Zhao, J.F.; He, S.M.; Xu, Q.Q.; Ma, X. Spatial econometric analysis of carbon emission intensity in Chinese provinces from the perspective of innovation-driven. *Environ. Sci. Pollut. Res.* **2019**, *26*, 13878–13895. [CrossRef]
58. Li, C.S.; Song, Y.; Kaza, N.; Burghardt, R. Explaining Spatial Variations in Residential Energy Usage Intensity in Chicago: The Role of Urban Form and Geomorphometry. *J. Plan. Educ. Res.* **2019**, *43*, 317–331. [CrossRef]
59. Zhang, P.; Li, W.; Zhao, K.; Zhao, S. Spatial Pattern and Driving Mechanism of Urban–Rural Income Gap in Gansu Province of China. *Land* **2021**, *10*, 1002. Available online: https://www.mdpi.com/2073-445X/10/10/1002 (accessed on 15 May 2024).
60. Whittemore, A.H.; BenDor, T.K. Rhetorical Framing in Planning: An Empirical Investigation of How Planners Discuss Density. *J. Plan. Educ. Res.* **2021**, *41*, 48–61. [CrossRef]
61. Hooper, M. Flatulence, Filth, and Urban Form: Do Primes for Hygiene Influence Perceptions of Urban Density? *J. Plan. Educ. Res.* **2022**, *42*, 185–198. [CrossRef]
62. Herdt, T.; Jonkman, A.R. The acceptance of density: Conflicts of public and private interests in public debate on urban densification. *Cities* **2023**, *140*, 104451. [CrossRef]
63. Moos, M. The Velocity of Density: Can We Build More Sustainable Cities Fast Enough? *Sustainability* **2017**, *9*, 2326. [CrossRef]
64. Schorcht, M.; Jehling, M.; Kruger, T. Where are cities under pressure?—An indicator for measuring the impact of building changes on urban density. *Ecol. Indic.* **2023**, *149*, 110142. [CrossRef]
65. Tabasi, S.F.; Rafizadeh, H.R.; Garmaroudi, A.A.; Banihashemi, S. Optimizing urban layouts through computational generative design: Density distribution and shape optimization. *Archit. Eng. Des. Manag.* **2023**. [CrossRef]
66. Fan, F.; Dai, S.Z.; Yang, B.; Ke, H.Q. Urban density, directed technological change, and carbon intensity: An empirical study based on Chinese cities. *Technol. Soc.* **2023**, *72*, 102151. [CrossRef]
67. Hong, A.Y.; Chakrabarti, S. Compact living or policy inaction? Effects of urban density and lockdown on the Covid-19 outbreak in the US. *Urban Stud.* **2022**, *60*, 1588–1609. [CrossRef]

68. Baics, G.; Meisterlin, L. Zoning Before Zoning: Land Use and Density in Mid-Nineteenth-Century New York City. *Ann. Am. Assoc. Geogr.* **2016**, *106*, 1152–1175. [CrossRef]
69. Cirianni, F.; Panuccio, P.; Rindone, C. A Comparison of Urban Planning Systems Between the UK And Italy: Commercial Development and City Logistic Plan. *Urban Transp. XIX* **2013**, *130*, 785–797. [CrossRef]
70. Ma, M.Y.; Rozema, J.; Gianoli, A.; Zhang, W.S. The Impacts of City Size and Density on CO_2 Emissions: Evidence from the Yangtze River Delta Urban Agglomeration. *Appl. Spat. Anal. Policy* **2022**, *15*, 529–555. [CrossRef]
71. Yi, Y.C.; Wang, Y.J.; Li, Y.Q.; Qi, J. Impact of urban density on carbon emissions in China. *Appl. Econ.* **2021**, *53*, 6153–6165. [CrossRef]
72. Norman, J.; MacLean, H.L.; Kennedy, C.A. Comparing high and low residential density: Life-cycle analysis of energy use and greenhouse gas emissions. *J. Urban Plan. Dev.* **2006**, *132*, 10–21. [CrossRef]
73. Ishii, S.; Tabushi, S.; Aramaki, T.; Hanaki, K. Impact of future urban form on the potential to reduce greenhouse gas emissions from residential, commercial and public buildings in Utsunomiya, Japan. *Energy Policy* **2010**, *38*, 4888–4896. [CrossRef]
74. Derudder, B.; Feng, X.; Shen, W.; Shao, R.; Taylor, P.J. Connections between Asian and European World Cities: Measurement, Analysis, and Evaluation. *Land* **2022**, *11*, 1574. [CrossRef]
75. Wang, Q.Z.; Zhang, Q. Foreign Direct Investment and Carbon Emission Efficiency: The Role of Direct and Indirect Channels. *Sustainability* **2022**, *14*, 13484. [CrossRef]
76. Wang, M.Q.; Zhang, L.Y.; Su, X.; Lei, Y.; Shen, Q.; Wei, W.; Wang, M.H. Assessing the technology impact for industry carbon density reduction in China based on C (3) IAM-Tice. *Nat. Hazards* **2019**, *99*, 1455–1468. [CrossRef]
77. Ghadami, M.; Dittmann, A.; Safarrad, T. Lack of Spatial Approach in Urban Density Policies: The Case of the Master Plan of Tehran. *Sustainability* **2020**, *12*, 7285. [CrossRef]
78. Chokhachian, A.; Perini, K.; Giulini, S.; Auer, T. Urban performance and density: Generative study on interdependencies of urban form and environmental measures. *Sustain. Cities Soc.* **2020**, *53*, 101952. [CrossRef]

International Journal of
Geo-Information

MDPI

Article

Spatial and Temporal Variation of GPP and Its Response to Urban Environmental Changes in Beijing

Le Chen [1,2,†], Simin Yu [1,2,†], Shi Shen [1,2,*], You Wan [3] and Changqing Song [1,2]

1 State Key Laboratory of Earth Surface Processes and Resource Ecology, Beijing Normal University, Beijing 100875, China; 202321051079@mail.bnu.edu.cn (L.C.); 202221051126@mail.bnu.edu.cn (S.Y.); songcq@bnu.edu.cn (C.S.)

2 Faculty of Geographical Science, Beijing Normal University, Beijing 100875, China

3 School of Resource and Environmental Sciences, Wuhan University, Wuhan 430079, China; wanyou@whu.edu.cn

* Correspondence: shens@bnu.edu.cn

† These authors contributed equally to this work.

Abstract: The carbon sequestration capacity of vegetation is the key to the carbon cycle in terrestrial ecosystems. It is significant to analyze the spatiotemporal variation and influencing factors of vegetation carbon sequestration ability to improve territorial carbon sink and optimize its spatial pattern. However, there is a lack of understanding of the impact of environmental conditions and human activity on the vegetation's carbon sequestration ability, especially in highly urbanized areas. For example, effective vegetation management methods can enhance vegetation Gross Primary Productivity, while emissions of air pollutants like O_3, CO, NO_2, and PM2.5 can suppress it. This paper mainly explores the factors influencing vegetation carbon sequestration capacity across different regions of Beijing. Based on remote sensing data and site observation data, this paper analyzed the spatiotemporal variation trend of Annual Gross Primary Production (AGPP) and the influence of environmental factors and human activity factors on GPP in Beijing from 2000 to 2020 by using the Theil–Sen's slope estimator, Mann–Kendall trend test, and comparing Geographically Weighted Regression method (GWR) and Geographically and Temporally Weighted Regression method (GTWR). GWR is a localized multiple regression technique used to estimate variable relationships that vary spatially. GTWR extends GWR by adding temporal analysis, enabling a comprehensive examination of spatiotemporal data variations. Besides, we used land use cover data to discuss the influence of land use cover change on AGPP. The results showed that the spatial distribution pattern of GPP in Beijing was higher in the northwest and lower in the southeast, and it showed an overall upward trend from 2000 to 2020, with an average annual growth rate of 14.39 g $C{\cdot}m^{-2}{\cdot}a^{-1}$. From 2000 to 2020, excluding the core urban areas, the GPP of 95.8% of Beijing increased, and 10.6% of Beijing showed a trend of significant increase, concentrated in Mentougou, Changping, and Miyun. GPP decreased in 4.1% of the regions in Beijing and decreased significantly in 1.4% of the areas within the sixth ring. The areas where AGPP significantly decreased were concentrated in those where land use types were converted to Residential land (impervious land), while AGPP showed an upward trend in other areas. CO and NO_2 are the main driving forces of GPP change in Beijing. O_3 and land surface temperature (LST) also exert certain influences, while the impact of precipitation (PRE) is relatively minor. O_3 and CO have a positive impact on AGPP as a whole, while LST and NO_2 generally exhibit negative impacts. PRE has a positive impact in the central area of Beijing, while it has a negative impact in the peripheral areas. This study further discusses opinions on future urbanization and environmental management policies in Beijing, which will promote the carbon peak and carbon neutrality process of ecological space management in Beijing. Besides, this study was conducted at the urban scale rather than at ecological sites, encompassing a variety of factors that influence vegetation AGPP. Consequently, the results also offer fresh insights into the intricate nexus between human activities, pollutants, and the GPP of vegetation.

Keywords: urbanization; environmental factors; vegetation carbon sequestration; Beijing; gross primary productivity

1. Introduction

The carbon sequestration capacity of vegetation is a crucial component of the carbon cycle in terrestrial ecosystems. Investigating the drivers of GPP spatiotemporal variations and their underlying mechanisms benefits revealing the main forces of GPP variations, thus promoting ecosystem management aiming at increasing carbon sequestration of terrestrial ecosystems [1–3]. Gross primary production (GPP) is the total amount of organic carbon fixed by plants per unit of time through photosynthesis, which reflects the vegetation productivity status of terrestrial ecosystems. It is one of the key indicators for assessing the carbon sequestration capacity of vegetation and understanding the carbon cycle of the terrestrial ecosystem [4]. Therefore, quantifying GPP and its spatiotemporal variation is of priority in carbon cycle studies [5]. Research indicates that the spatiotemporal variation in vegetation's ability to capture carbon is influenced by multiple factors such as climatic conditions [6], vegetation cover [7], land use patterns [8], and soil properties [9].

The current studies on factors affecting GPP changes mainly focus on meteorological conditions and land use changes, etc. [10,11]. However, in urban areas, the impact of urbanization on GPP is complex and multifaceted. On the one hand, urban expansion leads to changes in regional land cover and a decrease in vegetation coverage, which may directly reduce the productivity of vegetation [12]. On the other hand, human activities brought about by urbanization also have indirect positive impacts [13], such as the urban heat island effect, increased air pollution in the atmosphere, and more effective vegetation management and planning [14], which collectively promote the photosynthesis and carbon fixation capacity of vegetation. These indirect effects can offset the direct negative effects in some cases and even promote the growth of vegetation productivity in certain areas [15]. It has also been found that the effect of air pollution on GPP cannot be ignored. For example, O_3, which has strong oxidizing properties, is absorbed by plants and damages their cell membrane tissues, thus reducing vegetation GPP [16,17], and aerosol particles can increase GPP by increasing scattered light [18], too much particulate matter can cause plant stomatal blockage instead of causing GPP reduction. For example, Yue & Unger [19], in their analysis of the effects of O_3 and aerosols on GPP during 2002–2011, found that O_3 led to a global GPP decrease of 0.91 ± 0.44 PgCyr-1 and aerosols led to a global GPP increase of 0.05 ± 0.30 PgCyr-1. Therefore, in this study, we used four types of air pollution, O_3, NO_2, CO, and PM2.5, as the influencing factors of human activities caused by urbanization. We used Land Surface Temperature (LST), Air Temperature (T), and Preparation (PRE) as environmental factors. With these factors considered, we can further our understanding of the responses of the carbon cycle to environmental change and human activities brought about by urbanization, especially for those fast-growing cities.

Beijing, as a highly urbanized city in China and a pilot area for sustainable and low-carbon development, has seen the majority of GPP research focus on ecologically sensitive areas such as the Mentougou district. This focus has led to a neglect of the spatial and temporal dynamics of GPP and the factors influencing it. When assessing the impact of environmental factors on GPP, it is imperative to consider the spatial dependence characteristic to prevent biased evaluation outcomes. The urban environment and its suburban counterpart interact through wind patterns, suggesting that GPP in suburban areas may be influenced by urban various factors [10,20,21]. The Geographically Weighted Regression (GWR) model proposed by Brunsdon et al. [22] is a local multiple regression technique used to estimate variable relationships that vary spatially and can effectively explore the impact of multiple factors on GPP. Furthermore, studies have indicated varying impacts of air pollutants and meteorological factors on vegetation growth at different stages of urbanization. The Geographically and Temporally Weighted Regression (GTWR) model introduced by Huang et al. [23] takes into account the temporal dimension to address spatiotemporal heterogeneity, which has considerable applicability in this problem. Although

GWR and GTWR have been widely used in various studies, they are less frequently utilized in examining the determinants of GPP.

In this study, we used Theil Sen's slope estimation, Mann Kendall (M-K) trend test, GWR, and GTWR methods to investigate the spatiotemporal changes of GPP in Beijing and explore the impact of environmental and human activity factors on GPP using AGPP data from 2000 to 2020 and meteorological and air pollutant data from 2014 to 2020. Additionally, we also employed land use/land cover (LULC) data to analyze the impact of LULC transitions on Annual Gross Primary Productivity (AGPP). Through this comprehensive approach, we aim to optimize the spatial structure and facilitate the city's goal of achieving carbon neutrality.

2. Data and Methods

2.1. Study Area

Beijing, located at 39°56′ N latitude and 116°20′ E longitude, extends geographically from 115°25′ E to 117°30′ E and from 39°26′ N to 41°03′ N. The city experiences a temperate monsoon climate, with hot, wet summers and cold, dry winters. Surrounded by the Taihang and Yanshan mountain ranges, Beijing is positioned near the Bohai Sea to the east and the Yellow River to the south. This climate and location support a rich plant life and contribute to the city's agricultural potential. Beijing, extending over 16,410.54 square kilometers and comprising 16 districts, boasts a varied landscape that includes both mountainous regions and plains. The mountainous areas, which constitute around 10,200 square kilometers or 62% of the city's total area, and the plains, spanning 6200 square kilometers, together create Beijing's rich geographical diversity. The city's average elevation is 43.5 m; the plains typically range from 20 to 60 m in elevation, while the mountains peak over 1000 m. In the context of this study, areas above 500 m are classified as mountains, and those below are classified as plains. The plains of Beijing are mainly used for farming and urban development, whereas the mountainous areas are largely covered by forests and grasslands. Over the past twenty years, Beijing has seen a significant increase in urbanization, with the rate increasing from 77.52% in 2000 to 86.6% in 2020, and it is expected to reach 99.46% by 2030 [24].

2.2. Data Sets

In this paper, we choose the dataset of annual gross primary productivity in China's terrestrial ecosystems during 2000–2020, published by Fan et al. [25]. This dataset was constructed based on the annual GPP observation dataset obtained from ChinaFLUX [26] long-term observation data and publicly available data, and an annual GPP(AGPP) assessment model constructed by simulating annual GPP per unit leaf area using a random forest regression tree model, thereby accurately depicting the geographical and temporal fluctuations in GPP across China. We utilized the CNLUCC dataset, a high-precision national-scale multi-temporal land use/land cover database derived from manual interpretation of Landsat imagery, as our source for surface coverage data. Additionally, we incorporated Land Surface Temperature (LST) data, monthly precipitation (PRE) and temperature records (T), and Chinese air pollutant data. Table 1 delineates the specifics of these data sources, including their spatial and temporal resolutions. Due to the restricted period for which some pollutant data are available, this paper only examines the effects of seven environmental factors on AGPP from 2014 to 2020.

The AGPP data and LST data underwent preliminary processing in the ArcGIS 10.2 software environment, through the process including band extraction, reprojection, splicing, cropping, invalid value removal, unit conversion, etc., to finally obtain the annual average data of GPP and surface temperature in Beijing. At the same time, annual data were obtained for temperature and precipitation data using the image element statistics tool, and annual air pollutant data were obtained for air pollutant data in Beijing using the mask extraction and raster calculator tools.

Table 1. Research data in this study.

Data	Spatial Resolution	Temporal Resolution	Period	Data Source
AGPP	500 m	Year	2000–2020	https://www.scidb.cn/en/detail?dataSetId=b496b208f51e44fcaf326e8b0f792c34, accessed on 9 February 2024
LST	1 km	8-day	2014–2020	https://ladsweb.modaps.eosdis.nasa.gov/, accessed on 9 February 2024
LUCC	30 m	5-year	2000–2020	https://www.resdc.cn/DOI/doi.aspx?DOIid=54, accessed on 7 March 2024
Air Temperature	1 km	Month	2014–2020	http://loess.geodata.cn, accessed on 11 February 2024
Precipitation	1 km	Month	2014–2020	http://loess.geodata.cn, accessed on 11 February 2024
$PM_{2.5}$	1 km	Year	2014–2020	https://weijing-rs.github.io/product.html, accessed on 11 February 2024
CO	10 km	Year	2014–2020	https://weijing-rs.github.io/product.html, accessed on 11 February 2024
NO_2	10 km	Year	2014–2020	https://weijing-rs.github.io/product.html, accessed on 11 February 2024
O_3	10 km	Year	2014–2020	https://weijing-rs.github.io/product.html, accessed on 11 February 2024

To resolve the discrepancies in data resolution within our study, we implemented a systematic approach to standardize the data. Initially, we utilized ArcGIS to resample the data for the seven influential factors to a 500-m resolution, aligning it with the resolution of the AGPP data. Next, we determined the dataset's minimum spatial resolution, finding it to be 10 km, and adopted this as our sampling unit. For each unit, we identified the pixel location with the median AGPP value and extracted the corresponding data for the seven influencing factors from that pixel. These values were then used as the representative dataset for each unit, enabling subsequent analysis and processing to be conducted at the unit level. This approach effectively addressed the issue of inconsistent data resolutions and minimized the potential distortions in GWR and GTWR analyses that can arise from resampling.

2.3. Methods

2.3.1. Theil–Sen's Slope Estimator

When a time series exhibits a linear trend, the actual slope can be determined through a non-parametric method [27]. The slope estimation of the data is calculated using the following formula:

$$\beta = \text{mean}\left(\frac{x_j - x_i}{j - i}\right), \forall j > i \quad (1)$$

where x_j and x_i represent data points at times j and i respectively, with ($j > i$). The median of the N values of Q_i constitutes a Sen's slope estimator. A positive β signifies an ascending trend in the time series data, while a negative β points to a descending trend. This method, grounded in non-parametric statistics, is a robust approach for trend calculation. It is known for its computational efficiency, resilience to measurement errors, and resistance to outliers, making it a preferred choice for analyzing trends in extensive time series datasets.

2.3.2. Mann–Kendall Trend Test

The Mann-Kendall trend test evaluates time-series data for the presence of monotonic ascends or descends in the Y-axis values [28]. In this research, we employ the M-K test to identify such trends within the AGPP data series. The Mann–Kendall test for a time series is supposed to compute as follows:

$$\operatorname{sgn}(x_i - x_j) = \begin{cases} 1, & x_i - x_j > 0 \\ 0, & x_i - x_j > 0 \\ -1, & x_i - x_j > 0 \end{cases} \tag{2}$$

$$S = \sum_{i=1}^{N-1} \sum_{j=i+1}^{N} \operatorname{sgn}(x_i - x_j) \tag{3}$$

where N represents the dataset's size, and x_i and x_j are the data values at times i and j. A negative S suggests a downward trend, while a positive S suggests an upward trend. The trend's statistical significance is evaluated using the variance of the MK statistic, applicable when $N > 10$, and is computed as follows:

$$\operatorname{Var}(s) = \frac{n(n-1)(2n+5) - \sum_{i=1}^{p} q_i(q_i - 1)(2q_i + 5)}{18} \tag{4}$$

where p indicates the number of concurrent groups within the dataset, and qi refers to the count of data points in the kth group. If the grouping variable is missing, the summation is excluded from the equation. The test statistic's standard Z-value of the test statistic can be calculated using the subsequent formula:

$$Z = \begin{cases} \frac{S-1}{\sqrt{\operatorname{var}(S)}}\ S > 0 \\ 0; S = 0 \\ \frac{S-1}{\sqrt{\operatorname{var}(S)}} S < 0 \end{cases} \tag{5}$$

If the Z-value surpasses the critical value of significance level (α), the trend is identified as statistically significant.

This approach is advantageous as it is not contingent on data adhering to a normal distribution, nor does it necessitate linearity in trends. It remains unaffected by missing values and outliers, which has led to its extensive application in assessing trend significance within long-term time series. For this study, we utilize the "pyMann-Kendall" Python package developed by Hussain et al. [29] to conduct the M-K trend test.

2.3.3. Mann–Kendall (M-K) Mutation Test

The advantage of this method is that it is not only simple to calculate but also can determine the time when the mutation starts and point out the mutation area [30].

Assuming random independence in the time series, the statistics are calculated as below:

$$U_{Fk} = \frac{S_k - E(S_k)}{\sqrt{\operatorname{var}(S_k)}} (k = 1, 2, \cdots, n) \tag{6}$$

where $E(S_k)$ is the mean of S_k and $\operatorname{var}(S_k)$ is the variance of S_k.

The above process is repeated according to the reverse order $x_n, x_{n-1}, \ldots, x_1$ of the time series x, and the statistical variable UB_k ($k = n, n-1, \ldots, 1$) is obtained. At the same time: $UB_k = -UF_k$. The UB_k and UF_k are plotted as curves. If the two curves intersect and the U value at the intersection is less than 1.96, that point is identified as the sequence's mutation point, and the test confidence level is $\alpha = 0.05$.

2.3.4. Geographically Weighted Regression Method

Geographically Weighted Regression enhances the traditional regression framework [22] by allowing the model's parameters to vary spatially rather than remaining constant across the entire dataset. So that the model can be expressed as

$$Y_i = \beta_0(u_i, v_i) + \sum_k \beta_k(u_i, v_i) X_{ik} + \varepsilon_i \qquad i = 1, \ldots, n \tag{7}$$

where (u_i, v_i) signifies the spatial coordinates of point i, $\beta_0(u_i, v_i)$ is the intercept value, and $\beta_k(u_i, v_i)$ refers to the set of parameter values at point i. In contrast to the global model, which employs constant coefficients across different regions, this model permits spatial variation in parameter estimates, thus potentially enabling a more accurate representation of regional-specific impacts.

In order to calibrate the model, data near point i are assumed to more significantly influence the estimation of parameters $\beta_k(u_i, v_i)$ than data that are more distant from the observation point i:

$$\hat{\beta}(u_i, v_i) = \left[X^T W(u_i, v_i) X\right]^{-1} X^T W(u_i, v_i) Y \tag{8}$$

where $W(u_i, v_i)$ is an $n \times n$ matrix with diagonal elements representing the geographic weights of the observation data for point i, while the off-diagonal elements are zero. This weight matrix is determined for every point i where parameters are calculated.

The weight matrix in Geographically Weighted Regression signifies the varying significance of each data observation for parameter estimation at location i, with closer observations typically carrying more weight. Consequently, each point i's estimate is associated with a distinct weight matrix.

$$W_{ij} = \exp\left(-\frac{d_{ij}^2}{h^2}\right) \tag{9}$$

$$d_{ij} = \sqrt{(x_i - x_j)^2 + (y_i - y_j)^2} \tag{10}$$

where h is a non-negative parameter called the bandwidth, which causes the influence to decay with distance. We utilized the Corrected Akaike information criterion [31] to identify its optimal value. d_{ij} is the distance between locations i and j, and (x_i, y_i) and (x_j, y_j) denote the coordinates of the respective points.

2.3.5. Geographically and Temporally Weighted Regression

The GTWR model offers a robust estimation approach by incorporating time and space data into its weight matrices, effectively capturing non-stationary spatial and temporal dynamics. This feature addresses the GWR model's limitation of handling only cross-sectional data [23]. Thus, the model can be formulated as:

$$Y_{\mathrm{i}} = \beta_0(u_{\mathrm{i}}, v_{\mathrm{i}}, t_{\mathrm{i}}) + \sum_{k=1}^{d} \beta_{\mathrm{k}}(u_{\mathrm{i}}, v_{\mathrm{i}}, t_{\mathrm{i}}) x_{\mathrm{ik}} + \varepsilon_{\mathrm{i}} \tag{11}$$

where i ranging from 1 to n. Y_{i} denotes an $\mathrm{n} \times 1$ vector of explanatory variables, β_0 is the constant coefficient, and $(u_{\mathrm{i}}, v_{\mathrm{i}}, t_{\mathrm{i}})$ specifies the geographic and temporal coordinates of the observation. $\beta_{\mathrm{k}}(u_{\mathrm{i}}, v_{\mathrm{i}}, t_{\mathrm{i}})$ is the unknown parameter for the kth factor at these coordinates, with x_{ik} representing the $\mathrm{n} \times \mathrm{k}$ matrix of explanatory variables. Parameters are estimated using a locally weighted least squares approach, where nearby observations for a given point are assigned higher weights, and the difference between observed and fitted values is used to assign lower weights. The method minimizes the sum of squared differences to derive the parameter estimates.

Constructing the GTWR model hinges on defining the spatial weight matrix. Typically, the space-time weight matrix is formulated as $W(u_{\mathrm{i}}, v_{\mathrm{i}}, t_{\mathrm{i}}) = diag(w_{\mathrm{i1}}, w_{\mathrm{i2}}, \ldots, w_{\mathrm{in}})$, with diagonal element W_{ij} representing the spatiotemporal decay function. In this paper, according to the research of Huang et al. [23], a Gaussian function is utilized as the weighting function.

$$W_{ij} = \exp\left(-\left(\frac{d_{ij}}{h}\right)^2\right) \tag{12}$$

where h signifies the bandwidth that modulates the influence decay with distance, with its optimal value identified using the Corrected Akaike information criterion [31]. d_{ij} denotes the space-time gap between points i and j. To normalize disparities across dimensions, the GTWR model automatically adjusts the spatial scale parameter λ and the temporal scale parameter μ. The spatial distance d^S and the temporal distance d^T are integrated to form a composite spatiotemporal distance d^{ST}, thereby enabling the formulation of a spatiotemporal distance function.

$$d_{ij}^{ST} = \sqrt{\lambda\left[\left(u_i - u_j\right)^2 + \left(v_i - v_j\right)^2\right] + \mu\left(t_i - t_j\right)^2} \tag{13}$$

where $\lambda = 0$ implies the absence of spatial effects, reducing the space-time distance to a function proportional to the time distance, thus converting the GTWR model into a basic time-weighted regression. Similarly, if μ is zero, the model disregards temporal effects, effectively making it a GWR model. However, when both λ and μ are non-zero, the GTWR model constructs a space-time weight matrix as $(u_i, v_i, t_i) = \text{diag}\,(w_{i1}, w_{i2}, \ldots, w_{in})$, and W_{ij} can be calculated as below:

$$W_{ij} = \exp\left\{-\left(\frac{\lambda\left[\left(u_i - u_j\right)^2 + \left(v_i - v_j\right)^2\right] + \mu\left(t_i - t_j\right)^2}{h^2}\right)\right\} \tag{14}$$

The GTWR model applies local regression to each observed unit, estimating parameters at different time points as geographical positions change. This approach better captures the spatial dependence and spatiotemporal differences of each driving factor. Compared with GWR, GTWR accounts for temporal non-stationarity and the integrated effects of space and time, rendering it apt for analyzing complex spatiotemporal data that requires consideration of both spatial and temporal factors. Due to the short time frame of our data, ranging from 2014 to 2020, and constraints in data availability, we will conduct GWR and GTWR analyses. This will allow us to compare the outcomes and choose the most appropriate method for our study.

In this study, we function the GTWR ADDIN by Huang et al. [23] in ArcGIS 10.8 to get and compare the results of GWR and GTWR models.

3. Results

3.1. Spatiotemporal Distribution and Variation of AGPP in Beijing

This study used AGPP data from SCIDB to analyze the spatiotemporal distribution and variation of AGPP in Beijing. This AGPP data contains some no-data areas and some invalid data areas that are less than zero. The result is shown in Figure 1. The geographic distribution of AGPP across Beijing reveals a discernible trend, marked by higher values in the northwestern areas and lower values in the southeastern areas. From 2000 to 2020, AGPP in Beijing ranges from below 0 to 3000 g $C \cdot m^{-2} \cdot a^{-1}$, and the multi-year annual mean GPP varies from -116.18 to 2527.71 g $C \cdot m^{-2} \cdot a^{-1}$ (Figure 1). The mean AGPP in mountainous areas was significantly larger than the mean AGPP in the plains. In the northwestern mountainous areas, such as the Jundu Mountains, AGPP generally exceeded 1000 g $C \cdot m^{-2} \cdot a^{-1}$, while in the plain areas in the southeast, AGPP was generally lower, mostly below 1000 g $C \cdot m^{-2} \cdot a^{-1}$.

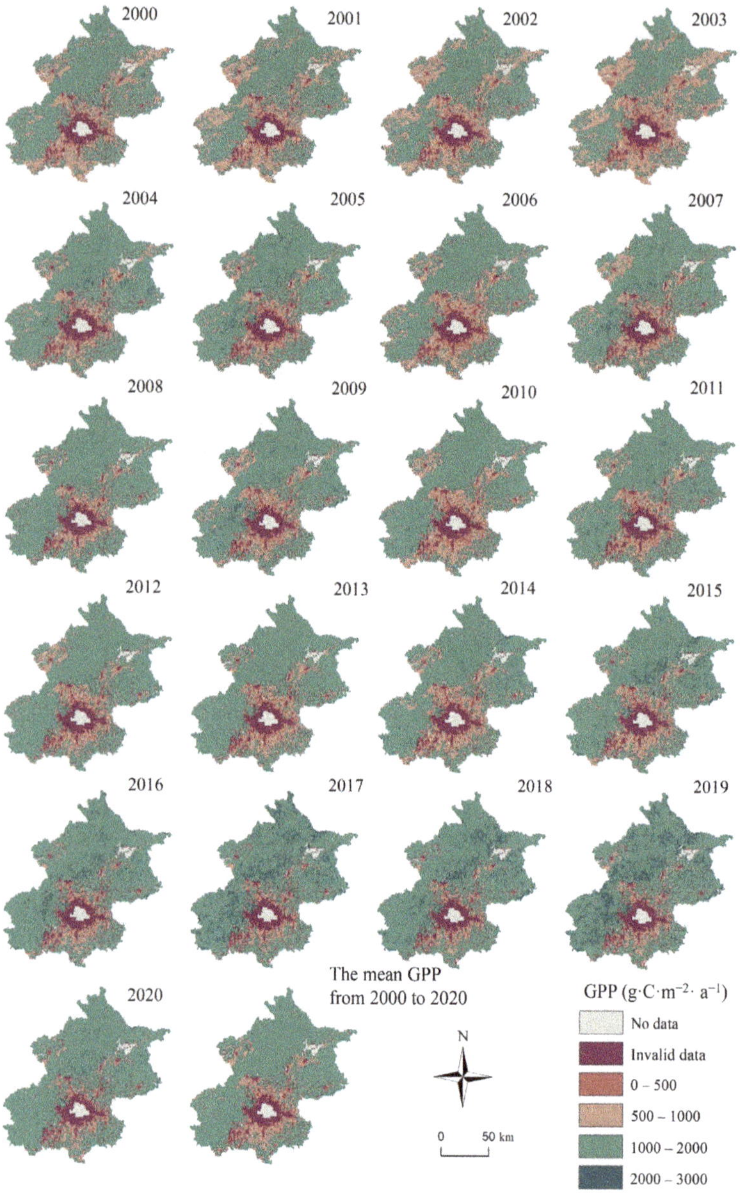

Figure 1. Spatiotemporal variation of AGPP in Beijing from 2000 to 2020.

Figure 2 shows spatial and temporal variation trends of AGPP, AGPP increased in most regions in Beijing, with a growth rate ranging from −55.35 and 71.98 g $C{\cdot}m^{-2}{\cdot}a^{-1}$ (Figure 2a,b). Significantly decreasing areas are mainly concentrated within Beijing's sixth ring road (Figure 2b). Figure 2c shows the 21-year time series of annual mean GPP. Figure 2c presents the 21-year time series of annual mean GPP, showing an overall increase from 2000 to 2020 at a rate of 17.41 g $C{\cdot}m^{-2}{\cdot}a^{-1}$. The apex of AGPP values was registered in the year 2019. Conversely, the nadir was noted in 2003.

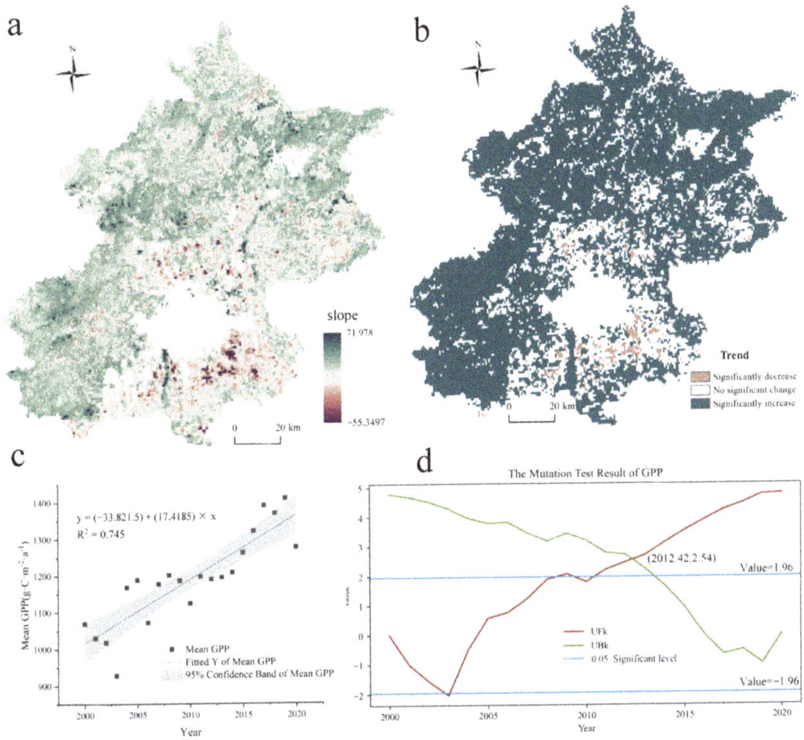

Figure 2. Spatial variation trend of AGPP from 2000 to 2020. (**a**) shows the slope of AGPP interannual variation, (**b**) shows the results of the significance test of AGPP change, (**c**) shows the time series of AGPP from 2000 to 2020, and (**d**) shows the mutation test result of AGPP.

In the M-K curve of the AGPP mutation test in Figure 2d, the UF curve is greater than 0 except for the case of less than 0 in 2000–2004, indicating that the average annual Gross Primary Productivity has shown a steady upward trend since 2004, and the upward trend has been significant since 2011. The UF and UB curves had only one crossing point in 2013, and this crossing point was outside the critical line with a critical value of 1.96, so it was hard to judge that 2013 was the time when the mutation began.

3.2. The Influence of Urban Environmental Factors on AGPP

To analyze the influence of urban environmental factors on AGPP, GWR and GTWR models were tested with the original data sets. Before fitting GWR and GTWR, a multicollinearity test is necessary to eliminate redundant indicators. This study used the Variance Inflation Factor (VIF) as an indicator and screened redundant factors with a threshold of 5 [32]. The results of the multicollinearity test are shown in Table 2. We removed the redundant factor with the highest VIF value step by step until the VIF value of all factors is not greater than 5. The results of the multicollinearity test for the remaining factors are shown in Table 2.

According to the results in Table 2, we removed PM2.5 and T. Then we fitted GWR and GTWR, and the results are reported in Table 3. The detailed results of GWR model can be seen in the Supplementary Materials.

Table 2. Variable Inflation Factor (VIF) Analysis of Remaining Factors.

Factors	VIF
CO	1.416183
LST	2.419883
NO_2	3.076362
O_3	1.932922
PRE	1.646616

Table 3. Comparison results of GTWR and GWR models.

	GTWR	GWR						
Year	2014–2020	2014	2015	2016	2017	2018	2019	2020
Bandwidth (°)	0.114996	0.121176	0.124996	0.16544	0.237059	0.124996	0.124996	0.314818
ResidualSquares	84,427,900	7,762,450	8,123,490	9,794,280	15,555,900	8,693,000	9,987,170	14,828,000
AICc	13,191.3	1926.78	1930.96	1904.79	1935.11	1938.39	1953.52	1914.67
R^2	0.602178	0.727502	0.731239	0.655626	0.526772	0.674349	0.724807	0.397604
R^2 Adjusted	0.599978	0.716514	0.720401	0.64174	0.507691	0.661218	0.713711	0.373314

Compared to the GTWR fitting results, the GWR fitting results exhibit notable year-to-year variations, which significantly diminishes their representativeness and reliability for in-depth analysis. Although the AICc value associated with GWR is considerably smaller than that of GTWR, this discrepancy is primarily due to the distinct data complexities inherent to each model. GTWR, which incorporates higher data complexity and greater model sophistication, naturally results in a larger AICc value. Therefore, the AICc values produced by GWR and GTWR should not be directly compared as they reflect different levels of model intricacy. A smaller AICc value for GWR does not necessarily imply that its fitting results are superior to those of GTWR. Consequently, the GTWR fitting results are deemed more appropriate and reliable for conducting a thorough analysis. Besides, we evaluated the significance of the GTWR fit using a p-value threshold of 0.05. The fitting results, along with their significance, are presented in Figure 3.

Figure 3 presents the GTWR fitting results derived from the original dataset, offering a comprehensive overview of how urban environmental factors influence AGPP in the Beijing area. The analysis reveals that CO and NO_2 are the most critical factors determining AGPP, with O_3 and LST also exerting certain influences, while the impact of precipitation (PRE) is relatively minor. Specifically, CO and O_3 generally have a positive effect on AGPP across most areas of Beijing, with their negative impacts being limited to a few regions in the northeastern and southwestern areas. In the northwest of the city, the positive effects of CO and O_3 are more pronounced. NO_2 and LST generally exhibit negative impacts on AGPP. LST only shows a positive effect on AGPP in a few areas of the southwest and east, with negative effects in the remaining regions and a more obvious negative impact in the southeast. NO_2 shows a positive effect on AGPP in a few areas of the northeast, southwest, and northwest, with negative effects in other regions and a more pronounced negative impact in the northwest. PRE has the least impact on AGPP, and its impact varies in different regions of Beijing. It has a positive effect on AGPP in most central areas of Beijing while having a negative effect in some northern and southwestern areas. Overall, the study results emphasize that CO and NO_2 play the most significant role in shaping the spatial pattern of AGPP in Beijing, with O_3 and LST also having certain influences. The impact of PRE is minimal and generally negligible. In terms of significance, the results indicate that the Geographically and Temporally Weighted Regression (GTWR) fits all factors with statistical significance across all domains, providing strong credibility for the model's predictive power.

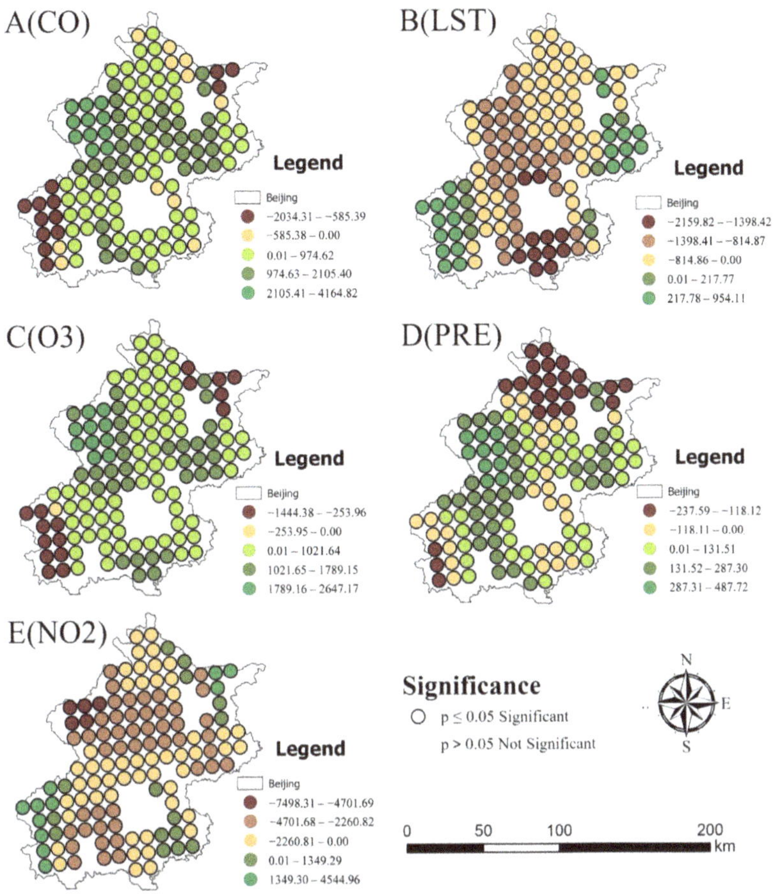

Figure 3. GTWR results from 2014 to 2020. (**A**) shows the fitting result of CO, (**B**) shows the fitting result of LST, (**C**) shows the fitting result of O_3, (**D**) shows the fitting result of PRE, and (**E**) shows the fitting result of NO_2.

4. Discussion

4.1. Spatiotemporal Distribution of AGPP in Beijing

In this paper, we observed that the spatial distribution pattern of AGPP in Beijing is generally higher in the northwest and lower in the southeast, which is consistent with the findings of Gao et al. [33]. There may be two factors contributing to this phenomenon: elevation and human activities. On the one hand, the elevation may cause differences in vegetation types, material metabolism, functional structure, and other vegetation itself, and on the other hand, human activities in the mountainous areas in northwestern Beijing and the plain areas in southeastern Beijing are not equally active, and human activities and urban development in the plain areas are relatively more, which may lead to the shrinkage of vegetation area and the degradation of vegetation growth environment. For example, a study by Lü et al. [34] demonstrated that from 2000 to 2010, 979.74 km^2 of vegetation in Beijing experienced severe degradation, 725.90 km^2 experienced moderate degradation, and 301.65 km^2 was mildly degraded. These degradation phenomena are directly related to the increase of urban development land. The phenomenon of lower AGPP in the area around the urban center is also consistent with the findings of Li [35] and Liu [36], and this result is mostly related to a large number of human construction activities in urban development. In addition, this paper also found that most of the AGPP in key ecological protection areas in Beijing showed a significant upward trend, which suggests that the increase of AGPP in Beijing may be related to the implementation of national and urban policies of returning farmland to forest and closing mountains for forestry.

In terms of temporal variation, this paper found that the AGPP in Beijing exhibited a general upward trend from 2000 to 2020, which is consistent with the findings of Zhang Xinzhu [37] and Du Wenli et al. [38] Previous studies have shown that the combined effects of climate change, CO_2 fertilization effects, nitrogen deposition, and ecological restoration efforts promote an increase in AGPP, while factors such as drought events and urban development land expansion may cause a decrease in AGPP. In the case of Beijing, major events and ecological restoration work may be the main reasons for the fluctuating rise in AGPP from 2000 to 2020. The years 2004 and 2008 had the greatest ups and downs in AGPP in Beijing and were the two years when ecological work was vigorously promoted to host the Beijing Olympics under the Beijing Urban Master Plan (2004–2020). Under the guidance of the Beijing Urban Master Plan (2004–2020) and the government's attention, the conditions for vegetation growth in Beijing were greatly improved, and AGPP surged.

4.2. Influence of LUCC on AGPP in Beijing

Land Use and Land Cover Change (LUCC) can exert an obvious influence on AGPP. For example, the conversion of vegetated areas into impervious surfaces can lead to a reduction in AGPP. Consequently, we also examined the effects of LUCC on AGPP using the CNLUCC dataset. To facilitate analysis, we have consolidated the original CNLUCC classification system into five primary categories: Cropland, Greenland, Water area, Residential land (impervious land), and Unused land. We have compiled land use/land cover data for Beijing spanning the years 2000 to 2020. The result is shown in Figure 4.

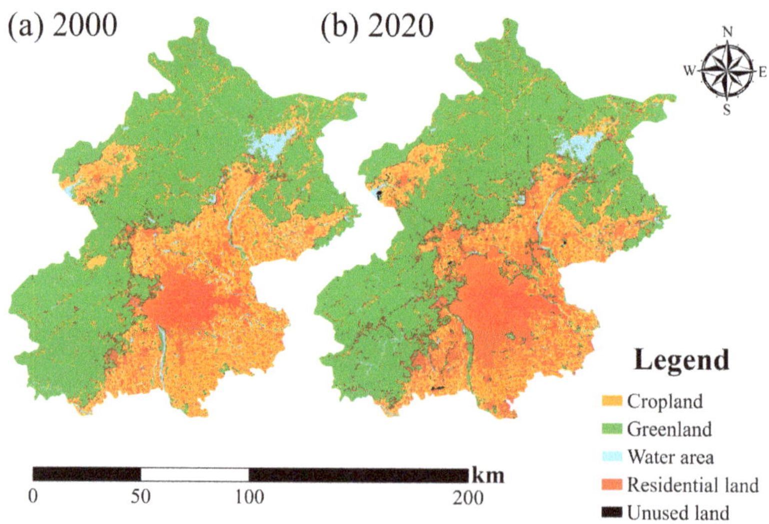

Figure 4. Land Use/Land Cover of Beijing in 2000 and 2020. (**a**) shows the Land Use/Land Cover of Beijing in 2000, and (**b**) shows the Land Use/Land Cover of Beijing in 2020.

As depicted in Figure 4, it is evident that there has been a significant conversion of various kinds of land into impervious land within the central region of Beijing, whereas transformations among other kinds of land cover are relatively minor. Building on this observation, we conducted an analysis of land use alterations in Beijing over the period from 2000 to 2020 and compared the result with the trend of AGPP during the same period. The comparative outcomes are illustrated in Figure 5.

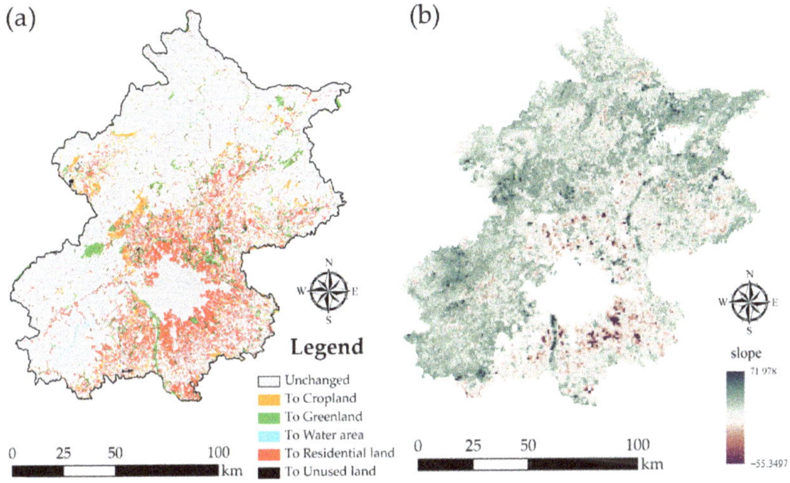

Figure 5. Spatial variation trend of AGPP and LUCC of Beijing from 2000 to 2020. (**a**) shows the results of LUCC in Beijing, and (**b**) shows the slope of AGPP interannual variation.

The result shows that areas with a negative trend in AGPP are predominantly found where the surface cover types have been converted to impervious surfaces, while AGPP remains stable or increases in other regions. The result indicates that urban sprawl exerts an obvious negative influence on AGPP. Therefore, for cities undergoing rapid expansion, the proper planning of urban growth areas and urban growth strategies is vital for the city's green transformation and sustainable development.

4.3. *Influence of Environmental Factors on AGPP in Beijing*

In terms of environmental factors, LST has significant and varying impacts on AGPP in Beijing. LST exerts a negative influence in the northern, central, and southeastern regions, where high surface temperatures likely limit plant productivity. However, in certain areas, such as those extending eastward and westward from the central parts, LST's impact shifts to a more positive influence, suggesting that moderate temperatures may benefit plant growth in these regions. This result highlights temperature as a key driving factor for vegetation productivity in Beijing. Within an optimal range, higher temperatures promote photosynthesis and boost plant productivity. However, excessive heat increases respiration and transpiration, causing dehydration and reducing productivity. These findings align with existing research [39–41].

The impact of precipitation (PRE) on AGPP in Beijing is relatively less significant, and its impact varies in different regions of Beijing. The negative impact of precipitation on AGPP in Beijing is attributed to the region's temperate semi-humid continental monsoon climate, which is marked by summer precipitation and a dearth of precipitation in spring and winter. This seasonal precipitation pattern induces water stress in vegetation during early growth stages, consequently affecting AGPP [42,43]. Additionally, extreme weather events, such as rainstorms and droughts, can exacerbate the decline in vegetation productivity. Rainstorms can lead to soil erosion and nutrient depletion, while droughts impede plant water uptake and photosynthesis [43]. As shown in Figure 6, the PRE in the Beijing area fluctuates greatly, and the intense fluctuations in PRE have caused obvious disturbance to vegetation growth. In the central regions of Beijing, human activities are capable of regulating water resources. Therefore, PRE exerts a positive influence on the

AGPP. On the periphery, however, human activities exhibit a limited ability to manage water resources effectively. Consequently, erratic precipitation patterns and the occurrence of extreme weather phenomena result in a negative impact of PRE on AGPP. In essence, the suppressive effect of precipitation on Beijing's AGPP is likely due to the seasonal distribution of precipitation and the frequency of extreme weather events.

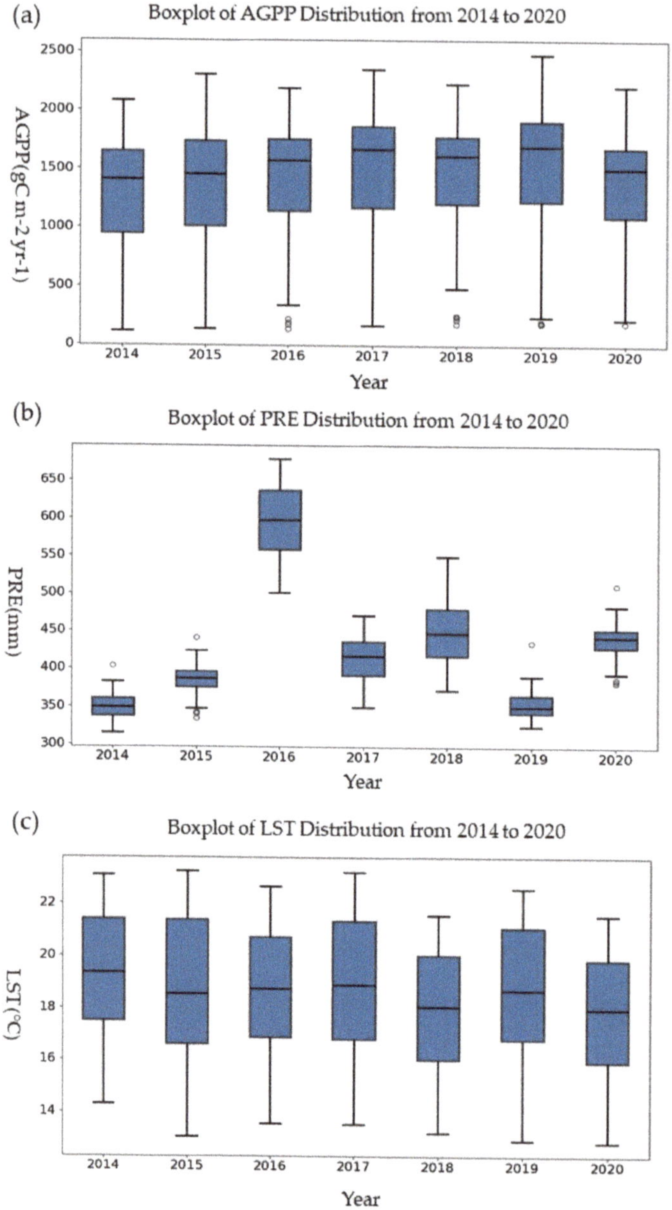

Figure 6. Boxplot of AGPP, PRE, and LST distribution. (**a**) shows the boxplot of AGPP distribution in Beijing, (**b**) shows the boxplot of PRE distribution in Beijing, and (**c**) shows the boxplot of LST distribution in Beijing.

Therefore, in the context of environmental considerations, the utilization of appropriate water resource management strategies to align the water-heat relationship, enhance urban water resource regulation capabilities, and mitigate extreme events such as floods and droughts can significantly enhance vegetation productivity in Beijing and contribute to the advancement of sustainable urban development.

4.4. Influence of Pollutant Factors on AGPP in Beijing

In terms of pollutant factors, CO and O_3 generally have a positive effect on AGPP throughout most of Beijing, with O_3 having a less obvious effect. However, in small areas in the northern, southwestern, and southeastern regions, CO and O_3 exhibit negative effects. NO_2 generally exhibits negative impacts on AGPP while showing a positive effect on AGPP in a few areas of the northeast, southwest, and southeast.

CO can affect vegetation AGPP by affecting the antioxidant and drought resistance of plants [44]. The strong oxidizing properties of O_3 and NO_2 also damage plant structure and have an inhibitory effect on AGPP [16,17]. At certain concentrations, NO_2 can stimulate plant growth, but at high concentrations, it becomes toxic, adversely affecting plant development [45]. There is a certain gap between the model fitting results and existing research results; the reason could be as follows: O_3 and CO levels can also indicate the intensity of human activities in this study, and vegetation may benefit from positive human interventions, such as more effective vegetation management strategies, which may offset some of the negative impacts of pollutants. This could result in a positive influence of CO and O_3 on AGPP across most areas. Conversely, areas exhibiting negative impacts are likely due to severe local pollution of O_3 and CO, where the detrimental effects of pollution on AGPP surpass the positive effects of human activities. In terms of NO_2, unlike O_3 and CO, the main source of NO_2 in China is industrial emissions [46]. Consequently, the correlation between NO_2 and human activities is relatively low, and the results fitted by GTWR are closer to the natural effects of NO_2. As a result, NO_2 generally has a negative impact on AGPP, with only a small fraction of areas exhibiting a promotional effect on AGPP.

Therefore, employing more effective vegetation management practices can significantly mitigate the impact of pollutants on AGPP. Focusing on key polluted areas and implementing strategies to reduce pollutant emissions, coupled with appropriate vegetation management, can contribute to the enhancement of AGPP in the Beijing region and facilitate the achievement of sustainable urban development. At present, Beijing has implemented comprehensive initiatives in pollution control, yielding substantial outcomes. As depicted in Figure 7, while O_3 concentrations are anticipated to persist in their upward trajectory due to the rising number of motor vehicles, both CO and NO_2 exhibited a pronounced decreasing trend over the period from 2014 to 2020. This trend underscores the substantial success of Beijing's environmental management strategies.

Moreover, Beijing has executed a multitude of strategies focused on the realms of urban greening and forest conservation in pursuit of sustainable urban development. For example, by implementing projects to enhance the ecological functions of urban green spaces, improve the quality and efficiency of plain forests, and accurately improve the quality of mountainous forests, the resilience of urban ecosystems has been strengthened. Besides, the Saihanba Mechanical Forest Farm's desertification control has significantly raised forest coverage from 11.4% to 82%, offering critical ecological services to the Beijing Tianjin Hebei region. These measures and achievements reflect Beijing's emphasis and commitment to the construction of ecological civilization.

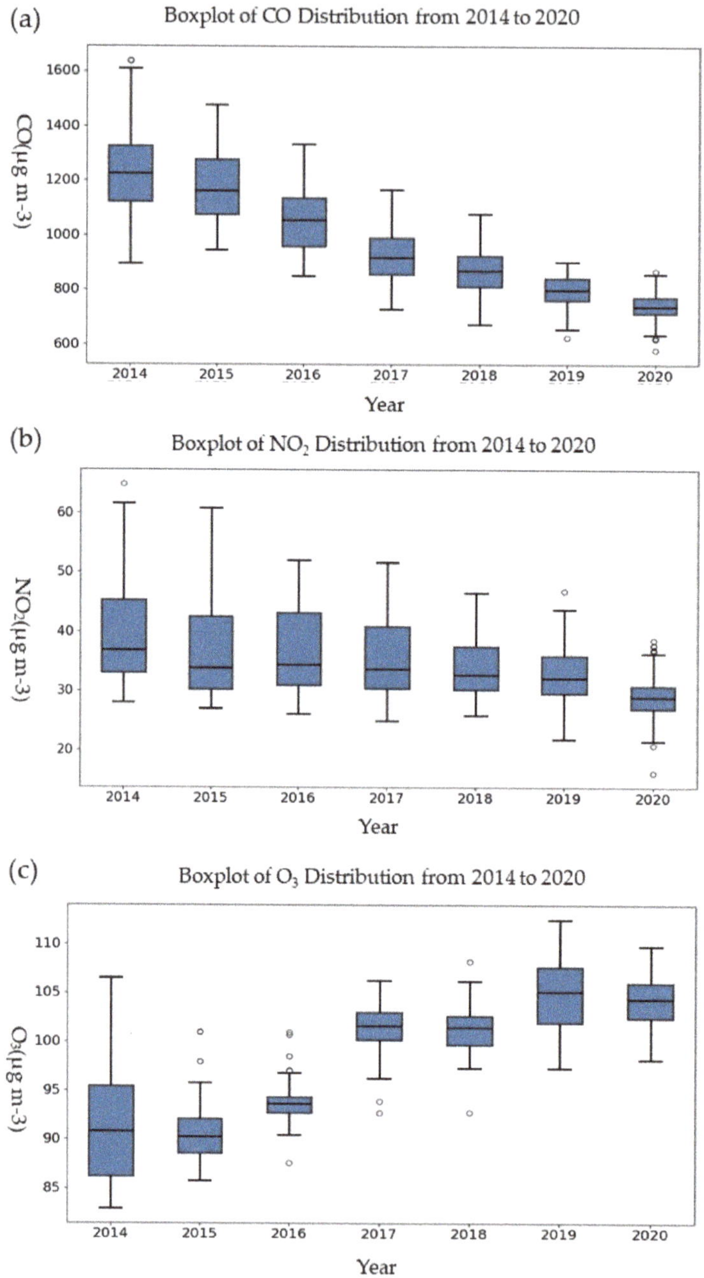

Figure 7. Boxplot of CO, NO_2, and O_3 distribution. (**a**) shows the boxplot of CO distribution in Beijing, (**b**) shows the boxplot of NO_2 distribution in Beijing, and (**c**) shows the boxplot of O_3 distribution in Beijing.

5. Conclusions

In conclusion, this study highlights the significant spatial and temporal patterns of AGPP in Beijing, with higher values in the northwest and lower values in the southeast.

Elevation and human activities, particularly urban development, play crucial roles in shaping these patterns. In terms of LUCC, the expansion of impervious land will have a significant negative impact on AGPP. Therefore, for cities experiencing rapid growth, the strategic planning of urban expansion is essential for facilitating a green transition and ensuring sustainable urban development.

Among the factors analyzed, CO and NO_2 are the most critical factors determining AGPP, with O_3 and LST also exerting certain influences, while the impact of precipitation (PRE) is relatively minor. Land Surface Temperature (LST) generally impacts AGPP negatively, whereas, in certain regions characterized by favorable temperatures, it enhances plant growth. The effect of PRE varies in different areas of Beijing, but it is not obvious. Irregular precipitation and extreme situations can both lead to negative impacts of precipitation on AGPP. Pollutants like CO and O_3 have a positive effect on AGPP in most areas, suggesting that effective vegetation management can offset their negative impacts. However, in severely polluted areas, the detrimental effects of these pollutants surpass the benefits of human activities. Unlike O_3 and CO, NO_2 exerts a detrimental effect on the AGPP due to its origins in industrial processes, which exhibit a low correlation with human activities. In conclusion, integrated water resource management, pollution control, and ecological restoration are crucial for enhancing AGPP and sustaining urban development in Beijing. These strategies are essential for fostering vegetation growth, achieving carbon neutrality, and ensuring environmental sustainability in the city. However, the impact of O_3 on AGPP in this study shows some differences from existing research findings. This discrepancy has not been fully examined and may be influenced by other environmental factors in Beijing. Further research and exploration are needed to understand this issue better.

Supplementary Materials: The following supporting information can be downloaded at: https://www.mdpi.com/article/10.3390/ijgi13110396/s1.

Author Contributions: All authors have conceptualized and discussed this work. Simin Yu and Le Chen processed various environmental data and discussed it online. The initial draft was written by Simin Yu and Le Chen. Shi Shen is responsible for supervising this paper. You Wan and Changqing Song contributed to the determination of the theme of the paper. All authors were equally reviewed and edited. All authors have read and agreed to the published version of the manuscript.

Funding: This research was funded by the National Natural Science Foundation of China, grant number 42230106, 42201498.

Data Availability Statement: All data used in this paper can be accessed for free. The AGPP data can be obtained from Science Data Bank (https://www.scidb.cn/, accessed on 9 February 2024). The LST data can be obtained from NASA-Earth-Data (https://ladsweb.modaps.eosdis.nasa.gov/, accessed on 9 February 2024). The Temperature data and Precipitation data can be obtained from National-Earth-System-Science-Data-Center (http://loess.geodata.cn, accessed on 11 February 2024). The $PM_{2.5}$, CO, NO_2, O_3 data can be obtained from an open-source dataset (https://weijing-rs.github.io/product.html, accessed on 11 February 2024). The Land Use/Land Cover data can be obtained from RESDC (https://www.resdc.cn/DOI/doi.aspx?DOIid=54, accessed on 7 March 2024).

Acknowledgments: The authors express their gratitude to the editor and the anonymous reviewers for their constructive feedback.

Conflicts of Interest: The authors declare no conflicts of interest.

References

1. Nuarsa, I.; As-Syakur, A.; Gunadi, I.; Sukewijaya, I. Changes in Gross Primary Production (GPP) over the Past Two Decades Due to Land Use Conversion in a Tourism City. *ISPRS Int. J. Geo-Inf.* **2018**, *7*, 57. [CrossRef]
2. Cui, Y.; Xiao, X.; Zhang, Y.; Dong, J.; Qin, Y.; Doughty, R.B.; Zhang, G.; Wang, J.; Wu, X.; Qin, Y.; et al. Temporal Consistency between Gross Primary Production and Solar-Induced Chlorophyll Fluorescence in the Ten Most Populous Megacity Areas over Years. *Sci. Rep.* **2017**, *7*, 14963. [CrossRef] [PubMed]
3. Fernandez, H.M.; Granja-Martins, F.M.; Pedras, C.M.G.; Fernandes, P.; Isidoro, J.M.G.P. An Assessment of Forest Fires and CO_2 Gross Primary Production from 1991 to 2019 in Mação (Portugal). *Sustainability* **2021**, *13*, 5816. [CrossRef]

4. Friedlingstein, P.; O'Sullivan, M.; Jones, M.W.; Andrew, R.M.; Hauck, J.; Olsen, A.; Peters, G.P.; Peters, W.; Pongratz, J.; Sitch, S.; et al. Global Carbon Budget 2020. *Earth Syst. Sci. Data* **2020**, *12*, 3269–3340. [CrossRef]
5. Wagle, P.; Xiao, X.; Torn, M.S.; Cook, D.R.; Matamala, R.; Fischer, M.L.; Jin, C.; Dong, J.; Biradar, C. Sensitivity of Vegetation Indices and Gross Primary Production of Tallgrass Prairie to Severe Drought. *Remote Sens. Environ.* **2014**, *152*, 1–14. [CrossRef]
6. Xu, B.; Li, J.; Luo, Z.; Wu, J.; Liu, Y.; Yang, H.; Pei, X. Analyzing the Spatiotemporal Vegetation Dynamics and Their Responses to Climate Change along the Ya'an–Linzhi Section of the Sichuan–Tibet Railway. *Remote Sens.* **2022**, *14*, 3584. [CrossRef]
7. He, Z.; Yue, T.; Chen, Y.; Mu, W.; Xi, M.; Qin, F. Analysis of Spatial and Temporal Changes in Vegetation Cover and Driving Forces in the Yan River Basin, Loess Plateau. *Remote Sens.* **2023**, *15*, 4240. [CrossRef]
8. Zheng, H.; Zheng, H. Assessment and Prediction of Carbon Storage Based on Land Use/Land Cover Dynamics in the Coastal Area of Shandong Province. *Ecol. Indic.* **2023**, *153*, 110474. [CrossRef]
9. Wang, S.; Xu, L.; Zhuang, Q.; He, N. Investigating the Spatio-Temporal Variability of Soil Organic Carbon Stocks in Different Ecosystems of China. *Sci. Total Environ.* **2021**, *758*, 143644. [CrossRef]
10. Ziska, L.H.; Bunce, J.A.; Goins, E.W. Characterization of an Urban-Rural CO2/Temperature Gradient and Associated Changes in Initial Plant Productivity during Secondary Succession. *Oecologia* **2004**, *139*, 454–458. [CrossRef]
11. Wei, S.; Chen, Q.; Wu, W.; Ma, J. Quantifying the Indirect Effects of Urbanization on Urban Vegetation Carbon Uptake in the Megacity of Shanghai, China. *Environ. Res. Lett.* **2021**, *16*, 064088. [CrossRef]
12. Li, C.; Sun, G.; Cohen, E.; Zhang, Y.; Xiao, J.; McNulty, S.G.; Meentemeyer, R.K. Meentemeyer Modeling the Impacts of Urbanization on Watershed-Scale Gross Primary Productivity and Tradeoffs with Water Yield across the Conterminous United States. *J. Hydrol.* **2020**, *583*, 124581. [CrossRef]
13. Liu, X.; Cui, Y.; Xiao, X.; Shi, Z.; Li, M.; Li, N.; Dong, J. Multi-Scale Analysis of Urbanization and Gross Primary Productivity during 2000–2018 in Beijing, China. *Environ. Res. Lett.* **2023**, *19*, 014023. [CrossRef]
14. Chen, Y.; Huang, B.; Zeng, H. How Does Urbanization Affect Vegetation Productivity in the Coastal Cities of Eastern China? *Sci. Total Environ.* **2022**, *811*, 152356. [CrossRef] [PubMed]
15. Liu, X.; Cui, Y.; Li, W.; Li, M.; Li, N.; Shi, Z.; Dong, J.; Xiao, X. Urbanization Expands the Fluctuating Difference in Gross Primary Productivity between Urban and Rural Areas from 2000 to 2018 in China. *Sci. Total Environ.* **2023**, *901*, 166490. [CrossRef] [PubMed]
16. Ainsworth, E.A.; Yendrek, C.R.; Sitch, S.; Collins, W.J.; Emberson, L.D. The Effects of Tropospheric Ozone on Net Primary Productivity and Implications for Climate Change. *Annu. Rev. Plant Biol.* **2012**, *63*, 637–661. [CrossRef] [PubMed]
17. Lombardozzi, D.; Sparks, P.; Bonan, B. Integrating O_3 Influences on Terrestrial Processes: Photosynthetic and Stomatal Response Data Available for Regional and Global Modeling. *Biogeosciences* **2013**, *10*, 6815–6831. [CrossRef]
18. Gu, L.; Baldocchi, D.D.; Wofsy, S.C.; Munger, J.W.; Michalsky, J.J.; Urbanski, S.P.; Boden, T.A. Response of a Deciduous Forest to the Mount Pinatubo Eruption: Enhanced Photosynthesis. *Science* **2003**, *299*, 2035–2038. [CrossRef]
19. Yue, X.; Unger, N. Fire Air Pollution Reduces Global Terrestrial Productivity. *Nat. Commun.* **2018**, *9*, 5413. [CrossRef]
20. McDonnell, M.J.; Pickett, S.T.A.; Groffman, P.; Bohlen, P.; Pouyat, R.V.; Zipperer, W.C.; Parmelee, R.W.; Carreiro, M.M.; Medley, K. Ecosystem Processes Along an Urban-to-Rural Gradient. In *Urban Ecology*; Marzluff, J.M., Shulenberger, E., Endlicher, W., Alberti, M., Bradley, G., Ryan, C., Simon, U., ZumBrunnen, C., Eds.; Springer: Boston, MA, USA, 2008; pp. 299–313. ISBN 978-0-387-73411-8.
21. Wang, S.; Ju, W.; Peñuelas, J.; Cescatti, A.; Zhou, Y.; Fu, Y.; Huete, A.; Liu, M.; Zhang, Y. Urban−rural Gradients Reveal Joint Control of Elevated CO_2 and Temperature on Extended Photosynthetic Seasons. *Nat. Ecol. Evol.* **2019**, *3*, 1076–1085. [CrossRef]
22. Brunsdon, C.; Fotheringham, A.S.; Charlton, M.E. Charlton Geographically Weighted Regression: A Method for Exploring Spatial Nonstationarity. *Geogr. Anal.* **1996**, *28*, 281–298. [CrossRef]
23. Huang, B.; Wu, B.; Barry, M. Geographically and Temporally Weighted Regression for Modeling Spatio-Temporal Variation in House Prices. *Int. J. Geogr. Inf. Sci.* **2010**, *24*, 383–401. [CrossRef]
24. Sun, D.; Zhou, L.; Li, Y.; Liu, H.; Shen, X.; Wang, Z.; Wang, X. New-Type Urbanization in China: Predicted Trends and Investment Demand for 2015–2030. *J. Geogr. Sci.* **2017**, *27*, 943–966. [CrossRef]
25. Fan, R.; Zhu, X.; Chen, Z.; Yu, G.; Zhang, W.; Han, L.; Wang, Q.; Chen, S.; Liu, S.; Wang, H.; et al. A dataset of annual gross primary productivity in China’s terrestrial ecosystems during 2000–2020. *China Sci. Data* **2023**, *8*, 1–13. [CrossRef]
26. Dong, J.; Li, L.; Li, Y.; Yu, Q. Inter-Comparisons of Mean, Trend and Interannual Variability of Global Terrestrial Gross Primary Production Retrieved from Remote Sensing Approach. *Sci. Total Environ.* **2022**, *822*, 153343. [CrossRef] [PubMed]
27. Ohlson, J.A.; Kim, S. Linear Valuation without OLS: The Theil-Sen Estimation Approach. *Rev. Account. Stud.* **2015**, *20*, 395–435. [CrossRef]
28. Hamed, K.H.; Ramachandra Rao, A. A Modified Mann-Kendall Trend Test for Autocorrelated Data. *J. Hydrol.* **1998**, *204*, 182–196. [CrossRef]
29. Hussain, M.; Mahmud, I. pyMannKendall: A Python Package for Non Parametric Mann Kendall Family of Trend Tests. *J. Open Source Softw.* **2019**, *4*, 1556. [CrossRef]
30. Wang, J. Determining the Most Accurate Program for the Mann-Kendall Method in Detecting Climate Mutation. *Theor. Appl. Climatol.* **2020**, *142*, 847–854. [CrossRef]
31. Hurvich, C.M.; Tsai, C.-L. A Corrected Akaike Information Criterion for Vector Autoregressive Model Selection. *J. Time Ser. Anal.* **1993**, *14*, 271–279. [CrossRef]

32. Murray, L.; Nguyen, H.; Lee, Y.-F.; Remmenga, M.; Smith, D. Variance Inflation Factors in Regression Models with Dummy Variables. *Conf. Appl. Stat. Agric.* **2012**. [CrossRef]
33. Deng, G.; Gao, J.; Jiang, H.; Li, D.; Wang, X.; Wen, Y.; Sheng, L.; He, C. Response of Vegetation Variation to Climate Change and Human Activities in Semi-Arid Swamps. *Front. Plant Sci.* **2022**, *13*, 990592. [CrossRef] [PubMed]
34. Lü, Y.; Zhang, L.; Feng, X.; Zeng, Y.; Fu, B.; Yao, X.; Li, J.; Wu, B. Recent Ecological Transitions in China: Greening, Browning and Influential Factors. *Sci. Rep.* **2015**, *5*, 8732. [CrossRef] [PubMed]
35. Li, Q.; Li, T.; Yang, M.; Ying, W. Spatiotemporal Variation of Ecosystem Services Value Based on Gradient Analysis in Wuhan: 2000–2010. *Acta Ecol. Sin.* **2017**, *37*, 2118–2125.
36. Liu, M.; Jiao, J.; Pan, J.; Song, J.; Che, Y.; Li, L. Spatial and Temporal Patterns of Planting NPP and Its Driving Factors in Qinghai Province. *Acta Ecol. Sin.* **2020**, *40*, 5306–5317.
37. Zhang, X.; Wang, H.; Yan, H.; Ai, J. Analysis of Spatio-Temporal Changes of Gross Primary Productivity in China from 2001 to 2018 Based on Romote Sensing. *Acta Ecol. Sin.* **2021**, *41*, 6351–6362.
38. Du Wen-li, S.; Wu, Y.; Song, Z. The Responses of Gross Primary Production to Drought in Terrestrial Ecosystems of China during 1980–2013. *Chin. J. Ecol.* **2020**, *39*, 23.
39. Liang, W.; Yang, Y.; Fan, D.; Guan, H.; Zhang, T.; Long, D.; Zhou, Y.; Bai, D. Analysis of Spatial and Temporal Patterns of Net Primary Production and Their Climate Controls in China from 1982 to 2010. *Agric. For. Meteorol.* **2015**, *204*, 22–36. [CrossRef]
40. Fu, Y.; Lu, X.; Zhao, Y.; Zeng, X.; Xia, L. Assessment Impacts of Weather and Land Use/Land Cover (LULC) Change on Urban Vegetation Net Primary Productivity (NPP): A Case Study in Guangzhou, China. *Remote Sens.* **2013**, *5*, 4125–4144. [CrossRef]
41. Wang, J.; Zhao, A.; Hu, X. Spatiotemporal Distribution of Vegetation Net Primary Productivity in Beijing-Tianjin-Hebei and Natural Driving Factors. *Ecol. Environ.* **2021**, *30*, 1158.
42. Lukas, E.; Jonathan, A. Sherratt Effects of Precipitation Intermittency on Vegetation Patterns in Semi-Arid Landscapes. *Phys. D Nonlinear Phenom.* **2020**, *405*, 132396. [CrossRef]
43. Xu, C.; McDowell, N.G.; Fisher, R.A.; Wei, L.; Sevanto, S.; Christoffersen, B.O.; Weng, E.; Middleton, R.S. Middleton Increasing Impacts of Extreme Droughts on Vegetation Productivity under Climate Change. *Nat. Clim. Change* **2019**, *9*, 948–953. [CrossRef]
44. Huang, B.-K.; Xu, S.; Xuan, W.; Li, M.; Cao, Z.-Y.; Liu, K.-L.; Ling, T.-F.; Shen, W.-B. Carbon Monoxide Alleviates Salt-Induced Oxidative Damage in Wheat Seedling Leaves. *J. Integr. Plant Biol.* **2006**, *48*, 249–254. [CrossRef]
45. Wang, J.; Wang, Y.; Zhang, H.; Guo, D.; Sun, G. Atmospheric Nitrogen Dioxide at Different Concentrations Levels Regulates Growth and Photosynthesis of Tobacco Plants. *J. Plant Interact.* **2021**, *16*, 422–431. [CrossRef]
46. Jion, M.M.M.F.; Jannat, J.N.; Mia, M.Y.; Ali, M.A.; Islam, M.S.; Ibrahim, S.M.; Pal, S.C.; Islam, A.; Sarker, A.; Malafaia, G.; et al. A Critical Review and Prospect of NO_2 and SO_2 Pollution over Asia: Hotspots, Trends, and Sources. *Sci. Total Environ.* **2023**, *876*, 162851. [CrossRef]

isprs International Journal of *Geo-Information*

MDPI

Article

Understanding the Carbon Footprint of Tile Transfer for Web Maps

Guillaume Touya [1,*], Azelle Courtial [1], Jérémy Kalsron [1,2], Justin Berli [1], Bérénice Le Mao [1] and Laura Wenclik [1]

[1] LASTIG, IGN, ENSG, University Gustave Eiffel, F-77420 Champs-sur-Marne, France; justin.berli@ign.fr (J.B.); berenice.le-mao@ign.fr (B.L.M.); laura.wenclik@ign.fr (L.W.)

[2] LIMOS, Clermont-Auvergne-INP, Mines de Saint-Étienne, Centre national de la Recherche Scientifique, Université Clermont Auvergne, F-63000 Clermont–Ferrand, France

* Correspondence: guillaume.touya@ign.fr

Abstract: As web maps are now extensively used by billions of users, the energy consumption of these maps is not marginal anymore. Green cartography seeks to reduce the energy consumption of maps to promote more sustainable digital tools. To reduce energy consumption, we first need to better understand the different sources of energy consumption for web maps. Among these sources, this paper focuses on the tiles that are stored on servers and then constantly transferred each time a user explores the map. This paper presents several experiments carried out with current web maps to assess this energy consumption. We first try to assess the number of map tiles that are loaded through the web when users explore web maps, and we determine which types of interaction are used with the maps, and a similar amount of tiles is loaded. Then, we try to assess which zoom levels are the most loaded by users; it appears that the medium–large scales are the most used (between zoom levels 11 and 17). Then, we explore the size of the map tiles and try to assess which ones are larger and thus require more energy to load over the web; we can find clear differences between zoom levels. Finally, we discuss how map generalization could be used to reduce energy consumption by creating lighter tiles. These experiments show that the current web maps are suboptimal regarding energy consumption, with many tiles loaded at zoom levels where the tiles are larger than necessary.

Keywords: green cartography; green computing; web mapping; tiling; map generalization

1. Introduction

As the direct consequences of climate change are constantly being unveiled, e.g., floods, storms, droughts, and a drop in biodiversity, we cannot design digital tools as we could 20 years ago due to the increasing impact of information and communication technologies on global greenhouse gas emissions [1]. As researchers, we are responsible for understanding how we can reduce the amount of energy required to run our digital map applications. Green computing research shows that we can design more sustainable digital tools [2].

Maps have now become mainly digital tools as they are mostly read on computer screens or smartphones, while printed paper maps have become marginal in use. Maps are extremely used as digital tools. For example, Google Maps had more than 1 billion users each month in 2019 [3] and the average use time per user was approximately 30 h per year during the same period [4]. These mapping applications consume energy in four manners [4–6]: by storing large volumes of data, by processing geographic data to generate up-to-date tiles, by constantly transferring map data (vectors and images) through networks, and by displaying the maps on digital screens.

Green cartography is a derivative of green computing that has recently emerged to find solutions to reduce the energy consumption of digital maps [4–6]. Although all sources of energy consumption from web maps are important to study, this paper specifically focuses on the transfer of map tiles through the network, as it is a repeated cost each time we interact with the map by zooming, panning, or flying to an address. In particular, we are interested in counting how many tiles are transferred when using such a map, and in finding which types of tiles are the most frequently transferred. The paper does not target a new web mapping architecture that reduces the amount of data transferred within each session but rather studies how we can assess this amount of data with practical experiments using those maps. From the assessment of the energy consumed by web maps when transferring or storing tiles, we will be able to derive an approximate amount of greenhouse gases emitted to generate this energy [7].

The article is structured as follows. Section 2 briefly describes previous research in cartography that proposes, deliberately or not, possible solutions to reduce the energy consumed by web mapping applications (including all types of energy consumption). Section 3 presents an experiment to assess the number of map tiles that current mapping applications transfer during map exploration. Section 4 presents a follow-up experiment that focuses on which map tiles are most transferred, i.e., which scales are more queried by the users of mapping applications. Section 5 presents a final experiment that compares the size of the tiles depending on the scales, the styles, and the landscapes that are depicted. Finally, Section 6 discusses the three experiments to derive a research agenda toward more energy-efficient maps.

2. Green Cartography and Green Computing

There are four major sources of energy consumption for web maps: screen display, storage, map computation, and map transfer. This section reviews past research projects that propose solutions that could reduce those four different sources of energy consumption, even though most of them were not initially proposed to reduce energy consumption.

The first source is the energy consumed by the screen of the phone or the computer. The energy used by OLED displays, which are now the most common type of display [4], can be estimated with models based on the pixel values of the image to be displayed [8]. With this sort of model, different map styles can be compared to assess which one consumes more energy on the screen. Cartographic style is a design variable that can be adjusted to adapt to several user needs [9,10], and style can also be adjusted to reduce the energy consumed by the screen rendering. For instance, from a color scheme defined in a map legend, it is possible to derive a dark version of the color scheme that reduces the energy used for rendering while preserving the semantic relations between the colors in the map [11]. It is also possible to define default energy-efficient color schemes, for choropleth maps, for instance [5]. Beyond style, all cartographic design decisions (e.g., generalization, typography, projection) can be constrained by a low-energy-consumption perspective [4]. For instance, font size for toponyms has a positive influence on energy consumption when the text is rendered with dark colors and the background is rendered with light colors, which is frequent in current cartographic styles.

The second source of energy consumption is the storage of map tiles on large servers in data centers. By reducing the amount of information that is stored, we can directly reduce the energy cost. The first proposition from the literature is to replace, as much as possible, raster tiles with vector tiling techniques, as storing vector tiles seems to be less volume-intensive than storing raster tiles for the same styles and spatial extent [12]. Vector tiling techniques are slowly replacing raster tiling for maps, but the literature suggests that it would be a faster way to reduce the energy cost. When using vector tiles, another solution to reduce the storage cost is to use map generalization to simplify the data to be used at smaller scales [12,13]. In this case, the data on the server side should be stored in a multi-scale or multi-representation database [14]. To go even further in the optimization of the storage space of this multi-scale database, some researchers propose database architectures that minimize redundancies between the information stored for the different scales [15]. Other rules can be applied to both vector and raster tiles to save storage space due to the multi-scalar nature of the maps. The scale of appearance in the map can be used as a new visual variable to convey hierarchies and avoid the redundant storage of recurring map elements [16]. For instance, when zooming into a named region (a country, city, or administrative region), it is not necessary to show the toponym of this region at larger scales, because the user knows they are inside the region; removing these redundant toponyms slightly reduces the size of the tiles at larger scales.

Several techniques are also proposed to reduce the third source of energy consumption, i.e., the energy used to compute map tiles while maintaining high-quality topographic maps. Once again, using vector tiles, it is not necessary to compute raster tiles on the server anymore, even though additional computations are required on the client side [12]. It may be beneficial if a large portion of the raster tiles that are re-computed after each update are rarely visualized by users. In this case, recent research on map generalization noticed that the generalization processes were often not used in current multi-scale web maps and rather propose to use them progressively at each zoom level [17,18]. With a constant reduction in and simplification of the data across scales, we might be able to reduce the complexity of the computation of the image/vector tiles for small scales.

The last source of energy consumption is the transfer of tiles (vector and raster) during the use of the map. Several solutions have been proposed to reduce the amount of information transferred this way, though these solutions were proposed to improve performance rather than to reduce the energy used. The first one lies in the data formats used to transfer data, in particular with vector tiles or, more generally, vector data [12]. For instance, GeoJSON is now widely used and it significantly reduces the amount of information transferred compared to older formats such as GML or KML. Cartographic generalization has been proposed to reduce the other sources of energy consumption. Still, it can also serve for the tile transfer energy cost, as good multi-scale generalization can reduce the amount of zooming in and out necessary to complete a task with the map [13,18] and, as a result, reduce the number of tiles transferred to the user's client.

Caching techniques can also be improved to reduce the energy consumed while transferring map tiles. If we minimize the number of tiles that are preventively cached without being used, unnecessary tiles are not loaded anymore. Alternatively, if we keep the tiles that might be reused later in the cache, tiles are not loaded several times during a session. The proposed solutions in the literature focus on the preventive caching of tiles: we can optimize this prefetching of tiles if the main interactions are pan and zoom [19], or by predicting the most popular tiles in a region (urban areas, density of points of interest) [20], or we can analzse the users' preferences to prefetch the tiles they are more likely to query [21].

Finally, in the beginning of the 2000s, when web cartography was emerging, several researchers studied the progressive transmission of geographic information [22–27]. Though progressive transmission was studied to render maps quickly, the same principles can also be used to transfer multi-scale data with a minimization of the redundancy between scales. For instance, important highways are usually displayed on the map at most scales. As a result, each time a new scale is queried from the client, the same highway vector is transferred with current architectures, which seems unnecessary because this vector was already transferred.

Figure 1 shows a mindmap that summarizes examples of good and bad practices for greener cartography, found in the papers reviewed in this section. There are four dimensions in which green cartography could reduce the impact of web maps: the architecture of the system, the content of the map, the display on the screen, and the use context. For each dimension, sub-dimensions are defined with examples that promote higher or lower energy consumption. It shows that there are already plenty of ideas to make greener maps, and they just need to be considered while designing web maps.

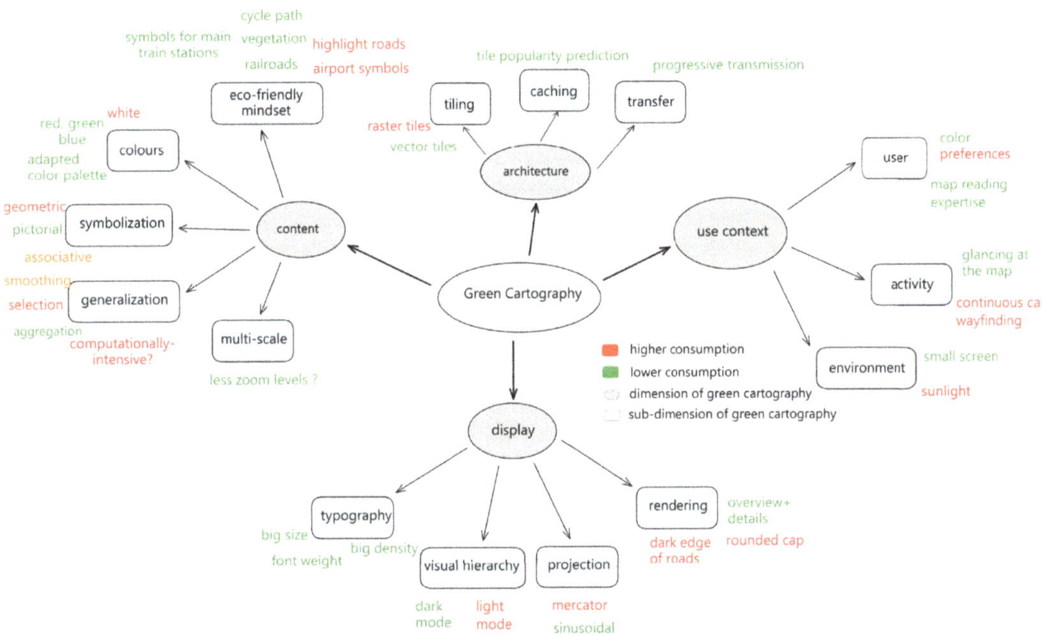

Figure 1. Mindmap that summarizes the different dimensions of green cartography and examples cited in the literature that increase or reduce the energy consumed by web maps.

The following sections present three small experiments to better understand the energetic cost of the tiling system of web maps. Each experiment covers a different aspect of this tiling system; Experiment 1 counts the tiles transferred when using the map Experiment 2 measures the size of the transferred tiles (Section 4), and Experiment 3 assesses the impact of map generalization on tile weight (Section 5).

3. How Many and Which Tiles Do Map Users Use?

3.1. Materials and Methods

Current web mapping applications usually follow standard architectures defined by the Open Geospatial Consortium (OGC) and use tiles to render maps on the client side. These tiles can be raster or vector depending on the architecture, but in both cases, the amount of energy used to transfer those tiles is directly related to (1) the size in bits of the transferred tiles and (2) the number of tiles that are transferred during a standard use session of the map. The issue of the weight of the tiles is addressed in Section 4, and this section addresses the second cause of energy consumption, the number of tiles that are transferred during a standard use of a web map.

The goal of this first experiment, called *Experiment 1* in the remainder of the paper, is to collect the tiles transferred during a standard use of a web map in order to check their number in relation to the duration of the session and their distribution over zoom levels, and to verify if these values vary with the types of interactions used. As the experiment was intended as a preliminary qualitative study, to better understand this issue of the number of tiles transferred while using the map, we selected only 9 participants who were expert users from our university, i.e., they were all trained in web cartography and were involved to some degree in research about cartography (master's students, PhD students, and post-docs). They ran the experiment in parallel and connected to a video chat application to communicate with each other.

The selected task was to find a target in the map based on textual indications. The task was repeated twice: both targets were shown at the same time at the beginning, and the participants were supposed to start looking at the second target as soon as the first one was found. As the aim was not to spend too much time on each task, additional indications were given to the participants who did not find one target after one minute of search. The task started at zoom level 6, centred on France, and the participants had to find the churches of two small towns: Rabastens and Fauverney. The indication for Rabastens was that it was located within a radius of 30 km around Toulouse, a large French city, but not in the direct conurbation of the city. The indication was similar for Fauverney, except that it was in a 30 km radius around Dijon, another French city (Figure 2). Those two regions were chosen because none of the participants had lived there in the past; they knew where the cities were, so they were able to find them easily while navigating in the map, but they did not know any town in the surroundings of these cities. The two parts of the task (Rabastens and Fauverney) were considered to be a single block, and not considered to be a repeated task because the starting point in the map was not the same before both parts (the second part starts in Rabastens and not at zoom level 6).

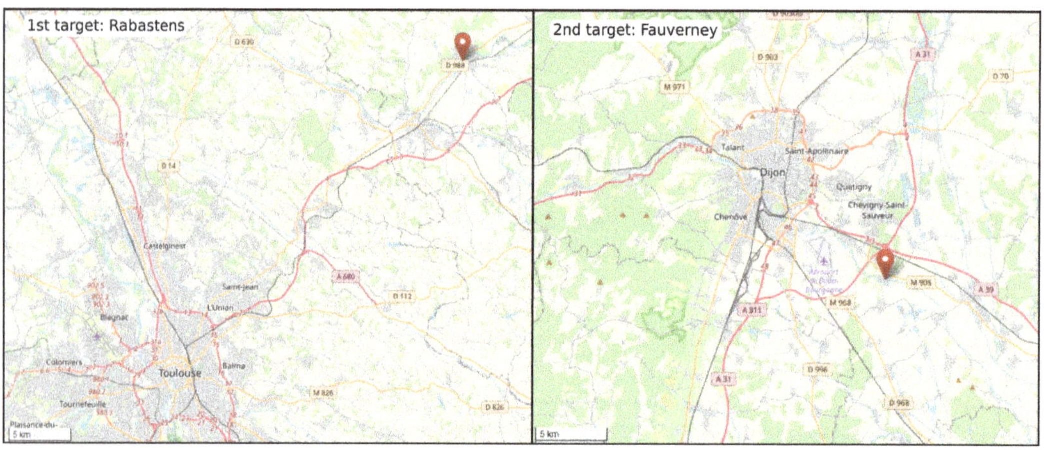

Figure 2. The two target towns of the first experiment and their location compared to the city cited in the textual indications.

As each participant was using a different computer/screen, we asked each participant to use the same internet browser, Firefox, and set the browser resolution to 1280×600 pixels. Each participant was asked to clear the browser cache to avoid using previously cached tiles. Comparing the number of tiles used in different mapping applications was not the goal of this experiment, so 8 of the 9 participants used the same application, a custom webpage, developed with OpenLayers, showing cartographic tiles from IGN, the French National Mapping Agency. The last participant used the default OpenStreetMap map browser (https://www.openstreetmap.org, accessed on 18 December 2024), to check if there was a noticeable difference. To determine the number of tiles transferred during the task, we asked the participants to open the network console of the web browser; each query from the client to retrieve the tiles was then printed in the console. At the end of the two tasks, the participants were asked to export the console logs as a .har file. This file was then parsed with a custom Python script to extract only the logs related to cartographic tiles. For each tile, we collected the following information: (1) the date, (2) the URL of the tile, (3) the name of the WMTS layer queried, (4) the size of the tile (in bytes), (5) the x coordinate of the tile in the tiling system, (6) the y coordinate of the tile in the tiling system, (7) the zoom level of the tile, and (8) the geographic coordinates of the tile as a WKT polygon. The output was then visualized in QGIS.

One of our hypotheses at the beginning of this experiment was that among the many possible browsing behaviors, some were more efficient in their consumption of tiles than others. To test this hypothesis, we defined different browsing behaviors (Table 1) and each participant adopted one of them while completing the task. To check the variability among the users adopting one of these behaviors, four participants adopted the same one. The first browsing behavior (BB1) was what we considered to be the standard use of a web map with a computer and a mouse: using the mouse wheel to zoom in and out slowly. We expected that this behavior would use many tiles as some unnecessary scales might be loaded on the way to the scales that were the most adapted to solve the task. The second behavior (BB2) aimed to mimic browsing with a touch screen on smartphones, but with a computer: the user only used the computer trackpad to browse the map. We also expected that this behavior would load many tiles due to clumsy and imprecise zooming gestures with the trackpad. The third browsing behavior (BB3) was to avoid panning as much as possible, and instead use zooming in and out to move across the map. We expected this behavior to generate more tiles due to unnecessary interactions, even though the unnecessary interactions may be mitigated by the tile cache if the user zooms out several times to the same tiles. The fourth behavior (BB4) was the opposite of the previous one, with a minimal use of zoom and an extensive use of panning. We expected this behavior to use many tiles due to the slower journey to the target. The next behavior (BB5) was called "center then zoom", which means that the user had to drag the location of interest to the center of the screen before zooming, to improve the precision of zooming and avoid unnecessary interactions. We expected this behavior to require fewer tiles than the previous behavior. This is the browsing behavior that was adopted by four participants. Finally, the last browsing behavior (BB6) required the use of a window zoom instead of the mouse wheel. A window zoom is possible with Open Layers-based mapping applications; the user presses the Ctrl key while defining a rectangle on the screen. We expected this behavior to require fewer tiles due to more precision in the zooming interaction, and the possibility of skipping some zoom levels.

Table 1. The six map browsing behaviors adopted by the experiment participants.

ID	Browsing Behavior	Number of Participants	Expected Impact on Tiles
BB1	small zoom with mouse wheel	1	many tiles loaded at intermediary scales
BB2	trackpad	1	more tiles loaded due to clumsy interactions
BB3	no panning	1	more tiles due to unnecessary interactions
BB4	mostly panning	1	more tiles loaded due to a slow journey to the target
BB5	center then zoom	4	fewer tiles due to a more precise zoom
BB6	window zoom	1	fewer tiles due to a more precise zoom

3.2. Results

The raw CSV files obtained during this experiment are available in a Zenodo repository (https://doi.org/10.5281/zenodo.14281428, accessed on 18 December 2024). Table 2 shows the completion time, the number of loaded tiles, and the number of tiles loaded per second for each of the participants in the experiment. As the task was easier for some participants than others due to luck (they directly zoomed close to the location of the targets), the completion time varied from 53 s to 333 s. As the number of tiles is dependent on the duration of the task and how much time the participants spent searching for the target in the map, we consider the number of tiles per second in the rest of the analysis, instead of the total number of loaded tiles. The number of tiles per second varied from 2.78 (third participant with BB5) to 12.2 (BB6).

Table 2. Number of tiles loaded during the experiment for each participant.

Browsing Behavior	Completion Time	Loaded Tiles	Tiles per Second
BB1	174	975	5.6
BB2	153	1383	9.0
BB3	195	1505	7.72
BB4	195	762	3.91
BB5-1	171	942	5.5
BB5-2	333	1908	5.73
BB5-3	222	620	2.78
BB5-4	90	993	11.03
BB6	53	648	12.2

The variability of 'tiles per second' values among the four participants adopting BB5 is as important as the variability among all participants (from 2.78 for participant BB5-3 to 11.03 for participant BB5-4). These results invalidate our hypothesis that there were browsing behaviors that were more energy-efficient than others, as the 'tiles per second' value seems more related to personal behaviors. To better understand the results, we interviewed the participants with extreme (low and high) values after the experiment. The participants with the lowest values all mentioned that they spent significant time reading the map, without any interaction. In other words, they spent a significant time in a static phase of map browsing as opposed to the transition phase [28] where the map changes due to zoom or pan. On the contrary, the participants with very high values reported that they were continuously in a transition phase, either panning or zooming, to find the target. It seems that what reduces the number of tiles used is to take the time to read maps with long static phases, instead of long transition phases. This result must be confirmed with a controlled user survey with enough participants for statistical significance.

Figure 3 shows the tiles loaded by the participant that adopted BB2 (use of the trackpad). This visualization confirms that the tiles collected during our experiment really correspond to the tiles visualized by this participant while completing the task, as the large-scale tiles are located around the cities of Toulouse and Dijon, the landmarks to find the targets. Analyzing this image, we can see that the participant searched longer for the first targets as most tiles at zoom level 14 around Toulouse were loaded, while the target around Dijon was directly found, resulting in fewer tiles loaded around the city.

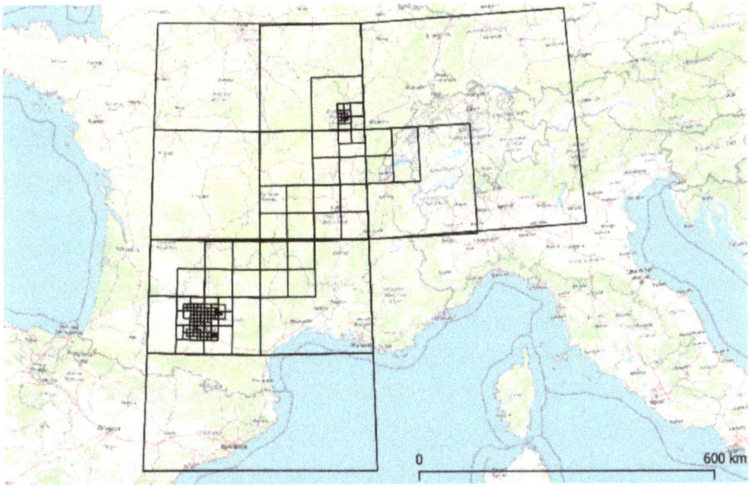

Figure 3. Visualization of the tiles loaded during the experiment by the participant with behavior BB2 (used the trackpad and not the mouse). ©OpenStreetMap contributors.

A question derived from the number of tiles transferred during the use of web maps is which types of tiles are mostly used and transferred, and more precisely, which zoom levels are more used than others. Knowing the zoom levels that are more frequently transferred on the client side could help designers focus on these zoom levels to reduce energy consumption. Figure 4 shows the mean number of tiles used by the participants of the experiment, from zoom level 1 to 20 (the tile from zoom level 0 was never loaded during the experiment). As the numbers visualized in this figure are related to only one session, each unique tile can be loaded only once, as it is stored in the cache afterwards. As a result, the numbers for the smallest zoom levels are limited by the small number of tiles at these scales.

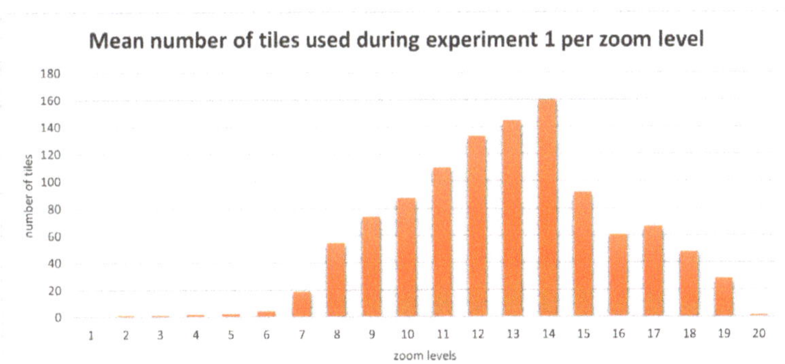

Figure 4. Mean number of tiles loaded per participant in Experiment 1 for each zoom level.

However, it can be noted that the peak in the number of tiles corresponds to zoom level 14. This result is surprising because the views at this zoom level cover a small part of the space the participant had to explore to complete the task (Figure 5). Furthermore, the tiles from zoom levels 12 and 13, where the names of the targets are visible, were also loaded a significant number of times.

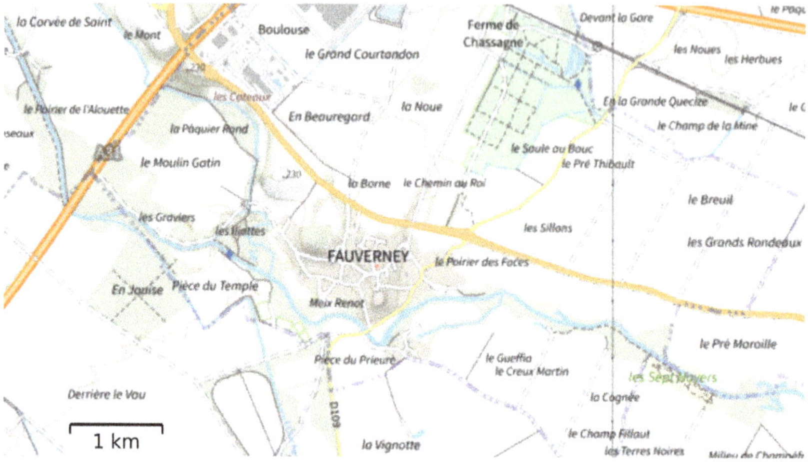

Figure 5. The small town of Fauverney, the second target of the experiment, visualized at zoom level 14 in the Plan IGN map (©IGN).

3.3. Comparison with Past Studies

As Experiment 1 only represents one of the many possible map uses, the number and the zoom level of the tiles used might be different when people are using the map for navigation or when they use the search bar to directly fly to an address or a point of interest. Thus, we searched the literature for other studies analyzing the tiles consumed by map users, to compare them with our results.

OpenStreetMap distributes the logs recording the use of their default tiles (https://planet.openstreetmap.org/tile_logs/, accessed on 18 December 2024), and we can compare the distribution of those tiles across zoom levels to the data we collected in our experiment, even if the information is slightly different: the logs from OpenStreetMap aggregate the access from many users to the same tiles, while our data only record one session at a time. Figure 6 shows the distribution of the tiles used during one week (between 13 November 2024 and 20 November 2024) and shows a prominent use of the medium–large zoom levels, between 13 and 17, with peak values at zoom levels 13 and 15. The curve is not very different from the one obtained with our experimental data, though it is skewed towards the largest zoom levels, with tiles at zoom levels 16 and 17 being highly accessed. Our interpretation of these figures is that the most loaded zoom levels are not the ones where we read the map during a static phase, but the ones that are used to navigate towards a target location and scale.

The results from an experiment with many users [29] exhibit a similar pattern after logging a large number of Google Maps sessions, but they are skewed even more towards larger scales. Zoom levels 15 and 17 are the most used zoom levels, which can be explained by the most prominent interaction used in Google Maps: the search bar. When typing an address or the name of a point of interest in the search bar, Google Maps flies to the location at zoom level 17.

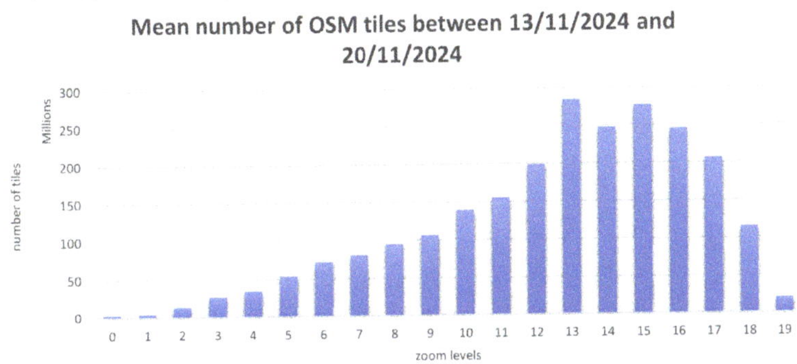

Figure 6. Mean number of OSM tiles loaded per day for each zoom level, for the week between 13 November 2024 and 20 November 2024.

In addition to the distribution of tiles across zoom levels, we can also analyze the spatial distribution of those tiles. Figure 7 shows a visualization of the tiles transferred by OSM users during a day (20 November 2024), using the OSM tile access logs viewer (https://osm-tile-access-log-viewer.raifer.tech/, accessed on 18 December 2024). At zoom level 11, we can see that the use is mainly centered in Western Europe, with peak values around large cities. At zoom level 16, only the urban tiles are transferred, and the large-scale rural tiles are never visualized. It appears that OpenStreetMap is mainly used to explore cities, where the data are usually more exhaustive.

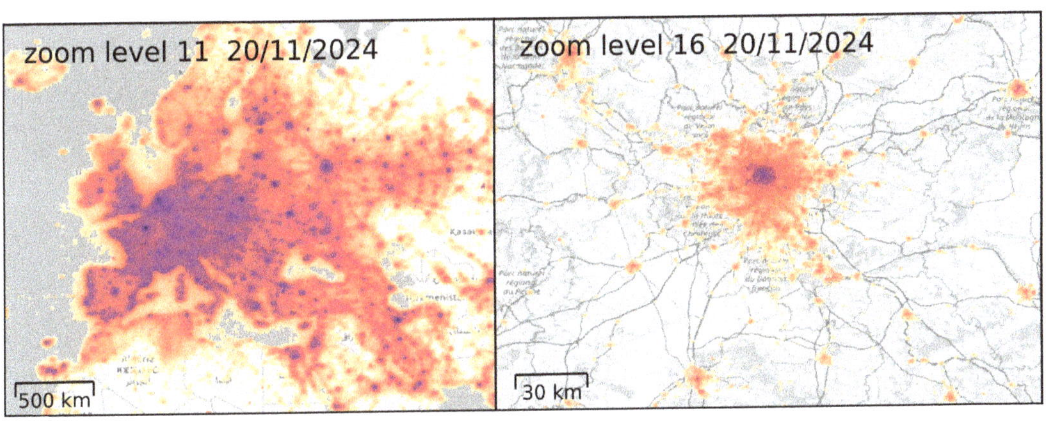

Figure 7. Visualization of the tiles loaded by OpenStreetMap users on 20 November 2024, from the OSM tile access logs viewer. Zoom level 11 tiles (on the **left**), zoom level 16 tiles (on the **right**).

4. Comparing the Weight of Map Tiles

Studying the number of tiles transferred during the use of web maps is not enough to quantify the energy consumption during this transfer because each tile has a different size in bits, even though the size in pixels is the same (256 × 256). Figure 8 shows three tiles from OpenStreetMap at the same zoom level (10). Tile (a) weighs 47 Kb, probably due to the numerous small forest patches. Tile (b), with a similar landscape but larger forest patches, weighs 38 Kb. Tile (c), with a large portion of sea, only weighs 5 Kb. To explore these size differences quantitatively, we designed another experiment, called Experiment 2 in the paper, to collect tiles at different zoom levels.

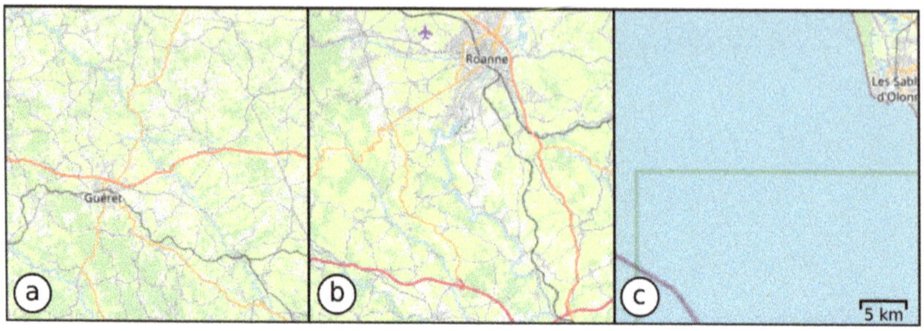

Figure 8. Three tiles from OpenStreetMap at zoom level 10. Tile (**a**) weighs 47 Kb, tile (**b**) Kb, and tile (**c**) 5 Kb. ©OpenStreetMap contributors.

4.1. Materials and Methods

We can see in the example in Figure 8 that size depends on the depicted landscape, and we designed a protocol to systematically collect tiles from different regions with large cities, small towns, rural areas, mountains, and coastal areas. We used the same environment as that in the previous experiment, with a Firefox browser opened and the network console printing the tile queries. This time, the zoom level was fixed and only panning was used. The task was to quickly pan for 1.5 min. After several beta tests, we decided that 1.5 min of quick pans were sufficient to collect a significant number of different tiles. The participants were given two additional instructions: (1) they were asked to avoid coming back to previously loaded tiles as much as possible, and (2) they were asked to scan as many different landscapes as possible. A tutorial was provided to obtain behaviors that were as consistent as possible among participants. At the end of the 1.5 min, the network console of the browser was saved as a .har file that was then parsed with the same Python script to obtain a CSV file of all the tiles loaded during the session. Figure 9 shows an example of the tiles collected at zoom level 16 following this protocol. The nine participants from Experiment 1 also formed the pool of participants in Experiment 2.

Following the discussion of the number of tiles used per zoom level in the previous section, we decided to collect tiles for all zoom levels between 10 and 17 (included). This protocol was reproduced for three different map styles: Plan IGN (https://www.geoportail.gouv.fr/carte, accessed on 18 December 2024) from IGN, the French National Mapping Agency; OpenCycleMap (https://www.opencyclemap.org/, accessed on 18 December 2024), which we considered to be a more complex topographic map with additional layers depicting cycling lanes and routes; and Positron Carto, which can be accessed with umap (https://umap.openstreetmap.fr, accessed on 18 December 2024), and is an example of a map with a faded topographic background, designed to include additional thematic layers on top. Without those additional layers, Positron Carto was expected to use lighter image tiles.

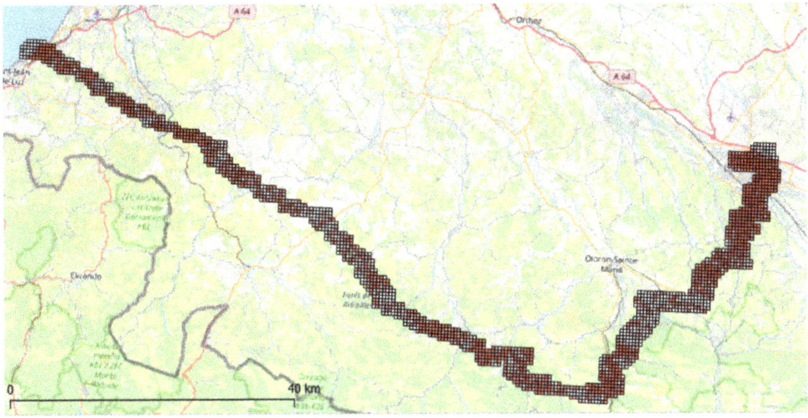

Figure 9. The tiles collected at zoom level 16 during the second experiment. The darker the red is, the larger the tile is (in Kb). The tiles are displayed over an OpenStreetMap background (©OpenStreetMap contributors).

4.2. Results

Figure 10 shows the distribution of the tile sizes in bytes for the eight tested zoom levels (from 10 to 17) from the Plan IGN map produced by the French national mapping agency. Zoom levels 11 and 14 contain the largest tiles, while zoom levels 16 and 17 contain the smallest tiles. When observing the map at the largest two zoom levels (11 and 14), the distribution of tile sizes is not surprising, as zoom level 11 corresponds to the appearance of the toponyms of all municipalities, while zoom level 14 corresponds to the appearance of the toponyms for hamlets and similar local places (Figure 11).

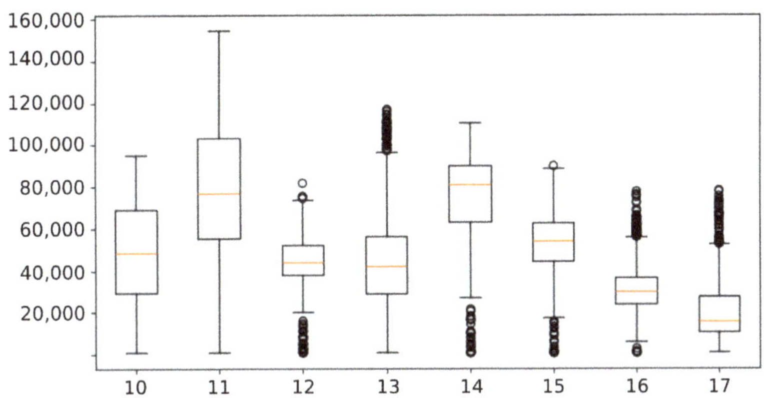

Figure 10. Box plot of the image tiles size in bytes across zoom levels 10 to 17 with the Plan IGN map style from IGN.

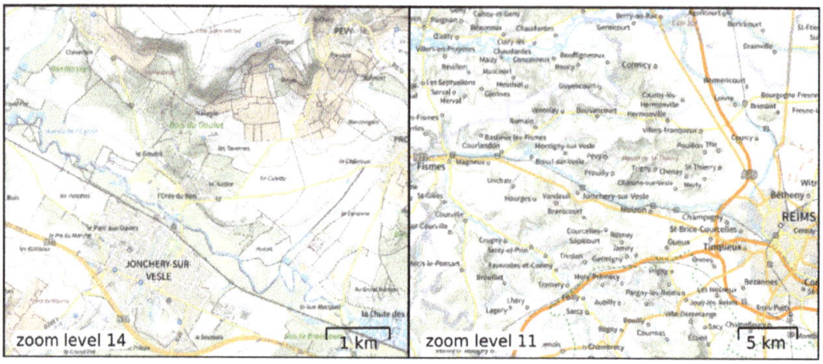

Figure 11. Two extracts of the Plan IGN map at zoom levels 11 and 14, the ones with the largest mean size values (Source: IGN).

Figure 12 shows the distribution of the tile sizes in bytes for the eight tested zoom levels (from 10 to 17), from the default OpenCycleMap style. It is first important to note that, contrary to our hypothesis, the tiles from OpenCycleMap are lighter than the ones from Plan IGN. As such, we can see that the tiles become lighter as the zoom level increases, zoom level 10 being the one with the largest tiles. This can be explained by the fact that as scale increases, there is no additional object depicted on the map, and almost all objects related to cycling routes are already visible at zoom level 10.

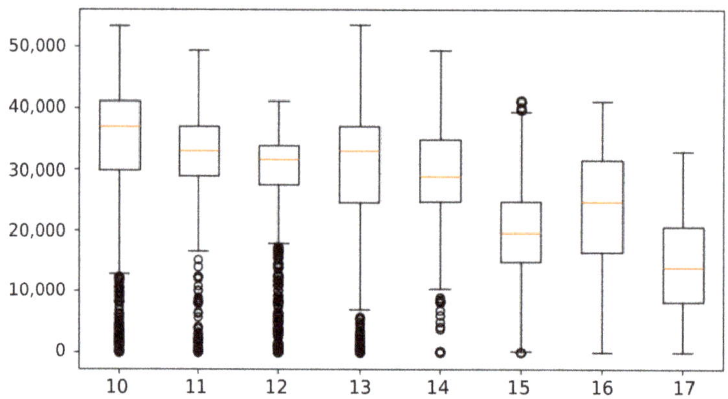

Figure 12. Box plot of the image tiles size in bytes across zoom levels 10 to 17 with the OpenCycleMap style.

Finally, Figure 13 shows the distribution of the tile sizes in bytes for the eight tested zoom levels (from 10 to 17) from the Positron Carto style produced by CartoDB, based on OpenStreetMap data. The curve of tile sizes across zoom levels is slightly different in this case. While the largest zoom levels are still the ones containing the lightest tiles, zoom levels 11 and 12 are the ones containing the largest tiles. Zoom level 12 corresponds to the appearance of the toponyms of most municipalities, which increases the visual complexity of the map, and thus the size of the tiles. It should be noted that, as expected, the tiles are much lighter than those of the two other maps tested in the experiment, with most zoom levels containing tiles with a mean size below 10 Kb.

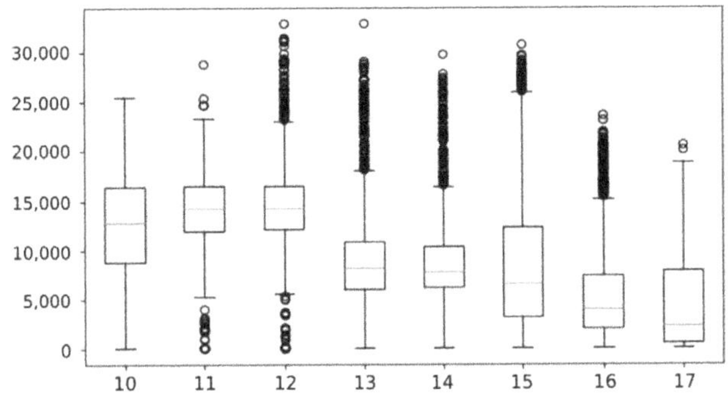

Figure 13. Box plot of the image tiles size in bytes across zoom levels 10 to 17 with the Positron map style from CartoDB.

5. The Role of Map Generalization

We noted in Section 2 that authors claim that web maps should rely as much as possible on vector tiles [12,13]. There is a general belief that image tiles are larger than vector tiles for the same spatial extent, but in reality, the sizes are quite similar. Experiment 3, presented in this section, explores the size differences between vector and raster tiles. Figure 14 shows a simple topographic map tile made from four vector layers (roads, buildings, vegetation, and water areas). The image is 502 × 501 pixels and weighs 120 Kb. If we convert the four layers into GeoJSON files (the usual data format to transfer vector data), the total size is 159.19 Kb (Table 3). To make sure the vector data are lighter (to reduce storage and transfer energy), we can transform the vector data. First, we reclassified the semantic data attached to the vectors as the attribute values were sometimes very long text strings (e.g., "Forêt fermée de feuillus" for the nature of vegetation, which means "forest with deciduous trees"). We converted all the text values into integer values (e.g., "Forêt fermée de feuillus" becomes '0'). This reclassification process reduces the vector file size to 151.9 (5% reduction). Then, we processed the four layers with generalization algorithms that reduce the number of vertices in the geometries without any visible change in the map at this scale. We used a Douglas–Peucker algorithm [30] for roads and buildings, and the Visvalingam–Whyatt algorithm [31] for water bodies and vegetation. The total size of the vector files drops to 103.55 Kb after generalization, below the 120 Kb of the raster tile. Of course, this experiment is not sufficient to demonstrate a general benefit from the use of map generalization for web maps with vector tiles, but it shows that further studies should be carried out to find generalization operations that reduce the size of vector tiles.

Table 3. Size in Kb of the GeoJSON files constituting the map from Figure 14.

Layer	Initial Size	After Attribute Simplification	After Generalization
roads	71.9	67.3	51.6
buildings	58.7	56.3	42.0
water	5.29	5.20	3.04
vegetation	23.3	23.1	6.91
complete map	159.19	151.9	103.55

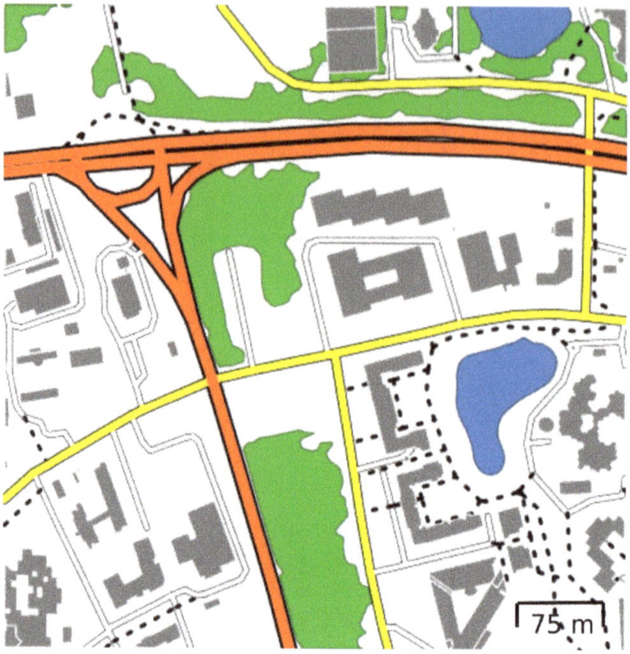

Figure 14. A raster map tile to illustrate the potential impact of map generalization on the size in Kb of the tiles.

In the first part of Experiment 3, presented above, generalization was used as a tool to reduce the size of a tile without changing the way the map is perceived by a human eye. In the second part of the experiment, we used generalization to visually simplify the map, to assess the impact of this transformation on the size of the tile, whether it is an image or a vector tile. On small datasets, we tested the impact of several types of map generalization operations on the size of the resulting images (Figure 15) and vector tiles. First, we processed four buildings from OpenStreetMap with the AGENT generalization model [32], which automatically chains enlargement, simplification, and squaring algorithms. In this case, generalization only results in a 2% reduction in image size (Figure 15a), while the reduction in terms of vertices is much more significant (from 48 to 29 vertices). The second example generalizes a road network using an automated selection process [33], this time resulting in a 23% reduction in image size (Figure 15b). The vector tile is much more reduced as the number of vertices is decreased by 55%. Finally, we tested the impact of a displacement algorithm [34,35] that moves the buildings that intersect the symbols of roads and rivers (Figure 15c). This time, the impact was negative on image size, and it increased by 6%, because more pixels are colored without overlaps. In this last case, the vector tile size remains unchanged as the number of vertices stays the same. Considering that map generalization usually mixes those operations (and others) to process a complete map, these measures show that the impact of map generalization on the weight of image and vector tiles is a complex problem that deserves further dedicated research.

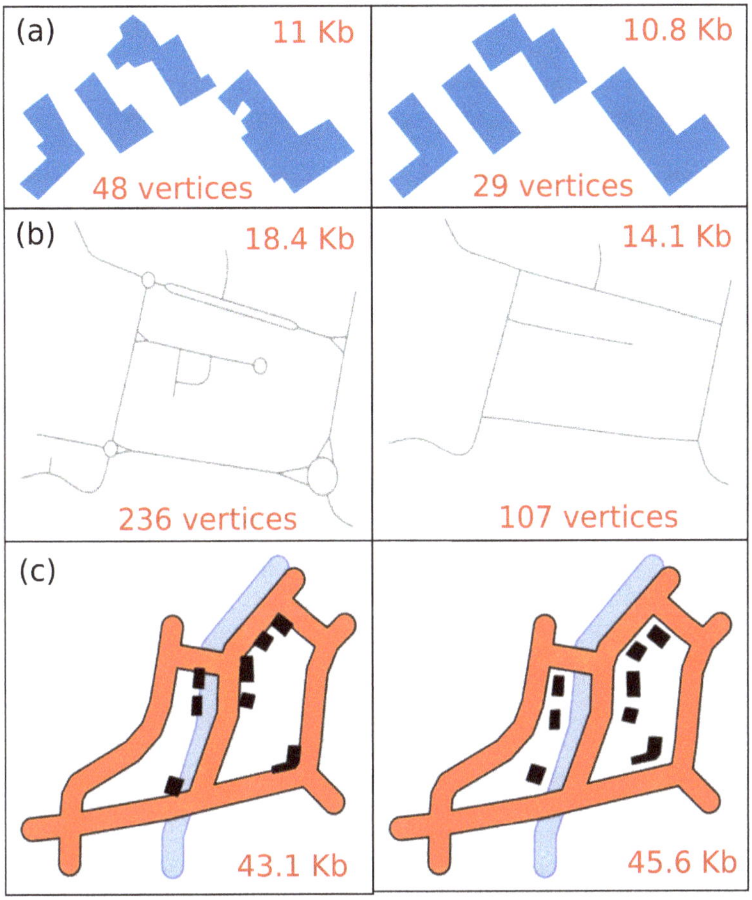

Figure 15. Three different types of generalization operations and their impact on the size of image tiles. (**a**) Building generalization (simplification, enlargement, and squaring) slightly reduces image size. (**b**) Road selection significantly reduces image size. (**c**) Displacement increases image size as the road and river symbol parts previously hidden under buildings are now visible.

6. Towards Energy-Efficient Maps

In our experiments, we have seen that many tiles are never visualized. Still, they are updated and recomputed frequently, a practice leading to increased energy consumption. Vector tiles would prevent some of this unnecessary computation, as styling is processed on the client side only when the tile is queried. Another solution would be to reinvestigate incremental update research [36–43]. Currently, when there is an update in the data somewhere (e.g., a new road or a new building), the complete tiles covering the new features are recomputed at all scales, sometimes including other tiles nearby, or even sometimes all the tiles across the world, because the process is global. Incremental updating proposes solutions to define the range of updates, and to propagate them across all scales/zoom levels with minimal data processing.

Experiment 1 shows that whatever the browsing behavior, map users query many tiles. But do we read all the tiles loaded during the use of a map? Or, to put it differently, do we really need all the tiles to be loaded during the use of the map? It might be interesting to study with eye tracking how much each of the loaded tiles is visualized by map users. In particular, the tiles that are preventively loaded in the cache might not all be useful. Such a study might also plead for the use of lighter tiles loaded during the transition phases because those tiles are only observed through peripheral vision [44].

We show in the results of Experiment 1 that all zoom levels are not equal in terms of use by map readers, and some levels are heavily used compared to others. If we want to make more energy-efficient maps, we need to focus on those specific zoom levels, to reduce their transfer cost, as well as the cost to render them on the screen. In our opinion, those heavily used zoom levels (zoom levels 10 to 17) are subject to improvement thanks to map generalization techniques, and we should study how much generalization can help reduce the energy cost at these zoom levels, following the short experiments presented in the previous section. We also assume that more map generalization could make each zoom level clearer and, as a result, reduce the number of zooms from the user [18]; this assumption could be tested with user surveys.

In recent years, zoom levels have become continuous values, which might be an incitement for changing the map between zoom levels to add more continuity to the zooming experience [45]. Recent Google Maps versions have slightly changed the map by adding text and points of interest at certain half-zoom levels (for example, some town names appear at zoom level 12.5). More changes between zoom levels imply more storage for these intermediate tiles. In the future, it will be necessary to find a balance between the number of zoom levels required for better usability of the map [17,45] and the energy cost of these additional levels.

Finally, even though our own experiment was not conclusive in the identification of eco-friendly map browsing behaviors, we still believe that there is an eco-friendly way to browse maps and that it should be taught to map users, or even enforced through interaction design choices. As a result, further experiments to better define eco-friendly map browsing behaviors are necessary. We could use trackers of map browsing behavior [29,46,47] on large populations of users to identify which ones generally use fewer tiles. We can also couple the tracking of interactions with more classical techniques to track the attention of the user during the exploration of the map. Eye-tracking is frequently used in cartography [48,49] and it could be used to better characterize what we identified as an efficient behavior, i.e., the use of long static phases and short transition phases. Mouse tracking [50] could also be useful to better characterize the browsing behaviors of users that tend to transfer fewer tiles than others for a similar task.

7. Conclusions

To conclude, this paper reported three short experiments to better understand the energy consumption of web maps related to tile storage and transfer. The first finding is that the number of tiles transferred during each use of a web map is significant. Second, the distribution of these tiles among zoom levels is not regular, with a peak between levels 13 and 16, though the distribution depends on the task accomplished with the map. The third finding is that the size of the tiles is irregular across zoom levels, with larger tiles often corresponding to the most heavily loaded tiles. Finally, we found that map generalization can play an important role in the mitigation of this energy consumption, but further studies are necessary to better define what its role can be in the design of greener web maps.

We are convinced that energy cost should become one of the key factors to consider when designing a map. In the future, researchers should provide evidence of good and bad practices in terms of energy consumption to help map designers move towards greener cartography.

Author Contributions: Conceptualization, Guillaume Touya and Azelle Courtial; methodology, Justin Berli, Azelle Courtial, Bérénice Le Mao, Guillaume Touya, and Laura Wenclik; software, Jérémy Kalsron; validation, Justin Berli, Azelle Courtial, Jérémy Kalsron, Bérénice Le Mao, Guillaume Touya, and Laura Wenclik; formal analysis, Guillaume Touya; data curation, Justin Berli, Azelle Courtial, Jérémy Kalsron, Bérénice Le Mao, Guillaume Touya, and Laura Wenclik; writing—original draft preparation, Guillaume Touya; writing—review and editing, Justin Berli, Azelle Courtial, Bérénice Le Mao, Jérémy Kalsron, and Laura Wenclik; visualization, Jérémy Kalsron and Guillaume Touya; supervision, Guillaume Touya; project administration, Guillaume Touya; funding acquisition, Guillaume Touya. All authors have read and agreed to the published version of the manuscript.

Funding: This project received funding from the European Research Council (ERC) under the European Union's Horizon 2020 research and innovation programme (grant agreement No. 101003012, LostInZoom).

Institutional Review Board Statement: Not applicable.

Informed Consent Statement: Not applicable.

Data Availability Statement: The dataset supporting the results presented in this article can be found in an open Zenodo repository (https://doi.org/10.5281/zenodo.14281428, accessed on 15 December 2024).

Conflicts of Interest: The authors declare no conflicts of interest.

References

1. Andrae, A.S.G.; Edler, T. On Global Electricity Usage of Communication Technology: Trends to 2030. *Challenges* **2015**, *6*, 117–157. [CrossRef]
2. Williams, J.; Curtis, L. Green: The New Computing Coat of Arms? *IT Prof.* **2008**, *10*, 12–16. [CrossRef]
3. Russell, E. 9 Things to Know About Google's Maps Data: Beyond the Map. 2019. Available online: https://mapsplatform.google.com/resources/blog/9-things-know-about-googles-maps-data-beyond-map/ (accessed on 18 December 2024).
4. Wu, M.; Lv, G.; Qiao, L.; Roth, R.E.; Zhu, A.X. Green Cartography: A research agenda towards sustainable development. *Ann. GIS* **2024**, *30*, 15–34. [CrossRef]
5. Han, Y.; Wu, M.; Roth, R. Toward green cartography & visualization: A semantically-enriched method of generating energy-aware color schemes for digital maps. *Cartogr. Geogr. Inf. Sci.* **2021**, *48*, 43–62. [CrossRef]
6. Wu, M.; Lv, G.; Yuan, L. Green Cartography and Energy-Aware Maps: Possible Research Opportunities. In *New Thinking in GIScience*; Li, B., Shi, X., Zhu, A.X., Wang, C., Lin, H., Eds.; Springer Nature: Singapore, 2022; pp. 197–205. [CrossRef]
7. Belkhir, L.; Elmeligi, A. Assessing ICT global emissions footprint: Trends to 2040 & recommendations. *J. Clean. Prod.* **2018**, *177*, 448–463. [CrossRef]
8. Dong, M.; Zhong, L. Power Modeling and Optimization for OLED Displays. *IEEE Trans. Mob. Comput.* **2012**, *11*, 1587–1599. [CrossRef]
9. Kent, A.J.; Vujakovic, P. Cartographic Language: Towards a New Paradigm for Understanding Stylistic Diversity in Topographic Maps. *Cartogr. J.* **2011**, *48*, 21–40. [CrossRef]
10. Christophe, S.; Hoarau, C. Expressive Map Design Based on Pop Art: Revisit of Semiology of Graphics? *Cartogr. Perspect.* **2013**, *73*, 61–74. . [CrossRef]
11. Hoarau, C. Reaching a Compromise between Contextual Constraints and Cartographic Rules: Application to Sustainable Maps. *Cartogr. Geogr. Inf. Sci.* **2011**, *38*, 79–88. [CrossRef]
12. Gaffuri, J. Toward Web Mapping with Vector Data. In *Geographic Information Science*; Xiao, N., Kwan, M.P., Goodchild, M.F., Shekhar, S., Eds.; Lecture Notes in Computer Science; Springer: Berlin/Heidelberg, Germany, 2012; Volume 7478, pp. 87–101. [CrossRef]
13. Gaffuri, J. Improving Web Mapping with Generalization. *Cartogr. Int. J. Geogr. Inf. Geovisualiz.* **2011**, *46*, 83–91. [CrossRef]
14. Balley, S.; Parent, C.; Spaccapietra, S. Modelling geographic data with multiple representations. *Int. J. Geogr. Inf. Sci.* **2004**, *18*, 327–352. [CrossRef]
15. van Oosterom, P. Variable-scale Topological Data Structures Suitable for Progressive Data Transfer: The GAP-face Tree and GAP-edge Forest. *Cartogr. Geogr. Inf. Sci.* **2005**, *32*, 331–346. [CrossRef]
16. Gruget, M.; Touya, G.; Muehlenhaus, I. Missing the city for buildings? A critical review of pan-scalar map generalization and design in contemporary zoomable maps. *Int. J. Cartogr.* **2023**, *9*, 255–285. [CrossRef]

17. Dumont, M.; Touya, G.; Duchêne, C. Designing multi-scale maps: Lessons learned from existing practices. *Int. J. Cartogr.* **2020**, *6*, 121–151. [CrossRef]
18. Courtial, A.; Touya, G. Does Generalisation Matters in Pan-Scalar Maps? In Proceedings of the 12th International Conference on Geographic Information Science (GIScience 2023), Leibniz International Proceedings in Informatics (LIPIcs), Leeds, UK, 12–15 September 2023; Volume 277, pp. 23:1–23:6. [CrossRef]
19. Vieira, M.R.; Bakalov, P.; Hoel, E.; Tsotras, V.J. A Spatial Caching Framework for Map Operations in Geographical Information Systems. In Proceedings of the 2012 IEEE 13th International Conference on Mobile Data Management, Bengaluru, India, 23–26 July 2012; pp. 89–98. [CrossRef]
20. Quinn, S.; Gahegan, M. A Predictive Model for Frequently Viewed Tiles in a Web Map. *Trans. GIS* **2010**, *14*, 193–216. [CrossRef]
21. Pan, S.; Chong, Y.; Zhang, H.; Tan, X. A Global User-Driven Model for Tile Prefetching in Web Geographical Information Systems. *PLoS ONE* **2017**, *12*, e0170195. [CrossRef]
22. Bertolotto, M.; Egenhofer, M. Progressive Transmission of Vector Map Data over the World Wide Web. *GeoInformatica* **2001**, *5*, 345–373. [CrossRef]
23. Buttenfield, B. Transmitting Vector Geospatial Data across the Internet. In *Geographic Information Science*; Egenhofer, M., Mark, D., Eds.; Lecture Notes in Computer Science; Springer: Berlin/Heidelberg, Germany, 2002; Volume 2478, pp. 51–64. [CrossRef]
24. Sester, M.; Brenner, C. A vocabulary for a multiscale process description for fast transmission and continuous visualization of spatial data. *Comput. Geosci.* **2009**, *35*, 2177–2184. [CrossRef]
25. Antoniou, V.; Morley, J.; Haklay, M. Tiled Vectors: A Method for Vector Transmission over the Web. In *Web and Wireless Geographical Information Systems*; Carswell, J.D., Fotheringham, A.S., McArdle, G., Eds.; Lecture Notes in Computer Science; Springer: Berlin/Heidelberg, Germany, 2009; Volume 5886, pp. 56–71. [CrossRef]
26. Corcoran, P.; Mooney, P. Topologically Consistent Selective Progressive Transmission. In *Advancing Geoinformation Science for a Changing World*; Geertman, S., Reinhardt, W., Toppen, F., Eds.; Lecture Notes in Geoinformation and Cartography; Springer: Berlin/Heidelberg, Germany, 2011; Volume 1, pp. 519–538. [CrossRef]
27. Glueck, M.; Khan, A.; Wigdor, D.J. Dive in! Enabling progressive loading for real-time navigation of data visualizations. In Proceedings of the SIGCHI Conference on Human Factors in Computing Systems, CHI'14, Toronto, ON, Canada, 26 April–1 May 2014; pp. 561–570. [CrossRef]
28. Potié, Q.; Touya, G.; Wenclik, L.; Mackaness, W. Browsing Map Browsers. *Abstr. ICA* **2024**, *7*, 130. [CrossRef]
29. Braga, V.G.; Corr, S.L.; Rodrigues, V.J.d.S.; Cardoso, K.V. Characterizing User Behavior on Web Mapping Systems Using Real-World Data. In Proceedings of the 2018 IEEE Symposium on Computers and Communications (ISCC), Natal, Brazil, 25–28 June 2018; pp. 01056–01061. [CrossRef]
30. Douglas, D.H.; Peucker, T.K. Algorithms for the Reduction of the Number of Points Required to Represent a Digitized Line or its Caricature. *Cartogr. Int. J. Geogr. Inf. Geovisualiz.* **1973**, *10*, 112–122. [CrossRef]
31. Visvalingam, M.; Wyatt, J.D. Line Generalization by Repeated Elimination of Points. *Cartogr. J.* **1993**, *30*, 46–51. [CrossRef]
32. Barrault, M.; Regnauld, N.; Duchêne, C.; Haire, K.; Baeijs, C.; Demazeau, Y.; Hardy, P.; Mackaness, W.A.; Ruas, A.; Weibel, R. Integrating multi-agent, object-oriented, and algorithmic techniques for improved automated map generalisation. In Proceedings of the 20th International Cartographic Conference, ICA, Beijing, China, 6–10 August 2001; Volume 3, pp. 2110–2116.
33. Touya, G. A Road Network Selection Process Based on Data Enrichment and Structure Detection. *Trans. GIS* **2010**, *14*, 595–614. [CrossRef]
34. Harrie, L.E. The Constraint Method for Solving Spatial Conflicts in Cartographic Generalization. *Cartogr. Geogr. Inf. Sci.* **1999**, *26*, 55–69. [CrossRef]
35. Sester, M. Generalization Based on Least Squares Adjustment. *Int. Arch. Photogramm. Remote. Sens.* **2000**, *33*, 931–938.
36. Kilpeläinen, L.T.; Sarjakoski, T. Incremental generalization for multiple representations of geographic objects. In *GIS and Generalization: Methodology and Practise*; Muller, J.C., Lagrange, J.P., Weibel, R., Eds.; Taylor & Francis: London, UK, 1995; pp. 209–218.
37. Harrie, L.E.; Hellström, A.K. A Prototype System for Propagating Updates Between Cartographic Data Sets. *Cartogr. J.* **1999**, *36*, 133–140. [CrossRef]
38. Skogan, D. Towards a Rule-based Incremental Generalization System. In Proceedings of the 5th AGILE Conference on Geographic Information Science, Palma, Spain, 25–27 April 2002.
39. Lecordix, F.; Lemarié, C. Managing Generalisation Updates in IGN Map Production. In *Generalisation of Geographic Information*; Mackaness, W.A., Ruas, A., Sarjakoski, L.T., Eds.; Elsevier: Amsterdam, The Netherlands, 2007; pp. 285–300. [CrossRef]
40. Anders, K.H.; Sester, M.; Bobrich, J. Incremental Update in a MRDB. In Proceedings of the 23rd International Cartographic Conference, ICA, Moscow, Russia, 4–10 August 2007.
41. Haunert, J.H.; Anders, K.H.; Sester, M. Hierarchical Structures for Rule-Based Incremental Generalisation. In Proceedings of the ISPRS Archives XXXVI Working Group II/2, Beijing, China, 3–11 July 2008.

42. Zhou, S.; Regnauld, N.; Roensdorf, C. Towards a Data Model for Update Propagation in MR-DLM. In Proceedings of the 11th ICA Workshop on Generalisation and Multiple Representation, ICA, Montpellier, France, 20–21 June 2008.
43. Straupis, T. Optimisation of generalisation re-calculation using partitioning. *Abstr. ICA* **2022**, *5*, 83. [CrossRef]
44. Rosenholtz, R. Capabilities and Limitations of Peripheral Vision. *Annu. Rev. Vis. Sci.* **2016**, *2*, 437–457. [CrossRef]
45. Dumont, M.; Touya, G.; Duchêne, C. More is less—Adding zoom levels in multi-scale maps to reduce the need for zooming interactions. *J. Spat. Inf. Sci.* **2023**, *27*. [CrossRef]
46. Popelka, S.; Vojtechovska, M.; Ruzicka, O.; Beitlova, M.; Stachon, Z. Adapting to the Mobile Majority: A New Approach to Interactive Map Usability Assessment. In Proceedings of the Online User Experiments: Seeing What Map Users See Without Seeing Them, Vienna, Austria, 9 September 2024.
47. Touya, G.; Brusq, R.; Campo, L.; Pech, V.; Wenclik, L. Recording the Use of Interactive Web Maps with ZoomTracker. *Abstr. ICA* **2024**, *7*, 176. [CrossRef]
48. Krassanakis, V.; Cybulski, P. Eye Tracking Research in Cartography: Looking into the Future. *ISPRS Int. J. Geo-Inf.* **2021**, *10*, 411. [CrossRef]
49. Fairbairn, D.; Hepburn, J. Eye-tracking in map use, map user and map usability research: What are we looking for? *Int. J. Cartogr.* **2023**, *9*, 231–254. [CrossRef]
50. Krassanakis, V.; Misthos, L.M. Mouse Tracking as a Method for Examining the Perception and Cognition of Digital Maps. *Digital* **2023**, *3*, 127–136. [CrossRef]

International Journal of
Geo-Information

MDPI

Article

Spatial Network Analysis of CO_2 Emissions in Major Cities in China: Regional Structures and Influencing Factors

Yue Zhao [1], Yuning Feng [1,*], Mingyi Du [1], Klaus Fraedrich [2] and Zehao Shen [1]

1 School of Geomatics and Urban Spatial Information, Beijing University of Civil Engineering and Architecture, Beijing 100044, China; 2108160222002@stu.bucea.edu.cn (Y.Z.); dumingyi@bucea.edu.cn (M.D.); 2108160223001@stu.bucea.edu.cn (Z.S.)

2 Max Planck Institute for Meteorology, 20146 Hamburg, Germany; klaus.fraedrich@mpimet.mpg.de

* Correspondence: fengyuning@bucea.edu.cn

Abstract: China's rapid industrialization and urbanization have led to significant and imbalanced CO_2 emissions, putting pressure on achieving sustainable development goals. This study analyzed the CO_2 emissions of 31 major cities in China from different sectors (total, power, industry, and transport) from 2019 to 2022. This study constructs a city-scale CO_2 emission correlation model to achieve nationwide and urban fine-scale research on CO_2 emission spatial networks from different sectors. This study revealed the following: (i) there is an increasing correlation among regions in China, and collaborative governance is crucial; (ii) there are differences in the structure, characteristics, and roles of CO_2 emission networks from different sectors; (iii) China's CO_2 emission network is mainly concentrated in the northern and eastern regions, which play an important role in emission reduction; and (iv) the impact factors have different effects on CO_2 emissions from different sectors, and we should actively contribute to promoting emission reduction. Correctly understanding the spatial characteristics and influencing factors of CO_2 emissions can help us formulate targeted and efficient emission reduction policies.

Keywords: CO_2 emissions; spatial network; sustainable development goals (SDGs); collaborative governance

1. Introduction

With the increasingly severe global climate change issue, CO_2 emissions have become a global focus of attention [1]. According to the United Nations Environment Programme (UNEP), global temperatures have been continuously rising in the past few decades, and one of the main causes is CO_2 emissions, which are produced by human activities [2]. In 2022, global CO_2 emissions reached a historic high, which has had a significant impact on global climate change, and fossil fuel combustion is the main source of CO_2 emissions [3]. This trend means the global ecological environment and sustainable development are seriously threatened. In order to achieve the long-term goals of sustainable development, countries worldwide have formulated a series of relevant policies and laws and proposed a series of goals, such as carbon neutrality, carbon peaking, sustainable development goals (SDGs), etc. [4,5].

The trend in CO_2 emissions in China presents a complex and severe situation, especially in the past few decades. China has become one of the world's largest CO_2 emitters, but the current trend has been curbed [6]. The Chinese government has recognized the seriousness of CO_2 emissions and has actively taken various measures to address it. They proposed the "dual-carbon" goal, which aims to achieve carbon peaking before 2030 and carbon neutrality before 2060 [7–9]. Due to the uneven level of regional economic development, there is a significant imbalance in the sectoral and spatial distribution of CO_2 emissions in China [10]. In addition, high-energy-consuming industries, especially power generation and manufacturing, account for over 80% of China's CO_2 emissions [11].

In recent years, due to the impact of the pandemic and the approaching deadline for SDGs, China has encountered new problems and challenges in controlling CO_2 emissions [12]. Considering the regional imbalance of CO_2 emissions in China, accurately and comprehensively understanding the distribution characteristics of CO_2 emissions can help the government formulate precise emission reduction policies, which is crucial for differentiated governance and coordinated development between different regions [13–18]. In previous studies, people explored the spatial characteristics of China's CO_2 emissions using the Kernel density estimation method, Markov chains, standard deviation ellipse, gravity models, etc. [19–22]. The results have shown that China's CO_2 emissions exhibit significant regional differentiation in different regions, with spatial correlations and spillover effects between regions [23]. In recent years, some studies have concluded that the spatial correlation of CO_2 emissions in China follows a network structure [24]. In addition, social network analysis has been applied in the field of geography research [25]. The network structure of CO_2 emissions has gradually become an important research direction. People not only pay attention to the regional heterogeneity of CO_2 emissions but also to the regional correlation of CO_2 emissions, which provides a new basis for China's emission reduction [26]. Early related studies were mostly conducted at the provincial level. For example, Fang et al. constructed a provincial CO_2 emission spatial network in China using a gravity model and analyzed its influencing factors [27]. However, Wang et al. pointed out that, at smaller study scales, the discrete nature of China's CO_2 emissions is more pronounced, which is conducive to reflecting the differences between regions [28]. In recent years, more research has explored the characteristic patterns of CO_2 emissions at the scale of cities. Luo et al. explored the changes in land use carbon budget and spatial correlation using 21 districts in Chongqing as the research scale [29]. Considering the regional imbalance in China, a study focusing on one province may not apply to other regions of China. Chen et al. conducted a gravity model and exploratory spatial data analysis of 274 cities at the prefecture level in China but lacked further analysis of network characteristics and influencing factors [30]. In addition, some studies only consider differences in CO_2 emissions and distance between regions when constructing spatial networks through the gravity model, ignoring the differences in economy, development level, etc., between regions [26,29]. Most of the research subjects are CO_2 emissions from total or one sector [10,19,21]. Considering the imbalance of CO_2 emissions among sectors in China [31], the lack of comparative analysis between sectors may lead us to overlook some results. There is still a research gap in nationwide, city fine scale, and sectoral CO_2 emission network studies.

This paper designed the city-scale CO_2 emission correlation model (CCCM) and further constructed spatial networks of CO_2 emissions between major cities in China, scientifically analyzed the structure and characteristics of the networks, and explored the influencing factors of CO_2 emissions from different sectors. The results revealed the spatial characteristics and influencing factors of China's CO_2 emissions and provided targeted emission reduction recommendations with a scientific basis. The main contributions of the research are (i) constructing a national spatial network at the city scale to more accurately display differences and features between regions and make fine results applicable to the whole country; (ii) designing CCCM, which is suitable for city-scale related research and can accurately reflect the correlation between CO_2 emissions from cities; and (iii) analyzing and comparing the impact of different influencing factors on CO_2 emissions from different sectors, to reveal the imbalance of CO_2 emissions among sectors in China from a new perspective, and provide a reference for relevant emission reduction policies.

2. Data and Methods

This study is structured around four major steps (Figure 1): (1) data collection and pre-processing of CO_2 emissions and relevant variables; (2) calculation and construction of the CO_2 emission correlation matrix through CCCM and calculation of the potential of each city through the potential model; (3) analyzing the characteristics of the CO_2 emission correlation network through social network analysis (SNA) to characterize the spatial pattern of China's CO_2 emissions; and (4) comparison and analysis of the influencing factors of CO_2 emissions from different sectors through the quadratic assignment procedure (QAP).

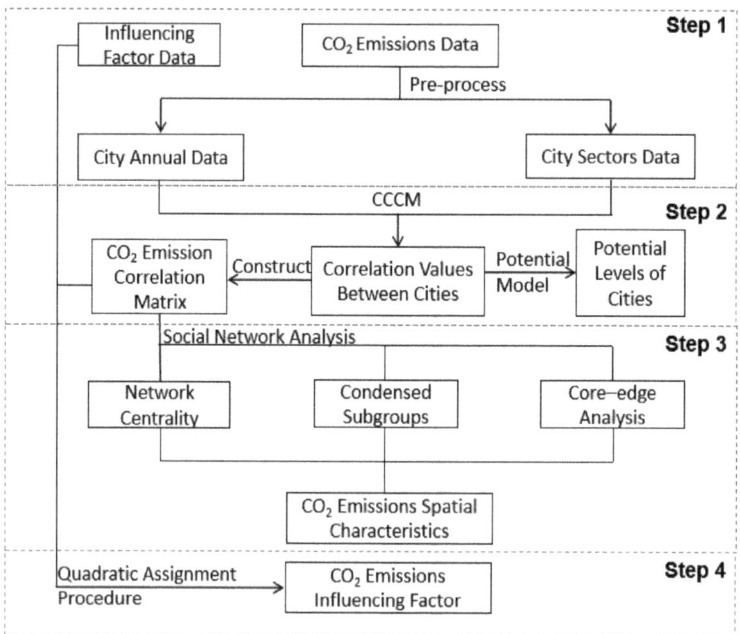

Figure 1. The process in this study.

2.1. Research Area and Data

Cities are key nodes in the CO_2 emission network. As the centers of development for their respective provinces, provincial capital cities often lead in terms of industrial activities, technological advancements, and environmental regulations, making them significant contributors to both economic growth and CO_2 emissions. Thus, they can, to some extent, represent the economic development and emission reduction situations within each province, providing a valuable snapshot of urban development and policy implementation at the provincial level. Consequently, studying these cities offers insights into the effectiveness of provincial-level policies and strategies to reduce emissions. Thus, we selected 31 provincial capital cities in China (excluding Hong Kong, Macau, and Taiwan) as study sites to represent the overall CO_2 emission pattern in China.

The types of CO_2 emission sources include the total, power, industry, and transport sectors, as well as other sectors. It should be noted that other sectors contribute little to CO_2 emissions and are therefore not included in the scope of this study. The study period was 4 years, from 2019 to 2022.

The raw data for this paper came from GRACED (Global gRidded dAily CO_2 Emissions Dataset). It provides near-real-time global gridded daily CO_2 emissions data and satisfies the demand for high-quality, high-precision, and near-real-time data on CO_2 emissions to support global emissions monitoring across various spatial scales [32–35]. We unified the coordinate system of all data, counted the grid data to the city scale, and resampled the data in the urban boundary area to reduce errors. We obtained annual data on CO_2 emissions from sectors (the total, power, industry, and transport sectors) from 2019 to 2022. The spatial distance between cities was obtained through ArcGIS; other data were obtained through the statistical yearbooks or statistical bulletins from the urban statistical bureau.

2.2. Construction of Spatial Correlation Network

The gravity model is a widely used diagnostic method for spatial interactions, based on the analogy of Newton's universal gravitation formula (as the product of the gravitational constant multiplied by the mass of two objects, divided by the square of their distance). It has been continuously extended and developed in many other fields and has gradually become the main tool for quantitatively analyzing spatial relationships and effects [36,37]. Based on the gravity model, we constructed the city-scale CO_2 emission correlation model (CCCM) to calculate the correlation between urban CO_2 emissions according to the requirements of this study and the characteristics of urban CO_2 emissions. At present, there is a significant difference in the economic level of cities in China, and the development of the economy is closely related to CO_2 emissions. Therefore, the difference in CO_2 emissions among cities is also enormous. In order to accurately reflect the CO_2 emissions under different levels of economic development, we used CO_2 emission intensity in the CCCM, which is defined as the amount of CO_2 emissions per unit of gross domestic product (GDP). In addition, we added the urbanization rate as a correction factor, which reflects the level of urban development and represents the proportion of CO_2 that plays a major role in the correlation between urban CO_2 emissions. In the CCCM, we used the product of urbanization rate and CO_2 emission intensity to represent mass, which reflects the differences in urban development, economic intensity, and CO_2 emissions levels. As for the distance in the model, we used the straight-line distance between cities obtained through Arcgis. In addition, we consider that the relationship between urban CO_2 emissions is directional, that is, influence or being influenced. Thus, we added the coefficient G to characterize the direction of the CO_2 emission impact. The CCCM defines r_{ij} as the value of the correlation of CO_2 emissions between city i and city j,

$$r_{ij} = G\frac{U_iF_iU_jF_j}{d_{ij}^2} \tag{1}$$

which reflects the strength and direction of their spatial correlation; d_{ij} defines the spatial distance between city i and city j. U_i and U_j measure the urbanization rate of the city in terms of weight and influential part of the city's CO_2 emissions; F_i and F_j introduce the CO_2 emissions per unit of GDP of a city. The coefficient G represents the direction of spatial correlation between urban CO_2 emissions defined as

$$G = \frac{F_i}{F_i + F_j} \tag{2}$$

We calculated the correlation values between each city to obtain an average value. If the correlation value between the two towns exceeded this average, we consider there to be a spatial correlation between them. We constructed spatial correlation matrices for each year and each sector based on whether there is a spatial correlation in CO_2 emissions between cities and finally constructed spatial correlation networks using UCINET software (Ver. 6) [38]. Note that the cities' CO_2 emissions spatial correlation matrix is a binary matrix where 1 indicates a correlation between the two cities, and 0 indicates none. It has also been used in social network analysis; a more detailed explanation can be found in the annotations of Section 2.4.

Based on the CCCM, we can calculate the spatial correlation value between two cities' CO_2 emissions. Further, we can achieve the intensity of CO_2 emission interaction from one city to all other cities by the CCCM, which reflects, to some extent, the city's current status in the CO_2 emission network. Moreover, due to the different status and speed of development of cities, the future development potential of their status in their network also varies. In order to study this potential and explore key cities for future policy implementation, we constructed a network model for the status potential of urban CO_2 emission:

$$\mathrm{T} = \frac{C}{\overline{r_i}} \tag{3}$$

$$\overline{r_i} = \frac{1}{n}\sum_{j=1}^{n} r_{ij} \tag{4}$$

T represents the potential for the future development of the city's status within the network; C is for the city's CO_2 emissions, which can represent the existing strength of a city and also measure its competitiveness in the network; and $\overline{r_i}$ represents the current status of a city in the CO_2 emission network in terms of an average "gravity" compared to all other cities. That is, the higher the current status of a city, the less space is available for future upward mobility. Therefore, T is directly proportional to strength C and inversely proportional to $\overline{r_i}$.

2.3. Analysis of Spatial Network Characteristics

After constructing the CO_2 emission spatial correlation network, this paper used social network analysis (SNA) to analyze the structure and characteristics of the network and its nodes. SNA is a method of quantifying the relationships between members embedded in complex and variable network structures. It can comprehensively characterize the complex relationships between individuals, between individuals and groups, and between groups, and thus provide a deep understanding of the network structure [39]. In addition, SNA uses mathematical and statistical tools for quantitative analysis, which makes the results more objective and accurate. In our research, the network represents the overall spatial scale pattern of CO_2 emissions in China, with nodes representing cities under study. The characteristic indicators of the network are shown in Table 1.

Table 1. Six indicators of network analysis.

Indicator	Definition	Relevance
Degree centrality	The number of other points directly connected to the point in the network.	Indicate the direct influence of cities in the CO_2 emission spatial correlation network.
Intermediate centrality	The degree to which the node is located on the shortest path of other nodes.	The degree of impact of the city on information flow and communication in the CO_2 emission network.
Approaching centrality	The sum of distances between the node and all other nodes in the network.	The degree to which a city's resources, influence, etc., are not controlled by other cities.
Condensed subgroups	Some nodes in the network are closely connected and form secondary groups.	Part of the cities under the overall CO_2 emission spatial network form a secondary network.
Core–edge analysis	Divided into core area and edge area based on the connections between nodes.	Cities occupy a central or peripheral position in the CO_2 emission network.
Core–edge density matrix	The connection between nodes in the core and edge areas of a network.	The degree of connectivity between cities in the core and peripheral areas.

2.4. Analysis of Network Influence Factors

The quadratic assignment procedure (QAP) is employed to study the factors that affect the spatial correlation of CO_2 emissions between cities. The QAP compares the similarity of elements in two matrices from a social network perspective [40]. It compares the elements in the matrices to obtain the correlation coefficient between the two matrices and performs nonparametric tests on the coefficients. It is based on the "relationship" between variables. Compared with traditional econometric methods, the QAP uses randomized testing to estimate the standard error of statistical measures, which can analyze the relationship between pairwise matrices.

According to existing research [41–43], we selected six factors (see Table 2): industrial level, electricity consumption, energy intensity, population size, technological level, and industrial structure, which represent influencing factors to explore their impact on the CO_2 emissions at the network scale. Each factor is quantified, as shown in Table 2, and a difference matrix is constructed characterizing the differences between all different cities. In addition, a correlation matrix is introduced, whose elements represent the spatial correlation between the CO_2 emissions of all cities (see Section 2.2). Now, the relationships between the CO_2 emission correlation matrix and the difference matrix of each influencing factor can be derived through the QAP to determine the individual impact of each factor on the CO_2 emission network.

Table 2. Six influencing factors and their definitions.

Influencing Factor	Definition
Industrial level	Differences in the value of the secondary industry between cities
Electricity consumption	Differences in overall electricity consumption between cities
Energy intensity	Differences in energy consumption per unit of GDP between cities
Population size	Differences in population between cities
Technological level	Differences in R&D expenditure relative to GDP ratio between cities
Industrial structure	Differences in the proportion of the secondary industry to GDP between cities

3. Results

3.1. CO_2 Emission Space Network

We construct CO_2 emission spatial networks based on periods and sectors. In Figures 2 and 3, cities with the same color have the same geographical orientation. Figure 2 shows the spatial network of total CO_2 emissions from 2019 to 2022. In recent years, the number of connections within the spatial network of CO_2 emissions in China has gradually increased, indicating that the network density is gradually increasing. Figure 3 shows the spatial network of CO_2 emissions from the total power, industry, and transport sectors in 2022. It can be seen that the CO_2 emission network of the industry sector is more closely connected, while the density of the power sector is lower, and the transportation sector has significant differences between cities. The network characteristic values are shown in Table 3. It can be seen that although the "gravity" between cities has decreased in recent years, the overall network density has gradually increased. We divided the calculation results of the potential model into five levels using natural breaks (which aim to divide numerical data into categories with minimum internal variance while maximizing the differences between the categories) [44], as shown in Table 4. It can be seen that Chengdu, Chongqing, Fuzhou, Urumqi, Guangzhou, and Lhasa have the highest potential for future development in the CO_2 emission network, which will be further discussed in Section 4.1.

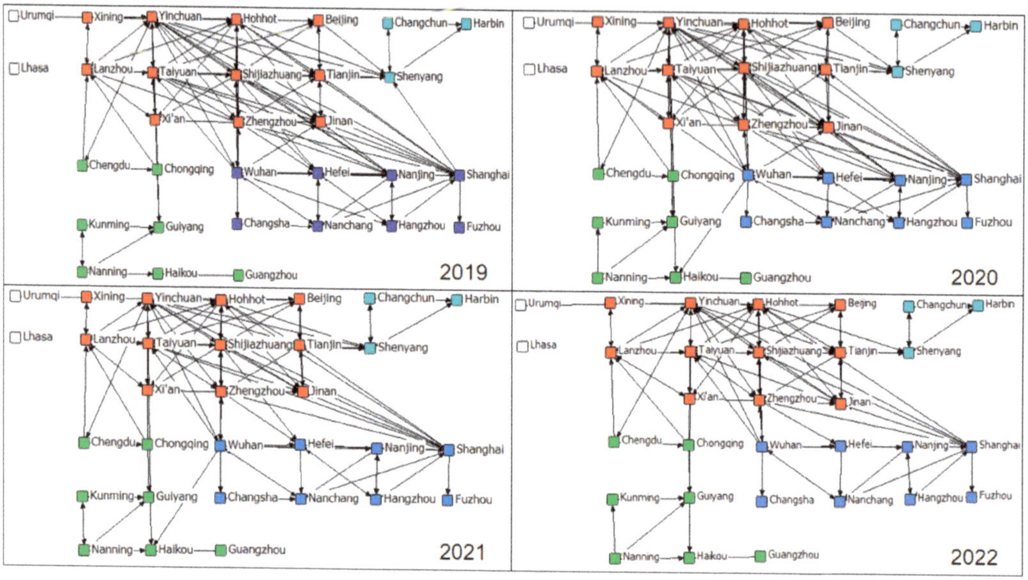

Figure 2. Spatial correlation networks of CO_2 emissions for each year (2019–2022).

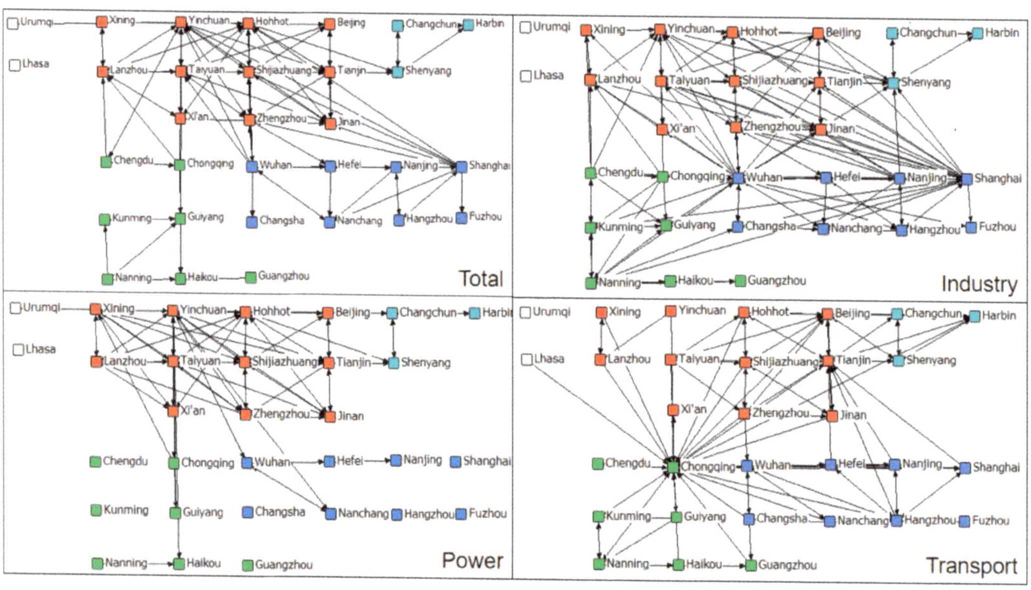

Figure 3. Spatial correlation networks of CO_2 emissions for each sector (total, power, industry, and transport) in 2022.

Table 3. Network characteristic values.

Network Characteristic	2019	2020	2021	2022	Power	Industry	Transport
Average of correlation	4.319	4.932	4.323	3.773	1.428	0.433	0.324
Network density	0.2172	0.2215	0.2194	0.2301	0.1290	0.2387	0.13333
Network standard deviation	0.4123	0.4153	0.4138	0.4209	0.3352	0.4263	0.3399
Maximum node network density	19	20	20	20	16	23	26
Average node network density	6.52	6.65	6.58	6.90	3.87	7.16	4.13

Table 4. Potential levels of each city.

Potential Level	Cities
5	Chengdu, Chongqing, Fuzhou, Urumqi, Guangzhou, and Lhasa
4	Guiyang, Hangzhou, Harbin, Kunming, Changsha, and Shanghai
3	Nanning, Nanchang, Shenyang, Changchun, Xi'an, and Wuhan
2	Hefei, Nanjing, Jinan, Haikou, Zhengzhou, and Beijing
1	Yinchuan, Taiyuan, Lanzhou, Shijiazhuang, Hohhot, Xining, and Tianjin

3.2. *Analysis of Network Characteristics*

We calculated various indicators of CO_2 emission network structure characteristics using SNA and noted the following results.

The network centrality for each city is shown in Table 5. The numerical range of degree centrality, which indicates the direct influence of cities in the CO_2 emission spatial correlation network, is relatively large, with a maximum value of 53.333 and a minimum value of 0. The numerical differences between the cities are significant, with most cities having degree centrality values lower than the average. With the exception of some cities with high values, the median centrality values of most cities are concentrated around the average or 0. Intermediate centrality, which shows the degree of impact of the city on information flow and communication in the CO_2 emission network, contains values between cities that also vary greatly. This reflects a considerable regional imbalance of urban CO_2 emissions in China. That is, there are great differences in the influence, status, and relationship of cities in the CO_2 emission network.

Approaching centrality reflects the degree to which a city is affected by other cities. In other words, the higher the value, the more susceptible the city is to the influence of other cities. The result shows that approaching the centrality of the major cities in China reveals obvious clustering characteristics, as almost all cities have values at a high level, and only a few cities have significantly lower values. This suggests that the connection and impact between urban CO_2 emissions appear to be a universal phenomenon.

To further explore the spatial relationship between urban CO_2 emissions, we carried out both cohesive subgroup analysis and core–edge analysis. The results of the cohesive subgroup analysis are shown in Figure 4. The connections between some cities in the overall network are relatively close, with four regions forming secondary networks. Among them, the north region has the most members and the most complex internal connections. The western and northeastern regions are classified together by cohesive subgroup analysis due to their low internal components and connections; that is, they have little or no connection with other cities.

Table 5. Network centrality of each city.

City	Degree Centrality		Intermediate Centrality	Approaching Centrality	
	Out Degree	In Degree		Out Degree	In Degree
Yinchuan	50	53.333	25.152	10.83	15.544
Shanghai	46.667	6.667	4.283	9.434	15.625
Shijiazhuang	40	33.333	5.433	10.526	15.306
Wuhan	36.667	20	10.376	10.417	15.152
Hohhot	36.667	36.667	3.842	10.638	14.925
Nanjing	33.333	16.667	5.447	10.067	15.152
Zhengzhou	33.333	30	1.514	10.49	14.925
Taiyuan	30	40	1.826	10.601	14.634
Tianjin	26.667	30	1.252	10.453	14.706
Beijing	23.333	13.333	0.069	10.067	14.634
Jinan	20	33.333	0.13	10.526	14.423
Shenyang	16.667	26.667	9.885	10.417	14.286
Lanzhou	16.667	30	0.958	10.345	14.151
Chongqing	13.333	3.333	0.057	9.434	14.085
Hefei	13.333	26.667	3.592	10.49	13.825
Hangzhou	10	13.333	1.317	9.868	14.019
Chengdu	10	6.667	2.414	10.101	14.019
Xi'an	10	20	0	10.239	14.019
Nanchang	10	16.667	3.027	9.934	13.636
Nanning	10	6.667	0.575	3.567	3.571
Kunming	6.667	3.333	0	3.559	3.567
Changchun	6.667	6.667	0	9.74	12.876
Harbin	6.667	6.667	0	9.74	12.876
Xining	6.667	10	0	10.135	13.889
Changsha	3.333	6.667	0	9.772	12.295
Haikou	3.333	6.667	0.345	3.567	3.563
Guangzhou	3.333	0	0	3.226	3.681
Guiyang	0	13.333	0	19.868	3.226
Fuzhou	0	3.333	0	9.677	3.226
Urumqi	0	3.333	0	11.152	3.226
Lhasa	0	0	0	3.226	3.226
Average	16.88	16.88	2.63	9.42	11.49
STDEV	14.52	13.49	4.93	3.16	4.82

The core–edge analysis (presented in Table 6) shows cities such as Beijing and Shanghai, which have always been in the core area, and most of these cities are located in North China. The core–edge density matrix (Table 7) reveals that the connections within the core area are much greater than those between the core–edge and edge–edge areas. The connections between different regions gradually increase, while the connections in the edge area are always low.

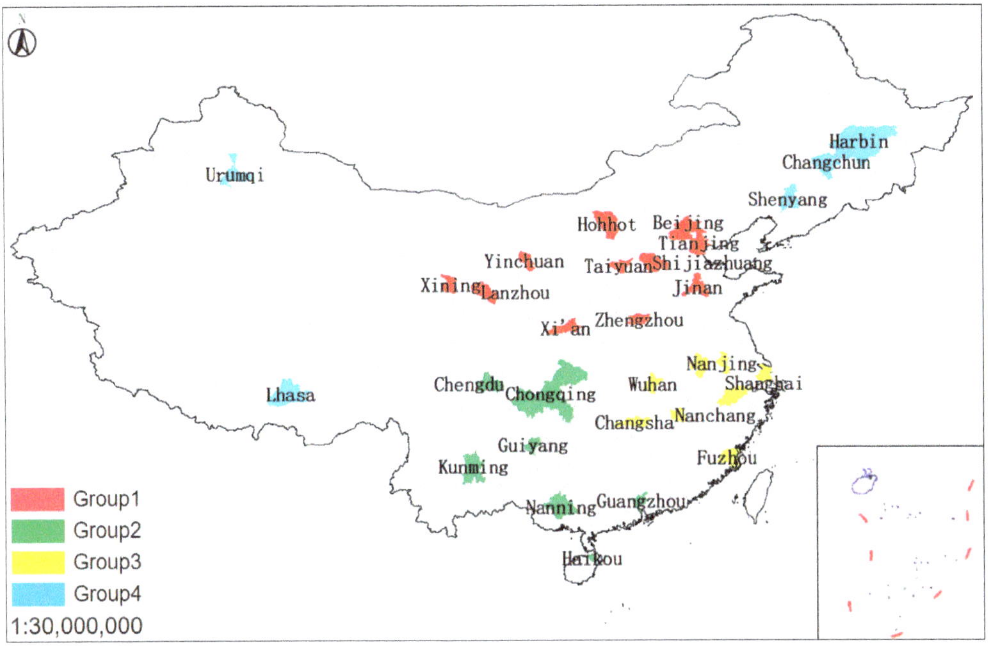

Figure 4. Results of cohesive subgroup analysis.

Table 6. Core–edge analysis.

Year	Core	Edge
2019	Beijing, Shanghai, Tianjin, Chongqing, Harbin, Changchun, Shenyang, Hohhot, Shijiazhuang, Taiyuan, Jinan, Zhengzhou, Xi'an, Lanzhou, Yinchuan, Xining, Hefei, Nanjing, Hangzhou, Nanchang, Wuhan, Chengdu, Guiyang	Urumqi, Changsha, Fuzhou, Guangzhou, Haikou, Nanning, Kunming, Lhasa
2020	Beijing, Shanghai, Tianjin, Nanjing, Zhengzhou, Xi'an, Shenyang, Hohhot, Shijiazhuang, Taiyuan, Jinan, Yinchuan, Wuhan, Lanzhou, Hefei	Changchun, Urumqi, Harbin, Chongqing, Hangzhou, Changsha, Nanchang, Xining, Chengdu, Guiyang, Fuzhou, Guangzhou, Haikou, Nanning, Kunming, Lhasa
2021	Beijing, Shanghai, Tianjin, Jinan, Zhengzhou, Xi'an, Shenyang, Hohhot, Shijiazhuang, Taiyuan, Lanzhou, Yinchuan, Wuhan	Chongqing, Harbin, Changchun, Urumqi, Hefei, Nanjing, Hangzhou, Changsha, Nanchang, Xining, Chengdu, Guiyang, Fuzhou, Guangzhou, Haikou, Nanning, Kunming, Lhasa
2022	Beijing, Shanghai, Tianjin, Shenyang, Hohhot, Shijiazhuang, Taiyuan, Jinan, Zhengzhou, Xi'an, Lanzhou, Yinchuan, Wuhan	Chongqing, Harbin, Changchun, Xining, Urumqi, Hefei, Nanjing, Hangzhou, Changsha, Nanchang, Chengdu, Guiyang, Fuzhou, Guangzhou, Haikou, Nanning, Kunming, Lhasa

Table 7. Core–edge density matrix.

		Core	Edge
2019	Core	0.287	0.022
	Edge	0.016	0.089
		Core	Edge
2020	Core	0.529	0.083
	Edge	0.050	0.054
		Core	Edge
2021	Core	0.56	0.101
	Edge	0.055	0.063
		Core	Edge
2022	Core	0.583	0.111
	Edge	0.064	0.092

3.3. Analysis of Influencing Factors

We analyzed the impact of each influencing factor on the CO_2 emission spatial network from each department using QAP, and the results are shown in Table 8. The correlation coefficient represents the value and direction of the impact, and the significance level (less than 0.1) indicates that the impact is significant. It can be seen that the impact of different factors on CO_2 emissions from different sectors is different and even opposite. The specific analysis and discussion of these impacts will be provided in Section 4.3.

Table 8. Results of the quadratic assignment procedure (Red indicates that the correlation coefficient is greater than 0, while blue indicates less than 0).

Influencing Factor	Total		Industry		Power		Transport	
	Correlation Coefficient	Significant Level	Correlation Coefficient	Significant Level	Correlation Coefficient	Significant Level	Correlation Coefficient	Significant Level
Industrial level	0.130	0.001	0.231	0.000	0.016	0.235	−0.137	0.015
Electricity consumption	0.144	0.002	0.200	0.004	0.024	0.199	−0.092	0.083
Energy intensity	0.024	0.186	−0.136	0.005	0.083	0.016	0.018	0.415
Population size	0.114	0.002	0.189	0.001	0.014	0.255	−0.183	0.001
Technological level	−0.05	0.084	−0.103	0.095	0.007	0.424	0.019	0.308
Industrial structure	−0.045	0.119	−0.063	0.134	0.028	0.113	0.001	0.447

4. Discussion

4.1. Overall Spatial Characteristics of CO_2 Emissions

This paper constructs a spatial correlation network of CO_2 emissions between provincial capitals in China based on CCCM and analyzes the characteristics of the CO_2 emission network through social network analysis. Overall, CO_2 emissions in various cities are generally spatially interconnected and exhibit networked structural characteristics, with not only spatial connections between neighboring cities but also complex spatial connections between non-neighboring cities. The mutual influence and interaction between cities make the spatial pattern structure of CO_2 emissions more complex and diverse. In recent years, due to the effective implementation of various energy-saving and emission-reduction policies, the intensity of CO_2 emissions has decreased, which is reflected in the decrease in the average correlation value. The overall spatial feature structure remains stable, and the network density is gradually increasing, which indicates that the correlation between cities is gradually increasing. Therefore, implementing energy conservation and emission reduction should not be limited to "local" areas but should develop towards "overall" areas. However, the gradual increase in network standard deviation indicates that the current CO_2 emission network structure in China is still unbalanced and requires optimization.

In the research on CO_2 emissions from sectors, total CO_2 emissions in China are concentrated in the North and East China regions, with more dense and frequent connections between cities. In the power sector, dense connections have been established between cities in North China, while cities in other regions have few connections. In the industrial sector, the correlation of CO_2 emissions covers almost all cities. It can be seen that CO_2 emissions from the industrial sector are the core of China's CO_2 emissions, driving connections between almost all cities and having a significant impact. As for the transportation sector, there are significant differences between cities. Some developed cities, such as Beijing and Tianjin, have established connections with more cities due to their developed economy and better transportation. Significantly, due to its geographical location as an intermediary point for north–south communication, Chongqing has established the densest connections with more cities. Compared with the previous studies on CO_2 emissions from the transport sector at the provincial level, which only pointed out the leading position of developed cities in eastern China [45], this study highlights the indispensable role of geographic location at the provincial level, a factor that has been underemphasized or overlooked in the existing literature. Our findings suggest that geographic context significantly shapes the spatial distribution of CO_2 emissions, especially in the transport sector, indicating that future research and practical applications should not neglect the impact of where events occur, or entities are situated. These results reflect the current imbalance in CO_2 emissions among sectors in China; therefore, the focus of energy-saving and emission-reduction tasks should be selected according to the actual situation of different sectors.

In addition, by calculating potential models, some cities have great potential for developing their position in the future CO_2 emission spatial pattern. Chengdu and Chongqing occupy an intermediary position in the north–south connection of CO_2 emissions, which has significant strategic significance. Lhasa and Urumqi, as relatively underdeveloped areas in the western region and targets for the country's western development, still have great potential for development. Guangzhou and Fuzhou are relatively developed coastal cities with weak connections to the mainland, which needs to be strengthened. It should also be noted that the influence of those cities is mainly reflected in international exchanges, which is not covered in our research. In fact, when the economic development of coastal cities reaches a bottleneck, they often migrate their industries inland, thereby driving the industrialization development of the central and western regions. That is to say, the impact of coastal cities on inland areas will continue to increase.

4.2. Analysis of Spatial Characteristics of CO_2 Emissions

4.2.1. Analysis of CO_2 Emission Network Center Cities

In the analysis of network centrality, we use the three centralities, degree centrality, intermediate centrality, and approaching centrality, to analyze the status and role of cities in the network.

The higher the value of the degree centrality, the more other nodes connected to that node indicate that the city occupies a central position in the CO_2 emission network. The degree centrality values, ranging from 0 to 50, denote significant differences in the status of each city in the network. Cities at a higher level are mostly located in the North or East China regions which have significant advantages in terms of their economic level and the scientific and technological impact, and their industrial industries are relatively more developed compared to the central and western regions. Therefore, their connections are more extensive, and most occupy central social network positions. The centrality of cities in the Western region has always been low, reflecting their disadvantaged location in China's CO_2 emission spatial network.

Intermediate centrality is the intermediary role nodes play in network connections, which reflect a city's control over resources and information. That is, the higher the value, the more information and resources are controlled by the city. Significant differences in the centrality of each city indicate an imbalance in their degree of control over resources and information among cities. Cities such as Shanghai have a high level of development and abundant resources. For Yinchuan and other cities located in a "middle" geographical position, it is more convenient to communicate with various regions.

Approaching centrality reflects the difficulty of spatially connecting with other cities, and the higher the value, the more likely the city is to experience spatial connectivity. The results show that the differences in approaching centrality among cities are relatively small, and the value range is relatively concentrated at a high level. It indicates that the spatial connection between cities appears to be common across the country. The proximity centrality of the vast majority of cities is higher than the average, and these cities have attained more developed technological levels, richer resources, or advantageous geographical locations for communication, making them more likely to establish spatial connections with other cities.

The calculation results of centrality indicate an increasing difference in the status of different cities in the CO_2 emission spatial network, and the uneven development between individual cities and regions is also evident. Therefore, collaborative governance and balanced development between different provinces and cities are crucial. Each province and city needs to combine their own actual situation, leverage their respective advantages in resources, technology, geography, etc., and strengthen regional cooperation. That is, correctly understanding one's own position and role in the network structure and considering the situation of all associated cities supports conceptualizing energy-saving and emission-reduction policies.

4.2.2. Analysis of Subgroups of Cities

Agglomeration subgroups can be used to reveal the secondary groups of some networks. Previous studies have employed block models to characterize the position and function of cities within networks by classifying cities based on their in and out roles in the network [46,47]. While this approach provides a structured way to understand the connectivity and interactions among cities, it has limitations, especially the oversimplified binary classification and the overlook of qualitative factors. The main factors (Geographical location, transportation, resources, industrial structure, economy, etc.) affecting their position in the network may differ for each city in the same category, which cannot be well demonstrated in the previous studies. However, cities form secondary networks because they have close geographical locations and primary influencing factors, so subgroup analysis in this study can better reflect the differences between different regions.

The results show that the overall spatial pattern of CO_2 emissions in China can be divided into four sub-blocks. Geographically, urban agglomerations can be divided into five regions: North China, South China, East China, Northeast China, and West China. Cities in Northeast China and West China are far from China's industrial centers, have a lower status and fewer connections in the CO_2 emission network, and are therefore divided into groups. Regarding the number of subgroup components, it is evident that there are more members and denser internal connections in the North and East China regions. The North China region is mostly plains with convenient transportation, and the East China region is mostly coastal, with frequent international trade and a booming economy. Due to their superior natural and socio-economic conditions and developed industries, the North and East China regions occupy a central position in the spatial network, making it easier to establish diverse connections with other provinces and control more information and resources. Due to its remote geographical location and relatively backward development level, the western region has a significant disadvantage in terms of network structure. Overall, the development level of the eastern and western regions is unbalanced, and the country needs to increase support for the western regions and improve their development level.

4.2.3. Core–Edge City Analysis

Core edge analysis in recent years has revealed significant fluctuations in the composition of the core and edge areas due to the impact of the epidemic and other factors. Overall, cities in East and North China are mostly concentrated in the core areas, while the peripheral areas are cities in the central and western regions. Due to their natural environment and socio-economic conditions, developed industries, and large and concentrated CO_2 emissions, the East and North China regions have become core areas. Therefore, they should bear more responsibility for emission reduction and provide support and guidance to the cities associated with them. The economic and development levels of the central and western regions are relatively low, and the peripheral areas are mostly cities in the central and western regions. Although their industrialization development is rapid, there is still a large gap compared to the eastern regions, and they still have high development potential. Accelerating the improvement of their development level and simultaneously considering environmental protection and controlling the increasing CO_2 emissions, appears to be an obvious and suitable policy.

From the density matrix, it can be concluded that there is a strong connection between the core area and its network density, which is much higher than between the edge areas and constantly increasing. Cities located in the core area often have good transportation and/or higher levels of development, and their industrial structures are more similar, resulting in closer connections. The network density between the core area and the edge area is also increasing, indicating that the high development of core area cities can provide guidance and support for edge area cities, driving the development of edge areas. The connections within the edge zone also show an upward trend, but still at a relatively low level, suggesting that the spatial network of the edge zone needs further optimization.

4.3. Analysis of Factors Affecting CO_2 Emissions

The QAP results show that different factors play different roles in different CO_2 emissions sectors. Technological level and industrial structure are negatively correlated with total CO_2 emissions. This indicates that developing green technology, optimizing industrial structure, and transforming high-energy-consuming industries into green, low-carbon industries are direct and effective strategies to achieve emission reduction goals. There is a positive correlation between energy intensity and total CO_2 emissions, which is consistent with previous research [48]. However, in the industry sector, this correlation is negative, indicating that China has achieved some results by improving energy utilization efficiency, enhancing energy intensity, and thereby controlling the growth of CO_2 emissions. As for the power sector, all influencing factors are positively correlated with CO_2 emissions. The power sector is an important support for urban development and economic growth [49], and the resulting increase in CO_2 emissions is inevitable. To achieve emission reduction, it is necessary to deeply explore effective influencing factors. Last, in the transport sector, industrial level and electricity consumption are negatively correlated with CO_2 emissions. With the development of industry, people have replaced transportation energy such as gasoline with electricity and some clean energy, thus achieving emission reduction. Furthermore, it is worth noting that population size is also negatively correlated with CO_2 emissions. However, most current research supports the conclusion that population growth promotes the increase of CO_2 emissions. For example, a study in a South Asian country showed that an increase in population density would have a negative impact on sustainable development [50]. However, for the Chinese provincial capital cities studied in this study, most of them have a high level of development and well-developed transportation facilities. The increase in population has improved the efficiency of infrastructure utilization and promoted the development of public transportation, thus no longer promoting an increase in transport CO_2 emissions. The impact of different factors on CO_2 emissions varies in different sectors. When implementing emission reduction policies in different sectors, we should select the focus of work in the light of specific conditions.

4.4. Study Limitations and Future Work

First, this study has mainly focused on the internal characteristics of China's CO_2 emission spatial network, and external influencing factors have not yet been included. Especially for some coastal cities (such as Guangzhou) with frequent international exchanges, their network status may be underestimated. This means that the conclusions of this study mainly support the interaction of domestic CO_2 emissions. Second, this study constructs China's CO_2 emission network based on cities as nodes. A total of 31 provincial capitals are selected as our research objects, which can, to some extent, represent the economic development and emission reduction situations within each province. However, it is important to acknowledge that although these capital cities serve as crucial indicators, they do not fully capture the national distribution of CO_2 emissions. To address this limitation, in future research, we can include more cities as research objects to build a more complete network, including secondary and tertiary urban centers and rural areas. This would allow for a more accurate representation of the national CO_2 emission landscape and facilitate the development of more targeted and effective mitigation strategies. Finally, connecting network analysis with concerted actions appears to be a requirement for future policymaking to achieve comprehensive, collaborative development between individual players and the community.

5. Conclusions

In order to study the spatial structure and characteristics of CO_2 emissions in China in recent years and explore more reasonable policies for energy conservation and emission reduction in the future, this paper took 31 provincial capital cities in China as objects, used the gravity model and SNA to study and analyze the structure of China's CO_2 emission network from 2019 to 2022, and explored its influencing factors by QAP. The following conclusions were drawn on the main characteristics of present CO_2 emissions in China:

(1) During most of the time in this study (2019–2022), the intensity of CO_2 emissions has decreased over time, indicating that China's emission reduction strategy has achieved certain results. The connections between regions are becoming stronger, while the differences between regions are also becoming more pronounced.

(2) CO_2 emissions mainly come from the industry sector. The industry sector has the greatest and most extensive influence, promoting connections between different regions of China. The power sector mainly promotes connections between cities in North China. The impact of the transport sector is mainly related to the geographical location and development level of the city.

(3) The location of the cities analyzed in this study can be grouped into five geographical regions: North, East, South, Northeast, and West China. The cities in the North are numerous, closely connected, and have great influence, and have long been in a core position, while the western, northeastern, and some other cities have weaker connections

and are in an edge position. The connections between different regions are strengthening, but still insufficient.

(4) In general, industrial level, electricity consumption, energy intensity, and population size promote an increase in CO_2 emissions, while Technological level and Industrial structure can control the increase in CO_2 emissions. In addition, due to the different characteristics of sectors, the impact of the factors on the CO_2 emissions from each sector varies.

Based on the above conclusions, future energy conservation, and emission reduction should achieve the following: (i) Correctly understanding the spatial network structure characteristics of CO_2 emissions, innovating environmental governance ideas, promoting energy conservation and emission reduction from "local" to "overall", and achieving regional collaborative governance. (ii) Formulating targeted and efficient emission reduction policies based on the differences in the structure, characteristics, and roles of CO_2 emission networks in different sectors. (iii) Core cities should play a central role, take on more emission reduction responsibilities, and guide emission reduction work in edge areas. While implementing emission reduction measures, the edge areas also need to improve their own development level. (iv) Promoting the development of innovation, energy-saving, and emission-reduction technologies and transforming high-energy-consuming industries into green and low-carbon industries. Develop targeted and efficient emission reduction policies based on the varying impacts of different factors on CO_2 emissions in different sectors. Explore more influencing factors of CO_2 emissions and leverage their positive effects to reduce CO_2 emissions.

Author Contributions: Methodology, Yue Zhao; Project administration, Mingyi Du; Software, Yue Zhao; Supervision, Yuning Feng and Mingyi Du; Writing—original draft, Yue Zhao; Writing—review and editing, Yuning Feng, Klaus Fraedrich and Zehao Shen. All authors have read and agreed to the published version of the manuscript.

Funding: The work was supported by the Beijing Natural Science Foundation (8254052), the Youth Research and Innovation Project—Young teachers research ability enhancement program (X23002), and the National Natural Science Foundation of China (41930650).

Data Availability Statement: The data presented in this study are openly available in SicienceDB at https://www.scidb.cn/en/anonymous/UmJ5ZXF1 (accessed on 24 October 2024). Derived data supporting the findings of this study are available from the corresponding author Y.F. upon request.

Conflicts of Interest: All authors have no conflicts of interest. On behalf of all authors, the corresponding author states that there are no conflicts of interest.

References

1. Li, Z.X.; Zhang, J.W. Visual analysis of hot research topics on economic development and carbon emissions. *Coop. Econ. Sci.* **2023**, *12*, 12–15. (In Chinese) [CrossRef]
2. UNEP. *Emissions Gap Report 2021: The Heat Is on—A World of Climate*; U.N. Environ. Programme UNEP DTU Partnersh: Nairobi, Kenya, 2021.
3. IPCC. *Climate Change 2022: Mitigation of Climate Change*; Cambridge University Press: Cambridge, UK, 2022; Available online: https://www.ipcc.ch/report/sixth-assessment-report-working-group-3/ (accessed on 20 October 2024).
4. Wei, Y.M.; Chen, K.; Kang, J.N.; Chen, W.; Wang, X.Y.; Zhang, X. Policy and management of carbon peaking and carbon neutrality: A literature review. *Engineering* **2022**, *14*, 52–63. [CrossRef]
5. Yang, G.L.; Zha, D.L.; Cao, D.Q.; Zhang, G.X. Time for a change: Rethinking the global renewable energy transition from the Sustainable Development Goals and the Paris Climate Agreement. *Innovation* **2024**, *5*, 100582. [CrossRef]
6. Zhao, X.; Ma, X.; Chen, B.; Shang, Y.; Song, M. Challenges toward carbon neutrality in China: Strategies and countermeasures. *Resour. Conserv. Recycl.* **2022**, *176*, 105959. [CrossRef]
7. Du, M.Z.; Zhang, J.N.; Hou, X.J. Decarbonization like China: How does green finance reform and innovation enhance carbon emission efficiency? *J. Environ. Manag.* **2025**, *376*, 124331. [CrossRef] [PubMed]

8. Wen, L.; Zhang, J.; Song, Q.Q. A scenario analysis of Chinese carbon neutral based on STIRPAT and system dynamics model. *Environ. Sci. Pollut. Res.* **2022**, *29*, 55105–55130. [CrossRef] [PubMed]
9. Wang, P.T.; Xu, Q.C.; Wang, F.Y.; Xu, M. Investigating the impacts of the Dual Carbon Targets on energy and carbon flows in China. *Energy* **2024**, *313*, 133778. [CrossRef]
10. Gao, Y.; Zhang, L.; Huang, A.S.; Kou, W.B.; Bo, X.; Cai, B.F.; Qu, J.B. Unveiling the spatial and sectoral characteristics of a high-resolution emission inventory of CO_2 and air pollutants in China. *Sci. Total Environ.* **2022**, *847*, 157623. [CrossRef] [PubMed]
11. Chen, S.Y.; Wang, C.; Guo, Y. Emission reduction path and strategic choice of China's industrial sector facing the goal of carbon neutrality. *J. Manag. Sci. China* **2024**, *27*, 1–20. (In Chinese) [CrossRef]
12. Zhang, Z.; Sharifi, A. Analysis of decoupling between CO_2 emissions and economic growth in China's provincial capital cities: A Tapio model approach. *Urban Clim.* **2024**, *55*, 101885. [CrossRef]
13. Wang, S.; Huang, Y.; Zhou, Y. Spatial spillover effect and driving forces of carbon emission intensity at the city level in China. *J. Geogr. Sci.* **2019**, *29*, 231–252. [CrossRef]
14. Huang, H.Q.; Zhou, J. Study on the Spatial and Temporal Differentiation Pattern of Carbon Emission and Carbon Compensation in China's Provincial Areas. *Sustainability* **2022**, *14*, 7627. [CrossRef]
15. Wang, Y.J.; Zhai, C.X.; Liu, C.Y.; Chen, Z.Y. Spatio temporal differences and influencing factors of carbon emissions from land use in Ningxia. *Environ. Sci.* **2024**, *45*, 5049–5059. (In Chinese) [CrossRef]
16. Li, Y.; Guo, W.; Li, P.; Zhao, X.; Liu, J. Exploring the Spatiotemporal Dynamics of CO_2 Emissions through a Combination of Nighttime Light and MODIS NDVI Data. *Sustainability* **2023**, *15*, 13143. [CrossRef]
17. Li, X.; Feng, D.; Li, J.; Zhang, Z. Research on the spatial network characteristics and synergetic abatement effect of the carbon emissions in Beijing–Tianjin–Hebei urban agglomeration. *Sustainability* **2019**, *11*, 1444. [CrossRef]
18. Cai, H.; Wang, Z.; Zhu, Y. Understanding the structure and determinants of intercity carbon emissions association network in China. *J. Clean. Prod.* **2022**, *352*, 131535. [CrossRef]
19. Teng, W.X.; Shi, C.Y.; Yu, Y.; Li, Q.Q.; Yang, J.B. Uncovering the spatiotemporal patterns of traffic-related CO_2 emission and carbon neutrality based on car-hailing trajectory data. *J. Clean. Prod.* **2024**, *467*, 142925. [CrossRef]
20. Guo, B.; Xie, T.T.; Zhang, W.C.; Wu, H.J.; Zhang, D.M.; Zhu, X.W.; Ma, X.Y.; Wu, M.; Luo, P.P. Rasterizing CO_2 emissions and characterizing their trends via an enhanced population-light index at multiple scales in China during 2013–2019. *Sci. Total Environ.* **2023**, *905*, 167309. [CrossRef] [PubMed]
21. Lu, C.; Du, Q. The heterogeneous effects of transportation infrastructure on trade-embodied CO_2 transfer: A multi-scale perspective. *Energy* **2024**, *302*, 131785. [CrossRef]
22. Gan, C.; Wang, K. The spatial network structure and driving factors of carbon emissions in China's inter provincial service industry. *Res. Environ. Sci.* **2022**, *35*, 2264–2272. (In Chinese) [CrossRef]
23. Zhang, W.; Zhang, T.; Wang, S.H. CO_2 emission spillover effects in the digital transformation of China's manufacturing industry and its inter-industry differences. *Energy Rep.* **2023**, *10*, 520–534. [CrossRef]
24. Sun, Y.N.; Liu, H.J.; Liu, C.M.; Cui, R. Research on spatial association of provinces carbon emissions and its effects in China. *Shanghai J. Econ.* **2016**, *28*, 82–92. (In Chinese) [CrossRef]
25. He, F.T.; Ji, L.Y.; Li, C.B. Grid disaster risk identification based on social network analysis. *Reliab. Eng. Syst. Safe* **2025**, *256*, 110756. [CrossRef]
26. Shen, W.R.; Liang, H.W.; Dong, L.; Ren, J.Z.; Wang, G.J. Synergistic CO_2 reduction effects in Chinese urban agglomerations: Perspectives from social network analysis. *Sci. Total Environ.* **2021**, *798*, 149352. [CrossRef] [PubMed]
27. Fang, D.C.; Wang, L.L. Research on the Network Characteristics and Influencing Factors of Spatial Correlation of Carbon Emissions in China. *Resour. Environ. Yangtze Basin* **2023**, *32*, 571–581. Available online: https://yangtzebasin.whlib.ac.cn/CN/10.11870/cjlyzyyhj202303012 (accessed on 21 October 2024). (In Chinese)
28. Wang, M.; Dai, X.; Zhang, H. Spatiotemporal Dynamics of CO_2 Emissions in China Based on Multivariate Spatial Statistics. *Atmosphere* **2024**, *15*, 538. [CrossRef]
29. Luo, H.R.; Zhou, Q.G.; Li, H.; Wu, L.J.; Mao, Y.F.; Xia, Y.S.; Chen, F.Y. Evolution of spatial correlation between land use carbon balance based on social network analysis method: Taking the urban area of Chongqing as an example. *Environ. Sci.* **2024**, *9*, 5003–5014. (In Chinese) [CrossRef]
30. Chen, J.D.; Xu, C.; Li, K.; Song, M.L. A gravity model and exploratory spatial data analysis of prefecture-scale pollutant and CO_2 emissions in China. *Ecol. Indic.* **2018**, *90*, 554–563. [CrossRef]
31. Wang, J.; Feng, L.; Palmer, P.I.; Liu, Y.; Fang, S.X.; Bösch, H.; O'Dell, C.W.; Tang, X.P.; Yang, D.X.; Liu, L.X.; et al. Large Chinese land carbon sink estimated from atmospheric carbon dioxide data. *Nature* **2020**, *586*, 720–723. [CrossRef] [PubMed]
32. Global gRidded dAily CO_2 Emission Dataset. Available online: https://carbonmonitor-graced.com/ (accessed on 13 March 2024).
33. Dou, X.Y.; Wang, Y.L.; Ciais, P.; Chevallier, F.; Davis, J.S.; Crippa, M.; Janssens-Maenhout, G.; Guizzardi, D.; Solazzo, E.; Yan, F.F.; et al. Near-real-time global gridded daily CO_2 emissions. *Innovation* **2022**, *3*, 100182. [CrossRef] [PubMed]

34. Liu, Z.; Ciais, P.; Deng, Z.; Lei, R.X.; Davis, S.J.; Feng, S.; Zheng, B.; Cui, D.; Dou, X.Y.; Zhu, B.Q.; et al. Near-real-time monitoring of global CO_2 emissions reveals the effects of the COVID-19 pandemic. *Nat. Commun.* **2020**, *11*, 5172. [CrossRef] [PubMed]
35. Liu, Z.; Ciais, P.; Deng, Z.; Davis, S.J.; Zheng, B.; Wang, Y.L.; Cui, D.; Zhu, B.Q.; Dou, X.Y.; Ke, P.Y.; et al. Carbon Monitor, a near-real-time daily dataset of global CO_2 emission from fossil fuel and cement production. *Sci. Data* **2020**, *7*, 392. [CrossRef] [PubMed]
36. Ren, X.S.; Li, Z.R. Evolution and influencing factors of spatial correlation network of building carbon emissions in China from the perspective of full life cycle. *Environ. Sci.* **2024**, *45*, 1243–1253. (In Chinese) [CrossRef]
37. Nie, X.Y.; Shi, P.J.; Lv, R. The spatiotemporal evolution of spatial interaction in oasis type towns in the Hexi Corridor. *Econ. Geogr.* **2016**, *36*, 76–83. (In Chinese) [CrossRef]
38. Liu, J. *Holistic Network Analysis: Practical Guide to UCINET Software*, 2nd ed.; Gezhi Publishing House: Shanghai, China, 2014.
39. Yue, Q.B.; He, J.H.; Liu, D.F. Identifying Restructuring Types of Rural Settlement Using Social Network Analysis: A Case Study of Ezhou City in Hubei Province of China. *Chin. Geogr. Sci.* **2021**, *31*, 1011–1028. [CrossRef]
40. Zhou, L.; Mao, J.W. Analysis on the influencing factors of China's regional foreign trade gap. *Stat. Decis.* **2021**, *37*, 157161. [CrossRef]
41. Cheng, Y.Q.; Wang, Z.Y.; Ye, X.Y.; Wei, Y.H. Spatiotemporal dynamics of carbon intensity from energy consumption in China. *J. Geogr. Sci.* **2014**, *24*, 631–650. [CrossRef]
42. Li, J.X.; Chen, Y.N.; Li, Z.; Liu, Z.H. Quantitative analysis of the impact factors of conventional energy carbon emissions in Kazakhstan based on LMDI decomposition and STIRPAT model. *J. Geogr. Sci.* **2018**, *28*, 1001–1019. [CrossRef]
43. Yang, X.; Ran, G. Forecasting China's CO_2 emissions and identifying key drivers: An application of the improved RFAGM model and LMDI decomposition methods. *Int. J. Sust. Dev. World* **2024**, *31*, 523–536. [CrossRef]
44. Huang, S.; Hu, Q.; Wang, S.; Li, H. Ecological Risk Assessment of World Heritage Sites Using RS and GIS: A Case Study of Huangshan Mountain, China. *Chin. Geogr. Sci.* **2022**, *32*, 808–823. [CrossRef]
45. Lei, T.; Wang, Y.Q.; Wang, C. Spatial correlation network and influencing factors of carbon emissions from inter provincial tourism and transportation in China. *Environ. Sci.* **2024**, *46*, 53–65. (In Chinese) [CrossRef]
46. Hai, X.H.; Zhan, X.X.; Wang, X. Research on the spatial correlation network characteristics of provincial carbon emissions under the background of the "dual carbon" target. *Stat. Deci.* **2024**, *40*, 85–89. (In Chinese) [CrossRef]
47. Dong, R.Y.; Zhou, X.W. The decoupling level of digital economy and carbon emissions and its spatial correlation network characteristics. *Stat. Deci.* **2023**, *39*, 129–133. (In Chinese) [CrossRef]
48. Xu, J. Study on spatiotemporal distribution characteristics and driving factors of carbon emission in Anhui Province. *Sci. Rep.* **2023**, *13*, 14400. [CrossRef] [PubMed]
49. Shi, X.; Li, W.B. Research on the Dynamic Relationship between Energy Consumption and Industrial Development. *Stat. Deci.* **2022**, *38*, 127–130. [CrossRef]
50. Ashraf, U.D.; Jian, M.; Imran, U.R.; Heesup, H.; Sunghoon, Y.; Rakan, R.A. Green road transportation management and environmental sustainability: The impact of population density. *Heliyon* **2023**, *9*, e19771. [CrossRef]

International Journal of
Geo-Information

MDPI

Article

Open Data for Transparency of Government Tenders: A State Analysis in Croatian Agriculture Land Lease

Larisa Hrustek [1], Karlo Kević [2,*] and Filip Varga [3,4]

[1] Department of Organization, Faculty of Organization and Informatics, University of Zagreb, Pavlinska ulica 2, 42000 Varaždin, Croatia; lhrustek@foi.hr
[2] Institute of Cartography and Photogrammetry, Faculty of Geodesy, University of Zagreb, Savska cesta 144A, 10000 Zagreb, Croatia
[3] Department of Plant Biodiversity, Faculty of Agriculture, University of Zagreb, Svetošimunska cesta 25, 10000 Zagreb, Croatia; fvarga@agr.hr
[4] Centre of Excellence for Biodiversity and Molecular Plant Breeding (CoE CroP-BioDiv), Svetošimunska cesta 25, 10000 Zagreb, Croatia
* Correspondence: karlo.kevic@geof.unizg.hr

Abstract: State-owned agricultural land is an asset that the state must manage in a responsible and transparent manner. Agricultural land is extremely important for farmers as it enables them to carry out agricultural activities. Due to its importance to farmers, it is often the subject of debate as stakeholders are often dissatisfied with the allocation and management of state-owned agricultural land. Qualitative research of the process of state agricultural land lease and the associated legislation in the Republic of Croatia enabled the analysis of the existing business model, with the results pointing to shortcomings in the Initial and Evaluation phases of the process. A steady rise in the number of tenders published in 2015–2022 was recorded. Local administrative units in the Continental region scored higher than those in the Adriatic region (both cities and municipalities) in terms of transparency. Unfortunately, the response rate from the local authorities was below 50% across both region and unit, further indicating low transparency. Based on the findings, a proposal of changes in the tendering process was made utilizing a digital platform as an environment for all stakeholders, which provides functionalities essential for the transparent implementation of tenders for the agricultural land lease in Croatia.

Keywords: transparency; accountability; agriculture land lease; transparency of tenders; tender process

1. Introduction

Government or public tenders are typically defined as formal invitations for interested parties to participate in the government procurement process by providing services, goods, or leasing state-owned assets. A core responsibility of governments is the transparent and accountable management of public assets and resources, such as state-owned agricultural land [1], especially in the digital age, when information on tenders, public calls, public procurement, budget management and reallocation, subsidies, but also visibility of all these processes can and should find its way to users who seek it [2,3]. To make this possible, numerous initiatives have been launched to support the development of digital society and public government [1,4]. One of these initiatives is the [1,4] open data (OD) initiative, which argues in favor of free and unrestricted access to (government) data [5]. Due to this, the OD initiative has the potential to transform business processes towards more transparent and accountable practices, which confirms it as one of the main pillars in achieving fully transparent and accountable open public government.

Transparency and accountability in public institution processes' is the basis for providing good quality services to society. From the user's point of view, this reflects the quality of public administration services in which digital technologies and open government data should be used to enable visibility and monitoring of activities, criteria, and results of a process from start to finish [6]. Further on, the creation of public value and value for the user is possible only through transparent and fair implementation of the process, and this requires understanding of public policies and ways of its implementation by stakeholders, i.e., government administration—counties, cities, and municipalities. Since government administration is responsible for the implementation of public tenders, it is on them to adopt an accountable, transparent, and fair approach to the management of state assets, e.g., state spatial capacities, financial resources, data capacities, infrastructure, and others, as well as the proper redistribution of all forms of state assets and resources [7,8].

State-owned agricultural land represents an important resource for a country, so its management must be done responsibly and transparently [9]. In many countries, e.g., Croatia, state-owned agricultural land is leased to private farmers based on an agricultural land lease tender conducted by the government administration. As part of the government procurement and in line with the principles of transparent and accountable state asset management, the process of land lease should be made fully transparent and accompanying data available as open data, but this is not always the case.

For several years there has been a debate about the (lack of) transparency of the tendering and allocation of state agricultural land in Croatia [10,11]. Numerous media articles [12–17] and scientific circles highlighted lack of transparency as one of the main problems of the Croatian agricultural sector. In one example, the author mentions leases won without all the necessary lease documents or contracts renewed without competitive bidding [14]. In some cases, a family connection was revealed between the beneficiaries of the tenders and local politicians [13], or the tendering process was canceled due to missing minutes of the lease commission where the most favorable bid was chosen [12].

The interest in agriculture in Croatia is decreasing, and more and more farmers are giving up this business out of great dissatisfaction caused by untransparent and potentially biased tendering processes. According to the Croatian Agency for Agriculture, Fisheries and Rural Development, the number of registered family farms has dropped by 25% from 2016 to 2023 [18]. Although Croatia continues to develop and refine its tendering frameworks, the issue of transparency appears to persist. Even the category-based scoring system introduced in the latest regulations reveals potential shortcomings.

Considering the aforementioned, the main goal of this paper is to provide insight into the current state of transparency of agricultural land leases in Croatia to detect barriers preventing a fully transparent land lease tendering process. To achieve this goal, authors propose the following research questions:

RQ1. *Are public calls for tenders of agricultural land lease publicly announced by the government administration?*

RQ2. *Is the decision on the most favorable offer publicly available as open data?*

RQ3. *To which extent are the offers from all the applicants in the tendering process available as open data?*

To answer the proposed research questions, the transparency issues of the current model of the state-owned agricultural lease tender, reconstructed from the binding laws, regulations, and practices, were analyzed. The transparency issues were analyzed based on available data on agricultural land lease tenders from the local level administration units' websites and through a survey with the local government.

2. Research Background

2.1. Research Theories and Practice Review

The Republic of Croatia has over 2.6 million hectares of agricultural land, of which about 33% is state-owned [19], and only 21% of state-owned agricultural land is offered through tenders for sale, lease, etc. [20]. Due to the fragmentation and unsettled ownership structures of agricultural land [21], many small agricultural producers rely on leasing state-owned land to ensure and stimulate the further development of agricultural production [19].

Inefficient management of state-owned agricultural land by local authorities in Croatia led to the establishment of the National Agency for Agricultural Land in 2009. The agency was responsible for leasing state-owned agricultural land [22]. With the establishment of the agency, responsibility for state-owned agricultural land was transferred from the local (regional) administrative level to the agency's centralized system. This should have helped speed up the land leasing process and make it more transparent. However, due to an even more inefficient land lease, the new 2018 Agricultural Land Act abolished the work of the Agency for Agricultural Land [23] and returned responsibility over state-owned land to the local administration.

Management of state agricultural land in Croatia is a topical issue that requires analysis and consideration of the issue from the perspective of legislation and implementation at lower levels. The stakeholders engaged in the tendering of state agricultural land leases include the local administrative unit, primarily responsible for issuing tenders for land leases, and farmers/agriculture producers, as applicants who constitute the demand side for agricultural land.

The evaluation of the practice of conducting public calls for tenders is a complex process that requires the understanding of various research theories. The starting point for the research is the analysis of the legislation and the responsible actors in its implementation, which is based on institutional theory.

Institutional theory focuses on institutions and their role in policy making, which can be interpreted through two policy cycle stages: agenda-setting stage as well as policy and implementation stage [24]. The institutional theory provides insight into how institutions shape behavior according to legislation rules on which they rely in their decision-making processes and how they coordinate the provision of goods and services [25]. Accordingly, the legislation for tenders of state agricultural land leases and the procedure for conducting the tenders were analyzed to identify problematic issues and shortcomings for stakeholders in the process. Agricultural land and its management in the Republic of Croatia is regulated by the Agricultural Land Act [26] and supplementary regulations and ordinances. According to the Act, state-owned agricultural land may be leased on a basis of a public tender conducted by the administration of a municipality or a city (Official Gazette 20/2018, Article 31) [23]. Although the act defines the conditions and provides for the possibility of leasing agricultural land, it does not provide guidelines for the implementation of the tender process. Public calls and the conduct of public tenders are regulated by the Ordinance on the Conduct of Public Tenders for the Lease of Agricultural Land and the Lease of Ponds Owned by the Republic of Croatia, which sets out in more detail the documentation, deadlines, and the required procedure (Official Gazette 104/2022). The act foresees the constitution of a Lease Commission whose task is to select the most favorable bid in the tender (Official Gazette 20/2018, Article 31), and the ordinance ensures a uniform tendering procedure at the country level. Thus, according to the ordinance, local-level administrative units, municipalities, or cities are obliged to conduct a public tender, i.e., to publicly announce a call for tender, giving all interested parties the opportunity to apply (Official Gazette 104/2022, Article 4). The ordinance also provides for the establishment of an information system for electronic tendering. Although it is in the early stages of implementation, the electronic conduct of tenders is expected to eliminate the possibility of manipulation in the tendering process. In addition to the act that provides for the possibility of leasing land and the ordinance that regulates the tendering procedure, the Program of Disposition of State-owned Agricultural Land must be in place for a land lease. Defined by the Act, the program includes all the agricultural land (and its characteristics) on the territory of the local administrative unit and ways of its administration, e.g., land lease. For a program to be binding, it must be approved by the competent ministry (Official Gazette 20/2018, Article 29). Without the adopted program, it is not possible to conduct a tender for lease. Acts and regulations on agricultural land in Croatia have been frequently amended in the last couple of years [27]. Often, these amendments created problems in practical implementation [28], but have also brought some improvements, like a more transparent land lease. The latest amendment of the act from 2022 introduced a scoring system in the tender for lease procedure. Now, tenderers score points based on different criteria, e.g., type of agricultural production (Official Gazette 57/2022, Article 18), and only the ones with enough points get land for lease. Scores per criteria are regulated by the act, so the scoring schema is fully transparent, but the evaluation of submitted bids according to the scoring criteria may not necessarily be transparent. Additionally, the complexity of the model, such as determining who benefits in specific cases, could be subject to misuse.

Another relevant research theory, on which the analysis of the implementation of the tender for the allocation of state agricultural land in the Republic of Croatia is based, is the prescriptive decision theory. To gain pertinent insights into applying prescriptive decision-making theory for enhancing and transforming existing practices transparently and responsibly, a review was conducted on practices in other countries and the accomplishments of extant research related to the tendering of state agriculture land leases. The 2019 European Open Data Directive plays a crucial role in transparency of government processes. While arguing for unrestricted and public access to government data, i.e., open government data, this Directive provides tools for the public to control the government and therefore to increase trust in public procurement processes [29]. The importance of regulating agricultural land leases is also recognized by the European Union, where in 2021 it issued a detailed report on the regulation of data and information on agricultural land [28]. However, the document does not clearly indicate the difference between the leasing of private and state land, and it is hard to determine the differences between countries related to state agricultural land leases, except by a detailed analysis of their legislation and the process of leasing state agricultural land. It is also important to examine practices in neighboring countries with similar historical heritage. For example, in Slovenia, the Fund for Agricultural and Forest Land manages the leasing of state-owned agricultural land and farms, regulated by the Regulation on the Leasing of Agricultural Land and Farms. The regulation defines ranking and preferential criteria among lessees but does not mandate publishing tender results [30]. In Serbia, recent changes in law were made to enable long-term leases (30 years) in order to stimulate potential investors [31]. In the Federation of Bosnia and Herzegovina, state agricultural land can be leased to residents who confirm at least ten years of agricultural activity and residency with a court-notarized declaration. Preference goes to families earning predominantly from agriculture with insufficient land for economic security [32]. The allocation of agricultural land is a significant issue, often critiqued as mismanaged [33,34].

Despite ongoing efforts, it is important to note that within scientific circles focused on related fields, this topic remains largely unexplored. Relevant research on the leasing of state agricultural land has not been conducted, likely due to the complexity of the issue. Existing research that dealt with a similar problem of tender implementation highlighted several key implications. Transparency and accountability in public administration are crucial for citizen trust. With the growing expectations of citizens [35], institutions need adequate resources and staff. Urban areas and cities have more resources (human, material, and financial), enabling them to act proactively and transparently compared to smaller rural environments [36,37]. Access to information on important issues is critical to building citizen trust. Spáč et al. [38] analyzed how local authorities (municipalities) in Slovakia operate and respond to requests for information. They also examined differences in municipalities based on ethnic group dominance, leadership by women, left-wing parties, or independents in providing information to citizens. The analysis shows that invoking the Freedom of Information Act increases the likelihood of obtaining information. The size of municipalities influences the processing of requests, with larger municipalities responding more efficiently [38]. More populous local governments receive more resources and operate more transparently [39]. The ethnicity, political orientation, and gender of local leaders also affect responses. Minorities may face language barriers, and independent mayors are more likely to respond. Municipalities with female mayors comply better with free access laws than those led by males [38].

The contribution of the mentioned research is important for the topic of this paper, given that this research investigates a specific area of local government activity, the allocation of state property (land), where access to information, transparency of the entire process, and public accountability are very important.

The prescriptive decision theory is the basis for decision making on guidelines and recommendations and drives the development of new models and recommendations to improve current practice. This theory includes clarification of how decisions should be made, what is desirable in the public sector, and how it can contribute to accountability and transparency [40].

2.2. *Current Agricultural Land Lease Model in Croatia*

The current agricultural land lease model in Croatia is shaped by the Agricultural Land Act and relevant ordinances. Although these do not formally define all the steps of the process, the crucial ones are covered and made uniform for all administrative units. Based on the binding agricultural regulations and knowledge from existing tendering procedures, the authors recreated a tendering model to serve as a starting point in transparency analysis. This authors' interpretation model considers two user perspectives, the perspective of the local administrative unit (municipality)—LAU and the perspective of applicants (farmers). We focus on these two user perspectives only because they represent the essential and sufficient stakeholders in the tendering process. Further on, the reconstructed model divides steps in the tendering process into five key phases: (1) initial phase, (2) tendering phase, (3) evaluation phase, (4) formalization phase, and (5) finalization phase. This structure of the model will allow authors to identify steps or phases in the process that are mostly related to transparency.

The initial phase gathers activities related to preparation and publication of the public tender. From the LAU perspective, it includes activities of preparing the tender and its announcement, and from the farmers' perspective, it considers the activity of considering the conditions and criteria for applying for the tender. The tendering phase refers to the time from the announcement of the call, that is, from the beginning to the end of the period in which the tender is open. This is the period during which the local administrative unit receives bids and farmers compile and complete the bidding documents. The evaluation phase includes the announcement and the event of the public opening of the bids and their evaluation by the commission formed by the local administrative unit, whereby farmers may take part in this public event. This phase also includes the publication of results. Although no regulation explicitly states when the commission should be formed, we consider it as one of the tasks in the evaluation phase. In the formalization phase, a contract is signed with the farmer who submitted the best bid, and in the finalization phase, the acceptance/non-acceptance of the bid is reported, and if the best bid is not selected, the public is informed that the tender has been canceled.

3. Materials and Methods

Based on the review of current practices in the tendering process of leasing state-owned agricultural land, the majority of transparency issues can be connected to several activities in two out of five phases of the tendering model. The first is the publication of the public tender (initial phase), while the other activities include the publication of the most favorable bid as well as all other bids in the call (evaluation phase). Thus, the following methodology focuses on collecting data on the aforementioned activities.

3.1. Data Collection

In order to evaluate the transparency of the current model, we collected data on the transparency of state agricultural land tendering and leasing from local administrative units in Croatia. Data are collected from two sources for the period 2015–2022 (the tendering process has been delegated from the national level to local government in 2015). First, we searched the official websites of all 556 units of local government (cities and municipalities) looking for information and documentation on state-owned land leases. If the websites were indexed and contained a website search option, we used key words in our search (e.g., "tender", "land lease", "agricultural land"). When the website search option was not present, we browsed the website under standard categories such as tenders or public calls, news, or documents, searching for information. Time needed to find the information was recorded, and a maximum of 15 min was devoted to searching each website.

The second approach was formally requesting the information on state-owned land leases (links to tenders, decisions on the selection of the most favorable offer and offers from all applicants) conducted in the 2015–2022 period from each local administrative unit via e-mail. In this e-mail, we introduced our research team and the purpose of the data that we are requesting. The elapsed time from inquiry to reply was also recorded (up to 60 days from the initial request). The responses were cross-checked with the data obtained in the first step, and a unified dataset was generated.

In addition, data on staff capacity [41] for all units of local government were collected for the most recent period available (2018) from the statistical reports of the Ministry of Justice and Public Administration. Data on population size of each unit of local government from the last national census [42] were also retrieved from the Croatian Bureau of Statistics. These variables have been shown to be related to the level of transparency in previous research.

3.2. Evaluating Local Government Transparency in State-Owned Land Lease

Each local administrative unit of government (municipality or city) was scored on their transparency when conducting state-owned land leasing based on three criteria:

(1) published tender for state-owned land lease (value 1—published tender on website and/or reply to inquiry with the information)

(2) published decision on the selection of the most favorable offer (value 1—published decision on the selection of the most favorable offer on website and/or reply to inquiry with the information)

(3) published offers from all applicants (value 1—published offers from all applicants on website and/or reply to inquiry with the information).

If no information on tenders was available on the official website and we received confirmation from the local administrative unit that they have not conducted tenders in the period 2015–2022, these units were excluded from further analysis of transparency. When no information on land-lease tenders was available on the website and we received no feedback from the unit, the unit scored zero on the transparency scale. If a unit conducted multiple tenders in the time period, each tender was scored separately, and an average score for the local government unit was calculated.

3.3. Statistical Analysis

Analysis of variance (ANOVA) was conducted on three levels (regional, county, and municipality/city) using the PROC GLM function in SAS statistical software version 9.4 [43]. Post hoc comparisons were conducted using Tukey's HSD test. Pearson's correlation coefficients were calculated between the average transparency scores of units of local government and staff capacity and total population size using the PROC CORR function in SAS statistical software version 9.4.

To measure spatial autocorrelation (spatial clustering) of the transparency scores in local government units, we calculated Moran's I using the spdep package v.1.3-6 [44] in R v.4.1 [45]. We tested the significance of the correlation coefficient using Monte Carlo simulation with 100 permutations. Based on the results of the analysis, a new and improved model of state agricultural land lease was proposed.

4. Results

After the first phase of data collection for this research, data on 335 tenders for leasing state-owned land in 204 units of local government were collected. The majority of units published only one tender in the period 2015–2022 (63%), with notable exceptions such as the city of Bjelovar, which published ten tenders in the same time period. Only tender documentation was published in the majority of cases (67%), followed by both tender and decision on the selection of the most favorable offer (30%). In only five cases was the complete tender documentation (including tender, decision, and applicant offers) published on the official website. In the second phase of the research, we received a relatively small response from local administrative units (only 36% of units responded to our inquiry). Out of 200 responses, 62% of them replied that they have not conducted a tender for state-owned land lease in the 2015–2022 period, excluding them from further transparency analysis.

When examining the 2015–2022 time period, we can see a slow but steady increase in the number of published tenders from 2015–2017 (71% increase from the initial 7 tenders in 2015 to 2016 and an additional 58% increase from 2016 to 2017), with a small decline in 2018. After 2018, a huge surge of tenders was recorded (150% increase in 2019 and an additional 217% increase in 2020). After 2019, the number of tenders fluctuates but is still above 75 tenders per year (Figure 1).

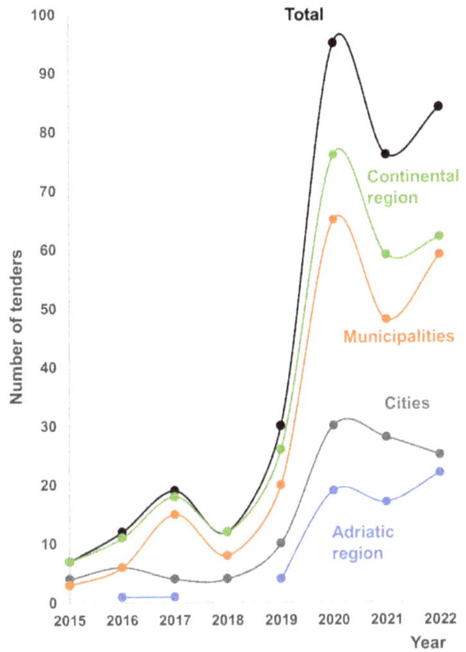

Figure 1. Number of published tenders for state-owned agricultural land lease per year by agricultural region (Continental, Adriatic) and type of local administrative unit (city, municipality).

Considering the unique peculiarities in the production of agricultural regions in Croatia and the geopolitical characteristics of the regions, the results of conducting tenders for the agricultural land lease are presented based on the Continental and Adriatic regions in order to identify possible differences (Figure 2). The Continental region has a total of 336 local administrative units, while the Adriatic region has 220 local administrative units.

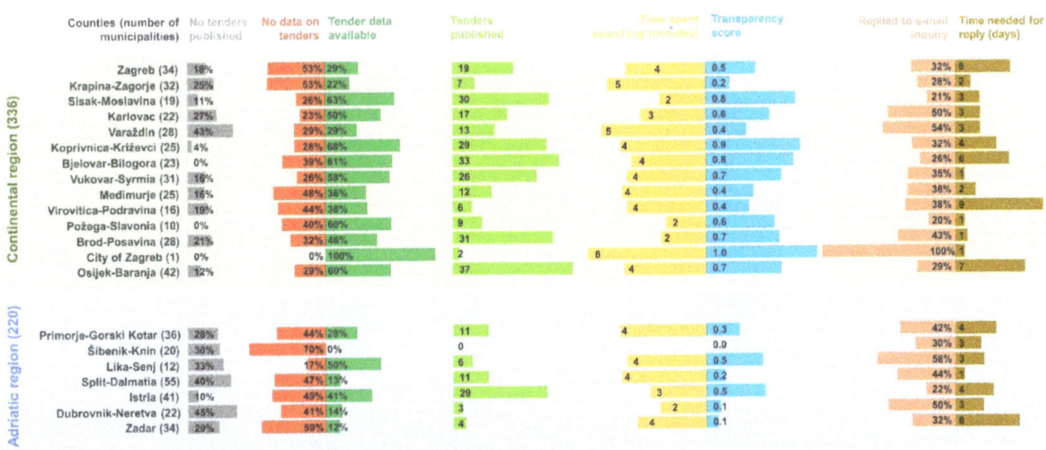

Figure 2. Tender for state owned agricultural land lease assessment results in Croatia on a county level divided into two agricultural regions.

In the Continental region, several counties had all of their local administrative units with at least one tender, in contrast to the Adriatic region, where the rate of counties with no tender published per local administrative unit was much higher. In the Continental region, there is a large percentage of local administrative units for which there is no information about tenders, with Zagreb and Krapina-Zagorje counties leading the way with 53%. In the Adriatic region, the situation is similar, with slightly higher percentages of local administrative units for which there is no data, with Šibenik-Knin county leading the way with 70% and Zadar with 59%.

In the Continental region, data are available for the city of Zagreb (100%), given that the only local administrative unit is the city itself. Koprivnica-Križevci with 68%, Sisak-Moslavina with 63%, and Bjelovar-Bilogora with 61% of the county where the most local administrative units announced a tender for the allocation of agricultural land. In the Adriatic region, Lika-Senj leads with 50% of them with a published tender. Šibenik-Knin does not have any available information about tenders.

It is noticeable that more tenders for agricultural land allocation were published in the Continental region than in the Adriatic region. Four counties had more than 30 tenders published in the Continental region, while in the Adriatic region, the county with the highest number of published tenders was Istria (29 tenders), while other counties had significantly fewer tenders for the allocation of agricultural land.

As the research was carried out by searching the official websites of local administrative units and contacting them, and these are also two important segments of transparent and accountable public government processes, the observed parameters for the implemented activities are presented. The time spent on the search in minutes ranges from 2 to 6 min. This is approximately how long it took the researchers to review the local administrative unit's website on average. It took the longest time to review the site of the city of Zagreb (6 min), which is not surprising considering that Zagreb is the capital of Croatia with a lot of published digital content. Analysis of variance showed no statistical differences in time needed to search and find data on tenders between regions (Continental and Adriatic) and type of local administrative unit (city and municipality).

When contacting local administrative units, data on responses and the time required for a response in days were recorded. In the Continental region, the city of Zagreb responded to the inquiry and was the only local administrative unit to provide insight into the data. Apart from them, only two other counties have more than 50% of collected responses from local administrative units, namely Varaždin (54%) and Karlovac (50%). In all other counties, less than 50% of responses from local administrative units to inquiries were collected, which is a worrisome fact because according to the Act on the Right of Access to Information, access to the requested data should have been enabled. In the Adriatic region, 58% of responses were received in Lika-Senj and 50% in Dubrovnik-Neretva. In all other counties, less than 50% of responses from local administrative units were collected. In the Continental region, local administrative units of four counties responded in a single day. At the same time, Virovitica-Podravina took the longest across both regions (9 days). In the Adriatic region, the responses came the fastest in Split-Dalmatia, on average in a single day. Analysis of variance once again showed no statistical differences in time needed to receive a response from local administrative units on tenders between regions (Continental and Adriatic) and type of local administrative unit (city and municipality).

Analysis of variance revealed statistical differences in transparency scores on both the regional level and between cities and municipalities ($p < 0.0001$ and $p = 0.0015$, respectively). Observing the Continental region, it is evident that cities have a higher transparency score compared to municipalities (Figure 3). The score for cities is 0.7, while for municipalities it is 0.6. In the Adriatic region, the transparency score is somewhat lower; for cities, it is 0.4, and for municipalities, 0.2. Namely, in municipalities, as rural areas, more tenders are published, but the transparency score is lower in both agricultural regions. The reason for this may be that rural areas do not have enough available human resources that provide the necessary information or publicly publish and update it on official websites.

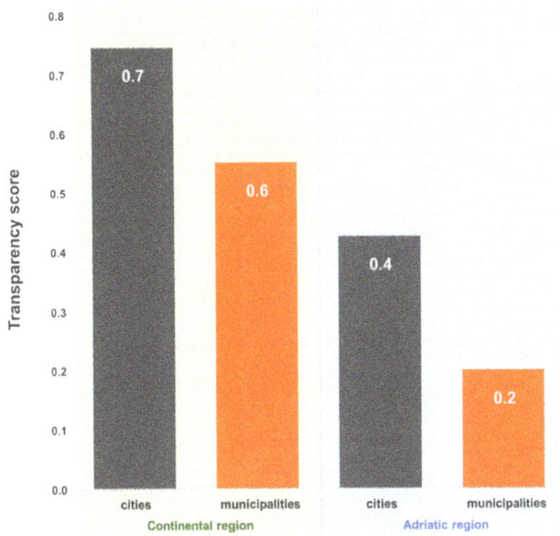

Figure 3. Average transparency scores of types of local administrative units (city, municipality) categorized by agricultural region.

Correlation analysis revealed a high correlation between the population size of units of local government and the staff number of cities/municipalities (r = 0.993, $p < 0.0001$), indicating that larger cities and municipalities in terms of population also have a greater number of staff. The results of the analysis between transparency scores and both population size and staff number showed no significant correlation (r = 0.02, $p > 0.68$) indicating that although larger cities/municipalities have higher resources, those are not invested into actions increasing the transparency level of the unit.

In total, 432 spatial units (cities/municipalities) were included in spatial autocorrelation analysis (units with no published tenders based on data collection were excluded). Out of those, 16 spatial units were neighborless, and the majority of units (66%) had 3–6 neighbors. The value of Moran's I was 0.186 and was highly significant ($p < 0.0001$). A positive value indicates that the transparency scores are spatially clustered to some extent. This can be seen in Figure 4, where small clusters of transparency score = 0 can be seen throughout the whole country.

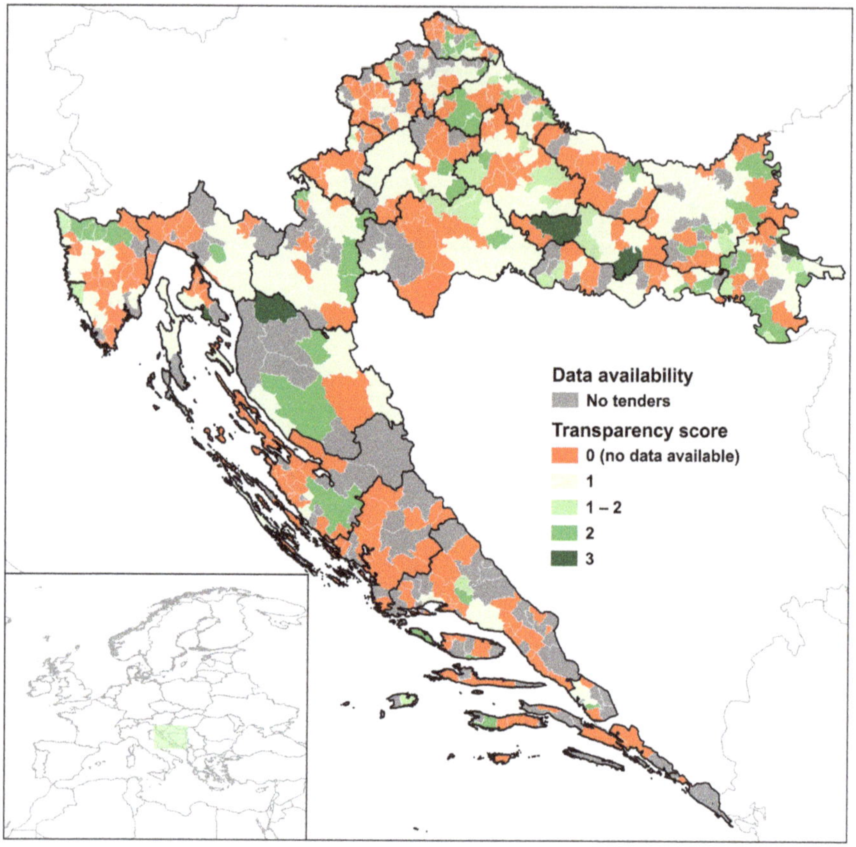

Figure 4. Tender transparency scores of local administrative units in the Republic of Croatia.

Small clusters of higher transparency scores can also be observed in Istria, along the border with Slovenia, the eastern part of the country (Slavonia), and central Croatia. Neighboring units with similar scores tend to belong to the same county, indicating that cities/municipalities adopt good (and bad) transparency policies from neighboring units that are pioneers in opening their data to the public according to EU and national regulations.

5. Discussion

5.1. Transparency in State-Owned Land Lease by Local Government

The analysis of the call for tender of the state agriculture land lease in Croatia has revealed open questions and opportunities for improvement. The analysis shows that more tenders were published in the Continental region than the Adriatic region, which was expected since the Continental region has more agricultural land and a long farming tradition in the country. The data collected on the level of transparency of local administrative units are worrying. Only in 5 counties, more than 50% of local administrative units provided data on the conducted tenders based on the request sent to them. Although the Law on the Right of Access to Information obliges them to provide the interested society with the requested information, this has not proven to be correct in practice. The question arises as to the reason for this practice and why the representatives of the local administrative units are shirking their responsibility and violating the transparency of the operations. Even though the obligation to publish the results and the most favorable decision is not defined in the current decision-making system, there is no obstacle to publishing them to raise and develop transparent practices. When conducting tenders, only 30% of units disclosed the decision on the most favorable tender.

The research carried out has shown that cities have better transparency compared to municipalities, although there are certain differences between cities depending on the region (Continental and Adriatic). Looking at the number of employees and inhabitants in the cities and municipalities, previous research states that cities and municipalities with a larger number of employees and inhabitants should be more transparent in business, but our research does not confirm this. No correlation was found to show that local administrative units with higher capacity have greater labor transparency.

The analyzed period from 2015 to 2022 was characterized by a weak but stable increase in the number of calls for tenders. Of interest is the year 2020, in which there was a significant increase in the number of tenders, and the question arises as to whether this is in connection with the COVID-19 pandemic and the pandemic-related situation on the market. According to the OECD [46], global production of key commodities such as wheat, corn, rice, and soybeans has increased. As there were numerous restrictions on the import and export of food, it is possible that national policies were geared towards this and gave their agricultural production a chance.

Spatially grouped data show that there is an influence between local administrative units, i.e., there is a high probability that when looking at a particular local administrative unit, the neighboring local administrative units have similar transparency scores. This suggests that neighboring local administrative units influence each other in their operational practices.

5.2. Potential Improvements of the Current Model

To ensure full transparency in the allocation of state-owned agricultural land through the tendering process, the implementation process and all related data should be open and publicly accessible to all interested stakeholders. To achieve this, we propose several improvements to the current model of leasing state agricultural land that should contribute to a higher level of transparency, primarily for the applicants in the tendering process, but also for the general public. The suggestions for improvement are considered to be simple but effective solutions and relate to different phases of the process (Figure 5).

The initial phase is considered an extremely important part of the process as it determines who can apply for the tender and under what conditions. This phase is managed by the local administrative unit (LAU), so the improvements are mainly focused on creating an unbiased and objective tendering environment. To increase transparency in this phase, the authors propose: (1) the establishment of a national digital tendering platform (TP), (2) the establishment of a national repository of state-owned agricultural land (SOAL) as part of the TP, and (3) various ways of disseminating tenders. The establishment of a tendering platform is expected to have three advantages. First, it will serve as a single access point for applicants to find all binding laws, regulations, obligations, and conditions in one place. Secondly, it will enable digital bid submission in the tendering process, and thirdly, it will serve as a dissemination platform for calls for tenders. In addition, the integrated part of the platform, the National Repository of SOAL, will provide an overview of vacant and occupied plots in local administrative units and their occupancy period in real time. This should help interested applicants better plan their future applications. Finally, the authors propose to publicize the call not only via the LAU website and the TP mentioned above, but also through additional online/offline advertising. This could contribute to a higher audience reach and more bids in the tendering process—transparency and healthy competition.

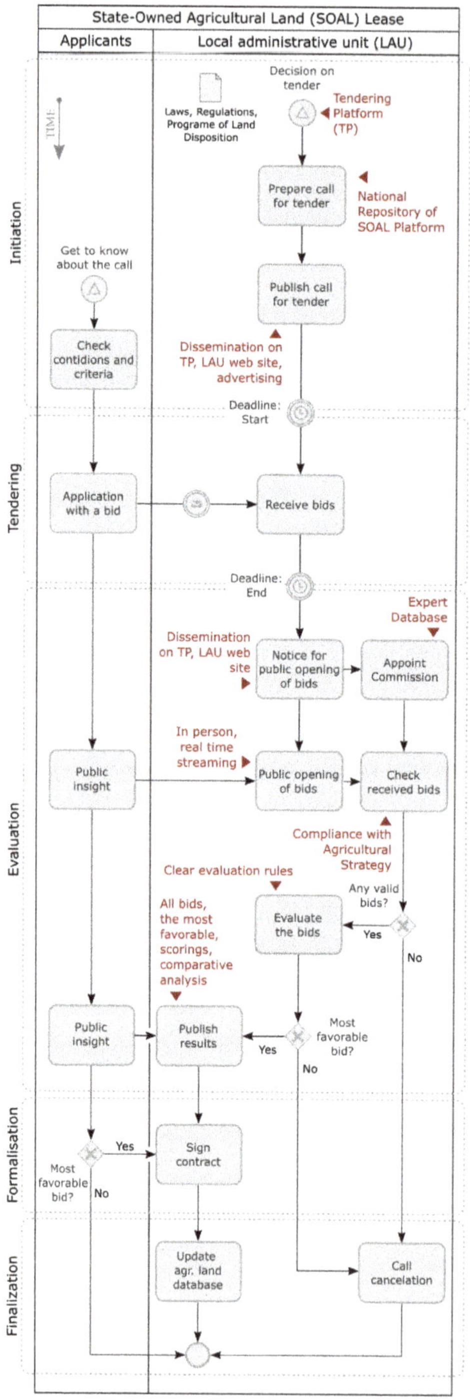

Figure 5. Proposed model for state agricultural land lease process in Croatia with improvements in initial and evaluation phases of the process.

The evaluation phase of the proposed model is where most of the improvements can be found, as most of the key tendering activities fall into this phase. These improvements relate to LAUs that have to decide on the most favorable bid and publish the results. As a first step, the new model proposes to announce the public opening of tenders via the tender platform and the LAU website to ensure that all applicants receive the information in time. A commission will then be appointed to review and evaluate the bids and decide on the most favorable bid, with the novelty being how the members of the commission will be selected. As the appointment of the commission requires the involvement of experts to evaluate the bids responsibly and fairly, a higher level of transparency could be achieved at this stage by randomly selecting the experts from the Expert Database, thus minimizing a priori subjective decisions. The public opening of bids is another activity in which all bids received are opened in front of the audience present. Instead of a traditional, face-to-face event, the new model proposes a hybrid event with face-to-face attendance and online streaming in real time (with video recording). In this way, all interested parties can be assured of a fair and transparent bid opening. The evaluation of the bids, the centerpiece of the tendering process, is a follow-up activity that requires the greatest attention and reasonably the highest level of transparency. Firstly, the applications must be in line with the objectives of the agricultural strategy in order for the bids to be qualified. However, the conditions for agricultural production set out in the strategy, and therefore the conditions that the interested parties must fulfil, should not be taken into account equally in all agricultural regions. If the applications do not match these objectives, the bid is dropped. For the bids that are in line with the strategy and therefore qualify, the evaluation should be based on clear evaluation rules and fair scoring. The evaluation rules should be publicized and include clearly defined criteria, sub-criteria, and possible points for each situation or status of the applicant. The final step in this phase is the publication of the results. The novelty in this activity is that all applications should be made publicly available on the tender platform together with the following (relevant) documents. This includes the list of all bids received, the selected—most favorable—bid, summarized scores per applicant according to criteria/sub-criteria, and a simplified comparative analysis of all applications (e.g., a table).

In the last year, there have been some changes related to the implementation of the call for tenders of agricultural land leases. Namely, since November 2022, the Ministry of Agriculture has started to transparently publish public calls for tenders on the official website, where all public tenders for leasing agricultural land owned by the Republic of Croatia can be found, but this is not an obligation. With this functional improvement, a positive shift in publishing tender is noticeable. In December 2023, the Ministry presented a new e-Lease application—a software solution for the disposal of state agricultural land whose goal is to digitize the process. The application is presented as a solution that will contribute to openness, transparency, and clarity for the entire public, but according to its functionalities, it does not enable this. Some of the functionalities offered by the newly developed application are guidelines when filling out the application, subsequent delivery of documentation, and drafting of contracts for agricultural land leases. Therefore, it can be concluded that the developed application solves the problem of an administrative nature, but not the deeper issue of the implementation of tenders for state agricultural land leases, and even less the evaluation procedure. Stakeholders in the process (local self-government units, the Ministry of Agriculture, or the applicant) log into the application, which means that it is not available and open to everyone, and thus transparent and open implementation of the tender is not ensured. Also, a large number of farmers are still dissatisfied and believe that there is still bias and rigging in the tenders that are conducted.

6. Conclusions and Policy Implications

Research on the implementation of tenders for agricultural land leases in Croatia has shown that it cannot be claimed that the process is transparent and even less that the allocation of land has been carried out responsibly. In the entire process, in most local administrative units, important data (evaluation and comparative analysis) for all stakeholders in the process are missing. Some shortcomings or loopholes in the current legislation were identified, which leave opportunities for manipulation in the implementation of tenders for agricultural land leases. In addition, the results of the evaluation and the selection of the most favorable offer should be made public and transparent. The fundamental change to the process relates to the introduction of a platform for the management of state agricultural land. These changes would enable an open and transparent procedure for all stakeholders and raise the management of state land to a higher level. Additionally, changes in legislature are needed as well, explicitly prescribing all stages in the tendering process. Without legal regulation of the tendering process, the responsibility will remain at the level of local administrative units, which does not guarantee uniform enforcement of the tendering conditions and full transparency. This could also have a positive impact on sustainable use of agricultural land in Croatia, since national policies include measures that ensure maintaining the long-term health of natural resources.

The limitations of this research relate to gaps in data collection. A small proportion of local administrative units did not respond to our request for data, nor did we find the data on their websites, limiting the analysis due to uncertainty of the status of tenders in said administrative units. Although linked to the Croatian agricultural land lease tendering process, proposed methods consider transparency aspects characteristic of any tendering process, which makes the procedure replicable no matter the model of land lease in background. Future research will be focused on the user-side of the agricultural land lease tendering process in Croatia in order to gain complete insights into the tendering process and to possibly identify additional barriers that were not observed in this research.

Supplementary Materials: The following supporting information can be downloaded at: https://www.mdpi.com/article/10.3390/ijgi13110401/s1.

Author Contributions: Conceptualization, Larisa Hrustek, Karlo Kević and Filip Varga; methodology, Larisa Hrustek, Karlo Kević and Filip Varga; validation, Larisa Hrustek, Karlo Kević and Filip Varga; formal analysis, Filip Varga; investigation, Larisa Hrustek, Karlo Kević and Filip Varga; writing—original draft preparation, Larisa Hrustek and Karlo Kević; writing—review and editing, Larisa Hrustek, Karlo Kević and Filip Varga. All authors have read and agreed to the published version of the manuscript.

Funding: This research received no external funding.

Data Availability Statement: The original contributions presented in the study are included in the Supplementary Materials, further inquiries can be directed to the corresponding author.

Acknowledgments: This research is part of the Twinning Open Data Operational project that has received funding from the European Union's Horizon 2020 research and innovation program under grant agreement no. 857592.

Conflicts of Interest: The authors declare no conflicts of interest.

References

1. Lnenicka, M.; Nikiforova, A. Transparency-by-Design: What Is the Role of Open Data Portals? *Telemat. Inform.* **2021**, *61*, 101605. [CrossRef]
2. Cahlikova, T.; Mabillard, V. Open Data and Transparency: Opportunities and Challenges in the Swiss Context. *Public Perform. Manag. Rev.* **2020**, *43*, 662–686. [CrossRef]

3. Varga, F.; Kević, K.; Hrustek, L. Transparency In Agricultural Land Lease By Local Government. In Proceedings of the International Conference on Open Data (ICOD 2022): Book of Abstracts; Varga, F., Đurman, P., Eds.; University of Zagreb, Faculty of Law: Zagreb, Croatia, 2023; pp. 44–49.
4. Park, S.; Gil-Garcia, J.R. Open Data Innovation: Visualizations and Process Redesign as a Way to Bridge the Transparency-Accountability Gap. *Gov. Inf. Q.* **2022**, *39*, 101456. [CrossRef]
5. Kević, K.; Kuveždić Divjak, A.; Welle Donker, F. Benchmarking Geospatial High-Value Data Openness Using GODI Plus Methodology: A Regional Level Case Study. *ISPRS Int. J. Geo-Inf.* **2023**, *12*, 222. [CrossRef]
6. Wiencierz, C.; Lünich, M. Trust in Open Data Applications through Transparency. *New Media Soc.* **2022**, *24*, 1751–1770. [CrossRef]
7. Croatian Ministry of Agriculture Public Tenders for the Agricultural Land Lease Owned by the Republic of Croatia. 2023. Available online: https://poljoprivreda.gov.hr/istaknute-teme/poljoprivreda-173/poljoprivredno-zemljiste/javni-natjecaji-za-zakup-poljoprivrednog-zemljista-u-vlasnistvu-rh/5806 (accessed on 18 August 2024).
8. van der Meer, J.; Hartmann, A.; van der Horst, A.; Dewulf, G. Raising Risk Awareness in Multi-Criteria Design Decisions for Integrated Design and Construction Tenders. *Constr. Manag. Econ.* **2022**, *40*, 296–312. [CrossRef]
9. Paresi, C.; Haile, S.; Permezel, M.; Asiama, S.; Kombe, W.; Gold, J.; Tuladhar, A.; Badiane, N.; Mongbo, R.; Bagré, A.; et al. *Tools to Support Transparency in Land Administration. Training Package: Trainers' Guide*; United Nations Human Settlements Programme (UN-Habitat): Nairobi, Kenya, 2013; ISBN 978-92-1-132549-2.
10. Petković, M. Vučković: OPG-Ovi Će Imati Prioritet Na Natječajima Za Državno Poljoprivredno Zemljište. Available online: https://www.agroklub.com/poljoprivredne-vijesti/vuckovic-opg-ovi-ce-imati-prioritet-na-natjecajima-za-drzavno-poljoprivredno-zemljiste/69501/ (accessed on 27 June 2023).
11. Gazdek, G. Upitna Dodjela Poljoprivredne Zemlje. Available online: https://gospodarski.hr/rubrike/ostalo/upitna-dodjela-poljoprivredne-zemlje/ (accessed on 27 June 2023).
12. Mandarić, A. U Velikoj Poništen Natječaj Za Zakup Poljoprivrednih Površina u Vlasništvu RH. Available online: https://www.034portal.hr/u-velikoj-ponisten-natjecaj-za-zakup-poljoprivrednih-povrsina-u-vlasnistvu-republike-hrvatske-752 (accessed on 27 June 2023).
13. Mandarić, A. Nećak Franje Lucića, Branko P. Ostvario Pravo Na Zakup Više Od 25 Hektara Zemlje u Pleternici Na 25 Godina?! Available online: https://www.034portal.hr/necak-franje-lucica--branko-p--dobio-na-zakup-vise-od-25-hektara-zemlje-u-pleternici---286 (accessed on 27 June 2023).
14. Šimić, Z. Malverzacije Sa Zemljištem s Onoga Svijeta I Oni Su Uredne Platiše: Zakup Državne Zemlje Plaća i 59 Pokojnika. Available online: https://www.jutarnji.hr/vijesti/hrvatska/malverzacije-sa-zemljistem-s-onoga-svijeta-i-oni-su-uredne-platise-zakup-drzavne-zemlje-placa-i-59-pokojnika-184009 (accessed on 27 June 2023).
15. Općina Lipovljani Sumnjiva Dodjela Zakupa Poljoprivrednog Zemljišta Na Području Općine Lipovljani. Available online: https://lipovljani.hr/vijesti/aktivnosti-opcine/sumnjiva-dodjela-zakupa-poljoprivrednog-zemljista-na-podrucju-opcine-lipovljani/ (accessed on 27 June 2023).
16. DNEVNIK.hr Prvo Nije Dobio Zemlju Pa Mu Je Naglo Dodijeljena Da Ga Se Ušutka? "Stranački Sam Nepodoban, a Na Vlasti Je HDZ". Available online: https://dnevnik.hr/vijesti/hrvatska/provjereno-sumnjiva-dodjela-zemlje-u-andrijasevcima---747877.html (accessed on 27 June 2023).
17. Gavranović, Ž. Slavonske Općine u Problemima. Morat Će Poništiti Natječaje? Available online: https://plusportal.hr/gospodarstvo/slavonske-opcine-u-problemima-morat-ce-ponistiti-natjecaje-47175 (accessed on 27 June 2023).
18. Croatian Agency for Agriculture Fisheries and Rural Development Register of Farmers. Available online: https://www.apprrr.hr/upisnik-poljoprivrednika/ (accessed on 14 October 2024).
19. Plješa, S. Država u Svom Vlasništvu Ima Skoro 900 Tisuća Hektara Poljoprivrednog Zemljišta. Koliko Brzo Ga Možemo Staviti u Funkciju? Available online: https://www.otvoreno.hr/gospodarstvo/drzava-u-svom-vlasnistvu-ima-skoro-900-tisuca-hektara-poljoprivrednog-zemljista-koliko-brzo-ga-mozemo-staviti-u-funkciju/264071 (accessed on 27 June 2023).
20. Svržnjak, K.; Franić, R. Developing Agrarian Structure through the Disposal of State-Owned Agricultural Land in Croatia. *Agroecon. Croat.* **2014**, *4*, 50–57.
21. Grgić, I. Dobitnici i Gubitnici Novog Zakona o Poljoprivrednom Zemljištu. Available online: https://gospodarski.hr/rubrike/ostalo/dobitnici-i-gubitnici-novog-zakona-o-poljoprivrednom-zemljistu/ (accessed on 27 June 2023).
22. Government of the Republic of Croatia Decree on the Establishment of the National Agency for Agricultural Land 2009. Available online: https://narodne-novine.nn.hr/clanci/sluzbeni/2009_03_39_878.html (accessed on 18 August 2024).
23. Croatian Parliment Agricultural Land Act 2018. Available online: https://www.zakon.hr/z/133/Zakon-o-poljoprivrednom-zemlji%C5%A1tu (accessed on 18 August 2024).
24. Mahmud, R. Understanding Institutional Theory in Public Policy. *Dyn. Public Adm.* **2017**, *34*, 135. [CrossRef]
25. Heikkila, T.; Isett, K.R. Modeling Operational Decision Making in Public Organizations. *Am. Rev. Public Adm.* **2004**, *34*, 3–19. [CrossRef]
26. Croatian Parliament. *Zakon o Poljoprivrednom Zemljištu*; Craotian Parliment: Zagreb, Croatia, 2022.

27. Croatian Ministry of Agriculture Objavljeno 229 Javnih Natječaja Za Zakup Poljoprivrednog Zemljišta i u Pripremi Još 64. 2024. Available online: https://poljoprivreda.gov.hr/vijesti/objavljeno-229-javnih-natjecaja-za-zakup-poljoprivrednog-zemljista-i-u-pripremi-jos-64/6018 (accessed on 18 August 2024).
28. Joint Research Centre (European Commission); Vranken, L.; Roebeling, P.; Tabeau, E.; Ciaian, P. *Agricultural Land Market Regulations in the EU Member States*; Publications Office of the European Union: Luxembourg, Luxembourg, 2021; ISBN 978-92-76-41990-7.
29. European Council. *Directive (EU) 2019/2024 of the European Parliament and of the Council of 20 June 2019 on Open Data and the Reuse of Public Sector Information*; European Council: Brussels, Belgium, 2019.
30. Government of Republic of Slovenia Agricultural Land Act Slovenia. Pravno-informacijski Sist. Republike Slov. 2023. Available online: http://pisrs.si (accessed on 18 August 2024).
31. Rajnović, L.; Kovačević, V. Lease of Agricultural Land in State Ownership-Possible Economic Effects. In *Sustainable Agriculture and Rural Development in Terms of the Republic of Serbia Strategic Goals Realization Within the Danube Region: Support Programs for the Improvement of Agricultural and Rural Development*; Subić, J., Kuzman, B., Vasile, A.Ž., Eds.; Institute of Agricultural Economics: Belgrade, Serbia, 2018; pp. 352–369, ISBN 978-86-6269-061-6.
32. Government of the Federation of Bosnia and Herzegovina Law on Agricultural Land BiH 2009 2023. Available online: https://fbihvlada.gov.ba/bosanski/zakoni/2009/zakoni/32bos.htm (accessed on 18 August 2024).
33. AGROklub Ima li Kriminala u Korištenju Državnog Poljoprivrednog Zemljišta? Available online: https://www.agroklub.ba/poljoprivredne-vijesti/ima-li-kriminala-u-koristenju-drzavnog-poljoprivrednog-zemljista/88501/ (accessed on 18 August 2024).
34. ONLINEstart Hrnjić, Topić i Ćulap o Dodjeli Državnog Poljoprivrednog Zemljišta Poljoprivrednim Proizvođačima @ StartBiH.ba. Available online: https://startbih.ba/clanak/hrnjic-topic-i-culap-o-dodjeli-drzavnog-poljoprivrednog-zemljista-poljoprivrednim-proizvodacima/218114 (accessed on 1 December 2023).
35. Raudeliuniene, J.; Meidute-Kavaliauskiene, I. Analysis of Factors Motivating Human Resources in Public Sector. *Procedia Soc. Behav. Sci.* **2014**, *10*, 719–726. [CrossRef]
36. Bearfield, D.; Bowman, A. Can You Find It on the Web? An Assessment of Municipal E-Government Transparency. *Am. Rev. Public Adm.* **2016**, *47*, 172–188. [CrossRef]
37. Emília, S.-B.; Martin, K.; Matúš, S. Exploring the Determinants of Transparency of Slovak Municipalities. *NISPAcee J. Public Adm. Policy* **2016**, *9*, 121–145.
38. Špáč, P.; Voda, P.; Zagrapan, J. Does the Freedom of Information Law Increase Transparency at the Local Level? Evidence from a Field Experiment. *Gov. Inf. Q.* **2018**, *35*, 408–417. [CrossRef]
39. Styles, A.K.; Tennyson, M. The Accessibility of Financial Reporting of u.s. Municipalities on the Internet. *J. Public Budg. Account. Financ. Manag.* **2007**, *19*, 56–92. [CrossRef]
40. Keeney, R.L. On the Foundations of Prescriptive Decision Analysis. In *Utility Theories: Measurements and Applications*; Edwards, W., Ed.; Studies in Risk and Uncertainty; Springer Netherlands: Dordrecht, The Netherlands, 1992; pp. 57–72, ISBN 978-94-011-2952-7.
41. Ministry of Justice and Public Administration Statistical Overview of the Ministry of Justice and Public Administration. mpu.gov.hr 2023. Available online: https://mpu.gov.hr/istaknute-teme/statisticki-prikaz/22205 (accessed on 18 August 2024).
42. Croatian Buro of Statistics Population Census. Državni zavod za Stat. 2023. Available online: https://podaci.dzs.hr/hr/podaci/stanovnistvo/popis-stanovnistva/ (accessed on 19 August 2024).
43. SAS Institute Inc. *Base SAS 9.4 Procedures Guide: Statistical Procedures*; SAS Institute: Cary, NC, USA, 2011; ISBN 978-1-60764-944-1.
44. Bivand, R.S.; Wong, D.W.S. Comparing Implementations of Global and Local Indicators of Spatial Association. *TEST* **2018**, *27*, 716–748. [CrossRef]
45. R Core Development Team A language and environment for statistical computing. R Foundation for Statistical Computing, Vienna, Austria. 2021.
46. Organisation for Economic Co-operation and Development Keep Calm and Carry on Feeding: Agriculture and Food Policy Responses to the COVID-19 Crisis 2024. Available online: https://www.oecd.org/coronavirus/policy-responses/keep-calm-and-carry-on-feeding-agriculture-and-food-policy-responses-to-the-covid-19-crisis-db1bf302/ (accessed on 19 August 2024).

International Journal of
Geo-Information

MDPI

Article

The Coordinated Development Characteristics of Rural Industry and Employment: A Case Study of Chongqing, China

Guoqin Ge, Yong Huang * and Qianting Chen

School of Architecture and Urban Planning, Chongqing University, No.174, Shazhengjie Street, Shapingba District, Chongqing 400030, China; 201815021041@cqu.edu.cn (G.G.); 20201501019g@cqu.edu.cn (Q.C.)
* Correspondence: hyong@cqu.edu.cn

Abstract: Developing industries and promoting employment are essential for rural revitalization. This study establishes a theoretical framework to support the coordinated development of rural industry and employment (RIE) with Chongqing, China as the study area. Methods include GIS spatial analysis, the entropy-weighted TOPSIS method, a coupled coordination degree model, and an optimal-parameter-based GeoDetector. The analysis examines the spatio-temporal evolution and driving mechanisms of the coordinated development of RIE. The main findings are as follows. (1) During the study period, Chongqing's RIE improved significantly overall, although rural industry is relatively lagging. (2) The evolution characteristics of the coordinated development of RIE exhibit "spatio-temporal ripple" and "spindle-shaped" patterns, and the spatial agglomeration has been enhanced. The growth of RIE is accompanied by the spatial diffusion of rural industry and the spatial echo of rural employment. (3) The primary driving mechanism for the coordinated development of RIE is "human-centered, natural resource-based socio-economic development." Finally, this study discusses employment-centered strategies for rural industrial development, providing a theoretical foundation for policy-making in rural industrial development.

Keywords: rural industry; rural employment; coordinated development; Chongqing

1. Introduction

With rapid industrialization and urbanization occurring globally, rural areas are experiencing a significant decline, making rural revitalization a pressing challenge. Promoting industrial development and productive employment has been widely recognized as a key strategy for rural revitalization [1,2]. However, the supply of industries in many rural regions remains highly limited. As such, achieving a balance between rural industries' "productive" and "welfare" functions is critical for realizing this path to revitalization. Understanding the evolution pattern and mechanism of the coordinated development of rural industry and rural employment is invaluable for shaping effective rural industrial development policies and maximizing the employment potential of rural industries.

Rural industry and employment (RIE) have traditionally been focal points of multidisciplinary research. In the context of rural industry, numerous studies have examined the phenomenon of rural industrial agglomeration [3,4] and the spatial characteristics of regional disequilibrium [5]. These studies suggest that industrial agglomeration can enhance rural incomes while reducing household income inequality. Other scholars have developed an evaluation framework to assess rural industrial integration, considering factors like agricultural value chain extension, multifunctionality, integrated service industry development, and the economic impact of industrial integration [6,7]. Additionally, rural industrial integration has been shown to improve agricultural production efficiency [8], strengthen community resilience [9], and enhance agricultural economic resilience [10]. Research on rural industrial development indicates that transportation infrastructure is a fundamental factor in promoting industrial integration, while fragmented land supply and tenure difficulties pose significant barriers to development [11]. Other scholars have examined the spatio-temporal characteristics of coordinated development among rural industry, settlement systems [12], the ecological environment [13], and population and land use [14], offering related development insights. Currently, there is a broad consensus on the need for urban–rural integrated development, with an increasing focus on advancing rural industry through urban–rural restructuring [15], the use of Internet technologies [16], intelligent innovations [17,18], inclusive financial instruments [19–21], and ecological governance models [22]. Research on rural employment encompasses population migration and rural livelihoods. In population migration, beyond the typical focus on rural-to-urban migration [23,24], numerous studies have examined the characteristics and influencing factors of urban-to-rural workforce migration, particularly concerning returning migrant workers [25,26]. Additional research has focused on the entrepreneurial activities of returning migrant workers [27–29]. The employment structure of the rural labor force is also a significant focus, especially in terms of off-farm employment characteristics and preferences [30,31]. Studies on rural youth indicate that off-farm employment predominantly occurs in peri-urban areas [32–34]. Factors like limited agricultural skills and a lower quality of life in rural areas have contributed to the reluctance of younger generations to engage in agriculture. Rural women are also a primary focus of employment research. Studies have shown that off-farm employment is increasingly the primary livelihood for rural women, although they remain at an overall disadvantage in employment opportunities compared to men [35–37]. Numerous studies examine rural residents' livelihood capital and strategies [38,39], including the impacts of rural tourism development [40–42], disaster shocks, like earthquakes [43,44], migrant relocation [45], and poverty [46,47].

In summary, rural industry is critical in rural revitalization, notably by expanding employment opportunities, raising farmers' incomes, and promoting social equity. Moreover, farmer participation positively impacts rural industry [8,48,49], enhances agricultural production efficiency [50], and mitigates potential ecological and cultural harms associated with industrial activities. With developed countries entering a "post-rural" era, current research on RIE has increasingly focused on the Global South. Regarding the research content, studies on RIE are often conducted separately. Rural industry research focuses on structural and sectoral elements, whereas rural employment research often centers on specific employment groups or behaviors. While some studies acknowledge the interaction between rural industry and employment, in-depth analyses of the characteristics and mechanisms behind their coordinated development remain limited. Additionally, due to data limitations, measurements of industry or employment levels are often conducted at the national level or based on micro-scale interviews with specific employment groups. Meso-scale research is relatively limited.

As the world's largest developing country, China has recently prioritized rural revitalization as central to high-quality development in the new era. Rural industrial development is seen as a crucial tool for promoting local employment among rural laborers and achieving rural revitalization [51,52]. These recent efforts have yielded notable results. Accordingly, this study establishes a theoretical framework for the coordinated development of RIE. It employs multivariate data and methodologies to assess spatio-temporal characteristics, influencing factors, and driving mechanisms of RIE in Chongqing, China. The study's findings can offer theoretical insights for developing countries in formulating rural industrial strategies.

2. Theoretical Framework

2.1. Connotation of RIE

RIE lacks a clear definition; however, the spatial and geographical attributes of the countryside influence its value characteristics. The discourse on rural industry traces back to debates on agricultural multifunctionality in the late 20th century. From the post-World War II period until the 1970s, Western countries prioritized maximizing agricultural output to ensure food security [53]. However, in the following two decades, an overemphasis on productivism led to issues like overproduction and land degradation [54]. Since the 1990s, developed countries have increasingly focused on the shift toward rural consumption. The role of rural spaces has expanded from purely agricultural production to include industrial production, services, tourism, and consumption [55]. The functional scope of agriculture now includes agricultural landscapes, biodiversity, rural employment, food quality and safety, agricultural cultural heritage, animal welfare, and more [56–58]. The National Rural Key Industry Guidance Catalog (2021) issued by China's Ministry of Agriculture and Rural Affairs classifies rural industries for national economic activities. Building on these foundations, this study defines rural industry as encompassing modern agricultural farming, agricultural product processing and distribution, and emerging service industries—such as leisure and tourism—centered around agricultural landscapes or production activities. Rural employment, a direct function of rural industry, includes self-employment and wage-based work within rural industries. Labor for rural employment comes primarily from rural residents, who have significant advantages in agricultural skills and proximity to workspaces. Conversely, urban residents serve as investors or technical guides within rural employment but constitute a small proportion of the workforce.

2.2. Coordinated Development of RIE

Coordinated development within the RIE system is essential to realizing rural industry's employment potential and achieving rural revitalization fully. Stakeholders in the RIE system—industry operators, employees, and the government—have distinct and sometimes conflicting interests. Industrial operators prioritize production efficiency and profitability, viewing employment promotion as a secondary outcome. Employees seek stable, well-paying jobs to maintain their quality of life. The government aims to balance rural industrial development that drives economic growth with employment initiatives that ensure an adequate labor force for industry. Achieving this balance hinges on the coordinated development of the RIE system (Figure 1). Rural industry generates diverse employment opportunities, reshaping livelihood options for residents [38,59,60]. When residents start their businesses, they strengthen rural industries, creating additional employment cycles [28]. Coordinated development of RIE also produces significant spatial effects. Transaction costs lead industries to cluster naturally [61]. However, considerations like distribution costs, the market environment, and organizational efficiency drive the need for industry decentralization [62]. Separation of labor processes promotes both spatial concentration and decentralization of industrial factors. As space becomes a means of production, disparities in spatial costs lead to its continuous reconfiguration alongside labor division [63]. Specialized clustering of industrial factors creates spatial inequality in job types and conditions, directly impacting rural employment stability. Each RIE subsystem has internal drivers, but coordinated development relies more on shared external factors. Chinese scholar Wu [64] categorized the rural habitat system into five subsystems: natural, human, social, residential, and support. This framework aids in identifying the external factors impacting the coordinated development of RIE, as RIE itself is a component of the habitat system.

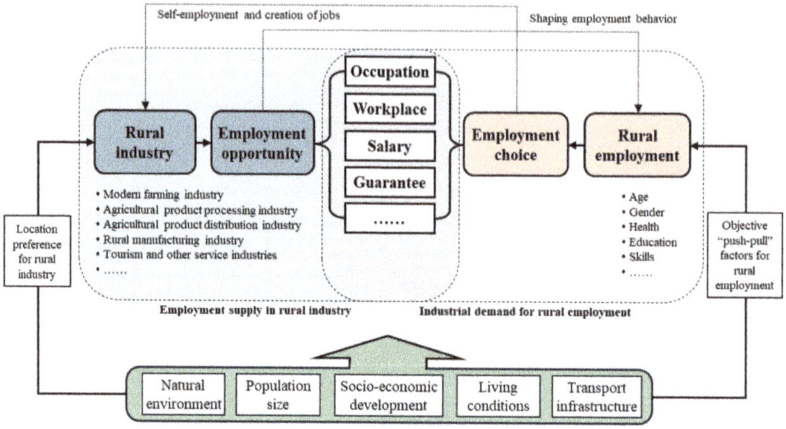

Figure 1. Coordinated development framework of RIE.

3. Materials and Methods

3.1. Study Area

Chongqing, located in the mountainous region of southwest China (Figure 2), is characterized by its "big city, big countryside, extensive mountainous, and reservoir areas" and was designated in 2007 as a pilot area for national urban–rural comprehensive reform. As the economic center of central and western China, Chongqing has a solid foundation for RIE development. Counties serve as the primary spatial unit for implementing RIE-related policies in China, including urban–rural integration and rural revitalization; thus, this study focuses on counties as the research unit. Because Yuzhong has been fully urbanized, the study focuses on the remaining 37 counties in Chongqing. This study examines 2013, 2018, and 2023 as part of its time series analysis. These years reflect fundamental policy shifts, including China's 2012 transition towards high-quality employment development and the 2017 introduction of the rural revitalization strategy.

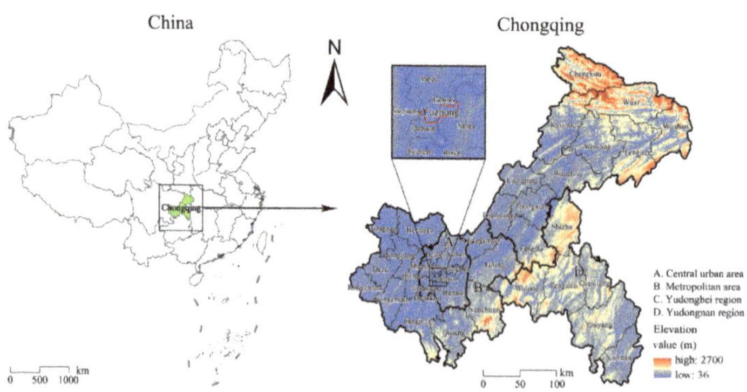

Figure 2. Location of the study area. These maps were drawn according to the standard map with drawing No. GS (2024) 0650, which was downloaded from the standard map service website of the Ministry of Natural Resources of the People's Republic of China. No modifications were made to the base map.

3.2. Methods

3.2.1. Coordinated Pattern Measure of RIE

- Evaluation indicator system

This study constructs a rating indicator system for rural industry and rural employment subsystems based on RIE characteristics and county-level data accessibility and representativeness (Table 1).

Table 1. RIE development level evaluation indicator system.

Subject	Target	Indicators	Indicator Description	Indicator Unit	Attribute	Weighting
Rural industry	Rural industry scale	Rural enterprise scale index	Sum of all enterprise size indices in the region	/	+	0.171
		Gross output value of agriculture	/	CNY 10,000	+	0.199
		Rural electricity consumption	Includes electricity consumption for agriculture, forestry, animal husbandry, and fisheries and electricity consumption by rural residents	10,000 kWh	+	0.234
	Rural industrial integration	Percentage of non-agricultural enterprise size indexes	Ratio of non-farm enterprise size index to total size index in the region	%	+	0.068
		Number of agricultural cooperatives	/	/	+	0.207
		Agricultural labor productivity	Gross value of agriculture/rural workers	CNY/person-year	+	0.121
Rural employment	Fuller employment	Rural employment numbers	/	10,000 people	+	0.116
		Rural population's permanent residence rate	Rural resident population/rural household population	%	+	0.087
		Index of rural employment opportunities	Calculated in the same way as the rural enterprise size index	/	+	0.144
		Government agricultural expenditure on agriculture, forestry, and water	/	CNY 10,000	+	0.081
		Government expenditure on social security and employment	/	CNY 10,000	+	0.143
	Higher-quality employment	Rural per capita disposable income	/	CNY	+	0.137
		Urban–rural income gap	Per capita disposable income of urban/rural permanent residents	%	-	0.070
		Proportion of non-farm employment	Calculated in the same way as the non-agricultural enterprise size index share	%	+	0.057
		Per capita rural consumption expenditure	/	CNY	+	0.165

Note: there are six categories of enterprise sizes: micro, small, medium, medium-high, large, and super-large. This study adopts the size determination criteria outlined in the "Statistical Approach to the Division of Large, Small, Medium, and Micro Enterprises" published by the National Bureau of Statistics of China. The size index weights are assigned as 1:3:9:18:36:72 to better reflect real-world conditions.

The evaluation of rural industry is based on previous research findings [5–7] and primarily considers industry scale and the level of industrial integration. Industrial scale forms the foundation of rural industrial development; this study evaluates it using three indicators reflecting rural industries' volume, input intensity, and output capacity: the rural enterprise scale index, the gross output value of agriculture, and rural electricity consumption. Industrial integration represents an advanced stage of rural industrial development, focusing on extending the industrial chain, expanding multifunctional roles, and enhancing production efficiency. This study thus uses three indicators to characterize integration: the percentage of non-agricultural enterprise size indexes, the number of agricultural cooperatives, and agricultural labor productivity.

China's "active employment policy", implemented in the early twenty-first century, is a basis for evaluating rural employment levels. The policy focuses on two core objectives: fuller employment and higher-quality employment. Fuller employment emphasizes the broader development and utilization of labor resources, accessible employment channels, and robust protection of workers' rights. This study uses five indicators to measure rural employment: rural employment numbers, rural populations' permanent residence rate, the index of rural employment opportunities, and government agricultural expenditure on agriculture, forestry, and water. Higher-quality employment emphasizes improving the employment environment, increasing labor remuneration, and enhancing social equity. Four indicators were selected to evaluate higher-quality employment: rural per capita disposable income, the urban–rural income gap, the proportion of non-farm employment, and per capita rural consumption expenditure.

- Indicator weighting and subsystem evaluation

This study uses the entropy weight TOPSIS method for indicator weighting and subsystem evaluation. The entropy weighting method is used for indicator weighting, and then TOPSIS (Technique for Order Preference by Similarity to an Ideal Solution) is used to measure the state values of the two subsystems of rural industry and rural employment. The calculation process is as follows.

(1) Entropy weights and comprehensive evaluation index

First, this study uses the min–max normalization method to normalize the original data matrix, and the entropy h_i and entropy weight w_i of the ith indicator in an assessment problem with m indicators and n evaluation objects are

$$h_i = -k\sum_{j=1}^{n} f_{ij}\ln f_{ij},\ f_{ij} = r_{ij}/\sum_{j=1}^{n} r_{ij},\ k = 1/\ln n\left(\text{when } f_{ij} = 0,\ f_{ij}\ln f_{ij} = 0\right) \quad (1)$$

$$w_i = \frac{1-h_i}{m-\sum_{i=1}^{m} h_i}\left(0 \le w_i \le 1,\ \sum_{1}^{m} w_i = 1\right) \quad (2)$$

Then, the comprehensive evaluation index is calculated as

$$G_i = \sum_{j=1}^{n} r_{ij}w_{ij} \quad (3)$$

(2) Relative proximity

In this study, relative proximity is used to reflect the relative level of RIE of a county in Chongqing among all spatial units. That is, the comprehensive evaluation index of a unit is elevated vertically (time series) but may still be declining in the horizontal (cross-sectional series) comparison. The relative proximity C_i is calculated as follows.

According to the weighting matrix Z, $Z_{ij} = w_{ij} \times r_{ij}$, the optimal solution Z^+ and the worst solution Z^- of each index are calculated. Then, determine the distance of each index to the ideal solution D_i^+ and the negative ideal solution D_i^-. Finally, the relative proximity of each research object to the optimal solution is determined, C_i. The calculation formula is

$$Z^+ = max_{i=1}^{m} Z_{ij},\ Z^- = min_{i=1}^{m} Z_{ij}(j = 1,2,\ldots,n) \quad (4)$$

$$D_i^+ = \sqrt{\sum_{j=1}^{n}\left(Z_{ij} - Z_i^+\right)^2},\ D_i^- = \sqrt{\sum_{j=1}^{n}\left(Z_{ij} - Z_i^-\right)^2}\ (i = 1,2,\ldots,m; j = 1,2,\ldots,n) \quad (5)$$

$$C_i = \frac{D_i^-}{D_i^+ + D_i^-} \quad (6)$$

where Z_{ij} are the elements in matrix Z and r_{ij} are the normalized data values. $C_i \in [0, 1]$; the larger the value of C_i, the higher the level of rural industry and employment development of the spatial unit among all county levels of that year, relatively.

- Coordination evaluation

This study uses the coupled coordination degree model (CCDM) to measure the coordinated development relationship between rural industries and rural employment in Chongqing. The calculation formula is as follows:

$$D = \sqrt{CI \times T},\ T = \alpha u_1 + \beta u_2 \tag{7}$$

$$CI = 2 \times \sqrt{\frac{u_1 \times u_2}{(u_1 + u_2)^2}} \tag{8}$$

where D is the coupling degree, CI is the coordination degree, T is the comprehensive coordination index, and u_1 and u_2 are the evaluation values of rural industry and employment level, respectively. α and β are the weights of the two, usually taking the value of 0.5. The coupling degree (D) indicates the coordination level of rural industry and employment, and the range of the value is [0, 1]; the larger the value, the higher the coordination level. According to the actual value range of the coupling degree D in this study, the level of coordinated development of RIE in Chongqing can be divided into 4 levels (Table 2). And, the hysteresis factors are categorized into industrial hysteresis and employment hysteresis according to the size contrast of u_1 and u_2.

Table 2. Classification of the level of coordinated development of RIE in Chongqing.

D	Types of Coordination	Development Features
$0.4 < D \leq 0.5$	Slight dissonance	RIE's coordinated development state is in a dysfunctional development stage of mutual games and conflicts
$0.5 < D \leq 0.6$	Barely coordinated	RIE's coordinated development state is in a less conflictual but still game-dominated phase
$0.6 < D \leq 0.7$	General coordination	RIE's coordinated development state is in a phase of transition to balanced harmonization
$0.7 < D \leq 0.8$	Better coordination	RIE's coordinated development state shows relatively good coupling

3.2.2. Coordinated Mechanism Analysis of RIE

- Variable Selection

In this study, the coupling degree of RIE's coordinated development is the dependent variable. Drawing on Wu's [64] framework for the rural habitat system, influencing factors were chosen across five dimensions: natural environment, population size, socio-economic development, living conditions, and transport infrastructure.

First, the natural environment is a crucial foundation for the coordinated development of RIE, which provides essential agricultural production conditions and fundamental livelihood capital. To characterize the natural resource conditions, this study selects four key indicators: the average slope, the disaster index, the arable land, and the forested land area. Second, a sufficient supply of labor resources is critical to the coordinated development of RIE. However, a huge population may lead to heightened competition for employment, potentially diminishing the level of RIE's coordination. In this context, the urban and rural resident populations represent the labor supply. Third, the coordinated development of RIE relies on investments from the government, investors, and farmers, which involve both productive construction and infrastructure investments. Additionally, a favorable consumer market, industrial land supply, and historical and cultural factors contribute positively to RIE. These elements are categorized under the socio-economic subsystem. Fourth, within the living conditions subsystem, rural settlement land may support informal land-use practices within the RIE system. At the same time, healthcare and education play significant roles in shaping the career choices of rural job seekers. Finally, transportation facilities influence RIE by facilitating the movement of raw materials and products for rural industries and improving work–life commuting. Based on the aforementioned analysis and the data's availability and representativeness, 16 specific variables have been selected to comprehensively reflect the factors influencing the coordinated development of RIE (Table 3).

Table 3. Variable settings for the coordinated mechanism analysis.

Type	Variable	Variable Interpretation	Variable Unit
Natural environment	Average slope (X_1)	Slope averages at 30 × 30 m^2 resolution	%
	Disaster index (X_2)	Percentage of area in low susceptibility zone × 1 + percentage of area in medium susceptibility zone × 2 + percentage of area in high susceptibility zone × 3	/
	Cropland area (X_3)	Sum of paddy and dryland areas	km^2
	Forest land area (X_4)	Area of natural and planted forests with >30% cover	km^2
	River and canal area (X_5)	Area of land below the perennial level of naturally occurring or artificially excavated rivers and their main stem	km^2
Population size	Number of permanent urban residents (X_6)	/	10,000 persons
	Number of permanent rural residents (X_7)	/	10,000 persons
Socio-economic development	Total investment in fixed assets (X_8)	/	CNY 100 million
	Total retail sales of consumer goods (X_9)	/	CNY 100 million
	Other construction land (X_{10})	Land area used for factories, mines, large industrial areas, oil fields, salt fields, quarries, etc.	km^2
	Rural historical and cultural resources index (X_{11})	Municipal historical and cultural village × 1 + municipal historical and cultural town × 5 + Chinese traditional village × 1 + Chinese historical village × 10 + Chinese historical and cultural town × 15	/
Living conditions	Number of health technicians (X_{12})	/	persons
	Number of basic education teachers (X_{13})	Combined number of full-time teachers in secondary and elementary school	persons
	Land for rural settlements (X_{14})	/	km^2
Transport infrastructure	Density of graded roads (X_{15})	Length of classified roads/area of administrative district	km/km^2
	Length of highways (X_{16})	Total length of transit highways	km

- Optimal-parameter-based GeoDetector (OPGD)

This study employs the factor detection and interaction detection functions of GeoDetector to identify critical factors and interactions affecting RIE coordinated development in Chongqing. Traditional GeoDetectors, however, depend largely on empirical decisions for discretizing continuous explanatory variables, which limits quantitative accuracy [65]. Thus, this study applies the optimal-parameter-based GeoDetector (OPGD) model [66]. To set the classification level of explanatory variables into 3–7 categories, this study uses equal interval, natural break, quantile, geometric interval, and standard deviation classifications. The parameter combination yielding the best q value was then chosen for discretization.

Factor detection is primarily used to quantify the explanatory power of various influencing factors on the coordinated development of RIE, calculated using the following formula:

$$q = 1 - \frac{1}{N\sigma^2}\sum_{h=1}^{L} N_h\sigma_h^2 \tag{9}$$

where q is the influence of the detection factor and the value of q is in [0, 1], where the larger the value, the greater the influence of the influence factor on the coordinated development level of RIE; L is the number of categories of the detection factor; N is the number of spatial units; and N_h is the number of county units of category h. σ^2 and σ_h^2 are the coordinated development level of RIE in the whole study area and the coordinated development level of RIE in the category h, respectively. Factor interaction detection refers to the enhancement or attenuation of RIE harmonization when two influencing factors work together or when both work together independently on RIE harmonization. The five types of interaction detections include nonlinearly weaken each other (Weaken, nonlinear), weaken each other uniformly (Weaken, uni), enhance each other bidirectionally (Enhance, bi), nonlinearly enhance each other (Enhance-nonlinear) and independent of each other (Independent).

3.3. Data Sources and Processing

This study relies on a comprehensive dataset, including economic and social statistics, enterprise business data, and geospatial data. The socio-economic data are sourced from the Chongqing Municipal Statistical Yearbook (http://tjj.cq.gov.cn) and other official government statistics and directories. Because rural electricity consumption and rural employment numbers of Chongqing in 2023 have yet to be available, this study uses data from the most recent available year, 2022, as a substitute. Rural enterprise data are obtained from the Tianyancha platform (http://www.tianyancha.com), one of China's largest online platforms for industrial and commercial data. A rural industrial and commercial enterprise directory was screened according to the National Rural Key Industry Guidance Catalog (2021 Edition) published by China's Ministry of Agriculture and Rural Affairs. The directory includes details on each enterprise's address, size, establishment and deregistration dates, and industry category. Geospatial data include topographic and land use information. The topographic data are derived from a 12.5 × 12.5 m^2 digital elevation model. Land use data are provided by the Data Center for Resources and Environmental Sciences, Chinese Academy of Sciences (RESDC) (http://www.resdc.cn), which offers 2013, 2018, and 2023 land cover vector data.

4. Results

4.1. Spatio-Temporal Characteristics of RIE's Coordinated Development

The results of the comprehensive evaluation index of the RIE system in Chongqing are shown in Figure 3. Overall, the RIE of most districts and counties has substantially improved over 2013–2023, with an average increase of 91.17% in the comprehensive evaluation index of rural industry and 88.45% for rural employment. Comparing the two periods, the rate of growth in the RIE system from 2013 to 2018 is more significant than that from 2018 to 2023. Some decline is also observed in the RIE system, notably in the rural industry systems of Nanan and Beibei from 2013 to 2018 and in Beibei and Qianjiang from 2018 to 2023; notably, all of these are municipal districts of Chongqing. In spatial terms, a marked spatial "ripple" effect emerges in the RIE system, characterized by two notable "peak" areas, one closer to the urban center and another in sub-marginal zones, where the comprehensive evaluation indexes of rural industry and rural employment are significantly higher than those of other regions. Regarding growth rates, the spatial trends of rural industry and rural employment are inversely related, with rural employment in peri-urban areas growing faster than in peripheral areas, while rural industry experiences relatively higher growth in Chongqing's peripheral areas.

In terms of the level of coordinated development (Figure 4a), the coupling degree of RIE in Chongqing counties ranges from 0.443 to 0.788 over the period 2013–2023, exhibiting an initial upward trend followed by a subsequent decline, with no indication of significant dysfunction within the RIE system. Notably, only two units are classified as slight dissonance, and this classification disappears by 2023. The number of barely coordinated counties was 12 in 2013, decreasing to 8 in 2018 and then increasing to 11 in 2023. The general coordination counties represent the largest group, with 20 in 2013, decreasing to 17 in 2018 and then increasing to 20 in 2023. Meanwhile, the number of counties with better coordination increased rapidly from 3 in 2013 to 18 in 2018, before decreasing to 6 in 2023. Between 2013 and 2023, 17 counties experienced a change in coordination type, with 9 experiencing an increase and 6 showing an initial increase followed by a decrease. Nanan and Jiangbei exhibited only decreases, declining from general to barely coordinated levels. Regarding lagging factors (Figure 4b), the relative lagging of rural industry remains the primary constraint on the coordinated development of RIE. However, this situation shows slight improvement by 2023, with the number of lagging counties decreasing from 31 to 28.

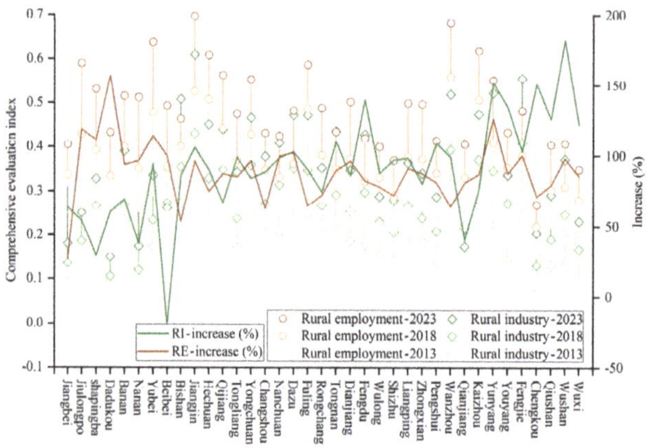

Figure 3. Comprehensive evaluation index for the subsystem of RIE. The X-axis represents the spatial distance from each county to the city hall, arranged in increasing order from left to right.

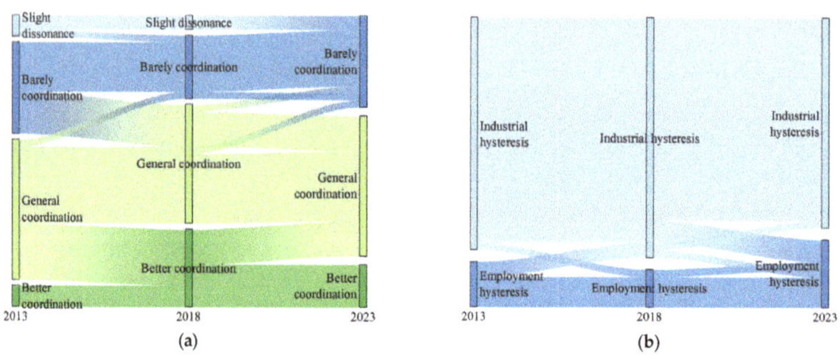

Figure 4. Types of and impediments to coordinated development of RIE. (**a**) Types of coordinated development; (**b**) impediments to coordinated development.

Regarding spatial distribution characteristics (Figure 5), regions with higher evaluation values are consistently concentrated around the Chongqing metropolitan area and in counties centered on Wanzhou. The range of extreme differences in the relative closeness of rural industry increased from 0.147 to 0.201, indicating a continued regional imbalance. In contrast, rural employment imbalance was even more prominent, with maximum relative proximity increasing from 0.611 to 0.663 and extreme differences expanding consistently from 0.348 to 0.400. In the spatial distribution of the RIE coupling degree, only Jiangjin, Fuling, and Wanzhou demonstrated better coordination in 2013. Fuling and Wanzhou are sub-centers in Chongqing's urban system, with this agglomeration pattern intensifying over time. Meanwhile, the coordinated development of RIE of countries in central urban area, such as Jiangbei and Nanan, continues to decline. Notably, the six countries—Qianjiang,

Wulong, Shizhu, Pengshui, Youyang, and Xiushan—in the Yudongnan region are administratively regarded as a single division within Chongqing. However, all remain consistently in a relatively underdeveloped position within the area's RIE system evaluation.

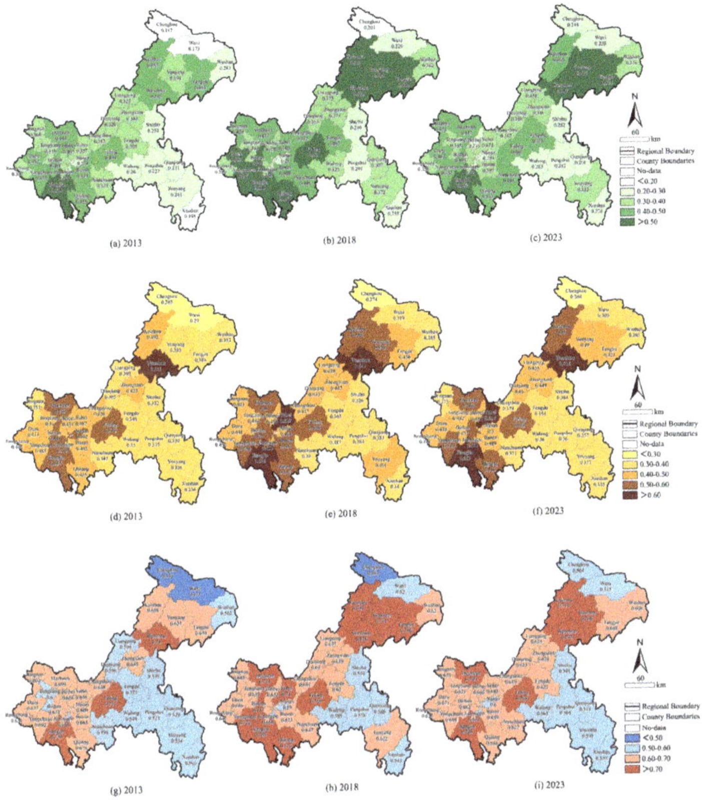

Figure 5. Spatio-temporal characteristics of RIE based on relative proximity. (**a–c**) Relative proximity of rural industry; (**b–f**) relative proximity of rural employment; (**g–i**) the coupling degree of RIE.

4.2. The Factors Influencing RIE's Coordinated Development

4.2.1. Factor Detection

The results of the factor detector (Table 4) indicate significant stage-specific variations in the factors affecting the coordinated development of RIE in Chongqing.

Regarding the natural environment, the average slope (X_1) has consistently failed the significance test, showing low and gradually diminishing explanatory power. Continuous improvements in Chongqing's rural infrastructure, including mechanized transformations of arable land, have reduced the limiting effect of terrain slope on the RIE system. In contrast, the explanatory power of the disaster index (X_2) has increased slightly, suggesting that disasters continue to limit the development of the RIE system significantly. In recent years, Chongqing has implemented proactive countermeasures, such as advancing migrant relocation initiatives in disaster-prone regions. The explanatory power of cropland area (X_3) has also increased slightly, indicating that agriculture remains a crucial foundation of the rural industry despite the de-farming trend. Forest land area (X_4) is the most critical natural factor, with Chongqing's forest coverage reaching 54.4% as of December 2021. The woodland area reflects ecological landscape richness and forest resources, which are essential foundations for RIE system development. Finally, while the river and canal area (X_5) has consistently failed the significance test, its explanatory power has notably improved from previous levels.

Table 4. The statistical significance of factor detection results.

Factors	Year		
	2013	2018	2023
X_1	0.473	0.365	0.347
X_2	0.393 **	0.357 *	0.430 *
X_3	0.467	0.467 **	0.500 ***
X_4	0.662 ***	0.626 ***	0.632 ***
X_5	0.285	0.392	0.38
X_6	0.755	0.591 ***	0.579 ***
X_7	0.386 *	0.519 ***	0.591 ***
X_8	0.731 **	0.645 ***	0.695 ***
X_9	0.799 ***	0.66	0.652 ***
X_{10}	0.457 **	0.384	0.475 *
X_{11}	0.14	0.235	0.274
X_{12}	0.725 ***	0.740 **	0.451 *
X_{13}	0.568	0.659	0.622 ***
X_{14}	0.663	0.557 ***	0.54
X_{15}	0.389	0.208	0.291
X_{16}	0.228	0.581	0.502 ***

Note: ***, **, and * indicate statistical significance at the 1%, 5%, and 10% levels, respectively.

The RIE system's average explanatory power of population size is relatively high, passing the significance test for all variables except for the permanent urban population in 2013. Notably, in 2013 and 2018, the explanatory power of the permanent urban residents (X_6) was significantly greater than that of the permanent rural residents (X_7). However, by 2023, the explanatory power of the rural resident population slightly surpasses that of the urban resident population.

The explanatory power of total investment in fixed assets (X_8) for the RIE system remains consistently high, indicating that investment in fixed assets plays a crucial role in advancing rural industrial development and employment growth. Similarly, the explanatory power of total retail sales of consumer goods (X_9) remains high, although it exhibits a notable decline over time. This suggests that while the local consumer market remains relevant to rural industrial development, recent initiatives to promote regional integration in the Chengdu–Chongqing area may have lessened the impact of the local market as the foreign market grows in influence. The explanatory power of construction land (X_{10}) is moderate, particularly in 2018, where it does not pass the significance test, likely due to the disorganized rural land supply from 2013 to 2018. However, implementing a more controlled land spatial planning system in 2019 curbed unregulated land expansion. In 2023, the explanatory power of construction land reached 0.475, showing a significant rise and passing the significance test once again. Meanwhile, the index of rural historical and cultural resources (X_{11}) consistently fails to pass the significance test, indicating that the contribution of rural industries based on historical and cultural resources is limited or that the level of development and utilization remains inadequate.

Regarding living conditions, the explanatory power of health technicians (X_{12}) is relatively high but decreases significantly by 39% in 2023 compared to 2018. Meanwhile, basic education teachers (X_{13}) passed the significance test in 2023 with an explanatory power of 0.622, suggesting that local employers in rural areas increasingly value education resources. Rural settlement land (X_{14}) only meets the significance threshold in 2018, with declining explanatory power over time.

Regarding transportation conditions, the density of graded roads (X_{15}) remains insignificant, with consistently low and declining explanatory power. In contrast, the length of highways (X_{16}) meets the significance threshold in 2023, showing a marked increase in influence. This increase is primarily attributed to Chongqing's "Village Passing" project, initiated in 2007, significantly enhancing rural road conditions. Highways are crucial in facilitating the direct movement of the county's goods, services, and people.

4.2.2. Interaction Detection

The results of interaction probes (Figure 6) reveal that the factors driving the coordinated development of the RIE system in Chongqing predominantly follow a bi-factor enhancement model. Among the interactions that passed the significance test, the top five factor pairs ranked by explanatory power in 2013 were $X_4 \cap X_{10}$ ($q = 0.923$), $X_2 \cap X_9$ ($q = 0.856$), $X_7 \cap X_9$ ($q = 0.854$), $X_7 \cap X_8$ ($q = 0.851$), and $X_7 \cap X_{12}$ ($q = 0.847$). In 2018, the top interactions were $X_6 \cap X_7$ ($q = 0.895$), $X_7 \cap X_8$ ($q = 0.893$), $X_4 \cap X_8$ ($q = 0.884$), $X_4 \cap X_6$ ($q = 0.874$), and $X_3 \cap X_6$($q = 0.872$). In 2023, the leading interactions included $X_3 \cap X_8$ ($q = 0.919$), $X_7 \cap X_8$ ($q = 0.911$), $X_7 \cap X_9$ ($q = 0.901$), $X_9 \cap X_{14}$ ($q = 0.899$), and $X_8 \cap X_{16}$ ($q = 0.890$). All of these factor interactions exhibit two-factor enhancement, with explanatory power exceeding that of individual factors. This suggests that the coordinated development of the RIE system in Chongqing results from the combined influence of multiple systemic factors. Among these interacting factors, those appearing most frequently are the number of permanent rural residents (X_7) and total investment in fixed assets (X_8). These results indicate that the RIE system in Chongqing remains in an investment-driven development phase, with rural residents at the center of both industrial activity and employment opportunities.

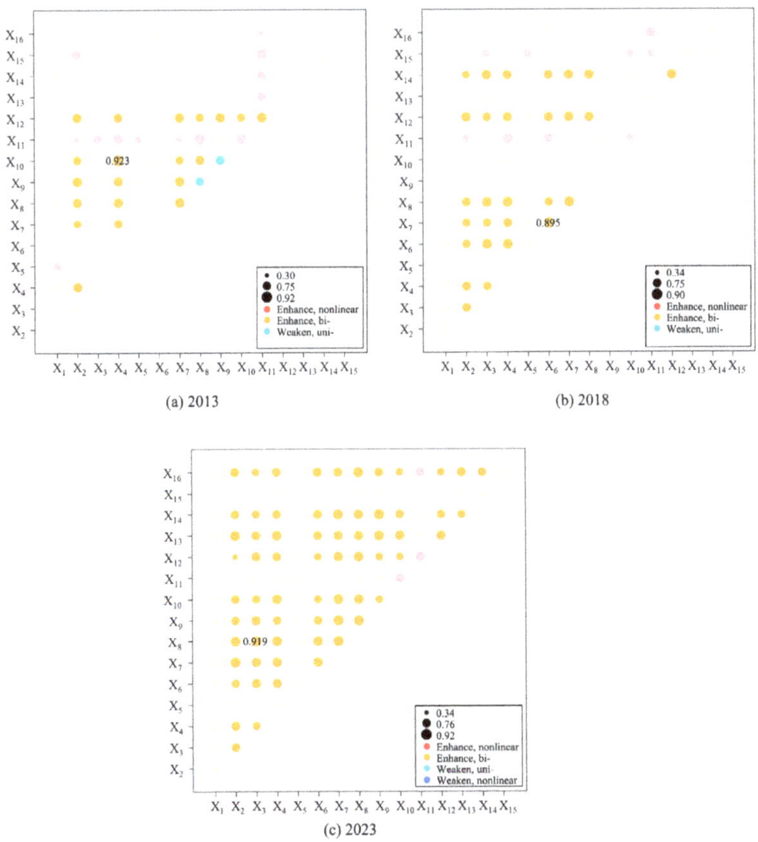

Figure 6. Interaction detection results of the factors influencing RIE's coordinated development.

5. Discussion

5.1. Evolution Pattern and Mechanism of the Coordinated Development of RIE

This study systematically analyzes the coordinated development of RIE in Chongqing, China and breaks through the limitations of traditional research that focuses only on the single-dimensional functions of rural industries, such as "productivity" or "welfare", to reveal the "non-equilibrium" spatial characteristics and the mechanism of coordinated development of RIE. This finding not only enhances the theoretical framework of the coordinated development of RIE but also provides a new perspective for the study of regional rural industry or employment development patterns.

Spatially, the spatial evolution of RIE's coordinated development displays distinct "Spatial-temporal ripples" characteristics, with dominant RIE countries concentrated around Chongqing's metropolitan areas and regional sub-centers. This clustering phenomenon has intensified, leading to a "spindle-shaped" spatial distribution pattern in RIE's coordinated development status. These findings are somewhat consistent with those of Shen [7] and Chen [67], who observed an increase in agglomeration and a polycentric distribution of rural industries in urban agglomerations in China, as well as Wang [68], who noted similar spatial processes in the study of rural areal functions in Xi'an's metropolitan area. Myrdal [69] proposed that a "diffusion effect" and an "echo effect" exist between core and peripheral areas. Similarly, the changes in the Chongqing RIE system are marked by a typical "echo effect" in employment and a "diffusion effect" in industry. Counties with higher rural employment growth are concentrated in the central urban, metropolitan, and regional sub-center areas, with average increases of 100.19%, 82.06%, and 64.41% between 2013 and 2023, respectively. In contrast, rural industry exhibits a different pattern, with average growth rates of 48.69%, 91.40%, and 99.18% in these areas, respectively.

In terms of mechanisms, a "human-centered, socio-economic development framework rooted in natural resources" broadly characterizes the driving mechanisms for the coordinated development of the RIE system in Chongqing (Figure 7). Population size consistently demonstrates high explanatory power in factor detection, alongside interactive solid effects with the natural environment, socio-economic development, and living conditions. The urban and rural populations provide a substantial labor force for industrial growth, which can reduce employment costs for enterprises while increasing both industrial scale and profitability. Among natural environment factors, cropland and forest land consistently show strong explanatory power as single factors and maintain high interaction strength with population size and socio-economic development. Favorable farming conditions and forest resources constitute an essential foundation for modern agricultural development, influencing the scale of rural households' livelihood capital. Prior research suggests that agricultural livelihood capital significantly impacts farmers' decisions to engage in agricultural activities [59]. Additionally, abundant natural resources are crucial for attracting investment. In socio-economic development, total investment in fixed assets and retail sales of consumer goods have consistently remained substantial. Fixed asset investment reflects the degree to which the industrial development environment and the living conditions for workers have been enhanced. It is considered one of the core driving forces behind the coordinated development of RIE. The interaction analysis between socio-economic development and the living conditions and transport infrastructure confirms that fixed asset investment is strongly correlated with improvements in both the living environment and transportation. Furthermore, it exerts a significant joint effect on the coordinated development of RIE. Transport infrastructure undoubtedly plays a central role in supporting rural industrial development [11]. However, internal transportation consistently fails to pass significance tests, possibly because its influence on RIE coordination primarily occurs through reduced commuting time. Moreover, increased commuting costs often arise from the spatial separation of employment and residence rather than from transportation facilities [12].

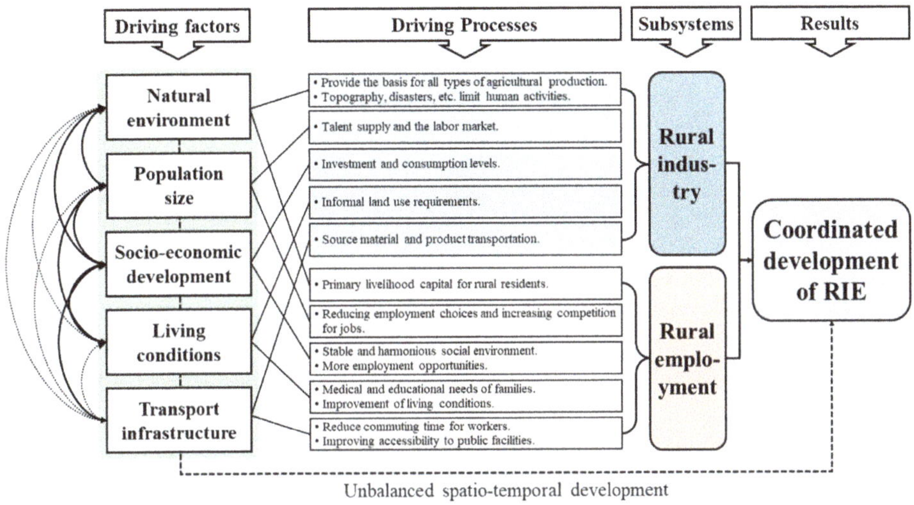

Figure 7. Mechanism framework for the evolution of the coordinated development of RIE. Solid lines between the five drivers indicate significant interactions, and dashed lines indicate minor interactions.

5.2. Employment-Centered Rural Industry: Fit Measures for the Coordinated Development of RIE

Industrial lag presents a significant constraint on the coordinated development of RIE (Figure 4), and this condition is likely to persist in the foreseeable future, particularly in the Global South. Rural industry primarily targets rural laborers, who are more vulnerable in the labor market and thus are more "welfare-oriented" than urban industry [70,71]. Promoting rural employment through rural industry is critical in fostering social equity and sustaining social stability. Thus, a more practical approach is to maximize the social function of rural industries in promoting employment despite the limitations in industrial supply.

Fully integrating the spatial and temporal characteristics and mechanisms of RIE is the prerequisite and foundation for establishing employment-oriented rural industrial development policies. At the spatial level, the findings of this study support the "polycentric" approach to RIE spatial development. They should encourage the creation of sub-centres in the peripheral areas of the region to form new growth centres and drive the development of RIE in the peripheral areas. The spatial imbalance within Chongqing's RIE is markedly pronounced (Figure 2). The Yudongbei region has established a favorable area for RIE's coordinated development centered around the regional sub-center of Wanzhou. In contrast, the Yudongnan region has consistently maintained a low RIE system status due to lacking a regional sub-center. Regional policymakers have recognized this issue and designated Qianjiang as a regional sub-center on par with Wanzhou in the Chongqing Territorial Spatial Master Plan (2021–2035) to stimulate growth in the Yudongnan region. Based on the phenomena of "spatial diffusion" of industries and "spatial return" of employment, the interventions of various influencing factors should adopt spatially differentiated strategies. Strengthening regional connectivity to promote the diffusion of industries to peripheral areas while simultaneously increasing investments in fixed assets, such as healthcare, education, and other public service facilities, will help improve the human environment and attract the return of rural employment. Furthermore, rural industry, as a subsystem of RIE that has historically lagged (Figure 4), requires concerted efforts from government agencies to guide the transfer of limited industrial resources to regions with significant gaps in RIE development within the scope of available basic conditions.

5.3. Limitations and Further Research

It is important to acknowledge that this study has certain limitations regarding constructing the indicator system and analyzing underlying mechanisms. First, the evaluation index system for the RIE system presented in this paper is not sufficiently comprehensive. Despite the availability of numerous existing studies to reference, the county-level scale and limitations in the availability and completeness of multivariate data constrained the extent to which this study could develop a more detailed and inclusive framework. Second, the factors influencing RIE often have multifaceted implications. While this paper focuses on quantifying the magnitude of influence exerted by these factors on the coordinated development of RIE, the mechanism of action is primarily explained through qualitative analysis. However, in practice, the modes of influence are likely to vary considerably depending on contextual factors, such as social dynamics and the specific intervention targets. Third, while this study emphasizes the coordinated development characteristics of RIE at the regional scale, some findings suggest significant multi-scale effects. For instance, the weak explanatory power of internal transportation at the county scale in influencing RIE development is noted. The impact of transportation facilities on RIE primarily pertains to improving commuting, and the rise in commuting costs resulting from the spatial separation of jobs and residences cannot be mitigated solely through transportation improvements. This issue warrants analysis at a more micro scale.

Looking ahead, further research should narrow the scope of study and enhance the theoretical framework for RIE's coordinated development by incorporating specific regional contexts and multiple stakeholder interests. Additionally, more diverse analytical methods (e.g., interviews and case comparisons) would help comprehensively explore the RIE coordination mechanism's multi-scale effects and dynamics.

6. Conclusions

This study aims to understand the evolution pattern and mechanisms of the coordinated development of rural industry and rural employment. Based on the systemic perspective, this study combines rural industry and rural employment to construct a theoretical framework for the coordinated development of RIE. This study employs various methods to analyze the spatio-temporal evolution pattern and mechanism of coordinated development of RIE across 37 counties in Chongqing from 2013 to 2023 and proposes corresponding development strategies. During the study period, the coordinated development of RIE in Chongqing has generally improved significantly, although the industrial subsystem has remained relatively lagging. The spatial-temporal characteristics of RIE exhibit clear non-equilibrium, with the spatial distribution being described as exhibiting "spatio-temporal ripples" and a "spindle shape". Agglomeration has been strengthened to a certain extent over time. The growth of RIE is accompanied by spatial diffusion of industries and spatial echo of employment. The "human-centered, natural-resource-based socio-economic development conditions" are the primary driving mechanism for the coordinated development of RIE in Chongqing.

Author Contributions: Conceptualization, Guoqin Ge; methodology, Guoqin Ge; software, Qianting Chen; validation, Guoqin Ge and Qianting Chen; formal analysis, Guoqin Ge; investigation, Guoqin Ge and Qianting Chen; resources, Yong Huang; data curation, Qianting Chen; writing—original draft preparation, Guoqin Ge; writing—review and editing, Guoqin Ge; visualization, Qianting Chen; supervision, Yong Huang; project administration, Yong Huang; funding acquisition, Yong Huang. All authors have read and agreed to the published version of the manuscript.

Funding: Project No. 2024CDJXY014. Supported by the Fundamental Research Funds for the Central Universities.

Data Availability Statement: The raw data supporting the conclusions of this article will be made available by the authors on request.

Conflicts of Interest: The authors declare no conflicts of interest.

References

1. Liu, Y.S.; Li, Y.H. Revitalize the World's Countryside. *Nature* **2017**, *548*, 275–277. [CrossRef] [PubMed]
2. Li, Y.H.; Westlund, H.; Liu, Y.S. Why Some Rural Areas Decline While Some Others Not: An Overview of Rural Evolution in the World. *J. Rural Stud.* **2019**, *68*, 135–143. [CrossRef]
3. Guo, D.; Jiang, K.; Xu, C.G.; Yang, X.Y. Industrial Clustering, Income and Inequality in Rural China. *World Dev.* **2022**, *154*, 105878. [CrossRef]
4. Wu, J.; Liu, X.; Ruan, J.; Qi, X.; Wang, C.; Fan, D. Space Power in Inclusive Development: Industrial Clusters and Rural Anti-Poverty. *Int. J. Environ. Res. Public Health* **2021**, *18*, 10943. [CrossRef] [PubMed]
5. Li, Z.T.; Yan, H.P.; Liu, X.X. Evaluation of China's Rural Industrial Integration Development Level, Regional Differences, and Development Direction. *Sustainability* **2023**, *15*, 2479. [CrossRef]
6. Wang, R.; Shi, J.W.; Hao, D.Q.; Liu, W.X. Spatial–Temporal Characteristics and Driving Mechanisms of Rural Industrial Integration in China. *Agriculture* **2023**, *13*, 747. [CrossRef]
7. Shen, Y.; Sargani, G.R.; Wang, R.; Jing, Y. Unveiling the Spatio-Temporal Dynamics and Driving Mechanism of Rural Industrial Integration Development: A Case of Chengdu–Chongqing Economic Circle, China. *Agriculture* **2024**, *14*, 884. [CrossRef]
8. Zhang, H.; Wu, D. The Impact of Rural Industrial Integration on Agricultural Green Productivity Based on the Contract Choice Perspective of Farmers. *Agriculture* **2023**, *13*, 1851. [CrossRef]
9. Zheng, Z.-Q.; Chou, R.-J. Rebuilding the resilience of mountainous rural communities by enhancing community capital through industrial transformation: A case study from rural Fujian, China. *Habitat Int.* **2024**, *149*, 13. [CrossRef]
10. Zhou, J.; Chen, H.P.; Bai, Q.Y.; Liu, L.X.; Li, G.H.; Shen, Q.L. Can the Integration of Rural Industries Help Strengthen China's Agricultural Economic Resilience? *Agriculture* **2023**, *13*, 1813. [CrossRef]
11. Zhang, H.; Wu, D. The Impact of Transport Infrastructure on Rural Industrial Integration: Spatial Spillover Effects and Spatio-Temporal Heterogeneity. *Land* **2022**, *11*, 1116. [CrossRef]
12. Gao, J.; Jiang, W.; Chen, J.; Liu, Y. Housing-industry symbiosis in rural China: A multi-scalar analysis through the lens of land use. *Appl. Geogr.* **2020**, *124*, 102281. [CrossRef]
13. Gu, J.; Zheng, J.Y.; Zhang, J. Research on the Coupling Coordination and Prediction of Industrial Convergence and Ecological Environment in Rural of China. *Front. Environ. Sci.* **2022**, *10*, 20. [CrossRef]
14. Li, X.; Liu, Y.; Guo, Y. The spatial pattern of population-land-industry coupling coordinated development and its influencing factor detection in rural China. *J. Geogr. Sci.* **2023**, *33*, 2257–2277. [CrossRef]
15. Deng, X.; Wang, G.; Song, W.; Chen, M.; Liu, Y.; Sun, Z.; Dong, J.; Yue, T.; Shi, W. An Analytical Framework on Utilizing Natural Resources and Promoting Urban–Rural Development for Increasing Farmers' Income Through Industrial Development in Rural China. *Front. Environ. Sci.* **2022**, *10*. [CrossRef]
16. Zhang, Z.; Sun, C.; Wang, J. How Can the Digital Economy Promote the Integration of Rural Industries—Taking China as an Example. *Agriculture* **2023**, *13*, 2023. [CrossRef]
17. Maja, P.W.; Meyer, J.; Von Solms, S. Development of Smart Rural Village Indicators in Line with Industry 4.0. *IEEE Access* **2020**, *8*, 152017–152033. [CrossRef]
18. Cowie, P.L. Townsend, and K. Salemink. Smart Rural Futures: Will Rural Areas Be Left Behind in the 4th Industrial Revolution? *J. Rural Stud.* **2020**, *79*, 169–176. [CrossRef]
19. Zhang, Y.; Ji, M.-X. From pixels to production: The impact of digital finance on the industrial fabric of rural China. *China Agric. Econ. Rev.* **2024**, *16*, 527–544. [CrossRef]
20. Ge, H.; Li, B.; Tang, D.; Xu, H.; Boamah, V. Research on Digital Inclusive Finance Promoting the Integration of Rural Three-Industry. *Int. J. Environ. Res. Public Health* **2022**, *19*, 3363. [CrossRef]
21. Tian, X.; Wu, M.; Ma, L.; Wang, N. Rural Finance, Scale Management and Rural Industrial Integration. *China Agric. Econ. Rev.* **2020**, *12*, 349–365. [CrossRef]
22. Dang, T.D.; Tran, T.A. Rural Industrialization and Environmental Governance Challenges in the Red River Delta, Vietnam. *J. Environ. Dev.* **2020**, *29*, 420–448. [CrossRef]
23. Ahmad, M.; Khan, Z.; Anser, M.K.; Jabeen, G. Do Rural-Urban Migration and Industrial Agglomeration Mitigate the Environmental Degradation across China's Regional Development Levels? *Sustain. Prod. Consum.* **2021**, *27*, 679–697. [CrossRef]

24. Sheng, Y.; Zhao, Y.; Zhang, Q.; Dong, W.; Huang, J. Boosting Rural Labor Off-Farm Employment through Urban Expansion in China. *World Dev.* **2022**, *151*, 15. [CrossRef]
25. Zhu, H.; Chen, Y.; Zhang, H.; Liu, Z. What Drives Migrants Back to Set up Firms? Return-home Entrepreneurial Intention of Rural Migrant Workers in China. *Chin. Geogr. Sci.* **2023**, *33*, 205–220. [CrossRef]
26. Gu, H.; Ling, Y.; Shen, T. Return or Not Return: Examining the Determinants of Return Intentions Among Migrant Workers in Chinese Cities. *Asian Popul. Stud.* **2021**, *17*, 51–70. [CrossRef]
27. Liu, J.F.; Yu, Y.W.; Qi, W.X.; Ma, X.; Han, Y. Innovation and Entrepreneurship of Chinese Returning Migrant Workers in Their Home Region. *Heliyon* **2024**, *10*, 16. [CrossRef] [PubMed]
28. Gustafsson, B.; Zhang, Y.D. Self-Employment in Rural China: Its Development, Chara Cteristics, and Relation to Income. *China World Econ.* **2022**, *30*, 136–165. [CrossRef]
29. Zhao, J.M. Internet Usage and Rural Self-Employment in China. *Asian Perspect.* **2020**, *44*, 77–101. [CrossRef]
30. Van Hoyweghen, K.; Broeck, G.V.D.; Maertens, M. Employment Dynamics and Linkages in the Rural Economy: Insights from Senegal. *J. Agric. Econ.* **2020**, *71*, 904–928. [CrossRef]
31. Zheng, H.; Ma, W.; Guo, Y.; Zhou, X. Interactive Relationship between Non-Farm Employment and Mechanization Service Expenditure in Rural China. *China Agric. Econ. Rev.* **2022**, *14*, 84–105. [CrossRef]
32. Dolislager, M.; Reardon, T.; Arslan, A.; Fox, L.; Liverpool-Tasie, S.; Sauer, C.; Tschirley, D.L. Youth and Adult Agrifood System Employment in Developing Regions: Rural (Peri-Urban to Hinterland) vs. Urban. *J. Dev. Stud.* **2021**, *57*, 571–593. [CrossRef]
33. White, B. Agriculture and the Generation Problem: Rural Youth, Employment and the Future of Farming. *IDS Bull. Inst. Dev. Stud.* **2012**, *43*, 9–19. [CrossRef]
34. Geza, W.; Ngidi, M.S.C.; Slotow, R.; Mabhaudhi, T. The Dynamics of Youth Employment and Empowerment in Agriculture and Rural Development in South Africa: A Scoping Review. *Sustainability* **2022**, *14*, 5041. [CrossRef] [PubMed]
35. Das, K.; Mahanta, A. Rural Non-Farm Employment Diversification in India: The Role of Gender, Education, Caste and Land Ownership. *Int. J. Soc. Econ.* **2023**, *50*, 741–765. [CrossRef]
36. Xie, F.; Liu, S.; Xu, D. Gender Difference in Time-Use of Off-Farm Employment in Rural Sichuan, China. *J. Rural Stud.* **2022**, *93*, 487–495. [CrossRef]
37. Koolwal, G.B. Improving the Measurement of Rural Women's Employment: Global Momentum and Survey Priorities. *World Dev.* **2021**, *147*, 16. [CrossRef]
38. Loison, S.A. Rural Livelihood Diversification in Sub-Saharan Africa: A Literature Review. *J. Dev. Stud.* **2015**, *51*, 1125–1138. [CrossRef]
39. Martin, S.M.; Lorenzen, K. Livelihood Diversification in Rural Laos. *World Dev.* **2016**, *83*, 231–243. [CrossRef]
40. Jing, Z.; Yu, Y.; Wang, Y.; Su, X.; Qiu, X.; Yang, X.; Xu, Y. Study on the Mechanism of Livelihood Behavior Decision of Rural Residents in Ethnic Tourism Villages in Western Sichuan. *Ecol. Indic.* **2024**, *166*, 15. [CrossRef]
41. Iorio, M.; Corsale, A. Rural Tourism and Livelihood Strategies in Romania. *J. Rural Stud.* **2010**, *26*, 152–162. [CrossRef]
42. Su, M.M.; Wall, G.; Wang, Y.A.; Jin, M. Livelihood Sustainability in a Rural Tourism Destination—Hetu Town, Anhui Province, China. *Tour. Manag.* **2019**, *71*, 272–281. [CrossRef]
43. Zhou, W.; Ma, Z.; Guo, S.; Deng, X.; Xu, D. Livelihood Capital, Evacuation and Relocation Willingness of Residents in Earthquake-Stricken Areas of Rural China. *Saf. Sci.* **2021**, *141*, 10. [CrossRef]
44. Fang, Y.P.; Zhu, F.B.; Qiu, X.P.; Zhao, S. Effects of Natural Disasters on Livelihood Resilience of Rural Residents in Sichuan. *Habitat Int.* **2018**, *76*, 19–28. [CrossRef]
45. Zhu, P.Y. Residential Segregation and Employment Outcomes of Rural Migrant Workers in China. *Urban Stud.* **2016**, *53*, 1635–1656. [CrossRef]
46. Gautam, Y.; Andersen, P. Rural Livelihood Diversification and Household Well-Being: Insights from Humla, Nepal. *J. Rural Stud.* **2016**, *44*, 239–249. [CrossRef]
47. Deng, Q.Q.; Li, E.L.; Zhang, P.Y. Livelihood Sustainability and Dynamic Mechanisms of Rural Households out of Poverty: An Empirical Analysis of Hua County, Henan Province, China. *Habitat Int.* **2020**, *99*, 102160. [CrossRef]
48. Qu, R.; Rhee, Z.; Bae, S.J.; Lee, S.H. Analysis of Industrial Diversification Level of Economic Development in Rural Areas Using Herfindahl Index and Two-Step Clustering. *Sustainability* **2022**, *14*, 6733. [CrossRef]
49. Wang, J.; Peng, L.; Chen, J.; Deng, X. Impact of Rural Industrial Integration on Farmers' Income: Evidence from Agricultural Counties in China. *J. Asian Econ.* **2024**, *93*, 14. [CrossRef]
50. Ouko, K.O.; Ogola, J.R.O.; Ng'on'ga, C.A.; Wairimu, J.R.; Serpa, S. Youth Involvement in Agripreneurship as Nexus for Poverty Reduction and Rural Employment in Kenya. *Cogent Soc. Sci.* **2022**, *8*, 20. [CrossRef]
51. Geng, Y.Q.; Liu, L.W.; Chen, L.Y. Rural Revitalization of China: A New Framework, Measurement and Forecast. *Socio-Econ. Plan. Sci.* **2023**, *89*, 16. [CrossRef]
52. Liu, Y.S.; Zang, Y.Z.; Yang, Y.Y. China's Rural Revitalization and Development: Theory, Technology and Management. *J. Geogr. Sci.* **2020**, *30*, 1923–1942. [CrossRef]

53. Marsden, T.K.; Murdoch, J.; Lowe, P.D.; Munton, R.J.; Flynn, A. Constructing the Countryside. *Trans. Inst. Br. Geogr.* **1994**, *19*, 120.
54. Cloke, P.; Goodwin, M. Conceptualizing Countryside Change: From Post-Fordism to Rural Structured Coherence. *Trans. Inst. Br. Geogr.* **1992**, *17*, 321–336. [CrossRef]
55. Holmes, J. Impulses Towards a Multifunctional Transition in Rural Australia: Gaps in the Research Agenda. *J. Rural Stud.* **2006**, *22*, 142–160. [CrossRef]
56. Marsden, T.; Sonnino, R. Rural Development and the Regional State: Denying Multifunctional Agriculture in the UK. *J. Rural Stud.* **2008**, *24*, 422–431. [CrossRef]
57. Zasada, I. Multifunctional Peri-Urban Agriculture—a Review of Societal Demands and the Provision of Goods and Services by Farming. *Land Use Policy* **2011**, *28*, 639–648. [CrossRef]
58. Renting, H.; Rossing, W.; Groot, J.; Van der Ploeg, J.; Laurent, C.; Perraud, D.; Stobbelaar, D.; Van Ittersum, M. Exploring Multifunctional Agriculture. A Review of Conceptual Approaches and Prospects for an Integrative Transitional Framework. *J. Environ. Manag.* **2009**, *90*, S112–S123. [CrossRef]
59. Bhandari, P.B. Rural Livelihood Change? Household Capital, Community Resources and Livelihood Transition. *J. Rural Stud.* **2013**, *32*, 126–136. [CrossRef]
60. Mbaiwa, J.E.; Stronza, A.L. The Effects of Tourism Development on Rural Livelihoods in the Okavango Delta, Botswana. *J. Sustain. Tour.* **2010**, *18*, 635–656. [CrossRef]
61. Castells, M.; Hall, P.G. Technopoles of the World: The Making of Twenty-First-Century Industrial Complexes. *Routledge* **1994**. [CrossRef]
62. Storper, M. The Transition to Flexible Specialisation in the U.S. Film Industry: External Economies, the Division of Labour, and the Crossing of Industrial Divides. *Camb. J. Econ.* **1989**, *13*, 273–305.
63. Massey, D. Spatial Divisions of Labour: Social Structures and the Geography of Production. *Macmillan* **1984**.
64. Wu, L.Y. *Introduction to Sciences of Human Settlements*; Architecture & Building Press: Beijing, China, 2001.
65. Wang, J.-F.; Li, X.-H.; George, C.; Liao, Y.-L.; Zhang, T.; Gu, X.; Zheng, X.Y. Geographical Detectors-Based Health Risk Assessment and Its Application in the Neural Tube Defects Study of the Heshun Region, China. *Int. J. Geogr. Inf. Sci.* **2010**, *24*, 107–127. [CrossRef]
66. Song, Y.; Wang, J.; Ge, Y.; Xu, C. An Optimal Parameters-Based Geographical Detector Model Enhances Geographic Characteristics of Explanatory Variables for Spatial Heterogeneity Analysis: Cases with Different Types of Spatial Data. *GIScience Remote Sens.* **2020**, *57*, 593–610. [CrossRef]
67. Chen, J.; Jiang, L.; Luo, J.; Tian, L.L.; Tian, Y.; Chen, G.L. Characteristics and Influencing Factors of Spatial Differentiation of Market Service Industries in Rural Areas around Metropolises—A Case Study of Wuhan City's New Urban Districts. *ISPRS Int. J. Geo-Inf.* **2022**, *11*, 170. [CrossRef]
68. Wang, X.; Liu, Y.; Shao, Y.; Li, S. Evolution Pattern and Mechanism of Rural Areal Functions in Xi'an Metropolitan Area, China. *Habitat Int.* **2024**, *148*, 103088. [CrossRef]
69. Myrdal, G. *Economic Theory and Underdeveloped Regions*; Duckworth: London, UK, 1957.
70. Bellemare, M.F.; Bloem, J.R. Does Contract Farming Improve Welfare? A Review. *World Dev.* **2018**, *112*, 259–271. [CrossRef]
71. Santika, T.; Wilson, K.A.; Meijaard, E.; Budiharta, S.; Law, E.E.; Sabri, M.; Struebig, M.; Ancrenaz, M.; Poh, T.M. Changing Landscapes, Livelihoods and Village Welfare in the Context of Oil Palm Development. *Land Use Policy* **2019**, *87*, 104073. [CrossRef]

International Journal of
Geo-Information

MDPI

Article

New Quality Productivity of Agriculture and Rural Areas at the Provincial Scale in China: Indicator Construction and Spatiotemporal Evolution

Xiangyang Cao [1], Jiahui Lei [1], Donghui Shi [2], Wenlong Yu [1,*], Tianhui Tao [3], Xun Zhang [4] and An Wang [1]

1 School of Civil Engineering, Shandong Jiaotong University, Jinan 250357, China; caoxiangyang@sdjtu.edu.cn (X.C.); cartopy@126.com (J.L.); cuifengkun@sdjtu.edu.cn (A.W.)
2 Institute of Geographic Sciences and Natural Resources Research, Chinese Academy of Sciences, Beijing 100101, China; shidonghui@igsnrr.ac.cn
3 College of Geomatics, Zhejiang University of Water Resources and Electric Power, Hangzhou 310018, China; taoth@zjweu.edu.cn
4 College of Science and Technology, Wenzhou-Kean University, Wenzhou 325060, China; zhanxun@kean.edu
* Correspondence: yuwenlong@sdjtu.edu.cn

Abstract: New quality productivity in agriculture and rural areas serves as a critical foundation for addressing the development needs of the times, advancing the comprehensive revitalization of rural regions, and overcoming the urban–rural dual structure. This paper studies the spatial changes and evolutionary trends of new quality productivity in agriculture and rural areas in 31 provinces of mainland China. The main conclusions are as follows: (1) At the provincial level, the development of China's new quality productivity exhibits a spatial gradient, with a decreasing trend from east to west. (2) At the national scale, while significant spatial autocorrelation exists in the new quality productivity of agriculture and rural areas, internal disparities are gradually narrowing. (3) The eastern region demonstrates significant advantages, the central region is making steady progress, the western region is rapidly catching up, and the northeastern region faces increasingly significant development pressures. This paper extends the research boundary of new quality productivity to the field of agriculture and rural areas, and we introduce a variety of spatial analysis methods to depict its distribution characteristics.

Keywords: new quality productivity; indicator construction; spatiotemporal analysis

1. Introduction

The current global economic situation is complex and volatile. Factors such as the adjustments in global trade patterns, the rapid advancement of technological revolutions, climate change, and geopolitical instability have introduced significant uncertainties to global economic growth [1]. Amid these profound changes, unseen in a century, nations worldwide are striving to identify new drivers of economic growth and development. While the Chinese economy has demonstrated a stable and positive trajectory, challenges such as lagging productivity and insufficient development momentum persist [2]. In September 2023, the concept of "New quality Productive Forces" was first introduced, aligning with the demands of high-quality development in the current era, and marking a new stage in the evolution of China's productive forces. Against the backdrop of the comprehensive rural revitalization strategy outlined in the No.1 Central Document of 2024, the concept of "New quality productivity of agriculture and rural areas" has emerged.

For decades, rural areas in China have predominantly relied on traditional farming practices, including conventional planting and breeding methods characterized by low technological content and production efficiency. This reliance has resulted in sluggish rural economic growth and has posed significant challenges to the sustainable development of agriculture. These issues have constrained the economic advancement of agriculture and rural areas, rendering them unable to meet the increasing demands of the population for an improved quality of life. Consequently, it is of critical importance to scientifically assess the emerging quality and productivity of agriculture and rural areas, thoroughly analyze their evolutionary characteristics and driving mechanisms, and explore effective development strategies. Such efforts are essential for achieving revolutionary breakthroughs in production technologies, fostering innovative allocation of production factors, and promoting the comprehensive transformation and upgrading of agricultural and rural industries.

1.1. Theoretical Research Related to the New Quality Productivity of Agriculture and Rural Areas

In the mid-to-late 18th century, Western countries entered the era of classical political economy, during which theories related to productivity began to attract significant attention. The French economist François Quesnay was the first to introduce the concept of "productivity", proposing the idea that "agriculture is the natural source of production", which sparked widespread discussions [3]. The British economist Adam Smith introduced the concept of "labor productivity", arguing that the division of labor and cooperative efforts could significantly enhance labor efficiency [4]. Jean-Baptiste Say and David Ricardo further integrated the concept of productivity into political economy, emphasizing its variability and adaptability [5]. Subsequent theoretical developments expanded the understanding of productivity from various perspectives. Karl Heinrich Marx's political economy, Franz Liszt's national economy, economic growth theory, and economic structural change theory discussed productivity from different perspectives. Marxist political economy emphasized the law of the contradictory movement between productive forces and production relations, becoming the basic theory for the study of productive forces [6,7]. Friedrich List's national economy emphasized the crucial role of national protection in the development of productive forces in late-developing countries [8]. The economic growth theory introduced the efficiency of factor combination into the analysis of productive forces [9]. Economic structural change theory focused on the issue of industrial forms in growth [10]. At this point, the theoretical system of productive forces was initially formed.

Building on these historic achievements, China has continuously deepened its understanding of the development patterns of productive forces. As the primary contradiction in the socialist stage has shifted from "the people's ever-growing material and cultural needs" to "the people's ever-growing needs for a better life", there is an urgent need to move away from dependence on traditional productive forces. In response, a new form of quality productivity, characterized by higher levels of integration, innovation, and technological advancement, has emerged [11]. This concept of new quality productivity advances Marxist theory on productive forces and deepens the understanding of the laws governing their development [12]. In the context of liberating and enhancing the productivity of agriculture and rural areas, scholars have conducted extensive research from multiple perspectives, including the role of information technology [13], system innovation [14], the elimination of the urban–rural dual structure [15], and moderate-scale operation. These efforts have yielded significant research outcomes. Furthermore, scholars have explored various aspects of new quality productivity, such as its connotation features [16], realization pathways [17], and practical applications [18].

In the theoretical framework of the new quality productivity of agriculture and rural areas, precision, digital, and smart agriculture technologies are the core elements of modern productive forces, reconstructing the traditional agricultural production model. Precision agriculture relies on the Internet of Things and other technologies to collect data, achieve precise fertilization and irrigation, and provide early warning of pests and disease, thus improving resource utilization efficiency [19]. Digital agriculture, supported by big data and cloud computing, builds a digital management platform for the entire industrial chain, optimizing every link from planting decisions to market circulation [20]. Smart agriculture integrates artificial intelligence and automated equipment, driving the transformation of agriculture from experience-driven to data-driven [21]. These technologies not only increase labor productivity but also contribute to the sustainable development of rural areas [22]. The evolution of the new quality productivity conforms to Marxist theory and provides a technological path for rural revitalization.

However, as a relatively new concept, limited research has been conducted on integrating new quality productivity with agricultural and rural development. In particular, systematic analyses of the connotations and characteristics of new quality productivity in the context of agriculture and rural areas remain scarce. This paper holds that the new quality productivity of agriculture and rural areas represents an advanced state of productive forces. It is characterized by a high-quality labor force as the main feature, disruptive innovation as the technological feature, the infiltration and integration of multiple factors as the allocation feature, the breakthrough and mutual integration of the boundaries between agriculture and rural areas as the structural feature, and digital intelligence and green development as the morphological features.

1.2. Research on the Index Construction and Spatiotemporal Evolution

Currently, numerous studies focus on the theoretical framework, connotations, and significance of new quality productivity. However, research involving quantitative analysis, such as the evaluation and spatiotemporal evolution of new quality productivity, remains relatively scarce. Some scholars used the improved entropy weight TOPSIS method to measure the new quality productivity in China from 2012 to 2021. The results showed that the level increased year by year, with an average annual growth rate of 21% [19]. Some other scholars constructed an index system to evaluate the new quality productivity of prefecture-level cities and found that the eastern region was leading in development, while the western and northeastern regions lagged behind, with significant differences between the regions [20]. In addition, some scholars used the Dagum Gini coefficient method to analyze new quality productivity, finding spatiotemporal differences between the north and south of China, among the four major regions and the five major economic belts, and that the development among provincial regions is unbalanced [21]. These studies provide important references for the quantitative analysis of new quality productivity. However, currently, there are relatively more studies on productivity at the provincial and urban scales, and these studies suffer from the problem of "overemphasizing theory and neglecting practice".

In conclusion, this study integrates multiple theoretical perspectives and employs an interdisciplinary approach to assess the new quality productivity of agriculture and rural areas while analyzing its spatiotemporal evolution characteristics. This paper has three research objectives: constructing a scientific and comprehensive evaluation index system suitable for the new quality productivity in agriculture and rural areas; accurately depicting its distribution characteristics from the dual perspectives of time and space; and, based on the actual conditions of various regions, proposing targeted and practical development measures tailored to local conditions to promote its efficient development. This study's innovations are reflected in three aspects: Firstly, it broadens the research scope of new quality productivity. On the one hand, it refines the research focus by applying the concept of new quality productivity specifically to the field of agriculture and rural areas. On the other hand, it moves beyond theoretical exploration by incorporating empirical analysis, thereby enriching the quantitative research on new quality productivity. Secondly, it advances the research methods for studying new quality productivity in agriculture and rural areas. This study employs a comprehensive approach, integrating hierarchical evaluation methods with various spatial analysis techniques across multiple disciplines, enhancing the methodological framework for this field of research. Finally, it contributes to the policy implications of new quality productivity in agriculture and rural areas. This study provides a rich body of research that offers scientific support for identifying key areas, primary models, and innovative strategies for the development of new quality productivity. These recommendations are tailored to regional and phased development needs and emphasize the importance of adapting to local conditions.

This paper is characterized by a rigorous logic, with its content presented in six parts that progress in a hierarchical manner. The Introduction analyzes the research background and motivation. The second part elaborates on the data sources and the construction of the index system. The third part introduces the research methodology. The fourth part presents the research findings. The fifth part conducts an in-depth discussion centered around the findings. Finally, the research achievements are summarized.

2. Data Sources and Construction of Index System

The data utilized in this study were sourced from the China Industrial Statistical Yearbook, China Energy Statistical Yearbook, China Environmental Statistical Yearbook, China Rural Statistical Yearbook, China Population and Employment Statistical Yearbook, China Urban and Rural Construction Statistical Yearbook, and the statistical yearbooks of 31 provinces in mainland China. The dataset spans the period from 2010 to 2022.

The index system is structured into four hierarchical levels, comprising a total of 23 detailed indicators. These indicators comprehensively and systematically reflect the performance and development trends of new quality productivity in agriculture and rural areas. Based on the principle of the three elements of productive forces, the new quality productivity in agriculture and rural areas is categorized into three dimensions: new quality laborers, new quality objects of labor, and new quality means of labor (Table 1). The dimension of new quality laborers is evaluated through laborer skills, labor productivity, and agricultural scientific and technological innovation capability [22,23].

Among them, the two indicators of forest coverage and urbanization rate were retrieved using 30 m resolution data obtained from the Landsat 8 satellite, with a timespan from 2010 to 2022. We integrated Band 5, Band 4, and Band 3 to obtain the standard false-color composite image. Then, the support vector machine method was used to conduct supervised classification on land-use types. Verification was carried out by establishing the confusion matrix for classification. It was found that the annual average value of the overall accuracy was 89%, and the annual average value of the kappa coefficient was 0.89, which met the requirements for the classification accuracy of the article. The standard false-color composite image of the original data is shown in Figure 1. In the figure, (a) is an example of Jiamusi, (b) is an example of Jinan, and (c) is an example of Tieling. According to the classification results, the corresponding indicators can be obtained by calculating the proportions of forest land and construction land in their respective provincial regions.

Table 1. Evaluation index system of new quality productivity in agriculture and rural areas.

Criterion Level	First-Level Indicator	Second-Level Indicator	Measurement Method
New quality laborers	Laborer's skills	Per capita education level	Worker's education level
		Rural cultural facilities	Number of rural cultural stations
	Labor productivity	Economic income	Net income
			Revenue growth rate
		Happiness index	Engel coefficient for rural residents
		Entrepreneurial concept	Farmers' entrepreneurial activity
	Agricultural innovation capability	Innovation achievements	Number of valid patent applications
New quality objects of labor	Agricultural development ability	Agricultural production capacity	Overall grain production capacity
		Agricultural production efficiency	Agricultural labor productivity
		Agricultural production value	Value added from agriculture, forestry, animal husbandry, and fisheries
	Ecological environment	Ecological environment governance	Area of soil erosion control
			Forest coverage
		Urbanization development	Urbanization rate
New quality means of labor	Tangible means of production	Agricultural development situation	Per capita power of agricultural machinery
			Irrigated area
			Application amount of pesticides and fertilizers
		Infrastructure	Rural digital base
			Fiber length
			Internet penetration
	Intangible means of production	Digital economy	Rural power consumption
			Agricultural fiscal expenditure
			Digital fixed assets investment
			Digital service fixed assets investment

(a) Jiamusi region (b) Jinan region (c) Tieling region

Figure 1. Examples of the standard false-color composite images of the original data; Jiamusi region (**a**), Jinan region (**b**) and Tieling region (**c**).

The dimension of new quality means of labor is divided into tangible and intangible means of production [24]. Tangible means of production encompass agricultural development and rural infrastructure construction. Specific indicators for agricultural development include the per capita total power of agricultural machinery, the effective irrigation area, and the application amount of pesticides and fertilizers. Rural infrastructure is characterized by indicators such as rural digital bases, fiber optic length, and Internet penetration rate. Intangible means of production primarily refer to the scale of digital elements, with the digital economy is measured by fixed asset investment in the digital industry and fixed asset investment in digital services.

3. Methodology

3.1. Index Weight Calculation Method

The weights of the indicators in this paper were calculated by using the entropy method. The entropy method completely determines the weight of indicators based on the degree of dispersion of the data themselves, avoiding subjective interference from human factors [25]. This method can objectively reflect the relative importance of each indicator, can fully utilize the information contained in each indicator, and is sensitive to reflecting the degree of variation in the indicator data. For comprehensive evaluation problems with multiple indicators and multiple objects, the entropy method can play a role in the fields of economy, management, or other fields [26]. Suppose the data are a matrix of m rows and n columns; the specific steps are as follows:

1 Data standardization:

$$\text{Positive indictors}: x'_{ij} = \frac{x_{ij} - x_{j(min)}}{x_{j(max)} - x_{j(min)}}\ j = 1,2\ldots n \tag{1}$$

$$\text{Negative indictors}: x'_{ij} = \frac{x_{j(max)} - x_{ij}}{x_{j(max)} - x_{j(min)}}\ j = 1,2\ldots n \tag{2}$$

where x_{ij} and x'_{ij} are the original data and normalized values, respectively; $x_{j(max)}$ represents the maximum value of the j-th indicator datum; and $x_{j(max)}$ represents the minimum value of the j-th indicator.

2 Calculate the proportion of the i-th datum in the j-th indicator w_{ij}:

$$w_{ij} = \frac{x'_{ij}}{\sum_{i=1}^{m} x'_{ij}}\ j = 1,2\ldots n \tag{3}$$

3 Calculate the entropy value (e_j) of each indicator:

$$e_j = -\frac{1}{ln(m)}\sum_{i=1}^{m}(w_{ij} * lnw_{ij})\ j = 1,2\ldots n \tag{4}$$

4 Calculate the information entropy residue d_j:

$$d_j = 1 - e_j \tag{5}$$

5 Calculate the weight W_j of the j-th indicator:

$$W_j = \frac{d_j}{\sum_{j=1} d_j} \tag{6}$$

3.2. Spatial Pattern Analysis Method

This paper uses Moran's I in exploratory analysis to study the spatial pattern mode of new quality productivity in agriculture and rural areas. It is mainly used to study whether the distribution of the research object in space has aggregated or dispersed characteristics [27]. If I is positive and significant, it indicates the existence of positive spatial autocorrelation; that is, similar values tend to cluster in space. If I is negative and significant, it means that there is negative spatial autocorrelation; that is, high values and low values tend to be dispersed in space. If I is close to zero, it indicates that the spatial distribution is random and there is no obvious spatial autocorrelation. When studying the new quality productivity of agriculture and rural areas across the country, Moran's I can detect whether there is obvious spatial clustering. Its calculation formula is as follows:

$$I = \frac{n\sum_{i=1}^{n}\sum_{j=1}^{n} w_{ij}(x_i - \overline{x})(x_j - \overline{x})}{\sum_{i=1}^{n}\sum_{j=1}^{n} w_{ij}\sum_{i=1}^{n}(x_i - \overline{x})^2} \tag{7}$$

where $\overline{x}$ is the expectation of the spatial unit, and w_{ij} represents the spatial weight matrix. Its significance level is usually determined by the p-value test. If the p-value is less than the given significance level α, then the null hypothesis is rejected. Otherwise, the null hypothesis is accepted. In the analysis of practical problems, the significance level α is usually 0.05.

3.3. Spatiotemporal Process Analysis Method

Standard deviational ellipse (SDE) is a method used to describe the directionality and degree of dispersion of the spatial distribution of geographical elements [28,29]. It intuitively shows the distribution characteristics of geographical elements in space by calculating parameters such as the centroid of the ellipse, the length of the x-axis, the length of the y-axis, and the direction angle. The centroid of the SDE represents the average central position of the distribution of geographical elements. The length of the x-axis and the y-axis represents the degree of dispersion in the horizontal and vertical directions, respectively. The longer the x-axis or y-axis, the more dispersed the distribution in that direction. The direction angle indicates the main trend direction of the distribution. The formulae are as follows:

$$SDE_x = \sqrt{\frac{\sum_{i=1}^{n}(x_i - \overline{X})^2}{n}} \tag{8}$$

$$SDE_y = \sqrt{\frac{\sum_{i=1}^{n}(y_i - \overline{Y})^2}{n}} \tag{9}$$

$$tan\theta = \left[\left(\sum_{i=1}^{n}\tilde{x}_i^2 - \sum_{i=1}^{n}\tilde{y}_i^2\right) + \sqrt{\left(\sum_{i=1}^{n}\tilde{x}_i^2 - \sum_{i=1}^{n}\tilde{y}_i^2\right)^2 + 4\left(\sum_{i-1}^{n}\tilde{x}_i\tilde{y}_i\right)^2}\right] + 2\sum_{i-1}^{n}\tilde{x}_i\tilde{y}_i \tag{10}$$

where SDE_x and SDE_y represent the lengths of the x-axis and y-axis of the ellipse, respectively, $tan\theta$ is the direction angle of the ellipse and $\tilde{x}_i$ and $\tilde{y}_i$ are the deviation values of the coordinates of element i from the centroid of the ellipse.

4. Results

4.1. Results of Spatial Pattern Analysis

At the provincial scale, significant disparities exist in the development levels of new quality productivity across China. Except for the northeast region, there is a general decreasing trend in new quality productivity from east to west. In 2022, the levels of new quality productivity in China ranged from 2.66 to 38.46, with an average value of 13.40 (Figure 2). Provinces with relatively high new quality productivity in agriculture and rural areas include Guangdong (38.46), Zhejiang (29.53), Shandong (23.49), Jiangsu (22.76), Hebei (18.70), Sichuan (18.15), and Henan (17.86). These demonstrate outstanding performances and possess advantages in aspects such as agricultural scientific and technological innovation, rural industry integration, and the cultivation of new agricultural business entities. Notably, Guangdong significantly outpaces other regions and serves as an important growth pole for the development of new quality productivity in China. Regions with medium levels of new quality productivity of agriculture and rural areas include Hubei (16.01), Hunan (15.17), Fujian (14.79), Heilongjiang (14.13), Shanghai (13.93), Anhui (13.49), Beijing (13.38), Guangxi (13.37), Chongqing (12.12), Yunnan (11.96), Jiangxi (11.68),

Shaanxi (10.99), Liaoning (10.76), Gansu (10.18), Xinjiang (10.12), Inner Mongolia (9.76), and Guizhou (9.65). Most of these regions are at a critical stage of agricultural modernization. There is a need to further strengthen agricultural infrastructure construction, the application of agricultural science and technology, and the cultivation of rural talents. For instance, as a major agricultural province, Heilongjiang has an advantage in the promotion of agricultural mechanization but still holds great potential for the development of smart agriculture. The remaining seven provinces, due to factors such as geographical limitations, weak economic foundations, and singular industrial structures, lack comprehensive agricultural infrastructure and advanced agricultural technology application conditions.

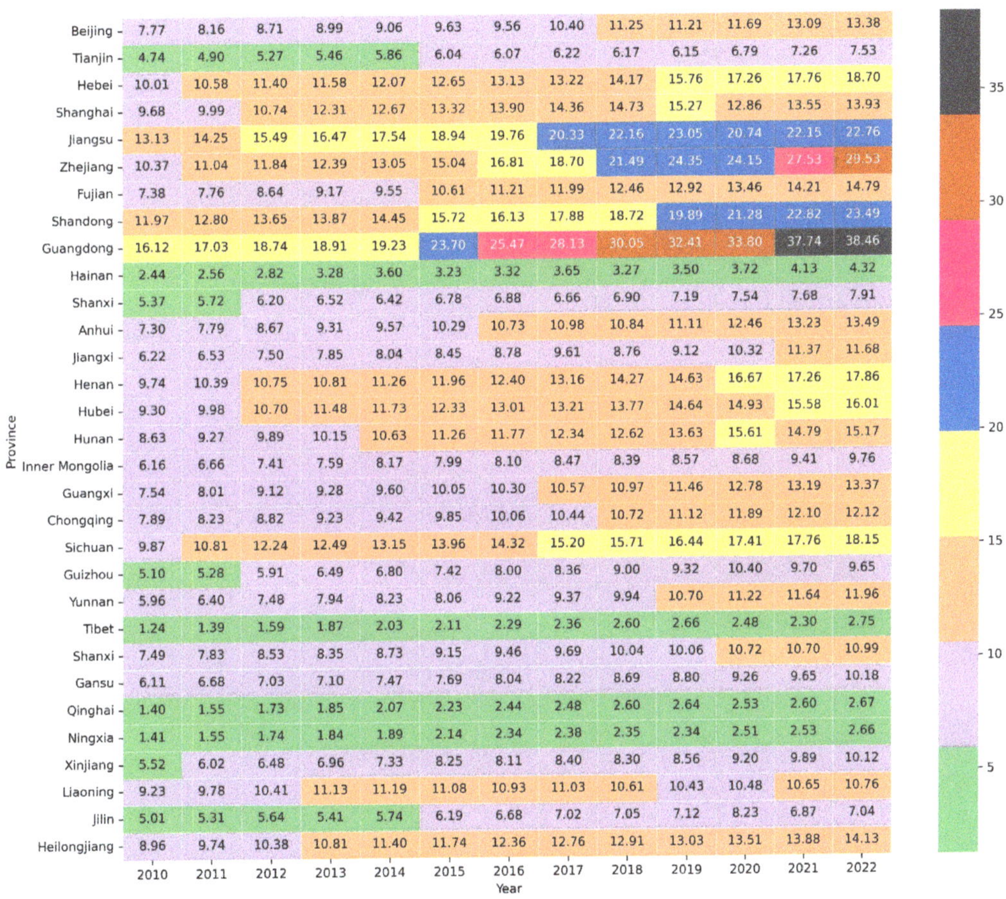

Figure 2. Calculation results of new quality productivity in each province.

There is a sharp contrast in the performance of new quality productivity between Sichuan and Tibet. In the western region, Sichuan performs relatively outstandingly. In terms of the industrial structure, its agricultural industry has developed diversely and formed a complete industrial chain. Meanwhile, its investment in scientific research is considerable. Numerous scientific research institutions and universities have facilitated the development of emerging models such as smart agriculture. In contrast, due to geographical constraints, Tibet has low agricultural production efficiency. Its relatively lagging economic development has led to insufficient investment in agriculture.

At the national scale, significant spatial autocorrelation exists in the new quality productivity of agriculture and rural areas; however, its significance is gradually weakening, and the internal disparities are diminishing. In 2010, the national Moran's I was 0.35, with a Z-score of 1.86 and a p-value of 0.049. In 2016, the national Moran's I decreased to 0.33, the Z-score to 1.51, and the p-value increased to 0.082. In 2022, the national Moran's I further declined to 0.30, the Z-score to 1.09, and the p-value rose to 0.134 (Figure 3). Moran's I passed the significance test at a 10% confidence level in 2010 and 2016, but not in 2022, indicating that while spatial autocorrelation has persisted, it has been weakening over time. The decreasing Moran's I and increasing p-value suggest a diminishing level of significance, which implies that the imbalance in new quality productivity among provinces is decreasing.

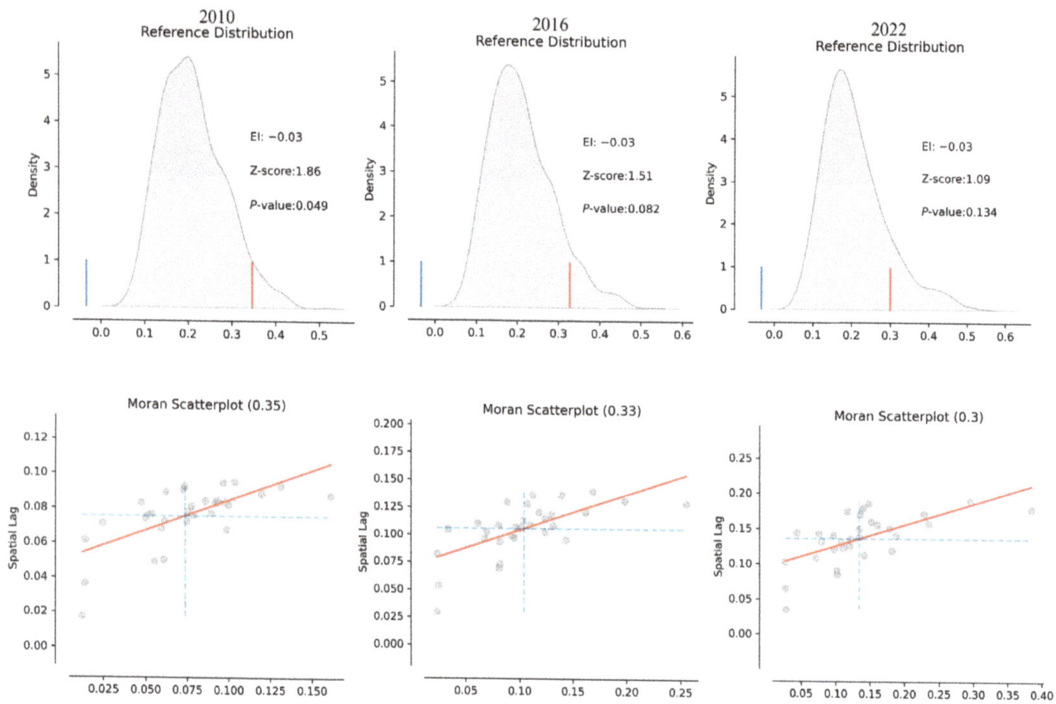

Figure 3. Spatial autocorrelation test.

4.2. Results of Time Trend Analysis

The average level of new quality productivity in agriculture and rural areas across China's provinces has shown an increasing trend; however, the growth rate has fluctuated over time. The highest growth rate occurred between 2011 and 2012, reaching 8.77%. In 2011, the central government allocated 988.45 billion RMB for expenditures on agriculture, rural areas, and farmers, an increase of 130.48 billion RMB over the previous year, representing a growth of 15.2%. Subsequently, the new quality productivity in agriculture and rural areas initially declined, then experienced fluctuations, and eventually tended toward stability (Figure 4).

Figure 4. Annual average value and growth rate of China's new quality productivity.

The level of new quality agricultural productivity in all four major regions of China has shown a yearly upward trend, with the northeast region exhibiting the slowest growth rate. Notably, the new quality productivity in the eastern region is significantly higher than the national average, rising from 9.36 in 2010 to 18.69 in 2022 (Figure 5). Regarding average annual growth rates, the eastern region leads, with 7.68%, followed by the western region at 6.67%, the central region at 6.27%, and the northeastern region at 3.19% (Table 2). The new quality productivity level in the central region is slightly above the national average, while the western region remains significantly below it. The development in the northeastern region has been relatively uneven, with its new quality productivity increasing from 7.73 to 10.04.

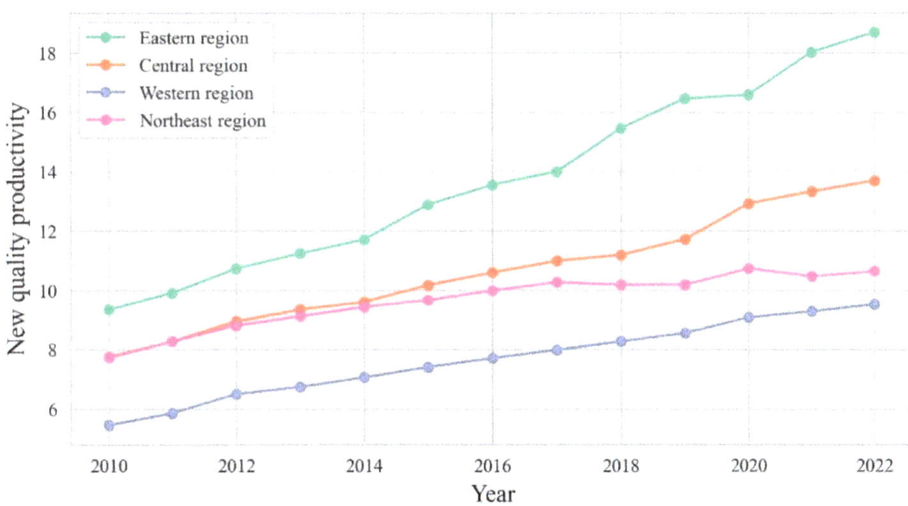

Figure 5. Development trend of new quality productivity in four major regions.

Table 2. Average growth rate of new quality productivity in major regions of China.

Area	Contained Cities	Average Growth Rate(%)
Eastern region	Beijing, Tianjin, Hebei, Shanghai, Jiangsu, Zhejiang, Fujian, Shandong, Guangdong, Hainan	7.68
Central region	Shanxi, Anhui, Jiangxi, Henan, Hubei, Hunan	6.27
Western region	Inner Mongolia, Guangxi, Chongqing, Sichuan, Guizhou, Yunnan, Tibet, Shaanxi, Gansu, Qinghai, Ningxia, Xinjiang	6.67
Northeast region	Liaoning, Jilin, Heilongjiang	3.19

The disparities in the levels of new quality productivity among the four major regions are relatively pronounced. The eastern region demonstrates significant advantages, the central region is making steady progress, the western region is continuously catching up, and the northeastern region faces greater development pressure. From 2010 to 2014, the level of new quality productivity in the northeastern region was slightly higher than the national level. However, after 2014, the gap between the northeastern region and the national level gradually widened, eventually approaching the development level of the western region (Figure 5). Although the western region started from a relatively low baseline, it has shown continuous growth. Before 2018, the growth in the western region was relatively slow. However, after 2018, the growth rate accelerated, with the level of new quality productivity increasing from 8.28 to 9.53 in 2022.

The growth of new quality productivity in the northeast region is slow, and it faces numerous difficulties. Most of the labor force engaged in agricultural production is elderly. They have difficulty adapting to the development of smart agriculture, which restricts the improvement of agricultural production efficiency. Recently, large-scale development has led to a decline in soil fertility in some areas, affecting the yield and quality of crops, and weakening the foundation of agricultural development. The northeast region has a high proportion of traditional industries. In recent years, it has faced great pressure in transformation and upgrading, and its driving effect on agricultural and rural development has weakened. At the same time, rural industries lack innovative vitality, further limiting the improvement of new quality productivity.

4.3. Results of Spatiotemporal Trend Analysis

The spatiotemporal changes observed in new quality productivity during the development process of the eastern region highlight the continuous optimization and transformation of its economic structure. The relative stability of the lengths of the x-axis and y-axis indicates that the core area and diffusion range of new quality productivity remain relatively stable. This stability can be attributed to the formation of a mature industrial cluster and economic development model in the eastern region (Table 3). The growth in new quality productivity relies more on technological innovation and industrial upgrading rather than large-scale spatial expansion. The change in the rotation angle reflects the directional adjustment of regional economic development, resulting in a certain degree of rotation in the spatial distribution of new quality productivity.

The stable spatial distribution pattern of new quality productivity in the central region reflects its important position and consistent role in the national economic development (Figure 6). Serving as a crucial hub connecting the eastern and western regions, the central region's economic development is relatively balanced, and its industrial structure remains stable. The constancy of the axis lengths and rotation angle indicates that there is no significant expansion or contraction trend in the spatial distribution of new quality productivity within the central region. This phenomenon suggests that the economic development of the central region relies primarily on the upgrading and transformation of traditional industries rather than large-scale spatial expansion.

Table 3. Changes in the properties of standard deviational ellipses in four major regions.

Region	Year	Centroid (x)	Centroid (y)	Axis (x)	Axis (y)	Rotation Angle
Eastern	2010	117.43	31.86	3.61	8.82	10.87
	2016	117.45	31.43	3.61	8.74	12.1
	2022	117.31	31.25	3.57	8.86	11.95
Central	2010	113.69	31.4	2.78	4.55	177.25
	2016	113.74	31.31	2.84	4.48	177.54
	2022	113.77	31.28	2.81	4.42	177.99
Western	2010	104.68	32.4	12.03	8.69	113.28
	2016	104.36	32.22	12.19	8.56	113
	2022	104.4	32.07	12.25	8.5	113.77
Northeastern	2010	125.37	44.34	0.79	5.23	38.1
	2016	125.53	44.53	0.8	5.2	37.78
	2022	125.68	44.72	0.79	5.18	37.46

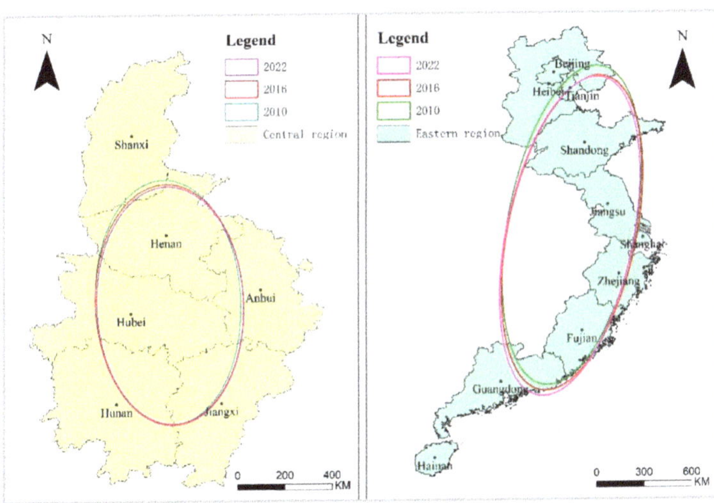

Figure 6. Spatiotemporal change trends in the eastern and central regions.

The spatial changes in new quality productivity within the western region are closely linked to resource exploitation and economic development in this region. The shift in the centroid is attributable to the continuous advancement of resource development and infrastructure construction, leading to a displacement of the center of gravity of agricultural and rural economic activities. Fluctuations in axis length and changes in rotation angle indicate that the spatial distribution of new quality productivity in the western region is constantly adjusting (Figure 7). This phenomenon occurs because the economic development of the western region is influenced by multiple factors, such as resource distribution, policy support, and market demand. Consequently, the growth of new quality productivity exhibits a certain degree of uncertainty.

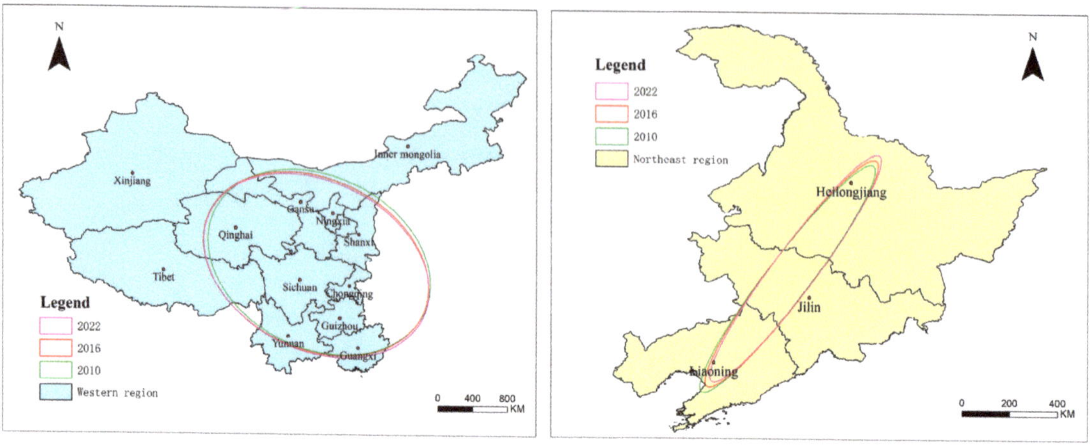

Figure 7. Spatiotemporal change trends in the western and northeastern regions.

The spatial variations in new quality productivity within the northeastern region reflect its efforts and exploration during industrial transformation. The relatively small change in axis lengths indicates that the core area of new quality productivity in this region remains stable. Moreover, the decrease in the rotation angle suggests that the economic development direction of the northeast region is gradually adjusting, placing greater emphasis on cooperation and coordinated development with neighboring regions.

5. Discussion

Research on new quality agricultural and rural productivity is vital for food security, rural economic development, and urban–rural integration. With rising food demand, traditional agriculture faces resource and environmental constraints [30–34]. New quality productivity can drive rural economic growth and narrow the urban–rural gap [35].

The utilization of entropy and spatial analysis methods allows for a comprehensive analysis of the spatiotemporal evolution trends of new quality productivity in agriculture and rural areas. The weights determined by the entropy method can help accurately measure the comprehensive levels of new quality productivity and identify key influencing factors. The spatial agglomeration characteristics revealed by exploratory spatial analysis provide a basis for formulating regional coordinated development policies and promoting the optimal allocation of resources [36]. Additionally, the distribution pattern shown by the standard deviational ellipse assists in planning infrastructure construction, optimizing the industrial layout, and promoting the balanced development of new quality productivity in agriculture and rural areas.

This study reveals that the development of new quality productivity in China's agriculture and rural areas is characterized by spatial heterogeneity and dynamic evolution. In combination with the rural revitalization policy, to promote its balanced development, in terms of policies, it is necessary to strengthen regional coordination to facilitate the cross-regional flow of technological factors. Differential investments should be made in infrastructure to address the development conditions of regions with weaknesses. By optimizing fiscal and financial policies, the endogenous innovation momentum can be stimulated. A two-way talent flow mechanism should be established to break the dilemma of the "siphon effect". Ecological protection and productivity improvements should be coordinated to avoid extensive development. The coordination of multi-dimensional policies can not only provide strong support for the implementation of the rural revitalization strategy but also achieve the coordinated development of new quality productivity.

Due to limitations in workload and time, this paper has not explored the spatial imbalances and its influencing mechanisms of new quality productivity in agriculture and rural areas. In future work, we intend to utilize the spatial Gini index and the geographically weighted regression model to further examine these characteristics.

6. Conclusions

This study evaluated the spatiotemporal evolution of new quality productivity in agriculture and rural areas across 31 Chinese provinces. Our key findings include the following:

At the provincial level, the development level of new quality productivity in China shows a decreasing trend from east to west. Economically, eastern provinces usually have a more developed economic foundation, advanced industrial structures, and comprehensive infrastructure, which provide favorable conditions for the development of new quality productivity. In contrast, western provinces have relatively low levels of economic development, more singular industrial structures, and less developed infrastructure construction, leading to numerous challenges in the advancement of new quality productivity.

There is significant spatial autocorrelation of new quality productivity in agriculture and rural areas across the country; however, this significance is gradually weakening, indicating that internal disparities are diminishing. With the continuous increase in policy support and proactive efforts in industrial transformation and upgrading, western provinces are expected to gradually narrow the gap with the eastern region in the development of new quality productivity.

Although the new quality productivity of agriculture and rural areas in China's four major regions has continued to grow, regional disparities remain significant. The eastern region has outstanding advantages, the central region is making steady progress, the western region is continuously catching up, and the northeastern region is facing greater development pressure.

Author Contributions: Conceptualization, Xiangyang Cao; methodology, Xiangyang Cao; software, Jiahui Lei and Donghui Shi; validation, Wenlong Yu; formal analysis, Tianhui Tao; investigation, Xun Zhang; resources, Xiangyang Cao; writing—original draft preparation, Xiangyang Cao; writing—review and editing, Wenlong Yu; supervision, Wenlong Yu and An Wang; funding acquisition, Xiangyang Cao and Wenlong Yu. All authors have read and agreed to the published version of the manuscript.

Funding: This research was supported by the General Program of the National Natural Science Foundation of China (Grant No. 42301349), the Shandong Provincial Natural Science Foundation Youth Project (Grant No. ZR2022QD146), and the National Statistical Science Research Project (No. 2023LY084).

Data Availability Statement: The datasets generated and analyzed during the current study are not publicly available due to data confidentiality and privacy concerns, but they are available from the corresponding author upon reasonable request.

Conflicts of Interest: All authors of this article declare that there is no conflict of interest.

References

1. Hickel, J.; Lemos, H.M.; Barbour, F. Unequal exchange of labor in the world economy. *Nat. Commun.* **2024**, *15*, 6298. [CrossRef] [PubMed]
2. Li, C. Study on the transformation of China's economic growth mode and growth sustainability. *Inf. Syst. Econ.* **2024**, *5*, 6132–6257.
3. Coffman, D.M. François Quesnay, Luigi Pasinetti and the historical contexts of economic theory. *Struct. Change Econ. D.* **2021**, *56*, 64–73. [CrossRef]

4. Matthew, S. Adam Smith on growth and economic development. *Hist. Econ. Rev.* **2023**, *86*, 2–15.
5. Sun, D.C. Ricardo and his "Principles of political economy and taxation". *Hist. Teach.* **1982**, 35–37.
6. Marshall, A. Karl Marx's Grundrisse: Foundations of the critique of political economy 150 years later. *Critique* **2010**, *38*, 329–331. [CrossRef]
7. Boyle, S.M. Performance and value: The work of theatre in Karl Marx's critique of political economy. *Theatre Surv.* **2017**, *58*, 3–23. [CrossRef]
8. Ince, U.O. Friedrich List and the Imperial origins of the national economy. *New Political Econ.* **2016**, *21*, 380–400. [CrossRef]
9. Amato, L.H.; Cebula, R.; Connaughton, E. State productivity and economic growth. *Reg. Stud. Reg. Sci.* **2022**, *9*, 180–203. [CrossRef]
10. Gao, F. The logic, multidimensional connotation, and contemporary significance of the proposal of "new quality productivity". *Political Econ.Rev.* **2023**, *14*, 127–145.
11. Li, Z.; Liao, X.D. The triple logic of theory, history, and reality in developing "new quality productivity". *Political Econ. Rev.* **2023**, *14*, 146–159.
12. Zhou, W.; Xu, L.Y. On new quality productivity: Connotation characteristics and important focus. *Reform* **2023**, *19*, 1–13.
13. Du, C.Z.; Shu, S.; Li, Z.H. Mechanism analysis and implementation path of promoting high-quality economic development through new quality productivity. *Econ. Rev.* **2023**, 20–28.
14. Wei, C.H. The basic meaning, historical evolution, and practical path of new quality productivity. *Theory Reform* **2023**, *6*, 25–38.
15. Wang, Q.M.; Yang, J.G. Research on digital new quality productivity and high quality development of agriculture in China. *J. Shanxi Norm. Univ.* **2023**, *52*, 61–72.
16. Jiang, Y.M.; Qiao, Z.Y. New quality productivity: Logic, Connotation, and Path. *Soc. Sci. Res.* **2024**, *1*, 10–18.
17. Jiang, C.H.; Jin, Z.W. Empowering new quality productivity with education: Theoretical logic and practical path. *Chongqing High. Educ. Res.* **2024**, *12*, 108–117.
18. Zhang, J.Y.; Xu, Z. Theoretical review, logical dialysis and practical path of new quality productivity from the perspective of Chinese path to modernization. *Xinjiang Soc. Sci.* **2024**, 34–45.
19. Bhakta, I.; Phadikar, S.; Majumder, K. State-of-the-art technologies in precision agriculture: A systematic review. *J. Sci. Food Agric.* **2019**, *99*, 4878–4888. [CrossRef]
20. Zhao, G.; Liu, S.; Lopez, C.; Lu, H.; Elgueta, S.; Chen, H.; Boshkoska, B.M. Blockchain technology in agri-food value chain management: A synthesis of applications, challenges and future research directions. *Comput. Ind.* **2019**, *109*, 83–99. [CrossRef]
21. Akkem, Y.; Biswas, S.K.; Varanasi, A. Smart farming using artificial intelligence: A review. *Eng. Appl. Artif. Intell.* **2023**, *120*, 105899. [CrossRef]
22. Zhao, Y.; Wang, Q. Research on the sustainable development of agricultural industry based on the pilot project of ecological circular agriculture. *Mod. Agric. Res.* **2024**, *30*, 36–39.
23. Lu, J.; Guo, Z.; Wang, Y.P. The development level, regional disparity and promotion path of new quality productivity. *J. Chongqing Univ.* **2024**, *30*, 1–17.
24. Zhu, F.X.; Li, R.X.; Xu, X.X.; Sun, J.C. Construction and spatiotemporal evolution of new quality productivity indicators in China. *J. Ind. Technol. Econ.* **2024**, *43*, 44–53.
25. Wang, Y.; Wang, R.J. New quality productivity: Index construction and spatiotemporal evolutio. *J. Xi'an Univ. Financ. Econ.* **2024**, *37*, 31–47.
26. Dong, X.Y.; Zhang, C.; Liao, M.Z. Theoretical connotation and empirical evaluation of new-quality productive forces. *Shenzhen Soc.Sci.* **2024**, *7*, 5–15.
27. Guo, X.M.; Lv, Z.F. The connotation characteristics, development blockages and practical paths of new-quality agricultural productivity. *Acad. J. Zhong Zhou* **2024**, 38–45.
28. Qi, W.H.; Zhao, C.; Su, Z. Research on the development path of new-quality productive forces based on the four "new" dimensions. *J. Lanzhou Univ.* **2024**, *52*, 15–24.
29. Yu, Z.; Qing, S.L.; Zhen, L. Research on gradual planning scheme of power grid based on AHP and entropy weight method. *Electron. Lett.* **2024**, *60*, 21–26. [CrossRef]
30. Bjerkebæk, I.; Toftaker, H. Reliability assessment combining importance resampling and the cross entropy method. *Electr. Power Syst. Res.* **2024**, *234*, 110722. [CrossRef]
31. Diawara, N.; Waller, L.; King, R.; Lorio, J. Simulations of local Moran's index in a spatio-temporal setting. *Commun. Stat. Simul.Comput.* **2019**, *48*, 1849–1859. [CrossRef]
32. Bostjan, P.; Robert, B.; Jozef, R. Evaluation of load frequency control performance based on standard deviational ellipses. *Trans. Power Syst.* **2017**, *32*, 2296–2304.
33. Rahmaniati, M.; Eryando, T.; Susanna, D.; Pratiwi, D.; Nugraha, F.; Riandi, U. The utilization of standard deviational ellipse (SDE) model for the analysis of dengue fever cases in Banjar City 2013. *Aspirator J. Vector Borne Dis. Stud.* **2014**, *6*, 21–28.

34. Zhou, J. Ensuring food security with new quality productivity: Internal logic, opportunities, challenges and countermeasures. *Econ. Rev. J.* **2024**, 31–40.
35. Chen, H.Q.; Zeng, F.S. Research on the impact of new quality agricultural productivity on the development of agricultural modernization. *Agric. Econ. Manag.* **2024**, 27–41.
36. Shi, X.T.; Yu, Z.Y. Measurement, structural decomposition and spatial convergence analysis of the regional new quality productivity level in China. *Ind. Technol. Econ.* **2024**, *43*, 90–99.

International Journal of
Geo-Information

MDPI

Article

Developing a Methodology for Assessing Visual and Environmental Sensitivity for Agrivoltaics Land Suitability Projects: The Case Study of Viterbo Province (Italy)

Ilaria Angelelli [1], Daniele Codato [2,*], Salvatore Eugenio Pappalardo [2] and Massimo De Marchi [2]

1 Independent Researcher, 00125 Rome, Italy; angelelli.ilaria@gmail.com
2 Department of Civil, Environmental and Architectural Engineering, University of Padua, 35131 Padua, Italy; salvatore.pappalardo@unipd.it (S.E.P.); massimo.de-marchi@unipd.it (M.D.M.)
* Correspondence: daniele.codato@unipd.it

Abstract: The transition to renewable energy is crucial for combating climate change but faces challenges like local socio-environmental impacts and territorial conflicts. Scientific research on mapping renewable energy suitability areas and identifying socio-culturally and environmentally sensitive zones is essential to guide project siting appropriately. This study proposes a replicable methodology to analyze scenarios and compare alternatives for agrivoltaics plant siting, using the province of Viterbo, Italy, as a case study. The methodology employs spatial data, thematic maps, and multi-criteria analysis in open-source GIS software to identify suitable solar belts, map environmental sensitivity through 14 criteria, and assess visual sensitivity based on proximity to landscape elements. The resulting workflow and customizable QGIS models provide a comprehensive, transparent decision-support tool to optimize agrivoltaics deployment while minimizing impacts and enhancing acceptance. Mapping multi-factor sensitivity offers crucial insights for sustainable planning and design. The Viterbo case study illustrates the 'conflict between green alternatives' where renewable energy development potential clashes with environmental and landscape protection needs. The analysis reveals significant spatial variability in suitability and sensitivity among the province's municipalities. The study highlights the importance of a nuanced approach to assessing suitability, moving beyond simple binary classification, and provides a tool adaptable to different regulations and contexts.

Keywords: agrivoltaics; sensitivity map; suitability map; spatial multi-criteria decision analysis; GIS

1. Introduction

Climate change fostered by anthropic activities is recognized worldwide as the main threat to human and non-human life, undermining the basis of human and natural systems [1]. The energy sector has emerged as the primary focal point for climate mitigation efforts [2], with national and international policies increasingly geared towards transitioning to a carbon-neutral global economy, including initiatives like the US Inflation Reduction Act, or the European Commission's REPowerEU plan [3–5]. This transition necessitates a substantial expansion of renewable energy sources, particularly wind and solar power [6]. Projections from the International Energy Agency suggest that by 2028, renewable energy will constitute over 42% of global electricity generation, with wind and photovoltaic energy production expected to double, reaching 25% [3]. However, the environmental urgency of scaling up renewable energies requires appropriate strategies for safeguarding the planet's biodiversity and ensuring sustainable human development.

Despite their benefits, renewable energy sources have raised several concerns, particularly regarding their impacts on ecosystems, communities, land occupation, and scenic landscapes at local and regional scales. Potential socio-environmental impacts include habitat fragmentation, soil degradation, water resource depletion, and disturbance to local ecosystems and wildlife. Additionally, large-scale solar panel system arrays on the landscape can be a source of concern for local communities and stakeholders involved in agriculture, tourism, and cultural heritage preservation [7–10].

Amid the development of sustainable energy alternatives to cope with these possible impacts of renewable energy production, agrivoltaics systems (hereafter agriPV) are gaining momentum worldwide as a promising solution to address the growing demand for renewable energy and food security. In these dual-purpose installations, photovoltaic panels are strategically elevated at a significant height above the ground, typically several meters, or sufficiently interspaced. This deliberate positioning creates a spacious area underneath or between the panels, allowing for the continuation of conventional farming practices while simultaneously reducing carbon footprint and increasing economic opportunities for farmers [11–14]. Yeligeti [15] recently estimated a global potential of 217 Terawatts production on 4.64 million km^2 of agricultural lands, according to the feasibility and profitability of several crops. However, the implementation of these projects, particularly in the form of large-scale industrial ones, could perpetuate most of the significant environmental and social impacts mentioned above, if not properly planned and managed. Although scientific investigations and currently implemented projects are promising in reducing competition for land and safeguarding food security, and while other potential benefits for crops such as reduced water evaporation, heat stress mitigation, and protection from extreme weather are encouraging [13,14,16], from a social perspective, the siting of agriPV is not exempt from conflicts with local communities if their concerns and values are not adequately considered. Issues such as land-use changes, potential impacts on traditional agricultural practices and agroecosystems or nearby natural and semi-natural ecosystems, and the perceived disruption of rural landscapes can generate opposition and hinder project acceptance [17,18].

Consequently, there has been a growing interest in developing suitability analyses and maps and identifying the best site locations for agriPV projects, together with the analysis and definition of areas socio-culturally and environmentally sensitive [19] to the potential impacts of these projects, with the aim of minimizing potential conflicts and impacts and addressing agriPV to the right places.

Recent research in this field has employed a combination of Geographic Information Systems (GIS) and Multi-Criteria Decision Analysis (MCDA, hereafter spatial MCDA or sMCDA) methodologies, by overlaying and weighting diverse geographical criteria: most investigations, related to ground-mounted PV or agriPV, combine technical and economic factors, such as solar energy potential, crop production, and closeness to power lines and/or socio-cultural and environmental aspects, such as protected areas, urban centers, and cultural heritage sites [11,20–23].

Singh Rathore and Kumar [24] highlighted the significant and yet underutilized potential of GIS, broadly speaking, in supporting all phases of Environmental Impact Assessment (EIA) studies, typically mandatory for agriPV projects, ranging from initial screening to the comprehensive evaluation of visual, environmental, and social impacts.

sMCDA can also incorporate experts' opinions, stakeholder engagement, and public participation processes, ensuring that the concerns and values of local communities, environmental organizations, and other relevant stakeholders are taken into account.

Within the European efforts for decarbonization, for Italy, one of the European nations that is consolidating itself as a major player in solar energy growth thanks to its geographical position, it is crucial to establish effective regulation for the renewable energy sector in a historical moment of exponential growth of project proposals, fueled by European funding. In 2021, Legislative Decree 199, transposing Directive (EU) 2018/2001 on the promotion of renewable energy resources (RES), introduced the concept of solar belts as part of Italy's strategy to accelerate the development of renewable energy. Solar belts are preferred areas identified by technical and environmental criteria for the installation of agriPV systems, which benefit from an expedited authorization process [25].

Recently, the issuance of two new key decrees between May and June 2024 [26,27] marks a fundamental step in the regulatory framework for RES installations, defining agriPV projects as the unique kind of photovoltaic RES permitted in agricultural lands and delegating sub-administrative units to define suitable areas. In particular, the Ministerial Decree of 21 June 2024, which entered into force on 3 July, set a deadline of 180 days for the adoption of regional laws on 'suitable areas', which expired in the first week of January 2025. However, as of the date of this study (September 2024), only five regions (Calabria, Lombardy, Piedmont, Apulia, Sardinia) of 22 had legislated on suitable areas, leaving the mapping incomplete and continuing to represent a significant obstacle [28].

Current regulations still only provide written texts without including official regional or sub-regional maps. In the absence of these, private operators are forced to independently prepare and submit their project plans, along with the necessary documentation for EIA, exposing the process to the risk of inaccuracies and divergent interpretations, causing delays.

Lately, Fattorusso [20] has proposed a methodology for agriPV land eligibility analysis using sMCDA at Italian regional scale, highlighting the necessity to investigate sMCDA methodologies and explore criteria useful to make agriPV sustainable not only as a food–energy nexus but also considering social and cultural aspects and the surrounding landscape.

In this framework, the main objective of this study is to develop and test a replicable methodology for assessing and mapping suitable areas for agriPV siting in the Italian territory, considering aspects of environmental and visual sensitivity. Specific objectives are to (1) develop a methodology for assessing suitability of a territory according to legislation; (2) develop a methodology for assessing environmental sensitivity to agriPV projects; (3) develop a methodology for assessing visual sensitivity to agriPV projects; and (4) test and validate the proposed methodologies using the Viterbo province (Lazio region) as a case study.

2. Materials and Methods

This research uses a combination of geospatial data and analysis, thematic maps, and sMCDA to identify areas suitable for the installation of agriPV projects. This is achieved by defining suitable belt areas based on Italian legislation (objective 1 in Figure 1) and integrates both environmental (objective 2 in Figure 1) and visual sensitivity (objective 3 in Figure 1) analysis.

The process begins with the collection and analysis of data, which are then integrated into the open-source GIS software QGIS (version 3.28) [29] to create spatial models that serve as the basis for further spatial analyses. The database for the project consists of an open-source spatial dataset, incorporating information available for most Italian regions. To ensure comparability, the datasets were reprojected to the EPSG: 32633–WGS 84/UTM Zone 33N reference system. Spatial overlay and distance analysis were performed using geoprocessing functions such as difference, buffering, and clipping, within the boundaries of the study case, i.e., the Viterbo province (objective 4).

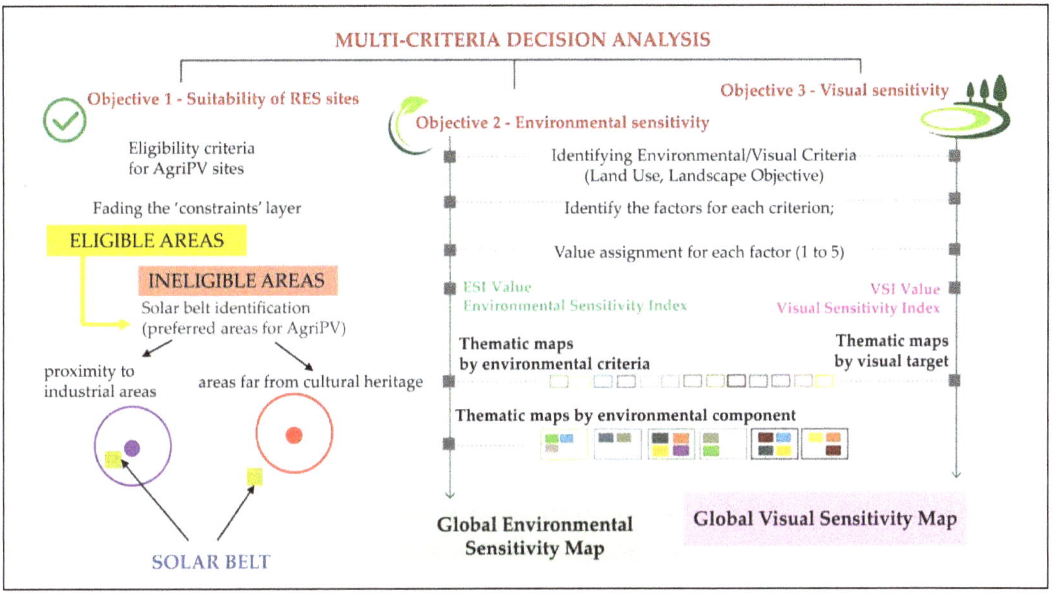

Figure 1. The figure shows the workflow concerning the three project objectives. The diagram illustrates the sequential stages that led to the final outputs.

2.1. *Workflow for Suitability Areas, Environmental and Visual Sensitivity Mapping*

Figure 1 shows the workflows diagram. We extensively used the QGIS model designer tool, which allows chaining several geo-operations in a unique model (hereafter GIS model) to simplify the workflows for criteria and index production and easily adapt the procedure to different contexts and changes in legislation.

To achieve the first objective, a workflow was developed to identify the solar belts for the installation of agriPV plants in compliance with Italian legislation, on the basis of key criteria such as distance from industrial zones and the absence of landscape constraints.

To assess the environmental sensitivity (second objective) of potential sites with the least impact for the construction of agriPV, a mapping process was conducted using a set of environmental criteria, including biodiversity protection, soil and subsoil quality, water resources, and landscape.

Each criterion was assigned a numerical value on a scale from 1 (less sensitive) to 5 (highly sensitive), known as the Environmental Sensitivity Index (ESI, see Table 1). This scale, calibrated on the basis of the experience of industry professionals and sectoral guidelines [30], facilitates the identification of the most sensitive areas, enabling the planning of appropriate mitigation measures to minimize potential environmental impacts. Additionally, progressive buffers were established around the most sensitive resources, ensuring further protection from disturbances related to the installation of renewable energy systems.

The assessment of the visual impact (objective 3) of an agriPV system was conducted in accordance with the Ministry of Culture's guidelines [31]. The analysis focused on key elements of visual sensitivity, identifying and mapping three main components: scenic roads, historic centers, and urban areas. Geoprocessing with progressive buffers were applied to these elements. A Visual Sensitivity Index (VSI) on a scale from 1 to 5 (see

Table 1) was assigned to each component to quantify the landscape value and the potential visual impact of the proposed intervention.

To standardize and quantify the data, each vector factor was then converted into raster maps of the same spatial resolution (100 × 100 m), based on the values from 1 to 5 assigned to each raster cell. This process identifies homogeneous areas, measurable in hectares (the standard commercial unit in the energy sector). Finally, criteria were combined using the linear weighted sum method, one of the most used methods in sMCDA, which combines multiple criteria scores by assigning weights to each criterion and summing the products, providing a straightforward approach to evaluate and rank alternatives based on their overall performance across various factors [32].

Table 1. Classification of ESI and VSI values.

Definition	ESI	Definition	VSI
Negligible environmental sensitivity	1	Negligible visual sensitivity (background level)	1
Low environmental sensitivity	2	Low visual sensitivity, where outlines and shapes of large masses are predominant	2
Medium environmental sensitivity	3	Medium visual sensitivity, where texture, color, and chiaroscuro effects are the most important	3
High environmental sensitivity	4	High visual sensitivity, where changes in structure and individual elements can be seen against a background	4
Very high environmental sensitivity	5	Very high visual sensitivity, distinguishing individual elements and perceiving multisensory factors	5

To conclude the process, a global assessment tool was designed to facilitate the comparison of results for decisionmakers. A single database was created in QGIS by calculating the average value of all pixel values for each factor and for EVI and ESI and the percentage of suitable or not suitable land for agriPV plant installation within each municipality of Viterbo Province. To preserve the nuances associated with local specificities, it was decided not to round the values but to retain the decimal points. This approach enables the comparison and functional ranking of the results among the 60 municipalities.

2.2. Study Area and Data Input

To test the methodology outlined, the province of Viterbo (Lazio region, see Figure 2, while the geographical framework of the province is available in Figure S1 of the Supplementary Materials) was selected as a case study (objective 4), as it serves as an exemplary case of territorial conflict between environmental protection and the goals of agriPV projects.

The Province of Viterbo, currently a central hub for renewable energy development in Italy, ranks first among Italian provinces in terms of the highest installed power per capita, with 2.49 kW per person. Additionally, it accounts for 80% of the renewable energy measures approved in the Lazio region in 2023 [33].

The potential associated with the area's high solar radiation can be found in Figure S31 of the Supplementary Materials related to the latest research from the Italian National Agency for New Technologies, Energy and Sustainable Economic Development (ENEA) [34].

The province covers an area of 3615 km^2 with a population of 308,187 and is characterized by low population density (85 inhabitants/km^2), one of the lowest in the region (300 inhabitants/sq km) and in the country (200 inhabitants/km^2) [35]. The distribution of land use in the Province of Viterbo is characterized by the predominance of agricultural areas (about 70%), followed by woods and forests (about 25% of the province and 21% of Lazio). Built-up areas and infrastructure occupy a smaller percentage of the territory, about 5%, according to land use land cover (LULC) updated in 2018 (see Table 2).

Figure 2. The figure shows the location of the province of Viterbo, highlighted with red fill, considered as the case study, within the Italian territory.

Table 2. Input data for the configuration of the Territorial Database.

Target	Source	Format	Year	Scale
Constraints (Table S2)	Map portal of the Lazio region/ Regional Landscape Plan of Lazio	Vector data in shapefile format	2021	1:25,000
Geomorphological and hydraulic hazards Aspects (Table S4)	District Basin Authority of the Central Apennines (Italy)	Vector data in shapefile format	2021	1:25,000
Protected natural areas (constraints) Natura 2000 network areas, protected national parks SCI (Site of Community Importance) and SPA (Special Protection Area)	Map portal of the Italian Ministry of the Environment	Vector data in shapefile format	2023	1:25,000
Landscape areas, Landscape roads Archaeological heritage, Historic centers (Table S5)	Regional Landscape Plan of Lazio	Vector data in shapefile format	2021	1:5000
Historical and architectural heritage (Table S3)	WebGIS of the Italian Ministry of Culture ('Vincoli in rete')	WMS	2024	1:25,000
Land Use Land Cover (LULC)	Map portal of the Lazio region	Vector data in shapefile format	2018	1:25,000

Table 2. *Cont.*

Target	Source	Format	Year	Scale
Hydrographic network	Map portal of the Lazio region	Vector data in shapefile format	2019	1:125,000
Fauna data	Map portal of the Lazio region	Vector data in shapefile format	2002–2016	1:25,000
DTM (Digital Terrain Model)	National Institute of Geophysics and Volcanology (INGV)	Raster—GeoTIFF	2023	10 m/pixel

While this situation highlights the great potential of the area, it has also led to growing tensions between institutions, citizens, and operators in the sector. The high potential of the territory for renewable energy production is in conflict with the need to protect valuable environmental and landscape heritage. In the case under consideration, approximately 68% of the area is constrained, creating a complex situation where different interests are pitted against each other.

Table 2 summarizes the main dataset used for the construction of the geodatabase of the study area, including sources, format, publication year, and the level of detail of the original data. For a detailed presentation of the dataset used, please refers to the Tables S2–S5 contained in the section 'Detailed information of Table 2', 'Input data for the configuration of the Territorial Database' in the Supplementary Materials.

3. Results

This section has been divided into three parts, corresponding to the three objectives of the project, using the case study of Viterbo province (obj. 4) as an example of the possible results produced by the analysis.

3.1. Suitability of AgriPV Sites and Solar Belt

To achieve the first objective, two GIS models were created. The first one is presented in Figure 3, while the second one is available in Figure S30 of the Supplementary Materials.

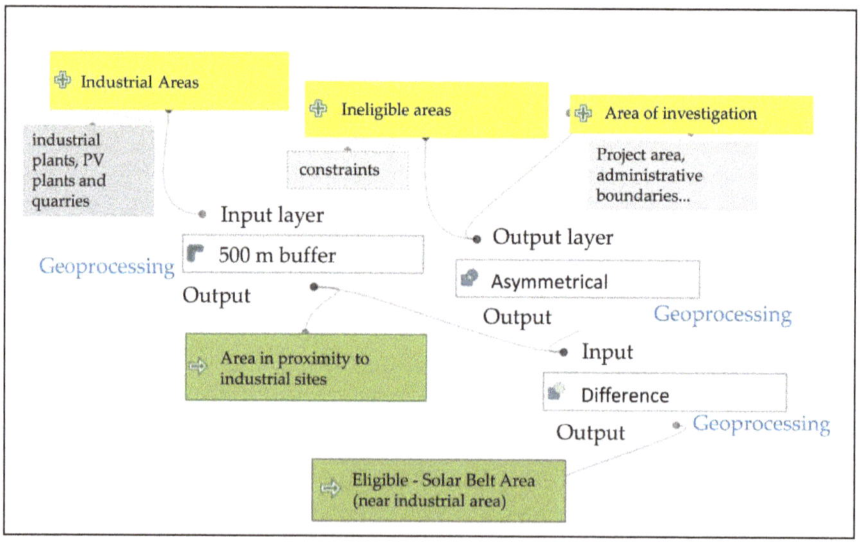

Figure 3. The figure illustrates the QGIS GIS model and the geoprocessing tools used to identify the solar belts, preferred areas due to their proximity to industrial zones.

The first model identifies suitable land located in priority areas related to the proximity of industrial installations. The main steps of the process are as follows:

Step 1. Identification of spatial constraints (unsuitable areas), including protected areas, nature parks, nature reserves, and other restricted areas. The constraints were merged into a single dataset representing the 'unsuitable areas' (Figure S3 of the Supplementary Materials).

Step 2. Data extraction (industrial zones): data on industrial plants were extracted from the productive plants category of the LULC of the Lazio region. These data constitute the starting point for the creation of the buffer and the subsequent spatial analysis (Figure S1 of the Supplementary Materials).

Step 3. Buffer creation: a 500 m buffer was created around the industrial plants. This buffer helps identify areas around industrial plants that, according to the regulations, are suitable for the installation of agriPV systems.

Step 4. Subtraction of constraints from the study area: once the dataset of non-suitable areas was created, a subtraction process was carried out with the entire study area. This resulted in the identification of suitable areas for the installation of agriPV systems (Figure S2 of the Supplementary Materials).

The second graphic model refers to another criterion established by the Italian legal framework, which identifies preferential areas for agriPV systems in the absence of cultural assets. To this end, data on cultural assets from the Regional Territorial Landscape Plan and the Ministry of Culture's Webgis were integrated.

The resulting areas, which have no constraints and are located within the industrial plants buffer or sufficiently far away from historical elements, were named the solar belts.

A map of areas potentially suitable for the installation of the agriPV System for Viterbo province was then drawn up by applying the two graphic models and on the basis of the legislation in force at the time of the analysis (an extract of the map is shown in Figure 4; the whole map is presented in Figure S4 of the Supplementary Materials). It is important to note that the RES legislation in Italy is constantly evolving, and the use of models allow the adaptability and replicability in different contexts and according to future changes in Italian regulations. In fact, for example, the decree (Ministerial Decree of 21 June 2024) requires regions to define their own buffer zone (ranging from 500 m to 7 km) around cultural heritage sites, with the current standard set at 500 m.

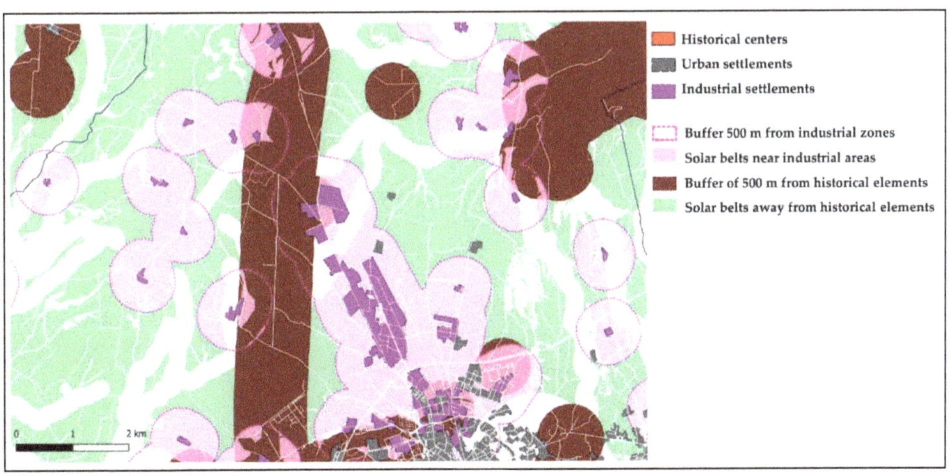

Figure 4. The figure shows an extract of the map produced by the application of the two GIS models. It illustrates the solar belts for the installation of the agriPV systems, with an example in the

municipality of Viterbo, considering the presence of historical road networks and other historical features, as well as a large industrial zone. The complete map is shown in the Supplementary Materials as Figure S4.

Maps such as the one shown in Figure 4 can provide a useful tool for various stakeholders: from investors who can quickly visualize where to prioritize investments, to environmental consultants who can clearly identify areas with simplified authorization procedures, to local public authorities for their planning activities. The solar belts close to industrial zones offer a unique combination of logistical benefits such as utilizing existing grid connections or repurposing brownfield sites.

3.2. Environmental Sensitivity

To assess environmental sensitivity (Objective 2), 14 thematic maps were developed using specific criteria to analyze the different environmental components that contribute to environmental impact assessment in the European context.

The factors and criteria were defined based on two main approaches:

1. the progressive spatial distance from a specific target
2. the intrinsic characteristics of landscape areas or land use, as identified in regional or national planning maps.

In the first approach, and in the absence of specific regulations defining protection and attention buffers for areas adjacent to sensitive features such as protected zones, this study has opted to adopt the 500 m buffer measure and its multiples and submultiples, in line with the provisions of Legislative Decree 199/2021 and the Decree on areas suitable for RES plants. These sectoral regulations have established that this minimum distance represents an optimal compromise in social, functional, environmental, ecological, and visual terms to ensure a balanced coexistence between the construction of agriPV installations and the protection of the analyzed features [31].

Regarding the second approach, even if agriPV projects deal with agricultural areas, in the analysis we considered some non-agricultural land covers, such as urban or semi-natural areas, as factors and not as constraints. This is necessary because of the spatial scale used to produce land-use and land-cover data or landscape categories by public administrations, which usually range between 1:10,000 and 1:100,000, and could include smaller agricultural spaces in other dominant classes, where agriPV projects could be proposed.

The following paragraphs present the workflows for the production of the individual thematic maps and their ESI value, which were then combined to produce the global maps, providing a more comprehensive overview. The datasets refer to the details provided in Table 2 and Tables S2–S5 in the Supplementary Materials.

3.2.1. Proximity to Protected Areas

The installation of agriPV systems requires careful consideration of proximity to protected areas. Although Italian regulations prohibit installation within these areas, they do not provide specific guidelines for the surrounding zones, making case-by-case analysis necessary. Specifically, for the case study, it is highlighted that the Viterbo province has an environmental and ecological heritage of great value, characterized by a variety of landscapes and a rich biodiversity. The province is home to 22 protected natural areas, including regional parks, nature reserves, and wetlands of international importance.

These areas protect a rich variety of habitats, including forests, lakes, volcanoes, coastal areas, and marshes. Sensitive areas and their ESI values were identified through the creation of three buffer zones from protected areas:

- 1500 m–1001 m (ESI value = 1): Areas located between 1501 m and 1001 m from a protected area are assigned a lower ESI value of 1. These areas are considered less sensitive in terms of their proximity to protected zones, likely because they are farther from areas under direct conservation management and may be less impacted by protection measures.
- 1000 m–501 m (ESI value = 3): As the distance decreases to between 1000 m and 501 m, the ESI value increases to 3. Areas within this range are more sensitive because they are closer to protected areas, meaning they are more likely to be influenced by the conservation measures in place. These zones could experience higher ecological pressure or benefits from the adjacent protected areas.
- 500 m–1 m (ESI value = 5): Finally, areas located within 500 m of a protected area are assigned the highest ESI value of 5. These areas are considered the most sensitive, as they are immediately adjacent to or within very close proximity to areas that are actively managed for conservation. These areas may experience the most direct ecological impacts from the protected zones, both positive (e.g., biodiversity conservation) and negative (e.g., human encroachment or disturbance from adjacent areas).

This gradation of ESI values reflects increasing environmental sensitivity as one approaches protected areas. It highlights the importance of considering proximity to conservation zones when evaluating the environmental impact or risk of human activities in a given area. The map for this criterion is presented in Figure 5a and is included in the Supplementary Materials as Figure S5.

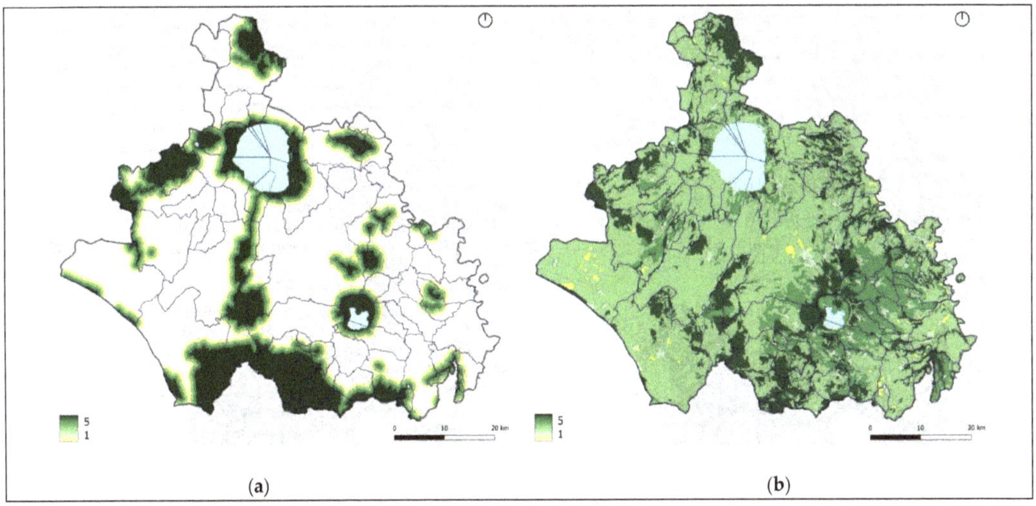

Figure 5. The figures show two examples of the thematic maps produced. The first map (**a**) shows the variation of a given parameter as a function of distance from protected areas, while the second map (**b**) shows the ESI values related to the spatial distribution of different types of LULC.

3.2.2. Proximity to the Forest and Natural System

In Italy, the installation of agriPV systems is prohibited within forested areas, though no regulations exist for the surrounding areas. Therefore, it is crucial to establish an ESI to determine the design parameters for agriPV systems based on the progressive distance from forested areas.

The forest system in the Viterbo province is a complex ecosystem that includes woods, forests, wooded areas, and arable land with trees. It is predominantly covered by oak, beech, and chestnut forests.

Most of these forests are protected by the Ministry of Culture due to their intrinsic environmental and landscape value. The implementation of renewable energy projects near wooded areas is a potential environmental impact that must be carefully managed. This can be achieved by preparing specific measures during the construction phase, as well as through targeted landscape design when selecting mitigation strategies. These measures ensure alignment with the tree and shrub species native to the area.

The assignment of the ESI index was carried out according to the procedures outlined in the previous paragraph for protected areas, based on progressive distance from the original LULC data representing the forest system Figure S6 of the Supplementary Materials).

3.2.3. Hydrogeological and Geomorphological Structure

For the characterization of the hydrogeological and geomorphological risk factors, vector data from the Central Apennine Basin Authority website in Italy were used. The classification of risk and vulnerability in the different basins of Lazio region proved to be challenging (a detailed overview is provided in Table S4 in the Supplementary Materials). The ESI values for various environmental factors are assigned as follows:

- Geomorphological Hazard–Areas to be Attended: These areas are assigned an ESI value of 3, indicating a moderate level of sensitivity that requires attention in terms of environmental management or mitigation strategies.
- Medium Hydraulic Hazard: Areas identified as having a medium hydraulic hazard are assigned an ESI value of 4, reflecting a higher sensitivity that calls for careful consideration and measures to minimize potential environmental impacts.
- High Geomorphological and Hydraulic Hazard: Areas categorized as high geomorphological or hydraulic hazard are classified as constraints, meaning they are highly sensitive and require stringent protection or management measures to prevent environmental degradation. The map can be consulted in Figure S7 of the Supplementary Materials.

3.2.4. Land-Use Classification

The assessment is based on the dataset of the LULC of 2018. The criterion used to assign a different sensitivity value to each land-use category is essentially based on the value of each area. Particular attention has been paid to orchards because, despite the fact that agriPV projects allow for the relocation of tree elements, orchards, together with wooded areas and natural environments, represent an important resource to be protected. The ESI indices have been identified as indicated in Table 3 (map available in Figure 5b and in Figure S8 of the Supplementary Materials).

Table 3. ESI value–Land use.

Factors	ESI Value
Artificial surfaces–industrial areas, mining areas, landfills	1
Artificial surfaces (residential areas, green areas)	2
Agricultural land (arable land, permanent pasture and meadow, other agricultural land)	3
Utilized agricultural area (permanent crops–vineyards–orchards–olive groves)	4
Areas with shrubs and/or herbaceous vegetation	5
Wooded and semi-natural areas	5

3.2.5. Land Capability Class

This classification, which takes into account the physical and chemical characteristics of soils and the environmental context [36], was used to identify areas unsuitable for hosting agriPV installations, in line with regional guidelines. This map is based on the Land Capability Map of the Lazio region, which divides soils into eight classes, where class I has the highest quality and class VIII the lowest. ESI values have been assigned in a progressive manner, starting with classes V, VI, VII, and VIII of lower quality with ESI 1, class IV with ESI 2, class III with ESI 4, and class II with ESI 5, considering class I as the constraint factor (Figure S9 of the Supplementary Materials).

3.2.6. Proximity to the Water System

The sensitivity of the area was considered proportional to the distance from water bodies. The dataset was created by combining layers from the regional technical map of the Lazio region. The ESI index was assigned progressively from 1 to 5, with 100 m intervals up to a distance of 500 m. The original data were derived from the hydrological grid (including primary and secondary watercourses) available within the shapefiles of the Lazio region's Landscape Plan (Figure S10 of the Supplementary Materials). Sensitivity buffers started from the end of the constrained zones around water courses established by law. In fact, the regulations establish an inviolable buffer zone of 150 m from main rivers and 10 m from minor watercourses.

3.2.7. Ecological Network

The areas of highest ecological value have been identified based on the regional ecological network plan of 2021. The network is an essential part of the regional plan of protected natural areas. The main objective is to highlight the areas of greatest naturalness and the connections between them for the purpose of establishing new protected areas and environmental assessments. The three main elements of the regional ecological network, as identified in the official documents, are as follows.

- Primary central areas: these are large natural areas with high functional and qualitative value for maintaining the vitality of target populations. They form the backbone of the ecological network. Primary core areas are characterized by centrality and are typically large in size, capable of supporting populations with high biodiversity and significant numbers. This helps reduce the risk of extinction for the populations, while also serving as a crucial source of dispersal for mobile individuals that can colonize (or recolonize) new external habitats.
- Secondary central areas: these areas, while still important for the overall ecological network, tend to be smaller or less central than primary core areas. They support biodiversity and contribute to the resilience of the system, but they may not have the same extensive functional role in supporting large populations. Secondary core areas play a supplementary role in linking and reinforcing primary core areas, contributing to habitat connectivity and acting as refuges for species, especially during periods of environmental stress or change.
- The connecting areas refer to the identified zones linking primary and secondary core areas, aligned with the specific ecosystem functions of areas important for forest species and open habitat species.

With reference to the hierarchy established by the ecological network plan of the Lazio region, the following evaluations have been assigned: the connecting areas were assigned an ESI value of 3, the central secondary areas (core) a value of 4, and the primary central areas (core) a value of 5. The choices are in line with the regional indications of the importance of the roles of the areas considered (Figure S11 of the Supplementary Materials).

3.2.8. Landscape Areas

The evaluation of this factor is based on the landscape areas defined in the Lazio landscape plan. The landscape system has been established according to the type, significance, and integrity of the landscape values present. These landscape areas form, through their morphological and geographical continuity, systems of homogeneous and recognizable elementary units within the territorial context. The map is available in Figure S12 of the Supplementary Materials. Table 4 shows the ESI values to the landscape components.

Table 4. ESI value–Landscape areas.

Landscape Areas	Valore ESI
Urban settlement landscape	1
Agricultural landscape of continuity	2
Agricultural landscape (significant value, agricultural value)	3
Natural landscape of continuity	4
Natural landscape	5

It is worth briefly introducing a discussion on high-value protected agricultural products, such as PDO (Protected Designation of Origin), PGI (Protected Geographical Indication), and DOC (Controlled Designation of Origin), among others. These designations should be included as additional criteria in the ESI analysis. However, the dataset available for the Lazio region is provided at a scale of 1:250,000, but for the specific needs of this study, which involves provincial-level analysis, a maximum scale of 1:25,000 is required. In the guidelines for agriPV plants, the Lazio region specifies that, for areas of high value, which have been analyzed at a scale of 1:250,000, detailed studies are required for an accurate assessment of individual cases.

3.2.9. Exposure

Exposure, understood as orientation to the points of the compass, is a key factor for soils. It includes a number of elements that influence the amount of solar radiation a soil receives, the microclimate that is created and, consequently, the fertility and productivity of the soil itself. Exposure acts in terms of degradation of soil structure, drought, or excessive soil erosion and affects the environmental sensitivity of a soil.

The exposure map is based on the Digital Terrain Model (DTM) at 10 m of spatial resolution. The ESI values have been assigned according to orientation (Table 5).

Table 5. Exposure map–ESI value.

Factors	ESI Value
North	1
North-West	2
East, West	3
South-West	4
South	5

In the case of agriPV systems, it is noted that south-facing exposure would offer an advantage for the system's productivity. At the same time, increased exposure results in more sensitive soils. In this context, a high sensitivity index was assigned, as these areas require careful consideration during the design phase. The thematic map for the exposure factor is included in the Supplementary Materials (Figure S13 of the Supplementary Materials).

3.2.10. Slope

Slope is an influential factor in the identification of areas of special ecological and environmental suitability, as the degree of slope depends on the function of water regulation and the risk of soil erosion.

Table 6 shows the five ESI classes with increasing values as slope increases, in proportion to the environmental sensitivity of the terrain. This environmental sensitivity is also inversely proportional to the complexity of implementing RES installations, especially in the case of large-scale installations such as agriPV. The classification is based on a combination of industry bibliography and professional experience. The thematic map for this factor is included in the Supplementary Materials in Figure S14.

Table 6. Slope–ESI value.

Factors	ESI Value
degrees 0–12°	1
degrees 13–20°	2
degrees 31–40°	3
degrees 41–50°	4
degrees 51–88°	5

3.2.11. Home Range and Core Areas of Mammals as a Function of Protection Level

Nesting sites and the presence of sensitive species must be protected, with designs carefully planned to minimize disturbance to wildlife.

The evaluation of the presence of mammals was performed processing the point datasets obtained from the Lazio mammal atlas project [37], a valuable resource providing information on the distribution and conservation status of species in the region. To create area layers, we performed the Kernel Density Estimation (KDE) method [38]. This method allows for estimating the probability of an animal's use of an area. In the KDE analysis, two percentage thresholds, 95% and 50%, are used to distinguish two areas within the territory frequented by an animal: the core area and the home range.

The core area represents the zone of maximum use by the animal, where the majority of its activity takes place, while the home range includes the entire area used by the animal for its movements and activities. The home range encompasses all areas where there is even a minimal probability of finding the animal. This condition is influenced by factors such as food, water, shelter availability, and the presence of predators or competitors. The application of the KDE method was carried out using a model builder available on sector-specific websites, updated in 2022 [39].

The ESI values assigned to home range and core areas are shown in Table 7. In this context, even non-protected species were assigned a level 1 of ESI.

Table 7. Home range and core areas of mammals as a function of protection level–ESI value.

Factors	ESI Value
Home range–No-protected species	1
Home range–Medium protection level species	2
Home range–High protection level species	3
Core area–Medium protection level species	3
Core area–High protection level species	4
Core area–Highest protection level species	5

The map produced with the application of the criteria presented in this paragraph is available in Figure 6a and in Figure S15 of the Supplementary Materials.

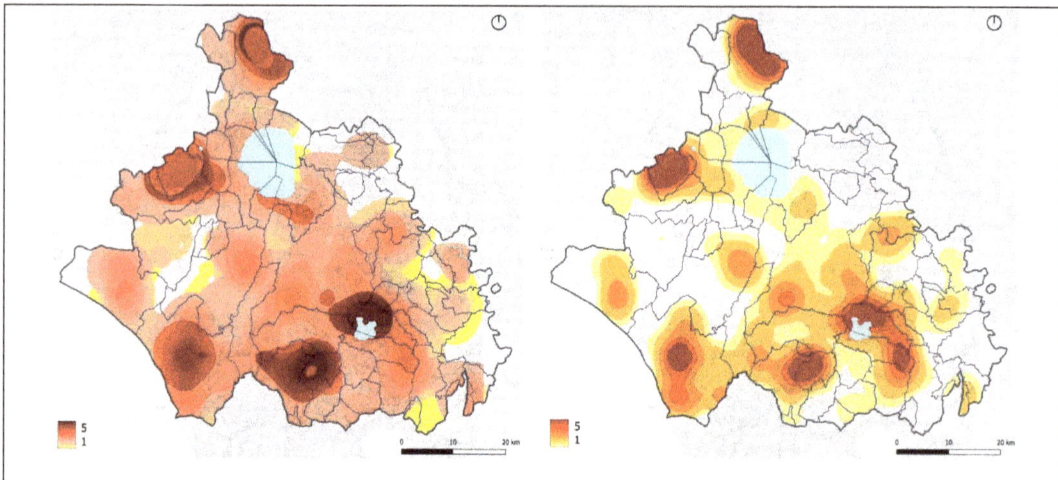

(**a**) (**b**)

Figure 6. These maps illustrate the results of the home range and core area analysis of medium-sized mammals (**a**) and micromammals (**b**), obtained using kernel density estimation.

3.2.12. Home Range and Core Areas of Micromammals

The management of a renewable energy facility requires a careful approach to wildlife conservation, particularly microfauna. Although areas designated for agriPV systems are often already fenced off, limiting the impact on larger species, microfauna, such as small mammals and insects, are highly mobile and more vulnerable to changes in land use. It is therefore essential to identify the main preferred corridors of microfauna so that the links to be protected can be considered at the design stage.

Starting from the data and the procedure described in the previous paragraph, the data that could be attributed to the home range of the micromammals were selected and processed into a new dataset to produce the map in Figure 6b and Figure S16 of the Supplementary Materials.

3.2.13. Bird-Nesting Probability

Birds play a crucial role in agricultural ecosystems, contributing to pest control and seed dispersal, but they may also be at risk of colliding with photovoltaic panels or being disturbed by environmental changes caused by the system's installation.

The map was produced using data from the Atlas of Breeding Birds of Lazio. In particular, data on the nesting probability of the various species were analyzed, obtaining a density map showing the areas with the greatest concentration of breeding birds. The ESI value was assigned based on this map, considering both the density of sightings and the probability of nesting. For the definition of home range and core, please refer to the paragraphs above. The map is shown in Figure S17 of the Supplementary Materials, while Table 8 presents the ESI values.

Table 8. Bird-nesting probability–ESI value.

Factors	ESI Value
Core–Possible nesting	1
Home range–Probable nesting	2
Core–Probable nesting	3
Home range–Nesting certain	4
Nesting certain	5

3.2.14. Probability of Birds of Prey Nesting

Following on from the above, raptor nesting has been included in the factors, based on data available on the Lazio region website. The dataset shows the distribution of birds of prey on a 10 × 10 km grid. Raptors are sensitive species and, as such, it is necessary to maintain an approximate range during the mapping phases. The assignment of the ESI value is a function of the probability of nesting (certain, probable, and possible) in Table 8. The map is available in Figure S18 of the Supplementary Materials.

3.2.15. Thematic Maps of Environmental Sensitivity

The 14 thematic maps were overlaid and their values summed to produce six thematic summary maps that group the criteria into homogeneous environmental components (soil, water, biodiversity/fauna, biodiversity/avifauna, biodiversity/flora, and landscape; see Table 9 and the maps in Figures S19–S24). In the final phase, these thematic summary maps were combined, resulting in an overall global ESI map (Figure 7). This map highlights the most sensitive areas and those most suitable for the installation of an agriPV system, considering the overlay of all components.

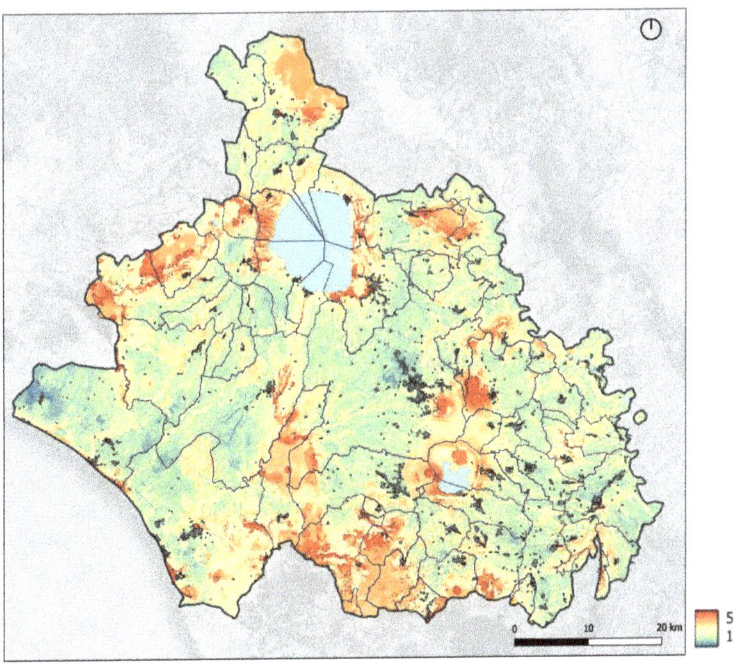

Figure 7. The map represents the global ESI map produced by overlaying the 14 thematic maps.

Table 9. Overview of the combination of thematic and global maps.

			Thematic summary maps (Environmental Components)						
			Soil	Water	Biodiversity (Flora)	Biodiversity (Fauna)	Biodiversity (Avifauna)	Landscape	Global ESI Map
Thematic maps Factors	1	Proximity to protected areas			x	x		x	x
	2	Proximity to forest systems	x		x		x		x
	3	Hydrological and geomorphological setting	x	x					x
	4	Land-use classification	x		x				x
	5	Soil capacity class	x						x
	6	Proximity to hydrographic system	x	x					x
	7	Ecological network	x		x				x
	8	Landscape areas	x		x	x	x	x	x
	9	Exposure	x						x
	10	Slope	x						x
	11	Probability of bird nesting					x		x
	12	Probability of raptor nesting					x		x
	13	Home range and core of mammals (medium size)				x			x
	14	Home range and core of micromammals				x			x

For this case study, the relative weight between criteria is set as equal, assuming all factors have the same importance. However, it is important to note that the methodology is flexible, and weights between criteria can be changed to reflect the opinions of stakeholders or to take into account different regional contexts that reflect varying territorial complexities. For example, in areas with a predominantly agricultural focus, the agricultural component may carry more weight. Conversely, in regions with significant natural habitats and ecological networks, the component related to protected areas may have a greater influence.

It is evident that biodiversity is a cross-cutting factor across all components, as it encompasses the assessment and cataloging of plant species present in a given territory, with particular focus on characterizing vegetation and plant species of ecological or conservation importance. This includes the identification and mapping of significant plant communities, as well as the analysis of vulnerability conditions arising from various pressure factors and the overall environmental degradation status. Furthermore, both the direct and indirect effects, both short term and long term, that may impact floral components and natural ecosystems are carefully examined, including potential irreversible consequences.

Evaluating environmental sensitivity through the combination of these factors is essential to ensure that agriPV projects are developed sustainably, minimizing impacts on local ecosystems and enhancing coexistence with nature. Using an approach based on the analysis of these indices allows for the identification of the most suitable areas for plant construction and the adoption of mitigation strategies to minimize negative effects on the environment.

3.3. *Visual Sensitivity*

Following the same detailed approach used for environmental sensitivity, three factors of visual sensitivity were identified and measured at receptors sensitive to the landscape component (scenic roads, historic and urban areas) according to a Visual Sensitivity Index (VSI) scale.

The VSI ranges from 1 to 5, reflecting the ability to discern landscape details at different distances. A high VSI (5) indicates very high sensitivity, allowing for the perception of individual elements and multisensory factors up to 500 m. As the distance increases, sensitivity diminishes, becoming negligible beyond 5000 m, where only a background plane is visible.

Each dataset produced for every criterion was then rasterized, assigning the VSI index to each cell to generate the specific criteria maps and subsequently the global map.

In constructing the visual sensitivity factors, in addition to the 500 m reference distance established by current regulations as a social compromise between the distance of sensitive elements and the impact of agriPV systems, it was deemed appropriate to use the visual analysis techniques outlined in the Guidelines for the Analysis, Protection, and Enhancement of the Scenic–Perceptual Aspects of the Landscape, developed in 2014 [40].

The aforementioned guidelines provide an analysis based on visibility bands. In the first plane, where the observation area is between 0 and 500 m, individual elements are distinguishable, and multisensory factors such as sounds and smells can be perceived. In the middle plane, where the observation range is between 500 and 1200 m, changes in structure and individual elements can be perceived against a background. In the next plane, known as the 'Second Plane,' with an observation range between 1200 and 2500 m, the effects of texture, color, and chiaroscuro become apparent. In the final plane, in the background, between 2500 and 5000 m, the contours and shapes of large masses are predominantly distinguishable. The distance to the horizon depends on the observer's height and the Earth's curvature. For an average person at sea level, the horizon is usually about 5 km away, and the dip of the horizon increases with altitude.

The evaluations are based on these factors related to the progressive distance between the target and the observer, and they are detailed below for each factor.

3.3.1. Proximity to Visual Systems

The landscape regulations of the Lazio region require the identification of a system of views, including vantage points, scenic routes, and visual cones. The original data are sourced from the landscape plan of the Lazio region. The dataset identifies precautionary distances from visual areas based on visual sensitivity. From these input data, a 500 m multi-buffer was generated for five iterations. Table 10 shows the generated ranges and the assigned indices. The map was generated based on panoramic views and referring to what has been presented, and is available in the Supplementary Materials as Figure S25.

Table 10. Table of distance ranges from the target criteria objects for VSI.

Range	VSI
2001–5000 m	1
1501–2000 m	2
1001–1500 m	3
501–1000 m	4
0–500 m	5

3.3.2. Proximity to Historic Centers

Historic centers are identified as privileged visual elements due to their conformation and location. The factor is based on the precautionary distances from historic centers according to visual sensitivity. The precautionary visual sensitivity zones have been assigned a VSI measured within 5 km, in accordance with what has been explained for visual systems in the previous paragraph (Table 10). The thematic criterion map is available in the Supplementary Materials as Figure S26.

3.3.3. Proximity to Urban Centers

A major challenge is to implement agriPV projects close to urban centers in a way that maximizes benefits and minimises negative impacts. This requires a careful planning approach that takes into account all environmental, social, and urban planning aspects. agriPV projects may face resistance from citizens living in the affected areas, who may perceive a negative impact on their quality of life or a change in their familiar environment.

The factor is based on precautionary distances from urban centers according to visual sensitivity. The precautionary visual sensitivity zones have been assigned a VSI based on the 500 m module identified as the basic module for measuring the minimum visual and social impact of agriPV installations on sensitive receptors. From these input data, a 100 m multi-buffer was generated for 5 circles. The VSI was determined by assigning a value from 1 to 5 to sub-multiples of 500 m (100–200–300–400–500). Refer to Figure S27 in the Supplementary Materials.

3.3.4. Global Map of Visual Sensitivity

The combination of the three criteria-based maps generated the global VSI map shown in Figure 8b and in Figure S28 of the Supplementary Materials.

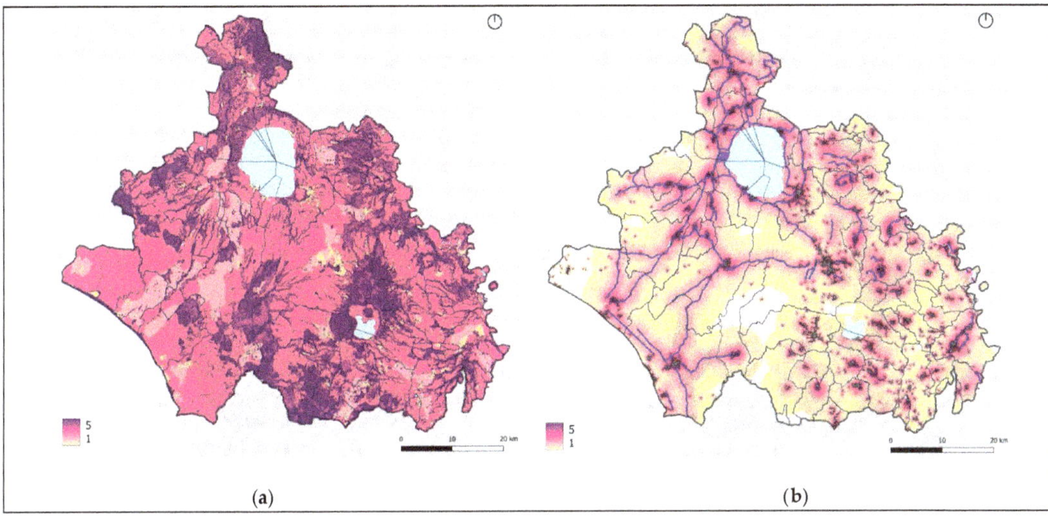

Figure 8. The figures represent a comparison between the thematic map produced according to the landscape component in terms of environmental sensitivity (**a**), see Section 3.2.8, landscape areas, and the global map produced according to the visual sensitivity criteria (**b**).

If the analysis reveals that certain portions of the territory are excessively exposed and visible, those areas can be excluded from the project, thereby reducing the overall visual impact and allowing for more suitable design solutions in those zones. Furthermore, the VSI guides the design of mitigation measures that, in cases of high visual sensitivity, can effectively minimize visual intrusion.

It is also interesting to compare the global VSI map with the landscape areas ESI map generated according to the approach outlined in Section 3.2.8 (Figure 8a). In some specific cases, even though the ESI of a landscape area may be high, if the visual sensitivity factor is lower, it may be possible to focus less on long-range mitigation measures and instead target more significant details at the local scale.

3.4. *A Global Assessment Tool for Decision-Makers*

Figure 9a for global ESI and Figure 9b for VSI index show two output map examples of the global assessment tool, i.e., the GIS project that displays the percentage of suitable or unsuitable areas and the average pixel values for EVI and ESI and their factors for each municipality in the Viterbo province.

This tool is designed to assist decisionmakers in interpreting and comparing results, allowing an analysis of the factors, helping to interpret the information, understand how the scenario evolves, and identify areas with the greatest potential for the installation of sustainable AgriPV plants. This method provides a clear and detailed overview of territorial suitability and environmental and visual sensitivity in each municipality of the Viterbo province.

By changing the symbology in QGIS, it is possible to assess each type of index in relation to each factor, as well as the global summaries for suitability, environmental sensitivity, and visual sensitivity.

All output maps can be consulted in Figure S29 of the Supplementary Materials and the values for each municipality in Table S1 of the Supplementary Materials (for the spatial distribution of municipalities with their names and codes, please refer to Figure S1 of the Supplementary Materials).

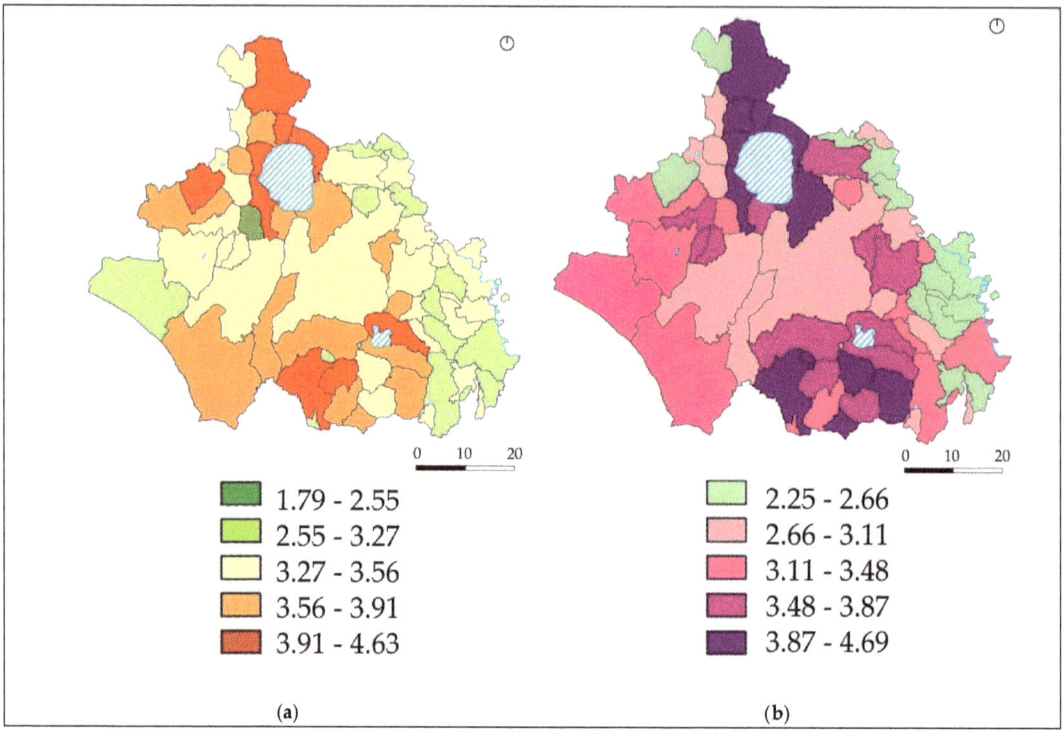

Figure 9. The maps show the global maps, respectively at ESI (**a**) and VSI (**b**) values, with two color ramps symbology that highlight municipalities with higher and lower values.

Beginning with the suitability data, an interesting pattern emerges regarding the geographical distribution of areas for agriPV systems. The analysis reveals a concentration of suitable areas in the northern part of the province, while the southern regions are less favorable. This discrepancy may be attributed to factors such as the territory's morphology, the presence of protected areas, or the varying agricultural characteristics of the municipalities.

The top five municipalities in terms of percentage of suitability areas include Piansano, with the highest value, followed by Vignanello, Fabrica di Roma, Corchiano, and Celleno (chart in Figure 10). The combination of the agricultural vocation of these areas with their low population density makes them particularly suitable for the installation of large-scale AgriPV plants. On the other hand, Calcata and Castel Sant'Elia have no suitable areas, while Gradoli and Faleria, despite having some favorable areas, are limited to less than 250 hectares.

The ESI analysis highlights areas with significant differences in terms of environmental vulnerability (chart in Figure 11). Gradoli, Bolsena, Caprarola, Farnese, Capodimonte, Blera, San Lorenzo, and Barbano Romano municipalities have a high overall ESI—higher than 4, confirming their environmental value, also linked to the presence of Lake Bolsena and the volcanic origins of the towns. Piansano, Montalto di Castro, and Civita Castellana have a medium/low ESI sensitivity, indicating a lower level of criticality compared to other municipalities.

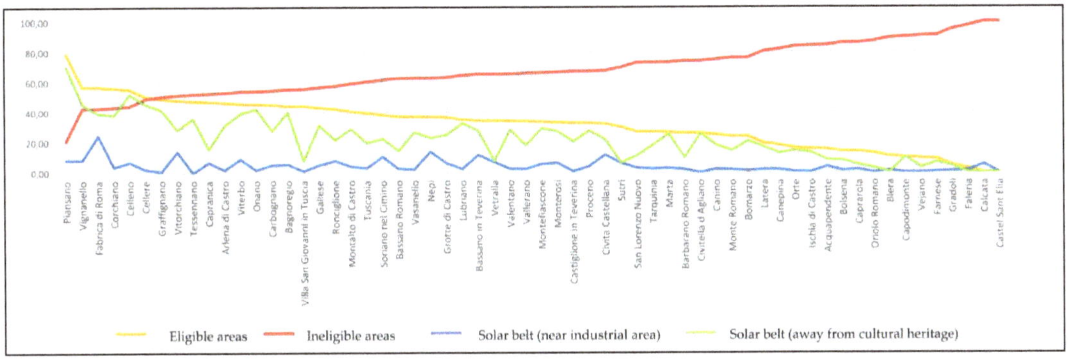

Figure 10. The graph shows the different scenarios according to the criteria and regulations applied for each municipality. The vertical axis represents the percentage of areas relative to the total surface area of the municipality. In red are the percentage of ineligible areas according to the constraint system. In orange are the percentage of eligible areas according to the regulations. In green and cyan are percentage area of solar belts, preferential areas for the installation of agriPV plants, in the presence of industrial areas or in the absence of cultural heritage.

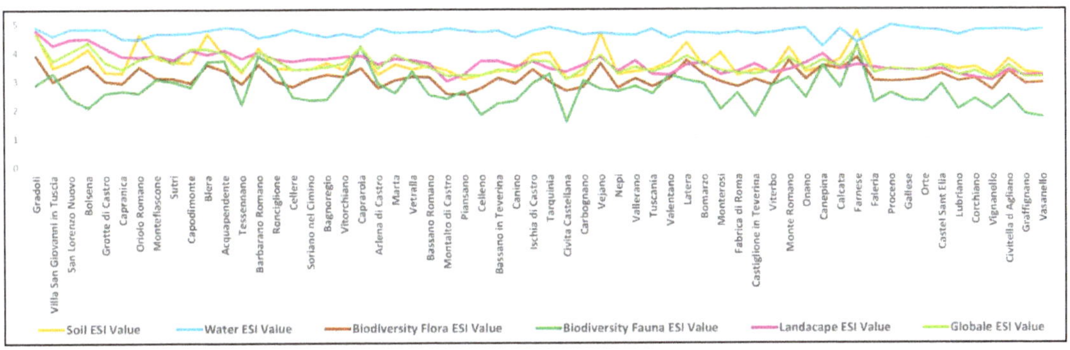

Figure 11. The graph shows the results of the comparison of the different average ESI values (the vertical axis), according to environmental components, for municipalities in the province of Viterbo. The green line shows the environmental sensitivity represented by the average of all environmental components.

4. Discussion

The proposed workflows and GIS models developed offer a flexible, scalable, and adaptable approach to evaluating suitable areas and sensitive territories for agriPV installation siting. The calibration of indices, based on suitability, environmental, and visual sensitivity criteria, can be easily modified to meet the specific characteristics of each Italian region and adapt to different territorial contexts. Moreover, it can be extended to other types of renewable energy production systems, reflecting the unique aspects of various renewable energy technologies.

The sMCDA conducted in this case study reveals that site selection for agriPV systems cannot rely solely on a binary logic of suitability versus non-suitability. Such a binary division based on predefined criteria could be limiting and insufficient for capturing the nuances of each territory. Instead, classifying territories based on environmental and visual sensitivity offers a more informed approach to design decisions and provides a common evaluation framework for decision-making processes, even though legal suitability criteria primarily dictate the project's feasibility. By identifying the most sensitive areas that are still suitable for agriPV projects, it becomes possible to align projects more closely with sustainability objectives.

The analysis of territorial suitability for agriPV system installation in the province of Viterbo, in light of recent regulatory updates in Italy, becomes increasingly important with the 2024 Suitable Areas Decree [27]. This decree introduces a new paradigm for identifying suitable areas for renewable energy installations, with a particular focus on solar belts. The mapping of suitable areas, required to be drafted by each region and autonomous province, excludes other areas that were previously deemed suitable.

The analysis highlights a clear geographical distribution of suitable areas. While the northern part of the province is more suited for agriPV system installation, the southern region shows lower suitability, possibly due to land morphology, the presence of protected areas, or differing agricultural land use. This geographical disparity underscores the necessity for an integrated approach that considers the territorial and environmental specifics of each municipality.

Analyzing various environmental and visual components in terms of their sensitivity allows the design process to be guided effectively using the indices derived from the study. For example, regarding biodiversity and wildlife, several strategies can be implemented to mitigate environmental impacts. In areas where significant movements of microfauna have been observed, measures such as installing permeable fences with a 30 cm-high base can facilitate the passage of microfauna while preventing larger species from entering. Additionally, creating ecological corridors composed of natural and cultivated herbaceous strips can help connect fragmented habitats, enhancing biodiversity and promoting the movement of microfauna. In the presence of significant visual impacts, it may be necessary to shift the location of the plant, modify its design, or adopt solutions like vegetation screening to reduce the impact on the landscape. In essence, a complete sustainable implementation of agriPV projects could be guaranteed by balancing environmental and visual sensitivity, by implementing sMCDA approaches that allow the combination of criteria from both spheres. For instance, in areas with high biodiversity but low visual impact, environmental criteria might be weighted more heavily. Conversely, in culturally significant landscapes, visual sensitivity may take precedence. Mitigation strategies can also play a role in this balance: innovative panel designs or strategic placement can reduce visual impacts, while biodiversity-friendly management practices can enhance environmental outcomes. Additionally, participatory planning processes involving local communities, environmental experts, and landscape architects can help identify optimal solutions that respect both sensitivities.

More generally, this article contributes to practices and reflections on the use of sensitivity mapping (SM) on environmental assessment.

SM for more than 50 years has accompanied the beginning, development, and consolidation of the environmental assessment and, although it is recognized as a key best practice in environmental assessment, it has not become a common practice. Despite the development and availability of open-source GIS tools, SM has not yet been a standard way of operating for the identification, description, and assessment of impacts and a systemic and territorial procedure to define significance, prevent undesirable impacts, and promote sustainability [41].

The lack of structural implementation of SM could be a combination of two elements. From one side, the discontinuity of requirements and references for environmental assessment by the environmental authorities: many examples of web GIS tools implemented by the environmental authorities are created and then abandoned; the last example is the web Sensitivity Maps created by the EPA Ireland [19]. From the proponent side (or the environmental consulting in the EIA market), the GIS approach on environmental assessment and the requirement to prepare a SM of each project to be evaluated is compared with other less time-consuming methodologies [24].

SM alone or combined with sMCDA contributes to the implementation of appropriate ecological economics approaches to recognize the values attributed by different territorial actors to ecosystems, beyond state or market logics [42–44].

The case presented is about an SM dealing with environmental and visual sensitivity applied to environmental assessment of projects related to agriPV. However, SM has many potentialities, from one side, in the articulated tiered ex ante evaluations from policies to operations; on the other side, including in the GIS platforms, different components of sensitivity: environmental, cultural, and social, in a perspective-integrated sustainability.

The climate emergency requiring a growing supply of renewable energy is offering an unexpected 'bonanza context' to agriPV from policies and science, with enthusiastic endorsement as 'the best win–win solution.' This apparently easy solution conceals other potential for solar energy production in urban and industrial areas [45].

However, in the framework of ecological economics, SM can help for a more complete mapping and disclosure of political economy and political ecology dimensions to implement just transition processes, not just energy transition solutions [46–49].

Comprehensive SM mapping can offer a key contribution to the transparency of decisionmaking processes, facilitating the multiple function of environmental ex ante evaluation and supporting the different dimensions of increasing requirement of impact assessment: climate change, biodiversity, and just transition.

Supplementary Materials: The following supporting information can be downloaded at: https://www.mdpi.com/article/10.3390/ijgi14030115/s1.

Author Contributions: Conceptualization, Ilaria Angelelli, Daniele Codato and Massimo De Marchi; methodology, Ilaria Angelelli, Daniele Codato and Massimo De Marchi; validation, Ilaria Angelelli, Daniele Codato and Massimo De Marchi; formal analysis, Ilaria Angelelli; investigation, Ilaria Angelelli, Daniele Codato and Massimo De Marchi; data curation, Ilaria Angelelli; writing—original draft preparation, Ilaria Angelelli, Daniele Codato and Massimo De Marchi; writing—review and editing, Ilaria Angelelli, Daniele Codato, Salvatore Eugenio Pappalardo and Massimo De Marchi; visualization, Ilaria Angelelli and Daniele Codato; supervision, Salvatore Eugenio Pappalardo and Massimo De Marchi; project administration, Massimo de Marchi; funding acquisition, Massimo De Marchi. All authors have read and agreed to the published version of the manuscript.

Funding: This article is part of the research activities of the Jean Monnet Center of Excellence on Just Fossil Fuel Transition, by the support of Erasmus+ Programme of the European Union, call for proposals ERASMUS-JMO-2024-HEI-TCH-RSCH; Project number: 101175896.

Data Availability Statement: No ethical approval was required due to the use of publically available and/or aggregate secondary data.

Conflicts of Interest: The authors declare no conflict of interest.

References

1. Intergovernmental Panel on Climate Change (IPCC) Summary for Policymakers. In *Climate Change 2023: Synthesis Report. Contribution of Working Groups I, II and III to the Sixth Assessment Report of the Intergovernmental Panel on Climate Change*; Lee, H., Romero, J., Eds.; IPCC: Geneva, Switzerland, 2023; pp. 1–34.
2. IEA. *World Energy Outlook 2024*; IEA: Paris, France, 2024.
3. IEA. *IEA Renewables 2023—Analysis*; IEA: Paris, France, 2024.
4. REN21. *REN21 Renewables 2023 Global Status Report Collection, Global Overview*; REN21: Paris, France, 2023; ISBN 9783948393113.
5. European Commission. *European Commission REPowerEU: Affordable, Secure and Sustainable Energy for Europe*; European Commission: Brussels, Belgium, 2022.

6. Intergovernmental Panel on Climate Change (IPCC) Summary for Policymakers. In *IPCC Special Report on Renewable Energy Sources and Climate Change Mitigation*; Edenhofer, O., Arvizu, D., Bruckner, T., Christensen, J., Logan, J., Eds.; Cambridge University Press: Cambridge, UK, 2011; p. 26.
7. Choi, C.S.; Cagle, A.E.; Macknick, J.; Bloom, D.E.; Caplan, J.S.; Ravi, S. Effects of Revegetation on Soil Physical and Chemical Properties in Solar Photovoltaic Infrastructure. *Front. Environ. Sci.* **2020**, *8*, 140. [CrossRef]
8. Pimentel Da Silva, G.D.; Magrini, A.; Branco, D.A.C. A Multicriteria Proposal for Large-Scale Solar Photovoltaic Impact Assessment. *Impact Assess. Proj. Apprais.* **2020**, *38*, 3–15. [CrossRef]
9. Hilker, J.M.; Busse, M.; Müller, K.; Zscheischler, J. Photovoltaics in Agricultural Landscapes: "Industrial Land Use" or a "Real Compromise" between Renewable Energy and Biodiversity? Perspectives of German Nature Conservation Associations. *Energy Sustain. Soc.* **2024**, *14*, 6. [CrossRef]
10. Wüstenhagen, R.; Wolsink, M.; Bürer, M.J. Social Acceptance of Renewable Energy Innovation: An Introduction to the Concept. *Energy Policy* **2007**, *35*, 2683–2691. [CrossRef]
11. Dere, S.; Elçin Günay, E.; Kula, U.; Kremer, G.E. Assessing Agrivoltaics Potential in Türkiye—A Geographical Information System (GIS)-Based Fuzzy Multi-Criteria Decision Making (MCDM) Approach. *Comput. Ind. Eng.* **2024**, *197*, 110598. [CrossRef]
12. Agostini, A.; Colauzzi, M.; Amaducci, S. Innovative Agrivoltaic Systems to Produce Sustainable Energy: An Economic and Environmental Assessment. *Appl. Energy* **2021**, *281*, 116102. [CrossRef]
13. Toledo, C.; Scognamiglio, A. Agrivoltaic Systems Design and Assessment: A Critical Review, and a Descriptive Model towards a Sustainable Landscape Vision (Three-Dimensional Agrivoltaic Patterns). *Sustainability* **2021**, *13*, 6871. [CrossRef]
14. Libra, M.; Kozelka, M.; Šafránková, J.; Belza, R.; Poulek, V.; Beránek, V.; Sedláček, J.; Zholobov, M.; Šubrt, T.; Severová, L. Agrivoltaics: Dual Usage of Agricultural Land for Sustainable Development**. *Int. Agrophys* **2024**, *38*, 121–126. [CrossRef]
15. Yeligeti, M.; Hu, W.; Scholz, Y.; Stegen, R.; von Krbek, K. Cropland and Rooftops: The Global Undertapped Potential for Solar Photovoltaics. *Environ. Res. Lett.* **2023**, *18*, 054027. [CrossRef]
16. Ali Abaker Omer, A.; Liu, W.; Li, M.; Zheng, J.; Zhang, F.; Zhang, X.; Osman Hamid Mohammed, S.; Fan, L.; Liu, Z.; Chen, F.; et al. Water Evaporation Reduction by the Agrivoltaic Systems Development. *Solar Energy* **2022**, *247*, 13–23. [CrossRef]
17. Kim, T.H.; Chun, K.S.; Yang, S.R. Analyzing the Impact of Agrophotovoltaic Power Plants on the Amenity Value of Agricultural Landscape: The Case of the Republic of Korea. *Sustainability* **2021**, *13*, 11325. [CrossRef]
18. Torma, G.; Aschemann-Witzel, J. Social Acceptance of Dual Land Use Approaches: Stakeholders' Perceptions of the Drivers and Barriers Confronting Agrivoltaics Diffusion. *J. Rural. Stud.* **2023**, *97*, 610–625. [CrossRef]
19. González Del Campo, A. Mapping Environmental Sensitivity: A Systematic Online Approach to Support Environmental Assessment and Planning. *Environ. Impact Assess. Rev.* **2017**, *66*, 86–98. [CrossRef]
20. Fattoruso, G.; Toscano, D.; Venturo, A.; Scognamiglio, A.; Fabricino, M.; Di Francia, G. A Spatial Multicriteria Analysis for a Regional Assessment of Eligible Areas for Sustainable Agrivoltaic Systems in Italy. *Sustainability* **2024**, *16*, 911. [CrossRef]
21. Adjiski, V.; Serafimovski, D. GIS-and AHP-Based Decision Systems for Evaluating Optimal Locations of Photovoltaic Power Plants: Case Study of Republic of North Macedonia. *Geomat. Environ. Eng.* **2024**, *18*, 51–82. [CrossRef]
22. Khan, A.; Ali, Y.; Pamucar, D. Solar PV Power Plant Site Selection Using a GIS-Based Non-Linear Multi-Criteria Optimization Technique. *Environ. Sci. Pollut. Res.* **2023**, *30*, 57378–57397. [CrossRef]
23. Murgante, B.; Rahmani, S.; Scorzelli, R.; Fattoruso, G.; Annunziata, A. Optimizing Agrivoltaic Integration: Spatial Analysis and AHP Assessment. In *Computational Science and Its Applications–ICCSA 2024 Workshops*; Gervasi, O., Murgante, B., Garau, C., Taniar, D., Rocha, A.M.A.C., Faginas Lago, M.N., Eds.; ICCSA 2024; Lecture Notes in Computer Science; Springer: Cham, Switzerland, 2024; Volume 14825. [CrossRef]
24. Singh Rathore, G.; Kumar, K. The Need and Use of Geographic Information Systems for Environmental Impact Assessment. *Int. J. Sci. Res.* **2021**, *10*, 206–213. [CrossRef]
25. Decreto Legislativo 8 Novembre 2021 n. 199; Italy. 2021. Available online: https://www.gazzettaufficiale.it/eli/id/2021/11/30/21G00214/sg (accessed on 14 January 2025).
26. DECRETO-LEGGE 15 Maggio 2024, n. 63; Italy. 2024. Available online: https://www.gazzettaufficiale.it/eli/id/2024/07/13/24A03724/SG (accessed on 14 January 2025).
27. DECRETO 21 Giugno 2024; Italy. 2024. Available online: https://www.gazzettaufficiale.it/atto/vediMenuHTML?atto.dataPubblicazioneGazzetta=2024-07-02&atto.codiceRedazionale=24A03360&tipoSerie=serie_generale&tipoVigenza=originario (accessed on 14 January 2025).
28. Legambiente Regioni e Aree Idonee. 2024. Available online: https://www.legambiente.it/wp-content/uploads/2024/11/Regione-e-aree-idonee_2024.pdf (accessed on 14 January 2025).
29. Open Source Geospatial Fundation QGIS. 2024. Available online: https://www.osgeo.org/ (accessed on 5 January 2025).
30. Ministero della Transizione Ecologica—Dipartimento per l'energia. *Linee Guida in Materia Di Impianti Agrivoltaici*; Ministero della Transizione Ecologica—Dipartimento per l'energia: Rome, Italy, 2022.

31. Decreto Del Presidente Del Consiglio Dei Ministri 12 Dicembre 2005; Italy. 2005. Available online: https://www.gazzettaufficiale.it/eli/id/2006/01/31/06A00800/sg (accessed on 14 January 2025).
32. Malczewski, J.; Rinner, C. *Multicriteria Decision Analysis in Geographic Information Science*; Advances in Geographic Information Science; Springer: Berlin/Heidelberg, Germany, 2015; ISBN 978-3-540-74756-7.
33. La Repubblica Fotovoltaico: Viterbo, La Città Italiana Con La Maggiore Potenza Installata per Abitante. Available online: https://www.repubblica.it/economia/rapporti/energitalia/trasformazione/2024/07/25/news/fotovoltaico_viterbo_la_citta_italiana_con_la_maggiore_potenza_installata_per_abitante-423413463/ (accessed on 7 November 2024).
34. ENEA Carta Della Radiazione Solare. Available online: http://www.solaritaly.enea.it/Radiazione/Mappe/DNI_2006%C3%B72022.png (accessed on 20 March 2024).
35. ISTAT Demografia in Cifre. Available online: https://demo.istat.it/ (accessed on 23 October 2024).
36. Klingebiel, A.A.; Montgomery, P.H. Land-Capability Classification. Agricultural Handbook. No. 210. Soil Conservation Service. 1961. Available online: https://www.govinfo.gov/content/pkg/GOVPUB-A-PURL-gpo20777/pdf/GOVPUB-A-PURL-gpo20777.pdf (accessed on 14 January 2025).
37. Capizzi, D.; Mortelliti, A.; Amori, G.; Colangelo, P.; Rondinini, C. (Eds.) *Mammiferi Del Lazio. Ecologia, Distribuzione e Conservazione*; ARP: Rome, Italy, 2012; ISBN 9788895213040.
38. Worton, B.J. Kernel Methods for Estimating the Utilization Distribution in Home-Range Studies. *Ecology* **1989**, *70*, 164–168. [CrossRef]
39. Home Range Analysis By Kernel Density Estimation. Available online: https://plugins.qgis.org/models/19/ (accessed on 15 February 2024).
40. MiBACT Ministero dei Beni e delle Attività Culturali e del Turismo Direzione Regionale per i beni culturali e paesaggistici del Piemonte, Regione Piemonte Direzione Programmazione strategica, politiche territoriali ed edilizia; Dipartimento Interateneo di Scienze, Progetto e Politiche del Territorio (DIST), P. e U. di T. *Linee Guida per l'analisi, La. Tutela e La. Valorizzazione Degli Aspetti Scenico-Percettivi Del. Paesaggio*. 2014. Available online: https://www.regione.piemonte.it/web/sites/default/files/media/documenti/2018-12/lineeguida.pdf (accessed on 14 January 2025).
41. Mariano, J.B.; de Oliveira Vasconcelos, A.; Mamede, P.; Carrascal, M.H.; La Rovere, E.L.; de Almeida, J.R.; Landau, L. GIS-Based Modeling of the Environmental Vulnerability of the Amazon Region to the Upstream Oil and Gas Activities. *Impact Assess. Proj. Apprais.* **2024**, *42*, 522–543. [CrossRef]
42. Martínez-Alier, J.; Muradian, R. *Handbook of Ecological Economics*; Edward Elgar Publishing: Cheltenham, UK, 2015; ISBN 9781783471416.
43. Splash, C.L. *Routledge Handbook of Ecological Economics*; Splash, C.L., Ed.; Routledge: London, UK, 2017.
44. Villamayor-Tomas, S.; Roy, B.; Muradian, R. The Barcelona School of Ecological Economics and Political Ecology: Building Bridges Between Moving Shores. In *Studies in Ecological Economics*; Villamayor-Tomas, S., Muradian, R., Eds.; Springer International Publishing: Berlin/Heidelberg, Germany, 2023; Volume 8, pp. 17–33, ISBN 978-3-031-22565-9.
45. Jing, R.; Liu, J.; Zhang, H.; Zhong, F.; Liu, Y.; Lin, J. Unlock the Hidden Potential of Urban Rooftop Agrivoltaics Energy-Food-Nexus. *Energy* **2022**, *256*, 124626. [CrossRef]
46. Hu, Z. Towards Solar Extractivism? A Political Ecology Understanding of the Solar Energy and Agriculture Boom in Rural China. *Energy Res. Soc. Sci.* **2023**, *98*, 102988. [CrossRef]
47. McGowan, K.; Antadze, N. Recognizing the Dark Side of Sustainability Transitions. *J. Environ. Stud. Sci.* **2023**, *13*, 344–349. [CrossRef]
48. Adelhardt, N.; Berneiser, J. Risk Analysis for Agrivoltaic Projects in Rural Farming Communities in SSA. *Appl. Energy* **2024**, *362*, 122933. [CrossRef]
49. Vezzoni, R. Farming the Sun: The Political Economy of Agrivoltaics in the European Union. *Sustain. Sci.* **2025**, 1–16. [CrossRef]

International Journal of
Geo-Information

MDPI

Article

A Comprehensive Survey on High-Definition Map Generation and Maintenance

Kaleab Taye Asrat and Hyung-Ju Cho *

Department of Software, Kyungpook National University, 2559 Gyeongsang-daero, Sangju-si 37224, Republic of Korea; kaleabtaye@knu.ac.kr
* Correspondence: hyungju@knu.ac.kr

Abstract: The automotive industry has experienced remarkable growth in recent decades, with a significant focus on advancements in autonomous driving technology. While still in its early stages, the field of autonomous driving has generated substantial research interest, fueled by the promise of achieving fully automated vehicles in the foreseeable future. High-definition (HD) maps are central to this endeavor, offering centimeter-level accuracy in mapping the environment and enabling precise localization. Unlike conventional maps, these highly detailed HD maps are critical for autonomous vehicle decision-making, ensuring safe and accurate navigation. Compiled before testing and regularly updated, HD maps meticulously capture environmental data through various methods. This study explores the vital role of HD maps in autonomous driving, delving into their creation, updating processes, and the challenges and future directions in this rapidly evolving field.

Keywords: autonomous vehicles; high-definition maps; HD map generation; point cloud generation; sensor fusion; SLAM

1. Introduction

The origins of autonomous driving can be traced back to earlier pioneering efforts, notably the research conducted by Ernst Dickmanns in the 1980s. Widely regarded as the father of autonomous vehicles, Dickmanns transformed a Mercedes-Benz van into an autonomous vehicle guided by an integrated computer system [1]. This groundbreaking work laid the foundation for subsequent advancements in the field.

At the onset of the 21st century, numerous competitions were instrumental in promoting research and development in autonomous driving (AD). The challenges held by the US Defense Advanced Research Projects Agency (DARPA) to evaluate the capabilities of intelligent vehicles in harsh and complex environments were particularly influential in driving advancements in this field [2].

High-definition (HD) maps have emerged as a crucial part of advanced driver assistance systems (ADASs) and autonomous vehicles (AVs). These highly detailed and precise digital representations of the physical world provide the essential contextual awareness and environmental understanding required for safe and reliable navigation. The concept of HD maps gained significant interest with the Bertha Benz Memorial Route project in 2013, where a Mercedes-Benz S500 successfully completed the historic route in a fully autonomous manner, leveraging an HD map that contained detailed information about road geometry, lane positions, and topological connections [3]. The precise map-relative localization facilitated by HD maps demonstrated their significance in enabling reliable autonomous navigation.

The importance of HD maps in the autonomous driving industry has grown exponentially in recent years, primarily due to their ability to define the environment with exceptional precision and capture even occluded areas [4]. HD maps provide crucial redundancy and complementary information to the sensor suite (LiDAR, cameras, and radars) employed by AVs for constructing the perception module of the autonomous system.

In recent years, significant advancements have been made in the generation and maintenance of HD maps, driven by the rapid development of sensor technologies, computational power, and machine learning techniques [5–7]. Researchers and industry leaders have proposed novel approaches to streamline data collection, improve mapping accuracy, and enable efficient map updates, paving the way for more reliable and scalable HD mapping solutions [2,8].

Moreover, HD maps play a vital role in addressing long-tail scenarios—rare, complex, and unpredictable events that pose significant challenges for AVs. These scenarios require sophisticated handling to ensure safety and reliability. HD maps offer centimeter-level accuracy in localization, which is essential in areas with poor GPS signals, such as urban canyons or tunnels. They enhance the vehicle's perception system by providing prior knowledge of road geometry, lane markings, and static objects, which is particularly useful in adverse weather conditions like heavy snow or fog. For instance, in dense fog, where camera visibility is reduced, an AV can rely on HD maps which are supported by multiple sensors to navigate by using reliable environmental data [9].

HD maps enable predictive awareness by allowing AVs to anticipate road features and potential hazards, such as sharp curves, steep gradients, or pedestrian crossings. This proactive approach improves safety; for example, an AV can slow down before reaching a known pedestrian crossing. HD maps also integrate real-time data on traffic conditions, construction zones, and temporary road closures, allowing AVs to adapt to changing environments. For instance, an AV can reroute to avoid a newly established construction zone. Furthermore, HD maps facilitate long-tail scenario training by providing a detailed and realistic virtual environment for extensive simulations. This allows AVs to train and test various scenarios without the risks associated with real-world testing, ultimately improving safety and reliability in autonomous driving.

The purpose of this research is to provide a comprehensive survey of recent advancements in HD map creation and update techniques by reviewing the state-of-the-art techniques. Additionally, the study discusses the challenges inherent in HD maps and examines how current systems address these issues. Furthermore, we explore the future direction of the autonomous driving industry. The paper is structured as follows: Section 2 covers the basics of the HD map, Section 3 delves into creating HD maps, Section 4 discusses updating HD maps, Section 5 addresses challenges and solutions, and Section 6 explores future directions which can further transform the AV industry.

2. Basics of HD Maps

Autonomous vehicles (AVs) require highly precise maps to navigate roads and lanes. Regular digital maps do not apply to AVs because they lack information about traffic lights and signs, traffic lanes, the height of pole-like objects, and the accurate size of curves. Designed specifically for human use, these maps cannot be used for AV localization, where the vehicle itself needs to understand the map and locate its position relative to the surroundings.

Though HD maps are crucial, there exists no unique standard HD map structure in the autonomous driving industry [5]. Different companies adopt various architectures and methods to create HD maps, highlighting the lack of a unified approach. Google's Waymo, a leader in autonomous taxis, uses high-resolution sensors, but specific HD map layers are not publicly known. Baidu's open-source Apollo platform supports accurate localization, precise path planning, and real-time updates for navigating diverse environments by continually enhancing its mapping technologies and integrating advanced sensors like LiDAR and 4D millimeter-wave radar [10].

However, we can look at some available architectures to understand common practices in the industry. HD maps from HERE, TomTom, and BerthaDerive (Lanelet) share a similar three-layer architecture but have slightly different functionalities in each layer. Figure 1 shows an example of HD map from HERE and Figure 2 illustrates HERE's HD map localization layer lane data. Lyft Level 5 has a different HD map structure with five layers [8,11,12]. The comparison between these architectures [5,13] is shown in Table 1. Although different companies have their own HD map structure, in this paper, we focus on the main categories of mapping information that an autonomous driving system uses: topological, geometric, semantic, and dynamic elements/real-time updates, and feature-based layers [4].

Figure 1. Example of the HD map from HERE [14].

Table 1. HD map layer structures from different companies involved in AD.

Layer	BERTHA Drive	TomTom	HERE	Lyft Level 5
1	Road network	Navigation data	HD road	Base map
2	Lane level map	Planning data	HD lanes	Geometric map
3	Landmarks/Road marking map	HD localization	HD localization	Semantic map
4	-	-	-	Map Priors layer
5	-	-	-	Real-time layer

2.1. Topological Representation

The fundamental component of a high-definition road map is the topology representing a road network [4,15]. In HD maps, topology captures how roads, lanes, intersections, and other features are connected. This topological information not only benefits the HD map to drive autonomously but also facilitates route optimization based on factors such as traffic speed, road quality, traffic delays, and traffic rules. The topological representation typically comprises nodes and edges, where nodes represent intersections, interchanges, or specific points of interest, while edges represent the road segments connecting these nodes. Edges are often associated with attributes such as road classifications, speed limits, and turn restrictions. They provide valuable context for path planning and decision-making. By accurately capturing the intricate relationships and interconnections within the road network, the topological representation of HD maps plays a crucial role in enabling safe and efficient autonomous driving, as well as facilitating advanced applications like intelligent transportation systems and fleet management [15].

2.2. Geometric Representation

The accurate representation of features and objects in HD maps is another crucial task. The geometric representation of HD maps refers to how the spatial features and layout of the environment are captured and represented in the digital map data. Geometric features include the shapes and positions of roads, lanes, sidewalks, buildings, and terrain. These features are typically represented using a vector data structure, which describes simplified geometric shapes such as points, lines, curves, circles, and polygons [8]. One notable example of a geometric representation layer is the Lyft Level 5 geometric map, which contains highly detailed 3D information about the environment, organized to support precise calculations and simulations [11,12]. This level of detail is essential for enabling accurate localization, path planning, and decision-making for AVs.

2.3. Semantic Representation

The semantic representation of HD maps builds on the geometric and topological representations by providing "semantic meaning" for features [4,11,12]. Semantic representation includes various traffic 2D and 3D objects such as lane boundaries, intersections, crosswalks, parking spots, stop signs, traffic lights, road speed limits, lane information, and road classification [4,11]. This semantic understanding is crucial for autonomous driving, where vehicles need to comprehend the surroundings to make smart, safe, and informed decisions. The semantic representation acts as a bridge between the raw geometric data and the higher-level understanding required for intelligent decision-making and path planning. By assigning semantic labels and attributes to the geometric features, it enables autonomous vehicles to interpret the environment in a more meaningful and contextual way. Notable examples of semantic representation layers include the HD map localization model layer from HERE [16] and the RoadDNA layer from TomTom [17]. These layers hold object-level semantic features, which help accurately estimate the vehicle's position using object locations and contextual information [6,16,17]. In general terms, the semantic representation assigns semantic labels and attributes to road features and objects defined in the geometric representation, providing a richer and more comprehensive understanding of the environment for autonomous driving systems [6].

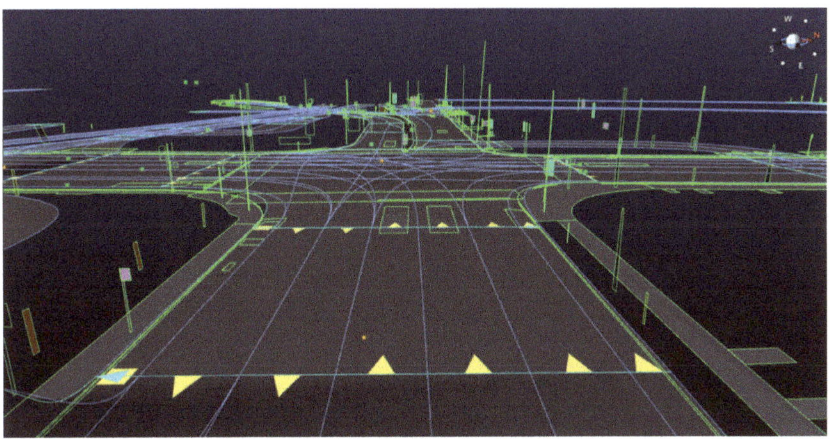

Figure 2. Lane data from HERE's HD map Localization layer [16].

2.4. Dynamic Elements

Dynamic elements are features, objects, or conditions within the environment that are susceptible to change over time. These features require continuous monitoring and updating to provide up-to-date information for autonomous vehicles. Dynamic elements like pedestrians, obstacles, and vehicles need to be updated for the HD map to be always precise and accurate [5]. The dynamic element layer of HD maps captures and represents these time-varying aspects of the environment, which are essential for safe and efficient path planning and decision-making. This layer includes features such as traffic conditions, construction zones, temporary road closures, and the positions and movements of other road users, such as vehicles, pedestrians, and cyclists. Accurate representation and frequent updating of dynamic elements are critical for autonomous vehicles to anticipate and respond to changing situations promptly. This information can be combined with the static elements of the HD map, such as road geometry and traffic rules, to enable more comprehensive and reliable path planning and decision-making. Proper management of dynamic elements, which will be explained in-depth in Section 4, is essential for planning a safe and efficient path for autonomous vehicles in dynamic and complex environments.

2.5. Feature-Based Map Layers

HD maps rely heavily on advanced feature-based map layers for accurate localization and navigation. These layers use various techniques to identify and match features in the environment to ensure precise vehicle positioning. One state-of-the-art approach in feature-based mapping is visual place recognition (VPR). Recent advancements in VPR, such as SeqNet, have improved performance through learning powerful visual features and employing sequential matching processes [18].

Another significant advancement is the use of 3D LiDAR maps for vehicle localization. The SeqPolar method introduces a polarized LiDAR map (PLM) and a second-order hidden Markov model (HMM2) for map matching, which offers a concise and structured representation of 3D LiDAR clouds. This method enhances localization accuracy significantly [19].

The integration of these techniques into HD map systems significantly enhances the accuracy and reliability of feature-based map layers, which is crucial for improving the localization and navigation capabilities of autonomous vehicles, especially in dynamic and complex environments.

In the autonomous driving industry, apart from layer structures, formats should be considered as well. Table 2 summarizes the key HD map formats and their respective representations.

Table 2. Common HD map formats and representations.

Format	Description	File Format	Representations
OpenDRIVE	Standard for logical description of road networks	XML	Geometric primitives
Lanelet2	HD map format for autonomous driving	XML, OSM	Points
NDS	Global standard for automotive map data	Binary	Vector data

One common format is Lanelet2 [20], derived from liblanelet [21]. It uses an XML-based OSM format and organizes maps into three layers: physical (points and line strings for physical elements), relational (lanelets, areas, and regulatory elements), and topological. Lanelet2 focuses on precise lane-level navigation and traffic regulation.

Another common format is OpenDRIVE [22] by ASAM. It uses an XML format to describe road networks with three layers: the reference line (geometric primitives for road shapes), lane (drivable paths), and features (traffic signals and signs). OpenDRIVE emphasizes static map features and road structure.

Apollo OpenDRIVE, owned by Baidu, utilizes a modified OpenDRIVE format [5,10,23]. The Apollo framework is an open-source project developed by Baidu, a company based in China. Apollo uses points, in contrast to the original OpenDRIVE format, which uses geometric primitives.

3. HD Map Generation Techniques

As mentioned in the previous sections, high precision is required for HD maps to be effective for autonomous driving. Initially, HD maps are generated offline; however, to ensure their relevance and accuracy, they must undergo real-time updates once created. The general steps taken to generate HD maps will be discussed in this section. The process contains data collection from multiple sources, sensor fusion, point cloud registration, and feature extraction.

3.1. Data Collection

The first step in the HD map generation process is data collection. A vehicle mounted with high-precision and well-calibrated sensors is sent to survey and collect comprehensive data about the environment. Mapping vehicles are equipped with mobile mapping systems (MMSs), which typically comprise various sensors. The sensors used in an MMS setup may vary depending on the requirements and the details needed for mapping. LiDAR is one of the common sensors used in MMS. LiDAR sensors provide highly accurate 3D point cloud data of the environment. Apart from LiDAR, cameras are used to capture high-resolution images of the surroundings. These high-resolution images provide additional information about road markings, traffic signs, and other features. The global navigation satellite system (GNSS) is another common sensor found in MMS. GNSS receivers are capable of receiving signals from multiple satellite constellations simultaneously, which increases accuracy and precision. GNSS sensors are usually coupled with an inertial measurement unit (IMU) sensor to help estimate the vehicle's motion, including position, velocity, and orientation. In addition, radar sensors could be used to detect and track moving objects. Odometer sensors can also be used to measure the distance traveled by the vehicle's wheels. Other sensors can be equipped based on the specific need of the mapping task. The above sensors can also be bought separately and configured to do the data collection. However this approach might not be simple and it is also a time-consuming task [6,24]. An MMS equipped with LiDAR, GNSS, IMU, and camera is shown in Figure 3.

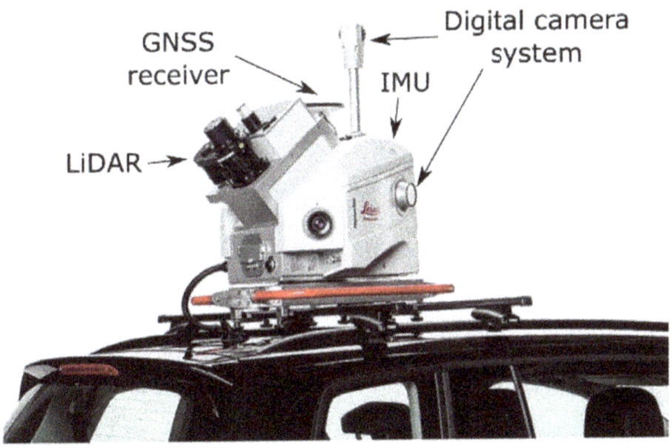

Figure 3. Mobile mapping system with various positioning and data collection sensors, designed to produce a precise georeferenced 3D map of the environment [24].

Typically, there are three primary approaches used to gather data for HD maps. Table 3 compares these approaches. The first involves sourcing data from publicly available datasets; however, these are not as plentiful for autonomous driving. Nonetheless, researchers can tap into a handful of datasets like the Level5 Lyft dataset [25], the KITTI dataset [26], the nuScenes dataset [27], and the Argoverse dataset [28]. These open-source datasets contain pre-labeled traffic data, 3D point clouds, images, and other sensor data, facilitating experimentation and map generation. The second method entails collecting data firsthand. Researchers may opt for this method when focusing on specific regions or features within HD maps, though resulting datasets tend to be geographically limited. Lastly, crowdsourcing emerges as a powerful third option. Leveraging contributions from numerous vehicles, this method yields vast and diverse datasets covering expansive areas [5]. Moreover, the collective nature of crowdsourced data captures a wide array of scenarios, offering invaluable insights into the challenges AVs might encounter. For instance, recent research proposes a crowd-sourcing framework to continuously update point cloud maps (PCM) for HD maps, integrating LiDAR and vehicle communication technologies to detect and incorporate environment changes in real time [29].

Table 3. Data collection methods for HD map.

Approach	Advantages	Disadvantages
Open-source dataset	Easily accessible Saves time	Dataset might be specific to some area Not actively updated
Self-collected dataset	Well-fitted for specific research Time-consuming	Costly compared with using open-source datasets
Crowdsourcing	Covers a wide area Data continuously updated	High cost Data preprocessing and labeling could be a difficult task

3.2. Sensor Fusion

Sensor fusion plays an important role in HD map generation by combining data from multiple sensors to create reliable and accurate HD maps. Sensor fusion integrates data from multiple sensors, such as LiDAR, camera, IMU, GPS, and radar in order to capture various aspects of the environment. It leverages data from various sensing modalities to minimize detection uncertainties and address the limitations of individual sensors operating independently [9,30–32]. Table 4 presents a comparison between LiDAR, camera, and radar sensors based on various factors to be considered in autonomous driving.

Integration of multiple sensors to provide redundancy and achieve a more accurate HD map requires careful calibration of sensors. Combinations of sensors widely used for autonomous driving are based on LiDAR, camera, and radar. Although other sensor integrations exist, the more widely researched combinations are among those three sensors [33]. These widely used sensor combinations are LiDAR–camera–radar, LiDAR–camera, and camera–radar.

Table 4. Comparison of LiDAR, camera, and radar sensors for the HD map.

Factors	LiDAR	Camera	Radar
Accuracy	Higher accuracy in measuring distances and capturing 3D point cloud data	Lower accuracy in depth perception as compared to LiDAR	Good accuracy in detecting objects even in adverse weather conditions
Range	Shorter range	Limited range	Longer range
Resolution	High spatial resolution	High resolution images	Lower resolution
Environmental Suitability	Affected by adverse weather conditions such as rain, snow, or fog	Affected by changes in lighting conditions, such as glare and low visibility, reflection	Less affected by environmental factors
Cost	Expensive	Affordable compared to LiDAR and Radar	Relatively cost-effective compared to LiDAR
Power Consumption	Higher power consumption	Lower power consumption	Moderate power consumption
Lane Detection	Less contribution	Very effective	Less contribution

The fusion of LiDAR and camera sensors yields high-resolution images and higher accuracy in distance measurements. LiDAR–camera fusion has proven to show better performance than the use of these sensors separately [34–38]. One approach used to couple these two sensors is to project the point cloud data from LiDAR onto the image obtained from the camera sensor [38]. Another way to implement LiDAR–camera fusion is to construct a representation of the camera sensor's 2D detections into the LiDAR point cloud data, as presented in FrustrumNet [39]. FrustrumNet's architecture, illustrated in Figure 4, involves projecting 2D object detections from the camera into 3D FrustrumNet candidates in the LiDAR point cloud. This fusion of complementary sensor data leverages the strengths of both modalities, enhancing the robustness and accuracy of perception in autonomous driving applications.

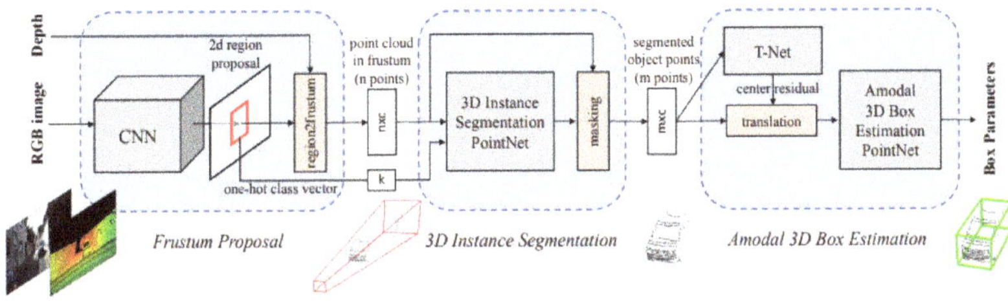

Figure 4. FrustrumNet architecture—a two-stage approach where, first, a 2D CNN detects objects, and then the point cloud within the corresponding 3D frustum is processed by PointNets to segment the object and estimate its amodal 3D bounding box [39].

Utilizing LiDAR–camera–radar fusion, as discussed in LRVFNet [40], involves the generation of distinct features from various sensor modalities (LiDAR, mmWave radar, and visual sensors) using a deep multi-scale attention-based architecture. This fusion method effectively integrates complementary information from these modalities to enhance accuracy and robustness in object detection. Similarly, the fusion mechanism utilized in CLR-BNN [41] employs a Bayesian neural network to fuse data from the three sensors. This approach results in improvements in detection accuracy and reduced uncertainties in diverse driving situations. Figure 5 illustrates the highlevel architecture of CLR-BNN.

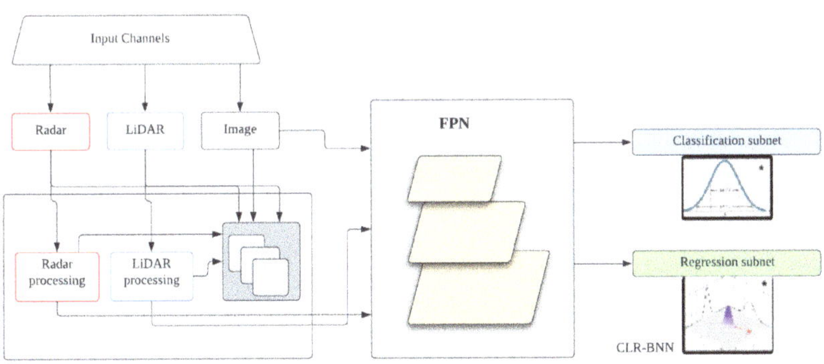

Figure 5. CLR-BNN high-level architecture for sensor fusion [41].

The integration of camera–radar sensors offers significant advantages for AVs, as it combines the high-resolution visual data from cameras, capturing intricate environmental details such as color, texture, and shape, with the long-range detection capabilities of radar sensors, which operate reliably even in challenging weather conditions and low visibility scenarios. CameraRadarFusionNet (CRF-Net) [42] shown in Figure 6 enhances

2D object detection networks by integrating camera data and sparse radar data within network layers, which autonomously determines the optimal fusion level, outperforming image-only networks across diverse datasets.

Sensor fusion integrates data from various sensors via early, middle, or late fusion strategies. These approaches differ when integration occurs: before processing (early), at the feature level (middle), or at the decision level (late). Figure 7 shows the basic working principle of these three fusion strategies.

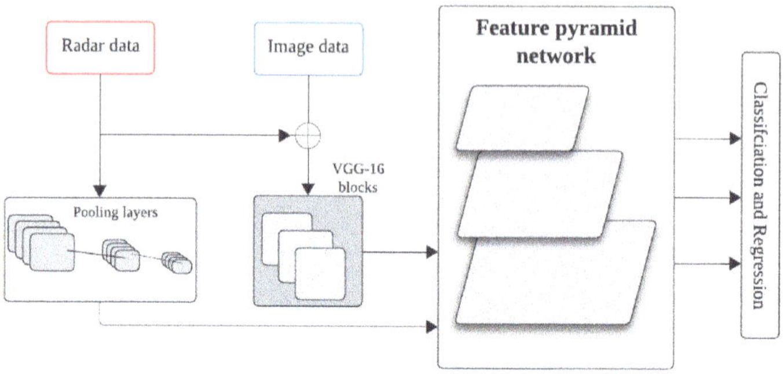

Figure 6. High-level architecture of CameraRadarFusionNet (CRF-Net) [42].

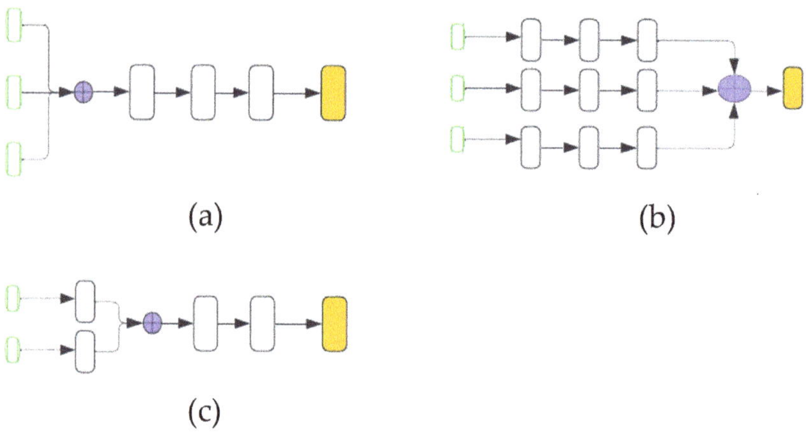

Figure 7. Three kinds of fusion strategies: (**a**) early fusion, (**b**) middle fusion, and (**c**) late fusion. Green blocks represent input sensors (camera, LiDAR, and radar), whereas yellow blocks represent output.

Among the most widely adopted algorithms used to implement sensor fusion strategies are object detection techniques such as YOLO [43] and SSD [44] for camera data, as well as 3D object detection methods like VoxelNet [45] and PointNet [46] for processing point cloud data from LiDAR or radar sensors [33].

3.3. Map Generation Algorithms

Within HD map generation techniques, this section specifically addresses the algorithms used in the map generation process. This section focuses on the methods and techniques employed to process sensor data and construct HD maps of the environment.

Point Cloud Registration Approaches

Algorithms play a critical role in transforming raw sensor data into meaningful map representations. After data collection, very powerful algorithms are crucial for the generation of HD maps. A process known as point cloud registration is performed to align several overlapping and non-overlapping point cloud data obtained from the data collection phase. The point cloud data gathered from different sensors needs to be aligned to obtain a comprehensive map of the environment. The alignment that needs to take place might not necessarily be from different sensors. This means that we can have point cloud data from the same sensor but for different times or the data could have been from a different viewpoint. Therefore, this condition also requires point cloud registration to align the acquired point clouds.

Generally, four techniques are used to perform point cloud registration: optimization-based approaches, probabilistic-based approaches, feature-based approaches, and deep learning techniques [47,48]. Table 5 compares and contrasts these approaches.

1. Optimization-based approaches—Optimization-based approaches involve two steps, which are correspondence and transformation. The correspondence step finds corresponding points between the point clouds that are being registered. Establishing accurate correspondence is crucial for determining how the point clouds should be aligned. Common techniques for establishing correspondences include the nearest neighbor search, which finds the closest point in the target cloud for each point in the source cloud based on distance metrics such as Euclidean distance. The second step, the transformation step, estimates the optimal transformation and aligns the source point cloud with the target point cloud. Typically, the transformation consists of translation, rotation, and also scaling. Optimization-based approaches aim to minimize the value of a cost function that quantifies the discrepancy between corresponding points in the two point clouds. The iterative closest point (ICP) is one of the most widely used optimization-based methods [49]. ICP iteratively refines the transformation parameters until they converge to an optimum solution. The goal of the ICP algorithm is to align the point clouds optimally [50]. Over the years, numerous variants and improvements to the ICP algorithm have been proposed [51,52], contributing significantly to the advancement of point cloud registration techniques. These algorithms have had a profound impact on the creation of highly accurate HD maps, which are essential for various applications, including autonomous driving and robotics [48].
2. Probabilistic-based approaches—While optimization-based approaches like ICP have proven beneficial for point cloud registration, they can encounter limitations when the point clouds have significant overlap or occlusions. In such scenarios, ICP derivative algorithms may converge to local optima, leading to registration failure. To address these limitations, probabilistic registration algorithms based on statistical models have gained traction in recent years, particularly those leveraging Gaussian mixture models (GMMs) [49] and the coherent point drift (CPD) algorithm [53].
   These probabilistic methods incorporate uncertainty into the registration process by estimating the posterior distribution of the transformation parameters. This approach is valuable for handling uncertain or noisy data, offering probabilistic estimates of the registration outcomes rather than deterministic solutions. GMM and CPD, along with their respective variants [54,55], are particularly useful when dealing with noisy or uncertain data, as they can provide probabilistic estimates of the registration results, accounting for the inherent uncertainties in the data.

Table 5. Comparison of point cloud registration approaches: advantages and limitations.

Approaches	Advantages	Limitations
Optimization-based	Well-established methods Generally computationally efficient Effective for rigid and affine transformations	Sensitive to initial alignment errors May struggle with noise and outliers May require manual parameter tuning
Probabilistic-based	Can handle uncertain or noisy data Provides probabilistic estimates of registration results	Computational complexity may be high May require significant training data Performance may degrade in unfamiliar scenes
Feature-based	Robust feature correspondence searching Facilitates precise registration outcomes	Dependence on feature extraction quality Requires extensive training data Performance may degrade in unfamiliar scenes with significant distribution differences
Deep learning-based	Capable of learning complex patterns and relationships Can improve robustness and accuracy Can handle non-linear transformations	May require large amounts of training data Training and inference may be computationally expensive

3. Feature-based approaches—Feature-based approaches for point cloud registration depart from traditional optimization-based methods by leveraging deep learning techniques to extract robust feature correspondences. These approaches focus on identifying distinctive key points or features from the point clouds, which serve as landmarks for establishing correspondences and facilitating registration. As highlighted in [47], these methodologies utilize deep neural networks as a powerful feature extraction mechanism, aiming to derive highly distinctive features that enable precise registration outcomes such as the SeqPolar method [19]. While feature-based approaches offer advantages in providing accurate correspondence searching and facilitating precise registrations, they also come with certain challenges. These methods typically require extensive training data to effectively learn the feature representations, and their performance may degrade when encountering unfamiliar scenes or environments that significantly deviate from the training data distribution.
4. Deep learning techniques—Deep learning techniques, particularly convolutional neural networks (CNNs), attention-based models, and other advanced architectures, have emerged as powerful tools for point cloud registration tasks. These methods leverage the ability of deep networks to learn feature representations directly from the point cloud data and capture complex patterns and relationships, enabling robust registration. Deep learning-based registration approaches can be classified into two main categories: end-to-end learning methods, where the entire registration pipeline is learned from data and hybrid approaches that combine deep learning with traditional techniques. Examples of deep learning-based registration methods include the deep closest point (DCP) [56] and DeepVCP [57]. DeepVCP shown in Figure 8, in particular, is an end-to-end deep learning framework for 3D point cloud registration that has demonstrated superior performance compared to both optimization-based approaches [49,58] and probabilistic-based approaches [53]. By leveraging the representational power of deep networks, these methods have shown promising results in addressing the challenges of point cloud registration, outperforming traditional techniques in various scenarios.

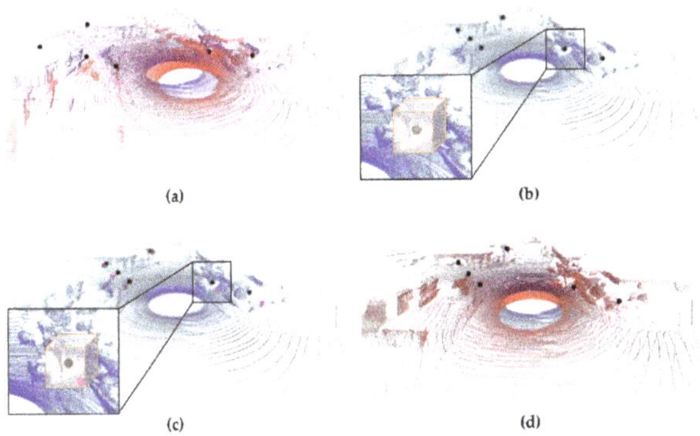

Figure 8. An illustration of the DeepVCP approach; the source point clouds and target point clouds are matched by the registration process. (**a**) The source (red) and target (blue) point clouds and the keypoints (black) detected by the point weighting layer. (**b**) A search region is generated for each keypoint and represented by grid voxels. (**c**) The matched points (magenta) generated by the corresponding point generation layer. (**d**) The final registration result computed by performing SVD given the matched keypoint pairs [57].

While significant advancements have been made in the algorithms employed for point cloud registration, several challenges and limitations persist, hindering the widespread adoption of these techniques. The most prominent challenges are as follows:

- Point cloud data collected from sensors usually contain outliers, noise, and missing data, which can affect the registration process.
- Point cloud registration algorithms often involve computationally intensive operations and require substantial computational resources, such as memory and processing power.
- In situations where point clouds are non-overlapping or have limited overlap, it is difficult for the registration process to find the correspondence and visualize the environment.
- Imprecise calibration of sensors could also be introduced to the registration process. Sensors involved in the data collection need to be properly calibrated for increased accuracy.
- Density differences in point clouds could result either due to sensor differences or there could be variations across different regions of the scene. For example, in regions where the point cloud is sparse, it could be difficult to find correspondence between points leading to errors in the point cloud registration.

3.4. Feature Extraction for HD Map

After the point cloud registration process, feature extraction becomes a crucial component in generating informative and high-quality HD maps for reliable use across various environments. Feature extraction involves identifying and extracting relevant features from sensor data to create a detailed representation of the environment. Features refer to distinct elements of the environment that are extracted from sensor data and used to create detailed representations of the surroundings. These features provide valuable information for navigation, vehicle localization, and various other applications in AVs and systems. Some common features found in HD maps include [5,59] road and lane markings, road networks, traffic signs, landmarks, and pole-like objects. Traditionally feature extraction was done manually, with human operators identifying and annotating features of interest in sensor data. This process is not cost-effective and it is a time-consuming task [5]. Recently, machine learning approaches, particularly deep learning techniques, have been employed to automate feature extraction from point cloud data [60].

Several deep learning methods have been proposed for extracting specific features, such as lane markings. LMPNet [59] is based on the feature pyramid network (FPN) proposed a method to extract the lane markings in images, which are then projected from a perspective space into a three-dimensional space in accordance with the position data. The authors of [61] used StixelNet, which employs deep convolutional networks, to obtain outstanding results in lane detection and object detection on the KITTI dataset [26]. In addition, deep deconvolutional networks with convolutional networks were used in [62], whereas EL-GAN [63] used generative adversarial networks (GANs) for feature extraction. DAGMapper [60] utilizes inference within a directed acyclic graphical model (DAG), integrating deep neural networks for conditional probabilities and employing a greedy algorithm for graph estimation. Deep learning algorithms for extracting traffic signs are also presented in [64]. Additionally, YOLOv3 architecture [65] has been employed for traffic sign detection [63]. For extracting road networks, aerial images are usually utilized since they offer a bird's-eye view of the terrain. Deep learning techniques like CNNs have been used in DeepRoadMapper [66], RoadTracer [67], and [68], to extract the road topology from aerial images. In addition, road networks can be extracted from 3D point clouds, sensor fusion, and camera images [5]. Figure 9 shows road network extraction methods.

Figure 9. Comparison of inferred road networks in Chicago (**top**), Boston, Salt Lake City, and Toronto (**bottom**). The yellow is the inferred graph (yellow) over ground truth from OSM (blue). RoadTracer (**Left**), RoadTracer-segmentation approach (**Middle**), and DeepRoadMapper (**Right**) [67].

4. Updating HD Maps

4.1. Importance of Regular Updates

Ensuring accurate and reliable navigation for AVs necessitates regular updates to HD maps. As AVs navigate through various environments, changes such as road infrastructure modifications, variations in traffic patterns, and dynamic events like road closures, construction sites, and accidents are inevitable [5,69]. Regularly updating HD maps is crucial to reflect these changes and support safe and efficient navigation strategies [7].

Moreover, frequent map updates play a vital role in enhancing the quality of route planning capabilities. By incorporating the latest information on road conditions, traffic patterns, and potential disruptions, AVs can dynamically adjust their planned routes, enabling timely responses to dynamic events [3,6]. This adaptability not only optimizes navigation efficiency but also contributes to improved safety by accounting for real-time changes in the environment. Therefore, maintaining up-to-date HD maps through regular updates is a critical aspect of autonomous vehicle operation, ensuring that the navigation system has access to the most accurate and current information, allowing for informed decision-making and seamless adaptation to the ever-changing conditions of the road network.

4.2. Real-Time Data Sources

Real-time data sources provide crucial information for updating HD maps. These sources include vehicle sensors, such as cameras, LiDAR, and radar, which capture data about the surrounding environment in real time. Additionally, connected vehicles (vehicle-to-vehicle (V2V)) and infrastructures (vehicle-to-infrastructure (V2I)), such as traffic monitoring systems and road sensors, contribute to real-time data collection. Other sources, such as crowd-sourced data from navigation apps and social media, could provide valuable insights into traffic conditions and road incidents if the sources are well-trusted.

4.3. Automated Update Process

To ensure efficient and accurate updates to HD maps, automated processes have been employed to streamline the integration of real-time data streams with map databases. These processes leverage algorithms to detect and incorporate changes into the maps, reducing the need for manual intervention and minimizing the risk of errors. In recent years, deep learning techniques have emerged as powerful tools for automating the detection of changes to be integrated into HD map updates [70,71]. In a comprehensive survey paper [6], various approaches employed in automated map update processes are outlined, including probabilistic and geometric-based approaches. Techniques such as Kalman smoothing and Dempster–Shafer's theory of beliefs, as demonstrated in [72], leverage probabilistic models to incorporate uncertainties and belief systems in the update process. Geometric-based approaches rely on geometric calculations and analyses to detect and incorporate changes in the environment, complementing deep learning techniques. By leveraging these automated techniques, HD maps can remain up-to-date with the latest information, reflecting changes in the environment with minimal manual effort. This automation not only improves the efficiency of the update process but also enhances the overall accuracy and reliability of the HD maps, ensuring seamless and safe navigation for AVs.

Simultaneous localization and mapping (SLAM) algorithms are another crucial component in the automated update process for HD maps. SLAM techniques enable AVs to generate and maintain maps of their environment while simultaneously determining their location within the mapped environment. This capability is particularly valuable for updating HD maps in real-time as the vehicle navigates through dynamic environments. LiDAR-based SLAM algorithms, such as LOAM (lidar odometry and mapping) [73], LeGO-LOAM [74], and SuMa++ [75], leverage point cloud data from LiDAR sensors to simultaneously estimate the vehicle's pose and create 3D maps of the environment. These algorithms can detect changes in the surroundings, such as new construction, obstacles, or road alterations, and automatically incorporate these updates into the HD maps. Visual SLAM techniques, like ORB-SLAM [76] and DSO (direct sparse odometry) [77], utilize visual odometry to perform localization and mapping. By analyzing the visual data and detecting changes in the environment, these algorithms can update the HD maps with new or modified features, such as road markings, traffic signs, or landmarks. Additionally, multi-sensor fusion approaches that combine data from LiDAR, cameras, IMU, and other sensors have been developed for SLAM-based HD map updates. Examples include LIO-SAM [78], R3-Live [79], FAST-LIVO [80], FAST-LIO [81], and FAST-LIO2 [82], which leverage the complementary strengths of different sensor modalities for robust and accurate map updates. By integrating SLAM algorithms into the automated update process, HD maps can continuously adapt to the dynamic changes in the environment, ensuring up-to-date and reliable information for autonomous vehicle navigation.

5. Challenges and Solutions

Over the years, HD maps have gained a lot of popularity and have experienced major breakthroughs. However, there are still areas where HD maps need improvement. These challenges will be discussed in this section.

5.1. Environmental Challenges

While keeping HD maps accurate helps mitigate the impact of environmental challenges [6], maintaining accurate HD maps and ensuring they function reliably under all environmental conditions is a significant challenge [4]. Handling construction zones, performing reliably in harsh weather conditions, and effectively mapping complex urban landscapes require constant updates and significant budgetary investments.

5.2. Technical Challenges

Universally scalable HD maps are essential for efficient autonomous driving systems in any city, region, or road. However, scalability is difficult to achieve due to factors such as varying traffic rules, signage, and road geometry across regions [6]. More efficient mapping algorithms are required to handle these discrepancies. Additionally, the lack of standardization in HD maps is a concern. Different companies use proprietary formats, making data exchange and interoperability challenging. The complexity of data collection presents another technical hurdle. After collecting terabytes of data, preprocessing it for just an hour of driving requires significant computing power, particularly for tasks like identifying and classifying relevant information [13]. Finally, delivering updated map data to AVs with minimal latency is crucial for safe operation. This is especially important in areas with poor internet infrastructure where slow network speeds can compromise safety.

5.3. Data Privacy and Security Concerns

In the technology and data-driven era, privacy is paramount. For AD systems, the requirement for centimeter-level precision necessitates recording the user's location and movement. This information can be leveraged to provide services like live updates on traffic congestion and road blockages [4]. Nevertheless, if not handled properly, these sensitive data could potentially be exploited for malicious purposes. The issue of security should not be taken lightly; it must be addressed comprehensively at every level by HD map providers.

6. Future Direction

6.1. Advances in Sensor Technologies

Although the future remains uncertain, and the technology we currently utilize may eventually be superseded, continued advancements in existing sensor technologies, such as LiDAR, cameras, radar, and IMUs, are anticipated to play a pivotal role in enhancing the capabilities of HD map creation and updating processes. Emerging sensor technologies, including solid-state LiDAR, high-resolution cameras, and multi-modal sensor fusion systems, offer improved accuracy, reliability, and cost-effectiveness for capturing and processing environmental data, further boosting the development of robust and up-to-date HD maps.

6.2. Integration of 5G and Edge Computing for Real-Time Updates

The integration of 5G cellular networks and edge computing infrastructure will facilitate real-time updates and seamless distribution of HD map data to vehicles and other connected devices. The high bandwidth and low latency offered by 5G networks, coupled with edge computing capabilities at roadside infrastructure and onboard vehicle systems, will enable dynamic map updates, efficient traffic management, and enhanced situational awareness for AVs. Furthermore, as 5G networks become available in the future, HD mapping systems will benefit from improved robustness, reliability, and potentially new functionalities, driven by the advancements in network performance and computing power.

6.3. AI in HD Map Creation and Updating

Artificial intelligence (AI) and machine learning (ML) algorithms will continue to play a critical role in HD map generation and updating processes. Deep learning, reinforcement learning, and generative adversarial networks (GANs) enable automated feature extraction, anomaly detection, change detection, and predictive modeling from sensor data. AI-driven approaches streamline map creation workflows, improve accuracy, and adapt to changing environments more effectively.

6.4. Standardized HD Mapping Ecosystem

Currently, most companies involved in AD produce their proprietary HD maps. However, collaboration among industry stakeholders, including automotive manufacturers, technology companies, government agencies, and standards organizations like the ASAM OpenDRIVE format [22], is essential for standardizing HD map formats, data exchange protocols, and quality assurance processes. Lanelet2, Apollo (utilizing a modified OpenDRIVE format), and NDS (Global standard for automotive map data) represent significant steps toward this standardization. Critical steps include fostering collaboration among stakeholders, adopting established formats such as OpenDRIVE, Lanelet2, and NDS for compatibility and interoperability, and developing unified specifications. Concerted standardization efforts promote seamless data sharing, and innovation across the entire ecosystem, driving continuous improvement in HD map creation, updating, and utilization for various applications. This collaborative approach fosters a unified and consistent HD mapping infrastructure and accelerates the development and deployment of advanced autonomous driving solutions, benefiting the industry as a whole. These future directions represent ongoing trends and areas of focus in the development and advancement of HD maps, with the potential to reshape the landscape of future transportation, navigation, and urban mobility in the years to come.

6.5. Autonomous Vehicles for Sustainability

The advancement of autonomous vehicles (AVs) represents a pivotal opportunity to foster sustainability in transportation. AVs have the potential to significantly reduce greenhouse gas emissions by optimizing driving patterns and minimizing energy consumption through the use of high-definition (HD) maps. These maps provide precise geographic data that enable AVs to navigate efficiently, thereby reducing fuel usage and emissions. Studies suggest that AVs can cut fuel consumption by 10–20% compared to conventional vehicles, particularly in urban areas where traffic congestion is prevalent [83]. Furthermore, AVs equipped with HD maps can enhance traffic flow by communicating with infrastructure and other vehicles, mitigating congestion, and improving overall air quality.

Autonomous driving technology also promotes sustainability through the promotion of shared mobility services and the potential reduction in vehicle ownership. By integrating with ride-sharing platforms and optimizing routes based on real-time data from HD maps, AVs can support a shift toward shared mobility models, thereby reducing the number of vehicles on the road and decreasing urban sprawl. Additionally, AVs contribute to safety improvements by leveraging HD maps for precise navigation and obstacle detection, thereby reducing traffic accidents and their associated environmental and economic costs. As cities and transportation systems evolve, the integration of AV technology with sustainable urban planning initiatives facilitated by HD maps holds promise for creating more efficient, livable, and environmentally friendly urban environments.

7. Conclusions

This comprehensive survey underscores the critical role of HD maps in propelling the advancement of autonomous driving technologies. It delves into various facets of HD map generation and maintenance, exploring the intricacies of map layers, data collection methodologies, sensor fusion strategies, point cloud registration techniques, and feature extraction processes essential for creating detailed and accurate representations of the environment. Moreover, the significance of regular updates in ensuring the currency and reliability of HD maps is emphasized. Furthermore, this review addresses the challenges inherent in HD map development and maintenance, while also exploring potential solutions and future perspectives that promise to enhance the effectiveness and utility of HD maps in driving the evolution of autonomous driving systems.

Author Contributions: Conceptualization, Kaleab Taye Asrat and Hyung-Ju Cho; methodology, Kaleab Taye Asrat; software, Kaleab Taye Asrat; validation, Kaleab Taye Asrat; formal analysis, Kaleab Taye Asrat; investigation, Kaleab Taye Asrat; resources, Kaleab Taye Asrat; data curation, Kaleab Taye Asrat; writing—original draft preparation, Kaleab Taye Asrat; writing—review and editing, Hyung-Ju Cho; visualization, Kaleab Taye Asrat; supervision, Hyung-Ju Cho; project administration, Hyung-Ju Cho; funding acquisition, Hyung-Ju Cho. All authors have read and agreed to the published version of the manuscript.

Funding: This research was supported by the Basic Science Research Program through the National Research Foundation of Korea (NRF), funded by the Ministry of Education (NRF-2020R1I1A3052713).

Data Availability Statement: No new data were created or analyzed in this study.

Conflicts of Interest: The authors declare no conflicts of interest.

References

1. Delcker, J. The Man Who Invented the Self-Driving Car (in 1986). Available online: https://www.politico.eu/article/delf-driving-car-born-1986-ernst-dickmanns-mercedes/ (accessed on 14 February 2024).
2. Chen, L.; Li, Y.; Huang, C.; Li, B.; Xing, Y.; Tian, D.; Li, L.; Hu, Z.; Na, X.; Li, Z.; et al. Milestones in Autonomous Driving and Intelligent Vehicles: Survey of Surveys. *IEEE Trans. Intell. Veh.* **2023**, *8*, 1046–1056. [CrossRef]
3. Ziegler, J.; Bender, P.; Schreiber, M.; Lategahn, H.; Strauss, T.; Stiller, C.; Dang, T.; Franke, U.; Appenrodt, N.; Keller, C.G.; et al. Making Bertha Drive—An Autonomous Journey on a Historic Route. *IEEE Intell. Transp. Syst. Mag.* **2014**, *6*, 8–20. [CrossRef]
4. Wong, K.; Gu, Y.; Kamijo, S. Mapping for Autonomous Driving: Opportunities and Challenges. *IEEE Intell. Transp. Syst. Mag.* **2021**, *13*, 91–106. [CrossRef]
5. Bao, Z.; Hossain, S.; Lang, H.; Lin, X. A review of high-definition map creation methods for autonomous driving. *Eng. Appl. Artif. Intell.* **2023**, *122*, 106125. [CrossRef]
6. Elghazaly, G.; Frank, R.; Harvey, S.; Safko, S. High-Definition Maps: Comprehensive Survey, Challenges, and Future Perspectives. *IEEE Open J. Intell. Transp. Syst.* **2023**, *4*, 527–550. [CrossRef]
7. Dahlström, T. The Road to Everywhere: Are HD Maps for Autonomous Driving Sustainable? Available online: https://www.autonomousvehicleinternational.com/features/the-road-to-everywhere-are-hd-maps-for-autonomous-driving-sustainable.html (accessed on 14 February 2024).
8. Ebrahimi Soorchaei, B.; Razzaghpour, M.; Valiente, R.; Raftari, A.; Fallah, Y.P. High-Definition Map Representation Techniques for Automated Vehicles. *Electronics* **2022**, *11*, 3374. [CrossRef]
9. Bijelic, M.; Gruber, T.; Mannan, F.; Kraus, F.; Ritter, W.; Dietmayer, K.; Heide, F. Seeing through fog without seeing fog: Deep multimodal sensor fusion in unseen adverse weather. In Proceedings of the IEEE/CVF Conference on Computer Vision and Pattern Recognition, Seattle, WA, USA, 14–19 June 2020; pp. 11682–11692.
10. Apollo. Available online: https://github.com/ApolloAuto/apollo (accessed on 6 June 2024).
11. Level5, W.P. Rethinking Maps for Self-Driving. Available online: https://medium.com/wovenplanetlevel5/https-medium-com-lyftlevel5-rethinking-maps-for-self-driving-a147c24758d6 (accessed on 14 February 2024).
12. Bučko, B.; Zábovská, K.; Ristvej, J.; Jánošíková, M. HD Maps and Usage of Laser Scanned Data as a Potential Map Layer. In Proceedings of the 2021 44th International Convention on Information, Communication and Electronic Technology (MIPRO), Opatija, Croatia, 27 September–1 October 2021; pp. 1670–1675. [CrossRef]
13. Seif, H.G.; Hu, X. Autonomous Driving in the iCity—HD Maps as a Key Challenge of the Automotive Industry. *Engineering* **2016**, *2*, 159–162. [CrossRef]
14. Bonetti, P. HERE Introduces HD LIVE Map to Show the Path to Highly Automated Driving. Available online: https://www.here.com/learn/blog/here-introduces-hd-live-map-to-show-the-path-to-highly-automated-driving (accessed on 1 March 2024).
15. Lógó, J.M.; Barsi, A. The Role of Topology in High-Definition Maps for Autonomous Driving. *Int. Arch. Photogramm. Remote Sens. Spat. Inf. Sci.* **2022**, *XLIII-B4-2022*, 383–388. [CrossRef]
16. HERE. Introduction to Mapping Concepts—Localization Model. Available online: https://www.here.com/docs/bundle/introduction-to-mapping-concepts-user-guide/page/topics/model-local.html (accessed on 14 February 2024).
17. TomTom. TomTom HD Maps. Available online: https://www.tomtom.com/products/hd-map/ (accessed on 14 February 2024).

18. Garg, S.; Milford, M. Seqnet: Learning descriptors for sequence-based hierarchical place recognition. *IEEE Robot. Autom. Lett.* **2021**, *6*, 4305–4312. [CrossRef]
19. Tao, Q.; Hu, Z.; Zhou, Z.; Xiao, H.; Zhang, J. SeqPolar: Sequence matching of polarized LiDAR map with HMM for intelligent vehicle localization. *IEEE Trans. Veh. Technol.* **2022**, *71*, 7071–7083. [CrossRef]
20. Poggenhans, F.; Pauls, J.H.; Janosovits, J.; Orf, S.; Naumann, M.; Kuhnt, F.; Mayr, M. Lanelet2: A high-definition map framework for the future of automated driving. In Proceedings of the 2018 21st International Conference on Intelligent Transportation Systems (ITSC), Maui, HI, USA, 4–7 November 2018; IEEE: Piscataway, NJ, USA, 2018; pp. 1672–1679.
21. Bender, P.; Ziegler, J.; Stiller, C. Lanelets: Efficient map representation for autonomous driving. In Proceedings of the 2014 IEEE Intelligent Vehicles Symposium Proceedings, Ypsilanti, MI, USA, 8–11 June 2014; IEEE: Piscataway, NJ, USA, 2014; pp. 420–425.
22. ASAM OpenDRIVE. Available online: https://www.asam.net/standards/detail/opendrive/ (accessed on 14 February 2024).
23. Gran, C.W. HD-Maps in Autonomous Driving. Master's Thesis, Norwegian University of Science and Technology, Trondheim, Norway, 2019.
24. Elhashash, M.; Albanwan, H.; Qin, R. A Review of Mobile Mapping Systems: From Sensors to Applications. *Sensors* **2022**, *22*, 4262. [CrossRef] [PubMed]
25. Houston, J.L.; Zuidhof, G.C.A.; Bergamini, L.; Ye, Y.; Jain, A.; Omari, S.; Iglovikov, V.I.; Ondruska, P. One Thousand and One Hours: Self-driving Motion Prediction Dataset. In Proceedings of the 4th Conference on Robot Learning, Cambridge, MA, USA, 16–18 November 2020; pp. 409–418.
26. Geiger, A.; Lenz, P.; Stiller, C.; Urtasun, R. Vision meets robotics: The kitti dataset. *Int. J. Robot. Res.* **2013**, *32*, 1231–1237. [CrossRef]
27. Caesar, H.; Bankiti, V.; Lang, A.H.; Vora, S.; Liong, V.E.; Xu, Q.; Krishnan, A.; Pan, Y.; Baldan, G.; Beijbom, O. nuScenes: A multimodal dataset for autonomous driving. In Proceedings of the IEEE/CVF Conference on Computer Vision and Pattern Recognition, Seattle, WA, USA, 14–19 June 2020; pp. 11621–11631.
28. Chang, M.F.; Lambert, J.; Sangkloy, P.; Singh, J.; Bak, S.; Hartnett, A.; Wang, D.; Carr, P.; Lucey, S.; Ramanan, D.; et al. Argoverse: 3D Tracking and Forecasting With Rich Maps. In Proceedings of the 2019 IEEE/CVF Conference on Computer Vision and Pattern Recognition (CVPR), Long Beach, CA, USA, 15–20 June 2019; pp. 8740–8749. [CrossRef]
29. Kim, C.; Cho, S.; Sunwoo, M.; Resende, P.; Bradaï, B.; Jo, K. Updating point cloud layer of high definition (hd) map based on crowd-sourcing of multiple vehicles installed lidar. *IEEE Access* **2021**, *9*, 8028–8046. [CrossRef]
30. Fayyad, J.; Jaradat, M.A.; Gruyer, D.; Najjaran, H. Deep Learning Sensor Fusion for Autonomous Vehicle Perception and Localization: A Review. *Sensors* **2020**, *20*, 4220. [CrossRef] [PubMed]
31. Wang, Z.; Wu, Y.; Niu, Q. Multi-Sensor Fusion in Automated Driving: A Survey. *IEEE Access* **2020**, *8*, 2847–2868. [CrossRef]
32. Wang, X.; Li, K.; Chehri, A. Multi-Sensor Fusion Technology for 3D Object Detection in Autonomous Driving: A Review. *IEEE Trans. Intell. Transp. Syst.* **2024**, *25*, 1148–1165. [CrossRef]
33. Yeong, D.J.; Velasco-Hernandez, G.; Barry, J.; Walsh, J. Sensor and Sensor Fusion Technology in Autonomous Vehicles: A Review. *Sensors* **2021**, *21*, 2140. [CrossRef] [PubMed]
34. Zhong, H.; Wang, H.; Wu, Z.; Zhang, C.; Zheng, Y.; Tang, T. A survey of LiDAR and camera fusion enhancement. *Procedia Comput. Sci.* **2021**, *183*, 579–588. [CrossRef]
35. Chen, X.; Ma, H.; Wan, J.; Li, B.; Xia, T. Multi-view 3d object detection network for autonomous driving. In Proceedings of the IEEE conference on Computer Vision and Pattern Recognition, Honolulu, HI, USA, 21–26 July 2017; pp. 1907–1915.
36. Shahian Jahromi, B.; Tulabandhula, T.; Cetin, S. Real-Time Hybrid Multi-Sensor Fusion Framework for Perception in Autonomous Vehicles. *Sensors* **2019**, *19*, 4357. [CrossRef]
37. Liang, M.; Yang, B.; Chen, Y.; Hu, R.; Urtasun, R. Multi-task multi-sensor fusion for 3d object detection. In Proceedings of the IEEE/CVF Conference on Computer Vision and Pattern Recognition, Long Beach, CA, USA, 16–20 June 2019; pp. 7337–7345.
38. Thakur, A.; Rajalakshmi, P. LiDAR and Camera Raw Data Sensor Fusion in Real-Time for Obstacle Detection. In Proceedings of the 2023 IEEE Sensors Applications Symposium (SAS), Ottawa, ON, Canada, 18–20 July 2023; pp. 1–6. [CrossRef]
39. Qi, C.R.; Liu, W.; Wu, C.; Su, H.; Guibas, L.J. Frustum pointnets for 3d object detection from rgb-d data. In Proceedings of the IEEE Conference on Computer Vision and Pattern Recognition, Salt Lake City, UT, USA, 18–22 June 2018; pp. 918–927.
40. Xiao, Y.; Liu, Y.; Luan, K.; Cheng, Y.; Chen, X.; Lu, H. Deep LiDAR-Radar-Visual Fusion for Object Detection in Urban Environments. *Remote Sens.* **2023**, *15*, 4433. [CrossRef]
41. Ravindran, R.; Santora, M.J.; Jamali, M.M. Camera, LiDAR, and Radar Sensor Fusion Based on Bayesian Neural Network (CLR-BNN). *IEEE Sens. J.* **2022**, *22*, 6964–6974. [CrossRef]
42. Nobis, F.; Geisslinger, M.; Weber, M.; Betz, J.; Lienkamp, M. A Deep Learning-based Radar and Camera Sensor Fusion Architecture for Object Detection. In Proceedings of the 2019 Sensor Data Fusion: Trends, Solutions, Applications (SDF), Bonn, Germany, 15–17 October 2019; pp. 1–7. [CrossRef]
43. Redmon, J.; Divvala, S.; Girshick, R.; Farhadi, A. You Only Look Once: Unified, Real-Time Object Detection. In Proceedings of the 2016 IEEE Conference on Computer Vision and Pattern Recognition, Las Vegas, NV, USA, 26 June–1 July 2016; pp. 779–788. [CrossRef]
44. Liu, W.; Anguelov, D.; Erhan, D.; Szegedy, C.; Reed, S.; Fu, C.Y.; Berg, A.C., SSD: Single Shot MultiBox Detector. In *Lecture Notes in Computer Science*; Springer International Publishing: Berlin/Heidelberg, Germany, 2016; pp. 21–37. [CrossRef]

45. Zhou, Y.; Tuzel, O. VoxelNet: End-to-End Learning for Point Cloud Based 3D Object Detection. *arXiv* **2017**, arXiv:1711.06396. Available online: http://arxiv.org/abs/1711.06396 (accessed on 13 April 2024).
46. Qi, C.R.; Su, H.; Mo, K.; Guibas, L.J. PointNet: Deep Learning on Point Sets for 3D Classification and Segmentation. *arXiv* **2017**, arXiv:1612.00593. Available online: http://arxiv.org/abs/1612.00593 (accessed on 13 April 2024).
47. Huang, X.; Mei, G.; Zhang, J.; Abbas, R. A comprehensive survey on point cloud registration. *arXiv* **2021**, arXiv:2103.02690. Available online: http://arxiv.org/abs/2103.02690 (accessed on 13 April 2024).
48. Chang, M.F.; Dong, W.; Mangelson, J.; Kaess, M.; Lucey, S. Map Compressibility Assessment for LiDAR Registration. In Proceedings of the 2021 IEEE/RSJ International Conference on Intelligent Robots and Systems (IROS), Prague, Czech Republic, 27 September–1 October 2021; pp. 5560–5567. [CrossRef]
49. Besl, P.; McKay, N.D. A method for registration of 3-D shapes. *IEEE Trans. Pattern Anal. Mach. Intell.* **1992**, *14*, 239–256. [CrossRef]
50. Zhang, Z. Iterative point matching for registration of free-form curves and surfaces. *Int. J. Comput. Vis.* **1994**, *13*, 119–152. [CrossRef]
51. Zhang, J.; Yao, Y.; Deng, B. Fast and Robust Iterative Closest Point. *IEEE Trans. Pattern Anal. Mach. Intell.* **2022**, *44*, 3450–3466. [CrossRef]
52. Wang, F.; Zhao, Z. A survey of iterative closest point algorithm. In Proceedings of the 2017 Chinese Automation Congress (CAC), Jinan, China, 20–22 October 2017; pp. 4395–4399. [CrossRef]
53. Myronenko, A.; Song, X. Point Set Registration: Coherent Point Drift. *IEEE Trans. Pattern Anal. Mach. Intell.* **2010**, *32*, 2262–2275. [CrossRef]
54. Liu, Z.; Yu, C.; Qiu, F.; Liu, Y. A Fast Coherent Point Drift Method for Rigid 3D Point Cloud Registration. In Proceedings of the 2023 42nd Chinese Control Conference (CCC), Tianjin, China, 24–26 July 2023; pp. 7776–7781. [CrossRef]
55. Gao, W.; Tedrake, R. Filterreg: Robust and efficient probabilistic point-set registration using gaussian filter and twist parameterization. In Proceedings of the IEEE/CVF Conference on Computer Vision and Pattern Recognition, Long Beach, CA, USA, 15–20 June 2019; pp. 11095–11104.
56. Wang, Y.; Solomon, J. Deep Closest Point: Learning Representations for Point Cloud Registration. In Proceedings of the 2019 IEEE/CVF International Conference on Computer Vision (ICCV), Seoul, Republic of Korea, 27 October–2 November 2019; pp. 3522–3531. [CrossRef]
57. Lu, W.; Wan, G.; Zhou, Y.; Fu, X.; Yuan, P.; Song, S. DeepVCP: An End-to-End Deep Neural Network for Point Cloud Registration. In Proceedings of the 2019 IEEE/CVF International Conference on Computer Vision (ICCV), Seoul, Republic of Korea, 27 October–2 November 2019; pp. 12–21. [CrossRef]
58. Segal, A.; Haehnel, D.; Thrun, S. Generalized-icp. In Proceedings of the Robotics: Science and Systems V, Seattle, WA, USA, 28 June–1 July 2009; pp. 21–31. [CrossRef]
59. Zhou, J.; Guo, Y.; Bian, Y.; Huang, Y.; Li, B. Lane Information Extraction for High Definition Maps Using Crowdsourced Data. *IEEE Trans. Intell. Transp. Syst.* **2023**, *24*, 7780–7790. [CrossRef]
60. Homayounfar, N.; Liang, J.; Ma, W.C.; Fan, J.; Wu, X.; Urtasun, R. DAGMapper: Learning to Map by Discovering Lane Topology. In Proceedings of the 2019 IEEE/CVF International Conference on Computer Vision (ICCV), Seoul, Republic of Korea, 27 October–2 November 2019; pp. 2911–2920. [CrossRef]
61. Levi, D.; Garnett, N.; Fetaya, E.; Herzlyia, I. Stixelnet: A deep convolutional network for obstacle detection and road segmentation. In Proceedings of the BMVC, Swansea, UK, 7–10 September 2015; Volume 1, p. 4.
62. Mohan, R. Deep Deconvolutional Networks for Scene Parsing. *arXiv* **2014**, arXiv:1411.4101. Available online: http://arxiv.org/abs/1411.4101 (accessed on 13 April 2024).
63. Ghafoorian, M.; Nugteren, C.; Baka, N.; Booij, O.; Hofmann, M. EL-GAN: Embedding Loss Driven Generative Adversarial Networks for Lane Detection. *arXiv* **2018**, arXiv:1806.05525. Available online: http://arxiv.org/abs/1806.05525 (accessed on 13 April 2024).
64. Plachetka, C.; Sertolli, B.; Fricke, J.; Klingner, M.; Fingscheidt, T. DNN-Based Map Deviation Detection in LiDAR Point Clouds. *IEEE Open J. Intell. Transp. Syst.* **2023**, *4*, 580–601. [CrossRef]
65. Redmon, J.; Farhadi, A. YOLOv3: An Incremental Improvement. *arXiv* **2018**, arXiv:1804.02767. Available online: http://arxiv.org/abs/1804.02767 (accessed on 13 April 2024).
66. Máttyus, G.; Luo, W.; Urtasun, R. DeepRoadMapper: Extracting Road Topology from Aerial Images. In Proceedings of the 2017 IEEE International Conference on Computer Vision (ICCV), Venice, Italy, 22–29 October 2017; pp. 3458–3466. [CrossRef]
67. Bastani, F.; He, S.; Abbar, S.; Alizadeh, M.; Balakrishnan, H.; Chawla, S.; Madden, S.; DeWitt, D. RoadTracer: Automatic Extraction of Road Networks from Aerial Images. *arXiv* **2018**, arXiv:1802.03680. Available online: http://arxiv.org/abs/1802.03680 (accessed on 13 April 2024).
68. Ventura, C.; Pont-Tuset, J.; Caelles, S.; Maninis, K.K.; Gool, L.V. Iterative Deep Learning for Road Topology Extraction. *arXiv* **2018**, arXiv:1808.09814. Available online: http://arxiv.org/abs/1808.09814 (accessed on 13 April 2024).
69. Liu, R.; Wang, J.; Zhang, B. High Definition Map for Automated Driving: Overview and Analysis. *J. Navig.* **2020**, *73*, 324–341. [CrossRef]
70. TaŞ, M.O.; Yavuz, S.; Yazici, A. Updating HD-Maps for Autonomous Transfer Vehicles in Smart Factories. In Proceedings of the 2018 6th International Conference on Control Engineering & Information Technology (CEIT), Istanbul, Turkey, 25–27 October 2018; pp. 1–5. [CrossRef]

71. Heo, M.; Kim, J.; Kim, S. HD Map Change Detection with Cross-Domain Deep Metric Learning. In Proceedings of the 2020 IEEE/RSJ International Conference on Intelligent Robots and Systems (IROS), Las Vegas, NV, USA, 25–29 October 2020; pp. 10218–10224. [CrossRef]
72. Jo, K.; Kim, C.; Sunwoo, M. Simultaneous Localization and Map Change Update for the High Definition Map-Based Autonomous Driving Car. *Sensors* **2018**, *18*, 3145. [CrossRef]
73. Zhang, J.; Singh, S. LOAM: Lidar odometry and mapping in real-time. In Proceedings of the Robotics: Science and Systems, Berkeley, CA, USA, 12–16 July 2014; Volume 2, pp. 1–9.
74. Shan, T.; Englot, B. LeGO-LOAM: Lightweight and Ground-Optimized Lidar Odometry and Mapping on Variable Terrain. In Proceedings of the 2018 IEEE/RSJ International Conference on Intelligent Robots and Systems (IROS), Madrid, Spain, 1–5 October 2018; pp. 4758–4765. [CrossRef]
75. Chen, X.; Milioto, A.; Palazzolo, E.; Giguere, P.; Behley, J.; Stachniss, C. Suma++: Efficient lidar-based semantic slam. In Proceedings of the 2019 IEEE/RSJ International Conference on Intelligent Robots and Systems (IROS), Macau, China, 3–8 November 2019; IEEE: Piscataway, NJ, USA, 2019; pp. 4530–4537.
76. Mur-Artal, R.; Montiel, J.M.; Tardos, J.D. ORB-SLAM: A Versatile and Accurate Monocular SLAM System Mur-Artal, Raul and Montiel, Jose Maria Martinez and Tardos. *IEEE Trans. Robot.* **2015**, *31*, 1147–1163.
77. Engel, J.; Koltun, V.; Cremers, D. Direct sparse odometry. *IEEE Trans. Pattern Anal. Mach. Intell.* **2017**, *40*, 611–625. [CrossRef]
78. Shan, T.; Englot, B.; Meyers, D.; Wang, W.; Ratti, C.; Rus, D. LIO-SAM: Tightly-coupled Lidar Inertial Odometry via Smoothing and Mapping. *arXiv* **2020**, arXiv:2007.00258. Available online: http://arxiv.org/abs/2007.00258 (accessed on 13 April 2024).
79. Lin, J.; Zhang, F. R 3 LIVE: A Robust, Real-time, RGB-colored, LiDAR-Inertial-Visual tightly-coupled state Estimation and mapping package. In Proceedings of the 2022 International Conference on Robotics and Automation (ICRA), Philadelphia, PA, USA, 23–27 May 2022; IEEE: Piscataway, NJ, USA, 2022; pp. 10672–10678.
80. Zheng, C.; Zhu, Q.; Xu, W.; Liu, X.; Guo, Q.; Zhang, F. FAST-LIVO: Fast and tightly-coupled sparse-direct LiDAR-inertial-visual odometry. In Proceedings of the 2022 IEEE/RSJ International Conference on Intelligent Robots and Systems (IROS), Kyoto, Japan, 23–27 October 2022; IEEE: Piscataway, NJ, USA, 2022; pp. 4003–4009.
81. Xu, W.; Zhang, F. FAST-LIO: A Fast, Robust LiDAR-Inertial Odometry Package by Tightly-Coupled Iterated Kalman Filter. *IEEE Robot. Autom. Lett.* **2021**, *6*, 3317–3324. [CrossRef]
82. Xu, W.; Cai, Y.; He, D.; Lin, J.; Zhang, F. Fast-lio2: Fast direct lidar-inertial odometry. *IEEE Trans. Robot.* **2022**, *38*, 2053–2073. [CrossRef]
83. Le Hong, Z.; Zimmerman, N. Air quality and greenhouse gas implications of autonomous vehicles in Vancouver, Canada. *Transp. Res. Part D Transp. Environ.* **2021**, *90*, 102676. [CrossRef]

International Journal of
Geo-Information

MDPI

Article

Analysis of Decoupling Effects and Influence Factors in Transportation: Evidence from Guangdong Province, China

Hualing Bi, Shiying Zhang and Fuqiang Lu *

School of Economic and Management, Yanshan University, Qinhuangdao 066004, China; bihualing081@126.com (H.B.)
* Correspondence: fuqiang_lu@126.com

Abstract: In recent years, global environmental issues have become increasingly prominent. The transportation industry, as the fundamental sector of national economic development, is also characterized by high energy consumption and carbon emissions. Therefore, it is imperative to conduct research on the carbon emission problem within this industry. In light of the Tapio decoupling model, an analysis of the correlation between traffic carbon emissions and economic development in Guangdong province during 1999–2019 was carried out. With the aim of encouraging Guangdong province's low-carbon transportation development, the factors affecting the transportation industry are analyzed utilizing the generalized Divisia index model (GDIM). We also introduced passenger and freight turnover as an influencing factor for analysis. The findings indicate that (1) Guangdong province's traffic carbon emissions increased from 1999 to 2019; (2) the traffic carbon emissions' decoupling effect is mainly "weakly decoupled", and the overall decoupling effect is not strong in Guangdong province; (3) among the traffic carbon emissions' factors, the effects of the production value of traffic and the turnover volume are at the forefront, and the effect of turnover volume has gradually exceeded the production value of traffic in recent years. The suppression of the intensity of carbon emissions is relatively large, while the suppression of the intensity of energy consumption and transport is relatively weak. Based on this, strategies were proposed to promote a cleaner energy mix, improve energy use efficiency, create energy savings, develop green technologies, and foster the restructuring of transportation.

Keywords: transportation; carbon emissions; Tapio decoupling model; generalized Divisia index model; decomposition of influencing factors

1. Introduction

The steady rise in the standard of living has been accompanied by a significant increase in the consumption of mineral fuels like coal and oil, which not only causes global climate change but also impacts energy, resources, and the ecological environment. The growing "greenhouse effect" has become a major barrier to achieving sustainable social and economic progress.

In 2015, the 193 member states of the United Nations formally adopted 17 Sustainable Development Goals at the summit. These include the goal of Sustainable Cities and Communities, which aims to render cities and human settlements inclusive, safe, resilient, and sustainable, and the goal of Climate Action, which intends to undertake urgent actions to combat climate change and its impacts. Numerous nations have begun to focus on reducing their carbon emissions in the context of global warming. In July 2021, the European Commission published a set of recommendations known as "Fit for 55", which put forward a wide range of initiatives in energy, industry, transportation, and buildings [1]. In March 2023, the EU and its member states reached a provisional agreement on the revision of the Renewable Energy Directive, reaffirming the EU's determination to achieve energy independence [2]. The United Kingdom has released its Net Zero strategy, which comprehensively sets out how the United Kingdom will achieve its net zero emissions commitment to climate change by 2050 [3]. The U.S. government released the first blueprint for decarbonizing the transportation sector in January 2023, aiming to decarbonize transportation by 2050 [4]. In February 2023, the Cabinet of Ministers approved the "Basic Policy for Realizing the Green Transition", which aims to realize the green transition and simultaneously decarbonize the country, stabilizing the energy supply [5]. In March 2023, the Korean government issued an outline of a basic plan for carbon neutral and green growth and a concrete implementation plan to achieve the 2030 greenhouse gas reduction target and 2050 carbon neutrality [6]. China is also concerned about the negative effects of carbon emissions. The "Fourteenth Five-Year Plan" and the Central Economic Work Conference have made it fairly evident that carbon neutrality objectives should be met before 2060 and carbon peaking targets by 2030 [7]. The 20th National People's Congress also shows that it is necessary to "aggressively and consistently advance carbon peaking and carbon neutrality", enhance the system of market-oriented resource and ecological component allocation, and cultivate low-carbon and green industries.

Transport is an indispensable basic industry for the national economy and people's lives. However, its energy consumption is huge, especially the elevated CO_2 emissions represented by fossil fuels. In this context, transport is a key driver for achieving the objectives of carbon peak and carbon neutrality.

Guangdong is situated at the hub of the maritime transportation routes between the Pacific, Indian, and Atlantic Oceans, with its back leaning on the vast hinterland of China and its front facing Southeast Asia. It is one of China's important maritime traffic chokepoints, with its communication channels to the outside world and major ports for foreign trade. The length of expressways open to traffic in Guangdong is about 11,500 km, ranking first in China for the 10th consecutive year. The operating mileage of the high-speed railway is about 2838 km, and the operating mileage of the urban rail transit system is about 1373 km. China has built about 400 10,000-ton berths and seven 10,000-ton ports. It has opened more than 400 international container liner routes, covering the world's major trading ports. Its inland waterways cover 1479 km of high-grade waterways, ranking first in China. Nine airports, including Guangzhou Baiyun and Shenzhen Baoan, have been built, and 36 general aviation airports have been registered, forming the initial system for the general aviation industry. The geographical location of Guangdong province is shown in Figure 1.

Figure 1. The geographical position of Guangdong province.

At present, Guangdong is still in the midst of a period of high-speed development, with a growing population and increased demand for travel and cargo transportation by residents, posing a serious challenge to China's goals of reducing emissions and conserving energy. By calculating China's transportation-related carbon emissions across all provinces, Guangdong ranked first in transportation emissions, with 84,645,700 metric tons, surpassing second-ranked Shanghai by about 48 percent. The total amounts of the import and export of goods, passenger and freight transportation, and turnover volume of Guangdong province are all ranked first in China, and it produces more carbon emissions while developing continuously. Thus, it is imperative to investigate the factors impacting Guangdong province's traffic carbon emissions.

The following are the fundamental contributions of this paper:

(1) The "up-bottom" method is applied to calculate each province's carbon emissions from traffic. Guangdong's transport carbon emissions are then measured from 1999 to 2019.
(2) Utilizing the Tapio decoupling model as a basis, the research investigates traffic carbon emissions' decoupling status in Guangdong province.
(3) The GDIM model is integrated with the passenger and freight turnover data to conduct a thorough analysis of the current Guangdong province traffic carbon emission issue, and corresponding countermeasures are proposed to lessen carbon emissions.

The reminder of this paper is organized as follows. Section 2 includes a literature review, Section 3 provides a methodology, Section 4 shows the data and results, and Section 5 presents conclusions and recommendations.

2. Literature Review

The Organization for Economic Co-operation and Development (OECD) formulated the decoupling theory first, which expects to effectively solve the problems of resource depletion and environmental contamination. Carbon emission decoupling is an ideal process in which the relationship between economic growth and greenhouse gas emissions constantly weakens and even disappears. That is, there is a gradual reduction of energy consumption on the basis of economic growth. At present, the methods used by scholars are the OECD or Tapio decoupling models. Tapio developed a novel decoupling model that was derived from the OECD decoupling model, investigating the European region's decoupling state [8]. Wang and Zhang investigated trade liberalization's impact, with the results indicating that wealthy countries gain from trade liberalization as it helps to separate economic expansion from carbon emissions. However, it negatively affects developing nations [9]. Wang et al. studied information and communication technology investment and the connection with carbon emissions and found that its decoupling state gradually improved [10]. Shi et al. assessed carbon emissions' decoupling status [11]. Nima used the Tapio model to ascertain the decoupling situation in the Middle East [12]. Zhao et al. revealed that different countries' decoupling states at different times differ in different periods. The U.S. was the best performer [13]. Lu et al. introduced fair preference into the distribution and improved the use requirements [14]. The Tapio decoupling model's benefit is that it does not need to choose the base period when calculating, and it will not be limited by the statistical outline over the OECD decoupling model. It also has a complete decoupling indicator system. As a result, the Tapio decoupling model was implemented to examine Guangdong province's traffic decoupling situation.

The influential factors for the identification and decomposition of transport carbon emissions are useful for encouraging the implementation of related policies and specific measures for subsequent low-carbon transportation development. At present, most scholars combine STIRPAT models, Kaya constant equations, and their expansion forms using index decomposition methods or factor decomposition to decompose factors. Shahbaz et al. examined Pakistan through the application of the STIRPAT model, finding that energy use and urbanization are causally related in a one-way manner [15]. Wang et al. broke down the carbon emissions factors in Chinese passenger and cargo industries between 1990 and 2015 by using the LMDI [16]. Solaymani determined the primary causes that impact carbon emissions by applying the average Divisia index to use seven carbon dioxide emissions countries as a test case [17]. Mine and Kemal evaluated the variables that impact the carbon emissions from Turkish traffic in 2000–2017 and determined that the population and economic factors are the primary drivers [18]. Li et al. integrated the threshold effect model with the decoupling index to assess the transportation departments in 30 provinces in China from 2005 to 2019 [19]. Zhang et al. examined the variables driving urban carbon emissions using the LMDI, which verified that the positive drivers included population growth and GDP per capita, with energy intensity being a negative driver [20]. Lu et al. introduced risk factors into the auction mechanism to improve the higher utility [21]. In terms of decomposition and the influencing factors of traffic carbon emissions, most scholars therefore decompose the factors affecting traffic carbon emissions by index decomposition, factor decomposition, and other methods, combined with the STIRPAT model, Kaya identity, and their extended forms. The exponential decomposition method can only consider one absolute factor at most, which may lead to unrealistic decomposition results. Consequently, this paper uses the GDIM model to discover the variables contributing to Guangdong province's traffic carbon emissions.

Currently, the GDIM model is widely used in the manufacturing industry, manufacturing's high-emission subsectors [22], the construction industry [23], the mining sector [24], and other fields, and used less in the transportation industry. This paper uses the GDIM model to study the transportation industry, which not only helps to expand the application field of the model but also provides new ideas and methods for the sustainable development of the transportation industry. While previous studies mainly approached the issue from the aspects of energy, the economy, and a population change analysis of carbon emissions, Tang and Jiang introduced the urbanization effect, however passenger and freight turnover rarely factor into the influences of the transportation sector [25]. This paper introduces passenger and freight turnover as influencing factors, contributing to a more comprehensive understanding of the transportation industry's carbon decoupling target.

3. Methodology

3.1. Carbon Emissions Calculation

According to the "Publications 2006 IPCC Guidelines for National Greenhouse Gas Inventories", there are two methods to calculate carbon emissions, including the "up-bottom" approach based on traffic fuel consumption and the "bottom-up" approach centered around traffic mileage [26]. The essence of the "up-bottom" approach is an excellent method based on fuel consumption statistics, which can avoid the difficulty of obtaining data. Consequently, this paper uses the "up-bottom" approach to calculate Guangdong province's transport carbon emissions. The formula of the calculation is as follows:

$$C = \sum_{i=1}^{n} (E_i \times F_i) \tag{1}$$

$$F_i = N_i \times M_i \tag{2}$$

where C is traffic carbon emissions, i is a specific kind of energy, E_i is the entire transportation's class i usage of energy, F_i is the carbon emissions coefficient (ton carbon/ton) of sector i, N_i is the average lower heating value (TJ/10,000 tons), and M_i is the carbon emissions factor of sector i (ton carbon/TJ).

The calculation formula for the carbon emissions of electricity is as follows:

$$E = \sum_{i}^{n} A \times B \times J \tag{3}$$

where E is the amount of carbon dioxide produced by electricity, A is the amount of electricity used, B is the standard coal coefficient of converting power, and J is the carbon emissions coefficient of power.

Carbon emissions in transportation refer to the traffic energy consumption, warehousing, and postal industry. This paper mainly uses energy, including the eight categories of electricity, natural gas, liquefied petroleum gas, fuel oil, diesel, kerosene, gasoline, and raw coal.

3.2. Tapio Decoupling Model

Decoupling reveals the regional relation of carbon emissions to economic growth. The Tapio decoupling model was chosen because of its less limited base period and clear judgment criteria, which can eliminate errors due to the base selection. Therefore, this paper builds on the Tapio decoupling model, measuring Guangdong province's transportation decoupling index. The following is the computation procedure:

$$t = \frac{\%\Delta C}{\%\Delta G} = \frac{\Delta C/C}{\Delta G/G} \tag{4}$$

where t represents the decoupling elastic index of the gross value of transportation and carbon emissions, ΔC symbolizes the change in traffic carbon emissions, and ΔG represents the change in the total value of transportation. According to different elastic index values, the decoupling state is divided into eight types [8], as shown in Table 1.

Table 1. Decoupling state classification.

Decoupling State Classification	Decoupling State	△C/C	△G/G	Decoupling Elastic
Decoupling	weak decoupling	>0	>0	0 < t < 0.8
	strong decoupling	<0	>0	t < 0
	recessive decoupling	<0	<0	t > 1.2
Coupling	expansive coupling	>0	>0	0.8 < t < 1.2
	recessive coupling	<0	<0	0.8 < t < 1.2
Negative decoupling	weak negative decoupling	<0	<0	0 < t < 0.8
	strong negative decoupling	>0	<0	t < 0
	expansive negative decoupling	>0	>0	t > 1.2

3.3. The Generalized Divisia Index

Japanese scholar Yoichi Kaya proposed a model for the quantitative analysis of carbon dioxide emissions [27], which linked carbon emissions with the population, economy, energy, and other factors, called the Kaya identity, and the expression is as follows:

$$C = \frac{C}{E} \times \frac{E}{G} \times \frac{G}{P} \times P \tag{5}$$

where C is carbon emissions, E is energy consumption, C/E is the energy carbon emissions factor, G is the GDP in China, E/G is energy strength, G/P is the economy, and P is the population.

This paper expands the influencing factors of the Kaya constant equation and decomposes the variables. It also constructs a factor decomposition model for Guangdong province's traffic carbon emissions. According to the GDIM [28], the following is a description of carbon emissions and related variables, and the specific meaning can be obtained from Table 2:

$$C = \frac{C}{G} \times G = \frac{C}{E} \times E = \frac{C}{T} \times T \tag{6}$$

According to $\Phi(X) = 0$:

$$C = \frac{C}{G} \times G, \frac{C}{G} \times G - \frac{C}{E} \times E = 0, \frac{C}{G} \times G - \frac{T}{C} \times C = 0, T - \frac{T}{G} \times G = 0, E - \frac{E}{G} \times G = 0 \tag{7}$$

For convenience, eight factors in turn are replaced by $X_1, X_2, \ldots, X_8$:

$$C = X_1 \times X_2, X_1 \times X_2 - X_3 \times X_4 = 0, X_1 \times X_2 - X_5 \times X_6 = 0, X_5 - X_1 \times X_7 = 0, X_3 - X_1 \times X_8 = 0 \tag{8}$$

Table 2. The specific meaning of various factors.

Factor	Symbol	Name	Meaning
X_1	G	the added value of transportation	A standard for measuring the level of the transportation industry in a region
X_2	C/G	carbon emission intensity	The level of reflecting economic development, technological progress, and the efficiency of energy utilization
X_3	E	energy consumption	Considering how energy affects carbon emissions
X_4	C/E	carbon intensity from energy consumption	Considering how clean energy affects carbon emissions
X_5	T	turnover volume	The product of the tonnage of cargo or the number of passengers transported and the distance they traveled
X_6	C/T	carbon emission efficiency	Reflecting the low-carbon level of the transportation method
X_7	T/G	transportation intensity	Reflecting a high or low transportation efficiency
X_8	E/G	energy consumption intensity	Representing the reliance of economic growth on energy

The factor X is written as the function C(X), and a Jacobian matrix composed of the various factors is constructed based on the above equation Φ_X:

$$\Phi_X = \begin{pmatrix} X_2 & X_1 & -X_4 & -X_3 & 0 & 0 & 0 & 0 \\ X_2 & X_1 & 0 & 0 & -X_6 & -X_5 & 0 & 0 \\ -X_1 & 0 & 0 & 0 & 1 & 0 & -X_1 & 0 \\ -X_8 & 0 & 1 & 0 & 0 & 0 & 0 & -X_1 \end{pmatrix} \tag{9}$$

According to the principle of GDIM, ΔC is decomposed into the form of the contribution sum of each influencing factor:

$$\Delta C[X|\Phi] = \int_L \nabla C^T (I - \Phi_X \Phi_X^+) dX, \nabla C^T = (X2, X1, 0, 0, 0, 0, 0, 0)^T \tag{10}$$

For the absolute quantity factors X_1, X_3, and X_5, the exponential function is defined as follows:

$$Q(t) = (Q_1/Q_0)^t \tag{11}$$

4. Data and Results

4.1. Analysis of Carbon Emissions Results in Transportation

4.1.1. Data

The data are given in Tables 3 and 4, including the carbon emission coefficient, power conversion standard coal coefficient, and carbon emission coefficient of petrochemical energy:

Table 3. Net heating value, carbon content, and emission coefficient of various energy sources.

Energy Type	Raw Coal	Gasoline	Kerosene	Diesel Oil	Fuel Oil	Liquefied Petroleum Gas	Natural Gas
Net heating value/(TJ.(10^3 t)$^{-1}$)	20.9	44.3	43.8	43.0	40.4	47.3	44.2
Carbon content /t.(TJ)$^{-1}$	26.37	18.9	19.6	20.2	21.2	17.2	17.5
Carbon oxidation rate/%	0.94	0.98	0.98	0.98	0.98	0.98	0.99
Emission coefficient of unit energy CO_2/t.t^{-1}	1.9	2.925	3.017	3.096	3.171	3.101	2.667

Table 4. Electricity coal folding coefficient and carbon emission coefficient.

Energy	Conversion of the standard coal coefficient (kg equivalent of coal. (kW·h)$^{-1}$)	Carbon emission coefficient (carbon. (t equivalent of coal)$^{-1}$)
Electricity	0.1229	8.12

The carbon emission coefficient is calculated according to the average low calorific of various energy sources given in the China Energy Statistical Yearbook and the carbon emission factor in the IPCC [26]. To completely mitigate the epidemic's effect on traffic carbon emissions, the 2019 data were used to calculate the carbon emissions in 30 provinces and cities.

4.1.2. Results

The "top-down" method was adopted to calculate the carbon emissions of various provinces in China, and the results are shown in Figure 2 below. From the figure, it is obvious that the carbon emissions of the transportation industry in Guangdong province rank first, reaching 84.6457 million tons, about 48% more than those of Shanghai, which is located in second place. Guangdong province's total import and export of goods, passenger and freight traffic, turnover, etc. are ranked first in China, while development has produced more carbon emissions, so it is urgent to study the influencing factors of Guangdong province's traffic carbon emissions.

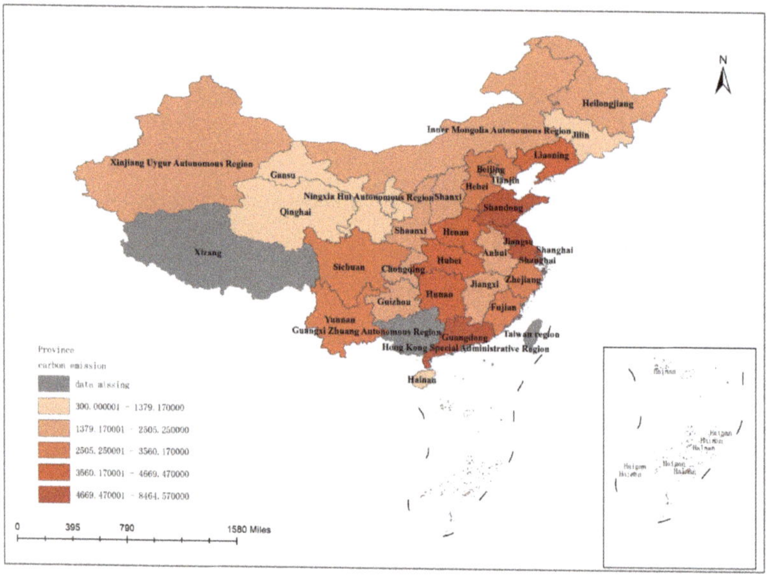

Figure 2. Carbon emissions of transport industry in China by province.

As a result, this paper uses the "up-bottom" method to calculate Guangdong province's carbon from 1999 to 2019, and Table 5 displays the outcomes. It is evident that Guangdong's traffic carbon emissions indicate a rising tendency. In 2003, the traffic carbon emissions increased significantly, mainly because the growth of port transportation demand this year increased the amount of water transportation, which led to the transportation's carbon emissions rising noticeably. In 2013, the traffic carbon emissions prematurely decreased first and then rose annually.

Table 5. Traffic carbon emissions in Guangdong province from 1999 to 2019.

Year	Carbon Emissions	Year	Carbon Emissions
1999	1787.91	2010	5815.64
2000	2057.68	2011	6063.79
2001	2254.24	2012	6346.04
2002	2457.78	2013	6157.36
2003	2940.33	2014	6446.25
2004	3422.06	2015	7044.92
2005	4066.47	2016	7562.99
2006	4192.19	2017	7827.27
2007	4635.13	2018	8111.29
2008	4990.32	2019	8464.57
2009	5267.83	—	—

4.2. Empirical Analysis of Carbon Emissions Decoupling Effects

On the basis of Guangdong province's economic development and carbon emission decoupling elastic index, the result is shown in Table 6, and the changing trend of the decoupling elastic index is shown in Figure 3.

Table 6. 1999–2019 Guangdong province's decoupling index and status judgment form.

Year	$\Delta C/C$	$\Delta G/G$	Decoupling Index	Decoupling Status
1999–2000	0.15	0.26	0.59	weak decoupling
2000–2001	0.10	0.22	0.43	weak decoupling
2001–2002	0.09	0.10	0.86	expansive negative decoupling
2002–2003	0.20	0.03	6.47	expansive negative decoupling
2003–2004	0.16	0.12	1.32	expansive negative decoupling
2004–2005	0.19	0.19	0.99	expansive coupling
2005–2006	0.03	0.16	0.19	weak decoupling
2006–2007	0.11	0.16	0.66	weak decoupling
2007–2008	0.08	0.14	0.55	weak decoupling
2008–2009	0.06	−0.04	−1.57	strong negative decoupling
2009–2010	0.10	0.12	0.84	expansive coupling
2010–2011	0.04	0.13	0.32	weak decoupling
2011–2012	0.05	0.11	0.42	weak decoupling
2012–2013	−0.03	0.06	−0.46	strong decoupling
2013–2014	0.05	0.11	0.42	weak decoupling
2014–2015	0.09	0.07	1.34	expansive negative decoupling
2015–2016	0.07	0.08	0.91	expansive coupling
2016–2017	0.03	0.10	0.35	weak decoupling
2017–2018	0.04	0.06	0.58	weak decoupling
2018–2019	0.04	0.09	0.50	weak decoupling

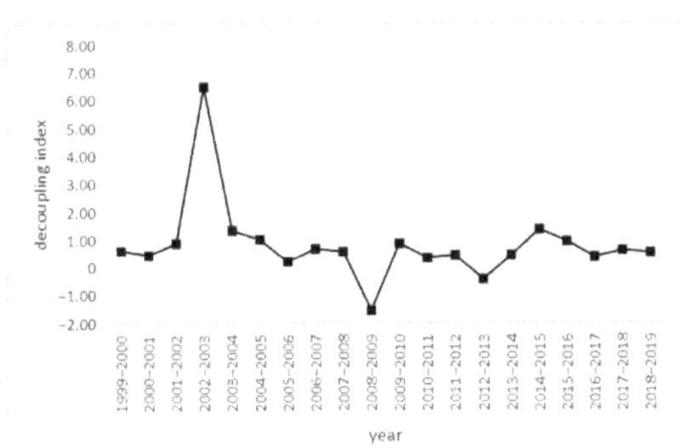

Figure 3. Guangdong province's transportation decoupling index change tracker.

From 1999 to 2005, Guangdong province's traffic decoupling status appeared as expansive coupling, expansive negative decoupling, and weak decoupling. The decoupling situation of the last few years is not optimal. In a great period of economic development, the quick expansion of various industries has increased the demand for logistics and transportation in Guangdong province. Moreover, because of the weakness of transportation infrastructure and the weak awareness of emissions reductions, it has caused a lot of consumption of transportation energy.

From 2006 to 2010, weak decoupling was the outstanding characteristic at this stage. During this period, the economy outpaced the rise in carbon emissions primarily because Guangdong attached great importance to infrastructure construction, forming a relatively complete comprehensive transportation system. Railway construction has achieved historic leaps, high-speed railway roads have considerably developed, and freight capacity has been considerably improved. In addition, Guangdong province has a lengthy coastal line on the coast. The rapid development of ports has led Guangdong's container throughput to increase to almost one-third of the country's.

From 2011 to 2014, the traffic carbon emissions growth rate during this period was lower than the rate of economic expansion, and Guangdong's economy was in a sustainable development state. Guangdong province attaches great importance to low-carbon development. Followed by Shenzhen in 2011, it was identified as a low-carbon transportation system construction pilot, and green transportation gradually entered into people's thoughts. In 2013, there was a state of strong decoupling, which is an ideal state. Guangdong province issued transportation energy conservation projects and kept pushing for energy saving, carrying out pilots, demonstration projects, and a transportation month and increasing the transportation of energy saving and technology applications, as well as creating and applying sustainable renewable energy, which has received excellent results.

From 2015 to 2019, there were three states where the traffic carbon emissions' decoupling state occurred, suggesting that it is still unstable. In 2015, the number of Guangdong civilian cars increased from 7.835 million in 2010 to 14.7233 million; by 2019, this had reached 23.2695 million vehicles, with an average growth rate of about 73%. With the swift expansion of the number of urban automobiles and the swift advancement of passenger transportation, freight business and the transportation industry's "carbon reduction" pressure is increasingly great; in the field of transportation, the situation is still grim.

On the whole, the economic development of China's transportation from 1999 to 2019 has not entirely rid itself from dependence on carbon emissions, and the environmental problems are still extremely serious. As China's economy continues to grow quickly, the road to low-carbon transformation is still lengthy. For this reason, it is essential to research the reasons that affect traffic carbon emissions.

4.3. *Carbon Emissions' Influencing Factors Decomposition Results*

4.3.1. Data

This paper selects the overall data on Guangdong's transportation from 1999 to 2019 as a research sample. Related data mainly come from the Chinese Energy Statistical Yearbook [29] and the Guangdong Statistical Yearbook [30,31].

In order to facilitate the calculations, the passenger turnover volume is converted into a freight turnover calculation through a certain transformation. The conversion coefficient is shown in Table 7 below:

Table 7. The conversion coefficient of each transportation method.

Transportation	Railway	Highway	Waterway	Aviation
conversion coefficient	1	0.2	0.3	0.83

4.3.2. Results

This paper breaks down the influential factors of the GDIM with R language. The contribution rate and contribution of different factors are presented in Tables 8 and 9.

Table 8. The influencing factors' contribution rate.

Year	G	C/G	E	C/E	T	C/T	T/G	E/G
2000	0.0854	−0.0298	0.0496	0.0000	0.0149	0.0363	−0.0049	−0.0007
2001	0.0739	−0.0369	0.0286	0.0027	0.0194	0.0123	−0.0028	−0.0017
2002	0.0353	−0.0045	0.0275	0.0025	0.0071	0.0233	−0.0010	0.0000
2003	0.0105	0.0565	0.0500	0.0162	0.0381	0.0274	−0.0007	−0.0018
2004	0.0418	0.0126	0.0519	0.0030	0.0485	0.0062	0.0000	−0.0001
2005	0.0640	−0.0005	0.0976	−0.0310	0.0002	0.0656	−0.0055	−0.0021
2006	0.0532	−0.0394	0.0097	0.0004	0.0228	−0.0130	−0.0008	−0.0021
2007	0.0534	−0.0168	0.0342	0.0007	0.0328	0.0021	−0.0004	−0.0003
2008	0.0467	−0.0192	0.0256	−0.0003	0.0075	0.0183	−0.0017	−0.0003
2009	−0.0119	0.0322	0.0170	0.0018	0.0270	−0.0076	−0.0020	−0.0009
2010	0.0405	−0.0063	0.0352	−0.0008	0.0623	−0.0263	−0.0007	−0.0001
2011	0.0424	−0.0275	0.0090	0.0052	0.0573	−0.0415	−0.0005	−0.0017
2012	0.0353	−0.0211	0.0124	0.0030	0.0970	−0.0726	−0.0061	−0.0014
2013	0.0205	−0.0309	−0.0099	0.0003	0.0617	−0.0663	−0.0032	−0.0020
2014	0.0363	−0.0211	0.0159	−0.0005	0.0678	−0.0490	−0.0017	−0.0009
2015	0.0236	0.0075	0.0321	−0.0008	−0.0053	0.0372	−0.0013	−0.0002
2016	0.0253	−0.0024	0.0221	0.0022	0.1351	−0.0906	−0.0178	−0.0002
2017	0.0321	−0.0212	0.0065	0.0052	0.0807	−0.0629	−0.0040	−0.0015
2018	0.0207	−0.0082	0.0080	0.0041	0.0083	0.0038	−0.0002	−0.0002
2019	0.0292	−0.0137	0.0063	0.0083	0.0087	0.0058	−0.0004	−0.0006

Table 9. The influencing factors' contribution.

Year	G	C/G	E	C/E	T	C/T	T/G	E/G
2000	152.68	−53.25	88.73	0.06	26.70	64.90	−8.72	−1.31
2001	152.03	−75.88	58.89	5.58	39.91	25.33	−5.77	−3.51
2002	79.57	138.94	122.80	39.91	93.72	67.43	−1.68	−4.42
2003	25.87	16.01	119.51	41.02	91.47	69.23	−1.31	−0.18
2004	122.92	36.98	152.53	8.80	142.71	18.22	−0.11	−0.32
2005	219.06	−1.54	333.89	−106.21	0.77	224.63	−18.93	−7.24
2006	216.24	−160.02	39.31	1.75	92.90	−52.99	−3.13	−8.34
2007	306.00	−141.68	142.11	3.12	136.25	9.09	−6.24	−5.68
2008	216.61	−88.97	118.62	−1.22	34.96	84.70	−8.05	−1.45
2009	−59.16	160.67	84.79	8.86	134.91	−38.00	−10.06	−4.50
2010	213.45	−32.95	185.55	−4.00	328.18	−138.31	−3.67	−0.45
2011	246.39	−160.08	52.49	30.04	333.51	−241.12	−2.98	−10.08
2012	214.19	−128.16	74.90	18.43	588.43	−440.40	−36.90	−8.23
2013	130.23	−196.00	−63.06	1.85	391.58	−420.75	−20.06	−12.47
2014	223.47	−129.63	97.67	−2.97	417.58	−301.61	−10.35	−5.26
2015	152.23	48.49	207.03	−5.13	−34.33	239.73	−8.09	−1.23
2016	177.92	−17.22	155.58	15.46	951.69	−638.25	−125.56	−1.54
2017	242.44	−160.28	49.20	39.24	610.52	−475.83	−29.96	−11.04
2018	162.37	−64.53	62.31	32.17	64.78	29.67	−1.31	−1.41
2019	237.24	−111.33	50.83	67.17	70.73	46.80	−3.45	−4.67

From the data in the table, carbon emissions have risen tremendously due to the production value of traffic (G). Although they have gradually decreased from the overall point of view, their contribution value is still increasing. Energy consumption (E) has an enhancive effect and is also at the forefront, and there is the overall presence of a rising first then moving downward trend. The turnover volume (T) promotes the increase in carbon emissions, with its impact gradually increasing first and then gradually weakening.

Moreover, the maximum value in 2016 reached 13.51%, mainly because the waterway freight remained rapidly developing in 2016, accounting for 22.7% of the total cargo volume and 82.4% of the cargo turnover.

In most cases, carbon emission intensity (C/G) has suppressed carbon emissions. Moreover, its degree of suppression is relatively large, indicating that the level of energy utilization has improved due to the ongoing advancements in technology development, thus playing an inhibitory role. The impact of carbon intensity from energy consumption (C/E) is mostly promoted, and the inhibitory effect is minor. The promotion and suppression effect of carbon emission efficiency (C/T) is equivalent, which shows that the low carbonization of energy in transportation is low.

Although the transportation intensity (T/G) and energy consumption intensity (E/G) have an inhibition effect, it was not more than 1%, and the inhibitory effect is weak. Therefore, transportation efficiency still needs to be continuously improved, and there should be a progressive reduction in the extent to which the economy depends on energy.

Divided from 1999 to 2019 into four sub-stages—from 1999 to 2004, 2004 to 2009, 2009 to 2014, and 2014 to 2019—the driving factor decomposition results were calculated, as shown in Figures 4 and 5.

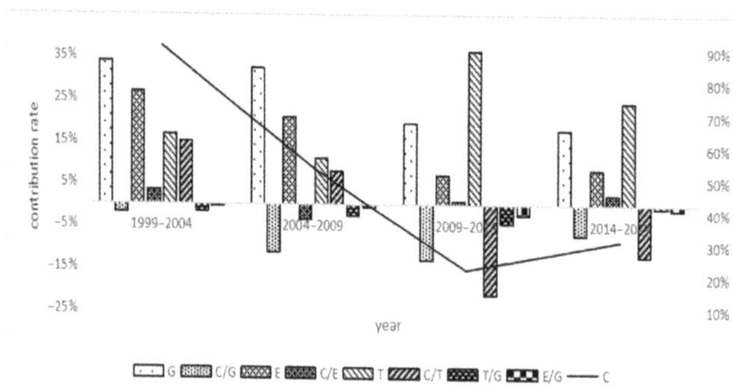

Figure 4. The various factors' contribution rates.

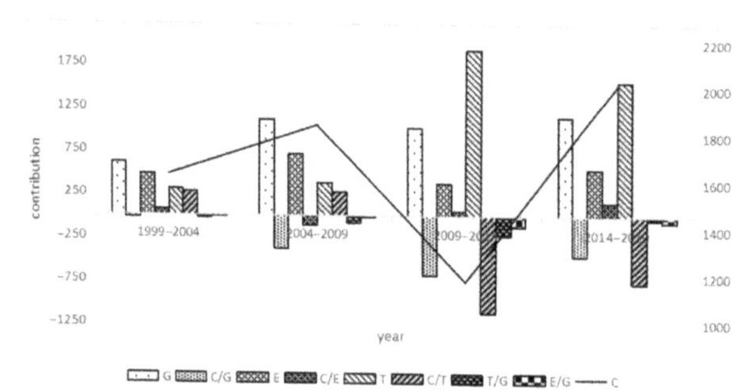

Figure 5. The various factors' contributions.

(1) The production value of traffic (G) is the end result of the production activities undertaken by the transportation industry throughout a specific time. It is evident that from 1999 to 2009, the effect of G was extremely obvious, and the contribution rate and contribution volume always ranked first. Owing to the national economy's explosive growth throughout this period, Guangdong province has also begun to attach importance to the transportation industry's development and investments. Various highways have been completed and used. Various transportation networks, such as railways, water transportation, civil aviation, and pipelines, are continuously improved. Motor vehicles

and transportation demands continue to increase, and numerous nonrenewable energies are consumed, resulting in increasing carbon emissions.

(2) Carbon emission intensity (C/G) is the economic output of the carbon emissions per unit. It continually acts as a brake on carbon emissions. In November 2010, Guangdong officially launched its national low-carbon provincial pilot work and established a number of programs and initiatives. In 2011, the low-carbon concept gradually permeated the transportation sector, from the construction of infrastructure to its promotion and application. It is clear that the contribution increased between 1999 and 2014. This suggests that the creation and efficient application of diverse low-carbon technologies during this period has improved the degree of low carbonization and the energy utilization efficiency of transportation. As a result, the suppression of the intensity of carbon emissions is relatively large. However, from 2014 to 2015, its contribution rate and contribution value decreased, indicating that the effect of carbon intensity gradually softened due to technical constraints. Between 1999 and 2019, energy consumption (E) made carbon emissions grow. The province was becoming less of an emitter of carbon from 26.6% to 8.4%, but the change in its contribution was relatively small. Between 1999–2004 and 2014–2019, carbon emissions increased less strongly than energy consumption, which shows that carbon emissions are still greatly influenced by energy consumption, and the reduction in its contribution rate is associated with other influencing factors. When other factors' contributions greatly increased, the energy consumption's contribution rate could decrease. As a result, the weakening of the positive driving effect cannot be explained only by the decrease in the contribution rate.

(3) Carbon intensity from energy consumption (C/E) has more of a promoting effect on carbon emissions than an inhibitory effect. It hardly slowed the increase in carbon emissions in 2004–2009, and its contribution rate accounts for only 3.45%, so the inhibitory effect is relatively minor. During this period, clean energy in Guangdong received unprecedented attention and development. Not only did it pay great attention to developing renewable energy sources, such as wind and water, but it also actively promoted pilot projects of solar energy and biomass energy utilization, so that the carbon intensity from energy consumption played a certain inhibitory role. Its facilitative role in 1999–2004 was due to the weak low-carbon awareness and underdeveloped technology at that time. Although its facilitative role was weak, the C/E transitioned from inhibition in 2004–2009 to growth in 2009–2019. This shows that, in the previous period, Guangdong province optimized the energy structure successfully, but the energy consumption structure is irrational in later periods, with unstable low-carbon degrees of energy and minor amounts of clean energy.

(4) Turnover volume (T) is the product of the tonnage of goods or the actual number of passengers and its transport distance, which increases carbon emissions. Its contribution rate and contribution volume first increased and then decreased, reaching a maximum of 36.41% and 19,181,010 tonnes in 2009–2014. This is a result of the increased demand for water transport during this period, with turnover increasing from 29,937.94 billion tonnes-km to 11,407.8 billion tonnes-km, resulting in a sudden increase in its contribution to carbon emissions. As a result of advancements in social and economic growth, as well as higher income levels among individuals and the rise in logistics and tourism, the demand for transportation in all walks of life was increasing. From 1999 to 2019, passenger and cargo turnover in Guangdong's rail, air, highways, and waterways increased substantially. Its impact on carbon emissions also continued to grow, and it has grown to become a significant driver of the rise in transport carbon emissions.

(5) There is less of a facilitative role due to carbon emission efficiency (C/T) on carbon emissions, which has gradually changed to being an inhibitory effect. Its contribution rate reduced from 14.9% to −12.2%, and the contribution value reduced from 18.4578 million tons to −7.867 million tons. From being the factor promoting carbon emissions' growth

to the primary obstacle, it is indicated that the increasing efficiency of carbon emissions suppresses carbon emissions. From 2009 to 2014, the greatest inhibition of carbon emission efficiency was 21.38%. Along with the development and promotion of shared bikes, Guangdong province has been trying to innovate and apply energy-saving technologies, promote current energy vehicles, strengthen public transportation, and speed up the development of a comprehensive three-dimensional transport network. Guangdong has actively developed various forms of transportation methods and increased the proportion of rail and waterways in integrated transportation, making efficiency in carbon emissions the main restraining factor.

(6) During 1999–2019, the rise in carbon emissions was restrained by transportation intensity (T/G). Transportation intensity is the ratio of transport turnover volume to the production value of traffic in a certain period of time, and low transport intensity represents an elevated transport efficiency. The contribution rate of transportation intensity is maintained at around −1% most of the time, and the overall inhibitory effect is weak. Its contribution rate reached a maximum of 5 percent from 2009 to 2014, indicating that Guangdong has strengthened its transportation infrastructure, developed advanced technological equipment, and changed its transportation structure to make it more efficient, so the inhibitory effect has become stronger. Although transportation intensity has a restraining effect, its inhibitory effect is relatively low compared with the influencing factors of promoting the growth effect, so it is still necessary to continuously improve transportation efficiency to enhance the inhibitory effect.

(7) Energy consumption intensity (E/G) influences carbon emissions in a positive way. From 1999 to 2019, energy consumption intensity prevented carbon emissions from rising, but the overall inhibitory effect was relatively tiny. Its contribution rate gradually increased to 2.22% from 2009 to 2014. During this period, the energy consumption intensity's inhibitory effect became more pronounced with Guangdong province's increased percentage of renewable energy sources. During 2014–2019, its contribution rate decreased to 0.64%; although it has a certain inhibitory effect, it is weak. Transportation development's dependence on energy is still large, and the growth of it will inevitably lead to increased energy use. There is still a need to build up the quantity of technological advancement and the implementation of clean energy.

To more accurately depict every variable's dynamic influence from 1999 to 2019, this paper adds up the contribution rate and contribution value of each factor from 1999 to calculate the cumulative contribution. In Figures 6 and 7, we can see that cumulative traffic carbon emissions rose by 69.56 million tons from 1999 to 2019. The production value of traffic (G) was a significant positive contributor. Because of the increase from 1,526,800 tons in 2000 to 2,372,400 tons in 2019, the cumulative contribution rate to the increase in carbon emissions was 75.83%. One of the key elements contributing to the rise in carbon emissions was turnover volume (T), and its contribution increased and increased dramatically, due to the increase in water transport turnover volume in 2016. From 1999 to 2019, the turnover volume increased by 45.1696 million tons of carbon emissions, and the final contribution rate reached 79.21% and exceeded the production value of traffic. After 2005, the effect of energy consumption (E) gradually became apparent, but its growth was gradual, with a 21.3365 million ton increase in total by 2019. Carbon intensity from energy consumption (C/E) also promoted carbon emissions, but the effect was tiny and less obvious, with only 1,939,500 tons by 2019.

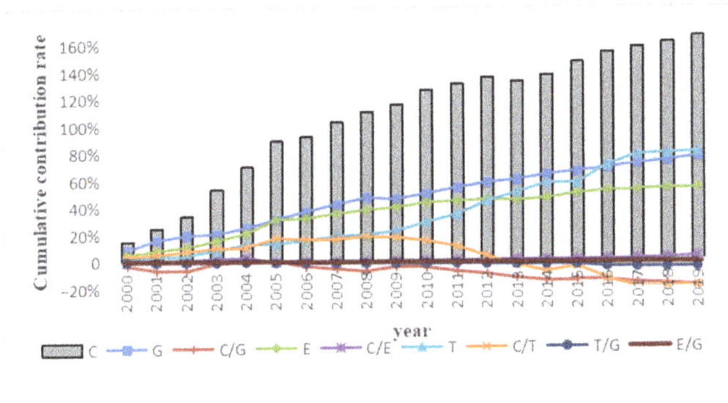

Figure 6. Cumulative contribution rate amount of each influencing factor.

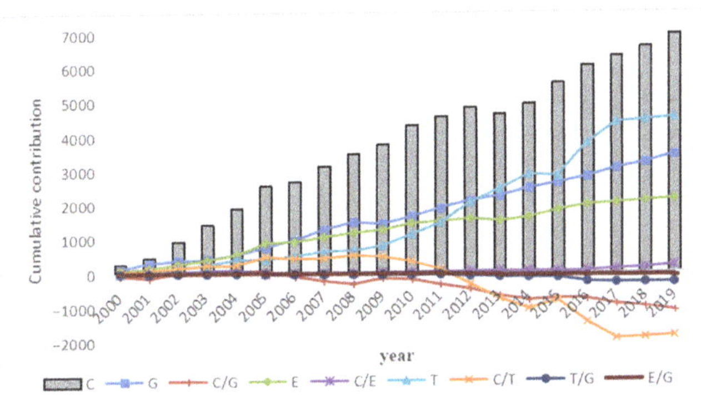

Figure 7. Cumulative contribution amount of each influencing factor.

Carbon emission intensity (C/G) and carbon emission efficiency (C/T) are the principal inhibitors, preventing an increase of 11.2044 million tons and 18.6751 million tons, respectively. However, both factors are not stable and occasionally promote carbon emissions. Carbon emission efficiency's promoting effect was slowly weakened in 1999–2011, then changed to an inhibitory effect in 2012, and gradually became an essential factor to restrain carbon emissions. The role that carbon emission intensity (C/G) inhibits is relatively tiny. There has been a gradual increase in inhibitory effects since 2011, but development has been limited. Energy consumption intensity (E/G) and transportation intensity (T/G) have relatively weak effects on carbon emissions. By 2019, carbon emissions were reduced by 933,300 tons and 3.0634 million tons, respectively. According to the needs of the "double-carbon" target, starting from the present situation and the development tendency of the transport industry, the carbon emission reduction policy of transportation should be implemented mainly by enhancing the energy economy and supporting the clean energy system.

5. Conclusions and Recommendations

5.1. Conclusions

This paper explores the traffic carbon emission's decoupling state in Guangdong province from 1999 to 2019 by using the Tapio decoupling index model. In conclusion, the decoupling state is poor, and low-carbon transformation still faces great challenges and needs long-term and arduous efforts.

(1) Through the breakdown of the contributing elements that can be obtained, one of the main causes of the rise in carbon emissions in the early stages was the production value of traffic, and its promoting effect was first enhanced and then weakened. In 2005, energy consumption's influence ranked top and then gradually weakened. However, turnover volume's steadily increasing effect grew year by year and gradually became the biggest factor. Although the carbon emission efficiency and energy consumption intensity showed an inhibitory effect for several years, they changed into a positive impact later on, indicating that the low-energy carbon degree of transportation is not stable enough and should be continuously strengthened.

(2) Among the inhibitory factors, the role of carbon emission intensity is relatively obvious. However, its inhibitory effect is unstable, which may be the bottleneck in technical aspects. Although energy consumption intensity and transportation intensity play an inhibitory role, their role is weak, indicating that the current transportation efficiency and energy structure cannot effectively inhibit carbon emissions and needs to be further improved.

5.2. Recommendations

(1) Rely on science and technology. Guangdong province should actively pursue innovation and research, as well as the promotion of cutting-edge technologies. Because of the marginal suppression of energy consumption and transportation intensity, it should give full play to its own technical advantages and talent advantages, aggressively advance environmentally friendly technologies, and enhance energy effectiveness. It is imperative to hasten clean energy technology's adoption and improve its use efficiency to promote clean production and the clean cycle of fossil energy. This can be achieved through developing energy-saving technology and innovating, introducing talents related to low-carbon transportation technology, and establishing professional talent teams. Intelligent and information management methods should be used to alleviate traffic congestion and to improve urban traffic efficiency and the degree of interconnection of highways, railways, and waterways, thereby reducing the operation time of transportation equipment. Additionally, high-performance battery technology could be developed to improve the energy density and life cycle of batteries and to reduce their cost. The research and development of battery systems for electric vehicles and hydrogen vehicles could also be promoted to improve the energy conversion efficiency and durability of batteries. Integration technology of solar and wind energy in transportation infrastructure should be strengthened to reduce the dependence on traditional energy sources. Carbon capture and sequestration technology could be developed to capture, compress, transport, and sequester the carbon dioxide generated in the transportation sector, thereby reducing carbon emissions.

(2) Accelerate the use of renewable energy. The installation of solar panels at transportation hubs, depots, and along major highways to harness solar energy for operational needs should be encouraged. There should also be the active development and use of biomass for the production of transport equipment, the strengthening of the management of transportation equipment and facilities, and the stimulation of the adoption of hybrid transportation equipment. Pilot projects showcasing the benefits of hybrid and electric transportation equipment should be developed and promoted. Charging infrastructure for electric vehicles and hydrogen fueling stations to facilitate the transition should be expanded, and the usage of clean energy vehicles to decrease fuel-powered vehicles should be aggressively encouraged, along with promoting the use of electricity-, hydrogen fuel-, and liquefied natural gas-powered heavy freight vehicles; improving the level of transport's own emissions technology; and implementing continuous control to reduce energy consumption.

(3) Optimize the transportation structure, promotion of transformation, and upgrading. There should be the improvement of the transportation structure, the optimization of urban transportation networks, and a decrease in freight turnover. Logistics hubs and

distribution centers that are strategically located to minimize transportation distances should be developed. Public transport and rail transit should be vigorously developed, along with promoting green low-carbon travel methods and raising the percentage of urban public transit, while gradually lowering the percentage of private vehicles. The development of multi-modal transport such as "railway to rail" and "railway to water could be promoted. The integration and development of green travel methods, such as rail, bus, and slow travel should be vigorously promoted, along with strengthening the construction of comprehensive transportation hubs and transportation facilities, optimizing the layout of the urban public transport line network, and promoting the coordinated development of various transportation methods. Various passenger and freight organizations should be optimized to reduce the empty load rate of each passenger transport. Partnerships between freight companies to share vehicles and reduce empty trips could be developed. The use of technology, such as load matching platforms, should be encouraged to improve the utilization of transportation assets.

(4) Increase policy support. A mechanism for rewarding energy conservation could be created. Special funds could be established, and there could be an increase in investment in research, developing improved techniques and new energy sources. The city's vehicles could be guided and encouraged to use fresh energy, replacing fuel vehicles with fresh energy vehicles, actively promoting green and low-carbon technologies, improving policies and measures, and promoting the application of new energy vehicles. For instance, a points-based system could be introduced where individuals and businesses would earn points for reducing energy consumption, which could be redeemed for various benefits such as discounts, vouchers, or even cash incentives. Green travel should be advocated, including advocating for the public to implement the concept of green travel, forming a good social style of low-carbon travel, and cultivating the public's low-carbon travel awareness. People should be guided towards using sustainable modes of transportation such as walking, cycling, and public buses. Infrastructure could be improved to make these options more convenient and appealing, such as building more bike lanes, pedestrian pathways, and efficient public transit systems.

Author Contributions: Conceptualization, Shiying Zhang; methodology, Shiying Zhang, Hualing Bi, and Fuqiang Lu; software, Shiying Zhang; validation, Shiying Zhang; formal analysis, Shiying Zhang; investigation, Shiying Zhang; resources, Shiying Zhang; data curation, Shiying Zhang; writing—original draft preparation, Shiying Zhang; writing—review and editing, Hualing Bi and Fuqiang Lu; visualization, Shiying Zhang; supervision, Hualing Bi and Fuqiang Lu; project administration, Hualing Bi and Fuqiang Lu; and funding acquisition, Hualing Bi and Fuqiang Lu. All authors have read and agreed to the published version of the manuscript.

Funding: This work is supported by the National Key R&D Program of China under grant no. 2020YFB1712802; the Research Program of Social Science of Qinhuangdao under grant no. 2024LX026; the Science and Technology Research and Development Plan Project of Qinhuangdao under grant no. 202401A143; the Key Applied Research Program of Social Science of Qinhuangdao under grant no. 2023LX044; and the Teaching Research and Reform Cultivation Project of Yanshan University under grant no. 2023XJJG084.

Data Availability Statement: The data that support the findings of this study are available from the corresponding author upon reasonable request.

Acknowledgments: The authors would like to gratefully acknowledge the anonymous reviewers and the members of the editorial team who helped to improve this paper through their thorough review.

Conflicts of Interest: The authors declare no conflicts of interest.

References

1. Marten, O.; Stef, P. Cost-effective reduction of fossil energy use in the European transport sector: An assessment of the Fit for 55 Package. *Energy Policy* **2022**, *168*, 113085.
2. Jessica, S.; Beatrice, G. Forest protection in the EU's renewable energy directive and nature conservation legislation in light of the climate and biodiversity crisis—Identifying legal shortcomings and solutions. *For. Policy Econ.* **2023**, *153*, 102996.

3. Daniel, S.; Aoife, H.; Manan, S.; Sushanth, B.; Ben, J.; Manish, D.; Ryan, S.; Yann, G.; Arild, G. Comparative review of international approaches to net-zero buildings: Knowledge-sharing initiative to develop design strategies for greenhouse gas emissions reduction. *Energy Sustain. Dev.* **2020**, *71*, 291–306.
4. Erin, M.; Jesse, J.; Eric, L.; Chris, G. Labor pathways to achieve net-zero emissions in the United States by mid-century. *Energy Policy* **2023**, *177*, 113516.
5. Hiroshi, O.; Brendan, F.D. Barrett Politics of climate change and energy policy in Japan: Is green transformation likely? *Earth Syst. Gov.* **2023**, *17*, 100187.
6. Xu, G.; Wang, Y.; Rehman, H. The future trajectory of carbon emissions in the process of carbon neutrality in South Korea. *J. Environ. Manag.* **2023**, *345*, 118588. [CrossRef] [PubMed]
7. Liu, Z.; Sun, T.; Yu, Y.; Ke, P.; Deng, Z.; Lu, C.; Huo, D.; Ding, X. Near-Real-Time Carbon Emission Accounting Technology Toward Carbon Neutrality. *Engineering* **2022**, *14*, 44–51. [CrossRef]
8. Tapio, P. Towards a theory of decoupling: Degrees of decoupling in the EU and the case of road traffic in Finland between 1970 and 2001. *Transp. Policy* **2005**, *12*, 137–151. [CrossRef]
9. Wang, Q.; Zhang, F. The effects of trade openness on decoupling carbon emissions from economic growth—Evidence from 182 countries. *J. Clean. Prod.* **2021**, *279*, 123838. [CrossRef]
10. Wang, J.; Jiang, Q.; Dong, X.; Dong, K. Decoupling and decomposition analysis of investments and CO_2 emissions in information and communication technology sector. *Appl. Energy* **2021**, *302*, 117618. [CrossRef]
11. Shi, B.; Xiang, W.; Bai, X.; Wang, Y.; Geng, G.; Zheng, J. District level decoupling analysis of energy-related carbon dioxide emissions from economic growth in Beijing, China. *Energy Rep.* **2022**, *8*, 2045–2051. [CrossRef]
12. Kouyakhi, N.R. CO_2 emissions in the Middle East: Decoupling and decomposition analysis of carbon emissions, and projection of its future trajectory. *Sci. Total Environ.* **2022**, *845*, 157182. [CrossRef]
13. Zhao, Y.; Su, Q.; Li, B.; Zhang, Y.; Wang, X.; Zhao, H.; Guo, S. Have those countries declaring "zero carbon" or "carbon neutral" climate goals achieved carbon emissions-economic growth decoupling? *J. Clean. Prod.* **2022**, *363*, 132450. [CrossRef]
14. Lu, F.; Wang, L.; Bi, H.; Du, Z.; Wang, S. An Improved Revenue Distribution Model for Logistics Service Supply Chain Considering Fairness Preference. *Sustainability* **2021**, *13*, 6711. [CrossRef]
15. Shahbaz, M.; Chaudhary, A.; Ozturk, I. Does urbanization cause increasing energy demand in Pakistan? Empirical evidence from STIRPAT model. *Energy* **2017**, *122*, 83–93. [CrossRef]
16. Wang, B.; Sun, Y.; Chen, Q.; Wang, Z. Determinants analysis of carbon dioxide emissions in passenger and freight transportation sectors in China. *Struct. Chang. Econ. Dyn.* **2018**, *47*, 127–132. [CrossRef]
17. Solaymani, S. CO_2 emissions patterns in 7 top carbon emitter economies: The case of transport sector. *Energy* **2018**, *168*, 989–1001. [CrossRef]
18. Isik, M.; Ari, I.; Sarica, K. Challenges in the CO_2 emissions of the Turkish power sector: Evidence from a two-level decomposition approach. *Util. Policy* **2021**, *70*, 101227. [CrossRef]
19. Li, R.; Li, L.; Wang, Q. The impact of energy efficiency on carbon emissions: Evidence from the transportation sector in Chinese 30 provinces. *Sustain. Cities Soc.* **2022**, *82*, 103880. [CrossRef]
20. Zhang, Y.; Zhang, Y.; Zhang, Y.; Gong, C.; Kong, Y. Analysis of the carbon emission driving factors and prediction of a carbon peak scenario—A case study of Xi'an city. *Heliyon* **2022**, *8*, e11753. [CrossRef]
21. Lu, F.; Hu, Y.; Bi, H.; Feng, W.; Wang, S.; Zhang, X. A Two-Stage Auction Mechanism for 3PL Supplier Selection under Risk Aversion. *Sustainability* **2021**, *13*, 9745. [CrossRef]
22. Liu, J.; Yang, Q.; Ou, S.; Liu, J. Factor decomposition and the decoupling effect of carbon emissions in China's manufacturing high-emission subsectors. *Energy* **2022**, *248*, 123568.
23. Liu, Y.; Gan, L.; Cai, W.; Li, R. Decomposition and decoupling analysis of carbon emissions in China's construction industry using the generalized Divisia index method. *Environ. Impact Assess. Rev.* **2024**, *104*, 107321.
24. Shao, S.; Liu, J.; Geng, Y.; Miao, Z.; Yang, Y. Uncovering driving factors of carbon emissions from China's mining sector. *Appl. Energy* **2016**, *166*, 220–238. [CrossRef]
25. Tang, Y.; Jiang, H. Analysis of the decoupling effect and driving factors of carbon emissions from the transportation sector in Guangdong Province. *Sci. Rep.* **2023**, *13*, 18744. [CrossRef]
26. Intergovernmental Panel on Climate Change (IPCC). 2006 IPCC Guidelines for National Greenhouse Gas Inventories. 2006. Available online: https://www.ipcc-nggip.iges.or.jp/public/2006gl/index.html (accessed on 1 May 2023).
27. Kaya, Y.; Yokobori, K. *Environment, Energy, and Economy: Strategies for Sustainability*; United Nations University Press: Tokyo, Japan, 1997.
28. Vaninsky, A. Factorial decomposition of CO_2 emissions: A generalized Divisia index approach. *Energy* **2014**, *45*, 389–400. [CrossRef]
29. National Bureau of Statistics of China. *China Energy Statistical Yearbook*; China Statistics Press: Beijing, China, 2000–2022.
30. Guangdong Provincial Bureau of Statistics, National Bureau of Statistics, Guangdong Survey Team. *Guangdong Statistical Yearbook*; China Statistics Press: Beijing, China, 2022.
31. Lu, F.; Hao, H.; Bi, H. Evaluation on the development of urban low-carbon passenger transportation structure in Tianjin. *Res. Transp. Bus. Manag.* **2024**, *55*, 101142.

International Journal of
Geo-Information

MDPI

Article

Evaluation of Cross-Border Transport Connectivity and Analysis of Spatial Patterns in Latin America

Changqi Miao, Yinbao Zhang, Xinjia Zhang *, Jianzhong Liu and Shike Zhang

The School of Geoscience and Technology, Zhengzhou University, Zhengzhou 450001, China; mcq@gs.zzu.edu.cn (C.M.); zhangyinbao@zzu.edu.cn (Y.Z.); liujianzhong@zzu.edu.cn (J.L.); shikezhang@gs.zzu.edu.cn (S.Z.)

* Correspondence: zhangxinjia@zzu.edu.cn

Abstract: The study of cross-border transport connectivity is significant for the development of regional integration and insight into global patterns. Comprehensive connectivity evaluations are lacking and insufficient attention has been paid to Latin American connectivity, so it is of great practical importance to comprehensively and rationally evaluate Latin American connectivity. In this article, based on the four modes of transport, namely, sea, road, air and railroad, and using the actual trade volume as a comparison, a connectivity evaluation index system with considerable reliability and generalization ability was constructed using the expert scoring method, QAP correlation analysis, QAP regression, and statistics, and the connectivity calculations of Latin America were obtained. Analyzing the connectivity structure of Latin America, it was found that cross-border passenger and cargo transport in the region was dominated by sea transport and supplemented by road and air transport, with railroads used the least. The overall connectivity of Latin America was low, and the overall development was unbalanced, with a strong law of spatial differentiation, which was mainly manifested in the strongest connectivity of the integrated coastal countries, followed by the island countries, and the lowest connectivity of the landlocked countries. Different countries assumed different roles in regional connectivity, which could be categorized into global hub type, local hub type and non-hub type based on the calculations. There was a spatial pattern of decreasing connectivity with distance in typical countries, but the rate of decline was closely related to their geographic location and the role they played in the connectivity network. This study can provide reference and inspiration for regional connectivity evaluation, improvement, and sustainable development.

Keywords: Latin America; cross-border transport connectivity; spatial patterns; evaluation system construction; QAP; sustainability

1. Introduction

Regional/inter-country connectivity is an essential condition for regional integration and economic globalization [1], of which transport infrastructure connectivity is an important component. A reasonable evaluation of the capacity of inter-country transport infrastructure connectivity is a reflection of the degree of regional economic integration. It can also provide recommendations for regional cross-border transport construction based on the evaluation results, which can provide enlightenment for regional and global sustainable development.

Connectivity is the ability of nodes in a network to interconnect and spatially interact with other nodes through various means at a given time [2,3]. The broader connectivity emphasizes the importance of availability of transport services and their capabilities within the framework of complex systems theory [4]. Cross-border transport, on the other hand, is defined as infrastructure used for the transport of passengers and goods across the administrative borders of two or more countries [5–11]. According to different modes of transport, it can be divided into sea, road, air and railroad. The level or capacity of spatial interactions (trade in goods, movement of people, etc.) between countries across administrative boundaries through various transport infrastructures can be considered as cross-border transport connectivity (hereinafter referred to as connectivity), which describes the degree of sophistication of the transport network and the ease of interaction between different countries.

At present, there are many results for cross-border transport connectivity. For example, Zong and Pan et al. analyzed the connectivity between global shipping routes and port connectivity based on complex networks [12,13]; Huang et al. studied the impact of "the Silk Road Economic Belt and the 21st-Century Maritime Silk Road" initiative on the containerized maritime transport connectivity of the countries along the route [14]; Zhang et al. analyzed the shipping connectivity characteristics of the "Asian Mediterranean" region from the dimensions of port functions and shipping routes [15]; Kanrak et al. analyzed the connectivity, classification and cluster structure of the cruise shipping network between Asia and Australia [16]; and Wang studied the development of transport corridors in border areas [17]. In addition, scholars such as Allroggen, del Rosal, and Zhang et al. studied shipping and aviation connectivity using gravity modeling, network analysis, and spatial analysis [4,18,19]. The above results mainly focused on a single mode of transport, and lacked comprehensive connectivity evaluation, in-depth study of its impact mechanism, and reliability verification of the evaluation method. At the same time, there is a lack of research on cross-border transport in Latin America, and even less on the study of connectivity in Latin America.

Among the existing studies on connectivity, the research foundation on its relationship with trade is relatively deep, and it can be argued that there is a generally significant positive correlation between connectivity and trade. For example, Calatayud et al. argued that improving connectivity was an increasingly important topic on the international trade and transport policy agenda [20]; Jiao et al. established a gravity model between multiple modes of transport and trade constant prices, and found that increased connectivity between China and the countries along the route positively promoted inter-country trade in goods, and that different modes of transport had different degrees of impact on trade [21]; Canbay, Hoffmann and Saeed et al. found a positive causal relationship between global shipping connectivity and trade, and assessed the extent of the impact of shipping connectivity on trade [22–24]; Oum et al. analyzed the impact of air connectivity on bilateral trade using the gravity model [25]; Bensassi et al. studied the relationship between transport infrastructure and trade using the gravity model [26]; and studies by Li, Matsumoto and Ng et al. also confirmed the positive correlation between connectivity and economy and trade, which were complementary and mutually reinforcing [27–29]. In addition, scholars such as Wang, Zhu, Guchang, Sharma, and Wong also analyzed and explained the linkages between connectivity, economy, and trade using the expert scoring method, modeling, complex networks, and double difference models [30–35].

Therefore, this paper used quantitative methods, based on the four modes of cross-border transport (sea, road, air and rail) and the actual amount of trade as a comparison, combined with expert scoring, statistics, QAP correlation analysis and regression analysis, to construct a comprehensive connectivity evaluation system that was reliable and able to respond to the reality to a considerable extent, and to evaluate the inter-country connectivity of Latin America. After that, the spatial pattern of connectivity in Latin America was analyzed by combining network analysis and visualization. The connectivity evaluation system established by this study is applicable to other regions of the world and has a certain degree of universality; at the same time, the evaluation and analysis of connectivity in Latin America fill a gap in this study in the region and can provide recommendations for improved regional connectivity and integration development.

2. Methods and Data

2.1. Technical Lines of Study

The technical line of research of this paper is shown in Figure 1.

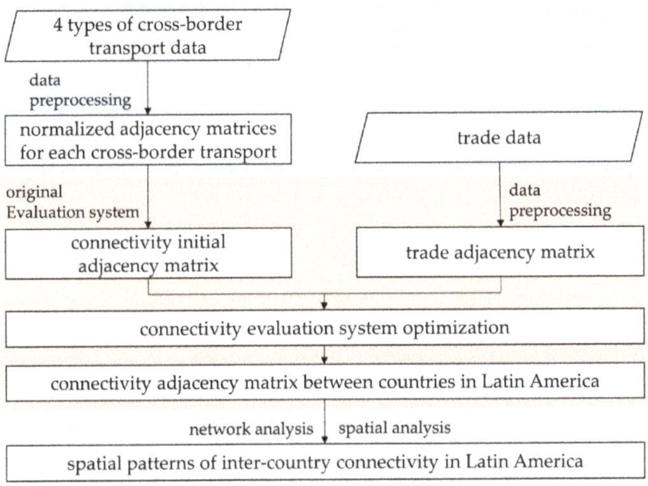

Figure 1. Technical lines of study.

This article used four types of cross-border transport data, namely, cross-border sea, road, air, and rail, for connectivity evaluation. Different cross-border transport types have different connectivity capacities and cannot be directly assigned using the data difference-based assignment method. Therefore, in this paper, the data were divided into two categories to be normalized separately: (i) transport infrastructure connected through physical materials. This form of connectivity is mainly reflected in the interaction between countries through land transport, such as roads, railroads, etc., so the connectivity measurement of this type is mainly based on the number and grade of line facilities. (ii) Transport infrastructure connected by routes. This form of connectivity is mainly reflected in the nodes between ports, airports and other stations, mainly manifested in the routes of transport means. Therefore, the connectivity measurement of sea and air transport is mainly carried out through the number and frequency of routes and flights. The Liner Bilateral Connectivity Index (LBCI) synthesized inter-country sea transport connectivity and could be used directly, while the air transport data were web-crawled to synthesize inter-country route counts and flight frequencies. The normalized adjacency matrices of each cross-border transport among Latin American countries were obtained by preprocessing the above four types of data.

The original connectivity evaluation system was determined by the expert scoring method, and the connectivity initial adjacency matrix of Latin America was calculated based on the normalized adjacency matrices of the above types of cross-border transport. However, the weights of the indicators in the original evaluation system did not fully correspond to the reality of inter-country connectivity in Latin America. Therefore, it was necessary to take the trade adjacency matrix as a comparison, based on the theoretical basis that connectivity and trade should show a strong positive correlation, combined with quantitative calculations to test the reliability of the initial adjacency matrix of connectivity; based on the test results of the original evaluation system of connectivity, it is necessary to optimize in order to obtain more reliable, more realistic results.

The QAP correlation analysis was used to calculate the connectivity initial adjacency matrix and trade adjacency matrix correlation to test the strength of correlation between two adjacency matrices and their significance level, the basic principle of which is shown in Figure 2a. If the calculated correlation coefficients reached a strong correlation ($r \geq 0.6$) and were within the significant level ($p < 0.01$), there was no need to adjust the weights of the indicators for the original evaluation system of connectivity. If it did not reach a strong correlation ($r < 0.6$) or was not within the level of significance ($p \geq 0.01$), the original evaluation system of connectivity needed to be adjusted to the weights of the indicators in order to reach a strong correlation and significance level.

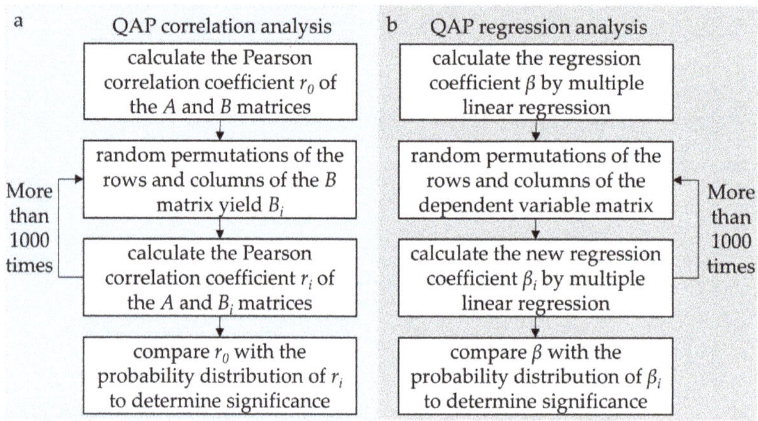

Figure 2. QAP basic principle.

QAP regression analysis was used to analyze and process the independent variable (normalized adjacency matrix of each cross-border transport) and the dependent variable (trade adjacency matrix), the basic principle of which is shown in Figure 2b. Based on the regression results, the impact of different cross-border transport modes on trade (connectivity) was determined, which was used to adjust the weights of the indicators in the original evaluation system of connectivity. Afterwards, the correlation between the adjusted connectivity adjacency matrix and the trade matrix was calculated, and the connectivity evaluation system with the largest r was selected as the optimized connectivity evaluation system.

The optimized connectivity evaluation system was used to calculate the connectivity among Latin American countries, and the spatial pattern of connectivity among Latin American countries was analyzed by combining network analysis methods and visualization tools. In this article, the overall spatial pattern of inter-country connectivity in Latin America was analyzed using overall connectivity and visualization; the spatial characteristics of typical country connectivity in Latin America were analyzed using country centrality, country meso-centrality, and individual country connectivity networks.

2.2. Data

2.2.1. Overview of the Study Area

Geographically, Latin America refers to the American region south of the United States, including Mexico, Central America, the Caribbean, and South America. Latin America has a vast territory, varied topography and landscape, and rich natural resources [36], but the economic development among countries is unbalanced, and the transport structure and level show large differences. This article evaluated the connectivity among 33 sovereign countries in Latin America and analyzed their spatial patterns. Based on the land and sea attributes of the Latin American countries, they were classified into three types: integrated, island, and landlocked. Integrated countries were defined as those connected to the continent and possessing a functional coastline; island countries were defined as those not connected to the continent and primarily consisting of islands; and landlocked countries were defined as those connected to the mainland but lacking a coastline. The names of Latin American countries, the types to which they belong and the corresponding ISO codes are shown in Table 1.

Table 1. Types of Latin American countries and their ISO codes.

Country	Code	Type	Country	Code	Type
Argentina	ARG	Integrated	Honduras	HND	Integrated
Antigua and Barbuda	ATG	Island	Haiti	HTI	Island
Bahamas	BHS	Island	Jamaica	JAM	Island
Belize	BLZ	Integrated	Saint Kitts and Nevis	KNA	Island
Bolivia	BOL	Landlocked	Saint Lucia	LCA	Island
Brazil	BRA	Integrated	Mexico	MEX	Integrated
Barbados	BRB	Island	Nicaragua	NIC	Integrated
Chile	CHL	Integrated	Panama	PAN	Integrated
Colombia	COL	Integrated	Peru	PER	Integrated
Costa Rica	CRI	Integrated	Paraguay	PRY	Landlocked
Cuba	CUB	Island	El Salvador	SLV	Integrated
Dominica	DMA	Island	Suriname	SUR	Integrated
Dominican	DOM	Island	Trinidad and Tobago	TTO	Island
Ecuador	ECU	Integrated	Uruguay	URY	Integrated
Grenada	GRD	Island	Saint Vincent and the Grenadines	VCT	Island
Guatemala	GTM	Integrated	Venezuela	VEN	Integrated
Guyana	GUY	Integrated			

2.2.2. Data Sources

This paper used cross-border sea, air, road, and railroad data and trade data among Latin American countries. The sources of the data and relevant information are presented in Table 2.

Table 2. Data sources and data information.

Data Classification	Data Sources	URL	Corresponding Time
sea	United Nations Conference on Trade and Development	https://unctad.org/ (accessed on 7 May 2024)	A.D.2020
road	natural earth	https://www.naturalearthdata.com/ (accessed on 12 May 2024)	A.D.2020
railroad	natural earth	https://www.naturalearthdata.com/ (accessed on 12 May 2024)	A.D.2020
air	VariFlight Map	https://map.variflight.com/ (accessed on 10 May 2024)	A.D.2020
trade	UN Comtrade	https://comtrade.un.org/ (accessed on 17 May 2024)	A.D.2020

Of these, cross-border roads and railroads were last updated in 2020, so all other data were queried and filtered according to 2020. Regarding sea transport data, the United

Nations Conference on Trade and Development (UNCTAD) database includes the LBCI. The LBCI is based on a combination of variables related to the number of trans-shipments, direct connections, public connections and the size of the largest vessel on the smallest route, and reflects the level of connectivity in the transport of goods by sea between countries.

2.2.3. Data Preprocessing

Using the attribute information contained in cross-border roads and railroads, cross-border roads and railroads were screened with the restriction of crossing the border line, and the cross-border road and railroad adjacency matrix between Latin American countries was constructed. For sea transport, the temporal resolution of the LBCI was quarterly, using 2020 data, filtered and averaged to obtain the sea transport adjacency matrix between Latin American countries. The air transport data mainly included the number of routes and the number of flights, and the air transport adjacency matrix between Latin American countries was obtained by crawling the data. The above adjacency matrices for sea, road, rail, and air transport were normalized and used as inputs for the connectivity calculations below using the expert scoring method. The trade adjacency matrix, on the other hand, serves as a comparison for the accuracy of the connectivity adjacency matrix. The data preprocessing steps are shown in Figure 3.

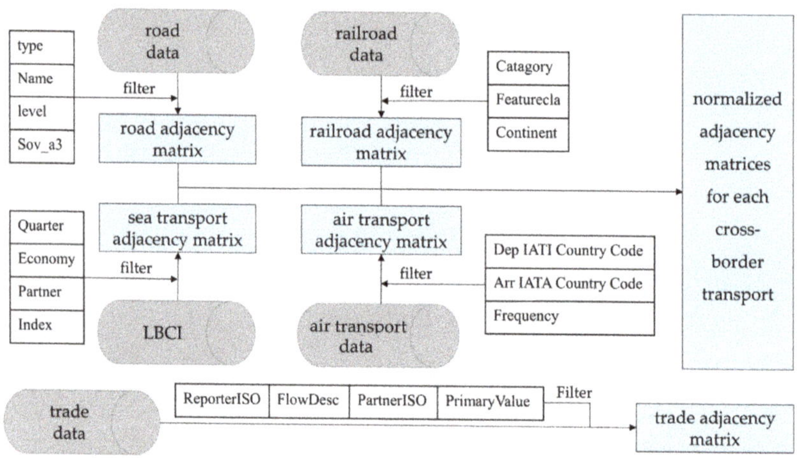

Figure 3. Data preprocessing.

3. Connectivity Evaluation System Construction

3.1. Original Evaluation System

The different indicators in the connectivity evaluation system could not directly adopt the assignment method based on data differences, which required judgment based on certain experience and cognition. Wang et al. [30] used the expert scoring method to assign weights for the connectivity evaluation indicators when evaluating the infrastructure connectivity of China–Silk Road countries. The experts came from the subject group of the third-party assessment report on the progress of the construction of the Silk Road Economic Belt and the 21st Century Maritime Silk Road; the indicators' weights had a certain authority and the calculation results were empirically verified in the analysis section. So, using this evaluation system for inter-country connectivity evaluation had a certain degree of reliability. Therefore, this paper referred to the connectivity evaluation system used by Wang et al. [30] to establish the original evaluation system for inter-country connectivity in Latin America, as shown in Table 3.

Table 3. Original evaluation system for inter-country connectivity in Latin America.

Connectivity Type	Score	Indicator	Score	Meaning
sea	35	LBCI	35	LBCI
railroad	29	Infrastructure connectivity	29	Number of infrastructures
road	18	Infrastructure connectivity	18	number of arterial road connections
air	18	direct route	14	number of routes
		frequency of flights	4	frequency of flights

Based on the normalized adjacency matrix of each cross-border transport, the connectivity initial adjacency matrix of Latin America was calculated using the weights of indicators from the original evaluation system of connectivity, i.e., the connectivity network, as shown in Table 4; using the trade data, the trade adjacency matrix among Latin American countries was screened and sorted out, i.e., the trade network, as shown in Table 5.

Table 4. Connectivity initial adjacency matrix among Latin American countries (2020).

	ARG	ATG	BHS	...	URY	VCT	VEN
ARG	0	13.61	22.45	...	42.61	14.06	14.51
ATG	13.61	0	13.61	...	13.61	14.64	14.73
BHS	22.45	13.61	0	...	22.28	14.02	14.66
⋮	⋮	⋮	⋮	0	⋮	⋮	⋮
URY	42.61	13.61	22.28	...	0	14.06	14.50
VCT	14.06	14.64	14.02	...	14.06+	0	20.41
VEN	14.51	14.73	14.66	...	14.50	20.41	0

Table 5. Trade adjacency matrix among Latin American countries (2020) (US$).

	ARG	ATG	BHS	...	URY	VCT	VEN
ARG	0	629 156	26 972 271	...	1 405 156 415	101 216	110 243 068
ATG	629 156	0	170 100	...	3 249	4 151 004	5 708
BHS	26 972 271	170 100	0	...	699 988	0	650 144
⋮	⋮	⋮	⋮	0	⋮	⋮	⋮
URY	1 405 156 415	3 249	699 988	...	0	0	4 808 796
VCT	101 216	4 151 004	0	...	0	0	0
VEN	110 243 068	5 708	650 144	...	4 808 796	0	0

However, since the realistic connectivity capacity of each transport mode varies in different regions, and although it could reflect the reality of the connectivity in the region to a certain extent, the connectivity in Latin America that was calculated using the original evaluation system was bound to have some bias. Using QAP correlation analysis and setting the number of permutations to 5000, the correlation between the initial adjacency matrix and the trade adjacency matrix for Latin American countries was calculated, and the results are shown in Table 6.

Table 6. Results of QAP correlation analysis between connectivity initial adjacency matrix and trade adjacency matrix among Latin American countries.

Indicator	Pearson Correlation	Significance	Average	Std Dev	N Obs
Value	0.469	<0.001	0.001	0.074	5000

In this result, the Pearson's correlation coefficient r = 0.469 between the connectivity initial adjacency matrix and the trade adjacency matrix proved that there was a moderate positive correlation between the two; the p-value (significance) was less than 0.001, which indicated that the correlation was statistically very significant, and that the event did not occur by chance. The above results proved that the comprehensive evaluation of inter-country connectivity using each cross-border transport as an indicator system was quite reliable. However, its correlation coefficient was $r < 0.6$; therefore, the degree of reflection of this connectivity initial adjacency matrix on the reality of connectivity (trade volume) among countries in Latin America needed to be improved and further optimized.

3.2. Connectivity Evaluation System Optimization

The connectivity evaluation system optimization process is shown in Figure 4.

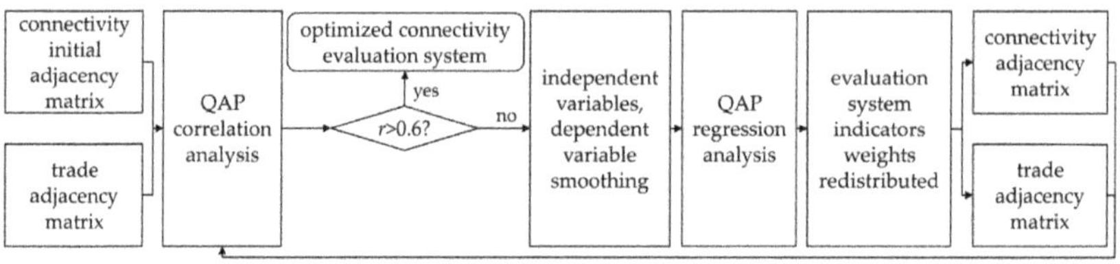

Figure 4. Connectivity evaluation system optimization process.

Regression analysis can measure the degree of influence of different independent variables on the dependent variable, and thus determine the appropriate allocation of weights for different indicators. Therefore, the impact of different modes of transport on trade was measured using the four cross-border transport normalized adjacency matrices as the independent variables and the trade adjacency matrix as the dependent variable. And the determination of the regression method required the analysis and processing of the data of the independent and dependent variables. The data structure analysis of the independent and dependent variables showed that each variable conformed to a normal distribution, but there were outliers (extremely large values, extremely small values, etc.). In order to exclude the effect of outliers, the independent and dependent variables were smoothed using bubble sort and interpolation.

The existence of multicollinearity between variables also affects the selection of regression methods. QAP analysis is a "method of measuring the relationship between relationships" [37], which can effectively solve the problems of multicollinearity and spurious correlation that exist in traditional measures of relational data processing [38]. A number of scholars have already used QAP regression analysis in the fields of transport, trade, tourism, and regional development to conduct research on the influencing factors and driving forces [39–44]. Therefore, in this article, QAP regression analysis was chosen to determine the degree to which the independent variables influence the dependent variable. The results of the QAP regression analysis are shown in Table 7.

Table 7. Results of QAP regression analysis.

Variables	Standardized Coefficient	Significance (p)	Number of Valid Values
sea	0.522	<0.001	1089
railroad	0.037	0.452	1089
road	0.208	<0.001	1089
air	0.202	<0.001	1089

In the above regression results, the larger the standardized coefficients and the stronger the significance (the smaller the p-value), meaning that this mode of cross-border transport had a greater impact on trade, the higher the degree of correlation. Observation of the data showed that sea, road and air transport all had a significant positive impact on trade; however, the degree of impact was different, with sea transport having the greatest impact, followed by road and air transport, while railroads had a non-significant impact on trade. According to the results of the regression analysis, the original evaluation system of connectivity was reallocated with reasonable weights with reference to the regression coefficient, and the allocation results are shown in Table 8.

Table 8. Evaluation system for inter-country connectivity in Latin America.

Connectivity Type	Score	Indicator	Score	Meaning
sea	55	LBCI	55	LBCI
railroad	5	Infrastructure connectivity	5	Number of infrastructures
road	20	Infrastructure connectivity	20	number of arterial road connections
air	20	direct route	16	number of routes
		frequency of flights	4	frequency of flights

Based on the four cross-border transport normalized adjacency matrices, combined with the connectivity evaluation system, the weight-adjusted connectivity adjacency matrix among Latin American countries was calculated, and the results are shown in Table 9.

Table 9. Connectivity adjacency matrix among Latin American countries (2020).

	ARG	ATG	BHS	...	URY	VCT	VEN
ARG	0	21.39	35.27	...	51.99	22.10	22.81
ATG	21.39	0	21.39	...	21.39	23.00	23.14
BHS	35.27	21.39	0	...	35.01	22.03	23.04
⋮	⋮	⋮	⋮	0	⋮	⋮	⋮
URY	51.99	21.39	35.01	...	0	22.09	22.78
VCT	22.10	23.00	22.03	...	22.09	0	31.72
VEN	22.81	23.14	23.04	...	22.78	31.72	0

The correlation coefficients between the above connectivity adjacency matrix and the trade adjacency matrix were calculated using QAP correlation analysis, and the results are shown in Table 10.

Table 10. Results of QAP correlation analysis between connectivity adjacency matrix and trade adjacency matrix among Latin American countries.

Indicator	Pearson Correlation	Significance	Average	Std Dev	N Obs
Value	0.678	<0.001	0.001	0.106	5000

Comparing Table 6 with Table 10, it could be seen that after weight adjustment, the correlation between the connectivity adjacency matrix and trade adjacency matrix was improved from 0.469 to 0.678, and a strong positive correlation level was reached between the two, and the significance was not decreased. Therefore, it could be argued that the adjustment of the indicator weights in the evaluation system of Latin American inter-country connectivity was reasonable and correct, and that the results calculated using this evaluation system were more in line with the reality of inter-country connectivity in Latin America.

The calculated inter-country connectivity results (adjacency matrix) for Latin America were visualized in conjunction with a map to obtain Figure 5, as shown below. The nodes in the figure are the 33 sovereign countries in Latin America, the center of the node is the geometric center of the country vector surface data, and the size of the node is the sum of the connectivity between the different countries and the rest of the countries in the study area; the edges indicate the inter-country connectivity, and the larger the value of the connectivity, the wider the edge.

Figure 5. Visualization of the results of the connectivity evaluation for Latin American countries.

3.3. *Comparisons with Existing Studies*

At present, there are many studies on transport connectivity based on a single mode of transport (e.g., sea transport, air transport, etc.), and the studies' perspectives are relatively simple. For example, Warnock-Smith et al. analyzed air transport connectivity between the EU and Latin America [45], but the study focused only on air transport connectivity and the development level of transcontinental air transport, which lacked comprehensiveness, and the scale of the study was based on continents and lacked detailed research. In addition, Wilmsmeier, Roach and Seabra et al. studied transport in Latin America focusing only on sea or land transport [46–48]. Meanwhile, Alejandra Saus and Velasco et al. evaluated and analyzed the internal transport of major Latin American cities, with the scale of the study limited to the city level [49,50]. In contrast, this study focused on cross-border transport among countries within the region, which was more complete as it covered the country-region levels at the spatial scale. In addition, this study was based on four modes of cross-border transport, sea, road, air and rail, which made the evaluation system more comprehensive.

Although the connectivity evaluation based on comprehensive indicators is not unique to this study, most of the existing studies have certain shortcomings. For example, Wang et al. constructed an infrastructure connectivity evaluation system by using a variety of cross-border transport data and the expert scoring method; they calculated the connectivity between China and other Silk Road countries and conducted related analyses [30]. However, this method was somewhat subjective and the reliability of the results obtained needed to be strengthened. In contrast, although the original evaluation system of this study was also obtained by the expert scoring method, the verification and correction links were added to enhance the reliability of the method and the credibility of the results. Based on the gravity model, Jiao and Li et al. used trade data to refer to connectivity among countries, and explored the impact of cross-border transport factors on connectivity, but did not really obtain a connectivity index among countries, and thus were unable to carry out further analyses, such as on the spatio-temporal evolution of connectivity [21,51]. The Data Envelopment Analysis (DEA) adopted by Netirith et al. had similar problems, i.e., it only evaluated the mechanisms by which multiple factors, including transport, affect trade [52]. In comparison, this study not only constructed a reliable connectivity evaluation system and explored the connectivity impact mechanism to some extent, but also constructed a connectivity network of Latin American countries and conducted an analysis of the spatial pattern of connectivity.

Above all, there was a lack of regional, cross-country and comprehensive transport connectivity evaluation in Latin America, and this paper constructed a scientific, reliable, comprehensive and expandable evaluation system to compensate for the lack of this aspect.

4. Spatial Pattern Analysis of Connectivity

In this article, network analysis was used to construct indicators measuring different perspectives on connectivity and combined with visual representations to analyze the spatial patterns of connectivity among Latin American countries, with the basic process shown in Figure 6.

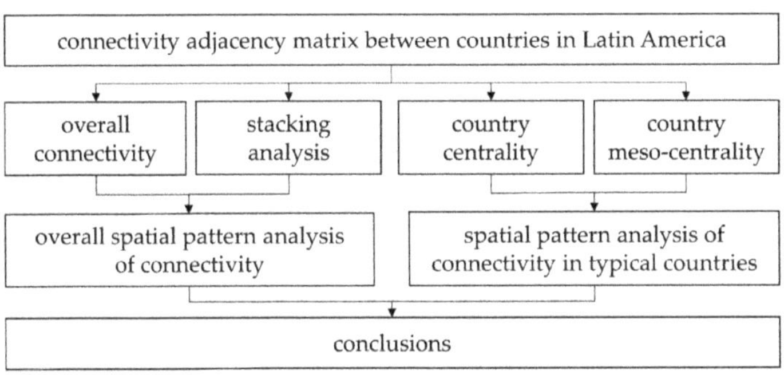

Figure 6. Process for analyzing spatial patterns of connectivity among Latin American countries.

The constructed indicators mainly contained overall connectivity, country centrality, and country meso-centrality:

(i) Overall connectivity

Overall connectivity reflects the overall level of connectivity between countries in the region, and it is solved by multiplying the overall density of the network by the sum of the weights of all edges in the network. Its calculation formula is as follows:

$$D = \frac{\sum_i^N M_i}{N(N-1)} * \sum_i^N Q_{ij} \quad (1)$$

where D denotes overall connectivity, i is a country in the network, M_i is the number of countries connected to country i, N is the total number of countries in the network, and Q_{ij}

is the corresponding inter-country connectivity. The calculated overall connectivity results range from 0 to 100, with higher values meaning a higher level of connectivity among Latin American countries; conversely, the closer to 0, the lower the level of connectivity.

(ii) Country centrality

Country centrality reflects the position of a country in the network, i.e., the higher the country centrality, the more central position the country has in the network; on the contrary, it is at the periphery [53]. Its calculation formula is as follows:

$$C_D(N_i) = \sum_{i=1}^{N} Q_{ij}/(N-1) \tag{2}$$

where $C_D(N_i)$ is country centrality, i is a country in the network, N is the number of countries in the network, Q_{ij} denotes the connectivity between two countries, and $N-1$ denotes the number of edges where country i is connected to all other countries.

(iii) Country meso-centrality

The country meso- centrality is the number of shortest paths passing through node (country) v_i divided by the number of all possible shortest paths in the shortest path from node (country) v_j to node (country) v_k [54,55]. Country meso-centrality reflects the ability of a country to act as the "hub" of a connectivity network. Its calculation formula is as follows:

$$C_B(N_i) = \sum_{j,k \in V} \frac{g_{jk}(i)}{g_{jk}} \tag{3}$$

where $C_B(N_i)$ is the country meso-centrality, g_{jk} denotes the number of shortest paths from node v_j to node v_k, and $g_{jk}(i)$ denotes the number of shortest paths passing through node v_i in the shortest path from node v_j to node v_k.

4.1. Overall Spatial Patterns Analysis

The overall connectivity among Latin American countries has been calculated to be 26.022. As shown by the box plot of connectivity among Latin American countries (Figure 7), the average value of connectivity between a Latin American country and other countries in the region was generally in the range of 20–30, with only a small number of countries having an average value of connectivity with other countries in the region exceeding 30, and individual countries having a connectivity greater than 50. Therefore, it could be argued that the overall connectivity among Latin American countries was low and the level of regional cross-border transport was underdeveloped.

Figure 5 can also represent a visualization of the connectivity network among Latin American countries. As can be seen from the performance of the nodes in Figure 5, the countries with high connectivity in Latin America were mainly those with medium volume (middle level of GDP) and above in the coastal region, such as COL, MEX, PAN, BRA, and DOM, followed by the Pacific Coastal countries. Among the above countries, BRA and MEX were the two largest countries in Latin America, and their level of economic development was also located in the top two; COL and PAN were important transport hubs in the region; other countries such as ARG, CHL, PER and DOM were medium-sized countries in the region, and had a certain degree of influence in the regional economy and politics. And from the side (inter-country connectivity) point of view, there were high levels of connectivity between BRA-ARG (URY), COL-PAN (MEX, DOM), ARG-CHL, CHL-PER, and MEX-PAN.

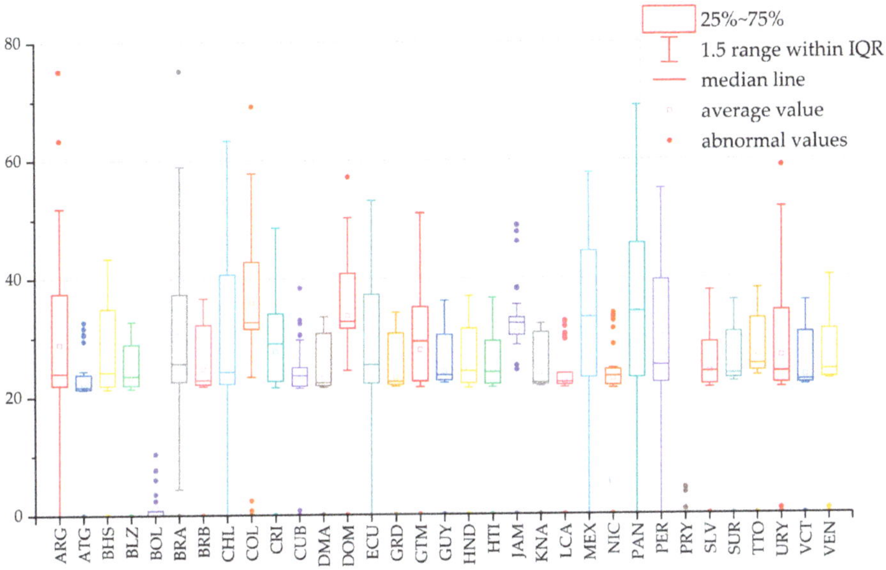

Figure 7. Box plot of connectivity among Latin American countries.

From the above results, it could be seen that the overall connectivity of Latin America was poor and the overall development was unbalanced, which was closely related to the overall physical and geographic characteristics and the level of economic development of the Latin American countries. The complex and varied terrain is the fundamental factor that leads to the low level of connectivity among Latin American countries. The region is bordered by the Atlantic Ocean in the east and the Pacific Ocean in the west; MEX is mainly a highland region, and the Central American peaks are predominantly uneven, which to a certain extent hampers the ease of regional connectivity; the countries of the Caribbean region are mainly island countries, which are convenient for sea transport, but the size of the countries is relatively small; the western part of South America is characterized by the Andes Mountains, and the eastern part of South America is dotted with a number of highland plateaus and plains, of which the best-known one is the Amazonian Plain; the Amazon rainforest and the Brazilian steppe have been shaped by climatic influences, and both mountainous isolation and climatic influences have somewhat hindered the improvement in regional connectivity.

The overall level of economic development in Latin America was low, and the economic growth rate has slowed down in recent years [56], which hindered the improvement in connectivity to a certain extent. The vast majority of Latin American countries are developing countries with a low level of economic development and a fragile economic structure; from 2014 to 2023, the annual growth rate of the Latin American economy was only 0.8%, which was less than half of the annual growth rate of the region in the "Lost Decade" of the 1980s; from 2015 to 2019, GDP per capita growth in Latin America was nearly stagnant. It was especially affected by the COVID-19 pandemic, declining sharply in 2020 [57], and by the end of 2023 had not recovered to the level of 2013–2014. Low, stagnant or even regressive economic growth and increased pressure on government debt were also key factors contributing to the low level of connectivity in Latin America. For example, the Colombian economy, which recovered rapidly after the shock of the epidemic and is growing slowly, has the highest level of connectivity in the region, but still suffers from poor rural roads and the absence of road connections in some cities. Brazil has a fluctuating level of economy, with a high level of connectivity in the region but slow growth. Argentina, whose economy has been on a downward trend in recent years, has a medium to high level of connectivity in the region, but does not have the capacity to invest significant financial support in transport infrastructure development.

The volatility and weak continuity of political systems were important reasons for the low level of connectivity among Latin American countries. Since 2015, many Latin American countries have held general elections, and party rotation and government change have continued to occur, with many of them ending in the ousting of the ruling party. This has led to a lack of continuity in the policies of many countries in Latin America and instability in their relations with neighboring countries, which makes it more difficult to realize long-term strategic investments, such as improvements in the level of transport connectivity. For example, as early as the 20th century, the United States and Latin American countries planned projects such as the Pan-American Highway to promote the construction of regional transport infrastructure. The construction of the Pan-American Highway has improved the road infrastructure of Latin American countries to a considerable extent and accelerated the realization of regional transport integration. However, the progress of the project has been stalled for decades at the Darien Gap (between COL and PAN) due to conflicts of interest, resulting in an imperfect Pan-American Highway [58,59].

Regional transport improvement projects are also being actively implemented within Latin American countries, such as the Iniciativa para la Integración de la Infraestructura Regional Suramericana (IIRSA) and the Proyecto de Integración y Desarrollo de Centroamérica (PM); the main part of these projects includes the construction of transport infrastructure. IIRSA began in 2000 and is now at its peak of project completion, with the construction of highway and railroad facilities being the most important part of the project [60]. PM began in 2001 and focuses on the construction of a highway network in terms of transport, which includes the Pacific Coast Highway and the Atlantic Coast Highway.

In addition, Brazil, Peru, Argentina and other countries have also proposed a number of railroad projects, such as the FIOL railroad in Brazil, the Peruvian railroad, the renovation of the San Martin Freight Railroad in Argentina, and the Two Oceans Railway. Although these projects are mainly domestic, small regional connectivity projects and lack intercontinental-scale connectivity, they can still make positive contributions to the enhancement of connectivity and regional integration in Latin America. At present, 22 countries in Latin America have signed cooperation documents with China to jointly build the Silk Road Economic Belt and the 21st Century Maritime Silk Road; the two sides will certainly continue to strengthen cooperation in infrastructure construction, which will help to further make up for the shortcomings of intercontinental connectivity and promote the integration of transport in Latin America, together with other agreements and projects [61].

4.2. Spatial Pattern Analysis of Typical Countries

Country centrality and country meso-centrality were calculated for Latin America, and the results are shown in Table 11.

As shown in Table 11, integrated countries showed the strongest connectivity with other countries, followed by island countries and finally landlocked countries. This was closely linked to the status of sea transport in international trade.

Based on the above results, the countries in Latin America were classified into three types: global hub type (country centrality > 35.000 and country meso-centrality = 1), local hub type (35.000 $\geq$ country centrality > 29.000 and country meso-centrality > 0.5) and non-hub type (country centrality < 29.000 or country meso-centrality < 0.5); the results are shown in Table 12.

Table 11. Country centrality, meso-centrality and its types in Latin America.

Country	Type	Country Centrality	Country Meso-Centrality	Country	Type	Country Centrality	Country Meso-Centrality
COL	Integrated	37.096	1.000	HND	Integrated	25.509	0.000
PAN	Integrated	35.262	1.000	SUR	Integrated	25.440	0.000
DOM	Island	34.805	0.000	GUY	Integrated	25.002	0.000
MEX	Integrated	34.291	0.000	HTI	Island	24.958	0.000
JAM	Island	31.739	0.000	SLV	Integrated	24.836	0.000
BRA	Integrated	31.389	1.000	VCT	Island	24.160	0.000
PER	Integrated	30.885	1.000	BLZ	Integrated	24.079	0.000
CHL	Integrated	30.496	1.000	GRD	Island	24.019	0.000
ARG	Integrated	29.887	1.000	KNA	Island	23.969	0.000
ECU	Integrated	29.797	0.419	DMA	Island	23.890	0.000
GTM	Integrated	28.755	0.000	CUB	Island	23.742	0.419
CRI	Integrated	28.610	0.000	NIC	Integrated	23.601	0.000
BHS	Island	27.795	0.000	LCA	Island	23.177	0.000
TTO	Island	27.786	0.000	ATG	Island	22.660	0.000
URY	Integrated	27.539	0.581	BOL	Landlocked	1.288	0.065
VEN	Integrated	26.172	0.419	PRY	Landlocked	0.511	0.017
BRB	Island	25.569	0.000				

Table 12. Latin American country types.

Type	Countries
global hub type	COL, PAN
local hub type	BRA, PER, CHL, ARG
non-hub type	MEX, DOM, ECU, JAM, GTM, CRI, BHS, TTO, URY, VEN, BRB, HND, SUR, GUY, HTI, SLV, VCT, BLZ, GRD, KNA, DMA, CUB, NIC, LCA, ATG, BOL, PRY

Colombia (global hub type), Brazil (local hub type) and Mexico (non-hub type) were the most representative of these three types of countries, as they had the largest land area and the largest economic volume. The spatial patterns and basic characteristics of their connectivity with other countries in the region will be analyzed below.

4.2.1. Colombia

COL was an important global hub for cross-border transport in the region. COL had the strongest country centrality in the region, which meant that it had the highest level of connectivity with the countries; at the same time, it also had the largest country meso-centrality and played an important role as a "bridge" in the region. COL connectivity with countries in the region was high and relatively homogeneous overall, but there was also some spatial heterogeneity (as shown in Figure 8): COL had the strongest connectivity with its land neighbor PAN, followed by its sea neighbors DOM and MEX, with the overall center of gravity of connectivity to the northwest. COL also had strong connectivity with landlocked neighboring countries such as PER, ECU, and BRA, and the connectivity tended to decline with increasing distance between countries. In addition, the size of the country was an important factor that significantly influenced the strength of its connectivity, e.g., it also maintained a high level of connectivity with countries that were geographically and spatially distant, such as CHL and ARG.

COL is strategically located, bordering VEN and BRA in the east, ECU and PER in the south, the Pacific Ocean in the west, PAN in the northwest and the Caribbean Sea in the north; together with PAN, it constitutes a transport hub between MEX, Central America, South America and the Caribbean Sea area. In addition, COL's GDP had long been the fourth highest in Latin America, which was among the higher levels in the region. This high economic level provided material support for high connectivity, and supported COL's well-established cross-border sea and air transport system. In terms of trade, PAN, MEX, BRA, ECU and other countries were closely connected to COL and had a high level of trade. In terms of exports, PAN was the largest source of COL's trade surplus, and was more dependent on COL for its domestic mineral resources; ECU, PER, BRA and MEX imported large quantities of chemicals and plastic and rubber products from COL. In terms of imports, MEX and BRA were the largest surplus countries for COL in the region, with COL importing large quantities of electromechanical products, transportation equipment, and base metals and products from both countries.

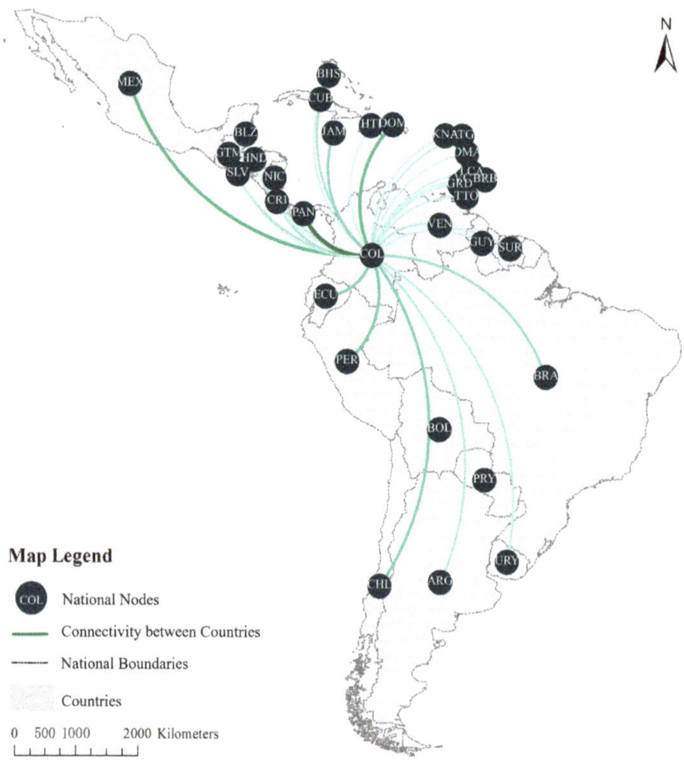

Figure 8. Spatial patterns of connectivity between Colombia and regional countries.

4.2.2. Brazil

BRA was a local hub for regional cross-border transport; in addition to fulfilling its own national cross-border transport needs, it also served as a "bridge" to a certain extent. BRA's country centrality was ranked sixth, representing a high level of regional connectivity; it also had the highest country median centrality, showing that BRA also served as a transit point to a certain extent. Overall, BRA's connectivity with regional countries had strong spatial heterogeneity (as shown in Figure 9): it was strongest with its land neighbors ARG and URY, with the overall center of gravity of connectivity skewed toward the south; the connectivity with other countries in the region decreased sharply with increasing distance; and although country size also affected connectivity between BRA and other countries to some extent, it was not as significant as COL.

Figure 9. Spatial patterns of connectivity between Brazil and regional countries.

BRA is the largest country in Latin America, bordering ten countries in South America. The relatively gentle terrain creates convenient geographic conditions for cross-border land transport; it is east of the Atlantic Ocean, with a coastline of about 7400 km. Economically, BRA had the highest GDP in Latin America. However, after experiencing the 2019–2020 COVID-19 pandemic shock, Brazil's economy fell back to the level of 10 years ago [62,63]. In terms of cross-border transport, BRA had relatively complete sea, air, and road transport systems, which were the basis for its high level of connectivity. After Lula was re-elected as president, he proposed policies to revitalize the economy and planned to invest BRL 1.7 trillion in infrastructure construction, which would help improve BRA's cross-border transport connectivity. In terms of trade, ARG was BRA's largest trading partner in Latin America, and the two countries cooperated closely in transportation equipment and electromechanical products; the port of Santos was BRA's largest port and also URY's trans-shipment port, which made an outstanding contribution to the connectivity of the two countries; CHL relied more on BRA's exports of minerals and electromechanical products; in addition to this, MEX, PAN, BER and COL were also important regional partners of BRA.

4.2.3. Mexico

MEX was a non-hub country for cross-border transport in the region and played a small role as a regional transit country. MEX ranked fourth in terms of country centrality and had a high level of overall connectivity in the region; however, it did not have a national meso-centrality, which indicated that it did not have a "hub" connectivity role in the region.

The connectivity between MEX and regional countries also showed spatial heterogeneity (as shown in Figure 10): it was most strongly connected to the neighboring hub countries PAN and COL, followed by the Pacific Coast countries, PER, CHL, and ECU, and other major countries in the South America, showing a strong tendency for the connectivity to decline with increasing distance.

Figure 10. Spatial patterns of connectivity between Mexico and regional countries.

MEX is bordered by the USA in the north, GTM and BLZ in the south, the Gulf of Mexico and the Caribbean Sea in the east, and the Pacific Ocean and the Gulf of California in the west, which makes it an essential place for land transport between North and South America. Economically, MEX was the second largest economy in Latin America, but the COVID-19 pandemic hit MEX's economy hard [64] and it only recovered to pre-pandemic levels by 2022, and then kept a low growth rate. In terms of cross-border transport, MEX had a relatively complete road system as well as more developed sea and air transport systems, which provided the foundation for its cross-border transport connectivity and trade. In terms of trade, BRA and COL were MEX's most important trading partners in the region. Of these, BRA was MEX's largest importer within Latin America in terms of transportation equipment and optical, horological, and medical equipment, while COL was MEX's largest exporter in Latin America in terms of trade in minerals.

5. Conclusions

This article constructed an inter-country connectivity evaluation system with considerable reliability and generalization ability based on various cross-border transport modes and trade data in a combination of methods. On this basis, an assessment of connectivity among the 33 countries of Latin America (2020) was carried out and the spatial patterns of connectivity were analyzed, leading to the following conclusions:

1. Cross-border transport in Latin America was mainly by sea, supplemented by road and air transport, while railroads rarely undertook cross-border transport.

2. Latin America's overall connectivity was low and development was unbalanced, with a strong law of spatial differentiation. The countries with higher connectivity were concentrated in coastal integrated countries of medium size and above, while island and landlocked countries generally had lower connectivity. This feature was closely related to the physical and geographic characteristics, the level of economic development and the stability of the political systems of the different countries.

3. Different countries in Latin America assumed different roles in regional connectivity, which could be categorized into three types: global hub type, local hub type and non-hub type. The countries represented by Colombia were the global hubs of regional cross-border transport, and they served as transit points for regional passenger and goods transport by virtue of their advantageous geographical location, high level of economic development and well-developed cross-border transport systems. Countries represented by Brazil were localized hubs for regional cross-border transport. They generally had a certain size of land area, a high level of economic development in the region and a relatively well-developed cross-border transport system, while undertaking transit tasks for some of the neighboring countries. Countries represented by Mexico did not undertake transit tasks in regional cross-border transport and mainly accomplished cross-border transport in their own countries.

4. Inter-country connectivity in Latin America generally declined with distance, but the rate of decline was closely related to the specific geographic location and the role assumed in the connectivity network. The closer the geographic location was to the regional center, the slower its connectivity declined with distance. And connectivity was less influenced by distance in global hub type countries compared to local hub type and non-hub type countries.

The connectivity evaluation method proposed in this paper has considerable universality and generalization ability, which is applicable in most regions of the world, and can provide references and ideas for other scholars' related research. Meanwhile, the calculation and analysis of connectivity in Latin America in this paper make up for the lack of connectivity studies in the region. But this study also has some limitations: the paper mainly explores connectivity and its spatial pattern among Latin American countries in 2020, and due to the limitation of data updating, it is not yet possible to assess and analyze the performance of connectivity across a longer time frame; the generalization of the connectivity evaluation method in this paper has certain constraints, which require a relatively complete transport infrastructure between the countries in the region, and a certain degree of accessibility of cross-border transport data that can support a reasonable assessment of connectivity. In subsequent studies, further research will be conducted on the performance of inter-Latin American connectivity and its evolutionary mechanisms on broader time scales. We will also attempt to predict future trends in connectivity.

Author Contributions: All authors made valuable contributions to this paper. Conceptualization: Changqi Miao, Xinjia Zhang, Yinbao Zhang and Jianzhong Liu; methodology: Changqi Miao and Shike Zhang; software: Changqi Miao; validation: Changqi Miao, Xinjia Zhang, Yinbao Zhang and Shike Zhang; formal analysis: Xinjia Zhang, Yinbao Zhang and Jianzhong Liu; resources: Changqi Miao, Xinjia Zhang and Shike Zhang; data curation: Changqi Miao, Xinjia Zhang and Yinbao Zhang; writing—original draft preparation: Changqi Miao, Xinjia Zhang and Shike Zhang; writing—review and editing: Changqi Miao, Yinbao Zhang and Jianzhong Liu; visualization: Changqi Miao and Xinjia Zhang. All authors have read and agreed to the published version of the manuscript.

Funding: This research was funded by the National Social Science Fund of China—"Theory and Method of Geo-Environmental Analysis in the Era of Big Data", under grant number 20&ZD138.

Data Availability Statement: Data are available in a publicly accessible repository.

Conflicts of Interest: The authors declare no conflicts of interest.

References

1. Gaens, B.; Sinkkonen, V.; Ruokamo, A. Introduction to the Special Issue: Theory and Practice of Connectivity in the Indo-Pacific-Spheres, Logics, and Regional Dynamics. *East Asia* **2023**, *40*, 195–207. [CrossRef] [PubMed]
2. Wang, G.W.Y.; Zeng, Q.H.; Li, K.; Yang, J.L. Port connectivity in a logistic network: The case of Bohai Bay, China. *Transp. Res. Part E-Logist. Transp. Rev.* **2016**, *95*, 341–354. [CrossRef]
3. Suwanwong, T.; Sopadang, A.; Hanaoka, S.; Rodbundith, T. Evaluation of air cargo connectivity and policy in Thailand. *Transp. Policy* **2018**, *72*, 24–33. [CrossRef]
4. Allroggen, F.; Wittman, M.D.; Malina, R. How air transport connects the world—A new metric of air connectivity and its evolution between 1990 and 2012. *Transp. Res. Part E-Logist. Transp. Rev.* **2015**, *80*, 184–201. [CrossRef]
5. Suau-Sanchez, P.; Burghouwt, G. Connectivity levels and the competitive position of Spanish airports and Iberia's network rationalization strategy, 2001–2007. *J. Air Transp. Manag.* **2012**, *18*, 47–53. [CrossRef]
6. Calatayud, A.; Mangan, J.; Palacin, R. Connectivity to international markets: A multi-layered network approach. *J. Transp. Geogr.* **2017**, *61*, 61–71. [CrossRef]
7. Fan, T. Improvements in intra-European inter-city flight connectivity: 1996–2004. *J. Transp. Geogr.* **2006**, *14*, 273–286. [CrossRef]
8. Jarumaneeroj, P.; Lawton, J.B.; Svindland, M. An evolution of the Global Container Shipping Network: Port connectivity and trading community structure (2011–2017). *Marit. Econ. Logist.* **2024**, *26*, 283–306. [CrossRef]
9. Lee, P.T.-W.; Lee, S.-W.; Hu, Z.-H.; Choi, K.-S.; Choi, N.Y.H.; Shin, S.-H. Promoting Korean international trade in the East Sea Economic Rim in the context of the Belt and Road Initiative. *J. Korea Trade* **2018**, *22*, 212–227. [CrossRef]
10. Zhang, Q.; Wang, B.; Xue, D. The Hub Competition in Delivering Air Connectivity between China and Oceania. *Sustainability* **2022**, *14*, 6482. [CrossRef]
11. Cheung, T.K.Y.; Wong, C.W.H.; Zhang, A. The evolution of aviation network: Global airport connectivity index 2006–2016. *Transp. Res. Part E-Logist. Transp. Rev.* **2020**, *133*, 101826. [CrossRef]
12. Zong, K.; Hu, Z. Research on the connectivity of shipping route network based on complex network. *J. Dalian Marit. Univ.* **2017**, *43*, 79–84.
13. Pan, J.; Wang, X. Port connectivity model based on complex network and its application. *J. Shenzhen Univ. Sci. Eng.* **2017**, *34*, 544–550. [CrossRef]
14. Huang, S.; Wang, B. The impact of the Belt and Road Initiative on container shipping connectivity enjoyed by the countries along the route. *J. Navig. China* **2024**, *47*, 97–102, 110.
15. Zhang, S.; Wang, L. A study on shipping connectivity in the Asian Mediterranean. *Acta Geogr. Sin.* **2022**, *77*, 2616–2632.
16. Kanrak, M.; Nguyen, H.-O. An analysis of connectivity, assortativity and cluster structure of the Asian-Australasian cruise shipping network. *Marit. Transp. Res.* **2022**, *3*, 100048. [CrossRef]
17. Wang, Q. Research on Development Strategy of Cross-Border Transport Corridor in Border Areas—A Case of Ruili in Yunnan. Master's Thesis, Chongqing Jiaotong University, Chongqing, China, 2013.
18. del Rosal, I.; Zanon Moura, T.G. The effect of shipping connectivity on seaborne containerised export flows. *Transp. Policy* **2022**, *118*, 143–151. [CrossRef]
19. Zhang, X.; Liu, C.; Peng, Y.; Lu, J. Connectivity-based spatial patterns and factors influencing international container multimodal hubs in China under the Belt and Road initiative. *Transp. Policy* **2023**, *143*, 10–24. [CrossRef]
20. Calatayud, A.; Palacin, R.; Mangan, J.; Jackson, E.; Ruiz-Rua, A. Understanding connectivity to international markets: A systematic review. *Transp. Rev.* **2016**, *36*, 713–736. [CrossRef]
21. Jiao, J.; Ma, W.; Li, H.; Bu, W.; Wang, J. Impact of connectivity on China's international trade: Taking countries along "the Belt and Road" as examples. *Sci. Geogr. Sin.* **2023**, *43*, 629–637.
22. Canbay, S. The analysis of relationships between global shipping networks and foreign trade volumes in developing countries. *Case Stud. Transp. Policy* **2024**, *17*, 101242. [CrossRef]
23. Hoffmann, J.; Saeed, N.; Sodal, S. Liner shipping bilateral connectivity and its impact on South Africa's bilateral trade flows. *Marit. Econ. Logist.* **2020**, *22*, 473–499. [CrossRef]
24. Saeed, N.; Cullinane, K.; Sodal, S. Exploring the relationships between maritime connectivity, international trade and domestic production. *Marit. Policy Manag.* **2021**, *48*, 497–511. [CrossRef]
25. Oum, T.H.; Wu, X.; Wang, K. Impact of air connectivity on bilateral service export and import trade: The case of China. *Transp. Policy* **2024**, *148*, 219–233. [CrossRef]

26. Bensassi, S.; Marquez-Ramos, L.; Martinez-Zarzoso, I.; Suarez-Burguet, C. Relationship between logistics infrastructure and trade: Evidence from Spanish regional exports. *Transp. Res. Part A-Policy Pract.* **2015**, *72*, 47–61. [CrossRef]
27. Li, W.; Bai, X.; Yang, D.; Hou, Y. Maritime connectivity, transport infrastructure expansion and economic growth: A global perspective. *Transp. Res. Part A-Policy Pract.* **2023**, *170*, 103609. [CrossRef]
28. Matsumoto, H.; Domae, K.; O'Connor, K. Business connectivity, air transport and the urban hierarchy: A case study in East Asia. *J. Transp. Geogr.* **2016**, *54*, 132–139. [CrossRef]
29. Ng, A.K.Y.; Jiang, C.; Li, X.; O'Connor, K.; Lee, P.T.-W. A conceptual overview on government initiatives and the transformation of transport and regional systems. *J. Transp. Geogr.* **2018**, *71*, 199–203. [CrossRef]
30. Wang, C.; Chen, P.; Wang, J.; Li, N. Evaluation method of infrastructure interconnection between China and other countries along the Belt and Road and its spatial pattern. *Geogr. Res.* **2020**, *39*, 2685–2704.
31. Zhu, Z.; Zhang, A.; Zhang, Y. Measuring multi-modal connections and connectivity radiations of transport infrastructure in China. *Transp. A-Transp. Sci.* **2019**, *15*, 1762–1790. [CrossRef]
32. Zhu, Z.; Zhang, A.; Zhang, Y.; Huang, Z.; Xu, S. Measuring air connectivity between China and Australia. *J. Transp. Geogr.* **2019**, *74*, 359–370. [CrossRef]
33. Guchang, A.O.; Yuanhua, J.I.A.; Huiling, Z. Research on Connectivity of Regional Road Network Based on the Capacity of Connectivity. *J. Beijing Jiaotong Univ.* **2009**, *33*, 42–46.
34. Sharma, S.; Ram, S. Investigation of Road Network Connectivity and Accessibility in Less Accessible Airport Regions: The Case of India. *Sustainability* **2023**, *15*, 5747. [CrossRef]
35. Wong, J.C.Y. Aviation Connectivity Impacts on Regional Economies in the United States. *Transp. Res. Rec.* **2018**, *2672*, 103–113. [CrossRef]
36. Carey, M. Latin American environmental history: Current trends, interdisciplinary insights, and future directions. *Environ. Hist.* **2009**, *14*, 221–252. [CrossRef]
37. Liu, J. QAP: A Unique Method of Measuring "Relationships" in Relational Data. *Chin. J. Sociol.* **2007**, *4*, 164–174+209. [CrossRef]
38. Chong, Z.; Qin, C. The Trade Network Structure of "One Belt One Road" and its Influence Factors: A Study Based on Analytic Network Process. *Int. Econ. Trade Res.* **2017**, *33*, 16–28. [CrossRef]
39. Byung-Hak, L.; Won, E.S. Impact of National Cultural Networks on Trade Network: QAP Analysis. *J. Int. Trade Commer.* **2021**, *17*, 11–20. [CrossRef]
40. Choi, J.H.; Barnett, G.A.; Chon, B.S. Comparing world city networks: A network analysis of Internet backbone and air transport intercity linkages. *Glob. Netw.-J. Transnatl. Aff.* **2006**, *6*, 81–99. [CrossRef]
41. Chong, Z.; Qin, C.; Pan, S. The Evolution of the Belt and Road Trade Network and Its Determinant Factors. *Emerg. Mark. Financ. Trade* **2019**, *55*, 3166–3177. [CrossRef]
42. Zhang, N.; Zhang, Y.; Chen, H. Spatial Correlation Network Structure of Carbon Emission Efficiency of Railway Transportation in China and Its Influencing Factors. *Sustainability* **2023**, *15*, 9393. [CrossRef]
43. Park, D.; Lee, G.; Kim, T. Intra-destination Travel Behaviors Analyses using Social Network Analysis Method among FITs from China, Japan, and North Americans to Seoul, South Korea. *J. Int. Trade Commer.* **2019**, *15*, 1–16. [CrossRef]
44. Song, J.; Zhu, Y.; Chu, X.; Yang, X. Research on the Vitality of Public Spaces in Tourist Villages through Social Network Analysis: A Case Study of Mochou Village in Hubei, China. *Land* **2024**, *13*, 359. [CrossRef]
45. Warnock-Smith, D.; Christidis, P. European Union-Latin America/Caribbean air transport connectivity and competitiveness in different air policy contexts. *J. Transp. Geogr.* **2021**, *92*, 102994. [CrossRef]
46. Wilmsmeier, G.; Cullinane, K.P.B.; Notteboom, T.; Sanchez, R.J. IAME 2011-contemporary studies on maritime transport in Latin America. *Marit. Policy Manag.* **2012**, *39*, 549–554. [CrossRef]
47. Roach, K.Q.; Kirton, R.M. Assessing the potential for increased and enhanced maritime transportation in Latin America and the Caribbean: A case study of Trinidad and Tobago and Venezuela. *World J. Sci. Technol. Sustain. Dev.* **2011**, *8*, 81–94. [CrossRef]
48. Seabra, F.; Patricio, T.; Bauer, M.M.; Gheller, L.; Saint Martin, J. Transport Infrastructure and Cross Border Connectivity Indices in Border Cities: The Case of Southern Brazil. *J. Lat. Am. Geogr.* **2022**, *21*, 112–126. [CrossRef]
49. Alejandra Saus, M. State of the art about urban development and transport: Reviewing tools for Latin America. *Econ. Soc. Y Territ.* **2023**, *23*, 991–1016. [CrossRef]
50. Velasco, A.; Gerike, R. A composite index for the evaluation of sustainability in Latin American public transport systems. *Transp. Res. Part A-Policy Pract.* **2024**, *179*, 103939. [CrossRef]
51. Li, J.Y.; Yang, Y.; Ji, X. Research on External Trade Effect of Cross-border Transport Corridor Based on Gravity Model: Taking Kunming-Bangkok Expressway as an Example. *J. Kunming Univ. Sci. Technol. (Soc. Sci.)* **2016**, *16*, 72–77. [CrossRef]
52. Netirith, N.; Ji, M. Analysis of the Efficiency of Transport Infrastructure Connectivity and Trade. *Sustainability* **2022**, *14*, 9613. [CrossRef]
53. Vidya, C.T.; Prabheesh, K.P. Implications of COVID-19 Pandemic on the Global Trade Networks. *Emerg. Mark. Financ. Trade* **2020**, *56*, 2408–2421. [CrossRef]

54. Freeman, L.C. A Set of Measures of Centrality Based on Betweenness. *Sociometry* **1977**, *40*, 35–41. [CrossRef]
55. Freeman, L.C. Centrality in social networks conceptual clarification. *Soc. Netw.* **1978**, *1*, 215–239. [CrossRef]
56. Yue, S. Analysis of Factors that Affect the Economic Growth in Developing Countries: A Case Study of Latin America. In Proceedings of the 7th International Conference on Applied Economics and Business (ICAEB), Copenhagen, Denmark, 24–26 August 2023; pp. 243–254.
57. Tejedor-Estupiñán, J.M. Impactos sociales y económicos de la pandemia de la COVID-19 en Latinoamérica. *Rev. Finanz. Y Política Económica* **2020**, *12*, 329–334. [CrossRef]
58. Orozco Cuello, C.; Sanandres Campis, E.; Molinares Guerrero, I. Colombia, Panamá y la Ruta Panamericana: Encuentros y Desencuentros. *Mem. Rev. Digit. Hist. Y Arqueol. Desde Caribe* **2012**, *15*, 101–130. [CrossRef]
59. Li, C. The Experience and Enlightenment of Pan-American Highway Construction. *J. Lat. Am. Stud.* **2019**, *41*, 133–153+158.
60. Kim, H. Analysis on the Characteristics of IIRSA: Focusing on the Construction of Infrastructure. *J. Assoc. Korean Geogr.* **2020**, *9*, 205–222. [CrossRef]
61. Xie, W. An Analysis of China-CELAC Joint Construction of the Belt and Road Initiative. *Pac. J.* **2018**, *26*, 80–90. [CrossRef]
62. Franz, T. Spatial fixes and switching crises in the times of COVID-19: Implications for commodity-producing economies in Latin America. *Can. J. Dev. Stud.-Rev. Can. D Etudes Du Dev.* **2021**, *42*, 109–121. [CrossRef]
63. Artiga-Purcell, J.A.; Chiasson-LeBel, T.; Leiva, F.; Watanabe-Farro, A. Disaster Extractivism: Latin America's Extractive Shock Therapy in the Age of COVID-19. *Lat. Am. Perspect.* **2023**, *50*, 172–192. [CrossRef]
64. Xiong, A. *Latin American Economy in 2023~2024: Recovery Slows Down*; Social Sciences Literature Publishing House: Beijing, China, 2024; pp. 137–151.

International Journal of
Geo-Information

MDPI

Article

Mapping the Dream: Designing Optimal E-Bike Routes in Valparaíso, Chile, Using a Multicriteria Analysis and an Experimental Study

Vicente Aprigliano [1,2,*], Catalina Toro [1], Gonzalo Rojas [1], Iván Bastías [3], Marcus Cardoso [4], Tálita Santos [5], Marcelino Aurélio Vieira da Silva [4], Emilio Bustos [1], Ualison Rébula de Oliveira [6] and Sebastian Seriani [1]

[1] Escuela de Ingeniería de Construcción y Transporte, Pontificia Universidad Católica de Valparaíso, Avda Brasil 2147, Valparaíso 2362804, Chile; catalina.toro@pucv.cl (C.T.); gonzalo.rojas.r@pucv.cl (G.R.); emilio.bustos@pucv.cl (E.B.); sebastian.seriani@pucv.cl (S.S.)

[2] Centro de Acción Climática, Pontificia Universidad Católica de Valparaíso, Avda Brasil 2147, Valparaíso 2362804, Chile

[3] Instituto de Geografía, Pontificia Universidad Católica de Valparaíso, Valparaíso 2362807, Chile; ivan.bastias.m@mail.pucv.cl

[4] Transport Engineering Program, Federal University of Rio de Janeiro, Av. Horácio Macedo 2030, Rio de Janeiro 21941-914, RJ, Brazil; cardoso@pet.coppe.ufrj.br (M.C.); marcelino@pet.coppe.ufrj.br (M.A.V.d.S.)

[5] Center for Sustainable Development, University of Brasília, Brasília 70910-900, DF, Brazil; ta.santos@unb.br

[6] Departamento de Administração e Administração Pública de Volta Redonda, Universidade Federal Fluminense (UFF)/MPA-ICHS, MSG-TCE and DSG-TCE, 783 Des. Ellis H. Figueira St Volta Redonda, Rio de Janeiro 27213-145, RJ, Brazil; ualisonrebula@id.uff.br

* Correspondence: vicente.aprigliano@pucv.cl

Abstract: The city of Valparaíso, Chile, faces significant mobility challenges due to its steep slopes, complex urban infrastructure, and socioeconomic conditions. In this direction, this study explores the potential promotion of E-bike uses by identifying the optimal routes that connect metro stations to strategic hilltop streets in the city. A hybrid methodology combining a multicriteria GIS-based analysis and an experimental study was used to evaluate potential routes and the possibility of increasing the power limitations for non-motorized mobility in Chile. Fifteen routes were assessed based on criteria including the slope, traffic safety, directionality, intersections, and travel distance. The results indicate that routes such as Cumming from Puerto and Bellavista stand out as the most viable for e-bike use given their favorable characteristics. The experimental study revealed that higher-powered E-bikes (500 W and 750 W) would be more able to overcome the steep slopes of Valparaíso, with an average speed of 5.36 km/h and 9.52 km/h on routes with a 10.88% average slope. These findings challenge the current regulatory limit of 250 W for non-motorized vehicles in Chile, highlighting the potential benefits of increasing their power limits to enhance sustainable mobility in the hilly urban contexts of this country. This study highlights the need to adapt urban mobility policies to the unique topographical conditions of each city. Future research should build upon more experimental studies, develop specific street-scale analyses using audit methods, incorporate climate-related variables, and evaluate the economic viability of e-bike infrastructure. Addressing these aspects could position Valparaíso as a leading example of sustainable urban mobility for cities facing comparable challenges.

Keywords: E-bikes; multicriteria analysis; experimental study; optimal routes; Valparaíso; Chile

1. Introduction

Great challenges lie ahead for cities around the world in integrating electric mobility (e-mobility), especially non-motorized solutions, into their urban dynamics. Some reasons for this relate to the higher speed of these modes, the lack of license requirements, and the constructive norms for cycleways, which are usually related to mechanic or regular bicycles' limitations. In this manner, studies developed by Niemann et al. [1] and Uluk et al. [2] evidence the growing accidents in urban environments involving e-mobility devices, also known as Electric Personal Mobility Devices (ePMDs).

Another important issue is that there is discussion in different countries about the power limits for non-motorized vehicles that do not require a specific driving license. In Germany and the Netherlands, for example, the law allows the use of E-bikes with motors of up to 250 watts and a maximum speed of 25 km/h without a specific driving license [3,4]. In the United States of America (USA), most states do not require a driving license for E-bikes with motors that do not exceed 750 watts and with a maximum speed of 45 km/h considering that they are pedal-assisted [5]. In some parts of Canada, for example, British Columbia, they allow the use of E-bikes with a maximum motor power of 500 watts that reaches up to 32 km/h without a driving license [6]. In Brazil, the maximum motor power and speed should not exceed 350 watts and 25 km/h [7].

In Chile, the regulation allows one to use an E-bike with a motor power of up to 250 watts with a maximum speed of 25 km/h without the need for a driving license. In this regard, it is important to understand that these E-bike characteristics generate limiting conditions that justify the implementation of road signs related to bicycle or E-bike use or specific cycleways in certain areas. For example, defining bicycle paths (whether shared with other modes or physically separated) is allowed and justified in sections with a maximum longitudinal slope of 6% [8]. In the city of Valparaíso, Chile, the normative slope criterion is a key barrier to the development of cycle-inclusive projects, although there is evidence of cyclists in regular and electric modes riding within the city and its surroundings [9,10].

As a starting point, it is important to understand the urban context of Valparaíso. As shown in Figure 1, most of the streets on the hills of this city, where approximately 94% of the population resides [11], present an average slope higher than 6%. Specifically, in Valparaíso, 57.29% of streets, representing 25.65 km^2, have slopes higher than 6%. The plain part of the city is the central business district, where most jobs are concentrated, and is mostly dedicated to commerce, education, health, and other services.

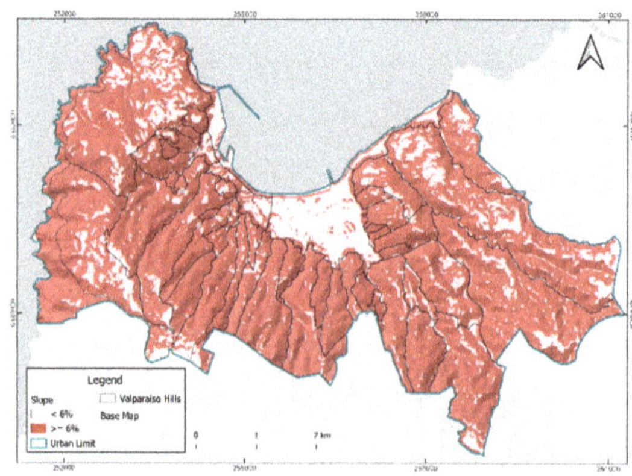

Figure 1. Slope map of the city of Valparaíso, indicating areas with slopes greater than and less than 6%. Source: elaborated by the authors.

In addition, it is a city with a complex urban infrastructure in the hills, with narrow streets, inadequate maintenance of the streets and sidewalks, inadequate use of public spaces, damaged street signs, and other problems. For example, in Figure 2a,b, it is possible to observe the use of both sides of the street for parking, which is not allowed and reduces mobility space significantly, while Figure 2c shows the extensive use of the street for parking and a pothole on the street.

(a) (b) (c)

Figure 2. Conditions of the streets in Valparaíso. (**a**) Templeman Street, between Lautaro Rosas and San Enrique, heading north; (**b**) Templeman, between Lautaro Rosas and San Enrique, heading south; (**c**) Urriola. Date: 25 April 2024. Source: elaborated by the authors.

Concerning the mobility context in Valparaíso, Fernandes and Tirado [12] indicate that the higher income of the population relates negatively to sustainable mobility patterns. Complementary to this, Aprigliano et al. [13] conclude that accessibility to work via public and private transportation is related to sociodemographic factors, such as household size, education, age, gender, and income. On the other hand, accessibility to leisure activities using all modes in Valparaíso is mostly influenced by urban factors, such as the street design, level of mixed land use, street safety, and distance to public transport stops and stations [13]. Interestingly, for both accessibility purposes mentioned, besides the previously noted influencing factors, the slope of the streets presented negative effects on non-motorized mobility [13]. Therefore, considering the complex urban structure of the city, which complicates the implementation of large-scale public transportation systems, and the socioeconomic inequality of the city, which limits citizens' equitable access to opportunities [12], Valparaíso requires improvements to non-motorized mobility solutions, such as the promotion of E-bikes and other similar ePMDs, which occupy less space and may reduce the accessibility gap in the city.

In this direction, this study aims to propose potential E-bike routes in the city of Valparaíso through a combination of multicriteria and spatial analyses, as well as experimental studies involving E-bikes on slopes similar to those found in the city's hills. Inspired by Aprigliano et al. [14], which identified built environment factors that influence metro ridership in Valparaíso, this study will analyze options for routes that connect metro stations to strategic streets and sectors on the hills of the city. An improved connectivity for E-bike users from the hills to the metro stations in Valparaíso would strengthen the accessibility to opportunities in neighboring cities. For the multicriteria analysis, the factors evaluated were topography, road safety, rider comfort, travel time, and potential cycling infrastructure. In addition, an analytic hierarchy process (AHP) was implemented, considering experts' perspectives to weigh the factors for the potential E-bike route analysis. As a complement, this research developed a preliminary experimental study with E-bikes that had different power characteristics to evaluate users' experience riding in steep slope conditions. More details can be found in the Materials and Methods section.

Proposing potential routes for e-bikes in the city of Valparaíso through multicriteria and spatial analyses is directly aligned with the United Nations Sustainable Development Goals (SDGs) [15]. Within the SDG 13 context concerning climate action, E-bikes, recognized as a sustainable means of transport, can play a significant role in reducing carbon emissions, contributing to climate change mitigation, and promoting responsible environmental practices [16,17]. Furthermore, by providing savings in transport costs, e-bikes favor economic sustainability, a determining factor for their increasing adoption, which aligns with the global objectives of promoting affordable and clean energy related to SDG 7 [17,18]. In this way, this study positions itself as a relevant contribution to the sustainable development agenda, encouraging sustainable and economically viable urban mobility, a concern of SDG 11 (Sustainable Cities and Communities). In addition, Chile has sought to promote the shared and safe use of streets through new laws, such as the incorporation norms in the national traffic law in 2018, related to the coexistence of different modes of transportation in the streets [19].

Besides its introduction, this paper is divided into five sections: a literature review focusing on identifying, from other studies, methods and variables for the definition of potential regular and electric bicycle routes; the Materials and Methods section, which describes the research process, variables, and analytical methods; a section dedicated to a descriptive analysis of the variables; the Results section, focusing on the analytical methods; and finally, the section on our conclusions and final considerations.

2. The Literature Review

2.1. Why Are Bicycles and E-Bikes Important?

The promotion of active mobility, with an emphasis on the use of bicycles, has been consolidated as an indispensable solution to the challenges faced by contemporary cities [20,21]. The benefits generated by policies that encourage the use of bicycles are mainly related to sustainability and improved urban quality of life [22,23]. Among the current cycling modalities, electric bicycles can greatly improve the accessibility to urban areas, reducing limitations related to age, physical condition, and certain disabilities [24]. This improvement in accessibility generates an increase in the number of daily bicycle users, promoting urban cohesion and sustainability [25].

The implementation of cycling networks that connect public transport stations and other urban attraction hubs not only optimizes the efficiency of transport systems but also makes it possible to reduce greenhouse gas emissions [24]. However, the benefits of using bicycles go beyond those arising from their integration with public transport systems. According to Brand et al. [26], the transition from car use to cycling can reduce daily CO_2 emissions, contributing to climate change mitigation and improving public health.

However, implementing infrastructure dedicated to the use of bicycles still faces barriers in many cities around the world. In large cities, the use of bicycles is often challenging due to their fragmented cycling infrastructure and/or complex terrain topology [27]. Furthermore, although the introduction of electric bicycles (E-bikes) in the context of urban mobility has proven to be an essential component of reducing the dependence on motorized vehicles [28], in many cities, public agents still tend to favor cars over cycling, which highlights the urgency of giving greater visibility to public policies and increasing investments in initiatives that promote the use of bicycles [29–31].

2.2. A Review of the Methods and Variables in Bicycle Route Planning

To maximize the benefits arising from the use of e-bikes, it is essential for cycling routes to be adequately planned, which requires consideration of multiple criteria, such as travel time, comfort, tranquility, and acceptable slope conditions [27,32,33]. In this sense, according to Lopez-Escolano et al. [34], effective urban planning and management are essential to overcome these challenges and fully integrate bicycles into the urban environment based on information-gathering. As Maltese et al. [35] state, a strategy to promote active mobility that includes citizen engagement, land use planning, and data collection is essential. After all, understanding the real needs of cyclists makes it possible to improve cycling infrastructure and adapt route planning to the actual behavior of users [36]. In this sense, based on revealed preference research, studies have shown that cyclists tend to prefer routes with dedicated infrastructure, such as segregated cycle paths, in addition to avoiding routes with heavy traffic or many intersections [37,38].

In the search for the optimization of cycling planning processes, the use of multicriteria tools to identify the best routes is essential so that they can be established according to the different preferences of cyclists [32]. In addition, advanced technologies, such as the use of GPS devices, smartphone applications, and the use of artificial neural networks, are increasingly being used in route planning for e-bikes. These technologies allow for the collection of detailed data on the behavior and real preferences of cyclists, enabling the development of more efficient and safe cycling infrastructure [36,39]. In this sense, even the incorporation of weather information can optimize the use of e-bikes further by recommending the ideal departure times to avoid adverse conditions and maximizing the energy savings of the equipment [28,40].

As demonstrated in the literature, the planning and implementation of cycle paths in urban environments have been explored using different methodological approaches in various parts of the world. In a study applied in the city of Taipei in Taiwan, Hsu and Lin [21] proposed the integration of multicriteria assessments with geographic information systems (GISs) for the planning of cycle networks in urban areas. From the perspective of cyclists, the authors developed a model for assessing the adequacy of streets, considering criteria such as traffic volume, lane width, and pavement quality. The model demonstrated effectiveness in identifying roads that needed improvements to become safer for cyclists.

In the same context of planning a cycling network, Milakis et al. [41], in a study carried out in the metropolitan region of Athens, Greece, sought to identify the ideal locations for its implementation. This study, whose objective was to create a network that simultaneously maximized connectivity and safety and minimized the negative impacts of bicycle traffic on motorized traffic, was based on a participatory multicriteria analysis integrated with a GIS, considering the perspectives of various stakeholders, such as cyclists and transport experts. Through the resulting methodological approach, it was possible to generate a network that was satisfactorily accepted by the local community, which highlighted the importance of public participation in urban planning.

In a study conducted in Cape Town, South Africa, Beukes et al. [42] proposed an alternative method of road infrastructure planning that promoted more livable neighborhoods and a more inclusive transportation network. Using a Spatial Multicriteria Assessment (SMCE) integrated into a geographic information system (GIS), this study sought to balance the needs of different transportation modes, including private vehicles and public transportation, as well as pedestrians and cyclists, with the contextual characteristics along the routes studied. This study demonstrated the potential of context-sensitive urban planning as a tool to improve the quality of life in cities, respecting the social, economic, environmental, and cultural aspects of the surroundings, promoting safety and mobility, and also integrating active transportation modes, such as cycling, into areas with diverse needs and characteristics.

In a case study applied in the city of Dublin, Ireland, Tal et al. [28] presented an innovative solution for planning electric bicycle routes that considered weather conditions to maximize the energy efficiency and improve the cyclist experience. To validate the proposal, the authors conducted a survey with 20 cyclists, which revealed that most of the participants were significantly affected by adverse weather conditions, thus demonstrating interest in the solution proposed by said study. This study highlights the importance of considering environmental factors in planning electric bicycle routes.

In Spain, Jiménez and Calatrava [43] studied the creation of a network of cycle paths that would facilitate daily commuting between the campuses of the Technical University and within the city of Cartagena, where the university is located. This study explored favorable geographical characteristics, such as mild climate and flat terrain, to promote the use of bicycles, reducing motorized traffic and CO_2 emissions. A multicriteria analysis of different cycle network projects developed based on a mobility survey with the academic community was carried out, considering factors such as the route characteristics, traffic segregation, accessibility, safety, and cost, resulting in the selection of the most suitable proposal for implementation.

In the city of Sparta, Turkey, Saplıoğlu and Aydın [38] combined the AHP method with GIS tools to identify cycling routes that would optimize cyclist safety. This study considered accident-prone areas and the presence of roadside parking as key variables, highlighting the need for segregated cycle paths to protect cyclists from car traffic. The authors also highlighted that cycling infrastructure should prioritize safety through adequate signage and regular maintenance, factors that are especially important in complex and dynamic urban contexts.

In Canada, Grisé and El-Geneidy [44] explored social equity in cycle path planning. In a study conducted in Quebec City and using a GIS-based prioritization index, the authors identified areas where the construction of new cycle paths could benefit socially disadvantaged groups by improving accessibility to active transportation. This approach highlights the importance of considering social equity as a central variable in urban planning, ensuring that all social groups have equal access to mobility opportunities.

In the town of Imotski, Croatia, Derek and Sikora [32] used a GIS-supported multicriteria analysis to plan cycling routes. They incorporated variables such as road type, slope, and proximity to emergency services, concluding that the integration of these variables would allow for the development of routes that balance safety, convenience, and leisure. This model is especially useful for e-bikes, where battery consumption management is critical, and routes with lower slopes and better road quality are preferable.

In suburban areas of Austria and in the city of San Francisco, USA, Perger and Auer [40] proposed a route planning system specifically designed for electric vehicles which took into account energy efficiency, travel time, and battery longevity, with special attention to the topography of the routes. The main objective of this study was to develop a multi-objective optimization algorithm that would allow users to prioritize variables such as energy consumption, travel time, weather conditions, and battery wear. The results showed that topography significantly influences energy consumption, especially for electric vehicles, where energy regeneration when going downhill can contribute to the overall efficiency. This approach is particularly relevant in cities with hilly terrain, where the slope of the roads can be a significant barrier to the use of bicycles, even electric ones.

In a study conducted in the city of Kaunas, Lithuania, Zagorskas and Turskis [45] developed a universal method for prioritizing the construction and renovation of cycling segments, using a hybrid multicriteria decision-making model (MCDM) integrated with a geographic information system (GIS). The criteria, divided into four main groups (utility, safety, convenience, and comfort), were developed based on surveys with local experts and cyclists and were weighted according to their importance. The analysis, visualized through a GIS, revealed that comfort and safety were the most influential criteria in determining the priorities. The authors concluded that the proposed method improved the efficiency in allocating resources for cycling infrastructure and offered a practical and adaptable solution for different urban contexts.

In Oslo, Norway, De Jong et al. [37] contributed an analysis based on GPS data that examined cyclists' route choices. Their study revealed that factors such as the presence of segregated cycling infrastructure and flat topography are determinants of the route definition and that cyclists tend to prefer safer and more pleasant routes, even if this means covering longer distances. This study reinforces the idea that safety and comfort are central variables that should be considered when planning cycling networks that encourage continued bicycle use.

In Portugal, based on an assessment of 1704 road segments in the city of Coimbra, Pais et al. [23] developed a multicriteria methodology for planning the maintenance of cycling infrastructure. Using the Electre Tri technique, these segments were classified into different performance categories based on criteria such as comfort, safety, lane width, and lighting and subdivided into specific subcriteria. The analysis revealed that safety was the most critical criterion, especially due to the lack of adequate separation between motorized traffic and cyclists. This methodology proved to be effective for planning maintenance interventions, allowing the authorities to prioritize the most critical segments for improvements.

Also in Portugal, Santos et al. [46] developed a decision-support tool to assess the suitability of urban road infrastructure for bicycle use, particularly in cities with hilly terrain. In a study applied in the city of Covilhã, the proposed assessment model was based on three main criteria, which were population density, proximity to main trip generation points, and road network characteristics. These criteria were combined using a weighted linear method within an MCDA approach, with the weights defined by a panel of urban mobility experts. This study identified that 23% of Covilhã's road network is suitable for the use of conventional bicycles, while 34% is suitable for e-bikes, with e-bikes demonstrating greater potential to overcome the city's topographical challenges. This study concluded that the methodological approach developed proved to be effective in assessing the suitability of urban road infrastructure for bicycle use, especially in cities with challenging topographical characteristics.

Overall, the reviewed studies show that integrating cycle paths with other modes of transport, taking into account local variables such as the topography and climate, and using advanced tools such as GISs and multicriteria analyses are essential to the success of urban cycle networks. Public participation and consideration of social equity are also very important aspects, as highlighted in studies such as those by Grisé and El-Geneidy [44] and Milakis and Athanasopoulos [47]. These studies suggest that creating well-planned cycle routes not only improves the efficiency of the transport system but also promotes public health, reduces carbon emissions, and improves the quality of life in cities.

As discussed previously, in order to fully realize the benefits associated with the use of electric or conventional bicycles, it is necessary for the planning of cycle routes to be carried out with a certain degree of rigor, which requires the analysis and careful consideration of several criteria. In Table 1, 13 essential factors that need to be observed when planning and implementing a cycle network are presented in light of the literature examined.

Table 1. Essential factors to consider when implementing and planning cycle paths. Source: elaborated by the authors.

Factors	Contextualization	Source
Connectivity and Accessibility	Ensure that cycle paths are connected to other modes of transport, maximizing accessibility, and facilitating intermodality.	[47,48]
Departure Time	Adjust the departure time to avoid heavy traffic and adverse weather conditions and optimize the energy consumption on electric bikes.	[28,40]
Topography	Assessment of road slopes, a critical factor in cities with uneven terrain, is especially important for the viability of electric bicycle routes.	[24,32,33,38,40,46]
Road Safety	Focus on reducing the risks for cyclists by assessing the intersection density, traffic volume and speed, and the presence of safe infrastructure and identifying areas with a high risk of accidents to avoid dangerous routes and ensure the safety of cyclists, both on regular and electric bikes.	[38,49,50]
Cycling Infrastructure	Emphasis is placed on the need for dedicated, high-quality cycle paths to ensure safety and comfort, which is essential for cyclists' route selection. Choosing main roads that offer better maintenance, reduced exposure to traffic conflicts (e.g., intersections), and safety is especially important for e-bikes due to their sensitivity to uneven surfaces.	[32,37,38,51,52]
Environmental and Urban Conditions	This includes consideration of green areas, landscaping, and air quality along the routes, improving the experience, and encouraging the use of bicycles.	[23,49,53]
Travel Time	Consider total travel time to select faster and more efficient routes, taking into account distance, obstacles, and elevation changes.	[27,33]
Multimodal Integration	This promotes the integration of cycle paths with public transport stations, facilitating the combined use of bicycles and public transport.	[43,48]
Weather Conditions and Climate	Consider the effects of weather, such as rain, wind, and temperature, which can affect the safety and comfort of cyclists, especially for e-bike users.	[28,40,52,54,55]
Demand and Potential Use	Assess projected demand and user profiles to ensure that cycle paths meet the needs of those who depend on them most.	[44,45]
Rider Comfort	Consider pavement quality, road width, directionality, and the presence of obstacles to ensure a safe and comfortable cycling experience.	[23,27,33,52]
Costs and Economic Viability	Analyze the cost of implementing and maintaining cycle paths to ensure that resources are used efficiently, balancing the benefits and investments.	[43,45]
Energy Cost	Assessment of the energy consumption along routes is required, considering energy regeneration during braking and energy efficiency in electric bicycles.	[18,56]

Active mobility should therefore not be seen as just a current trend but rather as an essential component of the sustainable development of cities. Ongoing research in the field of cycling planning, especially in the context of multimodal integration and adaptation to the local conditions, will continue to play a relevant role in transforming cities into more livable and resilient spaces.

3. Materials and Methods

3.1. Description of the Case Study

The municipality of Valparaíso, located in the region of the same name in central Chile, is part of the metropolitan area of Valparaíso (also known as Greater Valparaíso), which also includes Viña del Mar, Concón, Quilpué, and Villa Alemana, where around 900,000 people live [57]. Specifically, around 300,000 people live in Valparaíso itself [58], distributed in the hill and plain sectors, which are connected by approximately eight bus lines [59]. In addition, the commune of Valparaíso contains 6 of the 20 metro stations in the EFE Valparaíso network, located in the coastal area of the city (Figure 3). These metro stations were considered in selecting possible routes, taking them as the starting points of each route. For a better understanding of the location of the hills named in this article, Figure A1 in Appendix A presents the details of the location of each of these with their respective names.

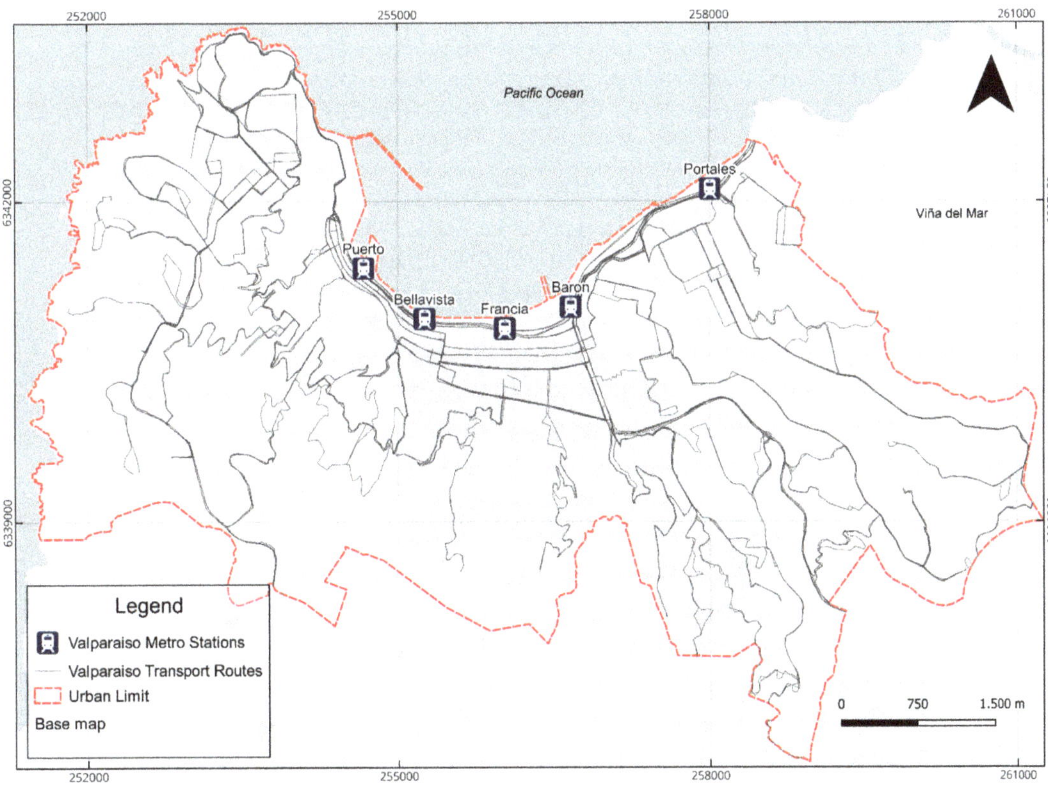

Figure 3. Transport contextualization of Valparaiso's urban limits. Source: elaborated by the authors.

Regarding the economic characteristics of the city, for the year 2022, according to the CASEN survey [58], Valparaíso had 18,589 companies, which represented 12.7% at the regional level and 1.23% at the national level. The majority were micro-enterprises in the sector of the wholesale and retail trade and repair of motor vehicles and motorcycles, which constituted 31.7% of the total. According to the Internal Revenue Service [58], there were

125,239 dependent workers, which corresponded to 17.8% at the regional level and 1.2% at the national level. Of these, 19,236 (15.4%) were employed in the transport and storage sector. On the other hand, the poverty rate in Valparaíso for 2022, according to CASEN, was 7.4%, compared to 6.5% at the national level. In December 2023, the rate of overcrowded households in Valparaíso was 6.7%, compared to 8.5% nationwide [58].

In the climatic context, the city of Valparaíso has a temperate Mediterranean coastal climate, where, thanks to the presence of the ocean, there are no major temperature variations throughout the year, averaging 14° per year; on the other hand, rainfall can reach 450 mm per year [60]. The heavy rains and their frequency can have a great negative impact on people's mobility since the city's topography favors the generation of rivers or canals in the steep streets, making pedestrian and vehicular traffic difficult. This was a tangible reality in the winter months of 2024 since several frontal systems occurred that left large amounts of precipitation in the city and forced local authorities to declare a red alert [61–63], causing floods that affected streets, stairs, funiculars, and homes, in addition to landslides.

3.2. Research Planning and Design

This study is divided into two approaches: a multicriteria analysis and an experimental study. Both of these approaches support the process of analyzing the possible routes for E-bike use in the city and the potential use of electric bicycles in Valparaíso.

The methodology for determining the optimal routes was based primarily on using geographic information system (GIS) tools and analyzing various characteristics of the streets of Valparaíso. The process was divided into five main stages. The first stage consisted of a literature review of studies in Google Scholar using keywords such as "Routes", "GIS", "Electric Bicycles" and "Multicriteria". The second stage, called "Route Design", focused on the identification of possible routes from the Valparaíso metro stations to the main cross street at an altitude of 100 m above sea level in the hills. The third stage, called "Design of Variables in GIS", involved considering factors such as the average slope, traffic accidents, directionality, route length, and the number and type of intersections, aiming to narrow down the route selection. In the fourth stage, called "Multicriteria Analysis in GIS", the routes were evaluated using a multicriteria approach to weigh the variables collected and determine the most suitable routes. Finally, the fifth stage involved analyzing the selected routes to validate their feasibility and optimization.

After the literature review, the routes were designed. The first step was to examine the routes registered on the Garmin Connect website [64], reviewing the "Training/Planning" section and the "Heatmap of popular routes" item. This map shows limited routes within Valparaíso used by cyclists who register their activity on Garmin. Since the routes available in Garmin were scarce, alternative routes were searched for using Google Maps, taking as the starting point a metro station in Valparaíso and as the destination Avenida Alemania, a street located at an average altitude of 115.5 m that connects several hills in the city, towards Plaza O'Higgins on the Washington route at an altitude of 126 m, towards Rodelillo Avenue in the case of Cerro Barón, and just over 200 m above the intersection between Manuel Antonio Matta and Javiera Carrera for the case of Cerro Placeres, with the latter located at 160 m and 170 m above sea level, respectively. From this search, the routes were digitized using QGIS software (v.3.40.2), creating vector files in line-type shapefile format, resulting in a total of 15 proposed routes within the city of Valparaíso, all with a slope greater than 6%, generating a division by the zones shown in Figure 4 for better visualization of the routes presented in the Results section, and the names of which are detailed in Table 2.

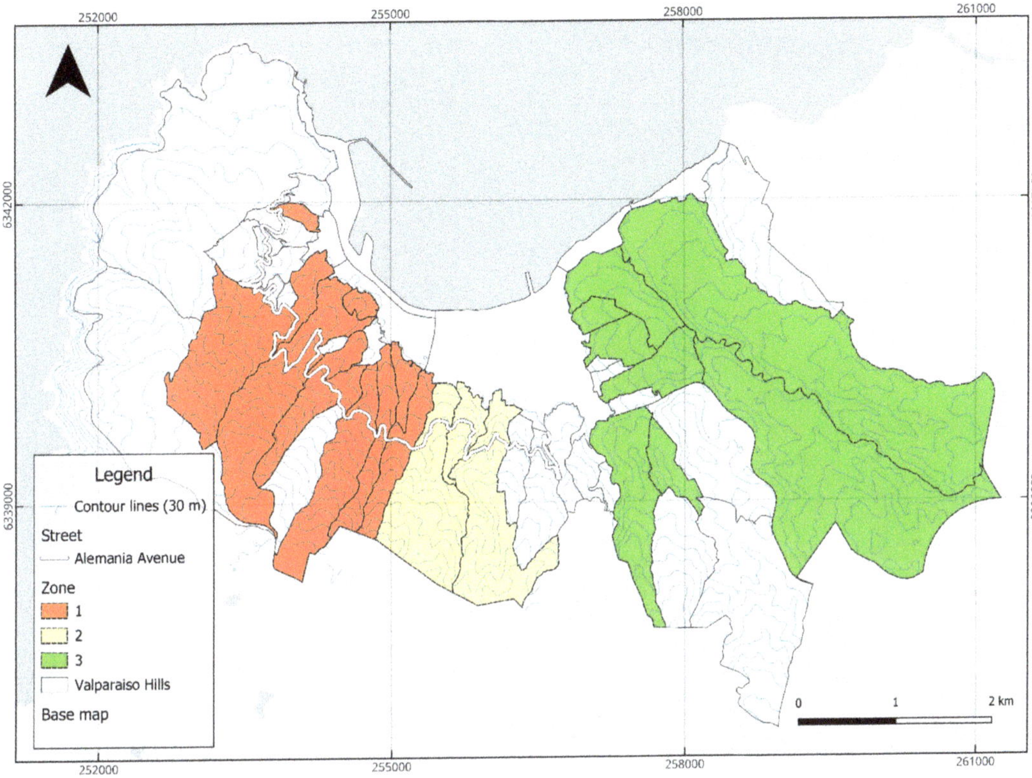

Figure 4. Zoning established to separate the routes. Source: elaborated by the authors.

Table 2. Names of the hills located in the areas designated for viewing the routes. Source: elaborated by the authors.

Zone	Hill Name	
1	Arrayán	Yungay
	Alegre	Florida
	Concepción	Cordillera
	Cárcel	Bellavista
	La Loma	San Juan de Dios
2	Mariposa	Monjas
	La Cruz	
3	Barón	Placeres
	Lecheros	Ramaditas
	Larraín	Rodelillo
	Polanco	O'Higgins

As for the experimental study, the objective was to evaluate the performance of the electric bicycles on a route that replicated the conditions of the proposed routes. The latter were designed for specific adaptations, which made conducting these experiments on such routes less feasible at the time of study due to the high pedestrian flow, as well as public and private traffic. Therefore, a remote test route was selected, with a low pedestrian and vehicular flow and similar characteristics to those of the evaluated routes. The purpose of this selection was to measure, in different individuals, both their physical capacity and their perception, as well as to evaluate the performance of different types of bicycles on slopes: conventional and 250 W, 500 W, and 750 W electric bikes. In addition, a stop was simulated halfway up the climb, where the participants had to stop in order to evaluate both their physical capacity and that of the bicycles to accelerate on slopes.

3.3. Description of the Variables for the Multicreteria Analysis

For the design of the GIS variables with the preliminary routes, criteria were established to reduce the number of routes and select the most viable ones based on the available information in the study area. The following factors were considered: topography, road safety, rider comfort, travel time, and cycling infrastructure. For each of the factors, there was a variable of measure. Respectively, this study applied the following variables: average slope; traffic accidents, favoring routes with a lower or no incidence of accidents; route length, preferring those with a shorter length; the number and type of intersections, selecting routes with a lower number and variety of intersections; and directionality, prioritizing one-way routes due to their greater safety. The measurement of each factor is described as follows:

- Topography [24,32,33,38,40,46]: A Digital Elevation Model of the Valparaíso region was used, obtained from the Spatial Data Infrastructure of Chile (IDE) [65]. This model was clipped to fit the area of Valparaíso. Next, the slope of each route was calculated using the "Zonal Statistics" tool in QGIS, obtaining the average of the slope from the beginning of the elevation to Avenida Alemania. The mask used was for routes from the foot of the hill with a buffer of a 5 m radius, similar to that stipulated in Decree 47 for the width of a local road corresponding to 11 m [66].
- Road safety [38,49,50]: The total number of traffic accidents along each proposed route was analyzed using the information available in the CONASET Geospatial Information Portal [67], which contains georeferenced accident data for the Valparaíso commune for the year 2022. To determine the number of accidents on each route, a polygon representing the width of the street was created based on the Google Maps satellite view and the QGIS base map. This polygon allowed the points where accidents occurred within each route to be cut out. Finally, the QGIS "Count Points in Polygon" vector tool was used to calculate the total number of accidents recorded along the routes. The mask used was for routes from the foot of the hill with a 5 m buffer [66].
- Rider comfort [23,27,33,52]: The directionality of each route was identified using Google Maps.
- Travel time [27,33]: From the digitization of the routes, the length of each route was calculated using the "Field Calculator" tool in QGIS. The "$length" option was selected to obtain the distance from the start point to the route's endpoint in meters.
- Cycling infrastructure [32,37,38,51,52]: Using the route lines and the base map in QGIS, all of the streets that intersected with the main route were digitized as points. The intersections were identified based on the number of streets that crossed an intersection, classifying them as intersections with one, two, or three streets. The mask used was for routes from the foot of the hill with a buffer of 5 m [66].

3.4. The Analysis Method for the Multicriteria Analysis

In the multicriteria analysis stage in the GIS, a score from 0 (very low) to 3 (very high) was assigned to each variable. It is relevant to highlight that the assessment of potential E-bike routes in Valparaíso was based on the reality in this city. This means that for some criteria, four categories were based on the quartiles of each variable in relation to the data on all of the assessed routes, such as distance, accidents, and the number of intersections (see Table A1 in Appendix A). In addition, this multicriteria analysis included the implementation of the analytic hierarchy process (AHP) [68] to weight the criteria differently based on students, academics, and professionals in the transport field. The combinations of the sections' average slopes were used to generate a final score for each route, as seen in Table A1 in Appendix A.

Regarding the slope conditions, Nematchoua et al. [25] evidenced E-bikes' advantages in overcoming steeper slopes compared to conventional bicycles. In this sense, adapting slope criteria to E-bike use on certain routes is interesting. Therefore, since a 5% or 6% slope is considered suitable and comfortable for the use of regular bicycles [8,69], for this criterion, the proposal was to increase the cycling potential for slopes by 2% from 6%, that is, optimizing routes with a slope of up to 8% and generating four slope ranges with their respective quartiles, with the first being less than 8%, thus penalizing streets that were too steep.

The distance criterion considered three indicators: the first being related to the total distance of the route from the metro station; the second being the total distance from the beginning of the hill's elevation; and the third being the average of the distances of the sections of the route. All of these indicators were conditioned by their quartiles, weighting the longest distances as unfavorable. Subsequently, traffic accidents were evaluated, considering the total number of accidents per route and an area of influence of 10 m in diameter [66], also weighting these according to their quartiles, and penalizing routes with more accidents.

The variable for type of intersection followed a similar logic to the slope variable; however, it sought to penalize intersections with complex dynamics (see the types of intersections in Figure 5).

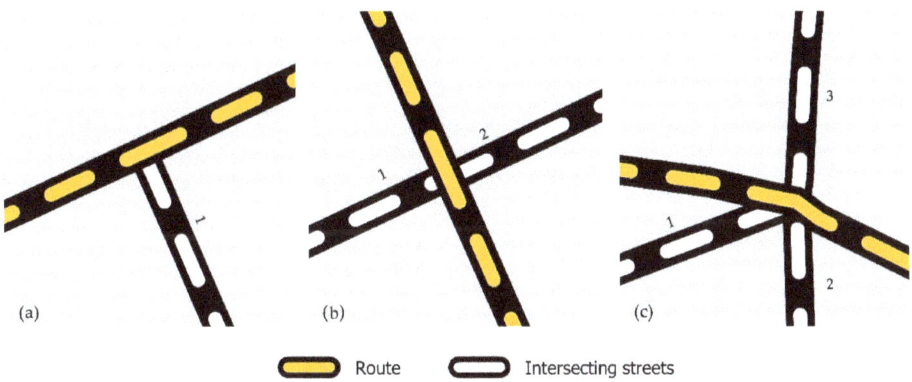

Figure 5. Type of intersections. (**a**) One intersection, for its type and weighting, is considered if it is unidirectional or bidirectional. (**b**) Two intersections, for their type and weighting, are considered if both are unidirectional or bidirectional. (**c**) Three intersections. Source: elaborated by the authors.

Finally, the speed criterion was considered, taking the average speed on the route as a variable. To this end, a 5 m influence area was used, from which coordinates corresponding to the people who start their journey in said area and who travel within Valparaíso were extracted, according to the data from the Origin and Destination Survey [70]. The distance and the declared time were considered, and the speed quartiles obtained were weighted, with the highest speeds classified as unfavorable.

Combinations of the variables that summed of their values and considered the weight of the criteria were used to rank the routes, from those with higher to less potential to be used as shared E-bike routes with other modes of transportation or even to be considered for further evaluation of building separated cycleways for E-bikes.

3.5. Description of the Variables for the Experimental Study

The experimental study was conducted using four types of bicycles: one regular bicycle and three electric bicycles with power ratings of 250 W, 500 W, and 750 W. The following variables were measured: the time it took the participant to complete the ride, the total distance traveled, and the average speed during each ride for each type of bicycle. In addition, the average heart rate, calories burned, and time in motion were recorded for each bicycle.

To ensure measurement accuracy, ride monitoring was performed using a Garmin EDGE 840 SENSOR BUNDLE GPS system, a VENU 2 GPS Smartwatch, and a Garmin cadence sensor and speed sensor. These devices provided accurate metrics on heart rate, calories, and movement time.

At the end of the routes, a survey was administered using Google Forms, in which each participant answered questions related to their socioeconomic data, mobility habits, level of physical activity, and experience using electric bicycles.

3.6. Analysis Methods for the Experimental Study

Different analyses were performed based on the data collected. First, the participants' profiles, socioeconomic status, and mobility patterns were obtained. Then, a performance analysis of the power of each bicycle was carried out, considering the variables of time, distance, and speed. Subsequently, a physical activity analysis was conducted, considering heart rate, calories burned, and time in motion. Finally, the participants' experience was analyzed, including factors such as comfort, safety, physical effort, and the ease of use of each type of bicycle.

4. Results

4.1. Multicriteria Analysis

A total of 15 routes were drawn from the five metro stations in Valparaíso: 4 from Puerto station, 6 from Bellavista station, 2 from Francia, 2 from Barón, and 1 from Portales. Only two of the routes go up and down the same road, while the others require a different route to return to the starting point. In addition, only four of the proposed routes utilize one-way streets: 1, 2, 4, and 5 (Table 3).

Table 3. Main characteristics of the routes.

Zone	ID	Street Name	Metro Station	Route type	Directionality
1	1	Almirante Montt	Puerto	Dual route	One-Way
	2	Almirante Montt	Bellavista	Dual route	One-Way
	3	Carampangue	Puerto	Dual route	Bidirectional
	4	Cumming	Bellavista	Dual route	One-Way
	5	Cumming	Puerto	Dual route	One-Way
	6	Ecuador/Cumming	Bellavista	Dual route	Bidirectional
	7	Ferrari	Bellavista	Dual route	Bidirectional
	8	Guillermo Rivera	Bellavista	Dual route	Bidirectional
	9	Tomás Ramos	Puerto	Dual route	Bidirectional
	10	Yerbas Buenas	Bellavista	Dual route	Bidirectional

Table 3. *Cont.*

Zone	ID	Street Name	Metro Station	Route type	Directionality
2	11	Baquedano	Francia	Round trip	Bidirectional
	12	Francia	Francia	Round trip	Bidirectional
3	13	Barón	Barón	Dual route	Bidirectional
	14	Placeres/Matta	Portales	Dual route	Bidirectional
	15	Washington	Barón	Dual route	Bidirectional

For better visualization of the routes, the routes were displayed in the three proposed hill areas (see Figures 6–8), accompanied by their respective characteristics (Tables A2–A4 in Appendix A). The first 10 routes were drawn from the Puerto and Bellavista stations, with all of them reaching Avenida Alemania (Figure 6).

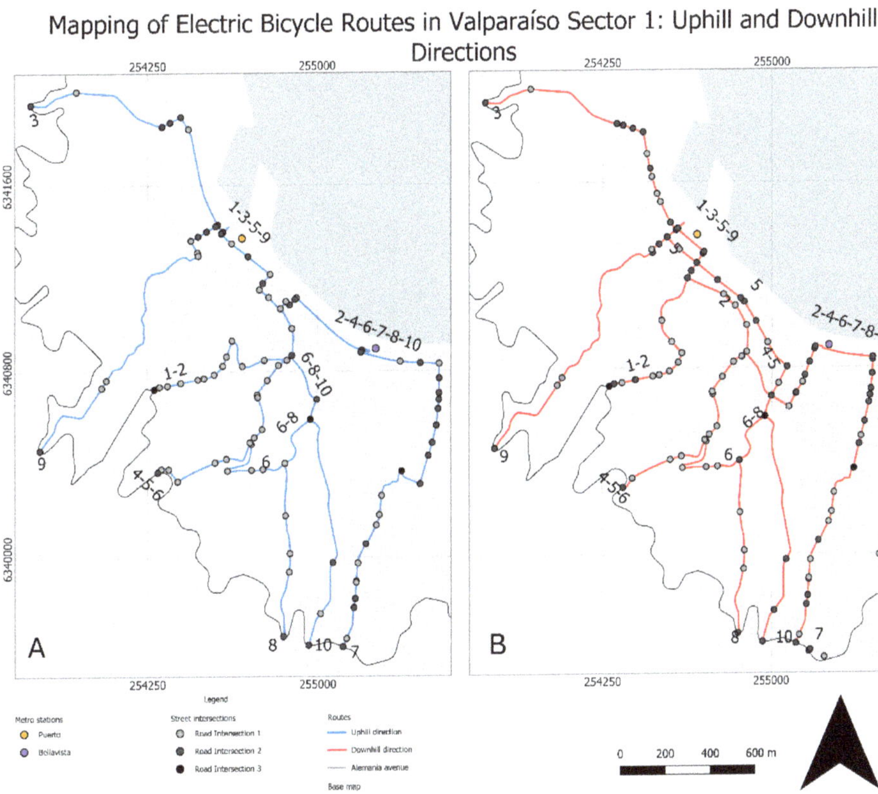

Figure 6. Uphill and downhill routes registered in zone 1. (**A**) Uphill direction; (**B**) downhill direction. Source: elaborated by the authors.

In the uphill section of zone 1, the average gradients vary from 10.85% in Ferrari (7) to 21.14% in Yerbas Buenas (10). The routes with the fewest traffic accidents are Tomás Ramos (9), Almirante Montt (1 and 2), and Cumming (4 and 5), while Ecuador/Cumming (6) and Yerbas Buenas (10) have the highest values, with 13 and 14 accidents, respectively. The distances from the initial elevation of the hill range between 1300 and 2200 m, with that for Carampangue (3) being the shortest. Most of the routes have more than 20 intersections, with Carampangue (3) standing out, with the fewest, at 8. In the downhill section, the results vary—for example, in terms of average slope, where the lowest slope is in Yerbas Buenas (10), at 10.25%, and the highest slope is 16.43%, in Almirante Montt (1). The distance from the initial elevation of the hill varies between 580 and 1680 m, with Carampangue (3) having the shortest distance, at 586.46 m (Table A2 in Appendix A).

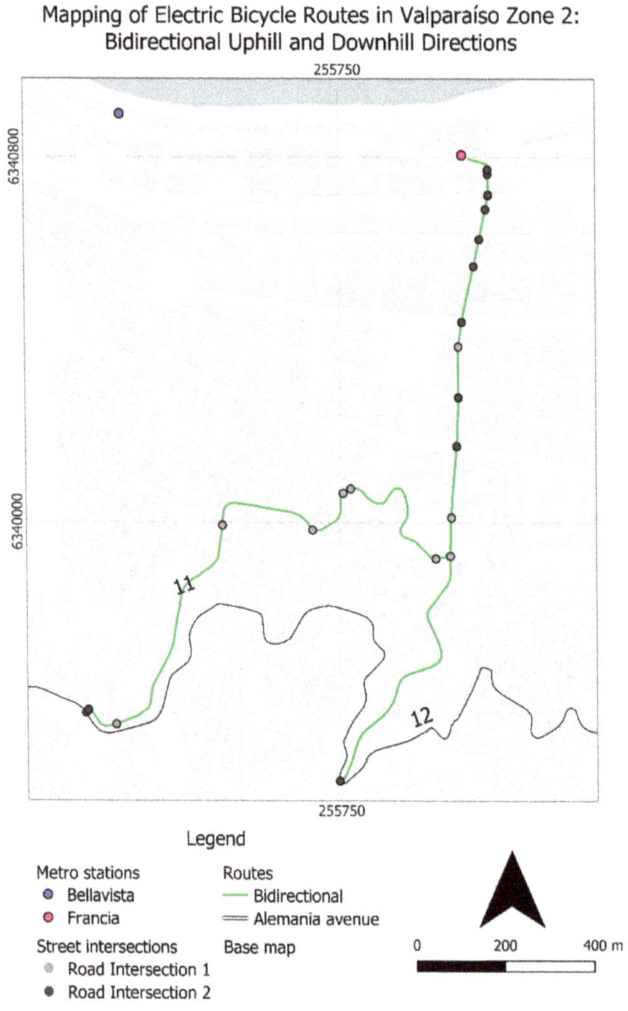

Figure 7. Bidirectional routes registered in zone 2. Source: elaborated by the authors.

In the second zone, only two routes were drawn, starting at Francia station, sharing a large part of their routes until a fork, and with both reaching Avenida Alemania (Figure 7). Although both are bidirectional and share a large part of their start, they have differences: the Baquedano route (11) registers 16 traffic accidents and has a steeper slope, a greater distance, and more intersections than the Francia route (12) (Table A3 in Appendix A).

In zone 3, three routes were laid out, with all of them being two-way routes, i.e., one uphill and one downhill. Routes 13 and 15 start from Barón station, while route 14 starts from Portales station, reaching altitudes of approximately 160 m above sea level. Although the routes are two-way, the changes between uphill and downhill are minor (Figure 8).

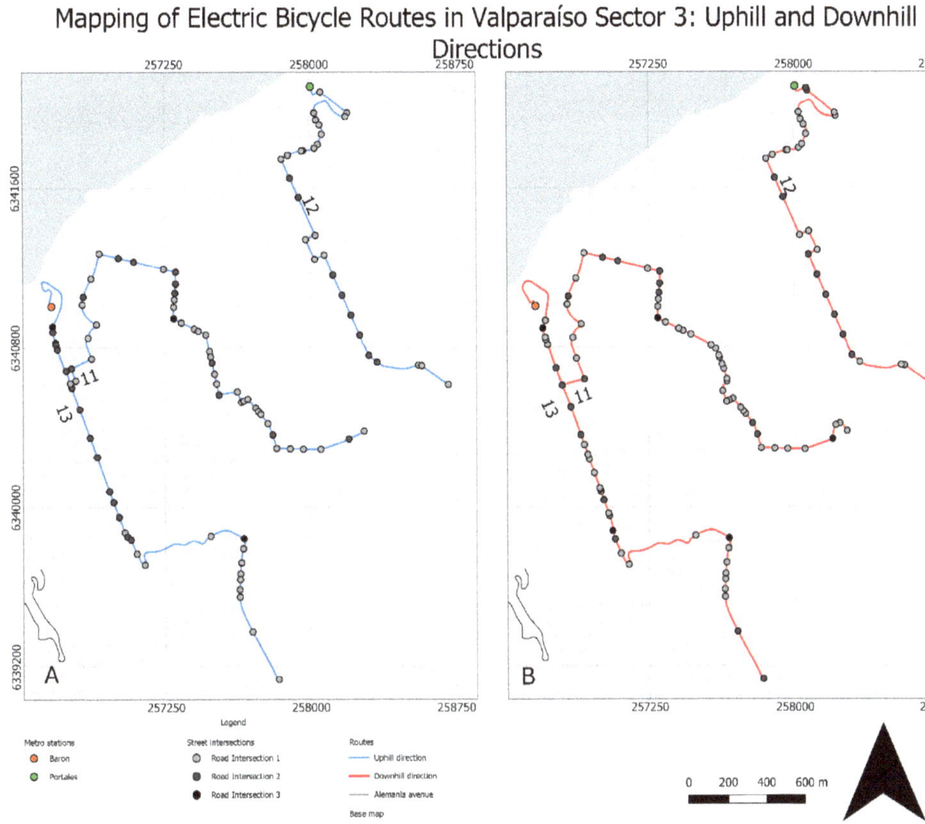

Figure 8. Uphill and downhill routes registered in zone 3. (**A**) Uphill direction; (**B**) downhill direction. Source: elaborated by the authors.

The average gradients of the three routes are similar between uphill and downhill, ranging between 9 and 13%. The number of traffic accidents ranges from 7 to 27 on the uphill route, while the downhill route has the highest number of accidents of all routes, with 43 in Washington (15). The distances are generally longer on these three routes, exceeding 2.5 km and reaching up to around 3.7 km from the metro stations, both uphill and downhill. Regarding intersections, it is observed that both uphill and downhill, the Baron route (13) is the one with the most intersections, with 50 and 53 (Table A4 in Appendix A).

In relation to the descriptive measures (see Table 4), the bidirectional routes have the lowest average slope (9.4%), reaching sections with a maximum of 30.64%, followed by the downhill routes (12.72%), with maximum sections of 42%, and finally, the uphill routes (15.1%), also registering a maximum slope of 42%. The shortest distance of the route to the nearest metro station is shorter in the bidirectional direction, at 1921.1 m, followed by downhill routes at 1935.23 m and uphill routes at 2080.54 m, while the distance of the routes from the beginning of the hill has the same distribution as the distance to the metro stations, at 1007.53 m, 1184.4 m, and 1190.07 m, respectively. On average, the distance of the route sections is greater for the bidirectional routes, at 109.81 m, also presenting the highest standard deviation (137.01 m) and highest maximum (620.63 m).

Table 4. Descriptive measures of downhill, uphill, and bidirectional routes. Source: elaborated by the authors.

Downhill							
Stats	**Mean**	**Std**	**Min**	**25%**	**50%**	**75%**	**Max**
Slope per section (%)	12.72	7.01	0.71	7.34	11.33	16.43	42
Total distance from metro station	1935.23	772	1105.71	1516.66	1560.03	2165.09	3713.58
Total distance from the initial elevation of the hill	1184.4	573.14	586.46	871.88	1021.86	1182.95	2612.7
Average distance of the route sections	83.06	89.02	0.6	35.79	60.4	97.6	788.27
Total number of accidents by area of influence of the route	12.92	10.79	0	6	12	14	43
Total intersections	22.38	11.41	9	14	21	26	53
Directionality of the route [Uni = 1]	6.42	3.37	1	4	6	8.25	12
Directionality of the route [Uni = 2]	5.08	2.29	1	3	6	6	9
Directionality of the route [Bi = 1]	7.67	7.69	1	2.75	4.5	9.5	27
Directionality of the route [Bi = 2]	2.83	2.37	1	1	2	3.25	8
Directionality of the route [Uni or Bi = 2 or 3]	2	1.61	1	1	1	2.5	5
Average route speed	9.52	4.34	4.19	8.22	9.17	9.92	20.04
Uphill							
Stats	**Mean**	**Std**	**Min**	**25%**	**50%**	**75%**	**Max**
Slope per section (%)	15.1	8.54	0	8.87	13.51	20.26	42
Total distance from metro station	2080.54	686.19	1302.68	1552.71	1871.34	2274.39	3648.27
Total distance from the initial elevation of the hill	1190.07	569.39	586.46	787.03	1021.81	1242.4	2573.13
Average distance of the route sections	91.41	97.97	0.6	35.59	62.2	106.95	788.27
Total number of accidents by area of influence of the route	10.23	6.5	0	7	8	13	27
Total intersections	22.46	10.5	8	15	22	25	50
Directionality of the route [Uni = 1]	7.23	4.85	1	3	8	11	15
Directionality of the route [Uni = 2]	4.23	2.83	1	2	4	4	11
Directionality of the route [Bi = 1]	6.92	6.49	1	2.75	4	7.75	22
Directionality of the route [Bi = 2]	3.4	2.5	1	1.25	2.5	5.5	8
Directionality of the route [Uni or Bi = 2 or 3]	2.5	1.08	1	2	2	3	5
Average route speed	8.68	2.79	4.29	7.36	8.46	9.62	14.52
Bidirectional							
Stats	**Mean**	**Std**	**Min**	**25%**	**50%**	**75%**	**Max**
Slope per section (%)	9.4	8.98	0	4	5.36	10	30.64
Total distance from metro station	1921.1	554.78	1528.82	1724.96	1921.1	2117.25	2313.4
Total distance from the initial elevation of the hill	1007.53	552.53	616.82	812.17	1007.53	1202.88	1398.23
Average distance of the route sections	109.81	137.01	1.44	38.99	66.87	117.4	620.63
Total number of accidents by area of influence of the route	15	1.41	14	14.5	15	15.5	16
Total intersections	16.5	4.95	13	14.75	16.5	18.25	20
Directionality of the route [Uni = 1]	3	-	3	3	3	3	3
Directionality of the route [Uni = 2]	5	0	5	5	5	5	5
Directionality of the route [Bi = 1]	5	2.83	3	4	5	6	7
Directionality of the route [Bi = 2]	5	0	5	5	5	5	5
Average route speed	6.74	0	6.74	6.74	6.74	6.74	6.74

Regarding accidents, the routes with the highest average number are the bidirectional ones, with 15, but these are the ones with the lowest standard deviation (1.41) and the lowest maximum value of accidents (16). On the other hand, the downhill routes are those with the highest number of accidents, with an average of 12.92, having the highest standard deviation (10.79) and the highest maximum (43). Uphill routes register an average of 10.23 accidents, with 6.5 standard deviations and 27 maximum accidents. Another relevant point is the speed, which is the highest on downhill routes, at 9.52 km/h on average, with a standard deviation of 4.34 km/h and a 20.04 km/h maximum.

Using Google Forms, a group of experts convened to answer a survey designed with comparative questions aligned with the AHP method. A total of 14 experts participated in this survey. Each participant's responses were entered into the AHP matrix and processed using the ahpy library in Python. In order to obtain the best Consistency Index (CI) and given the varied profiles of the experts, 1001 combinations of the participants' responses were made, considering subsets of 10 experts. This allowed for the best combination to be identified, reaching a CI of 0.156. This value is relevant considering the diversity of the participants' profiles and provides a first approximation for the weighting of the variables (see Table 5). Regarding the analytic hierarchy process (AHP), the output for weighting the criteria is presented in Table 5.

Table 5. Criteria and corresponding weights in the AHP model. Source: elaborated by the authors.

Criteria	Weight
Speed	0.4681
Directionality	0.1741
Intersections	0.159
Traffic accidents	0.0996
Distance	0.0534
Average slope	0.0458

In the AHP method applied in this study, the participants were six academics and eight professionals related to the field of urban transportation and mobility. The professional profiles of the participants are presented in Table 6.

Table 6. Professional profiles of the participants. Source: elaborated by the authors.

Professional Role	Average Work Experience (Years)		Total Experts
	Female	Male	
Academic	15.0	15.5	6 (43%)
Private System Professional	-	7.0	3 (21%)
Private System Professional, Postgraduate Student (Master's or Doctorate)	-	7.0	2 (14%)
Public System Professional	-	4.0	2 (14%)
Public System Professional, Postgraduate Student (Master's or Doctorate)	24.0	-	1 (7%)
Total Experts	3 (21%)	11 (79%)	14 (100%)

According to the analysis carried out of the identified routes, considering the variables described above, Table 7 ranks the routes from the most optimal to the least optimal. In first place, Cumming from Puerto (5), Cumming from Bellavista (4), and Almirante Montt from Puerto (1) stand out in the uphill routes. Meanwhile, the least optimal uphill routes are Yerbas Buenas (10), Carampangue (3), and Tomás Ramos (9). Moreover, of the downhill routes, Cumming from Puerto (5), Barón (13), and Almirante Montt (1) stand out. In contrast, the least optimal downhill routes are Placeres/Matta (14), Yerbas Buenas (10), and Tomás Ramos (9). Of the bidirectional routes, the most optimal route is Baquedano (11), and the least optimal is Francia (12); these routes are the only ones with bidirectional routing.

Table 7. Results of the routes' scores when evaluating them using the criteria established in Tables 5 and A1 in Appendix A. Source: elaborated by the authors.

ID	Street Name	Metro Station	Uphill	Downhill	Bidireccional
1	Almirante Montt	Puerto	7.06	6.56	-
2	Almirante Montt	Bellavista	6.03	5.75	-
3	Carampangue	Puerto	3.27	5.25	-
4	Cumming	Bellavista	8.03	6.47	-
5	Cumming	Puerto	9.41 (↑)	9.49 (↑)	-
6	Ecuador/Cumming	Bellavista	5.27	4.93	-
7	Ferrari	Francia	4.9	4.58	-
8	Guillermo Rivera	Bellavista	4.29	4.85	-
9	Tomás Ramos	Puerto	2.44 (↓)	2.17 (↓)	-
10	Yerbas Buenas	Bellavista	3.58	3.4	-
11	Baquedano	Francia	-	-	4.08 (↑)
12	Francia	Francia	-	-	3.87 (↓)
13	Barón	Barón	6.77	7.91	-
14	Placeres/Matta	Portales	4.31	4.02	-
15	Washington	Barón	5.17	5.35	-

Significance codes: (↑) weighted maximum, (↓) weighted minimum.

Of all the routes, Cumming from Puerto is the best, both uphill and downhill, reaching the highest scores of 9.41 and 9.49, respectively. On the other hand, among the less optimal routes, Yerbas Buenas and Tomás Ramos stand out, routes that are repeated as the least optimal both uphill and downhill. The Cumming from Puerto route stands out for its low values for the total distance from the foot of the hill, as well as for its total number of traffic accidents, total number of intersections, and speed, whose values are between the Q1 and Q2 quartiles. In addition, it stands out for having low values in terms of the type of intersections, with two that are bidirectional and others that are unidirectional or bidirectional in Q1. As for the Yerbas Buenas and Tomás Ramos routes, these stand out for their high values in the average distance of the sections and speed, with these values in Q3. For more details on the values of each route, see Appendix A Table A2, and to understand the context of the values in the quartiles, see Table 4.

4.2. The Experimental Study

The experiments were conducted on 1 October 2024 and 4 October 2024, from 9:00 a.m. to 2:00 p.m., under sunny weather conditions. They were carried out on the eastern roadway of Gastón Hamel Nieto Avenue, in Viña del Mar, with a total of 13 participants (6 in the first instance and 7 in the second). The proposed distance was approximately 550 m, with three control points—the start (14 m above sea level (masl)); the rest point (40 masl), which emulated an intersection, which tested the participants having to restart on a slope; and the end (68 masl)—reaching an average slope throughout the route of 10.88%. Figure 9 presents the topographic profile of the route.

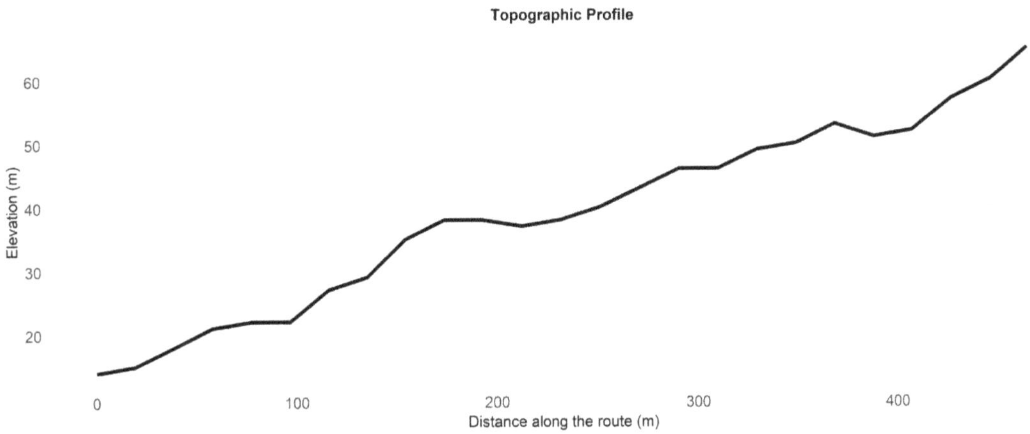

Figure 9. Topographic profile of the route on which the experiment was carried out. Source: elaborated by the authors.

During the activity, each participant used a regular bicycle, a 250 W electric bicycle, a 500 W electric bicycle, and a 750 W electric bicycle. Each participant completed the outbound (uphill) and return (downhill) legs on each bike, i.e., at a total of four rides per person. To ensure data accuracy, a monitoring system consisting of a Garmin EDGE 840 SENSOR BUNDLE GPS, a VENU 2 GPS Smartwatch, a Garmin pedaling cadence sensor, and a speed sensor was used. Personnel were assigned along the route to supervise the development of the activity.

The sample consisted of 13 participants, with most of them being men (77%), with an average age of 28 years, with 33% of them having an income that did not exceed the minimum wage, and with 22% of them having an income that exceeded 2 million Chilean pesos. The mobility patterns of the participants showed that the majority used minibuses, followed by walking and the metro. Their use of transport applications was less frequent, and it should be noted that none of the participants used a bicycle as a regular means of transport. In addition, 50% of the participants took between 45 min and 1 h on their usual journeys.

Regarding the results obtained in the experiment, Figure 10 shows the time it took each participant to travel the route using each bicycle, the distance traveled, and the average speed reached. The instruction given to the participants was to pedal with each bicycle as far as their physical condition and the slope allowed, resulting in some who completed the entire route on some bicycles and other participants who could not advance beyond the rest point. Based on the above, the time results shown in Figure 10 indicate that the shortest time taken to carry out the experiment was found for the regular bicycle (6 min) because it was also the bicycle on which the participants traveled the shortest distance (0.4 km), reaching an average speed of 4.09 km/h, where few participants managed to reach halfway during the climb. In terms of distance, the bike that reached the shortest distance after the conventional one was the 500 W e-bike, reaching an average of 0.68 km, because most of the participants reached the midpoint of the route since the bike model was heavier and could not provide assistance at 0 km/h but rather had to reach a certain speed for its assistance to be activated; as for the time for this bike, it averaged 7 min since the participants had a harder time going up the slope, with fairly low speeds that reached 5.36 km/h on average. On the other hand, the 250 W and 750 W bikes were those that had the best performance, recording 0.9 km and 0.93 km distances, respectively; that is, the vast majority of the participants reached the midpoint of the route and some reached the end of it, with the fastest bike being the 750 W bike.

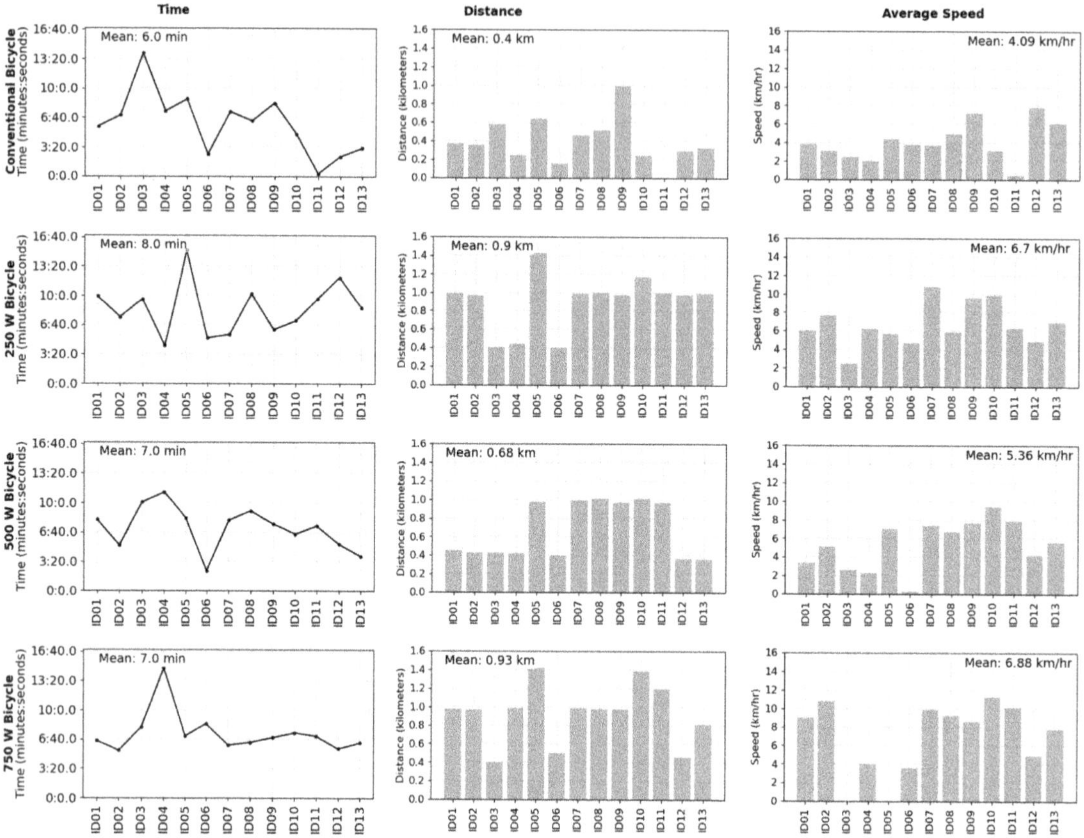

Figure 10. Metrics obtained from the sensors and instruments used in the experiments. Time and distance were obtained using the GPS and speed using a Garmin speed sensor. Source: elaborated by the authors.

Some of the results obtained from the survey carried out by the participants at the end of the experiment are summarized in Figures 11 and 12. Figure 11 shows the results related to the comfort and safety of the bicycles, highlighting the pattern of the low comfort and safety of the conventional bicycle. Regarding the 250 W bicycle, a medium level of comfort and safety predominate, with no responses for "very low". On the other hand, the 500 W and 750 W bicycles tend towards high comfort and safety, with some of the participants not considering them to be very safe, mainly due to the weight of these bicycles and the speeds they can reach.

In relation to physical effort and the ease of use of the bicycles, Figure 12 presents the results for these responses, where it is observed that the majority of the participants agree that the conventional bicycle is that with which the greatest physical effort must be applied to climb a significant slope, having a medium-level ease of use. On the other hand, the 250 W bicycle presents, according to the participants, a medium level of physical effort, considered by some to involve low and by a lesser amount to involve high physical effort, while, in terms of ease of use, the majority agree that it is medium to very high. Regarding the 500 W bicycle, the majority of the participants consider the physical effort used to be medium, highlighting one person who considered it to be very high; in terms of ease of use, it varies between very low and high, highlighting more responses at the medium level, with this being due to the fact that the bicycle was very heavy and the assistance option was only activated from a certain speed. Finally, the 750 W bike is the one that requires the least physical effort due to its power and ability to climb slopes easily, while its ease of use varies between low and very high.

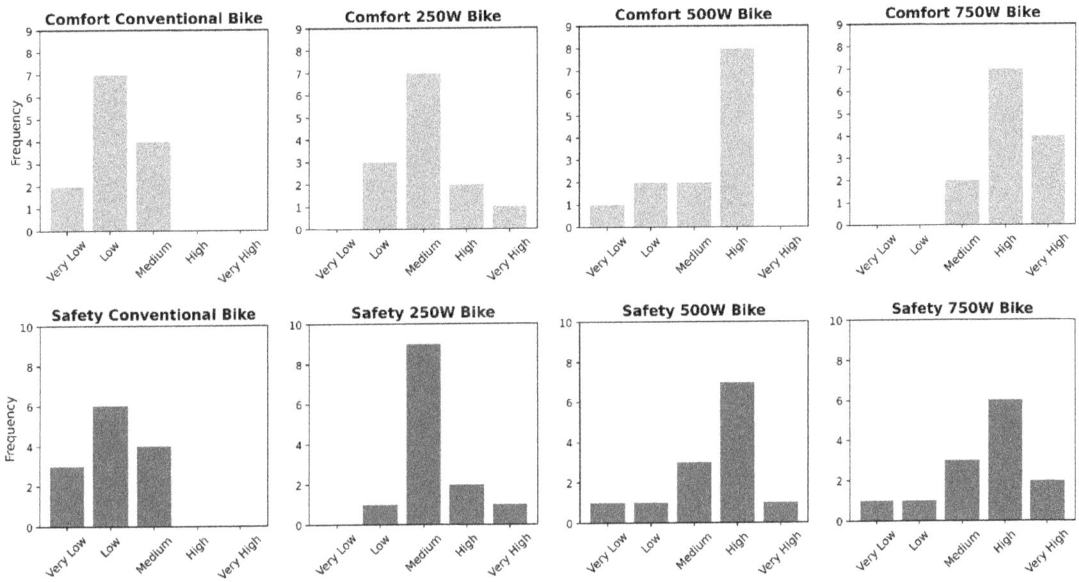

Figure 11. Respondents' responses to the survey section related to the perception of comfort and safety of each bicycle. Source: elaborated by the authors.

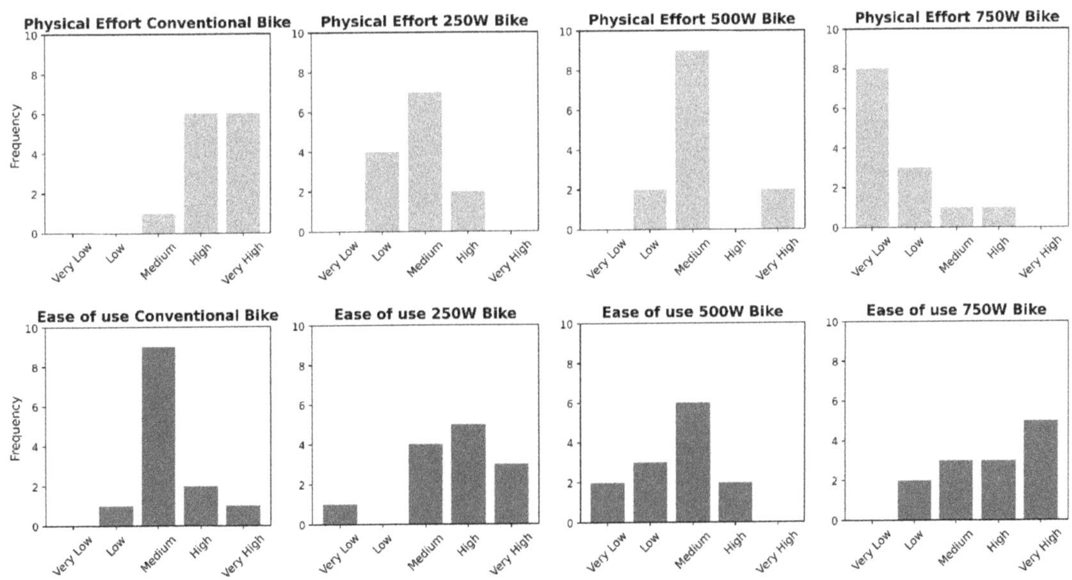

Figure 12. Respondents' responses to the survey section related to physical effort and the ease of use of each bike. Source: elaborated by the authors.

5. Conclusions

Aligned with the Sustainable Development Goals of the United Nations, this study identified and evaluated 15 potential E-bike routes in the city of Valparaíso that connected the city's metro stations with strategic streets in the hills, highlighting this as an important sustainable mobility solution in a challenging urban context. The routes with better conditions based on the criteria used were to Cumming from Puerto and Bellavista stations. These could be potential routes for promoting an E-bike-inclusive city. The results indicate that variables such as the slope, traffic accidents, intersections, and directionality of the street have a significant impact on the viability and safety of E-bike users, indicating the necessity to plan E-bike routes carefully, giving priority to safety and connectivity to promote a change in modal choices towards active and sustainable mobility, similar to the results found by De Jong et al. [37] and Saplıoğlu and Aydın [38]. As found by Santos et al. [46], the results of this study showed that E-bikes were capable of overcoming steep slope conditions, which was also noted by the experts that weighted the factors evaluated in this study.

Regarding the experimental approach, it was possible to demonstrate that the E-bikes with greater power (500 W and 750 W) allowed the cyclists to overcome the slopes encountered in Valparaíso with less physical effort and better comfort. However, the characteristics of the participants and their performance may have been affected depending on the design of the E-bikes (e.g., height) and their motorized settings, such as allowing motor assistance when the bicycle achieved a certain speed, which was the case for the 500 W E-bike. Different from the other studies reviewed in the literature review section, this research explored a case study, Valparaíso, that currently does not develop cycle-friendly infrastructure. One of the reasons for this is related to norms and regulations based on conventional bicycles for establishing cycleways and routes that limit the city from exploring and developing its sustainable mobility potential further.

The contributions of this study support the bettering of the definition of non-motorized vehicles in Chile, which currently is limited to E-bikes up to 250 W, supposedly because of the fear of non-motorized vehicles achieving speeds higher than 25 km/h. However, the experiments developed in this study showed that the 500 W and 750 W E-bikes achieved average speeds of 5.36 km/h and 6.88 km/h, respectively, and overcame slopes higher than the 6%, which is a limiting value for justifying cycleways in this country. Therefore, it was shown that in the case of Valparaíso, it could be possible to reduce the accessibility gap between less and more favored populations by increasing the limiting power for non-motorized mobility and the slope at which cycleways are justified.

Considering that this is a first approach to the city of Valparaíso regarding the potential use of E-bikes in this city, there are still challenges regarding the implementation and monitoring of E-bike cycleways on the proposed routes, the strengthening of public policies and investments to promote sustainable mobility, and education and awareness with respect to street safety and the environmental impact of transport systems. In this direction, this study seeks, in future opportunities, to pursue more detailed research regarding street scale analyses with the support of street audit tools to validate the global analysis, develop more experimental studies considering the limited number of participants to improve the validity of their results, and include climate variables, such as temperature and precipitation, as they also affect daily mobility. Furthermore, it would be interesting to explore an economic evaluation of cycleway projects in the city of Valparaíso in the future.

Author Contributions: Conceptualization: Vicente Aprigliano. Methodology: Vicente Aprigliano, Catalina Toro, Gonzalo Rojas, and Sebastian Seriani. Validation: Vicente Aprigliano and Sebastian Seriani. Formal analysis: Vicente Aprigliano, Catalina Toro, Gonzalo Rojas, and Sebastian Seriani.

Investigation: Vicente Aprigliano, Catalina Toro, Gonzalo Rojas, Emilio Bustos, Ivan Bastias, and Sebastian Seriani. Data curation: Catalina Toro, Gonzalo Rojas, Ivan Bastías, and Emilio Bustos. Writing—original draft preparation: Vicente Aprigliano, Catalina Toro, Gonzalo Rojas, Marcus Cardoso, Sebastian Seriani, Marcelino Aurélio Vieira da Silva, Talita Santos, Ualison Rébula de Oliveira, and Ivan Bastias. Writing—review and editing: Vicente Aprigliano, Catalina Toro, Gonzalo Rojas, Marcus Cardoso, Sebastian Seriani, Marcelino Aurélio Vieira da Silva, Talita Santos, and Ualison Rébula de Oliveira. Visualization: Vicente Aprigliano, Catalina Toro, Gonzalo Rojas, Emilio Bustos, and Sebastian Seriani. Supervision: Vicente Aprigliano. Project administration: Vicente Aprigliano. Funding acquisition: Vicente Aprigliano. All authors have read and agreed to the published version of the manuscript.

Funding: Agencia Nacional de Investigación y Desarrollo (ANID), Chile, through the FONDECYT Project 11230050.

Data Availability Statement: The raw data supporting the conclusions of this article will be made available by the authors on request.

Acknowledgments: The authors gratefully acknowledge the research support provided by Agencia Nacional de Investigación y Desarrollo (ANID), Chile.

Conflicts of Interest: The authors declare no conflicts of interest.

Appendix A

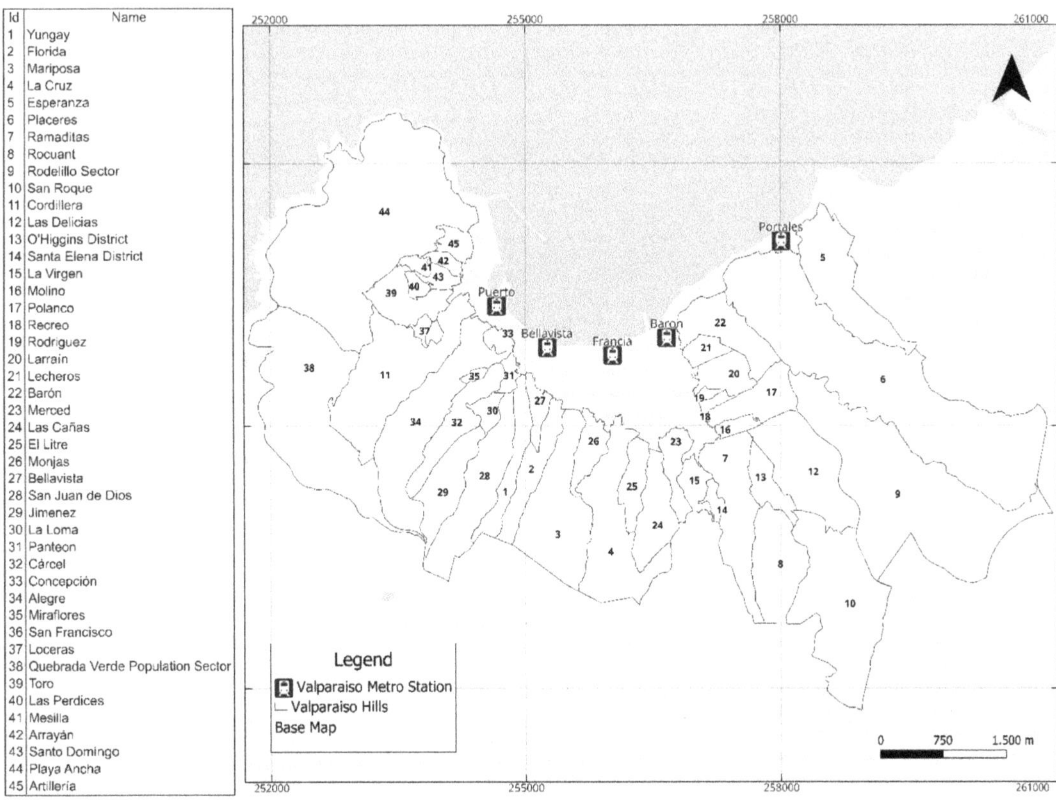

Figure A1. Names and locations of Valparaiso's hills. Source: elaborated by the authors.

Table A1. Criteria for establishing the scores for evaluating the most optimal route. Source: elaborated by the authors.

Criteria	Indicador	Description	Condition	Score	Formula
Avarege slope	Slope by proportion of the route section	For each section of the route, its gradient is calculated, a value that is weighted according to the specific conditions. This value is adjusted based on the proportion of the distance of the section within the total route.	<=8	3	$\textbf{Total}_x\ \textbf{Score} = \sum_{i=1}^{n} P_{i,x} * \left(\frac{D_{i,x}}{D_{total,\,x}}\right)$ $P_{i,x}$: $Score\ of\ section\ i\ in\ route\ x$ $D_{i,x}$: $Distance\ of\ section\ i\ in\ route\ x$ $D_{total,x}$: $Total\ distance\ of\ route\ x$
			>8 and <=10	2	
			>10 and <=12	1	
			>12	0	
Distance	Total distance from metro station	Total distance of the route from the metro station, weighted by the quartiles of each route.	<=Q1	3	$\mathbf{Q1} = D\left(\frac{(n+1)}{4}\right)$, $\mathbf{Q2} = D\left(\frac{(n+1)}{2}\right)$, $\mathbf{Q3} = D\left(\frac{3(n+1)}{4}\right)$ ***D:*** *Set of route distances from the plan* ***k:*** *Total number of route s(distances) in D.* ***n:*** *Total route distances from the plan, n = k.*
			>Q1 and <=Q2	2	
			>Q2 and <=Q3	1	
			>Q3	0	
	Total distance from the initial elevation of the hill	Total distance of the route from the start of the hill elevation, weighted by the quartiles of each route.	<=Q1	3	$\mathbf{Q1} = D\left(\frac{(n+1)}{4}\right)$, $\mathbf{Q2} = D\left(\frac{(n+1)}{2}\right)$, $\mathbf{Q3} = D\left(\frac{3(n+1)}{4}\right)$ ***D:*** *Set of route distances from the base of the hill.* ***k:*** *Total number of routes (distances) in D.* ***n:*** *Total distances of the route from the base of the hill, n = k.*
			>Q1 and <=Q2	2	
			>Q2 and <=Q3	1	
			>Q3	0	
	Average distance of the route sections	For each section of the route, its distance is calculated, and an average is obtained for each route, weighted by the quartiles of all the routes.	<= Q1	3	$\mathbf{Q1} = D\left(\frac{(n+1)}{4}\right)$, $\mathbf{Q2} = D\left(\frac{(n+1)}{2}\right)$, $\mathbf{Q3} = D\left(\frac{3(n+1)}{4}\right)$ ***D:*** *Set of distances of the route sections from the base of the hill.* ***k:*** *Total number of sections (distances) in D.* ***n:*** *Total average distance of the route sections from the base of the hill, n = k*
			>Q1 and <=Q2	2	
			>Q2 and <=Q3	1	
			>Q3	0	
Traffic accidents	Total number of accidents by area of influence of the route	For each route, an area of influence of 5 m was defined, within which the number of traffic accidents was calculated, weighted by the quartiles of all the routes.	<=Q1	3	$\mathbf{Q1} = A\left(\frac{(p+1)}{4}\right)$, $\mathbf{Q2} = A\left(\frac{(p+1)}{2}\right)$, $\mathbf{Q3} = A\left(\frac{3(p+1)}{4}\right)$ ***A:*** *Set of total number of accidents per area of influence.* ***m:*** *Total number of areas (total number of accidents) in A.* ***p:*** *Total accidents per area of influence, where p = m.*
			>Q1 and <=Q2	2	
			>Q2 and <=Q3	1	
			>Q3	0	

Table A1. *Cont.*

Criteria	Indicador	Description	Condition	Score	Formula
Intersections	Total intersections	For each route, a count of intersections with other streets was made, weighting them by the quartiles of all the routes.	<=Q1	3	$\mathbf{Q1} = D\left(\frac{(n+1)}{4}\right)$, $\mathbf{Q2} = D\left(\frac{(n+1)}{2}\right)$, $\mathbf{Q3} = D\left(\frac{3(n+1)}{4}\right)$ ***D:*** *Set of intersections on the routes.* ***k:*** *Total number of intersections in D.* ***n:*** *Total distances of the route from the base of the hill, n = k*
			>Q1 and <=Q2	2	
			>Q2 and <=Q3	1	
			>Q3	0	
	Type of intersections	For each intersection, its type of directionality was determined, favoring one-way directions and fewer intersections. The resulting weighting was adjusted according to the proportion of intersections of that type on the route.	Uni * = 1	3	$\boldsymbol{Total_x\ Score} = \sum_{i=1}^{n} P_{i,x} * \left(\frac{I_{i,x}}{I_{total,\ x}}\right)$ $\boldsymbol{P_{(i,x)}}$: *Weighted value of the intersection of route i for type x* $\boldsymbol{I_{(i,x)}}$: *Number of intersections of type x in route i.* $\boldsymbol{I_{total,x}}$: *Total intersections of type x in all routes.*
			Uni = 2	3	
			Bi = 1	2	
			Bi = 2	1	
			Uni or Bi = 2 or 3	0	
Directionality	Directionality of the route	Each route was identified in terms of its directionality.	Unidirectional	3	$\boldsymbol{D = 3} * I_{unidirectional} + 0 * I_{bidirectional}$ $\boldsymbol{I_{(unidirectional)}}$: *1 when the street is unidirectional and 0 when it is not.* $\boldsymbol{I_{(bidirectional)}}$: *1 when the street is bidirectional and 0 when it is not.*
			Bidirectional	0	
Speed	Average route speed	A 5 m area of influence was defined per route, within which EOD zones were identified. Car trips in Valparaíso were then analyzed to obtain an average distance and speed per zone, which allowed an estimated average speed per route to be calculated.	<=Q1	3	$\boldsymbol{Speed} = \frac{\sum_{j=1}^{m} D_{(j)}}{\sum_{k=1}^{p} T_{(k)}}$ $\boldsymbol{D_{(j)}}$: *Distance traveled by person j who travels by car.* $\boldsymbol{TI_{(k)}}$: *Time spent by person k traveling by car.* ***m:*** *Total number of people traveling by car (for the distances).* ***p:*** *Total number of people traveling by car (for thetime).*
			>Q1 and <=Q2	2	
			>Q2 and <=Q3	1	
			>Q3	0	

* Uni: unidirectional. Bi: bidirectional.

Table A2. Characteristics of the routes for zone 1. Source: elaborated by the authors.

Downhill									
ID	**Slope per Section (%)**	**Total Distance from Metro Station**	**Total Distance from the Initial Elevation of the Hill**	**Average Distance of the Route Sections**	**Total Number of Accidents by Area of Influence of the Route**	**Total Intersections**	**Type of Intersections**		**Average Route Speed**
1	16.43	1105.71	871.88	65.06	3	17	Uni = 1	6	4.62
							Uni = 2	6	
							Bi = 1	3	
							Bi = 2	1	
							Uni or Bi = 2 or 3	1	
2	15.4	1884.36	782.39	75.4	16	24	Uni = 1	8	9.04
							Uni = 2	6	
							Bi = 1	5	
							Bi = 2	2	
							Uni or Bi = 2 or 3	3	
3	11.22	1258.99	586.46	89.96	9	14	Uni = 1	6	15.51
							Uni = 2	5	
							Bi = 2	2	
							Uni or Bi = 2 or 3	1	
4	13.35	1560.03	894.02	70.93	12	21	Uni = 1	9	9.17
							Uni = 2	5	
							Bi = 1	4	
							Bi = 2	1	
							Uni or Bi = 2 or 3	2	
5	15.12	2165.09	893.74	83.3	12	26	Uni = 1	11	8.22
							Uni = 2	9	
							Bi = 1	4	
							Bi = 2	1	
							Uni or Bi = 2 or 3	1	

Table A2. *Cont.*

Downhill									
ID	Slope per Section (%)	Total Distance from Metro Station	Total Distance from the Initial Elevation of the Hill	Average Distance of the Route Sections	Total Number of Accidents by Area of Influence of the Route	Total Intersections	Type of Intersections		Average Route Speed
6	12.57	1677.74	1182.95	83.92	12	19	Uni = 1	5	9.92
							Uni = 2	7	
							Bi = 1	6	
							Uni or Bi = 2 or 3	1	
7	12.98	1559.32	787.02	70.9	6	21	Uni = 2	8	8.42
							Bi = 1	8	
							Bi = 2	4	
							Uni or Bi = 2 or 3	1	
8	10.56	1516.66	1021.86	101.15	12	14	Uni = 1	4	9.24
							Uni = 2	6	
							Bi = 1	2	
							Bi = 2	1	
							Uni or Bi = 2 or 3	1	
9	10.36	1447.93	1166.47	160.94	0	9	Uni = 1	1	20.04
							Uni = 2	3	
							Bi = 1	2	
							Bi = 2	3	
10	10.25	1537.23	1042.46	118.29	14	12	Uni = 1	2	9.92
							Uni = 2	6	
							Bi = 1	1	
							Bi = 2	2	
							Uni or Bi = 2 or 3	1	

Table A2. *Cont.*

Uphill									
ID	Slope per Section (%)	Total Distance from Metro Station	Total Distance from the Initial Elevation of the Hill	Average Distance of the Route Sections	Total Number of Accidents by Area of Influence of the Route	Total Intersections	Type of Intersections		Average Route Speed
1	16.47	1552.71	780.71	64.72	6	24	Uni = 1	12	7.36
							Uni = 2	4	
							Bi = 1	4	
							Bi = 2	1	
							Uni or Bi = 2 or 3	3	
2	19.77	1532.4	780.71	73	6	21	Uni = 1	11	9.62
							Uni = 2	3	
							Bi = 1	3	
							Bi = 2	1	
							Uni or Bi = 2 or 3	3	
3	12.52	1302.68	586.46	162.89	11	8	Uni = 1	2	11.98
							Uni = 2	4	
							Bi = 2	2	
4	18.77	1704.85	953.17	77.52	7	22	Uni = 1	14	8.17
							Uni = 2	2	
							Bi = 1	4	
							Uni or Bi = 2 or 3	2	
5	15.72	1725.2	953.19	69.03	7	25	Bi = 1	4	7.04
							Uni = 1	15	
							Uni = 2	4	
							Uni or Bi = 2 or 3	2	

Table A2. *Cont.*

Uphill									
ID	**Slope per Section (%)**	**Total Distance from Metro Station**	**Total Distance from the Initial Elevation of the Hill**	**Average Distance of the Route Sections**	**Total Number of Accidents by Area of Influence of the Route**	**Total Intersections**	**Type of Intersections**		**Average Route Speed**
6	19.14	2274.39	1242.4	103.42	13	22	Uni = 1	9	9.02
							Uni = 2	4	
							Bi = 1	7	
							Uni or Bi = 2 or 3	2	
7	10.85	1871.34	787.03	72	8	25	Uni = 1	2	8.46
							Uni = 2	9	
							Bi = 1	7	
							Bi = 2	6	
							Uni or Bi = 2 or 3	1	
8	19.99	2054.21	1021.81	136.99	12	15	Uni = 1	8	8.26
							Uni = 2	2	
							Bi = 1	2	
							Bi = 2	1	
							Uni or Bi = 2 or 3	2	
9	12.29	1536.81	1166.42	139.76	0	11	Uni = 1	3	14.52
							Uni = 2	2	
							Bi = 1	2	
							Bi = 2	4	
10	21.14	2136.59	1042.45	164.41	14	13	Uni = 1	4	9.17
							Uni = 2	4	
							Bi = 1	1	
							Bi = 2	2	
							Uni or Bi = 2 or 3	2	

Table A3. Characteristics of the routes for zone 2. Source: elaborated by the authors.

Bidirectional									
ID	Slope per Section (%)	Total Distance from Metro Station	Total Distance from the Initial Elevation of the Hill	Average Distance of the Route Sections	Total Number of Accidents by Area of Influence of the Route	Total Intersections	Type of Intersections		Average Route Speed
11	11.17	2313.4	1398.23	110.2	16	20	Uni = 1	3	6.74
							Uni = 2	5	
							Bi = 1	7	
							Bi = 2	5	
12	6.74	1528.82	616.82	109.24	14	13	Uni = 2	5	6.74
							Bi = 1	3	
							Bi = 2	5	

Table A4. Characteristics of the routes for zone 3. Source: elaborated by the authors.

Downhill									
ID	Slope per Section (%)	Total Distance from Metro Station	Total Distance from the Initial Elevation of the Hill	Average Distance of the Route Sections	Total Number of Accidents by Area of Influence of the Route	Total Intersections	Type of Intersections		Average Route Speed
13	13.06	3713.58	2612.7	67.54	23	53	Uni = 1	12	4.8
							Uni = 2	2	
							Bi = 1	27	
							Bi = 2	7	
							Uni or Bi = 2 or 3	5	
14	12.35	2603.41	2099.77	86.81	6	29	Uni = 1	4	10.67
							Uni = 2	1	
							Bi = 1	16	
							Bi = 2	8	
15	9.61	3127.91	1455.54	89.4	43	32	Uni = 1	8	4.19
							Uni = 2	3	
							Bi = 1	14	
							Bi = 2	2	
							Uni or Bi = 2 or 3	5	

Table A4. *Cont.*

Uphill									
ID	Slope per Section (%)	Total Distance from Metro Station	Total Distance from the Initial Elevation of the Hill	Average Distance of the Route Sections	Total Number of Accidents by Area of Influence of the Route	Total Intersections	Type of Intersections		Average Route Speed
13	13.16	3648.27	2573.13	71.56	15	50	Uni = 1	9	4.34
							Uni = 2	5	
							Bi = 1	22	
							Bi = 2	8	
							Uni or Bi = 2 or 3	6	
14	12.8	2567.07	2127.96	91.71	7	28	Uni = 1	4	10.67
							Uni = 2	1	
							Bi = 1	17	
							Bi = 2	6	
15	10.4	3140.45	1455.52	104.72	27	28	Uni = 1	1	4.29
							Uni = 2	11	
							Bi = 1	10	
							Bi = 2	3	
							Uni or Bi = 2 or 3	3	

References

1. Niemann, M.; Braun, K.F.; Otto, E.; Tiefenbrunner, M.; Wüster, J.; Stöckle, U.; Ahmad, S.S.; Märdian, S.; Graef, F. Dangers of E-Mobility: A Systematic Review and Meta-Analysis of Sustained Injury Patterns and Injury Severity. *Saf. Sci.* **2023**, *167*, 106283. [CrossRef]
2. Uluk, D.; Lindner, T.; Palmowski, Y.; Garritzmann, C.; Göncz, E.; Dahne, M.; Möckel, M.; Gerlach, U.A. E-Scooter: Erste Erkenntnisse über Unfallursachen und Verletzungsmuster. *Notf. Rettungsmed.* **2020**, *23*, 293–298. [CrossRef]
3. Germany Federal Ministry of Justice Road Traffic Act (Straßenverkehrsgesetz) (Section 7–20). Available online: https://www.gesetze-im-internet.de/englisch_stvg/englisch_stvg.html (accessed on 3 September 2024).
4. Ministerie van Binnenlandse Zaken en Koninkrijksrelaties Regeling Voertuigen. Available online: https://wetten.overheid.nl/BWBR0025798/2018-05-20 (accessed on 3 September 2024).
5. Consumer Product Safety Commission Electric Bicycles; Advance Notice of Proposed Rulemaking; Request for Comments and Information. Available online: https://www.federalregister.gov/documents/2024/03/15/2024-05472/electric-bicycles-advance-notice-of-proposed-rulemaking-request-for-comments-and-information (accessed on 3 September 2024).
6. Ministry of Transportation and Infrastructure. E-Bike Rules of the Road—Province of British Columbia. Available online: https://www2.gov.bc.ca/gov/content/transportation/driving-and-cycling/cycling/e-bike-rules-of-the-road (accessed on 3 September 2024).
7. Conselho Nacional de Trânsito. *Resolução N⁰ 465, de 27 de Novembro de 2013*; Contran: Brasília, Brasil, 2009.
8. Ministerio de Vivienda y Urbanismo. *Vialidad Ciclo-Inclusiva*; Serie Espacios Públicos Urbanos; Ministerio de Vivienda y Urbanismo: Santiago, Chile, 2015.
9. Bici Cosas Cicletada a Lo Vásquez, Todo lo Que Necesitas Saber—Bicicosas. Available online: https://www.bicicosas.cl/cicletada-a-lo-vasquez-todo-lo-que-necesitas-saber/ (accessed on 3 September 2024).
10. Estación Chile e-Bike Tour 7 Cerros de Valparaiso. Available online: https://www.estacionchile.cl/detalle-tour.php?tour=74 (accessed on 3 September 2024).
11. Instituto Nacional de Estadísticas Instituto Nacional de Estadísticas | REDATAM Procesamiento y Diseminación. Available online: https://redatam.org/ (accessed on 24 April 2024).
12. Fernandes, V.A.; Tirado, H.M. Sustainable Mobility in Valparaíso, Chile, and its Relationship with Topography and Socio-Spatial Conditions. *J. Sustain. Sci. Manag.* **2021**, *16*, 122–134. [CrossRef]
13. Aprigliano, V.; Toro, C.; Rojas, G.; Barros, G.; Fukushi, M.; Seriani, S. Effects of the Built Environment and Sociodemographic Factors on Accessibility in the City of Valparaíso, Chile. *J. Transp. Geogr.* **2024**, *118*, 103931. [CrossRef]
14. Aprigliano, V.; Seriani, S.; Toro, C.; Rojas, G.; Fukushi, M.; Cardoso, M.; Silva, M.A.V.D.; Cucumides, C.; De Oliveira, U.R.; Henríquez, C.; et al. Built Environment Effect on Metro Ridership in Metropolitan Area of Valparaíso, Chile, under Different Influence Area Approaches. *ISPRS Int. J. Geo-Inf.* **2024**, *13*, 266. [CrossRef]
15. United Nations. THE 17 GOALS | Sustainable Development. Available online: https://sdgs.un.org/goals (accessed on 3 September 2024).
16. Bunyamin, W.M.H.W.; Baharom, R.; Munim, W.N.W.A. Wireless Battery Charger with Power Factor Correction for Electric Bike. In Proceedings of the 2023 IEEE Industrial Electronics and Applications Conference (IEACon), Penang, Malaysia, 6 November 2023; IEEE: New York, NY, USA, 2023; pp. 139–144.
17. Yasir, A.; Hu, X.; Ahmad, M.; Alvarado, R.; Anser, M.K.; Işık, C.; Choo, A.; Ausaf, A.; Khan, I.A. Factors Affecting Electric Bike Adoption: Seeking an Energy-Efficient Solution for the Post-COVID Era. *Front. Energy Res.* **2022**, *9*, 817107. [CrossRef]
18. Salmeron-Manzano, E.; Manzano-Agugliaro, F. The Electric Bicycle: Worldwide Research Trends. *Energies* **2018**, *11*, 1894. [CrossRef]
19. Biblioteca del Congreso Nacional de Chile Ley 21088 | Modifica la Ley de Tránsito para Incorporar Disposiciones sobre Convivencia de los Distintos Medios de Transporte. Available online: https://www.bcn.cl/leychile/navegar?idNorma=1118358 (accessed on 26 November 2024).
20. Güldü, E.; Kuşçu Şimşek, Ç.; Selim, S. A Study on the Improvement of Bicycle Transportation in Sivas City Using Hybrid Multi-Criteria Model Based Network Analysis. *Env. Dev. Sustain.* **2024**. [CrossRef]
21. Hsu, T.P.; Lin, Y.T. A Model for Planning a Bicycle Network with Multi-Criteria Suitability Evaluation Using GIS. In *WIT Transactions on Ecology and the Environment*; WIT Press: Southampton, UK, 2011; Volume 148, pp. 243–252. [CrossRef]
22. Ryu, S.; Chen, A.; Su, J.; Choi, K. A Multi-Class, Multi-Criteria Bicycle Traffic Assignment Model. *Int. J. Sustain. Transp.* **2021**, *15*, 524–540. [CrossRef]
23. Pais, F.; Monteiro, J.; Sousa, N.; Coutinho-Rodrigues, J.; Natividade-Jesus, E. A Multicriteria Methodology for Maintenance Planning of Cycling Infrastructure. *Proc. Inst. Civ. Eng. Eng. Sustain.* **2022**, *175*, 248–264. [CrossRef]
24. Chondrogianni, D.; Stephanedes, Y.J.; Fatourou, P. Assessing Cycling Accessibility in Urban Areas through the Implementation of a New Cycling Scheme. *Sustainability* **2023**, *15*, 14472. [CrossRef]

25. Nematchoua, M.; Deuse, C.; Cools, M.; Reiter, S. Evaluation of the Potential of Classic and Electric Bicycle Commuting as an Impetus for the Transition towards Environmentally Sustainable Cities: A Case Study of the University Campuses in Liege, Belgium. *Renew. Sustain. Energy Rev.* **2020**, *119*, 109544. [CrossRef]
26. Brand, C.; Dons, E.; Anaya-Boig, E.; Avila-Palencia, I.; Clark, A.; De Nazelle, A.; Gascon, M.; Gaupp-Berghausen, M.; Gerike, R.; Götschi, T.; et al. The Climate Change Mitigation Effects of Daily Active Travel in Cities. *Transp. Res. Part D Transp. Environ.* **2021**, *93*, 102764. [CrossRef]
27. Hrncir, J.; Zilecky, P.; Song, Q.; Jakob, M. Practical Multicriteria Urban Bicycle Routing. *IEEE Trans. Intell. Transport. Syst.* **2017**, *18*, 493–504. [CrossRef]
28. Tal, I.; Olaru, A.; Muntean, G.-M. eWARPE—Energy-Efficient Weather-Aware Route Planner for Electric Bicycles. In Proceedings of the 2013 21st IEEE International Conference on Network Protocols (ICNP), Goettingen, Germany, 7–10 October 2013; IEEE: New York, NY, USA, 2013; pp. 1–6.
29. Behrendt, F. Cycling the Smart and Sustainable City: Analyzing EC Policy Documents on Internet of Things, Mobility and Transport, and Smart Cities. *Sustainability* **2019**, *11*, 763. [CrossRef]
30. Meireles, M.; Ribeiro, P.J.G. Digital Platform/Mobile App to Boost Cycling for the Promotion of Sustainable Mobility in Mid-Sized Starter Cycling Cities. *Sustainability* **2020**, *12*, 2064. [CrossRef]
31. Oliveira, F.; Nery, D.; Costa, D.G.; Silva, I.; Lima, L. A Survey of Technologies and Recent Developments for Sustainable Smart Cycling. *Sustainability* **2021**, *13*, 3422. [CrossRef]
32. Đerek, J.; Sikora, M. Bicycle Route Planning Using Multiple Criteria GIS Analysis. In Proceedings of the 2019 International Conference on Software, Telecommunications and Computer Networks (SoftCOM), Split, Croatia, 19–21 September 2019; IEEE: New York, NY, USA, 2019; pp. 1–5.
33. Hrncir, J.; Song, Q.; Zilecky, P.; Nemet, M.; Jakob, M. Bicycle Route Planning with Route Choice Preferences. *Front. Artif. Intell. Appl.* **2014**, *263*, 1149–1154. [CrossRef]
34. Lopez-Escolano, C.; Campos, A.P.; Pardos, S.V.; Nedeliakova, E.; Stefancova, V. Incorporating Bicycles into Urban Mobility: An Opportunity for Sustainable Development. *Komunikácie* **2017**, *19*, 68–73. [CrossRef]
35. Maltese, I.; Gatta, V.; Marcucci, E. Active Travel in Sustainable Urban Mobility Plans. An Italian Overview. *Res. Transp. Bus. Manag.* **2021**, *40*, 100621. [CrossRef]
36. Pritchard, R. Revealed Preference Methods for Studying Bicycle Route Choice—A Systematic Review. *IJERPH* **2018**, *15*, 470. [CrossRef] [PubMed]
37. De Jong, T.; Böcker, L.; Weber, C. Road Infrastructures, Spatial Surroundings, and the Demand and Route Choices for Cycling: Evidence from a GPS-Based Mode Detection Study from Oslo, Norway. *Environ. Plan. B Urban. Anal. City Sci.* **2023**, *50*, 2133–2150. [CrossRef]
38. Saplıoğlu, M.; Aydın, M.M. Choosing Safe and Suitable Bicycle Routes to Integrate Cycling and Public Transport Systems. *J. Transp. Health* **2018**, *10*, 236–252. [CrossRef]
39. Đerek, J.; Sikora, M.; Kraljević, L.; Russo, M. Using Neural Networks for Bicycle Route Planning. *Appl. Sci.* **2021**, *11*, 10065. [CrossRef]
40. Perger, T.; Auer, H. Energy Efficient Route Planning for Electric Vehicles with Special Consideration of the Topography and Battery Lifetime. *Energy Effic.* **2020**, *13*, 1705–1726. [CrossRef]
41. Milakis, D.; Athanasopoulos, K.; Vafeiadis, E.; Vasileiadis, K.; Vlastos, T. Planning of the Athens Metropolitan Cycle Network Using Participative Multicriteria Gis Analysis. *Procedia Soc. Behav. Sci.* **2012**, *48*, 816–826. [CrossRef]
42. Beukes, E.; Vanderschuren, M.; Zuidgeest, M.; Brussel, M. Creating Liveable Neighborhoods through Context-Sensitive Multimodal Road Planning. *Transp. Res. Rec.* **2011**, *2244*, 27–33. [CrossRef]
43. Jiménez, P.; Calatrava, M. Cycle Lane Network to Connect the Different Campuses of Technical University of Cartagena (Spain). In Proceedings of the 11th International Conferenceth on Urban Regeneration and Sustainability (SC 2016), Alicante, Spain, 12 July 2016; Volume 204, pp. 765–775.
44. Grisé, E.; El-Geneidy, A. If We Build It, Who Will Benefit? A Multi-Criteria Approach for the Prioritization of New Bicycle Lanes in Quebec City, Canada. *J. Transp. Land Use* **2018**, *11*, 217–235. [CrossRef]
45. Zagorskas, J.; Turskis, Z. Setting Priority List for Construction Works of Bicycle Path Segments Based on Eckenrode Rating and ARAS-F Decision Support Method Integrated in GIS. *Transport* **2020**, *35*, 179–192. [CrossRef]
46. Santos, B.; Passos, S.; Gonçalves, J.; Matias, I. Spatial Multi-Criteria Analysis for Road Segment Cycling Suitability Assessment. *Sustainability* **2022**, *14*, 9928. [CrossRef]
47. Milakis, D.; Athanasopoulos, K. What about People in Cycle Network Planning? Applying Participative Multicriteria GIS Analysis in the Case of the Athens Metropolitan Cycle Network. *J. Transp. Geogr.* **2014**, *35*, 120–129. [CrossRef]
48. Zuo, T.; Wei, H. Bikeway Prioritization to Increase Bicycle Network Connectivity and Bicycle-Transit Connection: A Multi-Criteria Decision Analysis Approach. *Transp. Res. Part A Policy Pract.* **2019**, *129*, 52–71. [CrossRef]

49. Fitch, D.T.; Handy, S.L. Road Environments and Bicyclist Route Choice: The Cases of Davis and San Francisco, CA. *J. Transp. Geogr.* **2020**, *85*, 102705. [CrossRef]
50. Hardinghaus, M.; Papantoniou, P. Evaluating Cyclists' Route Preferences with Respect to Infrastructure. *Sustainability* **2020**, *12*, 3375. [CrossRef]
51. Desjardins, E.; Higgins, C.D.; Scott, D.M.; Apatu, E.; Páez, A. Using Environmental Audits and Photo-Journeys to Compare Objective Attributes and Bicyclists' Perceptions of Bicycle Routes. *J. Transp. Health* **2021**, *22*, 101092. [CrossRef]
52. Liu, Y.; Yang, D.; Timmermans, H.J.P.; De Vries, B. Differences in Street-Scale Built Environment Preferences towards Biking: A Latent Class Analysis of Stated Choice Data. *Int. J. Urban. Sustain. Dev.* **2021**, *13*, 706–714. [CrossRef]
53. Özkan, S.P. Kentsel Alanda CBS Kullanarak Bisiklet Altyapısı Rota Planlaması: İzmir Örneği. *Planning* **2020**, *30*, 41275. [CrossRef]
54. Baeva, S.; Shterev, V.; Stoimenov, E.; Hinov, N. Risk Assessment and Route Optimization for Life and Health Self-Keeping during e-Cycling. In *AIP Conference Proceedings*; AIP Publishing: Sozopol, Bulgaria, 2019.
55. Baeva, S.; Shterev, V.; Yochev, K.; Hinov, N. Route Optimization for Long Durability Battery Life during E-Bike Cycling. In *AIP Conference Proceedings*; AIP Publishing: Sozopol, Bulgaria, 2019; p. 080017.
56. Bi, H.; Shang, W.-L.; Chen, Y.; Wang, K.; Yu, Q.; Sui, Y. GIS Aided Sustainable Urban Road Management with a Unifying Queueing and Neural Network Model. *Appl. Energy* **2021**, *291*, 116818. [CrossRef]
57. CEDEUS. Área Metropolitana de Valparaíso. Available online: https://indicadores.cedeus.cl/ciudad/16 (accessed on 12 September 2024).
58. Biblioteca del Congreso Nacional de Chile. Reportes Estadísticos 2024 de Valparaíso. Biblioteca Del Congreso Nacional de Chile. Available online: https://www.bcn.cl/siit/reportescomunales/comunas_v.html?anno=2024&idcom=5101 (accessed on 23 September 2024).
59. Ministerio de Transporte y Telecomunicaciones. Mapa de Recorrido de Buses, Trolebuses y Estaciones de Metro En Valparaíso 2014. Valparaíso, Chile. Available online: https://www.subtrans.gob.cl/region-de-valparaiso.html (accessed on 3 March 2024).
60. Biblioteca del Congreso Nacional de Chile. Clima y Vegetación Región de Valparaíso. Available online: https://www.bcn.cl/siit/nuestropais/region5/clima.htm (accessed on 23 September 2024).
61. BioBio Chile. Advierten Llegada de Sistema Frontal a la Región de Valparaíso: Se Esperan Hasta 100 mm de Agua. Available online: https://www.biobiochile.cl/noticias/nacional/region-de-valparaiso/2024/06/11/advierten-llegada-de-sistema-frontal-a-la-region-de-valparaiso-se-esperan-hasta-100-mm-de-agua.shtml (accessed on 23 September 2024).
62. BioBio Chile. Región de Valparaíso se Prepara Para Recibir dos Sistemas Frontales Esta Semana: Hay Alerta Preventiva. Available online: https://www.biobiochile.cl/noticias/nacional/region-de-valparaiso/2024/06/17/region-de-valparaiso-se-prepara-para-recibir-dos-sistemas-frontales-esta-semana-hay-alerta-preventiva.shtml (accessed on 23 September 2024).
63. Diario UChile Declaran Alerta Roja para las Provincias de Valparaíso y Marga Marga por Lluvias. Available online: https://radio.uchile.cl/2024/06/21/declaran-alerta-roja-para-las-provincias-de-valparaiso-y-marga-marga-por-lluvias/ (accessed on 23 September 2024).
64. Garmin Garmin Connect. Available online: https://connect.garmin.com/modern/home (accessed on 26 November 2024).
65. IDE Geoportal de Chile—DEM Alos Palsar Región de Valparaíso 2024. Ministerio de Bienes Nacionales. Santiago. Available online: https://www.geoportal.cl/catalog (accessed on 20 April 2024).
66. Ministerio de Vivienda y Urbanismo. *Decreto 47 FIJA Nuevo Texto de la Ordenanza General de la Ley General de Urbanismo y Construcciones*; MINVU: Santiago, Chile, 1992.
67. CONASET Portal de Información Geoespacial CONASET 2024. Santiago, Chile. Available online: https://mapas-conaset.opendata.arcgis.com (accessed on 25 May 2024).
68. Saaty, R.W. The Analytic Hierarchy Process—What It Is and How It Is Used. *Math. Model.* **1987**, *9*, 161–176. [CrossRef]
69. Matias, I.; Virtudes, A. Bicycle Solutions in Mountain Cities: CycloCable® in Trondheim-Norway. *IOP Conf. Ser. Mater. Sci. Eng.* **2019**, *603*, 022024. [CrossRef]
70. SECTRA Encuestas de Movilidad. Available online: https://www.sectra.gob.cl/encuestas_movilidad/encuestas_movilidad.htm (accessed on 30 May 2024).

International Journal of *Geo-Information*

MDPI

Article

Geospatial Analytics of Urban Bus Network Evolution Based on Multi-Source Spatiotemporal Data Fusion: A Case Study of Beijing, China

Xiao Li [1,2,3,4], Shaohua Wang [4,5,*], Liang Zhou [1,2,3], Yeran Sun [6], Jiayi Zheng [4], Chang Liu [4,5], Junyuan Zhou [1,2,3,4], Cheng Su [4,5] and Dachuan Xu [1,2,3,4]

[1] Faculty of Geomatics, Lanzhou Jiaotong University, Lanzhou 730070, China; 11220869@stu.lzjtu.edu.cn (X.L.); zhougeo@126.com (L.Z.); 11220851@stu.lzjtu.edu.cn (J.Z.); 12232104@stu.lzjtu.edu.cn (D.X.)

[2] National-Local Joint Engineering Research Center of Technologies and Applications for National Geographic State Monitoring, Lanzhou 730070, China

[3] Gansu Provincial Engineering Laboratory for National Geographic State Monitoring, Lanzhou 730070, China

[4] State Key Laboratory of Remote Sensing and Digital Earth, Aerospace Information Research Institute, Chinese Academy of Sciences, Beijing 100094, China; zhengjiayi@aircas.ac.cn (J.Z.); liuchang248@mails.ucas.ac.cn (C.L.); sucheng23@mails.ucas.ac.cn (C.S.)

[5] University of Chinese Academy of Sciences, Beijing 101408, China

[6] Department of Geography, School of Life and Environmental Sciences, University of Lincoln, Lincoln LN6 7TS, UK; yesun@lincoln.ac.uk

* Correspondence: wangshaohua@aircas.ac.cn; Tel.: +86-010-82178178

Abstract: Bus networks are a crucial support for urban commuting. By studying the evolutionary characteristics of bus networks, we can uncover their development patterns, coverage efficiency, and changes in regional balance, providing a scientific basis for sustainable urban development and the optimization of transportation resources. This study systematically analyzes the spatiotemporal evolution characteristics of the bus network in Beijing from 2006 to 2024 using specific spatial analysis tools to analyze spatiotemporal evolution characteristics. By analyzing spatial coverage rates of transit stations using road network and administrative division data, the study reveals the convenience of bus networks in different regions. By combining the research methodology of the Sustainable Development Goals (SDGs) report, a 500-m service radius for bus stops was assessed. A complex network model was used to extract the nodes and edges of the bus network, and the betweenness centrality (BC) characteristics were analyzed. The findings indicate that Beijing's bus network has gradually expanded from the central urban areas to peripheral regions, with notable expansion in Tongzhou and Yanqing, resulting in an improved balance in the distribution of stations and routes and the emergence of Tongzhou as a new bus hub. The diffusion characteristics of the bus network are significantly influenced by administrative boundaries and the layout of the ring roads. Bus routes and stops are highly concentrated in the central urban areas and within the Second Ring Road, while as the number of ring roads increases, various network indices gradually decrease. The distribution of bus stops shows notable clustering and an uneven directional development. Beijing's bus stop distribution exhibits significant clustering characteristics, and the areas with a high Population Conveniently Served by Buses (PCSB) are predominantly concentrated in the central urban areas, with a large gap compared to the outer suburban districts. These conclusions expand on the exploration of isolated and static characteristics of the bus network structure, revealing the dynamic mechanisms and evolution patterns of Beijing's bus network. They provide guidance and recommendations for improving the bus network and offer more comprehensive support for urban planning and resource allocation.

Keywords: bus network; evolutionary trends; spatiotemporal characteristics; geospatial analysis

1. Introduction

Under the framework of the United Nations Sustainable Development Goals (SDGs), SDG 11—"Make cities and human settlements inclusive, safe, resilient, and sustainable"—has become a central concern for nations worldwide [1]. Among its key components, building a safe, convenient, and sustainable public transportation system for all humanity is an essential aspect of achieving this goal [2]. At the same time, in the face of worsening global warming and the acceleration of urbanization [3–5], promoting public transportation modes can effectively reduce carbon emissions, alleviate traffic congestion, and improve urban environments [6]. This initiative aligns with the requirements of SDG 11 to enhance urban livability and sustainability [7], while also providing strong support for the development of more convenient and livable urban communities. Public transit stations and routes build the city's road network, constituting the core elements of urban transit systems. Urban public transit systems connect various parts of the city, strengthening interregional links [8,9]. A well-developed public service infrastructure improves the quality of life, narrows social disparities, and ensures sustainable social development [10]. A mature transit network system, as an efficient and sustainable transportation option, provides substantial convenience, relieving pressure on urban traffic congestion [11]. Compared to private vehicle travel, it significantly reduces air pollution and promotes carbon neutrality, playing a crucial role in urban social and economic development [12]. Therefore, given the indispensable role of bus networks in urban areas, a thorough analysis of the characteristics and evolutionary patterns of urban bus networks is of great significance. Such research supports the optimization of transportation infrastructure under the framework of sustainable development goals and plays a crucial role in building more livable and resilient cities.

Urban public transit network evolution is a complex process of system optimization and self-organization, involving temporal nodes, spatial distribution, and transit behavior patterns [13]. The structure, connectivity, and complexity of urban transit networks are constantly adjusted over time [14]. Dynamic studies on transit networks over multiple periods can reveal the intrinsic rules of urban bus network evolution, providing valuable guidance for urban transportation planning. This study adopts a spatiotemporal perspective to analyze bus network evolution across ring roads and administrative divisions, offering insights into the roles of stations and routes within the network and better explaining the development mechanisms of bus networks. Population variables were incorporated into the analysis to comprehensively reveal bus network development trends. Through statistical feature analysis, we derived results regarding the relationships among population, stations, and routes, captured bus network evolution trends, identified patterns and issues, and provided practically significant insights for real-world applications.

The main innovations of this study are as follows. First, it introduces a spatiotemporal perspective to uncover the dynamic evolution patterns of urban bus networks. By analyzing multiple periods and spatial scales, along with a comprehensive examination of different ring roads and administrative divisions, it delves into the dynamic evolution characteristics of urban bus networks. Compared to static network structure analysis, studying the dynamic mechanisms and evolution patterns of urban bus networks can reveal the system's adaptability and adjustment mechanisms in complex real-world scenarios, providing more comprehensive support for urban planning and resource allocation. Second, it integrates the SDG framework to assess the features of bus networks, revealing their impact on urban sustainability. Under the SDG framework, the study further refines the evaluation of bus network connectivity, complexity, and coverage characteristics, systematically analyzing their contribution to urban sustainability. By analyzing the spatiotemporal evolution characteristics of bus networks from the perspective of sustainable development, this research addresses the gap in studying the evolution processes and dynamic mechanisms of urban bus networks. The evaluation system developed through this approach can be applied to other cities to explore the evolution patterns of their bus networks and evaluate the network's balance and spatial layout rationality. The structure of this study is as follows: Section 2 reviews related work and summarizes the general research directions of bus networks; Section 3 introduces the study area and specific data for analyzing the evolution of Beijing's bus network; Section 4 describes the experimental process and specific analytical methods; Section 5 visualizes and provides a detailed description of the experimental results; Section 6 discusses the study as a whole, offering a summary and outlook; and Section 7 concludes the study by summarizing the overall research.

2. Related Work

Urban public transportation has various forms, and buses and rail transit constitute today's main modes in cities. Compared to rail transit, conventional buses require less investment and offer better route flexibility, playing an essential role in urban transportation systems [15,16]. Early research on bus networks primarily focused on the quality of bus services. Newell analyzed various bus scheduling policies, clarifying the relationships between trips, timetables, and passenger demand during scheduling. He proposed several scheduling models and evaluated their effectiveness in actual operations, providing theoretical support for the design of bus scheduling strategies and specific recommendations on balancing punctuality and service quality in bus systems [17]. Glaister explored the impact of removing capacity regulation on the service quality of bus routes through stochastic simulation [18]. Bly and Oldfield, using London as an example, developed a bus service demand model to assess optimal bus service sizes. They investigated the potential improvements in bus services after introducing smaller buses to optimize London's existing transportation system [19]. However, due to the limited development of early bus networks and the immature research techniques at the time, studies were more focused on passenger experience but paid less attention to the structural characteristics of bus network changes.

Geography, a discipline that examines spatial phenomena and their causal relationships, provides multi-layered and multi-faceted analytical perspectives [20,21]. With the advancement of GIS technology, integrating geographic knowledge and techniques into the study of bus networks can provide a more comprehensive and scientific foundation for understanding and optimizing urban bus systems. Research on transit networks focuses primarily on GIS technology and complex network theory [22,23]. For example, Li and Guo analyzed urban transit network topologies and fractal characteristics using GIS to quantitatively describe its spatial morphology [24]. Tung et al. used spatial Durbin models to analyze the relationship between dynamic transit network characteristics and spatial structures across 265 Chinese cities [25]. Sienkiewicz and Hołyst analyzed public transit systems in 22 Polish cities, revealing network structures and characteristics [26]. Gao et al. constructed and analyzed transportation flow networks based on traffic flow models, employing a modified ChSch model to simulate traffic flow states and examine how traffic dynamics impact network structure. Varying traffic conditions led to different connection distributions within the network, and three types of small-world networks emerged in traffic flow evolution [27]. Sui et al. hypothesized that urban transportation demand decreases concentrically from the city center outward and that the quantity of transportation demand follows a power-law distribution. Based on this, they constructed a spatial evolution model. The model describes the network structure by defining a set of candidate stations, routes, and node metrics. It is used to balance the impact of passenger and investor demands on the network structure. The model's validity was verified using actual bus network data from nine Chinese cities obtained from the 8684 website [28]. Wang studied Beijing's traffic network from the perspectives of network topology construction, evolution, and characteristic analysis using transportation network theory and complex network science [29]. Bin Asad and Yuan, using taxi trajectory data from Nanjing, applied texture analysis algorithms to identify human spatial movement patterns and analyze city traffic patterns [30]. Abenoza et al. conducted surveys on public transit service quality expectations and satisfaction among different market segments, revealing the significance of service quality attributes over time and across groups [31]. Yu et al. built a complex network model and evaluated network density, centrality, clustering coefficients, and average distance to analyze Nanjing's transit network evolution. This innovative approach described subway network spatial structure characteristics, proposing star and ring structures to depict subway network spatial patterns, and gradually illustrated the city's commercial spatial layout [32]. Wang et al. conducted spatial configuration studies of urban transit networks, proposing a framework based on geographic spatial network analysis. By constructing an Urban Bus Spatial Network (UBSN), which represents the spatial configuration and connectivity of the urban bus system, they conducted in-depth analyses on macro statistical characteristics, centrality metrics, and community detection. Findings showed evident spatial clustering characteristics in UBSN, revealing the spatial distribution and interconnections of transit stations and routes within urban spaces. Based on these findings, they proposed optimization recommendations for urban transport infrastructure planning [23]. By integrating GIS technology with complex networks, existing studies provide a multi-level and comprehensive analytical framework for bus network structures, uncovering the complexity of transportation networks and their interaction patterns with the surrounding spatial environment.

Bus networks serve urban residents, and good network performance characteristics contribute to the robust operation of cities. Additionally, these performance characteristics are key indicators for evaluating the structural features of public transport networks, encompassing reliability, vulnerability, and resilience [33]. Various factors, such as climate change, disease spread, and information technology, impact the resilience of urban transit networks to varying degrees, which can be evaluated through categorized indicators [34]. Xu and Chopra discussed how Hong Kong's transit system's resilience is reliant on interconnectedness and its contribution to network robustness [35]. Liu and Bardaka analyzed the spatial correlation between poverty rates and transit network characteristics by coupling suburbanization in impoverished areas with public transit accessibility [36]. Fan et al. focused on the three major economic circles in Shandong Province, incorporating space syntax theory into analyzing the spatial distribution of compactness and selectivity in road networks using a spatial design network analysis model, which presents its spatial pattern and optimization recommendations [37]. Lee et al. applied a multi-layer network model to model and visualize Seoul's public transit system, further evaluating urban travel patterns, complexity, and robustness [38]. These studies are no longer limited to the surface-level aspects of bus service efficiency and quality but instead focus on the deeper structure of networks and their interactions with the surrounding spatial environment, providing both theoretical foundations and practical guidance for evaluating and optimizing urban bus networks. Overall, existing research methods on transportation networks are diverse but primarily emphasize single-characteristic analyses, such as complex network theory and spatial analysis [39,40].

However, most of these studies are confined to static analyses of network states during a single period. Research that systematically addresses the dynamic mechanisms and evolution patterns of urban bus networks is relatively scarce, with a lack of exploration of the spatial patterns of bus networks from a spatiotemporal perspective. Compared to the study of static bus network structure characteristics, exploring the dynamic mechanisms and evolution patterns of urban bus networks can reveal the system's adaptability and adjustment mechanisms under different periods and scenarios. Studying the changes in the network during real-world operations, it allows for a deeper understanding of how the system responds to factors such as population movement, changes in traffic demand, and policy adjustments. This helps identify service gaps in transportation demand, provides support for urban planning and resource allocation, and, from a multi-temporal perspective, serves as an important complement to static network structure analysis.

3. Materials

3.1. Study Area

Beijing, one of the world's largest cities, is located on the North China Plain in eastern China, with a total area of approximately 16,410 square kilometers [41]. The city consists of 16 county-level administrative divisions aligned longitudinally and 327 streets of varying sizes [42]. According to the Seventh National Census, Beijing's permanent population reached 21.893 million, a net increase of 2.281 million over the decade since the Sixth Census in 2010 [43]. As an ancient capital of China, Beijing has a rich history and profound cultural heritage. Throughout its history, Beijing has served multiple times as the political and cultural center of China, witnessing the rise and fall of several dynasties [44]. The development and evolution of Beijing's bus network have paralleled these historical changes. Over a long historical evolution, Beijing's public transportation system has developed into a distinctive, mature, and complex network [45]. Drawing on both domestic and international advanced practices, it continuously adopts innovative technologies to enhance the quality and efficiency of transit services, making Beijing's transit system a model for urban transportation networks worldwide [46]. Given the complexity and maturity of its transit network, dense population, and intricate urban structure, Beijing was chosen as the study area, as shown in Figure 1.

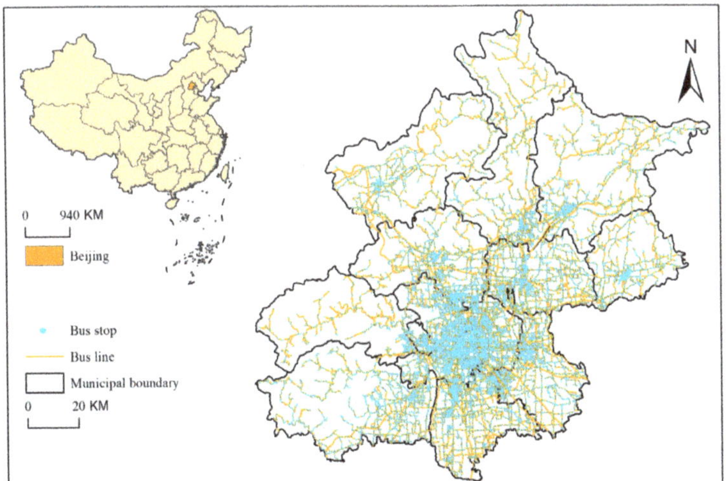

Figure 1. Study area.

3.2. Study Data

The urban bus network is an integral part of urban public facilities, serving the daily lives of city residents. Factors such as major urban events and city policies influence the distribution patterns of public service facilities [47]. Therefore, to better study the evolutionary characteristics of Beijing's bus network, we selected key temporal nodes corresponding to significant events in Beijing, and, considering the construction timeline of the bus network, we chose to analyze the bus network data from the year following each event. Based on this, we selected bus network data for four time points: 2006, 2009, 2017, and 2024. These datasets effectively capture the evolutionary process of the bus network. Utilizing data from the National Science and Technology Ministry's 863 Program project "Dynamic Peer-to-Peer Structured Open Navigation and Positioning Electronic Map Service Network Platform Construction and Demonstration", Wang combined with data from the Beijing Municipal Bureau of Statistics, the Beijing Bus Network, and digital map sources to compile the transit network data for Beijing in 2006 and 2009. He analyzed the spatiotemporal patterns of roads and transit networks over the years from the perspective of complex networks, employing various spatial analysis methods. This study focused on Beijing as the research area, revealing the evolutionary characteristics of its transit network and providing guidance for similar research on urban networks [29] (https://d.wanfangdata.com.cn/thesis/Y2430285 18 October 2023). Data for the 2017 and 2024 bus networks were obtained from Amap's website, a leading provider of digital map content, navigation, and location-based services in China. Amap is also one of the top ten mobile applications in China. The data obtained from Amap include information such as route names for both the inbound and outbound bus lines, the number of stops, latitude/longitude coordinates, and other relevant details (https://gaode.com/ accessed on 25 May 2024).

As a vital element in urban development, the population closely is linked with public transportation. This study used 2020 population data at a spatial resolution of 100 m × 100 m, shared on the Figshare platform. This dataset was generated using a downscaling method employing stacking ensemble learning, integrating random forest, XGBoost, and LightGBM models to segment residential areas from geospatial big data based on population data from the Seventh Census, thereby improving the accuracy of population grid data estimation. Compared to mainstream population datasets, this dataset has a coefficient of determination (R^2) of 0.8936, outperforming WorldPop (R^2 = 0.7427) and LandScan (R^2 = 0.7165) [48] (https://doi.org/10.5194/essd-16-3705-2024 accessed on 15 July 2024). To ensure the reliability of the population data, this study incorporated data from China's Seventh National Population Census to calibrate the initial population data (https://www.stats.gov.cn/sj/pcsj/ accessed on 15 July 2024). For this study, the 2024 road network data from OpenStreetMap (OSM) was used to segment Beijing's land parcels and analyze the distribution of stations. OSM is an open-source map data platform (https://www.openstreetmap.org/ accessed on 20 May 2024), where global users upload data on roads, waterways, buildings, and more. The downloaded road network data includes fields such as road names, types, and numbers [49]. Additionally, vector data for Beijing's administrative divisions were used to facilitate extraction, partitioning, and other analyses in subsequent experiments.

4. Research Methods

This study aims to analyze the spatial-temporal evolution of Beijing's transit network through data from multiple years, including transportation networks, road networks, and population data. Figure 2 illustrates the specific methodological workflow of this work. Given the heterogeneous forms of data sources, GIS software was employed for data preprocessing tasks such as coordinate calibration and range clipping to ensure data consistency. First, block units were segmented using road networks of different levels in conjunction with Beijing's administrative division data and overlay analysis was performed by intersecting segmented block units with transit stops to obtain coverage statistics within different spatial areas, illustrating the overall evolutionary trends. Second, a transit stop service radius of 500 m was established to create service area polygons for each stop. These were then combined with the population grid data, and spatial analysis tools were used to calculate the population proportion of convenient public transport services for each administrative district and ring road in different years, and the visualization charts were displayed [50]. Then, based on the bus network data, a complex network model was established to identify the corresponding relationships between nodes and edges in the route network, and Gephi was used to calculate the betweenness centrality of nodes and corresponding edges. Lastly, multi-distance spatial cluster analysis tools were applied to reveal anisotropic patterns in transit network density over time. From a spatiotemporal perspective, this study aims to reveal the spatiotemporal evolution of Beijing's bus network, through the coverage rate, density extension, and equity.

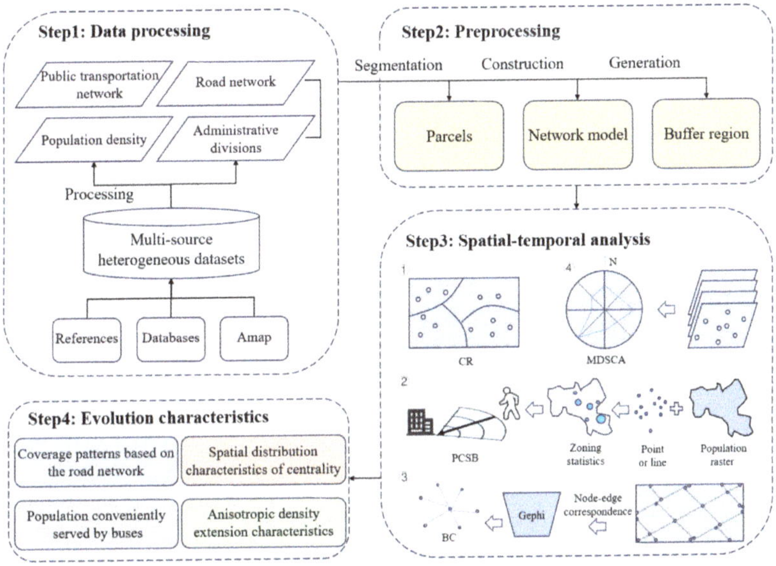

Figure 2. Research framework.

4.1. Coverage Rate (CR)

The coverage rate is an important metric for evaluating the efficiency of public service facilities. It refers to the extent of an area within a certain jurisdiction covered by a service facility, which directly reflects the service range and accessibility of the facility [51]. A higher coverage rate indicates better accessibility, meaning the facility can serve more people or regions [52]. Parcels, as fundamental units in urban research and spatial planning, are widely applied in coverage rate statistics for public service facilities, naturally segmented by road networks within urban areas [51]. This study selects parcels as the smallest unit of analysis and focuses on the coverage rate of bus stops in Beijing.

Beijing's road network is well-structured and rectangular, making parcels an ideal unit for analysis. Based on open-source map data (OSM), parcels are divided according to the administrative area of Beijing [53]. In the segmentation process, main roads, such as highways and arterials, are selected as dividing lines, while minor roads are excluded to prevent excessive segmentation, which could make statistics complicated and distort results. Based on the administrative area of Beijing, road network data are used to divide the area, and elongated parcels under 5000 square meters without significant socioeconomic functions are excluded [54]. The bus stops within each parcel are then counted, and the number is compared to the parcel area to obtain a score for coverage frequency within different parcels.

$$P_i = \frac{N_{covi}}{A_i} \tag{1}$$

P_i represents the coverage rate for a parcel, where i denotes the parcel number, N_{covi} represents the number of bus stops in parcel i, and A_i is the actual area of parcel i.

4.2. PCSB Spatial Statistics

Use software such as Supermap and GIS to create a 500-m buffer zone around bus stops in the public transport network data from 2006 to 2024 and determine the service coverage area for each bus stop. The buffers were overlaid with administrative division data and road network data, ensuring each buffer was assigned to the corresponding district and ring road for more detailed analysis, which facilitated the calculation of PCSB (Population Conveniently Served by Buses). Subsequently, the classified buffers were overlaid with 2020 population data at a 100 m × 100 m resolution to extract population data within each service area, thereby showcasing the PCSB for each region.

To ensure data accuracy, this study incorporates data from the seventh national population census to calibrate the initial population data, refines population details in each service area, and provides a more precise age distribution. By analyzing the calibrated population data, the proportion of PCSB in each service area is determined, areas for improvement are identified, and patterns and potential service gaps in bus accessibility are revealed.

4.3. Betweenness Centrality (BC)

Centrality is an indicator used to assess the importance or influence of nodes within a network. Common types of centralities used to measure network structural characteristics include Degree Centrality, Betweenness Centrality (BC), Closeness Centrality, and Eigenvector Centrality [55]. BC is a measure in complex network theory that identifies key nodes by calculating their frequency on the shortest paths within the network [56]. If a node is on the shortest paths between multiple adjacent nodes and it has a high BC, its significant influence on network efficiency and connectivity is indicated, serving as a crucial hub [46]. In this study, GIS software was first used to combine with bus route data to establish a network dataset spanning from 2006 to 2024. Subsequently, Gephi software was employed to calculate the Betweenness Centrality (BC) of the bus network. This method effectively demonstrates the importance of each node in a complex network.

$$B_i = \left(\sum_{j,k \in V, j \neq k \neq i} \frac{n_{jik}}{n_{jk}} \right) \times \frac{1}{(N-1)(N-2)} \tag{2}$$

B_i represents the BC of a bus route node, where n_{jik} is the number of shortest paths between nodes j and k passing through i, n_{jk} is the total number of shortest paths between route nodes j and k, and N denotes the total number of nodes in the entire bus network.

4.4. Multi-Distance Spatial Clustering Analysis (MDSCA)

Using the two-dimensional distribution information of Beijing's bus stops from 2006 to 2024, Ripley's $K(d)$ function analysis method can quantify the spatial distribution characteristics of the stops. The $K(d)$ function, as a covariance indicator, measures the distribution of distances between stops. As the spatial scale d changes, $K(d)$ reflects the distribution characteristics of stops across different distance ranges [57].

$$K(d) = \left(\frac{A}{n^2}\right)\sum_{i}^{n}\sum_{j}^{n}\frac{l_d(u_{ij})}{W_{ij}}(i \neq j) \tag{3}$$

In this formula, n represents the number of bus stops; d is the spatial distance scale; A is the area within the jurisdiction; u_{ij} is the distance between points i and j, where $u_{ij} > d$, $l_d(u_{ij}) = 0$, and where $u_{ij} \leq d$, $l_d(u_{ij}) = 1$; and W_{ij} represents the ratio of the arc length of a circle centered on i with radius u_{ij} to the circumference within A. The indicator function $l_d(u_{ij})$ is used to evaluate whether the distance between two bus stops falls within the specified threshold.

To balance variance stability, a square root linear transformation using $\frac{K(d)}{\pi}$ corrects the data [58], yielding $L(d)$, as follows:

$$L(d) = \sqrt{\frac{K(d)}{\pi}} - d \tag{4}$$

The clustering status of bus stops is determined by the values of $L(d)$, observed values, and expected values.

5. Results and Analysis

5.1. The Global-Scale Evolution Trend Based on Coverage

The road network density in Beijing shows a trend of gradual expansion from the center outward, with coverage rates positively correlated with the complexity of the road network. The complex topography of Beijing features mountainous areas in the northwest and flat terrain in the southeast. Due to natural terrain influences, the road coverage rate is lower in the northwest, where bus stops are relatively scarce, leading to lower bus coverage compared to the southeast (Figure 3). Especially before 2006, public transportation planning in Beijing focused on the city center, resulting in a dense layout of bus stops within the Second Ring Road, particularly around important commercial and administrative centers, as well as concentrated residential areas. During this period, the distribution of bus stops formed a dense pattern centered around the city, with the Second Ring Road as a boundary. The coverage rate in peripheral areas was significantly lower, particularly in the northwest, where the development of bus stops and roads progressed slowly due to terrain constraints.

As the city developed, bus coverage gradually expanded outward. In 2009 and 2017, Beijing's bus network exhibited a trend of incremental outward diffusion. Coverage rates weakened along the Second Ring, Third Ring, and beyond. By 2024, the bus stop coverage pattern in Beijing formed a concentrated trend centered around the Second Ring Road, and the coverage in peripheral areas remained low, but some regions like Yanqing District had local improvements. The convenience of transportation greatly affects residents' daily travel and has a profound impact on the sustainability of urban development.

5.2. PCSB Analysis

According to the calculated proportions of PCSB in various regions (as shown in Figures 4 and 5), there has been a significant increase from 2006 to 2024 in the proportions of PCSB across Beijing. Notably, areas with high population density, high bus network density, and high public transportation accessibility overlap significantly, indicating that the development of the bus network effectively serves the city's high-density population areas. Taking Shunyi District as an example, the number of PCSB was 239,850 in 2006, which only provided convenient transport for 19.54% of the district's residents. By 2024, this number rose to 915,220, with the proportion significantly increasing to 74.54%. The expansion of the bus network and the improvement in infrastructure greatly enhanced bus convenience in Shunyi District. This trend is also prominent in other districts such as Yanqing, Pinggu, and Fangshan, where bus convenience gradually improved, and population accessibility steadily increased. This growth can be attributed to ongoing improvements in transportation infrastructure, the expansion of bus networks, and the continued development of urban fringe areas. In core areas such as Dongcheng and Xicheng, changes in the proportion of PCSB have been relatively small. In Dongcheng, for example, the proportion was 96.94% in 2006, rising slightly to 98.79% by 2024. These core areas already had a very high coverage rate for bus early on, limiting the potential for further increases. Therefore, the further optimization of bus in core areas mainly focuses on improving service quality rather than expanding coverage.

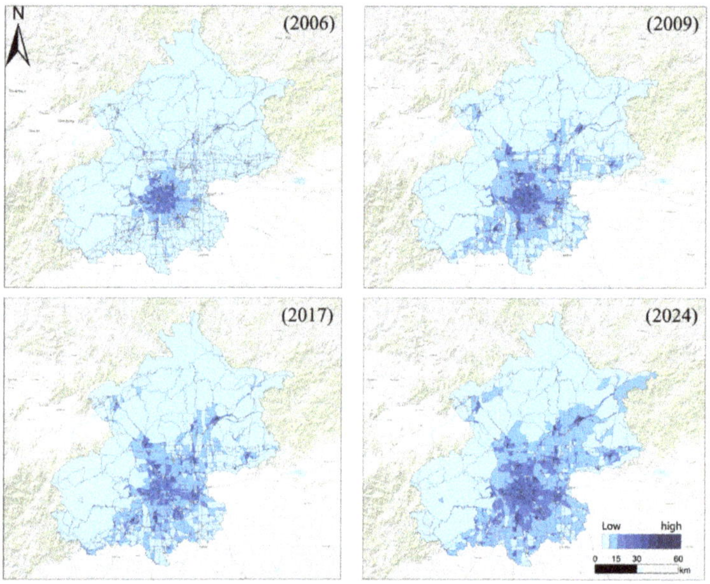

Figure 3. Coverage rate.

Between 2006 and 2009, the number of bus routes and their coverage significantly increased, with many new routes added to the outskirts, optimizing the transportation connections between the core urban areas and peripheral regions. In particular, the convenience of public transportation in the far suburban districts improved dramatically. For example, the proportion of the population with convenient access to public transportation in Changping District rose sharply from 36.33% in 2006 to 71.53% in 2009. This significant growth can be attributed to the acceleration of urbanization and the rapid population increase (according to the 2010 Sixth National Census, Beijing's population was approximately 19.61 million, a net increase of 6.04 million people, or about 44.6%, compared to the Fifth National Census), which drove an expansion of the city's bus capacity. Similar changes occurred in districts like Daxing and Shunyi, where the bus networks in these peripheral areas were notably improved between 2006 and 2009, making public transportation more accessible to a larger population. For core urban areas such as Dongcheng and Xicheng, where the proportion of PCSB was already very high in 2006 (96.94% and 98.66%, respectively), the increase in the PCSB ratio was more limited.

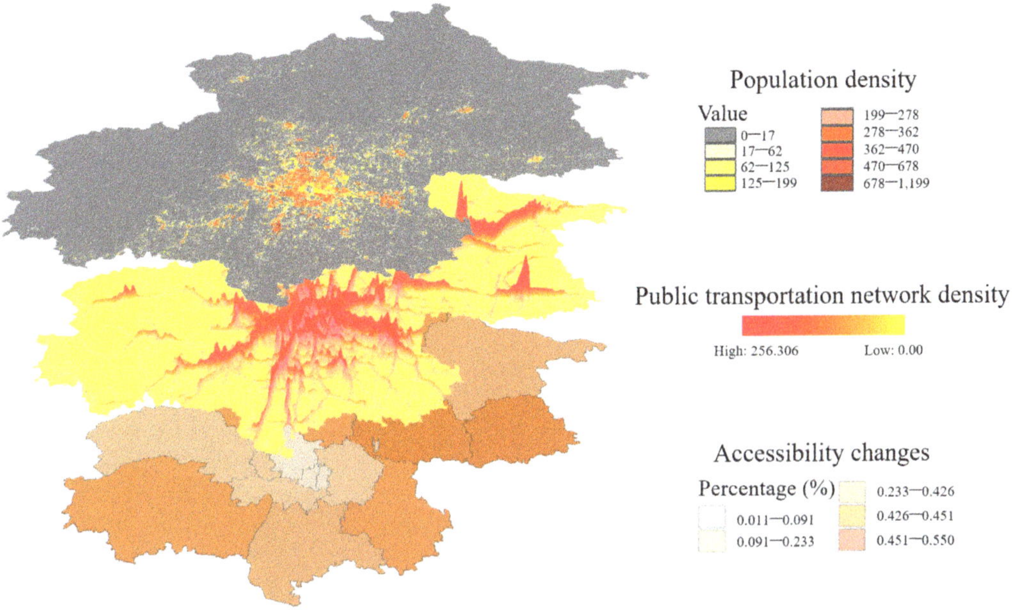

Figure 4. Overall characteristics of PCSB proportions.

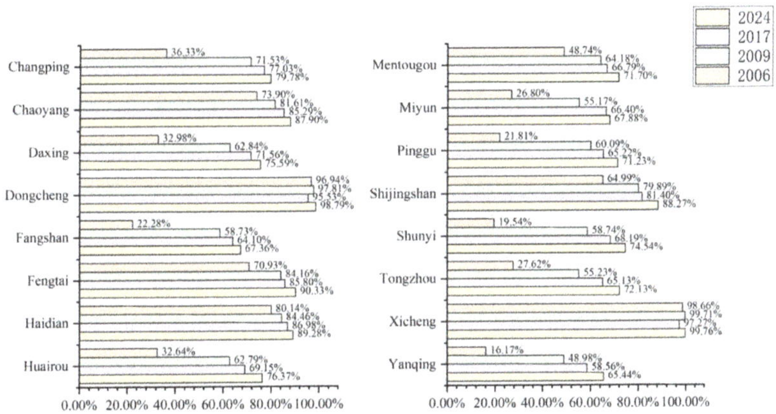

Figure 5. The changes in PCSB proportions across districts and counties.

Beijing's ring road traffic network has had a significant impact on bus convenience. As the city expands year by year, the ring road system gradually extends outward, and the PCSB changes accordingly in diverse areas. The specific changes are shown in Figure 6, where the bar chart on the left axis represents the specific number of PCSB, and the line chart on the right axis reflects the PCSB ratio. The x-axis divides the data by different ring roads, with multiple colors representing different years. From the changes in PCSB across the ring roads, it can be observed that as the ring roads expanded, the population within each ring road increased exponentially, and this growth was positively correlated with the expansion of the ring roads. This is because the addition of ring roads also brought an increase in the number of bus stops within each area, which in turn served more residents. In 2017, the PCSB values in the second to fourth ring roads were relatively low, and it is likely due to the Beijing Municipal Government's official proposal in 2016 for developing Tongzhou District as the city's sub-center. After that, development resources and bus network construction were no longer fully focused on the central urban areas, leading to a decrease in PCSB in that year. Looking at the PCSB ratio across the ring roads, in the city's center, the second ring road already had a highly developed bus network in 2006, in which 98.55% of the population can conveniently access to public transportation. With the continued development of the city and the optimization of public transportation services, this ratio further increased to 99.46% by 2024, almost reaching full coverage. The central urban areas have consistently maintained a high level of service. This reflects the fact that the transportation infrastructure in the core areas has long been well-developed, with a wide and dense bus network that meets the travel needs of the vast majority of residents. However, unlike the central urban areas, the proportion of PCSB decreases as one moves further from the city center. For example, in the Sixth Ring Road, the PCSB ratio was only 71.77% in 2006, much lower than that in the core urban areas. This gap was mainly due to the underdeveloped bus network in the far suburban districts, where bus stops were sparse and travel relied heavily on private vehicles. Over time, as Beijing's urbanization increased, the bus network within the Sixth Ring Road was significantly improved at several key points. By 2024, the PCSB ratio within the Sixth Ring Road had increased to 89.56%. This change reflects the gradual improvement in the bus network in the far suburbs, making public transport more accessible to a larger population and significantly enhancing transportation convenience. Despite these improvements, the issue of uneven public transportation development between the city's edge and the core urban areas persists.

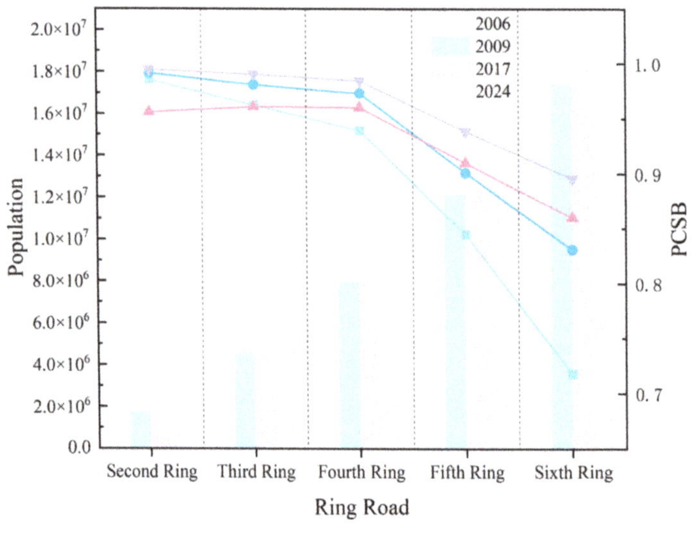

Figure 6. The changes in PCSB proportions across different ring roads.

Overall, the number and proportion of PCSB are negatively correlated. This is because, as the ring roads expand, the number of bus stops and routes within their coverage increases, which in turn serves a larger population and contributes to the PCSB count. However, the total population within the area also increases, and since the overall population is much larger than the PCSB count, the two factors are negatively correlated. The development of bus networks in Beijing's central urban areas and far suburban districts has improved. By 2024, the bus network has gradually expanded to the Sixth Ring Road, significantly enhancing bus convenience in the Sixth Ring and surrounding areas. This also reduced the gap slightly with the central urban areas and improved the travel quality for residents in these regions. This reflects Beijing's efforts to optimize transportation services in the outer districts through policy support and infrastructure development while expanding the city's boundaries. The number of bus routes, the density of bus stops, and the connections between the far suburban districts and the core urban areas have significantly improved, leading to notable enhancements in bus convenience and accessibility. Although the overall bus network in Beijing has developed positively, and the PCSB ratio in peripheral areas has gradually increased, there remains a certain gap between the far suburban districts and the core urban areas. The PCSB proportion in several far suburban districts is still around 60% (e.g., in districts like Miyun and Fangshan), while the core urban areas maintain a high ratio of over 90%. This uneven development somewhat limits the balanced growth of the city, impacting the quality of life and economic development potential in the far suburban districts. Given the noticeable spatial gradient in the distribution of bus networks and public service accessibility, further investment in transportation infrastructure is needed in the far suburbs to narrow the gap in transportation convenience with the city center.

5.3. BC Spatial Distribution Characteristics

As shown in Figure 7, panels a, b, c, and d represent the BC maps for the years 2006 to 2024, and panel e shows the topographic map of Beijing. In 2006, areas with high BC in Beijing's bus network were mainly concentrated in Chaoyang, Haidian, Xicheng, and Dongcheng. These areas not only serve as the city's commercial hubs but also handle significant commuting traffic, making them central nodes in the bus network. High BC means that these nodes are well-connected and accessible within the network, effectively supporting the transportation needs of other regions. Chaoyang and Haidian, as two of Beijing's main economic centers, are the core of numerous office buildings, commercial complexes, and educational institutions, attracting a high volume of people. Meanwhile, Xicheng and Dongcheng, as Beijing's traditional administrative centers, have a relatively well-developed bus system that plays a key role in the city's bus network. The high BC of these areas highlights their critical position within the network and reflects structural characteristics in Beijing's urban development.

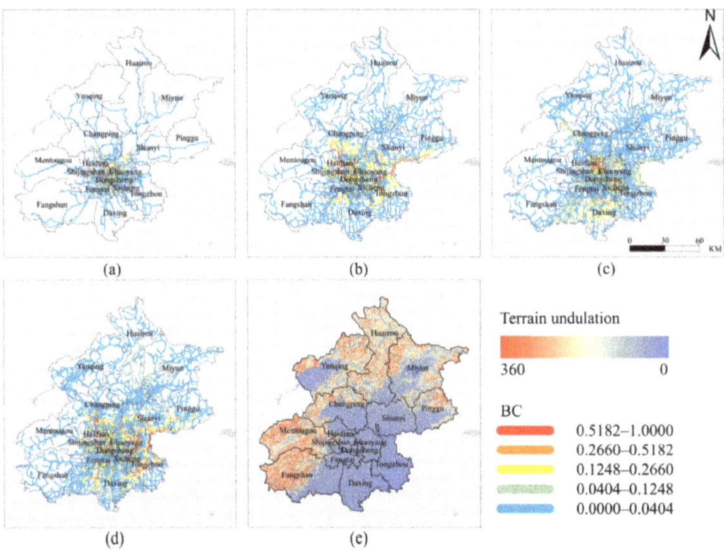

Figure 7. Betweenness centrality and Terrain Relief in Beijing (**a**) BC in 2006; (**b**) BC in 2009; (**c**) BC in 2017; (**d**) BC in 2024; (**e**) Terrain Relief.

By 2017, with the upgrading of public transportation facilities, Tongzhou District's BC steadily increased, indicating that the region's importance and influence within the bus network were growing. The strong connectivity between Tongzhou and core areas like Chaoyang and Dongcheng has not only enhanced commuting convenience but also boosted economic and population mobility, injecting new vitality into the eastern part of Beijing. In contrast, Fangshan District, located in Beijing's southwest, has relatively low BC in the bus network due to its geographic location and natural conditions. Fangshan's complex terrain—with plains in the south and mountainous areas in the west and northwest—presents challenges for infrastructure development, making road construction and bus route layout difficult. Furthermore, the district's relatively low population density and economic activity have limited the pace and extent of bus network expansion. Unlike the flatter southeastern areas, Fangshan's challenging topography and slower development have left it in a peripheral position within Beijing's bus network. Thus, the low BC in Fangshan's bus network indicates the area's limited capacity to serve as a multi-regional transportation hub.

The distribution and BC shifts within Beijing's bus network also reflect the optimization of urban spatial layout. In recent years, the Beijing municipal government has promoted spatial rebalancing of urban functions through policy guidance and resource allocation, gradually advancing the development of the sub-center and peripheral areas. With increasing BC in areas like Tongzhou, Beijing's overall transportation network is transitioning from a single-center structure reliant on core hubs to a network structure shared by multiple central hubs. This trend enhances inter-regional connectivity, promotes rational urban function distribution, and establishes a solid foundation for achieving urban planning goals aimed at "relieving pressure on central urban areas and fostering regional coordinated development".

5.4. Directional Density Extension Characteristics

By analyzing the directional density extension characteristics of bus stops in Beijing, a clustering pattern in the distribution of bus stops can be observed, obtaining variations across different time frames and distances. The red curve (Observed K) represents observed values, while the blue curve (Expected K) represents expected values (Figure 8a). Overall, the values of the Observed K curve are generally higher than those of the Expected K curve, indicating a clear clustering pattern rather than a random or dispersed distribution of bus stops in Beijing. This distribution pattern reflects a centralization trend in the layout of bus stops, influenced by urban planning and actual demand. The difference between observed and expected values increases with distance initially, then decreases. The degree of clustering of bus stops varies with distance: it is lower at shorter distances and gradually increases along with distance before decreasing slightly, showing distinctive characteristics across spatial scales.

In 2006, 2009, 2017, and 2024, the differences between observed and expected values reached their maximum at distances of 26,043 m, 36,026 m, 38,150 m, and 36,750 m, respectively, indicating the strongest clustering characteristics at these points. In 2009, 2017, and 2024, the maximum differences all occurred between 36,000 and 38,200 m, a range where the clustering characteristic was most pronounced, reflecting dynamic adjustments in bus stop layout in line with Beijing's urban expansion and development. From 2006 to 2024, bus stop density extended outward each year, gradually covering a wider area (Figure 8b). However, this expansion was uneven, with notable directional differences in density growth. The southern and southeastern directions became the main areas of bus stop expansion, while the western and northern directions had relatively slower network expansion. The density of bus stops in the southern direction increased from 0.95 in

2006 to 2.59 in 2024, marking it as the highest-density area, highlighting a concentrated trend in bus development in this region. In the southeast, density also rose significantly, from 0.25 to 1.54. In contrast, growth in the western and northern directions was slower, with densities rising from 0.05 and 0.06 to 0.51 and 0.48, respectively. This uneven spatial distribution reflects an imbalance in resource allocation during Beijing's urban expansion process. Rapid urbanization in the southeast has led to faster infrastructure development, which significantly increases the coverage and density of bus stops. However, the western and northern areas, limited by geographic and resource constraints, experienced relatively slower network expansion, resulting in lower clustering and accessibility of bus stops.

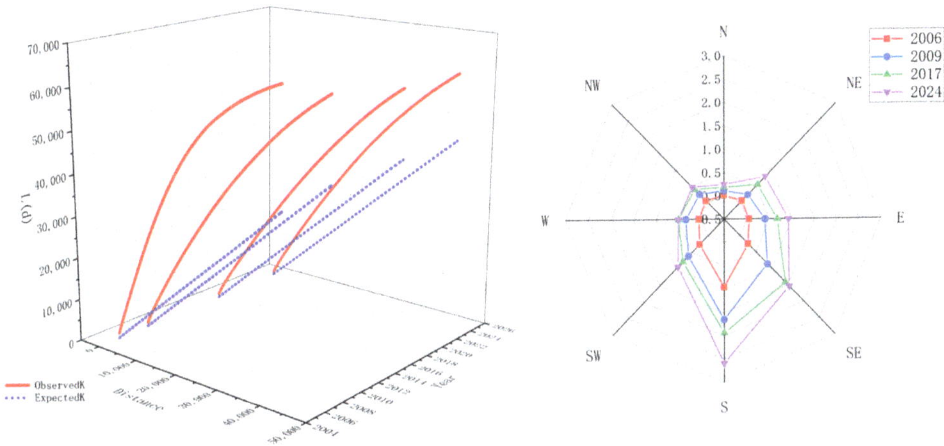

(a) (b)

Figure 8. Directional density extension characteristics (**a**) Change with increasing distance; (**b**) Change with direction.

6. Discussion

After years of urbanization, Beijing's bus network has become a uniquely complex network system, making it a typical case of rapid transportation network development. Although previous studies have delved into various aspects of Beijing's bus network, such as node characteristics, topological properties, and urban demand, which focus on its overall static structural features [46,59], this study further combines dynamic multi-period and multi-source data to explore the spatiotemporal evolution characteristics of Beijing's bus network from a spatiotemporal perspective. By analyzing data from 2006 to 2024, we found a clear clustering tendency of the bus network in central urban areas (Figure 3), while peripheral urban areas have shown improvement in recent years. However, compared to the central area, a significant disparity remains in the overall service level, and the network coverage and service levels across various peripheral regions are greatly different (Figure 5), particularly in areas like Tongzhou and Yanqing, which are influenced by key events.

As discussed earlier, the central areas of Beijing exhibit a high clustering of the bus network, with various indices reaching high values, and ring roads significantly impacting network clustering (Figure 7). This finding aligns with previous research. Dai et al. pointed out that central regions like Chaoyang, Haidian, Dongcheng, and Xicheng show high network BC, highlighting the core position of the central urban transportation network [60]. Liu et al. found an oversupply of transportation within the Second Ring Road area and an undersupply in areas beyond the Sixth Ring Road and suburban regions, reflecting an imbalance in transportation resource allocation [61]. Zhang et al., through analyzing the connectivity and service capacity of transportation nodes, identified key transportation network nodes in the Beijing-Tianjin-Hebei region, further highlighting the strong clustering characteristic of Beijing's bus network in the central region [62]. These findings reveal the complex interaction between the population, terrain, and evolution of the bus network, assessing the inter-regional imbalance in transportation resource allocation and providing a reference value and theoretical support for the sustainable development of public transportation in Beijing.

6.1. Suggestions

In light of the issues within Beijing's bus network, we propose the following recommendations. (1) Although the bus network in Beijing's central area is well-established, peripheral districts, especially those beyond the Sixth Ring Road, have relatively sparse bus stop layouts, leading to lower public transit convenience. Therefore, priority should be given to the development of public transportation in peripheral areas, reducing regional disparities, and improving the coverage and service level of bus networks in these areas. This can be achieved by increasing the number of bus lines and optimizing station distribution, particularly in areas with high residential density and proximity to important public facilities. Optimizing the connection with the city center and adding rapid bus lines can improve the travel efficiency of residents in peripheral districts, promoting the balanced development of transportation resources across the city and enhancing the overall livability of the city. (2) Bus network expansion and optimization should consider the demand generated by major events, such as the Winter Olympics and international conferences. Beijing's experience shows that infrastructure improvements from major events can significantly enhance regional bus networks. Similar opportunities in the future should be used to promote overall regional development while ensuring that the facilities established during these events continue to serve residents' travel needs in the long term. In future urban development, public transit planning and adjustments should be aligned with the development strategies of emerging areas like Beijing's sub-center in Tongzhou and new urban areas in Changping, ensuring that the bus network in these regions keeps pace with urban expansion. (3) In the northwestern mountainous and complex terrain areas of Beijing (such as Fangshan), the development of public transportation has been relatively lagging, but these areas also have significant ecological protection value. Due to the complex terrain, the coverage of public transportation is low. Therefore, when planning the bus network in these areas, it is recommended to incorporate green transportation concepts and prioritize the development of low-carbon, which is the environmentally friendly mode of transport. This could include the introduction of small electric buses, customized buses, and shared transport tools to improve transportation convenience while minimizing the environmental impact and promoting harmonious coexistence between the city and nature. Additionally, considering factors such as population, ecology, and terrain, the bus routes should be planned rationally to minimize construction costs and enhance the green accessibility of public transportation. (4) For core urban areas (e.g., Dongcheng and Xicheng), where bus stop coverage is nearly saturated, future efforts should focus more on improving the quality of bus services. Measures like increasing service frequency, improving vehicle comfort, and promoting new energy buses can enhance the service level of the bus system, reduce congestion, and improve the travel experience. Further optimization of connections with key commercial and administrative areas would provide efficient and convenient bus services to residents.

6.2. Limitations of This Study and Future Research

This study has some limitations. The study period is limited to data from 2006 to 2024, offering a relatively short period; it does not analyze multiple influencing factors to fully interpret the evolution characteristics of the bus network and lacks suggestions for optimization. Future research should consider bus network data over a longer time frame to explore the influencing factors and mechanisms behind the bus network's evolution in greater depth, providing better guidance for urban public transit planning and sustainable development. For example, spatial optimization could combine intelligent transportation and big data technology to further optimize the layout of bus routes and stops, which can improve the efficiency and service quality of public transportation and meet residents' travel needs. The more comprehensive research can offer a valuable empirical foundation for understanding the evolution and future trends of Beijing's bus network, contributing to sustainable urban transportation planning and management.

7. Conclusions

This study uses spatial analysis methods to examine the bus network across multiple time points, incorporating population distribution information and employing various spatial analysis techniques to investigate the evolution patterns and outcomes of Beijing's bus network. The evolution of the bus network is closely linked to the relationship between stops and routes, providing essential insights for urban transportation planning and public transit system optimization.

The main conclusions from our findings are as follows:

(1) The number of bus stops and routes has increased year by year, with the network structure becoming more complex, showing a trend of expansion from central to peripheral areas. The coverage of bus stops in Beijing has expanded outward with the increasing complexity of the road network. Road coverage is relatively low in the northwest, where bus stops are sparse, resulting in lower public transportation coverage than in the southeast. The proportion of PCSB in the central area remains relatively stable, while the suburbs indicate significant variation, with change rates exceeding 50% in some areas;

(2) The diffusion characteristics of the bus network are significantly influenced by administrative divisions and ring road layout, with notable expansion in areas like Tongzhou and Yanqing. Bus routes and stops are highly concentrated within the central urban area and the Second Ring Road, with various indices, gradually decreasing as ring roads increase. Over time, the betweenness centrality of routes and stops tends toward regional balance, with an increasing number of central hubs, and Tongzhou is gradually becoming a new transportation hub. The road network-based stop coverage in Yanqing has also markedly improved;

(3) The distribution of bus stations shows significant clustering characteristics, and the directional development is unbalanced. The distribution of bus stations in Beijing shows obvious clustering, particularly in the southeastern areas where urbanization has rapidly developed, leading to a sharp increase in bus station density. In contrast, due to geographical constraints, the expansion of the bus network in the western and northern areas is relatively lagging, which has a slow density growth. Compared to the outer suburban districts, high PCSB value areas are notably concentrated in the central urban areas, with a significant gap of around 30%. This spatial distribution discrepancy reflects the imbalance in resource allocation and urbanization processes in Beijing.

Author Contributions: Conceptualization, Xiao Li and Shaohua Wang; methodology, Xiao Li and Chang Liu; software, Shaohua Wang; validation, Liang Zhou and Shaohua Wang; formal analysis, Liang Zhou; investigation, Yeran Sun; resources, Yeran Sun; data curation, Cheng Su; writing—original draft preparation, Xiao Li; writing—review and editing, Xiao Li, Junyuan Zhou, and Dachuan Xu; supervision, Jiayi Zheng; project administration, Shaohua Wang; funding acquisition, Shaohua Wang. All authors have read and agreed to the published version of the manuscript.

Funding: This study was financially supported by the National Key R&D Program of China (Grant No. 2023YFF0805904), the Talent Introduction Program Youth Project of the Chinese Academy of Sciences (E2Z10501), and the innovation group project of the Key Laboratory of Remote Sensing and Digital Earth Chinese Academy of Sciences (E33D0201-5).

Data Availability Statement: The analysis datasets for the current study are available from the corresponding author upon reasonable request.

Conflicts of Interest: The authors declare no conflicts of interest.

References

1. Schmidt-Traub, G.; Kroll, C.; Teksoz, K.; Durand-Delacre, D.; Sachs, J.D. National baselines for the Sustainable Development Goals assessed in the SDG Index and Dashboards. *Nat. Geosci.* **2017**, *10*, 547–555. [CrossRef]
2. Caudill, C.M.; Pulsifer, P.L.; Thumbadoo, R.V.; Taylor, D.R.F. Meeting the Challenges of the UN Sustainable Development Goals through Holistic Systems Thinking and Applied Geospatial Ethics. *ISPRS Int. J. Geo-Inf.* **2024**, *13*, 110. [CrossRef]
3. Liu, X.; Pei, F.; Wen, Y.; Li, X.; Wang, S.; Wu, C.; Cai, Y.; Wu, J.; Chen, J.; Feng, K.; et al. Global urban expansion offsets climate-driven increases in terrestrial net primary productivity. *Nat. Commun.* **2019**, *10*, 5558. [CrossRef]
4. Melnikova, I.; Yokohata, T.; Ito, A.; Nishina, K.; Tachiiri, K.; Shiogama, H. Emergent constraints on future Amazon climate change-induced carbon loss using past global warming trends. *Nat. Commun.* **2024**, *15*, 7623. [CrossRef]
5. Wang, Y.; Sarkar, A.; Hasan, A.K.; Tian, Y.; Wu, Q.; Hossain, M.S.; Wei, F. The evaluation of temporal and spatial trends of global warming and extreme ocean surface temperatures: A case study of Canada. *ISPRS Int. J. Geo-Inf.* **2021**, *11*, 21. [CrossRef]
6. Ercan, T.; Onat, N.C.; Tatari, O. Investigating carbon footprint reduction potential of public transportation in United States: A system dynamics approach. *J. Clean. Prod.* **2016**, *133*, 1260–1276. [CrossRef]
7. Almulhim, A.I.; Sharifi, A.; Aina, Y.A.; Ahmad, S.; Mora, L.; Filho, W.L.; Abubakar, I.R. Charting sustainable urban development through a systematic review of SDG11 research. *Nat. Cities* **2024**, *1*, 677–685. [CrossRef]
8. Wang, W.; Wang, Y.; Correia, G.H.A.; Chen, Y. A Network—Based Model of Passenger Transfer Flow between Bus and Metro: An Application to the Public Transport System of Beijing. *J. Adv. Transp.* **2020**, *2020*, 6659931. [CrossRef]
9. Liu, L.; Yu, H.; Zhao, J.; Wu, H.; Peng, Z.; Wang, R. Multiscale effects of multimodal public facilities accessibility on housing prices based on MGWR: A case study of Wuhan, China. *ISPRS Int. J. Geo-Inf.* **2022**, *11*, 57. [CrossRef]
10. Xu, Y.; Olmos, L.E.; Abbar, S.; González, M.C. Deconstructing laws of accessibility and facility distribution in cities. *Sci. Adv.* **2020**, *6*, eabb4112. [CrossRef]
11. Zhang, L.; Yuan, J.; Gao, X.; Jiang, D. Public transportation development decision-making under public participation: A large-scale group decision-making method based on fuzzy preference relations. *Technol. Forecast. Soc. Change* **2021**, *172*, 121020. [CrossRef]
12. Bi, S.; Hu, J.; Shao, L.; Feng, T.; Appolloni, A. Can public transportation development improve urban air quality? Evidence from China. *Urban Clim.* **2024**, *54*, 101825. [CrossRef]
13. Tranos, E.; Mack, E. Big data: A new opportunity for transport geography? *J. Transp. Geogr.* **2019**, *76*, 232–234. [CrossRef]
14. Freudendal-Pedersen, M. Sustainable urban futures from transportation and planning to networked urban mobilities. *Transp. Res. Part D Transp. Environ.* **2020**, *82*, 102310. [CrossRef]
15. Vuchic, V.R. *Urban Public Transportation Systems;* University of Pennsylvania: Philadelphia, PA, USA, 2002; Volume 5, pp. 2532–2558.
16. Chen, N.; Liu, S.; Wang, S. Spatial Evolution Features of Transit Bus: An Empirical Study of Guangzhou. *Sci. Geogr. Sin.* **2017**, *37*, 55–63. [CrossRef]
17. Newell, G.F. Dispatching policies for a transportation route. *Transp. Sci.* **1971**, *5*, 91–105. [CrossRef]
18. Glaister, S. Competition on an urban bus route. *J. Transp. Econ. Policy* **1985**, *19*, 65–81.
19. Bly, P.H.; Oldfield, R.H. Competition between minibuses and regular bus services. *J. Transp. Econ. Policy* **1986**, *20*, 47–68.
20. Gao, B.; Yang, J.; Chen, Z.; Sugihara, G.; Li, M.; Stein, A.; Kwan, M.; Wang, J. Causal inference from cross-sectional earth system data with geographical convergent cross mapping. *Nat. Commun.* **2023**, *14*, 5875. [CrossRef] [PubMed]

21. Zhang, C.; Lin, H.; Chen, M.; Zhang, X.; Li, R.; Ding, Y. A modelling system with adjustable emission inventories for cross-boundary air quality management in Hong Kong and the Pearl River Delta, China. *Comput. Environ. Urban Syst.* **2017**, *62*, 222–232. [CrossRef]
22. Ma, J.; Yang, Y.; Guan, W.; Wang, F.; Liu, T.; Tu, W.; Song, C. Large-Scale Demand Driven Design of a Customized Bus Network: A Methodological Framework and Beijing Case Study. *J. Adv. Transp.* **2017**, *2017*, 3865701. [CrossRef]
23. Wang, Y.; Deng, Y.; Ren, F.; Zhu, R.; Wang, P.; Du, T.; Du, Q. Analysing the spatial configuration of urban bus networks based on the geospatial network analysis method. *Cities* **2020**, *96*, 102406. [CrossRef]
24. Li, J.; Guo, Q.S. Quantitative representing complexity of urban traffic network base on GIS. *J. Cent. China Norm. Univ. (Nat. Sci.)* **2002**, *4*, 534–537. [CrossRef]
25. Tung, C.L.; He, S.; Mei, L.; Zhang, H. Exploring the influence of transportation on urban spatial structure using the spatial Durbin model: Evidence from 265 prefecture-level cities in China. *Comput. Urban Sci.* **2024**, *4*, 5. [CrossRef]
26. Sienkiewicz, J.; Hołyst, J.A. Statistical analysis of 22 public transport networks in Poland. *Phys. Rev. E* **2005**, *72*, 046127. [CrossRef]
27. Gao, Z.; Li, K.; Li, X.; Huang, H.; Mao, B.H. Scaling laws of the network traffic flow. *Phys. A Stat. Mech. Its Appl.* **2007**, *380*, 577–584. [CrossRef]
28. Sui, Y.; Shao, F.; Sun, R.; Li, S. Space evolution model and empirical analysis of an urban public transport network. *Phys. A Stat. Mech. Its Appl.* **2012**, *391*, 3708–3717. [CrossRef]
29. Wang, S. Research on Evolution and Characteristics of Urban Traffic Network—Case of Beijing. 2013. Available online: https://webvpn.lzjtu.edu.cn/https/494a553139386968732a235e35546e2834614a7f1404eb0e2856ebc694a6fb2fdd4ab9ef/thesis/Y2430285 (accessed on 18 October 2023).
30. Bin Asad, K.M.; Yuan, Y. The impact of scale on extracting urban mobility patterns using texture analysis. *Comput. Urban Sci.* **2023**, *3*, 33. [CrossRef]
31. Abenoza, R.F.; Cats, O.; Susilo, Y.O. Travel satisfaction with public transport: Determinants, user classes, regional disparities and their evolution. *Transp. Res. Part A Policy Pract.* **2017**, *95*, 64–84. [CrossRef]
32. Yu, W.; Chen, J.; Yan, X. Space–time evolution analysis of the Nanjing metro network based on a complex net-work. *Sustainability* **2019**, *11*, 523. [CrossRef]
33. Gu, Y.; Fu, X.; Liu, Z.; Xu, X.; Chen, L. Performance of transportation network under perturbations: Reliability, vulnerability, and resilience. *Transp. Res. Part E Logist. Transp. Rev.* **2020**, *133*, 101809. [CrossRef]
34. Serdar, M.Z.; Koç, M.; Al-Ghamdi, S.G. Urban transportation networks resilience: Indicators, disturbances, and assessment methods. *Sustain. Cities Soc.* **2022**, *76*, 103452. [CrossRef]
35. Xu, Z.; Chopra, S.S. Interconnectedness enhances network resilience of multimodal public transportation systems for Safe-to-Fail urban mobility. *Nat. Commun.* **2023**, *14*, 4291. [CrossRef]
36. Liu, C.; Bardaka, E. The suburbanization of poverty and changes in access to public transportation in the Triangle Region, NC. *J. Transp. Geogr.* **2021**, *90*, 102930. [CrossRef]
37. Fan, W.; Zhang, J.; Wang, J.; Shan, B. Traffic Network Pattern and Optimization of the Three Economic Circles in Shandong Province from the Perspective of Space Syntax. *Geogr. Geo-Inf. Sci.* **2023**, *39*, 55–61. Available online: https://www.geores.com.cn/dlydlxxkx/CN/Y2023/V39/I01/55 (accessed on 1 August 2024).
38. Lee, S.; Kim, S.; Kim, J. Robustness Analysis of Public Transportation Systems in Seoul Using General Multilayer Network Models. *J. Supercomput.* **2024**, *80*, 26589–26613. [CrossRef]
39. De Bona, A.A.; de Oliveira Rosa, M.; Fonseca, K.V.O.; Lüders, R. A reduced model for complex network analysis of public transportation systems. *Phys. A Stat. Mech. Its Appl.* **2021**, *567*, 125715. [CrossRef]
40. Dadhich, P.N.; Hanaoka, S. Spatial investigation of the temporal urban form to assess impact on transit services and public transportation access. *Geo-Spat. Inf. Sci.* **2012**, *15*, 187–197. [CrossRef]
41. Song, X.; Zhang, J.; AghaKouchak, A.; Sen Roy, S.; Xuan, Y.; Wang, G.; He, R.; Wang, X.; Liu, C. Rapid urbanization and changes in spatiotemporal characteristics of precipitation in Beijing metropolitan area. *J. Geophys. Res. Atmos.* **2014**, *119*, 11250–11271. [CrossRef]
42. Yu, Q.; Gu, Y.; Yang, S.; Zhou, M. Discovering Spatiotemporal Patterns and Urban Facilities Determinants of Cycling Activities in Beijing. *J. Geovis. Spat. Anal.* **2021**, *5*, 16. [CrossRef]
43. Hu, R.; Zhang, C. Research on Urban Evolution in China from an Ecological Perspective: A Case of Beijing. *J. Urban Plan. Dev.* **2023**, *149*, 05022047. [CrossRef]
44. Kong, L. Cultural icons and urban development in Asia: Economic imperative, national identity, and global city status. *Political Geogr.* **2007**, *26*, 383–404. [CrossRef]
45. The People's Government of Beijing Municipality. 20 October 2024. The People's Government of Beijing Municipality. Available online: https://www.beijing.gov.cn/gate/big5/www.beijing.gov.cn/ywdt/gzdt/202410/t20241010_3915407.html (accessed on 28 October 2024).

46. Du, C.; Wang, J.; Liu, B.; Huang, D. Impacts of street and public transport network centralities on housing rent: A case study of Beijing. *Prog. Geogr.* **2019**, *38*, 1831–1842. [CrossRef]
47. Zhan, D.; Zhang, W.; Chen, L.; Yu, X.; Dang, Y. Research progress and its enlightenment of urban public service facilities allocation. *Prog. Geogr.* **2019**, *38*, 506–519. [CrossRef]
48. Chen, Y.; Xu, C.; Ge, Y.; Zhang, X.; Zhou, Y. A 100-m gridded population dataset of China's seventh census using ensemble learning and geospatial big data. *Earth Syst. Sci. Data Discuss.* **2024**, *16*, 3705–3718. [CrossRef]
49. Zhao, P.; Jia, T.; Qin, K.; Shan, J.; Jiao, C. Statistical analysis on the evolution of OpenStreetMap road networks in Beijing. *Phys. A Stat. Mech. Its Appl.* **2015**, *420*, 59–72. [CrossRef]
50. Zhong, E. Deep Mapping—A Critical Engagement of Cartography with Neuroscience. *Geomat. Inf. Sci. Wuhan Univ.* **2022**, *47*, 1988–2002. [CrossRef]
51. Murray, A.T.; O'Kelly, M.E.; Church, R.L. Regional service coverage modeling. *Comput. Oper. Res.* **2008**, *35*, 339–355. [CrossRef]
52. Yao, J.; Zhang, X.; Murray, A.T. Location optimization of urban fire stations: Access and service coverage. *Comput. Environ. Urban Syst.* **2019**, *73*, 184–190. [CrossRef]
53. Liu, X.; Long, Y. Automated identification and characterization of parcels with OpenStreetMap and points of interest. *Environ. Plan. B Plan. Des.* **2016**, *43*, 341–360. [CrossRef]
54. Chen, W.; Huang, H.; Dong, J.; Zhang, Y.; Tian, Y.; Yang, Z. Social functional mapping of urban green space using remote sensing and social sensing data. *ISPRS J. Photogramm. Remote Sens.* **2018**, *146*, 436–452. [CrossRef]
55. Chen, Y.; Chen, Y.; Yin, G.; Song, C.; Hou, Y. The Influence of Road Network Centrality on the Spatial Distribution of Catering: A Case Study of Qingdao Main Urban Area. *Sci. Geogr. Sin.* **2022**, *42*, 1609–1618. [CrossRef]
56. Zhang, B.; Cheng, S.; Wang, P.; Lu, F. Inferring freeway traffic volume with spatial interaction enhanced betweenness centrality. *Int. J. Appl. Earth Obs. Geoinf.* **2024**, *129*, 103818. [CrossRef]
57. Liao, Z.; Zhang, L. Spatial distribution characteristics and accessibility analysis of characteristic towns in Guangdong Province based on Ripley's K function. *J. Math.* **2022**, *2022*, 2873707. [CrossRef]
58. Ge, Y.; Liu, W.; Li, Y. Algorithm Modification and Estimation Comparison of Ripley's K-function. *Geomat. Inf. Sci. Wuhan Univ.* **2023**, *48*, 970–978. [CrossRef]
59. Zheng, X.; Chen, J.; Shao, J.; Bie, L. Analysis on topological properties of Beijing urban public transit based on complex network theory. *Acta Phys. Sin.* **2012**, *61*, 95–105. [CrossRef]
60. Dai, T.; Ding, T.; Liu, Q.; Liu, B. Node centrality comparison between bus line and passenger flow networks in Beijing. *Sustainability* **2022**, *14*, 15454. [CrossRef]
61. Liu, J.; Meng, B.; Xu, J.; Li, R. Exploring Public Transportation Supply–Demand Structure of Beijing from the Perspective of Spatial Interaction Network. *ISPRS Int. J. Geo-Inf.* **2023**, *12*, 213. [CrossRef]
62. Zhang, Y.; Zheng, S.; Chen, Y. Identification of key nodes in comprehensive transportation network: A case study in Beijing-Tianjin-Hebei urban agglomeration, china. *Transp. Res. Rec.* **2024**, *2678*, 827–840. [CrossRef]

International Journal of *Geo-Information*

MDPI

Article

RouteLAND: An Integrated Method and a Geoprocessing Tool for Characterizing the Dynamic Visual Landscape Along Highways

Loukas-Moysis Misthos * and Vassilios Krassanakis

Department of Surveying & Geoinformatics Engineering, University of West Attica, Egaleo Park Campus, Ag. Spyridonos Str., Egaleo, 12243 Athens, Greece; krasvas@uniwa.gr

* Correspondence: lmmisthos@uniwa.gr

Abstract: Moving away from a static concept for the landscape that surrounds us, in this research article, we approach the visual landscape as a dynamic concept. Moreover, we attempt to provide an interconnection between the domains of landscape and cartography by designing maps that are particularly suitable for characterizing the visible landscape and are potentially meaningful for overall landscape evaluation. Thus, the present work mainly focuses on the consecutive computation of vistas along highways, incorporating actual landscape composition—as the landscape is perceived from an egocentric perspective by observers moving along highway routes in peri-urban landscapes. To this end, we developed an integrated method and a Python (version 2.7.16) tool, named "RouteLAND", for implementing an algorithmic geoprocessing procedure; through this geoprocessing tool, sequences of composite dynamic geospatial analyses and geometric calculations are automatically implemented. The final outputs are interactive web maps, whereby the segments of highway routes are characterized according to the dominant element of the visible landscape by employing (spatial) aggregation techniques. The developed geoprocessing tool and the generated interactive map provide a cartographic exploratory tool for summarizing the landscape character of highways in any peri-urban landscape, while hypothetically moving in a vehicle. In addition, RouteLAND can potentially aid in the assessment of existing or future highways' scenic level and in the sustainable design of new highways based on the minimization of intrusive artificial structures' vistas; in this sense, RouteLAND can serve as a valuable tool for landscape evaluation and sustainable spatial planning and development.

Keywords: peri-urban visual landscape; dynamic field of view (*d*FoV); dynamic viewsheds; highway routes; visual significance; landscape composition; geoprocessing automation; web mapping; route characterization; sustainable spatial planning

1. Introduction

The visual landscape is a vital resource for human well-being. People's landscape preferences are crucially and positively affected by the presence of natural landscape elements, such as water (e.g., [1–5]); (woody) vegetation (e.g., [5–8]); land of particular/mountain morphology (e.g., [3,9]); and overall elements of wilderness (e.g., [2]). On the contrary, certain landscape elements associated with the urban settings, such as industrial units, paved roads, electric power lines, and, in general, urban sprawl, arouse negative impressions (e.g., [2,5,10–12]). Yet, not all built elements are negatively associated; for instance, well-preserved rural man-made elements such as villas of vernacular architecture and traditional farm buildings can cause positive impressions [2,13].

Urbanization has profoundly transformed prevailing rural landscapes into urban ones over the last decades, inducing a dramatic alteration of land cover/use composition and structure [14]. Urban and peri-urban landscapes are composed of a great amount of artificial (i.e., non-natural), built elements; thus, natural landscape elements are typically sparse and, therefore, are particularly coveted in such landscape areas. Environmental psychology has long ago shown the positive effects of natural elements on human health and psychological well-being (e.g., [15]). The beneficial, restorative effects of nature, emerging from natural landscapes [6,16,17], are diminished in urban settings; in such settings of the built environment, urban residents tend to be more prone to mental diseases, suffering from increased stress levels [18–20]. As an effect, sustained interaction with the non-natural, built environment leads to the depletion of cognitive, affective, and physiological human functions [19,21].

Since 1950, the rapid growth of the world urban population—from 0.8 billion (30%) to 4.2 bullion (55%) [22]—is an alarming indicator regarding the decreasing exposure of human population percentages to natural landscape elements. By 2050, approximately 6.7 billion people (68%) will be estimated to live in urban settlements [22]—a fact implying a further reduction in the world population percentage that is expected to be exposed to natural landscapes.

One important aspect when approximating the visual urban and, particularly, peri-urban landscapes refers to the fact that a large number of people move or drive in certain linear routes, i.e., roads and highways, on an everyday basis. The peri-urban landscapes that are being produced due to increasing urban sprawl tend to be increasingly "cluttered, unattractive, and monotonous" [12]. In this sense, highways can be considered as critical markers of landscape transformation in peri-urban regions, directly influencing viewers' visual perception and experience [12]. Given the dire necessity of urban residents to have contact with nature and natural elements in order to lead a healthy life [23], it is essential to study the visual experience of observers while moving in such routes.

The visual landscape can be modeled under two perspectives: either from the human (ground) standpoint, i.e., the way people experience landscape as they stand, move, etc., or in the way a landscape is cartographically represented, i.e., in 2D maps [24]. The first perspective constitutes the *egocentric*, while the second one constitutes the *exocentric* perspective (e.g., [24–29]). Furthermore, there is a crucial need to quantitatively describe not only the static but also the dynamic perspective when perceiving the visible landscape, especially along linear components, such as highways [12,30]. However, several issues are raised when it comes to the integration of the two available perspective modes of the visual landscape, especially when the dynamic visual landscape is to be modeled (see Sections 2.1 and 2.2 for a more thorough investigation).

In this research work, our endeavor focuses on how to overcome these challenges from a conceptual but chiefly technical point of view. Hence, this article deals with modeling the dynamic landscape along highway routes from a human perspective. In this sense, it mainly aims at the development of both a method and a software tool for automatically and objectively characterizing highway routes depending on the (vehicles') movement speed, the dynamic (human) field of view (*d*FoV) at all segments of the highway routes, and the actual relative contribution (in %) of each landscape element within the *d*FoV. Aside from developing and implementing this method, the article also aims at providing as final outputs interactive web maps visualizing the prevailing landscape elements within the *d*FoV attributed to all predefined segments of highway routes. In this context, a use case of the method and the tool is provided that refers to a selected study area. This case study does not lie at the core of this research work; it has been selected to display and showcase the usefulness and the usability of the developed geoprocessing tool in a peri-urban region.

2. Background and Related Work

2.1. Describing the Perceived Size of Landscape Elements from the Egocentric (Human) Perspective: Concepts and Issues

Describing the visual landscape experience as apperceived by humans, i.e., from the egocentric perspective, is a core matter in landscape studies [28,29,31–33]. Various methods, techniques, and technologies have been employed for registering/recording and measuring observers' landscape visual perception, such as eye tracking, EEG, and fMRI (see [24] for a list of research studies). However, these approaches, no matter how valid they indeed are, are 'plagued' with subjectivity, in that the apperceived landscape impressions—in both behavioral (i.e., gaze patterns), and neuronal/cognitive (i.e., brain activity) terms—may significantly vary across potential observers. On the other hand, typical exocentric perspectives objectively represent the landscape but do not involve the observer, since they constitute vertical views of an area from above.

Therefore, having as a starting point the egocentric perspective, while moving a step back, a prerequisite is to define what parts of the landscape and which landscape elements *can be* visually perceived within a potential observer's field of view (e.g., [28,31]). The quantitative description of the landscape elements that lie within the human field of view (hFoV) at any position (viewpoint) is a crucial prerequisite for further studying the visual perception and evaluation of the landscape, since this description provides an objective information of the landscape composition at the level of the hFoV.

In this line of thought, Brabyn [32] developed a GIS-based method for combining space, time, and landscape character: for a given track (i.e., route), they have calculated the contribution of different landscape character types (e.g., "High Hill, Indigenous Scrub") in certain distance zones (0–2, 2–5, 5–10, and 0–10 km) by introducing "experions", which are a new unit for measuring visual landscape experience. Despite the conceptual merits of this method, aspects related to the inadequacy of precisely calculating the actually perceived size of each landscape type and its relative contribution within the hFoV from the observer's viewpoint have limited the utility of the method. Moreover, the relative time during which each landscape type is exposed to an observer poses another challenge. Other innovative, GIS-based modeling methods such as those developed in the research work of Schirpke et al. [31] and Misthos and Menegaki [28] are also indirect approximations of the perceived landscape composition from the egocentric perspective.

Some other research studies have addressed the issue of miscalculating the landscape elements' perceived size by employing different means. For instance, Minelli et al. [34] calculated the perceived size of an object (i.e., wind turbines) as a proportion of its area in a varying field of view. Nutsford et al. [35] developed an index (VVI) to rectify standard viewshed analysis outputs by calculating the vertical angle (vertical angular size) between an observer's eye level and the top and bottom points of each visible cell within a viewshed output; by also indirectly correcting the horizontal angular size, the VVI became a determinant of the visual significance of each visible cell on a terrain from certain observers (viewpoints), providing an accurate representation of landscape visibility from a human (egocentric) perspective.

2.2. Describing the Dynamic Landscape Along Higways: Concepts and Issues

Some decades ago, in the influential work of J.J. Gibson, the visual perception of the environment has been acknowledged as a dynamic process—in that humans, like animals, perceive the world around them as a sequence of changing ambient optic arrays—that persists during long acts of locomotion [36]. In today's everyday life, and as urbanization keeps increasing, moving in vehicles is a common practice. So, people typically perceive a cityscape chiefly by means of the dynamic visual experience [37]. The dynamic visual landscape is a concept which specifies the landscape's visual experience from the perspective of moving observers (e.g., [27,38,39]) and/or vehicles. Roads and highways, in particular, have expanded in order to meet increased urban area sprawl [40]. Provided that many highways have been designed without adequately taking into consideration visual landscape preferences or quality, visual landscape degradation is anything but rare [40]. Since highways are crucial physical linkages enabling transportation [12,40] but also offering to large numbers of moving observers the potential to witness impressive scenery [12,30], the quantitative description of the dynamic visible landscape along such linear components is a key issue.

In the last decades, some previous research studies have focused on visualizing and quantitatively analyzing the visible parts of a terrain from the perspective of a moving observer along linear routes in different topographic feature types (ridgelines, valleys, pass-lines, etc.) [27,38,39]. In these research studies, viewsheds in different types of routes have been visualized in animated maps, while statistical correlations and inferences among elevation/topography and viewshed area (acreage) have been established. Nevertheless, these studies have not dealt with the qualitative aspect (i.e., landscape composition) of the area viewed from each viewpoint/observer and have not taken into consideration the actual size of visible/not visible cells within the hFoV. In another research study, a general framework is provided which, by combining various indicators (e.g., sky expanse, landmark distribution, etc.) at multiple dimensions and emphasizing the continuity of the linear urban landscape (LUL), serves to quantify dynamic visual perception [41]. This study has taken into consideration the continuous and dynamic features of visual perception, quantitatively approximating the continuous dynamic visual perception of consecutive viewpoints along linear routes (LULs) and also under different movement modes (walking, cycling, and boating). Nonetheless, the emphasis in this study has been placed on modeling the dynamic/changing views of the cityscape expressed/extracted in the form of photographs or perspective drawings and also on the simultaneous utilization of visual evaluation indicators in order to assess LULs based on distinct scenario preference criteria and patterns.

3. Materials and Methods

3.1. Overall Rationale

The overall approach of this research refers to the selection of initial highway routes (polyline features) in a peri-urban area in Attica region in order to derive the *d*FoV's landscape elements' composition information, which is ultimately attributed to each segment of the route and visualized in properly designed interactive web maps. The implementation of the computations is realized via the execution of an autonomous geoprocessing Python tool, introduced and developed within the presented work, which receives as inputs open geospatial data, while the tool's outputs are visualized using JavaScript and html technologies.

3.2. Input Data (Case Study)

For the implementation of the methodology analyzed below (Section 3.4), three different geospatial data inputs are required: (i) a polyline vector road network layer for creating the highway route (of the case study), (ii) a raster digital elevation/surface model (DEM/DSM) for deriving terrain information (elevation, slope, aspect), and (iii) a polygon vector land cover layer for deriving landscape elements' information.

The presented case study relies on the exclusive use of open geospatial data. More specifically, the road network vector features used in the case study were adapted from OpenStreetMap (OSM) platform (https://www.openstreetmap.org/, accessed on 15 October 2024) and downloaded freely from BBBike Extract Service (https://extract.bbbike.org/, accessed on 15 October 2024). The crowdsourcing platform of OSM provides information stored in different fields, among others, for the road type (field: fclass), for the maximum allowed speed (field: maxspeed), etc. Regarding the road type, seven main road type categories are present, including motorway, trunk, primary, secondary, tertiary, residential, and unclassified networks; the OSM also includes special road- and path-type categories (e.g., pedestrian, path, footway, cycleway, raceway, etc.). The utilized polyline layer was extracted both spatially (only segments within the study area (see Section 3.3 for the description of the study area) were retained) and thematically (only segments/records classified as 'trunk' were retained) in order to meet the aims of this research study.

Furthermore, the European Digital Elevation Model (EU-DEM), provided by OpenDEM portal (https://www.opendem.info/, accessed on 15 October 2024), was employed in the performed analysis. EU-DEM has been derived by the Shuttle Radar Topography Mission (SRTM) and the Advanced Spaceborne Thermal Emission and Reflection radiometer (ASTER) Global Digital Elevation Model (GDEM) having a ~30 m (1 arc per second) spatial resolution. EU-DEM has been preserved by the Copernicus Land Monitoring Service (CLMS) implemented by the European Environmental Agency (EEA) and Joint Research Center of the European Commission.

In addition, the Urban Atlas Land Cover/Land Use 2018 (LU/LC 2018) (https://land.copernicus.eu/en/products/urban-atlas/urban-atlas-2018, accessed on 15 October 2024), also provided by the CLMS, was used as an input for implementation of the developed software tool. Urban Atlas constitutes a pan-European dataset that involves seventeen (17) urban classes of land use/land cover (LU/LC) with the Minimum Mapping Unit (MMU) of 0.25 ha and ten (10) rural classes with 1ha of MMU (https://land.copernicus.eu/en/products/urban-atlas/urban-atlas-2018, accessed on 15 October 2024). Urban Atlas is a satellite product with six-year update cycle, which is distributed as a vector dataset.

Aside from the three initial input datasets, two other datasets were derived from the EU-DEM—also required for performing the tool, namely, the terrain slope and the terrain aspect products—having the same spatial resolution as the EU-DEM (~30 m). In total, the input layers that were utilized are presented in Table 1.

Table 1. Input datasets or layers accompanied by their formats and corresponding geometries, and by their source geospatial datasets.

Input Dataset (Layer)	Format—Geometry	Open Geospatial Dataset
highway routes	feature class/shapefile—polyline vector	OSM roads
terrain elevation	GeoTIFF—raster	EU-DEM
landscape elements	feature class/shapefile—polygon vector	Urban Atlas LU/LC
terrain slope *	GeoTIFF—raster	EU-DEM
terrain aspect *	GeoTIFF—raster	EU-DEM

* derived product.

3.3. Study Area

The area of concern is a peri-urban region in the southeastern Attica region, lying almost exclusively within the boundaries of the Lavreotiki municipality (whereas the selected route is exclusively within the Lavreotiki municipality). The study area consists of a high variety of LU/LC types (over 20 different types), including urban fabric (of varying density), roads, industrial areas, agricultural land, pastures, herbaceous vegetation, forests, water, sea, etc. This variety in LU/LC is directly connected with the landscape elements' diversity or complexity in the study area. This area, being a peri-urban region, displays a high landscape complexity, since it comprises a mixture of elements found in urban, agricultural, and forest areas. Other landscape regions, for instance rural/agricultural landscape regions, display a much lower complexity [42].

The selected route—which was characterized in terms of landscape composition within the *d*FoV—is part of a highway (i.e., main road but not motorway) with the generic name "Lavriou" and "Markopoulo-Lavrio". This route is an important dual carriageway road, i.e., divided road with two directions connecting Lavrio town and Eleochori settlement. The route length is approximately 5 km, while its general direction (from Lavrio toward Eleochori) is south–north; more precisely, for the first 1.5 km (from Lavrio), the general direction is southeast–northwest, while for the rest of the 3.5 km, the general direction is southwest–northeast. The reference map of the study area with the selected route is presented in Figure 1. As displayed in Figure 1, since the route has two directions, the vehicles can also move toward the opposite direction (from Eleochori toward Lavrio). In addition, points upon this route have been placed at a certain spacing interval (50 m) for the needs of the methodology implementation (see Figure 1 and Section 3.4.2).

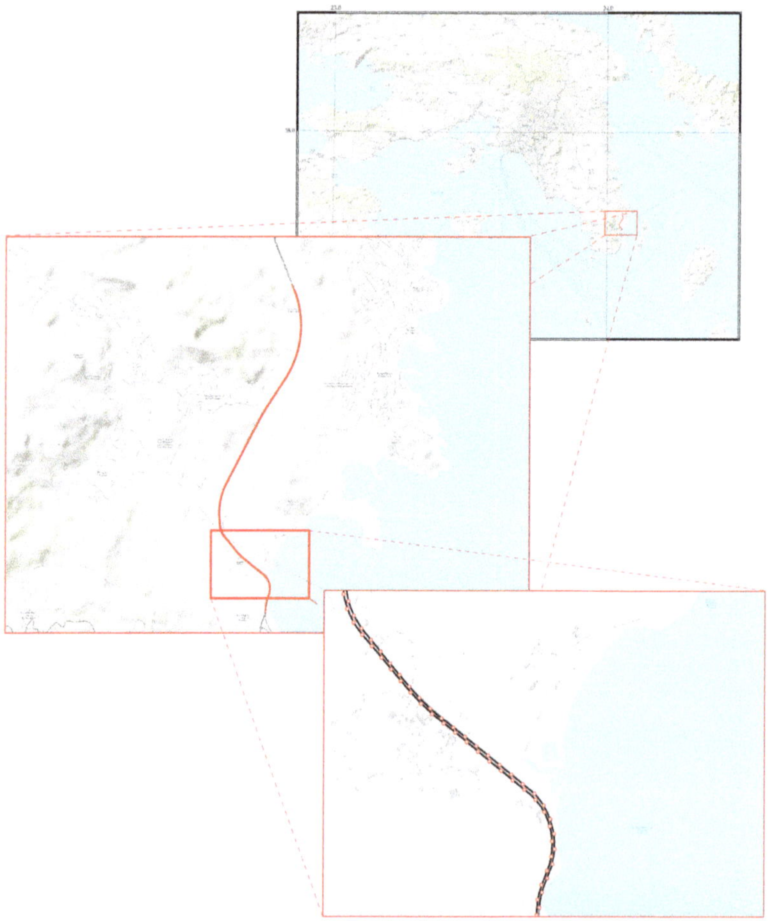

Figure 1. Reference map (top map), the selected highway route (middle map), and the points (red hollow circles) placed at a 50 m spacing interval on each direction of the route (bottom map).

Moreover, this route was divided in segments according to the maximum allowed speed. As will be shown in the following sections, the human field of view varies with movement speed. Therefore, the horizontal field of view for a hypothetical driver along this route was calculated based on the maximum allowed speed or speed limits, ranging from 50 to 80 km/h per segment of the route (and per movement direction) for the study area (harnessing the OSM field *maxspeed*).

3.4. Methodology

3.4.1. Overview

In order to attain the aims of the research study and address the issues and challenges pointed out in Sections 2.1 and 2.2, two complementary tasks were accomplished. At first, a new method was developed and introduced for integrating a sequence of geoprocessing computations and geometric calculations in order to quantify the dominant landscape elements within the selected highway route's *d*FoV. Secondly, an integrated algorithmic procedure, namely, a geoprocessing, Python-based software tool—named "RouteLAND"—was developed and implemented on the basis of the introduced method. Both the method development and the tool implementation can be considered to consist of certain semantically and sequentially distinct sub-routines ('sub-models'). In the rest of this section, the five 'sub-models' are described separately in the respective sub-sections because of their autonomous rationale within the method's and the geoprocessing tool's algorithmic sequence. The five sub-models are summarized as follows:

1. Sub-model 1: Generation of viewpoints along highway routes and computation of route segments' viewing directions;
2. Sub-model 2: Computation of (start and end) horizontal viewing directions based on the route segments' mean viewing directions and speed limits (max speed) for further viewshed analyses computation;
3. Sub-model 3: Iterative computation of viewsheds and other relationships or attributes (distance, elevation, slope, aspect, LU/LC) of the visible/viewshed cells (or points) associated with each respective viewpoint;
4. Sub-model 4: Computation of viewing significance and actual landscape composition, at the level of the highway route's viewpoints *d*FoV, based on the spatial/thematic relationships and analyses among the viewshed points' attributes and the associated viewpoints' positions;
5. Sub-model 5: Qualitative and quantitative characterization of the highway routes regarding the dominant landscape element 'occupying' the *d*FoV per route segment (Figure 2).

3.4.2. Method Development: Rationale and Computations (Geoprocesses and Calculations)

Before implementing the main procedure (divided in the aforementioned five sub-models), some pre-processing tasks were executed. First, a 6 km buffer zone around the highway route was produced in order to restrict both the computation process within certain boundaries and the volume of the input data. Hence, the DEM dataset, along with its derivatives—slope and aspect datasets—and the Urban Atlas LU/LC dataset were clipped within this 6 km buffer zone. Furthermore, the initial highway route was dissolved in order to generate a single-record feature.

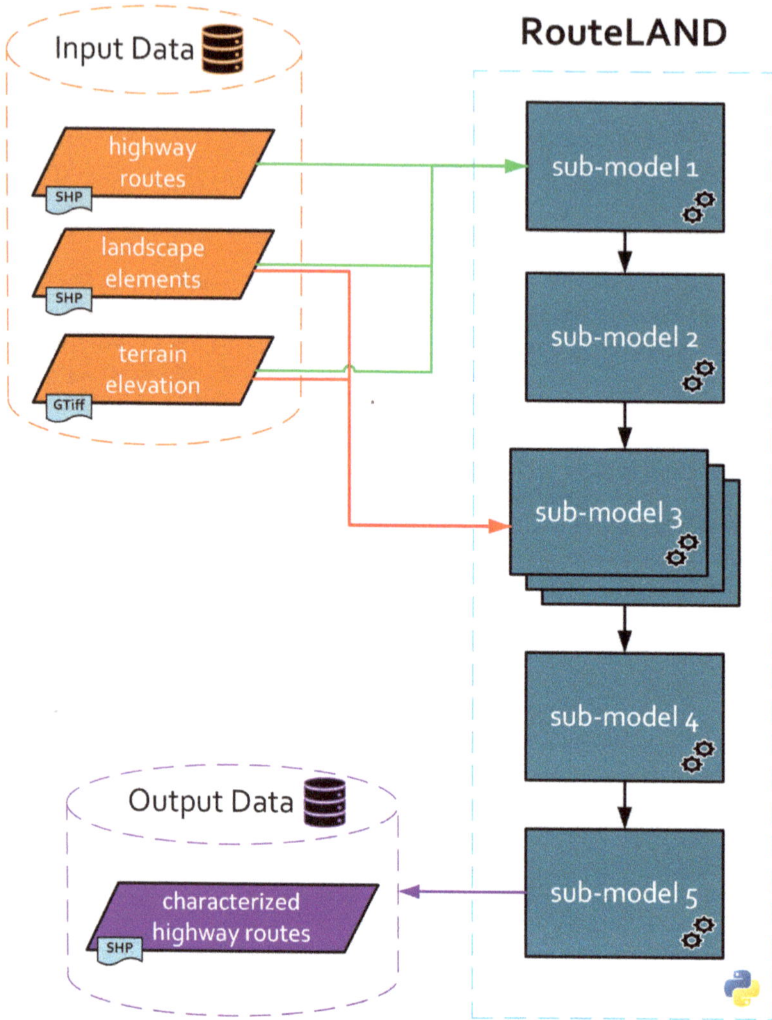

Figure 2. The five sequential 'sub-models' of RouteLAND method and tool.

For the first sub-model, the required input data were (a) (i) the initial highway route (OSM polyline feature layer) and (ii) the dissolved highway route (polyline feature layer) and (b) the DEM (raster layer). At the beginning, the dissolved polyline feature was segmented, and point features were generated at a 50 m interval. This interval was selected for approximating the dynamic change of the visual landscape along a route for observers moving upon vehicles with relatively high speeds (>50 km/h), since other research studies have suggested 10 to 40 m intervals for 'walking' observers (e.g., [38,39]), and since the DEM's resolution is approximately the same size (30 m). The previously generated points were ascribed (via spatial join) all the attributes (e.g., max speed, type, etc.) existing in each record of the initial OSM polyline feature layer. The dissolved highway route was further split at the 50 m interval points in order to adjust the geometry of the polyline route according to the generated points, while all the attributes of the points were transferred back (via spatial join) to the split highway route. A new field ("road_str") was created in the new split highway route to retain each FID corresponding to each 50 m highway segment, and the mean direction (azimuth) was calculated for each 50 m segment and stored in a new field ("COMPASS"). The vertices of this directed polyline were converted to point features by retaining only the starting vertex of each directed 50 m segment. In addition, the elevation from the DEM was attributed to each viewpoint ("RASTERVALU" field), and the initial information of the max speed (or speed limit) was also transferred to these point features. In total, among others, useful information about (i) the elevation, (ii) the max speed, and (iii) the mean azimuth (i.e., mean viewing direction) of each segment of the route was finally stored in the 50-meter-spaced viewpoints produced (upon the route).

The second sub-model received as inputs the previously produced viewpoints (point feature layer). The horizontal field of view was calculated under a certain condition, becoming narrower as the (max) speed increased—based on the relevant literature (e.g., [43,44]). So, the horizontal fields were set to: 120° (azimuth) for very low speeds, 110° for speeds up to 30 km/h, 100° for speeds greater than 30 km/h and up to 65 km/h, 60° for speeds greater than 65 km/h and up to 80 km/h, 40° for speeds greater than 80 km/h and up to 100 km/h, and 20° for speeds greater than 100 km/h (Table 2). In this respect, the left-/right-hand azimuth angles relative to the mean (viewing) direction ("COMPASS") were the half of the aforementioned values (horizontal half-angles), i.e., 60, 55, 50, 30, 20 and 10°, respectively, and were stored in a new field ("hlf_angle"). Two new fields were created ("st_angle" and "end_angle") to store the horizontal start/end azimuth angles for each one of the 50-meter-spaced viewpoints. The start azimuth angles ("st_angle") were calculated by *abstracting* the horizontal half-angles ("hlf_angle") from the mean (viewing) direction ("COMPASS") of each segment, while the end azimuth angles ("end_angle") were calculated by *adding* the half-angle left-/right-hand azimuth angles ("hlf_angle") to the mean (viewing) direction ("COMPASS") of each segment, correspondingly. It is worth noticing that the final left-/right-hand azimuth angles were recalculated in order to fall in the interval [0, 360]°, satisfying the following conditions. The calculations of "st_angle", "end_angle"—under conditions—are specified below as pseudocode:

if st_angle < 0°: $st\_angle = (360 - |st\_angle|)^\circ$;
else if 0° <= *st_angle* <= 360°: $st\_angle = (st\_angle)^\circ$;
else (if *st_angle* > 360°): $st\_angle = (360 - st\_angle)^\circ$;

$$st\_angle = Compass - hlf\_angle; \quad (1)$$

if *end_angle* < 0°: $st\_angle = (360 - |end\_angle|)^\circ$;
else if 0° <= end_angle <= 360°: $end\_angle = (end\_angle)^\circ$;
else (if *end_angle* > 360°): $end\_angle = (360 - end\_angle)^\circ$;

$$end\_angle = Compass + hlf\_angle; \quad (2)$$

Table 3 presents an illustration of calculating the start/end azimuth angles with regard to certain, indicative (max) speeds and mean directions along different highway segments.

Table 2. Speed limits per highway segment accompanied by their horizontal and horizontal half-(azimuth) angles.

Speed Limit (km/h)	Horizontal Angle (°)	Horizontal Half-Angle (°)
0	120	60
0 < sl <= 30	110	55
30 < sl <= 65	100	50
65 < sl <= 80	60	30
80 < sl <= 100	40	20
>100	20	10

Table 3. Start/end azimuth angles occurring by indicative (max) speeds and mean directions.

Max Speed (km/h)	Mean Direction (°)	Half-Angle (°)	Start Angle (°)	End Angle (°)
20	355	55	300	50
20	5	55	310	60
50	265	50	215	315
50	95	50	45	145
70	175	30	145	205
70	185	30	155	215
90	85	20	65	105
90	275	20	255	295
110	10	10	0/360	20
110	350	10	340	0/360

Within the third sub-model, the geoprocesses and calculations were executed in an iterative manner. The input data were (i) the LU/LC (polygon feature layer); (ii) the DEM (raster layer); the two other DEM/terrain derivatives, i.e., (iii) the slope (raster layer) and (iv) the aspect (raster layer); and (v) the viewpoints (point feature layer) produced in the previous sub-model (2). The fundamental procedure here was the iterative selection of each one of the viewpoints within a loop; in this loop, both the FID and the elevation of each viewpoint were stored (fields: "fid_value" and "elev_value"), and the visibility/viewshed analyses were executed for each viewpoint. The parameters for this iterative viewshed analyses were specified either from the information calculated in the previous models and stored in the fields of the viewpoints (e.g., horizontal start/end viewing angle) or from information set at will (manually) in the RouteLAND tool (e.g., observer offset: 1 m; outer analysis radius: 5 km). The horizontal field of view had been already calculated in sub-model 2; the observer offset was set to 1 m, taking into consideration a typical height of drivers' eyes in a conventional vehicle; the radius for computing the viewshed was confined to 5 km in order to diminish the extensive computation load, since landscape elements that are far away (>5 km) from a moving observer are considered to have a diminished or even negligible visual impact compared to elements that lie in an observer's foreground. The latter assumption is endorsed in other research works and technical reports [29,45–47]. Now, for every viewpoint of the route, a 'cloud' of *visible* terrain cells was produced; cells that were not visible were excluded from the rest of the geoprocessing to significantly reduce the subsequent computation load. Additionally, for every iteration, a unique code number (FID field: "fid_value") characterizing each viewpoint had already been stored in order to associate each one viewpoint with its 'cloud' of produced visible or viewshed cells—along with other outputs in subsequent geoprocesses. The several 'clouds' of viewshed terrain cells were converted into viewshed points (features) to enable the subsequent management of both descriptive and geometric information. In the next stage, the three DEM derived values, i.e., terrain elevation, terrain slope, and terrain aspect of the visible cells, were transferred into the viewshed points as separate fields, namely, "ELEV", "SLOPE", and "ASPECT". By utilizing the LU/LC, information about the type of the visible landscape elements (fields: "code_2018", "class_2018") was also transferred—in a similar manner—in the 'clouds' of viewshed points. Moreover, the distance and the horizontal angle/azimuth from each viewshed point of the cloud to each associated viewpoint upon the route were computed as well, while this information was stored and two new fields as "NEAR_DIST" and "NEAR_ANGLE", respectively. Finally, two attributes of every viewpoint of the route were joined with each associated 'cloud' of viewshed points, that is, the viewpoint FID ("VPoint" = "fid_value") and the viewpoint elevation ("VPElev" = "elev_value"). In doing all the previous tasks, the several 'clouds' of viewshed points were attributed to the information of (a) elevation (i), slope (ii), aspect (iii), land cover (iv), and distance from viewpoints (v)—attributes unique for each viewshed point in every viewshed point 'cloud'—as well as (b) FID (vi) and elevation (vii)—attributes shared/common in every viewshed point 'cloud'. This information was stored in different fields of the several 'clouds' of viewshed points. It should be noted that each viewshed point 'cloud' was descriptively (i.e., non-spatially) distinguished from any other viewshed point 'cloud' on the basis of its unique FID—'carrying' the FID of the associated viewpoint along the route. The entire algorithmic procedure of the third sub-model described above is depicted in the following summarized flowchart (Figure 3).

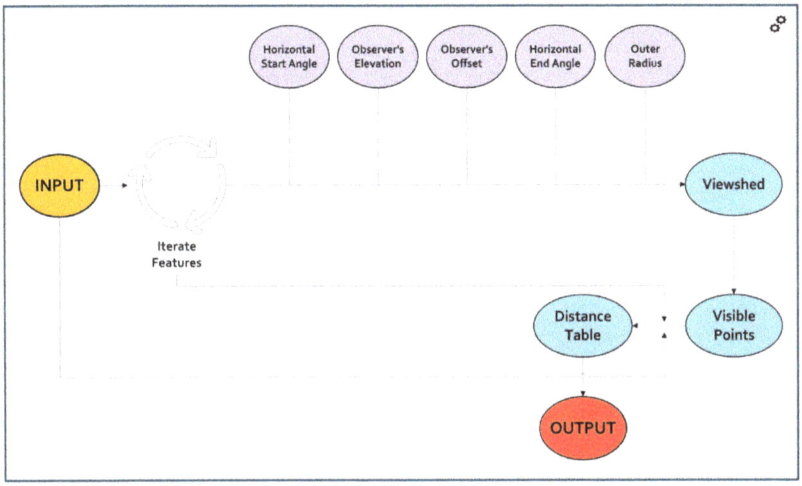

Figure 3. Summarized flowchart of the third sub-model. Note that the input data may enter or may not enter the iterative stage (loop). (i) The viewpoints produced from the second sub-model do enter the iterative selection, so (a) the viewsheds/visible points and (b) the distances (between viewpoints and visible points) are also iteratively computed. (ii) The DEM derivatives (elevation, slope, and aspect) and the LU/LC do not enter the iterative selection, so their attribute values are assigned on the iterated visible points. The output is multiple (see also Figure 2, sub-model 3).

In the fourth sub-model took place all the calculations within each viewshed point 'cloud' regarding the actually perceived size (perceived visual significance) of each visible landscape element and its contribution or percentage (landscape composition) within the field of view. The most important part of calculations refers to the *Perceived Visual Significance Index* (*PerVSI*). PVSI calculation can be subdivided to the calculations of (A) *Perceived Adjusted Vertical* (*Viewing*) *Angle* (*PAVA*) and (B) *Perceived Adjusted Horizontal* (*Viewing*) *Angle* (*PAHA*).

A. For the calculation of *PAVA*, several individual physical quantities were calculated:

1. The vertical (positive/negative) viewing angle (*VerAng*) at which each viewpoint of the route is connected to each viewshed point of the cloud (sharing a common FID) was calculated as a function of the viewpoint elevation (VPElev), viewshed elevation (ELEV), and horizontal distance among viewpoint and viewshed points (NEAR_DIST), which is defined as follows:

$$VerAng = \tan^{-1}\left(\frac{VPElev - ELEV}{NEAR\_DIST}\right); \quad (3)$$

2. The actual (i.e., not just horizontal) distance (*ActDist*) from each viewpoint of the route to each viewshed point of the cloud (sharing a common FID) was

also calculated as a function of the viewpoint elevation (VPElev), viewshed elevation (ELEV), and horizontal distance among viewpoint and viewshed points (NEAR_DIST), which is defined in the following:

$$ActDist = \sqrt{(VPElev - ELEV)^2 + NEAR\_DIST^2}; \quad (4)$$

3. The relative vertical viewing angle (*RelVAng*) that occurs between (i) the vertical angle at which each viewpoint of the route is connected to each viewshed point of the cloud (VerAng) and (ii) the slope of the surface of the viewshed cells (SLOPE), which is calculated as follows:

$$RelVAng = ||VerAng| - SLOPE|; \quad (5)$$

4. The actual height (*ActHei*)—due to the relative viewing vertical angle—of each viewshed cell of the cloud relative to each viewpoint of the route (sharing a common FID), which is herein computed as follows:

$$ActHei = \sin(RelVAng * cell\ size); \quad (6)$$

5. The perceived adjusted vertical viewing angle (*PAVA*) as a function of *ActHei* (Equation (6)) and *ActDist* (Equation (4)), which is based on the following formula:

$$PAVA = 2 * \tan^{-1}\left(\frac{ActHei}{2 * ActDist}\right); \quad (7)$$

B. On the other hand, for the calculation of *PAHA*, the following calculations took place:

1. The relative horizontal viewing angle (*RelHorAng*) that occurs between the following: (i) the horizontal angle (azimuth) at which each viewpoint of the route is connected to each viewshed point of the cloud (NEAR_ANGLE) and (ii) the aspect of the surface of the viewshed cells calculated under certain conditions in two stages (a, b):

    (a) In order to delimit the relative horizontal viewing angles within the [0–180] range, the following condition was enforced, obtaining the following two arguments (NEAR ANGLE and ASPECT): if $ASPECT == -1$ (flat surface): $ASPECT = 90°$,
    else if $0 <= ASPECT <= 180°$: $ASPECT = ||NEAR\_ANGLE| - ASPECT|$,
    else (if $ASPECT > 180°$): $ASPECT = ||NEAR\_ANGLE| - |ASPECT - 360°||$;
    and then an intermediate variable, *RelHorA*, was calculated as a function of the two arguments:

    $$RelHorA = f(NEAR\_ANGLE,\ ASPECT); \quad (8)$$

    (b) In order to delimit the relative horizontal viewing angles within the [0–90] range, the following condition was enforced:
    if $RelHorA <= 90°$: *RelHorA* remains unchanged;
    else (if $RelHorA > 90°$): $RelHorA = RelHorA - 90°$;
    and, so:

    $$RelHorAng = RelHorA; \quad (9)$$

2. The actual width (*ActWid*)—due to the relative viewing horizontal angle—of each viewshed cell of the cloud relative to each viewpoint of the route (sharing a common FID) was computed using the following equation:

$$ActWid = \sin(RelHorAng * cell\ size); \quad (10)$$

3. The perceived adjusted horizontal viewing angle (*PAHA*) as a function of *ActWid* (Equation (10)) and *ActDist* (Equation (4)) was calculated based on the following formula:

$$PAHA = 2 * \tan^{-1}\left(\frac{ActWid}{2 * ActDist}\right); \tag{11}$$

PAVA and *PAHA* indices measure the perceived angular sizes of visible/viewshed cells in the two (vertical and horizontal) dimensions, i.e., perceived angular height and width, while their angle values range in the interval $[0, 180)^{\circ}$. The Perceived Visual Significance Index (*PerVSI*) is a compound index that measures the two-dimensional (solid) angle (measured in square degrees) and occurs as the product of *PAVA* and *PAHA*, namely,

$$PerVSI = PAVA * PAHA \tag{12}$$

The fifth sub-model received as inputs all the viewshed points in each 'cloud'. Each one of these viewshed points has been ascribed several attributes from all the aforementioned geoprocesses, such as the previously calculated PerVSI, the LU/LC, the FID of the initial viewpoint (along the route), etc. Now, by spatially and thematically aggregating (dissolving) the PerVSI based on the same LU/LC type (field: "class_2018"), the total (sum) PerVSI per LU/LC type was calculated inside each 'cloud' of viewshed points and stored in a new field ("SUM_PerVSI"). By summing all the SUM_PerVSIs for all different LU/LC types existing in each 'cloud' (summing the PerVSI for each viewshed point is equivalent), a new field was created ("SUM_SUM_PerVSI"), defining the total sum of the two-dimensional angles of visible terrain within the *d*FoV. *LandComp*, which represents the relative contribution (percentage) of each LU/LC type or landscape element, i.e., the landscape composition normalized in the [0, 100] within the within the dFoV, was calculated as a division:

$$LandComp = \left(\frac{SUM\_PerVSI}{SUM\_SUM\_PerVSI}\right) * 100 \tag{13}$$

The dominant LU/LC or landscape element ("new_class_2018") and its respective relative contribution (percentage) ("MaxLand") were computed by further aggregating (dissolving) the "LandComp" values; by selecting the features/rows having the maximum "LandComp" value, for each viewshed points 'cloud', only one point–row was retained, creating a set of multipart point features, while the number of this set equaled the number of the initial viewpoints (along the route). After this aggregation, the multipart point features were joined to their associated initial viewpoints. As an effect, all the attributes of the multipart point features (e.g., "MaxLand", "LandComp", etc.) were transferred to the initial viewpoints. These viewpoints–features were merged into a single-part point feature, and so each row represented the viewpoints—having been attributed a variety of fields. In this last stage, all the attributes stored in the viewpoints were transferred to the segments of the highway route via spatial join processing. Therefore, in these polyline features (viewlines), multiple lines–rows were occupying the same highway segment, since more than one point was spatially joined to each segment. These lines–rows were aggregated (i.e., dissolved) into a unique ID under the following condition: if the "new_class_2018" (LU/LC) was the same, e.g.,: "Complex and mixed cultivation patterns", with "MaxLand" values, e.g., 54.67 and 62.85, respectively, then the LU/LC was retained ("Complex and mixed cultivation patterns"), and the "MaxLand" mean value was calculated by averaging the two values (58.76); if the "new_class_2018" (LU/LC) was different, e.g.,: "Complex and mixed cultivation patterns" and "Other roads and associated land", with "MaxLand" values, e.g., 44.31 and 52.25, then the LU/LC corresponding to the larger value would be retained ("Other roads and associated land"), with the respective "MaxLand" value

(52.25). So, in the final output, only the initial segments of the initial segments are retained, while two new fields are created, storing (i) the description of the dominant LU/LC ("final_class_2018") and the quantitative values of the dominant LU/LC ("DomLand"). One last field ("LandChar") was created suitable for storing an information fusion mixture of two fields, namely, the Urban Atlas LU/LC coding, "code_2018" and the "DomLand", by concatenating string values; thence, "LandChar" contains a verbal description of both qualitative and quantitative information about the dominant landscape character corresponding to each highway route segment (e.g., 24000: 58.76%; 12220: 52.25%, etc.).

3.4.3. Tool Implementation: Employing ArcPy Toolboxes, Toolsets, and Tools

The whole methodology was implemented via a software tool particularly developed for the needs of this research work. More specifically, a Python (version 2.7.16) script-based geoprocessing tool, named "RouteLAND", was developed and executed within the ArcMap 10.8, ESRI®, by making use of the ArcPy Python site package, provided by ESRI®, obtaining as inputs freely available, open geodata (described in Section 3.2). The input data were imported into the script, and then the path to the output data was set. The geoprocesses and calculations were implemented by employing several geoprocessing ArcPy toolboxes, toolsets, and tools. In Table 4, the employed tools are summarized and classified based on their corresponding toolsets and toolboxes, while the layers involved are also summarized based on the toolboxes, harnessing them as input data; moreover, the last row of the table constitutes a somewhat separate one, providing information about the "SearchCursor" element—a class enabling the iterative data access (selection) of each viewpoint separately for further proceeding to the computations per viewpoint. After leveraging all the displayed tools within the overall geoprocessing sequence, the final output was a polyline feature layer, whereby each segment–record was characterized by the dominant landscape element within the highway's *d*FoV.

Table 4. Employed (ArcPy) tools (column 4), classified per toolset (column 3) and toolbox (column 2), along with the datasets serving as inputs within each toolbox (column 1). The last row is differentiated in that it constitutes a class-function, enabling iterative selection (SearchCursor) of feature records.

Input Layer	Toolbox/Data Access Module—ArcPy Function	Toolset	Tool/Class-Function
1. Initial highway route (OSM roads); 2. Landscape elements (UrbanAtlas LU/LC); 3. Intermediate highway route	Analysis	Extract	Clip; Select
		Overlay	Identity; SpatialJoin
		Proximity	GenerateNearTable; Buffer
		Statistics	Statistics (Summary Statistics)
1. Initial highway route (OSM roads); 2. Landscape elements (UrbanAtlas LU/LC); 3. Terrain elevation; 4. Terrain slope; 5. Terrain aspect (EU-DEM); 6. Intermediate highway route	Management	Fields	AddField; CalculateField; DeleteField
		General	Merge
		Generalization	Dissolve
		Features	CopyFeatures; FeatureVerticesToPoints
		Joins	JoinField
		Layers and Table Views	MakeFeatureLayer; SelectLayerByAttribute
		Raster Processing	Clip
		Sampling	GeneratePointsAlongLines

Table 4. *Cont.*

Input Layer	Toolbox/Data Access Module—ArcPy Function	Toolset	Tool/Class-Function
1. Initial highway route (OSM roads); 2. Terrain elevation; 3. Terrain slope; 4. Terrain aspect (EU-DEM)	Spatial Analyst	Surface	Aspect; Slope; Viewshed2
		Extraction	ExtractMultiValuesToPoints
1. Initial highway route (OSM roads); 2. Terrain elevation; 3. Terrain slope; 4. Terrain aspect (EU-DEM)	Conversion	From Raster	RasterToPoint
1. Initial highway route	Spatial Statistics	Measuring Geographic Distributions	DirectionalMean
1. Initial highway route	Classes/Cursors	-	SearchCursor

3.4.4. Web Map Design

The final output of the presented case study was used to design an interactive web map that visualizes the prevailing landscape elements, within the *d*FoV, for each one of the predefined segments of the highway routes. More specifically, the web map portrays the segments of the highway routes, categorized by the dominant land uses/land covers, on the cartographic backgrounds of ESRI Gray (dark) and OpenStreetMap (OSM). The web map serves as an exploratory tool that allows users to reach the attributes of LU/LC, the DomLand values, and the LandChar descriptions for each route segment. Additionally, the users can filter the depicted routes' segments based on both DomLand values and LU/LC attributes, as well as measure path distances and areas on the map. Moreover, we have restricted the zoom levels of the web map from 13z to 16z to help potential users to efficiently explore the selected highway routes. The development of the interactive web map was based on the utilization of open-source geospatial software. In more detail, highway route symbolization was implemented in QGIS software (Version 3.34.12 LTR), utilizing the standard UrbanAtlas legend for the different LU/LCs (https://land.copernicus.eu/en/map-viewer?dataset=70903c20fc2a4a90ad200bc95a7557d4%5C accessed on 10 December 2024), while different cartographic backgrounds were imported from the corresponding web map services using the QuickMapServices QGIS plugin (https://plugins.qgis.org/plugins/quick_map_services/, accessed on 10 December 2024). Finally, for the production of the interactive web map, the qgis2web QGIS plugin (https://plugins.qgis.org/plugins/qgis2web/, accessed on 10 December 2024) and LeafLet JavaScript Library (https://leafletjs.com/, accessed on 10 December 2024) were utilized.

4. Results

4.1. Presentation

The results of this article are both the Python-based geoprocessing tool ("RouteLAND.py") and the web map. The source code of the developed tool has been freely distributed to the scientific community under the third version of the GNU General Public License (GPL v3) on the GitHub platform (https://github.com/), as presented below, in the Supplementary Materials' section. As regards the interactive web map, it is freely accessible on the web via the webpage https://routeland.geo.uniwa.gr/ (created on 29 November 2024). An indicative snapshot of the web map is illustrated in Figure 4. In the illustrated snapshot, a popup emerges, showing the produced attributes' information (stored in the respective fields)—i.e., LU/LC, DomLand, and LandChar—as the user interactively explores and hovers along different segments of the highway route (Eleochori–Lavrio).

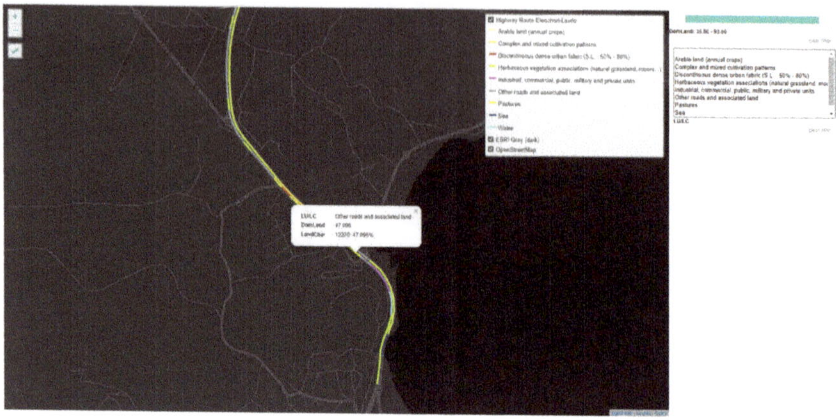

Figure 4. Web map snapshot of interactively exploring the highway route of the case study.

In Table 5, we further elaborate on the derived attributes' information. More precisely, the second column of the table presents the percentage area (%) covered by each LU/LC within the route's 6 km buffer zone—i.e., study area—while the third column displays the percentage of segments (%) dominated by the vista of each LU/LC along the highway route. Thence, the second column provides the landscape composition from an exocentric perspective, whereas the third column provides the route's dominant landscape element (LU/LC) composition from the egocentric perspective of an observer moving on a highway, implementing computations utilizing the RouteLAND tool. There are some striking cases where the percentages of these two columns/perspectives differ dramatically. For instance, while the study area is exocentrically dominated by the "Sea" LU/LC (35.84%), only in 1.03% of the totality of the route does the "Sea" LU/LC dominate moving observers' vista (i.e., from an egocentric perspective). Another very indicative case that refers to the "Other roads and associated land" LU/LC is the following: while within the study area, this LU/LC can be hardly found (2.74%), it constitutes one of the most prevailing segments (34.36%) of the route in terms of the LU/LC dominating the vistas of moving observers. In total, out of the 23 UrbanAtlas LU/LC types existing in the study area, only 9 LU/LC types have been computed to constitute dominant LU/LC types viewed from the highway route.

Table 5. UrbanAtlas LU/LC types expressed as landscape composition percentage (%) within the route's 6 km zone—exocentric perspective (column 2); route's composition percentage (%) mainly viewing a specific LU/LC type—egocentric perspective (column 3); mean/max DomLand values in each group of route's segments' mainly viewing a specific LU/LC type—egocentric perspective (column 4).

UrbanAtlas LU/LC Type	Area of LU/LC Within the Route's 6 km Zone (%)	Number of Segments with Dominant LU/LC in Route's dFoV (%)	Mean/Max DomLand for Each Segment (%)
Arable land (annual crops)	2.10	1.03	46.37/46.37
Complex and mixed cultivation patterns	6.61	9.23	61.75/88.39
Construction sites	0.01	-	-
Continuous urban fabric (S.L.: >80%)	0.29	-	-
Discontinuous dense urban fabric (S.L.: 50–80%)	2.02	0.51	62.08/62.08
Discontinuous low-density urban fabric (S.L.: 10–30%)	0.88	-	-
Discontinuous medium density urban fabric (S.L.: 30–50%)	2.35	-	-
Discontinuous very-low-density urban fabric (S.L.: <10%)	0.04	-	-

Table 5. *Cont.*

UrbanAtlas LU/LC Type	Area of LU/LC Within the Route's 6 km Zone (%)	Number of Segments with Dominant LU/LC in Route's dFoV (%)	Mean/Max DomLand for Each Segment (%)
Forests	1.45	-	-
Green urban areas	0.02	-	-
Herbaceous vegetation associations (natural grassland, moors...)	34.92	47.18	55.13/86.21
Industrial, commercial, public, military, and private units	1.47	1.54	78.67/80.09
Isolated structures	1.23	-	-
Land without current use	0.11	-	-
Mineral extraction and dump sites	0.09	-	-
Open spaces with little or no vegetation (beaches, dunes, bare rocks, glaciers)	0.01	-	-
Other roads and associated land	2.74	34.36	66.12/93.66
Pastures	3.22	4.10	41.90/49.15
Permanent crops (vineyards, fruit trees, olive groves)	1.35	-	-
Port areas	0.67	-	-
Sea	35.84	1.03	41.45/41.45
Port areas	0.67	-	-
Water	2.38	1.03	34.24/37.30
Total	100	100	-

In the fourth column of the table, the mean and maximum (max) values of the "DomLand" are summarized for the nine group segments dominated by each separate LU/LC type. By comparing the third and the fourth columns, it becomes clear that for some LU/LC types, such as "Complex and mixed cultivation patterns", their maximum "DomLand" values are very high (88.39%) but do not constitute segments that significantly contribute as LU/LC types dominating the moving observers' vistas in the totality of the route (9.23%). On the other hand, the "Herbaceous vegetation associations" LU/LC type is found to prevail along the route (47.18%), yet it does not present equally high max "DomLand" values (86.21%).

The previous deviations show the great magnitude of differences occurring between (i) the two perspectives and (ii) the spatial and arithmetic occurrences of the landscape composition.

The first deviation occurs because the landscape elements existing in the study area (exocentric perspective) may be present in very different percentages or may be absent within the *d*FoV of the moving viewers (exocentric perspective)—as computed with Route-LAND. Computing the number/percentage at which one LU/LC type prevails within the hFoV within a highway route is based on the calculation of the sum of each viewed cell that represents the particular LU/LC type, which is expressed as an angular size/diameter visually perceived by an observer. What is more, when dealing with the *d*FOV of a moving observer, the horizontal hFoV varies with the observer's motion speed along a highway. As a consequence, the percentage of highway segments principally viewing the particular LU/LC type may be very different compared to the percentage of this particular LU/LC type within the highway's 6 km buffer zone—especially in areas where the landscape element diversity is relatively high (e.g., peri-urban areas).

The second deviation arises because a large percentage of (segments of) a highway may be mainly viewing a specific LU/LC type, yet the magnitude of the percentage within the *d*FoV may be not very high (e.g., 55%)—albeit higher than any LU/LC type within the *d*FoV. This means that a specific LU/LC type (e.g., "Herbaceous vegetation associations") may be the most frequent type in spatial terms (traced in the majority of segments along the highway (e.g., 90%)), while at the same time (compared to other segments of the highway where their *d*FoVs are dominated by other LU/LC types) the mean and max values of this specific LU/LC type—along the segments where it is dominantly viewed—may be relatively low.

4.2. Evaluation

The results produced by our modeling approach cannot be evaluated in a direct fashion; hence, the evaluation can only be based on an indirect method. For this reason, we generated (3D) perspective geovisualizations at representative viewpoints, imitating the egocentric perspective of the visual experience. By doing so, we were able to visually compare the mean DomLand values (produced by the RouteLAND's main export) with the dominant LU/LC landscape elements (produced by the perspective geovisualizations) at specific points/segments of the highway route.

Specifically, we utilized two indicative instances from the RouteLAND's exocentric views; at these two segments, the route was characterized by DomLand values corresponding to "Herbaceous vegetation associations" (52.07%) and "Complex and mixed cultivation patterns" (44.515%), respectively (Figure 5). In the same Figure, we also exclusively display the LU/LC landscape elements dominating in the previous perspective (i.e., "Herbaceous vegetation associations" and "Complex and mixed cultivation patterns"), albeit this time, these elements have been visualized under the egocentric perspective—viewed from the same segments of the route (as in the exocentric perspective). The horizontal viewing angle for both points/segments is 100°; for the point/segment at the top position of the figure, the mean viewing direction (azimuth) is approximately 189°, whereas for the point/segment at the bottom position, it is approximately 310°.

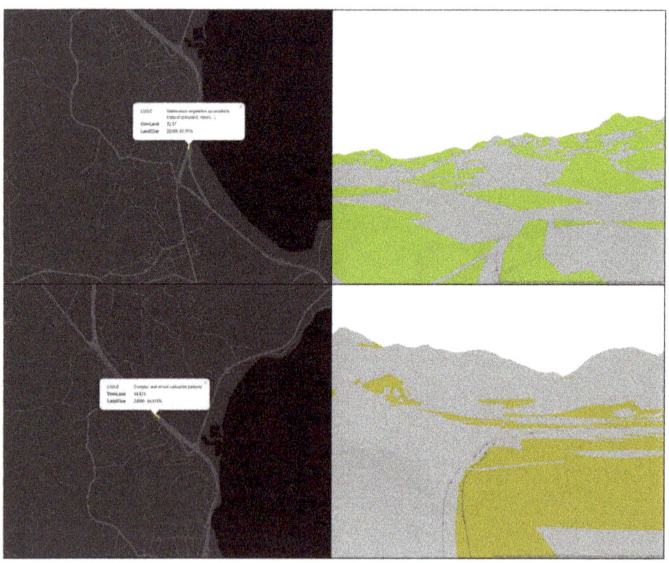

Figure 5. The results of the visible dominant landscape represented/characterized by RouteLAND' s main export (left side) and from the perspective geovisualizations (right side). The colored (i.e. not gray) parts of the landscape at the right side of the figure represent the two different LU/LC landscape elements ("Herbaceous vegetation associations": greenish hue, and "Complex and mixed cultivation patterns": yellow/brownish hue) dominating in the perspective geovisualization. The two geovisualizations placed at the top of the figure and the two geovisualizations placed at the bottom of the figure comprise the two alternative geovisualizations for the same point/segment.

From the visual inspection of these two alternative geovisualizations, the main output produced by the developed method/tool appears to significantly correspond to the results produced by the egocentrically based perspective. More precisely, according to the top-left geovisualization results (utilizing RouteLAND), the specific route segment's vista is mainly characterized as (i.e., 52.07%) "Herbaceous vegetation associations"; this characterization is visually validated by the top-right geovisualization, where the aforementioned LU/LC is indeed a dominant element of the visible landscape. In the same vein, the results depicted at the bottom-left geovisualization (utilizing RouteLAND) are also visually validated by the generated bottom-right geovisualization.

5. Discussion and Conclusions

5.1. Synopsis

The present article had as a main goal to provide a new, integrated method for characterizing highway routes based on landscape composition and visual significance (i.e. actually perceived sizes of landscape elements), along with a software tool (RouteLAND) through which the sequence of the required geocomputations and calculations were built and implemented. The application of the method through the tool produced specific results for a selected use case; these results were cartographically visualized in an interactive web map—also developed for the needs of this article. By summarizing the results of executing RouteLAND in a selected peri-urban area in Attica region, Greece, it was found that the most frequently occurring (within the route) prevailing landscape elements were "Herbaceous vegetation associations", while the quantitatively most dominant landscape elements—albeit less frequently occurring—were "Other roads and associated land". The web map generated provides an overall visualization of the results of the use case.

5.2. Assumptions and Limitations

The introduced method has been built upon previous attempts to quantify the landscape composition within the *d*FoV along highway routes. Toward this end, there have been some assumptions made referring either to the utilized datasets or to certain conceptualization issues. Some additional limitations are also associated with the developed software.

The UrbanAtlas LU/LC dataset—a categorized geodata layer—has been considered to appropriately represent the exocentric perspective of the visual landscape. It is apparent that using this dataset constitutes a simplification of the actual visual landscape (compared to, e.g., a high spatial and spectral resolution satellite image), yet this is a reasonable one, taking into account the purpose and scale of the produced maps. The elevation dataset used also constitutes a simplification of the actual morphological conditions: the 30 m EU-DEM is a medium-scaled approximation of the topographic relief compared to a very precise and detailed model of the actual surface morphology, including vegetation and buildings' heights (e.g., fine-scaled DSM). However, for the requirements of this landscape-based application and provided that the main focus was on the development of the method and the tool, the usage of this elevation geospatial dataset is assumed to be adequate. All the previous assumptions are supported by the pertinent literature (e.g., [28,31,39]).

Additionally, the maximum viewing distance, which was set to 5 km, poses a concern; however, it is a reasonable assumption that elements positioned in the far background of the viewing field are considered to have a negligible contribution within the hFoV. Moreover, the calculation of the horizontal field of view according to the maximum allowed speed of each segment of the highway, namely, the 'narrowing' of the horizontal hFoV with the increase in the movement speed, is also based on assumptions, yet they are plausible and empirically derived ones; what is more, the hFoV applied from each viewpoint is associated with a more simplified (rectangular) version of the actual hFoV. Another point of concern is the spacing intervals at which the highway route was 'sampled'; extracting viewpoints for every 50 m of the route and computing viewsheds from these viewpoints was considered an appropriate viewpoint interval for specifying the *d*FoV and executing all the associated computations. At this point, it should be noted that related research work also supports the abovementioned assumptions (e.g., [29,38,39,43–47]).

Regarding the software tool, the first released version of RouteLAND has two main limitations. First, although RouteLAND has been freely distributed to the scientific community, its functionality depends on the utilization of the commercial ArcPy Python site package provided by ESRI®. Second, the developed tool was tested and executed in Python version 2.7.16, which is compatible with the ArcMap 10.8 (ESRI®) environment in which the presented case study was implemented.

5.3. Future Research and Outlook

The present methodology executed through the RouteLAND tool has enabled the automatic characterization of any potential route according to the landscape element that dominates the *d*FoV. Thus, through this tool, we modeled the vistas apperceived from vehicles moving upon highway routes, hence managing to cartographically represent the actual composition of landscape elements that drivers and passengers are experiencing in their everyday transportation. Since highways are crucial transportation links involving large numbers of observers, this method/tool can be immediately embedded and utilized in cases of comparative landscape analyses in peri-urban regions, further feeding public decision support systems for landscape evaluation and sustainable spatial planning. This first endeavor can be further amended, ameliorated, and expanded to meet the needs and goals of relative applications and (scientific) fields.

As regards the software tool itself, there are some enhancements that can be made: In order to be more generic and less dependent on existing GIS software and commercial Python site packages, future versions of the tool could be based on the exclusive use of open-source libraries and/or modules. Future versions of the tool could be developed in Python version(s) 3.x, also taking advantage of the core improvements of the Python programming language. In addition, RouteLAND can be further extended by incorporating an automatic process for generating a web map that visualizes the final output and supports effective exploration of the produced results. Finally, the existing source code constitutes a solid groundwork toward developing a new standalone software or a QGIS plugin with an integrated graphical user interface (GUI).

Aside from the future work toward enhancing the tool's usability and functionality, the applications of this integrated method can be further extended and expanded. In the realm of landscape studies and research, this method/tool can be adopted in relation to Landscape Character Assessment (LCA). As LCA is concerned primarily with attributing the landscape character—i.e., the distinct and consistent pattern of elements in the landscape—to an area [48] viewed from the exocentric perspective [24], this method/tool can enable a shift for LCA from the exocentric to the egocentric perspective—that is, computing the LCA along routes of great importance. A step further could be the evaluation of landscape character according to certain criteria, such as naturalness, complexity, disturbance (see, e.g., [49] for such criteria), or the evaluation and rating of a landscape's visual quality/preferences along routes. Dramstad et al. [50] have long ago investigated the relationship between visual landscape preferences and map-based indicators (indices and metrics) of landscape composition and structure by encompassing viewshed analyses into their investigation. RouteLAND has the potential to accurately measure and calculate indicators of landscape composition and structure from the egocentric perspective and within the *d*FoV along routes.

The RouteLAND integrated method could also address issues related to visual impact assessment (VIA) and 'feed' alternative design scenarios toward the optimization of spatial/landscape planning. For instance, the visual impacts from intrusive surface mining activities (i.e., quarries) have already been quantified and mapped through a geospatial model (GEMMELIM) [29]. However, this model does not take into consideration the *d*FoV. RouteLAND could be adjusted to serve as a tool for the dynamic VIA of quarries or other technical works related to energy projects (e.g., wind farms, solar farms, etc.) along routes. From a somewhat different point of view, RouteLAND could be utilized as a tool for spatial planning supporting decisions about either the assessment of alternative routes or the design of new roads (routes) by creating outputs that 'feed' cost path/surface geoprocesses; the computation of the least cost path regarding the contribution of an intrusive energy project within the *d*FoV can support urban planners to qualify a certain route (which minimizes the visual nuisance) among a multitude of alternative ones. Even more fundamentally, the introduced method of this article could support existing legislation concerned with VIA, providing a scientifically sound basis for quantifying the impacts from existing or future technical projects (see also [51]). The other side of the coin of VIA is associated with the maximization of exposure to vistas of landscape elements that are either positively rated by people (e.g., lakes, forests, buildings of traditional architectures, etc.), or are of particular importance (e.g., monuments, mountain peaks, various landmarks, etc.). In this sense, RouteLAND can assess existing (land or sea) routes or prompt the design of new routes that maximize their vistas to such valuable landscape resources. As an effect, RouteLAND can support spatial planning decisions aiding in the amelioration of the touristic experience along routes and, more generally, in further promoting sustainable landscape design.

Finally, the introduction of this method/tool furnishes a new opportunity to fuse GISystems and GIScience with the fields of eye movement analysis and visual perception. This outlook has been introduced and empirically explored by Schirpke et al. [52] and revisited by Misthos et al. [24]. RouteLAND computes the visible landscape elements within the *d*FoV in both a quantitatively and a spatially explicit manner. Thence, the embedment of eye movement recordings and analyses appears to be a very promising ground for coupling map (2D) geospatial information with information derived from the visual perception of actual observers. This fusion becomes even more intriguing when it is synthesized in a dynamic sequence in cases of routes upon the ground and observers moving upon these routes.

Supplementary Materials: The source code of RouteLAND is freely available at https://github.com/lmmisthos/RouteLAND (accessed on 11 March 2025) under the third version of the GNU General Public License (GPL v3). The final output of the RouteLAND tool (for the selected use case) is available at https://routeland.geo.uniwa.gr/ (created on 29 November 2024).

Author Contributions: Conceptualization, Loukas-Moysis Misthos; methodology, Loukas-Moysis Misthos; software development, testing, and validation, Loukas-Moysis Misthos and Vassilios Krassanakis; formal analysis, Loukas-Moysis Misthos; resources, Loukas-Moysis Misthos; data curation, Loukas-Moysis Misthos and Vassilios Krassanakis; writing—original draft preparation, Loukas-Moysis Misthos; writing—review and editing, Loukas-Moysis Misthos and Vassilios Krassanakis; visualization, Loukas-Moysis Misthos and Vassilios Krassanakis All authors have read and agreed to the published version of the manuscript.

Funding: This research received no external funding.

Data Availability Statement: The data presented in this study are available upon request. Open data used in this study were extracted from the websites https://extract.bbbike.org/, https://www.

opendem.info/ and https://land.copernicus.eu/en/products/urban-atlas/urban-atlas-2018 (all accessed on 15 October 2024).

Conflicts of Interest: The authors declare no conflicts of interest.

References

1. Bergen, S.D.; Ulbricht, C.A.; Fridley, J.L.; Ganter, M.A. The Validity of Computer-Generated Graphic Images of Forest Landscape. *J. Environ. Psychol.* **1995**, *15*, 135–146. [CrossRef]
2. Arriaza, M.; Cañas-Ortega, J.F.; Cañas-Madueño, J.A.; Ruiz-Aviles, P. Assessing the Visual Quality of Rural Landscapes. *Landsc. Urban Plan.* **2004**, *69*, 115–125. [CrossRef]
3. Bulut, Z.; Yilmaz, H. Determination of Landscape Beauties through Visual Quality Assessment Method: A Case Study for Kemaliye (Erzincan/Turkey). *Environ. Monit. Assess.* **2008**, *141*, 121–129. [CrossRef]
4. Howley, P.; Donoghue, C.O.; Hynes, S. Landscape and Urban Planning Exploring Public Preferences for Traditional Farming Landscapes. *Landsc. Urban Plan.* **2012**, *104*, 66–74. [CrossRef]
5. López-martínez, F. Visual Landscape Preferences in Mediterranean Areas and Their Socio-Demographic Influences. *Ecol. Eng.* **2017**, *104*, 205–215. [CrossRef]
6. Ulrich, R.S. Human Responses to Vegetation and Landscapes. *Landsc. Urban Plan.* **1986**, *13*, 29–44. [CrossRef]
7. Misgav, A. Visual Preference of the Public for Vegetation Groups in Israel. *Landsc. Urban Plan.* **2000**, *48*, 143–159.
8. Han, K.-T. Responses to Six Major Terrestrial Biomes in Terms of Scenic Beauty, Preference, and Restorativeness. *Environ. Behav.* **2007**, *39*, 529–556. [CrossRef]
9. Hammitv, W.E.; Pattersonb, M.E.; Noe, F.P. Identifying and Predicting Visual Preference of Southern Appalachian Forest Recreation Vistas. *Landsc. Urban Plan.* **1994**, *29*, 171–183. [CrossRef]
10. Purcell, A.T.; Lamb, R.J.; Mainardi Peron, E.; Falchero, S. Preference or Preferences for Landscape? *J. Environ. Psychol.* **1994**, *14*, 195–209. [CrossRef]
11. Strumse, E. Perceptual Dimensions in the Visual Preferences for Agrarian Landscapes in Western Norway. *J. Environ. Psychol.* **1994**, *14*, 281–292.
12. Gao, H.; Abu Bakar, S.; Maulan, S.; Yusof, M.J.M.; Mundher, R.; Chen, B. How Highway Landscape Visual Qualities Are Being Studied: A Systematic Literature Review. *Land* **2024**, *13*, 431. [CrossRef]
13. Tempesta, T. The Perception of Agrarian Historical Landscapes: A Study of the Veneto Plain in Italy. *Landsc. Urban Plan.* **2010**, *97*, 258–272. [CrossRef]
14. Forman, R.T.T. Town Ecology: For the Land of Towns and Villages. *Landsc. Ecol.* **2019**, *34*, 2209–2211. [CrossRef]
15. Ulrich, R.S. Visual Landscapes and Psychological Well-Being. *Landsc. Res.* **1979**, *4*, 17–23.
16. Mcmahan, E.A.; Estes, D. The Effect of Contact with Natural Environments on Positive and Negative Affect: A Meta-Analysis. *J. Posit. Psychol.* **2015**, *9760*, 1–13. [CrossRef]
17. Kaplan, S. The Restorative Benefits of Nature: Toward an Integrative Framework. *J. Environ. Psychol.* **1995**, *15*, 169–182. [CrossRef]
18. Gullone, E. The Biophilia Hypothesis and Life in the 21st Century: Increasing Mental Health or Increasing Pathology? *J. Happiness Stud.* **2000**, *1*, 293–322. [CrossRef]
19. Ulrich, R.S.; Simons, R.F.; Losito, B.D.; Fiorito, E.; Miles, M.A.; Zelson, M. Stress Recovery during Exposure to Natural and Urban Environments. *J. Environ. Psychol.* **1991**, *11*, 201–230. [CrossRef]
20. Grinde, B.; Patil, G.G. Biophilia: Does Visual Contact with Nature Impact on Health and Well-Being? *Int. J. Environ. Res. Public Health* **2009**, *6*, 2332–2343. [CrossRef] [PubMed]
21. Ulrich, R.S. Aesthetic and Affective Response to Natural Environment. In *Behavior and the Natural Environment*; Altman, I., Wohlwill, J.F., Eds.; Springer: Boston, MA, USA, 1983; pp. 85–125. ISBN 978-1-4613-3539-9.
22. UN Department of Economic and Social Affairs. *World Urbanization Prospects*; United Nations: New York, NY, USA, 2019; Volume 12, ISBN 9789211483192.
23. Franěk, M. Landscape Preference: The Role of Attractiveness and Spatial Openness of the Environment. *Behav. Sci.* **2023**, *13*, 666. [CrossRef]
24. Misthos, L.-M.; Krassanakis, V.; Merlemis, N.; Kesidis, A.L. Modeling the Visual Landscape: A Review on Approaches, Methods and Techniques. *Sensors* **2023**, *23*, 8135. [CrossRef]
25. Aznar-Casanova, J.; Matsushima, E. Interaction between Egocentric and Exocentric Frames of Reference Assessed by Perceptual Constancy Parameters. *Cogn. Stud.* **2008**, *15*, 22–37.
26. Harrower, M.; Sheesley, B. Moving beyond Novelty: Creating Effective 3D Fly-over Maps. In Proceedings of the Anais: 22th International Cartographic Conference Mapping Approaches into a Changing World, A Coruña, Spain, 9–16 July 2005; pp. 9–16.
27. Misthos, L.-M. *Mountainous Landscape Exploration Visualizing Viewshed Changes in Animated Maps*; National Technical University of Athens: Zografou, Greece, 2014.

28. Misthos, L.-M.; Menegaki, M. Identifying Vistas of Increased Visual Impact in Mining Landscapes. In Proceedings of the 6th International Conference on Computer Applications in the Minerals Industries (CAMI 2016), Istanbul, Turkey, 5–7 October 2016.
29. Misthos, L.M. *Development of a Geospatial, Multiparametric Model for Assessing Landscape Impacts from Mining*; National Technical University of Athens: Zografou, Greece, 2022.
30. Kent, R.L. Determining Scenic Quality along Highways: A Cognitive Approach. *Landsc. Urban Plan.* **1993**, *27*, 29–45. [CrossRef]
31. Schirpke, U.; Tasser, E.; Tappeiner, U. Landscape and Urban Planning Predicting Scenic Beauty of Mountain Regions. *Landsc. Urban Plan.* **2013**, *111*, 1–12. [CrossRef]
32. Brabyn, L. Modelling Landscape Experience Using "Experions". *Appl. Geogr.* **2015**, *62*, 210–216. [CrossRef]
33. Council of Europe. *The European Landscape Convention—Firenze, 20X. 2000 (ETS No. 176)*; Official Text in English; Council of Europe: Strasbourg, France, 2000.
34. Minelli, A.; Marchesini, I.; Taylor, F.E.; De Rosa, P.; Casagrande, L.; Cenci, M. An Open Source GIS Tool to Quantify the Visual Impact of Wind Turbines and Photovoltaic Panels. *Environ. Impact Assess. Rev.* **2014**, *49*, 70–78. [CrossRef]
35. Nutsford, D.; Reitsma, F.; Pearson, A.L.; Kingham, S. Personalising the Viewshed: Visibility Analysis from the Human Perspective. *Appl. Geogr.* **2015**, *62*, 1–7. [CrossRef]
36. Gibson, J.J. *The Ecological Approach to Visual Perception*; Classic Edition; Psychology Press: London, UK, 2014.
37. Jin, X. A Review of Cityscape Research Based on Dynamic Visual Perception. *Land* **2023**, *12*, 1229. [CrossRef]
38. Misthos, L.M.; Nakos, B.; Mitropoulos, V.; Krassanakis, V.; Menegaki, M.; Batzakis, D.V. The Effectiveness of Propagating Viewsheds' Geovisualization from Topographically Prominent Viewroutes. In Proceedings of the 10th International Congress of the Hellenic Geographical Society, Thessaloniki, Greece, 22–24 October 2014; pp. 22–24.
39. Misthos, L.-M.; Nakos, B.; Krassanakis, V.; Menegaki, M. The Effect of Topography and Elevation on Viewsheds in Mountain Landscapes Using Geovisualization. *Int. J. Cartogr.* **2019**, *5*, 44–66. [CrossRef]
40. Jaal, Z.; Abdullah, J. User's Preferences of Highway Landscapes in Malaysia: A Review and Analysis of the Literature. *Procedia—Soc. Behav. Sci.* **2012**, *36*, 265–272. [CrossRef]
41. Jin, X.; Wang, J. ScienceDirect Assessing Linear Urban Landscape from Dynamic Visual Perception Based on Urban Morphology. *Front. Archit. Res.* **2021**, *10*, 202–219. [CrossRef]
42. Dupont, L.; Ooms, K.; Duchowski, A.T.; Antrop, M.; Van Eetvelde, V. Investigating the Visual Exploration of the Rural-Urban Gradient Using Eye-Tracking. *Spat. Cogn. Comput.* **2017**, *17*, 65–88. [CrossRef]
43. Barbu, D.M. Visual Field Evaluation Method Of The Automobile Drivers In Traffic. *Ann. Fac. Eng. Hunedoara* **2016**, *14*, 163.
44. Dharmasena, S.; Edirisooriya, S. Impact of Roadside Landscape to Driving Behaviour; Lessons from Southern Highway, Sri Lanka. *Cities People Places Int. J. Urban Environ.* **2018**, *3*, 66. [CrossRef]
45. Rodrigues, M.; Montañés, C.; Fueyo, N. A Method for the Assessment of the Visual Impact Caused by the Large-Scale Deployment of Renewable-Energy Facilities. *Environ. Impact Assess. Rev.* **2010**, *30*, 240–246. [CrossRef]
46. Menegaki, M.E.; Kaliampakos, D.C. Evaluating Mining Landscape: A Step Forward. *Ecol. Eng.* **2012**, *43*, 26–33. [CrossRef]
47. U.S. Government Printing Office. *USDA Forest Service National Forest Landscape Management, Vol. 1: Agriculture Handbook No 434*; U.S. Government Printing Office: Washington, DC, USA, 1973.
48. Swanwick, C. *Landscape Character Assessment: Guidance for England and Scotland*; The Countryside Agency and Scottish Natural Heritag: London, UK, 2002.
49. Tveit, M.; Ode, Å.; Fry, G. Key Concepts in a Framework for Analysing Visual Landscape Character. *Landsc. Res.* **2006**, *31*, 229–255. [CrossRef]
50. Dramstad, W.E.; Tveit, M.S.; Fjellstad, W.J.; Fry, G.L.A. Relationships between Visual Landscape Preferences and Map-Based Indicators of Landscape Structure. *Landsc. Urban Plan.* **2006**, *78*, 465–474. [CrossRef]
51. Palmer, J.F. The Contribution of a GIS-Based Landscape Assessment Model to a Scientifically Rigorous Approach to Visual Impact Assessment. *Landsc. Urban Plan.* **2019**, *189*, 80–90. [CrossRef]
52. Schirpke, U.; Tasser, E.; Lavdas, A.A. Potential of Eye-Tracking Simulation Software for Analyzing Landscape Preferences. *PLoS ONE* **2022**, *17*, e0273519. [CrossRef] [PubMed]

International Journal of
Geo-Information

MDPI

Article

Ecological Network Construction Based on Red, Green and Blue Space: A Case Study of Dali City, China

Rong Chen [1], Shunmin Zhang [1], Xiaoyuan Huang [2], Xiang Li [1] and Jiansong Peng [3,*]

[1] College of Soil and Water Conservation, Southwest Forestry University, Kunming 650224, China; cr13579@swfu.edu.cn (R.C.); zsm2021@swfu.edu.cn (S.Z.); lx330@swfu.edu.cn (X.L.)
[2] College of Economics and Management, Southwest Forestry University, Kunming 650224, China; hxy21cn@swfu.edu.cn
[3] College of Ecology and Environment, Southwest Forestry University, Kunming 650224, China
* Correspondence: pjs@swfu.edu.cn

Abstract: Rapid urbanization leads to fragmentation and reduced connectivity of urban landscapes, endangering regional biodiversity conservation and sustainable development. Constructing a red, green, and blue spatial ecological network is an effective way to alleviate ecological pressure and promote economic development. Using circuit theory, hydrological analysis, and suitability analysis, this study constructs a composite ecological network under urban–rural integration. The results show the following: (1) A total of 22 ecological corridors with a length of 349.20 km, 22 ecological pinch points, and 22 ecological barrier points are identified in the municipal area, mainly distributed in Haidong Town. There are 504 stormwater corridors, which are more evenly distributed, 502 riverfront landscape corridors, and 130 slow-moving landscape corridors. (2) A total of 20 ecological corridors, with a length of 99.23 km, 19 ecological pinch points, and 25 barrier points were identified in the main urban area, and most of them are located in the ecological corridors. There are 71 stormwater corridors, mainly located in the northwestern forest area, 71 riverfront recreation corridors, and 50 slow-moving recreation corridors. (3) Two scales of superimposed ecological source area of 3.65 km^2, and eleven ecological corridors, are primarily distributed between Erhai Lake and Xiaguan Town. There are two superimposed stormwater corridors and fourteen recreational corridors. The eco-nodes are mostly distributed in the east and south of Dali City; wetland nodes are mainly situated in the eighteen streams of Cangshan Mountain; and landscape nodes are more balanced in spatial distribution. The study results can provide a reference for composite ecological network construction.

Keywords: complex ecological network; red–green–blue space; urban–rural integration; multi-scale; Dali City

1. Introduction

Urban–rural ecological contrasts, fragmentation of the landscape, and other ecological issues have been brought about by human activity and continuous urban expansion, which stress the ecosystem [1]. Because of this, nations all over the world have placed a strong emphasis on sustainable development, and the idea of an "eco-city" has been proposed [2]. An eco-city is a perfect city model that has a reasonable structure, stable function, and socioeconomic development that does not exceed the ecological environment's carrying capacity [3]. The relationship between China's urban and rural areas has entered a new phase of growth. This is due to the promotion of new urbanization, rural revitalization, and other national programs. The concept of "urban-rural integration and development" was first introduced in the 19th CPC National Congress Report in 2017. It emphasizes that urban and rural areas as a whole, rather than a single town or rural area, are the center of gravity of urban–rural development, focusing on multi-level penetration and integration [4]. It is centered on penetration and intermingling, which is multi-level, multi-domain, and all-around [5]. A key strategy for ecological integration of urban and rural areas is the construction of composite socio-ecological networks [6]. These networks link disparate habitat patches into a cohesive network of landscapes and habitats. This ensures regional ecological security on a limited amount of ecological land, enhances natural ecosystem health, and safeguards biodiversity.

Various administrative levels, the extent of ecological space, project management, planning concepts, and the meaning of ecological land are all significant considerations in eco-city design. Various ecological space levels provide distinct functions and need to be taken into account for various planning and value considerations. While regional differences are evident at the macro level, specific attention should be paid to the network relationship between nature, humans, and structures at the micro level, with the partitioning of ecological space based on the region as a whole [7]. In order to connect urban and rural areas organically and preserve urban ecosystem stability, it is necessary to integrate urban and rural ecological space in a way that is tailored to the local environment. This will also ensure ecological security.

Red, green, and blue space is a collection of water, green space, and recreational resources. Blue space includes natural waters, such as rivers, lakes, mudflats, and wetlands, as well as man-made reservoirs and ditches; green space refers to public green space, protective green space, and ancillary green space integrated with urban construction land [8]. Red space, on the other hand, is a tourism resource that provides slow walking and leisure recreation for the public. Currently, there is no uniform definition of ecological networks in the academic world; in general, ecological networks, in the eyes of different experts and scholars, help to maintain ecological processes' consistency [9]. In 1939, Troll [10], a leading German biogeographer, proposed the formal introduction of landscape into ecology as the totality of the environmental space in which human beings live and the whole of everything they can see visually, and emphasized that landscape ecology is an integrated study combined with recently developed aerial photogrammetry, ecology, and geography [11]. Ecological networks are network systems of natural ecological space and artificial green space with ecological significance within the city, connected by linear ecological corridors [12].

To date, a large number of scholars have examined ecological networks from the angles of geography, urban planning, and landscape ecology [13]. The main studies include the structure [14] and function [15,16] of ecological networks and the construction [17], evaluation [18], and optimization [19] of ecological networks. They have done this using techniques such as minimum cumulative resistance (MCR) [20], graph theory [21], morphology spatial pattern analysis (MSPA) [22], circuit theory [23], etc. to form the "comprehensive ecological source area identification, landscape resistance surface construction, ecological corridor extraction, ecological network." Formally, the main paradigm has been formed as "identification of ecological sources, construction of landscape resistance surfaces, extraction of ecological corridors, and construction of ecological networks" [24–26]. Researchers have studied several types of region at several scales, and conducted a number of optimization studies in recent years of the determination the width of ecological corridors [27,28], extraction of ecological nodes [29], and multi-objective [30,31] composite ecological networks. Nonetheless, the distinctive ecological requirements of cities and the objective of offering recreational services to citizens have been partially disregarded by the current research, which has also not adequately taken into account the nested relationships of ecological networks at various scales and the paucity of studies on the ecological abundance of low-county units. Given that ecological restoration and protection are major carriers of new urbanization, more focused and methodical research is required to increase the efficacy of ecological networks in these kinds of areas. Weighing the hierarchy of multi-objective demands in red, green, and blue networks, and optimizing the spatial coordination of multiple networks, are also difficult problems to solve in ecological network optimization.

As the only international wetland city in Yunnan Province, Dali City is a popular destination for tourists in China, the rapid urbanization of the region has harmed the ecological environment, causing the issue of urban and rural landscapes to become more serious. In view of this, this study takes Dali City as an example, and the research objectives include the following: (1) identifying the ecological source in the municipal and main urban areas in terms of biological processes, hydrological processes, and humanistic processes; (2) constructing the landscape resistance surfaces of the municipal and the main urban areas; (3) applying the circuit theory to identify the nested and connected ecological corridors, pinch points, and barrier points of the municipal and the main urban areas, and constructing the multi-scale composite ecological network of the red, green, and blue spaces; and (4) proposing the "point, line, and surface" combination of the optimization scheme to coordinate the ecological protection and restoration of natural elements, and make suggestions for further ecological management. The results of the study will alleviate the contradiction between ecological protection and economic development in highland wetland cities and provide the necessary basis and reference for exploring the ecological environment and construction mode of highland wetland cities.

2. Materials and Methods

2.1. Study Area

Dali City, located in the western part of Yunnan Province, in the heart of Dali Bai Autonomous Prefecture, is the political, economic, and cultural center of the state and also the capital of Dali Prefecture. Located between 99–58° to 100–27° east longitude and 25–25° to 26–02° north latitude, the city has jurisdiction over nine towns, one township, and three streets, as well as the Dali Economic Development Zone and Dali Haidong Development Committee. The terrain is high in the northwest and low in the southeast. It is situated in the low-latitude and high-altitude zones and the subtropical plateau monsoon climate zone, and the average annual temperature is 15.2 °C. Erhai Lake within the territory covers an area of 250 km^2, accounting for 13.7% of the total area. In recent years, due to urban development, habitat patches have been fragmented and ecological pressure has intensified, and there is an urgent need to resolve the conflict between economic development and the ecological environment. Therefore, ecological protection cannot be neglected. To ensure the ecological security of Dali City, the multi-scale network structure provides an effective solution. The overview of the study area is shown in Figure 1.

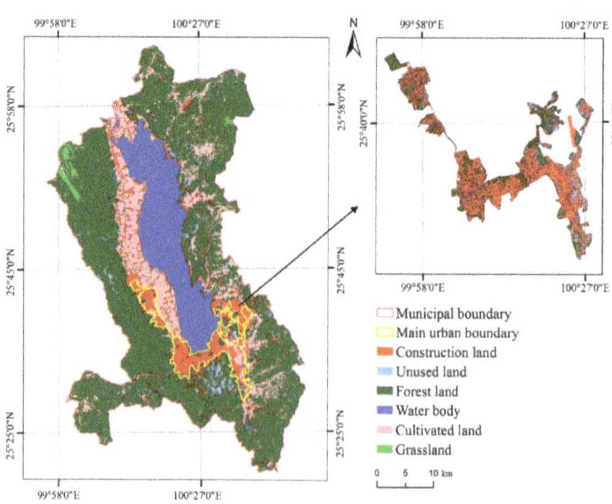

Figure 1. Overview of the study area.

2.2. Data Sources and Pre-Processing

The land use data and vegetation cover involved in this study were obtained from remote sensing image decoding; the data were obtained from the Geospatial Data Cloud

Platform "http://www.gscloud.cn (accessed on 21 March 2024)"; the slope was obtained from the (digital elevation grid) DEM through ArcGIS "https://desktop.arcgis.com/ (accessed on 21 March 2024)" slope analysis; the road network vector data were obtained from the 2006 and 2016 two-phase forest resources type II survey data; and the nighttime light data, (point of interest) POI data, and administrative boundary data were obtained from the BIGEMAP GIS Office "http://www.bigemap.com/ (accessed on 21 March 2024)". All images and data are unified in the WGS_1984_UTM_Zone_47N coordinate system.

2.3. Research Methodology

In this study, the main urban and Dali city municipal scales are used to form the ecological network. At the highest level of building, municipal works can support the stability of the surrounding ecological environment, thereby maintaining the overall ecological security of the city. In the "Dali City Provincial Forest City Construction Master Plan (2017–2026)," the town development boundary is defined as a small-scale study object, defined as the urban area, and Dali City is defined as a large-scale study object, defined as the municipal area, according to the "Dali City Territorial Spatial Master Plan (2021–2035)." When building an urban network, attention is paid to creating open ecological spaces that serve the needs of the local population while also enhancing and optimizing the network's structural integrity.

Therefore, according to the actual situation of the study area, this study considers the functional positioning and spatial needs of different scales; takes bio-hydrology–humanities as the goal orientation; adopts the methods of ecological source identification, landscape resistance surface construction, and ecological element identification to construct a habitat network; adopts the methods of river network extraction, stormwater corridor construction, and wetland node construction to construct a water–green network; and constructs recreation and leisure corridors according to the suitability analysis and accessibility analysis. The recreation corridor is constructed based on suitability analysis and accessibility analysis, and the multi-scale nesting of the network and the construction and optimization of the multi-objective ecological network are carried out. The specific research framework is shown in Figure 2.

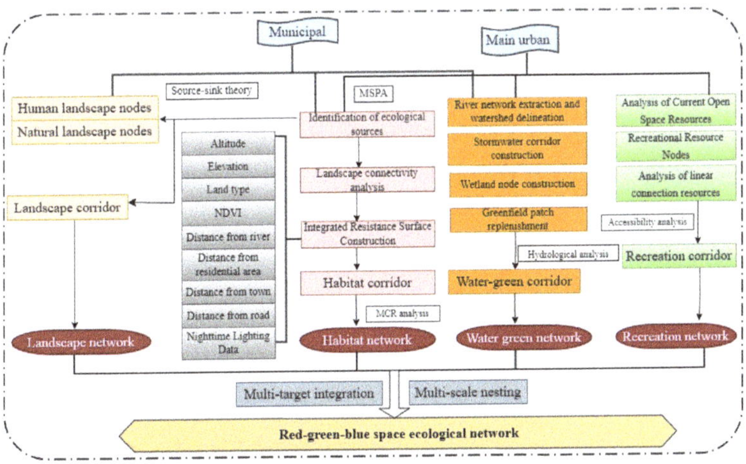

Figure 2. Research framework. Source–sink theory suggests that landscapes can be categorized as "sources" or "sinks", with "source" landscapes being those that facilitate processes and "sink" landscapes being those that prevent or retard processes. "Sinks" landscapes are those that prevent or retard the development of processes. Accessibility analysis is an important indicator of the effectiveness of an urban transport system, referring to the ability of residents to reach their destinations within a certain time and cost.

2.3.1. Habitat Network Construction

(1) Ecological source area identification

MSPA (Morphological Spatial Pattern Analysis) is a customized sequence of mathematical morphological operators targeted at the description of the geometry and connectivity of the image components. Based on geometric concepts only, this methodology can be applied at any scale and to any type of digital image in any application field [32]. MSPA constitutes an analytical method that can identify essential patches that have a significant influence on the ecological connectivity of the study area. In this study, by taking forest and grassland as the foreground and other land types as the background, seven landscape types (core, isolated, bridging, edge, pore, loop, and spur) were obtained by setting the width of the edge of the municipal area to 5 pixels and the urban area to 3 pixels using the eight-neighborhood method in the MSPA analysis module in Guidos Toolbox [33] "https://forest.jrc.ec.europa.eu/en/activities/lpa/gtb/ (accessed on 22 March 2024)".

As the selection range of the source region, 2 km and 1 km buffer zones were established for the municipal area and urban area, respectively, in order to maintain range continuity and lessen the influence of each administrative entity on ecosystem services [34,35]. Landscape connectedness illustrates the impact of many types of landscapes, modifying the dispersion resistance of a species [36]. Regional landscape connectivity and the importance of each ecological patch were measured by overall connectivity (IIC) and probable connectivity (PC) [37,38], and the larger the value of dPC, the more important the patch is to the overall connectivity of the landscape. According to the range of patch area in the core area ($\geq$2 km^2 in the municipal area and $\geq$0.1 km^2 in the urban area), Conefor software "www.conefor.org (accessed on 22 March 2024)" was used to set the connectivity distance threshold to 500 m [39], and the probability of connectivity to 0.5, and the patch connectivity index (dPC) was calculated so as to select the ecological source in the municipal area and the main urban area.

$$\text{IIC} = \frac{\sum_{i-1}^{n} \sum_{j=1}^{n} \frac{a_i a_j}{1+nl_{ij}}}{A_L^Z} \tag{1}$$

where n is the total number of patches in the landscape, a_i and a_j represent the areas of patch i and j, respectively, nl_{ij} is the number of connections between patch i and patch j, and A_L^Z is the area of the landscape. $0 \leq \text{IIC} \leq 1$; a value of IIC of 0 indicates that there is no connection between the habitat patches. Conversely, when the value of IIC is 1, it means that the entire area involved in the calculation is a habitat patch.

$$\text{PC} = \frac{\sum_{i=1}^{n} \sum_{j=1}^{n} 1 \neq jP_{IJa_i a_j}^{*}}{A_L^Z} \tag{2}$$

where n is the total number of patches in the landscape, P $\times$ ij represents the maximum probability of species spreading between patch i and patch j, a_i and a_j represent the areas of patch i and j respectively, and A_L^Z is the area of the landscape. $0 < \text{PC} < 1$.

$$\text{dPC} = \frac{PC - PC_{remove}}{PC} \times 100\% \tag{3}$$

where PC indicates the possible connectivity index of a patch in the landscape and PC_{remove} indicates the possible connectivity index of the landscape after the patch is removed.

(2) Ecological resistance surface construction

The resistance surface illustrates both the challenge of relocating between ecological patches of various habitats and the significance of landscape variety in the ecological flow process. The resistance factors to be constructed in this study are land use type, elevation, slope, normalized difference vegetation index (NDVI), distance from roads, distance from rivers, distance from settlements, and nighttime lighting index (the selection and weighting of specific indicators are shown in Table 1). These selections are based on previous research, as well as the actual conditions of the study area [40,41] . The integrated resistance surface of the municipal area and the main urban area was obtained using the hierarchical analysis method (AHP) "http://ahpman.metadecsn.com/ (accessed on 21 March 2024)" and the raster calculator tool (Figure 3). Each resistance factor was categorized according to a five-level system; the lower the level, the lower the resistance, and vice versa.

Table 1. Resistance factors and resistance values.

Scale	Drag Factor	Resistance Value Rating					Weight
		1	2	3	4	5	
Municipal	Land use type	Forest, waters	Grasslands	Cultivated land	Unutilized land	Building land	0.28
	Elevation (m)	<1500	1500–2000	2000–2500	2500–3000	>3000	0.19
	Slope (°)	0–6.20	6.20–14.63	14.63–23.20	23.20–33.33	33.33–70.99	0.14
	NDVI	−0.21	0.02–0.16	0.16–0.25	0.25–0.34	0.34–0.56	0.06
Main urban	Land use type	Forest	Water	Cultivated land	Unutilized land	Building land	0.24
	Elevation (m)	<2000	2000–2100	2100–2200	2200–2300	>2300	0.15
	Slope (°)	0–6.05	6.05–18.45	18.45–31.69	31.69–45.87	45.87–75.22	0.15
	NDVI	−0.14	0.01–0.11	0.11–0.20	0.20–0.29	0.29–0.55	0.06
	Distance from river (m)	<500	500–1000	1000–1500	1500–2000	>2000	0.07
	Distance from road (m)	<500	500–1000	1000–1500	1500–2000	>2000	0.07
	Distance from residential area (m)	<1000	1000–2000	2000–3000	3000–4000	>4000	0.12
	Nighttime Light index	<5	10	15	20	>20	0.14

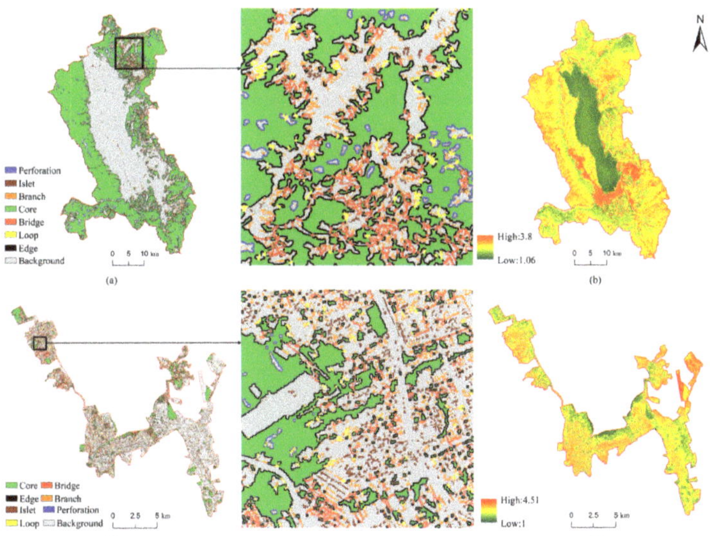

Figure 3. Spatial distribution of MSPA landscape type and ecological resistance surface. (**a**) Municipal MSPA landscape type. (**b**) Municipal ecological resistance surface. (**c**) Main urban MSPA landscape type. (**d**) Main urban ecological resistance surface.

(3) Ecological network construction and optimization

By giving various ecological significances to physical parameters like resistance, current (which indicates the probability of a species spreading along a given path during movement), and voltage (used to predict the likely probability that a species will leave any one focus and successfully reach the next given focus), circuit theory—which focuses on the random walk properties of wildlife—allows the discovery of ecological corridors [42]. The main principle of the MCR is that the spatial components need to overcome a certain landscape resistance when flowing; the stronger the landscape ecological service function, the more perfect the landscape function, and the smaller the resistance. It minimizes data redundancy and combines the benefits of graph theory and random migration theory in comparison to the MCR model [43]. Circuit theory was used to identify ecological corridors at both scales in Dali City, taking into account the variations in ecological sources and ecological resistance at various scales.

Based on the construction of ecological source and integrated resistance surfaces, the Linkage Mapper toolbox "https://linkagemapper.org/ (accessed on 22 March 2024)" in the ArcGIS platform was used to extract the ecological corridors at the municipal and main urban scales, i.e., ecological circulation paths, and the weighted cost distances of the municipal and main urban corridors were set to 20,000 m [44]. Using Circuitscape software "https://circuitscape.org/ (accessed on 22 March 2024)" in the many-to-one and pairwise modes, the cost-weighted distance of the corridor was determined to be 1000 m based on the construction of the corridor. Next, the ecological pinch points were identified by identifying the locations with greater current density (which indicates the likelihood that a species will migrate or spread along a particular path). The natural breakpoint method was utilized to classify them into five categories, and their maximum values were superimposed with the key ecological corridor to obtain key ecological pinch points, with the rest as general ecological pinch points. Ecological pinch points are critical nodes in the movement of organisms, while barrier points are areas of high resistance to the movement and migration of organisms [45]. The ecological barrier points were searched with 300 m and 100 m as two scales of radius, and the maximum value was used to obtain the simulation results of biological activity barrier areas. The results were then divided into five categories using the natural breakpoint method, and the maximum value was superimposed with the key corridor to extract key ecological barrier points; the rest were general ecological barrier points. The migration patterns of animals within a research area can affected by ecological corridors, pinch points, and barrier points. This information can be used to inform the design and optimization of multi-scale synergistic ecological networks.

Key corridors [46] can be identified and targeted for protection by creating an interaction matrix linking source sites. The gravity model, sometimes referred to as the gravitational model, can be utilized as a standard for determining the significance of corridors and the priority of protection. The formula for the gravity model is as follows:

$$G_{ij} = \frac{N_i N_j}{D_{ij}^m} = \frac{\left[\frac{1}{P} \times \ln(S_i)\right]\left[\frac{1}{P} \times \ln(S_j)\right]}{\left(\frac{L_{ij}}{L_{max}}\right)^2} = \frac{L^2_{max \ln(S_i S_j)}}{L_{ij}^2 P_i P_j} \tag{4}$$

where G_{ij} is the interaction force between source patch i and patch j; N_i and N_j are the weight values of patches i and j. D_{ij} is the standardized corridor resistance value between the two patches. P_i denotes the resistance value of patch i; P_j denotes the resistance value of patch j; S_i and S_j are the areas of the two patches; L_{ij} denotes the cumulative resistance value of the corridor between patch i and patch j. L_{max} refers to the potential maximum cumulative resistance value of the corridor.

2.3.2. Water Green Network Construction

The river network was extracted using the ArcGIS hydrologic analysis tool to study the stormwater catchment process in the study area. The level of the river network is different for different cumulative amounts of catchment, and when the cumulative amount is larger, the level is higher [47]. Referring to the related research and after many attempts at debugging, this study selected the thresholds of 1000 and 4000 for the two scales of municipal area and main urban area, respectively, and divided the river network through Strahler's grading method so as to obtain the stormwater corridor of this study. The wetland node at the catchment node can achieve the tasks of flood storage and flood abatement. The catchment node is the primary hydrological node of the water cycle in the stormwater ecological network [48]. By capturing the pouring point, the catchment node is obtained based on the extraction of the stormwater corridor.

Based on previous research [49], the municipal area chooses forest land and grassland areas larger than 2 km^2, as well as lakes and reservoirs larger than 0.1 km^2, in the land use data. Meanwhile, the main urban area extracts forest land larger than 0.002 km^2 and lakes and reservoirs larger than 0.01 km^2 as the origin of the water–green network. The multi-level stormwater corridors connect wetland nodes, greenland patches, and hilly patches to form the water–green ecological network.

2.3.3. Landscape Network Construction

The primary goal of constructing a landscape network is to improve the connection between natural and humanistic landscape (townships and characteristic villages) nodes [7]. Based on Dali City's spatial shape and resource distribution, this article uses the expert scoring method and the hierarchical analysis method for the hierarchical division and weight evaluation. Afterwards, using the three levels of accessibility, landscape, and service, it builds the appropriateness evaluation system. The landscape nodes are connected by the water system, banded green areas, historical and cultural neighborhoods, and featured highways as the main line and the traffic arteries as the link to form a three-dimensional landscape network. Table 2 demonstrates the indicators, evaluation methods, and weights selected for constructing the landscape network.

Table 2. Landscape network construction suitability evaluation system.

Category	Indicator Layer	Evaluation Criteria	Weight
Accessibility	Resource accessibility	Assign 4, 3, 2, 1, and 0 points for distance from landscape nodes	0.18
	Accessibility to transportation facilities	4, 3, 2, 1, 0 points for distance from important transportation terminals	0.29
	Existing linear landscape resources	Assigned 4, 3, 2, 1, and 0 points for distance from existing linear landscape resources	0.12
Scenic	Characteristic of the place	4, 3, 2, 1, 0 points according to whether it is a historical and cultural district under key protection and has good landscape and windscape	0.12
	River	Assign 4, 3, 2, 1, 0 points for distance from the river	0.08
	NDVI	4, 3, 2, 1, 0 points were assigned according to the vegetation cover class	0.05
Serviceability	Distribution density of visitor services	4, 3, 2, 1, 0 points according to space heat level	0.08
	Distribution density of civic service facilities	4, 3, 2, 1, 0 points based on space heat level	0.08

2.3.4. Recreation Network Construction

The goal of the recreation network is to meet the needs of human leisure and transportation [50]. Based on prior research and the actual conditions of the study area [51], 11 evaluation indicators were chosen from 4 categories in this study: ecological landscape, natural resources, humanistic landscape, and transport resources. The weights were then categorized and assessed using the AHP hierarchical analysis method (the selection and weighting of specific indicators are shown in Table 3). Natural and humanistic recreational resources are two types of resources that can be used to construct an ecological network for recreational purposes. In contrast to humanistic recreational resources, which mostly consist of museums, memorials, urban historic districts, and cultural relic preservation units, natural recreational resources in the city include river systems, urban parks, squares, and other natural landscape features.

Table 3. Appropriateness evaluation system for recreational network construction.

Level 1 Element		Secondary Element		Evaluation Methodology
Element	Weight	Element	Weight	
Ecological background	0.16	Elevation	0.1	4, 3, 2, 1, 0 points are assigned according to the slope classification
		NDVI	0.06	4, 3, 2, 1, 0 points were assigned according to the vegetation cover class
Natural resources	0.4	River	0.12	Assign 4, 3, 2, 1, 0 points for distance from the river
		City parks and squares	0.18	Distance from city parks and squares are given 4, 3, 2, 1, 0 points
		Forest parks, scenic areas	0.1	4, 3, 2, 1, 0 points for the popularity, grade and area of forest parks and scenic areas
Human resources	0.24	Unit of cultural relic protection	0.08	4, 3, 2, 1, 0 points for distance from the heritage conservation unit
		Characteristic Historical Neighborhoods	0.08	4, 3, 2, 1, and 0 points for distance from the Distinctive Historic District
		Museums	0.08	Assigned 4, 3, 2, 1, 0 points for distance from the museum respectively
Transportation facilities	0.2	Road traffic network	0.06	4, 3, 2, 1, 0 points given for distance from road traffic network
		Major transportation terminal	0.04	4, 3, 2, 1, 0 points for distance from important transportation terminals
		Slow-moving road network	0.1	Assign 4, 3, 2, 1, 0 points for distance from the slow-moving road network

Based on the findings of the suitability assessment, point and linear regions that are appropriate for the building of greenways are linked, while inappropriate regions are avoided throughout the linking process. The linear resources for connections are divided into rivers and roads. Rivers are natural ecological recreation corridors, so rivers are prioritized as the main recreation corridors in the city. The city road network is a potential recreational corridor in the leisure and recreational ecological network, and the road linear connectivity resources adopt the principle of being slow moving, in view of which, roads with high value and potential for slow-moving recreational transformation are selected as slow-moving recreational corridors in terms of accessibility and resource landscape.

2.3.5. Corridor Width Optimization

Corridor width directly affects organisms' migration and energy flow. Many scholars have also proposed the appropriate corridor width in terms of landscape structure and function [52]. Combined with the current situation of the study area, 30, 60, 90, 150, 300, and 600 m were selected for buffer analysis to obtain the area proportion change values of land type composition under different corridor widths, and the change inflection points were used to determine the appropriate width of each type of corridor.

2.3.6. Multi-Scale Nesting and Multi-Target Integration

The construction of a multi-scale ecological network focuses on the integration and expression of ecological elements at each scale. The multi-scale ecological environment is constructed through different indicator evaluation systems, and direct superposition will lead to a complex ecological environment with high construction costs and the inability to meet multiple objectives at a later stage. Therefore, it is necessary to use a multi-scale nested model to realize the organic connection between the municipal area and the main urban area to ensure that the large scale dominates and regulates the small scale, as well as to convey ecological validity by sharing the important elements across the scales.

Using the ArcGIS overlay analysis method, a "triple-networked" red, green, and blue spatial ecological network was built at the biological, hydrological, and humanistic levels. It consisted of a habitat network intended to protect biodiversity, a water green network intended to improve the ecological environment, and a recreation network intended to promote leisure and open space.

3. Results

3.1. Municipal Network Construction

(1) Habitat network

The core class has the greatest patch in the front background, with a municipal area of 807.6 km^2, or 46.44% of the entire research area, according to the MSPA identification results. Figure 3 illustrates how the municipal area's core is more widely distributed, with the largest patches mostly found in Xiaguan Town and the Cangshan Nature Reserve. The remaining patches are smaller, more dispersed areas, and the overall spatial distribution is fractured. Patch 1 (Cangshan Mountain Reserve) had the greatest impact on the connectivity of ecological networks in Dali City, followed by patches 2 and 3, which had importance indices greater than 10, and were primarily located in the mountainous areas or forested lands in the regions of Xizhou Town, Fengyi Town, Shuanglang, and Xiaguan Town. Based on the analysis of dPC values, 13 ecological sources were identified, with a total area of 759.78 km^2 and an evenly distributed but low overall connectivity. These ecological sources are mostly the study area's mountains, nature reserves, or forest parks, which are important habitats.

Figure 3 depicts the municipal's ecologically resistant surface. The aforementioned findings suggest that the south is mostly home to higher resistance levels, whereas Erhai Lake is home to lower resistance values. Using the Linkage Mapper tool, twenty-two ecological corridors totaling 349.20 km were located. According to the distribution of the corridors, there were corridors connecting each source patch, and they were primarily located in Haidong Town, Wase Town, Fengyi Town, and Shuanglang Ancient. In the backdrop of a karst fissure environment, the eastern region of the country has developed a unique ecosystem. This is the result of special geological formations. This ecosystem is extremely fragile, with a high degree of soil erosion sensitivity, making ecological recovery and reconstruction difficult. In the south, the original forest vegetation has been destroyed, and the functions of water conservation and soil conservation are weak, increasing soil erosion. The degradation and sanding of meadows brought about by rocky desertification are prominent, threatening Erhai Lake's ecological security. Therefore, the conservation and restoration of mountain forest vegetation is the focus of forest city construction in the region.

We numbered the 13 ecological sources and calculated the interaction strength between different ecological source locations in the study area using the gravity model (Figure 4), and the stronger the interaction force between ecological source locations, the more meaningful the construction of intersource corridors. In the municipal area, corridors with an interaction force greater than 200 were selected as key ecological corridors, and the remaining corridors were treated as general ecological corridors. This resulted in 13 key ecological corridors and 9 general ecological corridors. Among them, sources 5 and 6 have the largest interaction force of 3029.13, indicating that the spatial correlation between the two is the strongest, and the lower the resistance encountered by species when migrating and spreading between the two patches, the more favorable it is for regional ecological protection. The interchange of materials is challenging at sources 8 and 11, which also have the lowest contact force and the strongest resistance to organism migration between the two sites. Therefore, to encourage the interchange of materials and biological exchanges between the north and the south, the number and area of ecological nodes in the urban built-up region can be expanded during the planning of possible corridors based on the extracted corridors.

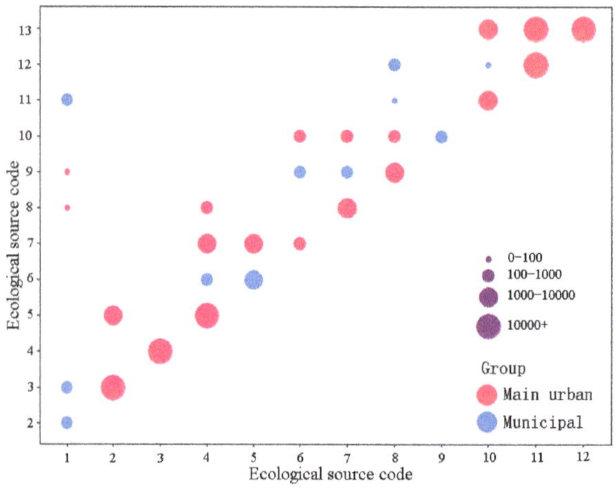

Figure 4. Bubble chart of interaction between ecological source areas based on a gravity model. In the figure, 1–12 represent the ecological source code, and the range of values represents the strength of interaction forces between different ecological sources; the larger the bubble, the stronger the interaction force.

The Pinchpoint Mapper program utilizes Circuitscape to identify ecological pinch points at both sizes. A total of 22 ecological pinch points were found by merging the two modes (Figure 5), of which 9 were general ecological pinch points and 13 were key ecological pinch points in the municipal territory. The eastern and southern sections are more thickly distributed and are primarily centered around the boundaries of the source region and corridor crossings, with the exception of the western and northern sections. With six ecological pinch points, the town of Haidong has the most, making up 27% of all the pinch points.

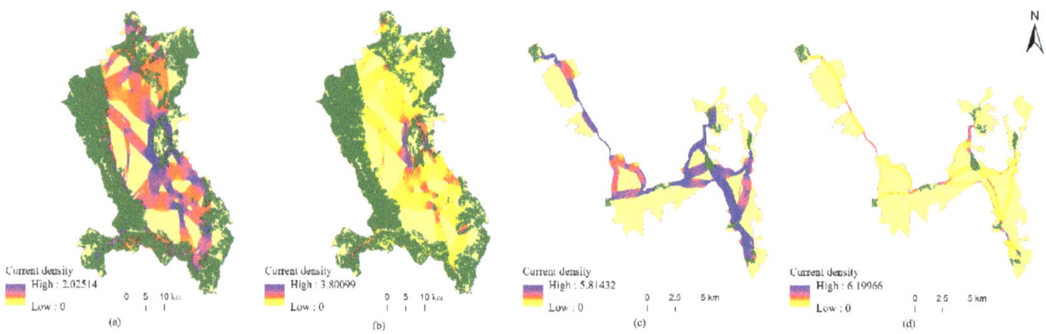

Figure 5. Eco-node analysis based on Pinchpoint Mapper. (**a**) Municipal "pairwise" model. (**b**) Municipal "all to one" model. (**c**) Main urban "pairwise" model. (**d**) Main urban "all to one" model.

The analysis of the barrier points using Linkage Mapper yielded two scenarios for the nodes: unselected improvement scores relative to (least-cost distance) LCD percentage and selected improvement scores relative to LCD percentage (Figure 6). Out of the 22 ecological barrier points extracted within the municipal area, 9 were general ecological barrier points and 13 were key ecological barrier points (Figure 7). The geographical distribution of barrier points is as follows: The majority of the eastern and western locations are found at the intersection of urban road networks and ecological corridors, where the ecological environment is more delicate. The southeast region is home to a concentrated number of barrier points; the area is part of Fengyi Township's development and construction, and the land used for transportation has impeded the flow of ecological elements.

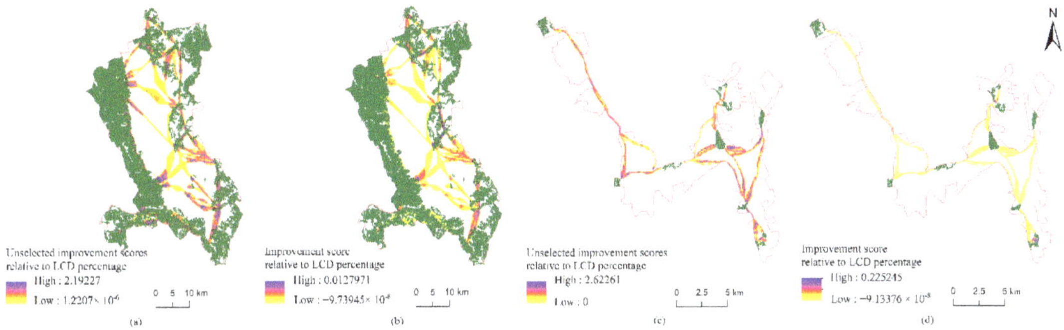

Figure 6. Eco-node analysis based on Barrier Mapper. (**a**) Municipal unselected improvement scores relative to LCD percentage. (**b**) Municipal improvement scores versus LCD percentage. (**c**) Main urban unselected improvement scores relative to LCD percentage. (**d**) Main urban improvement score versus LCD percentage.

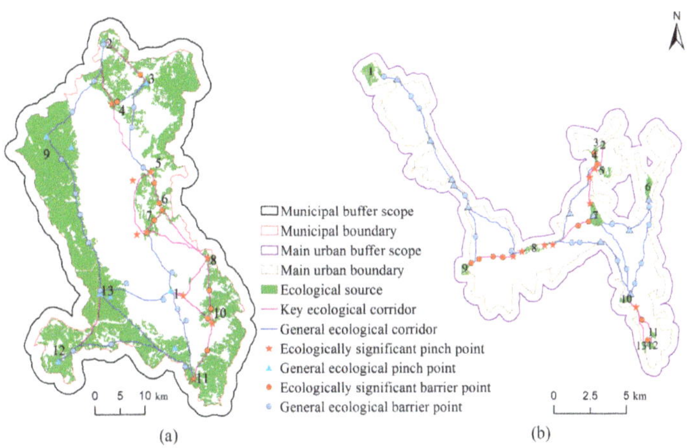

Figure 7. Spatial distribution of ecological corridors and nodes in (**a**) municipal and (**b**) main urban areas.

(2) Water Green Network

Based on the ArcGIS 10.8 hydrological analysis tool used to simulate the stormwater natural confluence process, the municipal's river network is divided into five levels according to the runoff level, and the river situation is extracted into three levels of stormwater corridors, of which the first level, second level, and third level of stormwater corridors are 40, 78, and 386, respectively. The stormwater corridors are evenly distributed, and the Cangshan area is densely distributed, forming an interconnected situation.

According to the analysis of catchment nodes, wetland node construction is carried out at river confluence nodes and watershed outlet nodes; 112 wetland nodes are identified, and 29 greenland patches are selected, including 14 lakes and reservoirs and 15 woodland meadows, with a total area of 1016.14 km^2, accounting for 58% of the total area of the region. The stormwater ecological network was supplemented to form a water and green ecological network combining points, lines, and surfaces (Figure 8).

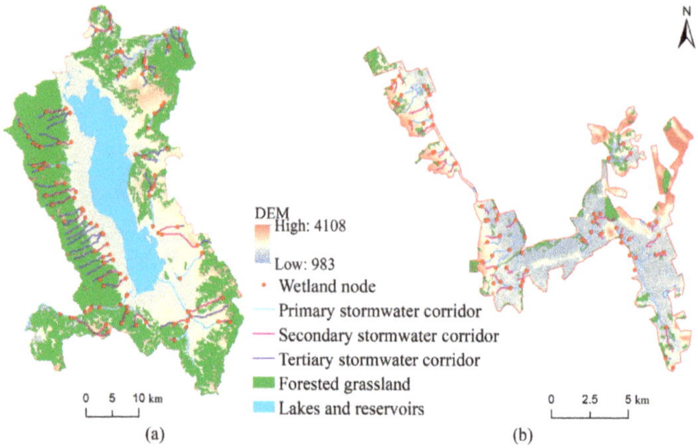

Figure 8. Spatial distribution of water green network in (**a**) municipal and (**b**) main urban areas.

(3) Landscape networks

Figure 9 illustrates that a high distribution in the center and a low dispersion around indicate the suitability of landscape network design in the study area. The northeast and southwest regions have few recreational opportunities and low landscape value, making them unsuitable for the establishment of landscape ecological networks. The highest areas are primarily found around Erhai Lake, Dali Ancient City, and Xiaguan Town, where selective placement is required.

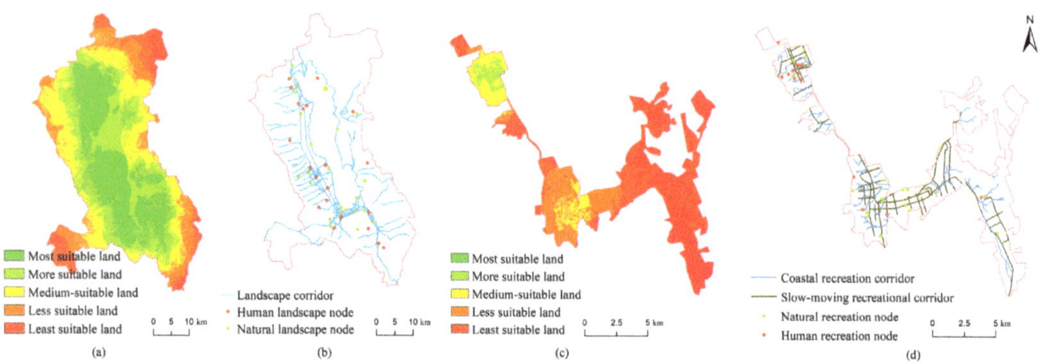

Figure 9. Suitability assessment and spatial distribution of landscape recreation networks. (**a**) Municipal suitability assessment. (**b**) Municipal landscape network. (**c**) Main urban suitability assessment. (**d**) Main urban recreation network.

The municipality has identified 130 slow-moving landscape routes and 502 riverfront landscape corridors. Landscape corridors generally have a distribution pattern "more in the south and less in the north". Erhai Lake flows from the northeast to the southwest, with the southwestern end being the outlet, forming a river convergence. The study area's northeastern region has fewer corridors due to its natural woodland landscape type, which is not conducive to recreational activities. Additionally, the region has fewer water systems and roads, which contribute to its low spatial accessibility. This makes it difficult to engage in experiential and recreational activities there. The study area's natural landscape resources include parks, rivers, and squares; its humanistic landscape resources are mostly museums, distinctive historical neighborhoods, and units for cultural relic safeguarding. The municipal region yielded 27 natural landscape nodes and 21 humanistic landscape nodes.

3.2. Network Construction in Main Urban

(1) Habitat networks

The landscape area of the main urban core category is 16.90 km^2, which is 15.69% of the total area of the study area. There is more room for improved ecological network construction in the main urban area because there are fewer core green space patches, fewer bridging-type patches than other types of patches, and insufficient spatial connectivity between the large and medium-sized core-type patches (Figure 3). With a combined size of 5.31 km^2, 13 patches with dPC > 1 were chosen as ecological source locations within the municipal area. Small and sparsely distributed peri-urban woodlands and villages, with low intensity of human development, are a good ecological source. The area is well connected and mostly situated on the outskirts of the city, and is primarily made up of low-intensity human development communities and forest land.

In the main urban area, twenty ecological corridors totaling 99.23 km in length were found. Of these, those with interaction forces larger than 1000 were designated as major ecological corridors (Figure 4), with the remaining corridors being considered general ecological corridors. Figure 7 displays the 7 general ecological corridors and the 13 key ecological corridors that were identified. Key ecological corridors are primarily found around the towns of Fengyi and Haidong, where the district has a straightforward forest structure and little to no forest cover. Building extensive forest ecological corridors and forest protection green spaces within towns, between urban districts, between towns and cities, on the outskirts of cities, and along transportation routes is a priority for urban construction in the district with the goal of improving the quality of the human environment on a large scale. Of these, patch 1 and patch 8 had the weakest interactions between patches because they straddled built-up land, had the greatest resistance to flow from the source site, made it more difficult for species to migrate, and thus had the weakest interactions between patches, while the greatest interactions were found between patches 12 and 13 in the main urban area, indicating the strongest connectivity in the source area. In order to improve the ecological corridor and maximize the ecological network, the ecological network should be built by adding ecological pinch points or altering the migration paths in the source area, where there is greater flow resistance and greater difficulty for species migrating.

In the main urban area, 19 ecological pinch points were extracted, of which 8 were key ecological pinch points and 11 were general ecological pinch points. There are 25 ecological barriers, of which 10 are key ecological barriers and 15 are general ecological barriers. The majority of ecological barriers are found inside ecological corridors or on the periphery of ecological sources at both scales. Some of the barriers can be restored by strengthening the creation of main urban green space and taking steps like converting farmland back to forests in order to improve and restore the isolation area and maximize the connectivity between ecological resources, as indicated by the remote sensing image map, which shows that some of the barriers are located on town roads.

(2) Water Green Network

The main urban river network was split into three tiers based on the hydrological analysis in order to identify the secondary stormwater corridors, of which 31 and 40 were primary and secondary stormwater corridors, respectively. A total of 80 wetland nodes and 140 greenland patches—along with 11 lakes and reservoirs and 333 forests and grasslands—were chosen based on the analysis of catchment nodes. The main urban water and green network was created by connecting the corridors, which were primarily located in the northwest section of the city, with one another via woods (Figure 8).

(3) Recreation Network

In the main urban region, Xiaguan Town and Dali Old Town are the most suited, while Fengyi Town is the least suited because of its lack of recreational opportunities and difficult access to transportation. Combining leisure, recreation, and planning into one linear space is known as a recreational corridor. Thirty-one slow-moving recreational routes and seven riverfront recreational corridors are identified based on the utilization of the available resources and the people' recreational demands. Nineteen natural recreation nodes and fifteen humanistic recreation nodes were extracted (Figure 9). The construction of the recreation ecological network should be close to and connect the recreation nodes with the existing corridors to ensure the spatial continuity of the recreation resources and improve the overall connectivity of the recreation network.

3.3. Complex Ecological Network Construction

3.3.1. Appropriate Widths for Different Corridors

In the municipal biological migration corridor (Figure 10), in the secondary corridor, the area of forest land and water area gradually decreased, the area of cultivated land, unutilized land, and constructed land continued to increase, and the inflection point of the rate of change was 150 m. In the main urban corridor, the area of forest land in the range of 30–90 m remained stable, the area of constructed land decreased and then increased, and the inflection point of the rate of change was 60 m. In the municipal stormwater corridor, the area of forest land, cultivated land, constructed land, and unutilized land decreased and then increased, and the inflection point of the change was 60 m. In the main urban corridor, the area of forest land, cultivated land, construction land, and unutilized land decreased and then increased, and the inflection point of the change rate was 90 m. The area of forest land in the corridor in the main urban remained stable within the range of 60–300 m, the construction land and water areas decreased, and the inflection point of the change rate was 60 m. According to the current land use characteristics of the city, the suitable widths of different biomigration corridors are 150 and 60 m, respectively; the suitable widths of stormwater corridors are 90 and 60 m; and the suitable widths of open space corridors are 150 and 60 m. Rivers and roads are selection lines, so the width is a fixed value and cannot be set.

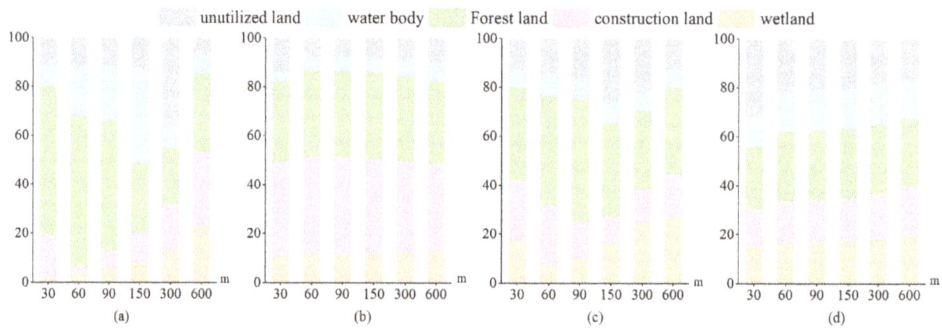

Figure 10. Percentage of area in land use types with different corridor widths. (**a**) Width of biological migration corridor in municipal area. (**b**) Width of biological migration corridor in main urban area. (**c**) Width of municipal stormwater corridor. (**d**) Width of main urban stormwater corridor.

3.3.2. Nested Multi-Scale Ecological Networks

Regarding overlapping and intersecting ecological sources, corridors, and nodes at both municipal and main urban scales (Figure 11), the number of common ecological sources at both scales was obtained to be 5, with an overlapping area of 3.65 km^2, and there were 11 overlapping ecological corridors with an overlapping length of 12.93 km, which accounted for 3.7% and 13.03% of the length of ecological corridors in the municipal area and the main urban area, respectively. The overlapping ecological corridors are mainly located around the built-up area. There are 3 overlapping stormwater corridors with a length of 2.73 km and a total of 2.73 km^2 of forested land, and 17 overlapping recreational corridors with a length of 38.39 km. These overlapping ecological sources and corridors support ecological processes between the two scales and need to be protected and optimized. The absolute contribution of the shared source sites and corridors in the regional ecological network is low, but the fact that the shared source sites overlap most of the source sites in the main urban area within the municipal suggests that the nested substrate of habitats between the two scales is good and the biological processes may be effective.

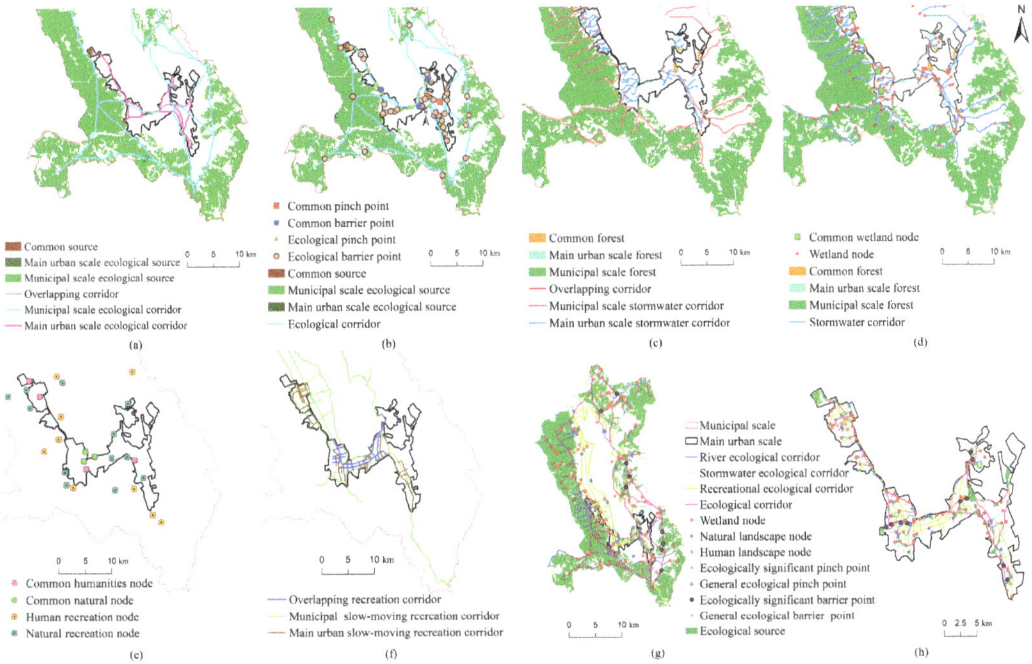

Figure 11. Spatial distribution of red–green–blue ecological networks with nested analysis. (**a**) Nesting source and ecological corridors. (**b**) Nesting of pinch points and barrier points. (**c**) Nesting of source and stormwater corridors. (**d**) Nesting of wetland nodes. (**e**) Nesting of recreational nodes. (**f**) Nesting of landscape and recreational corridors. (**g**) Municipal complex ecological network. (**h**) Main urban complex ecological network.

Overlapping ecological pinch points and barrier points at both scales (Figure 11) yielded two overlapping pinch points, accounting for 4.44% and 9.52% of the number of pinch points in the municipal and main urban areas, and a total of three overlapping barrier points, accounting for 13.64% and 12% of the total number of barrier points in the municipal and main urban areas, respectively. There are six overlapping wetlands nodes. These represent 5.36% and 7.59% of the number of wetland nodes in the municipal and main urban areas, respectively. There are seven overlapping recreation nodes, accounting for 14.58% and 20.59% of the number of recreation nodes in the municipal and main urban areas, respectively. The low degree of overlap of ecological nodes across scales suggests that future optimization of the ecological network needs to pay attention to the transfer of ecological processes between multiple scales to facilitate the flow of matter and energy. The control of the ecological security boundary and the solution of important ecological problems in Dali City need the support of the main urban area, while the protection and restoration of key ecological nodes in the main urban area is an important link in constructing and connecting the overall ecosystem.

3.3.3. Multi-Targeted Ecological Network Integration

Red–green–blue ecological networks, primarily defined by rivers, biology, stormwater, and recreation, are created by superimposing single-objective ecological networks (Figure 9). Protecting biodiversity, controlling urban runoff, enhancing the urban environment, and offering inhabitants leisure and recreation are just a few of the roles that the river biological corridor plays. After careful examination, it can be seen that the three functional corridors—which are oriented around Erhai Lake and Xiaguan Town—have a very balanced design. There are more recreational and water–green corridors in the southwest, of which the number of ecological corridors is the highest in the southeast, mainly due to the fact that some of the smaller source areas play the role of stepping stones to reduce the cost of migratory distances while enhancing the connectivity between the source areas, while a small number of stormwater corridors and ecological corridors are distributed in the north due to the topography and transportation. When comparing the nodes' spatial distribution, it can be seen that the ecological nodes are primarily found in the east and south. This is likely due to the city's highly developed south and developing east. The eighteen streams on Cangshan Mountain are home to the majority of wetland nodes, which can help with runoff rate reduction, flood storage, and water quality enhancement. The overall spatial distribution of the landscape nodes is fairly balanced, with multi-functional, multi-type recreational nodes like temples, commercial districts, and science and technology parks on the west side, and parks and green spaces such as Shangguan Flower Park, Erhai Forest Park, Erhai Jade Cabbage Wetland, Erhai Ancient City, and Xiubeishan Forest Park as participatory ecological spaces.

3.3.4. Multi-Functional Composite Ecological Network Pattern Optimization Strategy

The coverage of the ecological network in Dali City is centered on ecological sources, but the eastern part of the city lacks large habitat patches and has poor connectivity with the north and south, so it is necessary to focus on increasing the construction of ecological sources in the eastern part of the city to enhance the protection of regional ecological sources. Some of the stormwater corridors and biological corridors are side-by-side or intersect to enhance the vegetation richness of the composite corridors and strengthen the connection between the corridors, which can improve the potential utilization rate of the corridors. Recreational corridors and ecological corridors intersect to form ecological breakpoints in municipal and main urban areas, respectively. There are breakpoints 16, 15, mainly distributed on the national highways G214, G320, and S221; and ecological corridors at the intersection of the road can be restored through the installation of above-ground flyovers and underground tunnels. Other actions include reducing the interference of town development and construction on all types of corridors, strengthening the spatial coordination of composite sources and corridors, and promoting the optimal construction of a multi-level, multi-objective red, green, and blue ecological network.

Dali City has a great variety of wild animals and plants, and rich vegetation types, and has built various kinds of nature reserves, wetland parks, and geological parks. In recent years, Dali City has carried out the project of returning farmland to forests, closing the mountains to cultivate forests, and continuously strengthening ecological environment construction while protecting and making good use of the existing forests. Unique natural landscape conditions and rich natural resources have created favorable conditions for ecological construction in Dali City. The cost–benefit analysis of ecological network construction and its subsequent development is a key link in assessing its value for environmental protection and sustainable development. As a result, the risk of ecosystem degradation is effectively reduced by promoting biodiversity conservation and restoration, further reducing the costs of governance in the future. Secondly, it improves the efficiency of resource use, such as water conservation, soil conservation, and other ecological services, and indirectly reduces the cost of resource acquisition and utilization. Finally, the construction of ecological networks promotes the development of green industries such as eco-tourism, injects vitality into the local economy, and achieves a win–win situation for both economic and ecological benefits.

4. Discussion

Building a multi-objective composite ecological network is crucial to the development and integration of recreational, water–green, and regional habitat networks. Using MSPA, circuit theory, and the landscape connectivity approach, a multi-level and multi-objective composite ecological network was built in Dali City for this study. The two most critical steps in the process of building ecological networks are the extraction of ecological sources and the identification of ecological corridors. Nowadays, ecological sources are identified by landscape connectivity [37], ecological sensitivity [7], and ecosystem service importance [53]. In this study, MSPA and landscape connectivity were used to identify ecological sources [54]. The construction of resistance surfaces is a prerequisite for the identification of ecological corridors [39,55] and in the past, the construction of resistance surfaces usually used topography, surface cover type, and road network factors as the leading indicators to classify and assign values to them. However, these methods ignore the high density of socio-economic activities brought about by urbanization [41].

Circuit theory and the MCR model are used to build ecological corridors. The MCR model can better depict the trend of ecological corridors [22], but it cannot pinpoint important nodes, such as "pinch points" and "obstacle points," inside the corridors. In this study, circuit theory was used to construct the ecological corridors between the Dali municipal area and the main urban area [23]. The key ecological corridors were extracted using a gravity model. According to the results, which are in line with those of Li et al. [20] and Xu et al. [21], ecological sources are primarily distributed in Cangshan Mountain, Shangguan Town, and Fengyi Town. These are primarily core areas with widely dispersed woodland landscapes. The Cangshan region is home to the majority of forests, which are dispersed due to a variety of causes, including topography, climate, and human activity. As a result, there are few ecological sources found and there is a lot of edge fragmentation. The study region is centered on significant ecological corridors, which are areas of ecological connection and are crucial for the optimization and restoration of the entire landscape. The study identified 22 and 25 barrier points, respectively, in the municipal and main urban areas. These points were primarily spread inside the built-up area range and road network, indicating their location at the periphery of ecological sources or corridors. These findings are in line with prior research [56]. Furthermore, by examining the ecological corridor's width, we discovered that, within the 30 to 150 m range, there was minimal variation in the landscape structure. This was associated with the type of landscape, which was primarily forested and featured a sizable forested area with strong landscape connectivity. This was in line with earlier research [57]. Narrow ecological corridors are not good for species survival because they have a negative "edge effect" on species' movement and distribution. This is because, as a corridor's width declines, so does landscape connectivity. Thus, covering as many different types of environmental gradients as possible will improve environmental variability and ensure biodiversity.

At present, the majority of research works have constructed networks on a single scale, emphasizing the region's general average value at large scales while disregarding the variations at smaller ones [38]. The linkages between various ecological network levels and various administrative levels are also not widely studied [31]. Nor is the geographical variation in ecological landscape patterns at various scales taken into account [58]. It is important to take into account changes at various scales when building ecological networks and developing regional plans because of the scale effect, which makes ecological processes at different scales independent . The multi-scale and multi-level requirements of ecological restoration work in land space are combined in this study, and scale nesting is incorporated into the ecological network construction link in the study area, serving as an external support role and improving the effectiveness of regional ecological protection [59]. This not only builds the overall ecological security pattern at the mesoscale, but also takes into account Dali City's small-scale land use characteristics and landscape connections. Additionally, this study built a habitat network, a water–green network, and a recreation network in Dali City from the perspectives of biological, hydrological, and humanistic processes. In contrast, an ecological network driven by a single goal that solely considers recreational or ecological conservation would be different [24,60,61]. It was shown that applying this method helps to increase the connectivity of landscape patches, which is advantageous for preserving regional landscape patterns and protecting biodiversity.

In the process of constructing the composite ecological network system, this study comprehensively applied the quantitative analysis method. This improves the traditional method based on empirical judgment and qualitative analysis. This study can provide practical references for urban and rural ecological construction and is of great significance for the sustainable development of fragile ecosystems. Although this study provides new insights into the construction of a composite ecological network, the following research deficiencies still exist: Firstly, because the natural environment is an organic whole, this study only utilizes the county scale and the urban scale, and future research can consider the relationship between larger scales. Secondly, the study concentrates on the tension between the increase in construction land and the deterioration of the terrestrial biological environment, primarily because of the study area's relatively intact water network. Although certain significant water networks are taken into account, the significance of the water environment and the preservation of water systems need to be enhanced. Third, the ecological corridors created in this study are only useful for conserving biodiversity in the area; the requirements of certain species for corridor width have not been sufficiently taken into account. Future research should categorize ecological networks based on the requirements of various species. Ultimately, because of the data's limitations, the impact of variables like residents' travel patterns and spatial distribution on the recreation corridor was not fully taken into account. However, future research can optimize the suitability evaluation system by combining big data from multiple sources, including population travel patterns, settlement distribution, and holiday travel patterns.

5. Conclusions

Building a composite ecological network is an effective way to mitigate habitat fragmentation and biodiversity loss. This study used Landsat remote sensing data and forest resources type II survey data in Dali City for 2016, adopting source–sink theory, circuit theory, appropriateness analysis, and hydrological analysis in order to achieve ecologically sustainable development. The study was based on the principles of ecological protection and multi-functional coupling and coordination. The study constructed a composite ecological network of the municipal area and main urban area under the spatial perspective of red, green, and blue in Dali City. The conclusions are as follows:

(1) In the "source-corridor-key node" ecological network, 3.65 km^2 of overlapping ecological sources was identified, mainly concentrated in the main urban area of the study area. There are 11 overlapping ecological corridors, 3 stormwater corridors, and 17 recreational corridors. The overlapping areas are the key to achieving overall ecological sustainability and biodiversity in the region and should be protected as a priority.

(2) Two overlapping pinch points, three obstacle points, four wetland nodes, and seven open space nodes were identified between the municipal and the main urban area. The ecological pinch points and barrier points are areas where species migration and movement between source areas are strongly impeded, and ecological problems can be addressed through the nested integration of multi-scale spatial structures.

(3) The triple-function corridors are mainly distributed in the central part of the study area and around Erhai Lake, with stormwater corridors and recreational corridors dominating in the west and ecological corridors mostly distributed in Fengyi Town

and Haidong Town in the southeast. Therefore, in the future, the protection of Shuanglang Town and Xiaguan Town should be increased to build ecological sources and ecological corridors with stronger connectivity. At the same time, the protection of ecologically fragile areas in the west and south should be increased to reduce ecological damage and prohibit indiscriminate logging and deforestation to maintain ecological functions.

(4) This study emphasizes the importance of a composite ecological network for ecological sustainability, which can be applied to different levels of territorial space restoration. In terms of constructing low corridors, corridors can be constructed through measures such as restricting land development and implementing vegetation restoration. In addition, according to the actual situation, additional protective measures can be installed in the watersheds where the population is concentrated, at the policy level, in order to strictly control the pollution of the watersheds caused by agricultural production and the sewage discharge from domestic life. This study can provide a reference for the composite ecological network of similar cities.

Author Contributions: Data preparation and design, Rong Chen, Shunmin Zhang, and Xiang Li; formal analysis, writing—original draft preparation, Rong Chen; Methodology, validation, Rong Chen and Shunmin Zhang; Visualization, writing—review and editing, Rong Chen; Project administration, conceptualization and supervision, funding acquisition, Xiaoyuan Huang and Jiansong Peng. All authors have read and agreed to the published version of the manuscript.

Funding: This research received no external funding.

Institutional Review Board Statement: Not applicable.

Informed Consent Statement: Not applicable.

Data Availability Statement: Data will be made available upon a reasonable request.

Conflicts of Interest: The authors declare no conflicts of interest.

References

1. Jones, K.R.; Venter, O.; Fuller, R.A.; Allan, J.R.; Maxwell, S.L.; Negret, P.J.; Watson, J.E.M. One-third of global protected land is under intense human pressure. *Science* **2018**, *360*, 788–791. [CrossRef] [PubMed]
2. Heynen, N.; Perkins, H.A.; Roy, P. The political ecology of uneven urban green space: The impact of political economy on race and ethnicity in producing environmental inequality in Milwaukee. *Urban Aff. Rev.* **2006**, *42*, 3–25. [CrossRef]
3. Ma, C.; Deng, J.; Zhao, X.; Zhang, Y.C. Theory and practice of ecological city construction. *IOP Conf. Ser. Earth Environ. Sci.* **2018**, *186*, 012058. [CrossRef]
4. Cui, S.Q.; Zhou, G.H.; Dai, L.Y. Progress and prospect of research on urban-rural integration development based on geography perspective. *Econ. Geogr.* **2022**, *42*, 104–113. [CrossRef]
5. Dong, Y.; Jin, G.; Deng, X.; Wu, F. Multidimensional measurement of poverty and its spatio-temporal dynamics in China from the perspective of development geography. *Geogr. Sci.* **2021**, *31*, 130–148. [CrossRef]
6. Brose, U. Improving nature conservancy strategies by ecological network theory. *Basic Appl. Ecol.* **2010**, *11*, 1–5. [CrossRef]
7. Yang, Y.; Feng, Z.; Wu, K.; Lin, Q. How to construct a coordinated ecological network at different levels: A case from Ningbo city, China. *Ecol. Inform.* **2022**, *70*, 101742. [CrossRef]
8. Wang, S.F.; Liu, L.B. From Corridor to Whole Region: Construction of Urban and Rural Blue-Green Space Network Under the Guidance of Green Urban Design. *Landsc. Archit.* **2021**, *28*, 45–50. [CrossRef]
9. Jongman, R.H.G. Nature conservation planning in Europe: Developing ecological networks. *Landsc. Urban Plan.* **1995**, *32*, 169–183. [CrossRef]
10. Naveh, Z.; Lieberman, A.S. *Landscape Ecology: Theory and Application*; Springer: Berlin/Heidelberg, Germany, 1994; pp. 27–109. [CrossRef]
11. Wu, J.G. *Landscape Ecology: Pattern, Process, Scale and Levels*, 2nd ed.; Higher Education Press: Beijing, China, 2007; pp. 27–125.
12. Cunha, N.S.; Magalhães, M.R. Methodology for map** the national ecological network to mainland Portugal: A planning tool towards a green infrastructure. *Ecol Indic.* **2019**, *104*, 802–818. [CrossRef]
13. Zhou, D.; Song, W. Identifying ecological corridors and networks in mountainous areas. *Int. J. Environ. Res. Public Health* **2021**, *18*, 4797. [CrossRef] [PubMed]
14. Bascompte, J. Structure and dynamics of ecological networks. *Science* **2010**, *329*, 765–766. [CrossRef] [PubMed]
15. Baguette, M.; Blanchet, S.; Legrand, D.; Stevens, V.M.; Turlure, C. Individual dispersal, landscape connectivity and ecological networks. *Biol. Rev.* **2013**, *88*, 310–326. [CrossRef] [PubMed]

16. Säterberg, T.; Sellman, S.; Ebenman, B. High frequency of functional extinctions in ecological networks. *Nature* **2013**, *499*, 468–470. [CrossRef] [PubMed]
17. Gurrutxaga, M.; Lozano, P.J.; del Barrio, G. GIS-based approach for incorporating the connectivity of ecological networks into regional planning. *J. Nat. Conserv.* **2010**, *18*, 318–326. [CrossRef]
18. Pocock, M.J.O.; Evans, D.M.; Memmott, J. The robustness and restoration of a network of ecological networks. *Science* **2012**, *335*, 973–977. [CrossRef]
19. Geng, J.; Yu, K.; Sun, M.; Xie, Z.; Huang, R.; Wang, Y.; Zhao, Q.; Liu, J. Construction and Optimization of Ecological Networks in High-Density Central Urban Areas: The Case of Fuzhou City, China. *Remote Sens.* **2023**, *15*, 5666. [CrossRef]
20. Li, Q.Y.; Tang, L.N.; Qiu, Q.Y.; Li, S.T.; Xu, Y. Construction of urban ecological security pattern based MSPA and MCR Model: A case study of Xiamen. *Acta Ecol. Sin.* **2024**, *44*, 2284–2294. [CrossRef]
21. Xu, Z.Y.; Wang, S.H.; Gu, T.; Wang, S.Y.; Pei, C.Y.; Zhang, Q.F. Ecological network-based identification of critical areas for ecological restoration—A case study of Yan'an City. *J. Arid. Land.* **2024**, *47*, 1073–1083.
22. Chen, W.; Liu, H.; Wang, J. Construction and optimization of regional ecological security patterns based on MSPA-MCR-GA Model: A case study of Dongting Lake Basin in China. *Ecol. Indic.* **2024**, *165*, 112169. [CrossRef]
23. Huang, L.; Chen, Z.; Yang, Z.; Chen, M.; Chen, X.; Zhai, T.; Qiu, T. Integrating "quality-risk-demand" framework and circuit theory to identify spatial range and priority area of ecological security pattern in a rapidly urbanizing landscape. *Ecol. Inform.* **2024**, *82*, 102673. [CrossRef]
24. Zeng, W.; He, Z.; Bai, W.; He, L.; Chen, X.; Chen, J. Identification of ecological security patterns of alpine wetland grasslands based on landscape ecological risks: A study in Zoigê County. *Sci. Total Environ.* **2024**, *928*, 172302. [CrossRef]
25. Wang, M.; Liang, S.; Wang, Y.C. Comparison and Synthesis of Scenarios of Greenland Ecological Network Construction in the Context of Urban Shuangxiu—Based on the Perspective of "Socio-Ecological" Functional Composite. *South Archit.* **2019**, 1–8. [CrossRef]
26. Wang, B.; Liu, Z. Study on the optimization strategy of ecological network in Beijing central area under the background of urban renewal. *Urban Stud.* **2022**, *29*, 113–120.
27. Dong, J.; Peng, J.; Liu, Y.; Qiu, S.T.; Han, Y.N. Integrating spatial continuous wavelet transform and kernel density estimation to identify ecological corridors in megacities. *Landsc. Urban Plan.* **2020**, *199*, 103815. [CrossRef]
28. Shen, J.; Wang, Y. An improved method for the identification and setting of ecological corridors in urbanized areas. *Urban Ecosyst.* **2023**, *26*, 141–160. [CrossRef]
29. Yu, Q.; Yue, D.P.; Wang, Y.H.; Kai, S.; Fang, M.Z.; Ma, H.; Zhang, Q.B.; Huang, Y. Optimization of ecological node layout and stability analysis of ecological network in desert oasis: A typical case study of ecological fragile zone located at Deng Kou County (Inner Mongolia). *Ecol. Indic.* **2018**, *84*, 304–318. [CrossRef]
30. Zhou, Y.; Li, B.; Li, P.Y.; Yao, J.; Chen, M.K.; Tang, M.; Zhang, L.; Chen, J. Optimization of ecological network pattern of composite green space based on multi-functional coupling of "ecology-climate adaptation- recreation". *Acta Ecol. Sin.* **2024**, *44*, 5854–5866. [CrossRef]
31. Shen, Z.; Wu, W.; Tian, S.; Wang, J. A multi-scale analysis framework of different methods used in establishing ecological networks. *Landsc. Urban Plan.* **2022**, *228*, 104579. [CrossRef]
32. Wei, Q.; Halike, A.; Yao, K.; Chen, L.; Balati, M. Construction and optimization of ecological security pattern in Ebinur Lake Basin based on MSPA-MCR models. *Ecol. Indic.* **2022**, *138*, 108857. [CrossRef]
33. Wang, T.; Li, H.; Huang, Y. The complex ecological network's resilience of the Wuhan metropolitan area. *Ecol. Indic.* **2021**, *130*, 108101. [CrossRef]
34. Koen, E.L.; Bowman, J.; Sadowski, C.; Walpole, A.A. Landscape connectivity for wildlife: Development and validation of multispecies linkage maps. *Methods Ecol. Evol.* **2014**, *5*, 626–633. [CrossRef]
35. Liu, J.; Yin, H.W.; Kong, F.H.; Li, M.H. Structure optimization of circuit theory-based green infrastructure in Nanjing, China. *Acta Ecol. Sin.* **2018**, *38*, 4363–4372. [CrossRef]
36. Hu, C.; Wang, Z.; Wang, Y.; Sun, D.; Zhang, J. Combining MSPA-MCR model to evaluate the ecological network in Wuhan. China. *Land* **2022**, *11*, 213. [CrossRef]
37. Zhou, Y.; Tang, M.; Chen, J.; Li, B. Optimization of green space ecological network in Chengdu city based on morphological spatial pattern analysis and mapping theory. *Chin. J. Ecol.* **2023**, *42*, 1527–1536. [CrossRef]
38. Yu, H.; Wang, Y.; Eme, C.; Fan, X.; Rong, Y.; Zhang, Q.; Liang, Z. Construction of ecological network based on multi-scale conversion and nesting. *Water* **2021**, *13*, 1278. [CrossRef]
39. Fan, J.; Wang, Q.; Sun, M.Y.; Feng, Y.; Yang, F.; Zhang, Z. Ecological network construction and gradient zoning optimization strategy in urban-rural fringe: A case study of Licheng District, Jinan City. China. *Ecol. Indic.* **2023**, *150*, 110251. [CrossRef]
40. Hashemi, R.; Darabi, H. The review of ecological network indicators in graph theory context: 2014–2021. *Int. J. Environ. Res.* **2022**, *16*, 1–26. [CrossRef]
41. Lu, J.; Jiao, S.; Han, Z.W.; Yin, J.W. Promoting Ecological Restoration of Deeply Urbanized Hilly Areas: A Multi-Scale Ecological Networks Approach. *Ecol. Indic.* **2023**, *154*, 110655. [CrossRef]
42. McRae, B.H.; Dickson, B.G.; Keitt, T.H.; Shan, V.B. Using circuit theory to model connectivity in ecology, evolution, and conservation. *Ecology* **2008**, *89*, 2712–2724. [CrossRef]

43. Chen, J.; Mei, Z.; Wang, B.; Wei, J. Construction of Ecological Security Patterns Based on Circuit Theory under the Resistance Distance Principle. *Int. J. Envirn. Res. Public Health.* **2022**, *19*, 6298. [CrossRef] [PubMed]
44. Zhu, J.; Su, J.; Yin, H.W.; Kong, F.H. Construction of an ecological network in Xuzhou based on comprehensive identification of source sites and multiscale nesting. *J. Nat. Resour.* **2020**, *35*, 1986–2001. [CrossRef]
45. An, Y.; Liu, S.L.; Sun, Y.X.; Shi, F.N.; Beazley, R. Construction and optimization of an ecological network based on morphological spatial pattern analysis and circuit theory. *Landsc. Ecol.* **2021**, *36*, 2059–2076. [CrossRef]
46. Wei, B.J.; Su, J.; Hu, X.J.; Xu, K.H.; Zhu, M.L.; Lin, L.Y. Comprehensive identification of eco-corridors and eco-nodes based on principle of hydrological analysis and LinkageMapper. *Acta Ecol. Sin.* **2022**, *42*, 2995–3009. [CrossRef]
47. Ahmad, I. Digital elevation model (DEM) coupled with geographic information system (GIS): An approach towards erosion modeling of Gumara watershed, Ethiopia. *Environ. Monit. Assess.* **2018**, *190*, 568. [CrossRef] [PubMed]
48. Pan, Y.; Yuan, X.; Wang, F.; Zhou, L.; Zhang, D.; Yu, X.H. Ecological network construction and optimal regulation based on rural water ponds. *J. Hydroecol.* **2023**, *44*, 99–106. [CrossRef]
49. Shao, J.; Huang, K.; Tang, W.; Liang, X.; Wu, W.; Huang, G. Discussion on the Construction of Ecological Water Network in Guangxi Province of China. *Res. Ecol.* **2021**, *3*, 8–13. [CrossRef]
50. Ma, R.; Xie, M.; Yun, W.; Zhu, D. Evaluating responses of temperature regulating service to landscape pattern based on "Source-Sink" theory. *Isprs. Int. J. Geo-Inf.* **2020**, *9*, 295. [CrossRef]
51. Lu, J.; Wu, X. Research on Urban Greenway Alignment Selection Based on Multisource Data. *Sustainability* **2022**, *14*, 12382. [CrossRef]
52. Peng, J.; Zhao, H.; Liu, Y. Urban ecological corridors construction: A review. *Acta Ecol. Sin.* **2017**, *37*, 23–30. [CrossRef]
53. Zhou, G.; Huan, Y.; Wang, L.; Lan, Y.; Liang, T.; Shi, B.; Zhang, Q. Linking ecosystem services and circuit theory to identify priority conservation and restoration areas from an ecological network perspective. *Sci. Total Environ.* **2023**, *873*, 162261. [CrossRef]
54. Cui, L.; Wang, J.; Sun, L.; Lv, C.D. Construction and optimization of green space ecological networks in urban fringe areas: A case study with the urban fringe area of Tongzhou district in Beijing. *J. Clean. Prod.* **2020**, *276*, 124266. [CrossRef]
55. Shuai, N.; Hu, Y.; Gao, M.; Guo, Z.; Bai, Y. Construction and optimization of ecological networks in karst regions based on multi-scale nesting: A case study in Guangxi Hechi, China. *Ecol. Inform.* **2023**, *74*, 101963. [CrossRef]
56. Kang, J.; Qing, Y.; Lu, W. Construction and optimization of the Saihanba ecological network. *Ecol. Indic.* **2023**, *153*, 110401. [CrossRef]
57. Shi, F.; Liu, S.; Sun, Y.; An, Y.; Zhao, S.; Liu, Y.; Li, M.Q. Ecological network construction of the heterogeneous agro-pastoral areas in the upper Yellow River basin. *Agric. Ecosyst. Environ.* **2020**, *302*, 107069. [CrossRef]
58. Nie, W.; Shi, Y.; Siaw, M.J.; Yang, F.; Wu, R.; Wu, X.; Zheng, X.; Bao, Z. Constructing and optimizing ecological network at county and town Scale: The case of Anji County, China. *Ecol. Indic.* **2021**, *132*, 108294. [CrossRef]
59. Cui, Y.; Guo, B.; Li, W.; Kong, X.M. Assessment of urban blue-green space cooling effect linking maximum and accumulative perspectives in the Yangtze River Delta, China. *Environ. Sci. Pollut. Res.* **2023**, *30*, 121834–121850. [CrossRef]
60. Qian, Y.M.; Huang, Y.T.; Cao, Y.R.; Wu, J.Y.; Xiong, Y.M. Ecological network construction and optimization in Guangzhou from the perspective of biodiversity conservation. *J. Environ. Manag.* **2023**, *336*, 117692. [CrossRef]
61. Wang, T.; Huang, Y.; Cheng, J.; Xiong, H.; Yin, Y.; Feng, Y.; Wang, J. Construction and optimization of watershed-scale ecological network based on complex network method: A case study of Erhai Lake Basin in China. *Ecol. Indic.* **2024**, *160*, 111794. [CrossRef]

International Journal of
Geo-Information

MDPI

Article

The Coupling Coordination Relationship Between Urbanization and the Eco-Environment in Resource-Based Cities, Loess Plateau, China

Shuaizhi Kang [1,2,3], Xia Jia [1,2,3], Yonghua Zhao [1,2,3,*], Manya Luo [1,2,3], Huanyuan Wang [4] and Ming Zhao [1,2,3]

[1] School of Land Engineering, Chang'an University, Xi'an 710054, China; 2020035002@chd.edu.cn (S.K.); jiaxia@chd.edu.cn (X.J.); luomanya@chd.edu.cn (M.L.); zm199231@chd.edu.cn (M.Z.)
[2] Key Laboratory of Degraded and Unused Land Consolidation Engineering of the Ministry of Natural Resources, Chang'an University, Xi'an 710054, China
[3] Shaanxi Key Laboratory of Land Consolidation, Chang'an University, Xi'an 710054, China
[4] Shaanxi Provincial Land Engineering Construction Group Co., Ltd., Xi'an 710075, China; wanghyuan@chd.edu.cn
* Correspondence: yonghuaz@chd.edu.cn

Abstract: Resource-based cities face numerous sustainability challenges, making the coupled and coordinated relationship between urbanization and the eco-environment critical for sustainable development strategies. The Loess Plateau is an essential energy base and ecologically fragile area in China, holding unique and significant research value. This research employed the Remote Sensing Ecological Index (RSEI) and the Compound Night Light Index (CNLI), based on MODIS and night light data, to investigate the socio-economic development and eco-environmental changes across 25 resource-based cities on the Loess Plateau (LP) in China over the past 20 years. The Coupling Coordination Degree Model (CCDM) and Multi-Scale Geographically Weighted Regression (MGWR) were utilized to assess the relationship between urbanization and ecological factors. The average RSEI values for these cities ranged from 0.4524 to 0.4892 over the 20 years, reflecting an upward trend with a growth rate of 8.13%. Simultaneously, the average CNLI values ranged from 1.5700 to 6.0864, with a change of 4.5164. Over the past two decades, all cities in the study area experienced rapid urbanization and ecological development. The correlation between urbanization and ecological factors strengthened, alongside an increasing spatial heterogeneity. While the coupling coordination relationship in most cities showed improvement, many remained within the low to middle grades. These findings enhance the understanding of the intricate relationships between urbanization and ecology, offering valuable insights for policy-making aimed at creating sustainable and livable resource-based cities.

Keywords: resource-based cities; eco-environment; urbanization; RSEI; Loess Plateau; sustainability

1. Introduction

Since the launch of China's reform and opening-up policies in 1978, the country's economy and society have rapidly developed alongside the urbanization process [1]. The latest data from the National Bureau of Statistics (2021) indicate that China's urbanization level reached 63.89% at the end of 2020. Furthermore, this trend of rapid increase is likely to continue in the coming years [2]. In process of urbanization, the land use and land cover change (LUCC) caused by human activities, as a complex and diverse risk source, has sharply modified the structure and morphology of cities, meanwhile facing a gradually increasing demand for environment [3]. The development of urbanization has resulted in an increase in urban population density, which, in turn, has led to the overconsumption of resources [4,5], moreover, actual and potential problems in the ecological environment such as atmospheric pollution, urban heat islands, and water and soil pollution [6–9]. In addition, the degraded eco-environment acts as a hindrance to the continued expansion of urbanization [10]. This situation particularly occurs in resource-based cities, where dominant industries depend on natural resources such as minerals, gas, and oil [11].

The United Nations' 2030 Agenda introduced the Sustainable Development Goals (SDGs) in 2015, which have since provided a framework for both national institutions and international organizations to develop strategies for achieving sustainable development [12]. As sustainable development gains increasing attention, it becomes crucial to prioritize the coupling and coordination between urbanization and the eco-environment. This relationship is highly complicated, non-linear, and multifaceted [13]. Researchers have studied the correlation degree, coupling degree, and coordination degree of various regions, such as Chongqing, using CCD and GIS spatial methods [14]. Other scholars, like Zou et al., have explored the dynamic changes in and spatiotemporal heterogeneity of the coupling relationship between urbanization and the ecological environment in Shaanxi Province using CCDM, RDDM, and GTWR [15]. Fu et al. [16] used a comprehensive index system to quantitatively evaluate the evolution of the relationship between urbanization and the eco-environment development in Qingdao. Overall, there has been a growing scholarly interest in exploring the connection between urbanization and the eco-environment in recent years, indicating that relevant research remains a hot topic in urban ecology [17].

The concept of coupling, derived from physics, describes the interaction of two or more subsystems. Modeling the degree of coupled coordination evaluates the harmony between subsystems—a method widely used in resource and environmental studies. The coupling coordination degree is typically established through the evaluation of both the urbanization level and the eco-environmental quality, which has been accomplished in various studies through the use of different statistical indicators [18]. For instance, Yang et al. [19] characterized the ecological environment using ecological footprint, while Shi et al. [20] used primary productivity (NPP) and fractional vegetation cover (FVC) to evaluate the eco-environment. Zhang et al. [21] also employed different indicators to support indexes in cities. However, the subjectivity in constructing indicator systems and selecting specific indicators, as well as the challenge of obtaining and filtering large spatial- and temporal-scale data, pose difficulties in these studies. To address this situation, remote sensing can offer an efficient and convenient approach [22,23]. Traditional statistical data analysis reflects regional trends at the overall city level, but fails to capture changes in the internal spatial structure of cities. As a result, the consistency and reliability of long-term statistical data are often difficult to ensure. In contrast, remote sensing data objectively reflect regional changes without requiring weight parameters factitiously.

China boasts abundant mineral resources and relatively favorable natural conditions, making it the world's largest energy producer [24]. Resource-based cities in China play a critical role in supporting the improvement of people's standard of living and driving national economic development by providing a significant quantity of resources [25]. Nowadays, resource-based cities serve as crucial components of China's energy and resource security base, and the sustainability of these areas is essential for ensuring the stability of both the economy and the environment [26]. As urbanization continues to develop, it is crucial to prioritize environmental protection. Consequently, it becomes both urgent and necessary to transition resource-based cities from a focus on high-speed growth to a focus on high-quality development [27]. As a less developed inland region with both economic and ecological vulnerability [28], the Loess Plateau (LP) has 39 resource-based cities, accounting for 14.8% of China. The LP is characterized by a fragile local environment and a high concentration of resource-based cities, leading to an evident contradiction between the demand for economic development and ecological protection in the region [29]. As such, it presents a typical case for exploring the relationship between urbanization and the eco-environment, which could inform policies related to environment conservation and urban development. Therefore, as a class of important cities with special significance facing huge problems in sustainable development, to protect these resource-based cities, we urgently need to implement a sustainable development coupling analysis with SDGs as the basic guideline [30].

The LP is characterized by a fragile local environment and a high concentration of resource-based cities, leading to an evident contradiction between the demand for economic development and ecological protection in the region. Therefore, as a class of important cities with special significance facing huge problems in sustainable development in the LP, to protect these resource-based cities, we urgently need to implement a sustainable development coupling analysis with SDGs as the basic guideline. This study aimed to achieve the following three key objectives: (1) How have the eco-environments of resource-based cities located in the LP changed over the past two decades, and what were the underlying factors contributing to these changes? (2) What were the spatiotemporal patterns of the urbanization levels in the resource-based cities located in the LP over the past two decades, and how have they varied across different regions and time periods? (3) To what extent was there a coupling coordination between the eco-environment and urbanization levels in resource-based cities located in the LP, China, and how have the spatial relationships between urbanization and the environment varied across different regions and time periods? This research will provide more detailed and meaningful information about urbanization and the environment for both policy makers and scholars to achieve the sustainable development of resource-based cities.

2. Materials and Methods

2.1. Study Area

The LP has a total area of 6.487×10^5 km^2 and is rich in energy and mineral resources. Resource-based cities represent a critical component of both the national energy security base and ecological security barrier. According to the National Plan for Sustainable Development of Resource-Based Cities (2013–2020) issued by the State Council, there are 39 resource-based cities within this region, comprising 25 prefecture-level cities, 12 counties (cities), and 2 municipal districts. For data consistency and availability across administrative units, this study focuses on 25 resource-based cities within the LP (Figure 1). These cities, identified by the national development plan as resource-based, are distributed across the Loess Plateau, resulting in a non-continuous study area. Among these cities, growing and mature resource-based type cities account for a relatively high proportion; more than half of them are coal-based cities. In 2020, the GDP of all 25 resource-based cities in the Loess Plateau region was 4.2135×10^{12} RMB. The population is 7.148×10^7 people, including 4.086×10^7 people in urban areas.

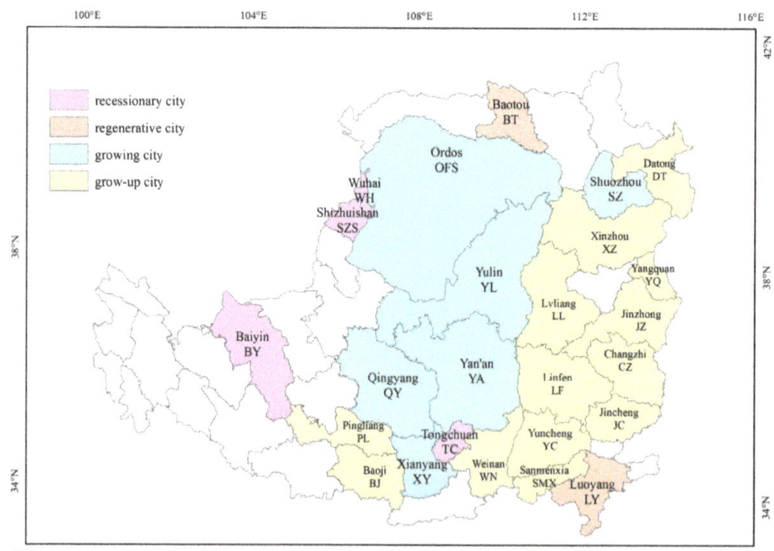

Figure 1. Development stage and location of resource-based cities in the Loess Plateau area.

2.2. Data Sources

This study primarily utilizes the following data: MODIS (Moderate-resolution Imaging Spectroradiometer) data, including the MOD09A1, MOD11A2, and MOD13A1 products from the GEE (https://earthengine.google.com/ (accessed on 30 September 2024)) dataset, which, respectively, provide terra surface spectral reflectance, land-surface temperature (LST), and Vegetation Index (VI). The satellite products collected in the years 2000, 2005, 2010, 2015, and 2020 were deemed suitable for the study, with a 20-year interval between each collection period; to enable better comparison across different years, image collection was limited to the period between June and September, while the population data were obtained from the WorldPop Global Project Population Data. Nighttime light data were obtained from NPP-VIIRS-like NTL Data from National Tibetan Plateau Data Center [31], generated via the NTLSTM network, and exhibit a high reliability with an RMSE of 0.73, R^2 of 0.95. Administrative and physical geographic boundary data were provided by Center of Resources and Environment Science, Chinese Academy of Sciences (https://www.resdc.cn/ (accessed on 1 October 2020)).

2.3. Data Methods

2.3.1. Calculation of the RSEI

RSEI is an aggregated index developed in recent years for the fast monitoring of ecological circumstances exclusively through remotely sensed data [32,33]. To represent the greenness index, which reflects vegetation growth and its distribution density in the study area, the Normalized Difference Vegetation Index (NDVI) was used [34]. Additionally, the Normalized Difference Impervious Surface Index (NDISI) was calculated using the Index-based Built-up Index (IBI) and Soil Index (SI), which reflects the spectral response of land desertification and urbanization, as reported by [35]. To obtain the Land Surface Temperature (LST) indicator of the study area, the MODIS LST product data were used, which can reflect the heat condition of the study area [36]. Lastly, the Water and Soil Erosion Index (WET) was calculated using the Tasseled Cap Transformation (TCT), a widely used method for analyzing remote sensing data, which was used to reflect the land degradation caused by the soil water content [37]. All of the aforementioned indexes were operationally processed on the Google Earth Engine platform. In this study, the RSEI was calculated using the following formulas.

$$\mathrm{NDVI} = (\rho_{\mathrm{NIR}} - \rho_{\mathrm{Red}})/(\rho_{\mathrm{NIR}} + \rho_{\mathrm{Red}}) \tag{1}$$

$$\mathrm{Wet} = 0.1147\mathrm{B}_1 + 0.2489\mathrm{B}_2 + 0.2408\mathrm{B}_3 + 0.3132\mathrm{B}_4 - 0.3122\mathrm{B}_5 - 0.6416\mathrm{B}_6 - 0.5087\mathrm{B}_7 \tag{2}$$

$$\mathrm{NDSI} = \tfrac{1}{2}(\rho_{\mathrm{Swir1}} + \rho_{\mathrm{Red}} - \rho_{\mathrm{Blue}} + \rho_{\mathrm{Nir}}) \times \left(\frac{\rho_{\mathrm{Swir1}}}{\rho_{\mathrm{Swir1}}+\rho_{\mathrm{Red}}+\rho_{\mathrm{Blue}}+\rho_{\mathrm{Nir}}} - \frac{\rho_{\mathrm{Nir}}}{2\rho_{\mathrm{Swir1}}+\rho_{\mathrm{Nir}}+\rho_{\mathrm{Red}}} + \frac{\rho_{\mathrm{Green}}}{\rho_{\mathrm{Green}}+\rho_{\mathrm{Swirl}}}\right) \tag{3}$$

$$\mathrm{RSEI}_0 = 1 - \{\mathrm{PC}_1[\mathrm{f}(\mathrm{Wet}, \mathrm{NDVI}, \mathrm{NDSI}, \mathrm{LST})]\} \tag{4}$$

$$\mathrm{NI}_i = \frac{\mathrm{I}_i - \mathrm{I}_{\min}}{\mathrm{I}_{\max} - \mathrm{I}_{\min}} \tag{5}$$

$$\mathrm{RSEI} = \frac{\mathrm{RSEI}_0 - \mathrm{RSEI}_{\min}}{\mathrm{RSEI}_{\max} - \mathrm{RSEI}_{\min}} \tag{6}$$

where ρ stands for the surface reflectance bands; ρ_{red}, ρ_{blue}, ρ_{green}, ρ_{nir1}, ρ_{nir2}, ρ_{swir1}, and ρ_{Swir2} are the surface reflectance of the MDOIS bands; PC_1 represents the first component of the spatial principal component analysis (SPCA); NI_i indicates the normalized result of each index; and RSEI means the normalized result of the $RSEI_0$.

2.3.2. Calculation of the Nighttime Light Index

The level of urbanization reflects a comprehensive integration of factors such as population, industrial structure, and regional spatial development, while the average intensity of night lighting represents the combined influence of these elements. Therefore, utilizing the average intensity of night lighting to analyze and evaluate urbanization levels is both appropriate and effective. The Compound Night Light Index (CNLI) was developed as a means of using nighttime light data to gauge the level of regional urbanization, offering a greater accuracy compared to previous studies and aligning closely with results derived from population and land indicators [38]. Hence, the CNLI was employed to assess the overall level of urbanization in each city using the following formula [39]:

$$CNLI = LAP \times MLI = \frac{\sum_{i=T}^{DN_M} (DN_i \times n_i)}{N_L \times DN_M} \times \frac{Area_N}{Area} \quad (7)$$

where LAP represents the average night light brightness of all pixels in the region, while MLI represents the proportion of urban areas to the total area of the region. DN_i is the DN value of the ith gray level; n_i is the number of pixels at the ith gray level; DN_M is the maximum DN value; N_L is the number of pixels with a DN value between T and DN_M; N is the number of total pixels of the region; $Area_N$ is the area of city region in study area; and Area is the total area of the study area.

2.3.3. Coupling Coordination Degree Model

The degree of coupling coordination between urbanization and the eco-environment, denoted as C (U, E), was calculated using the following formula [40]:

$$C = 2 \times \frac{\sqrt{UE}}{U + E}, 0 \leq C \leq 1 \quad (8)$$

The coupling degree (C) between urbanization and eco-environment was calculated as the value of the relationship between the Compound Night Light Index (CNLI) and the Remote Sensing Ecological Index (RSEI), where U represents CNLI and E represents RSEI.

The level of effect between urbanization and eco-environment (T) was defined as follows:

$$D = \sqrt{C \times T}, T = \alpha U + \beta E \quad (9)$$

where D is the coupling coordination degree and α and β are contribution comprehensive weight coefficients. In this study, assuming that the eco-environment is as important as urbanization, hence $\alpha = \beta = 1/2$ [41]. When D is between 0 and 0.4, the urban ecosystem is characterized by a low coupling coordination; a value of D between 0.4 and 0.5 indicates a moderate coordination, while a value of D between 0.5 and 0.8 indicates a high coordination. If D is between 0.8 and 1, the urban ecosystem is considered to have an extremely high coordination [42]. The coupling coordination degree classification between urbanization (U) and eco-environment(E) is shown in Table 1.

Table 1. The classification levels of the CCD.

Composite Category	Coordination Level	Subcategory	Systematic Exponential Comparison	Type
Coordinated development	$0.7 < D \leq 1$	High coordination	E-U > 0.1 U-E > 0.1 |E-U| < 0.1	Sluggish urbanization (IV-1) Ecological environment lag (IV-2) Coordinated development (IV-3)
Transformation development	$0.6 < D \leq 0.7$	Moderate coordination	E-U > 0.1 U-E > 0.1 |E-U| < 0.1	Sluggish urbanization (III-1) Ecological environment lag (III-2) Coordinated development (III-3)
Uncoordinated development	$0.4 < D \leq 0.6$	Reluctant coordination	E-U > 0.1 U-E > 0.1 |E-U| < 0.1	Sluggish urbanization (II-1) Ecological environment lag (II-2) Coordinated development (II-3)
	$0 < D \leq 0.4$	Serious imbalance	E-U > 0.1 U-E > 0.1 |E-U| < 0.1	Sluggish urbanization (I-1) Ecological environment lag (I-2) Coordinated development (I-3)

2.3.4. MGWR

Compared with traditional geographically weighted regression (GWR), which can only choose the same bandwidth for all independent variables for regression, multi-scale geographically weighted regression (MGWR) uses the respective optimal bandwidth for each independent variable, solving the problem of different variable scales and bandwidths [43]. Hence, we applied the MGWR model to identity the spatially varying relationships in the cities. The general structure of the MGWR is as follows [44]:

$$y_i = \sum_{j=1}^{k} \beta_{bwj}(u_i, v_i)x_{ij} + \varepsilon_i \quad (10)$$

where X_{ij} is the jth predictor variable, (u_i, v_i) is the center-of-mass coordinate of each block i, and β_{bwj} represents the bandwidth of the regression coefficient of the jth variable.

3. Results

3.1. The Spatio-Temporal Pattern of the Eco-Environment and Urbanization

3.1.1. Changes in Comprehensive Ecological Environment

Principal component analysis results showed that the average contribution rate of the index exceeded 75% in the data for the fifth phase, indicating that PC1 explained over 75% of the variance in the ecological index on average. This suggests that the coupled RSEI effectively represented the ecological environment quality. The RSEI values in the LP were categorized into the following five classes: Very Bad (0–0.2), Bad (0.2–0.4), Medium (0.4–0.6), Good (0.6–0.8), and Very Good (0.8–1.0) (Figure 2). The results indicated a significant upward trend in the RSEI of resource-based cities in China's LP from 2000 to 2020. The average RSEI values of resource-based cities in China's LP in the five periods were 0.4524, 0.4282, 0.4609, 0.4491, and 0.4892, which revealed a fluctuating up trend with approximately 8.13% growth rates. In the past 20 years, significant changes have occurred in the ecological status of resource-based cities in China's LP. The proportion of the "Very Bad" ecological status was reduced by 79.4%, decreasing from 4.71% in 2000 to 0.97% in 2020. Moreover, the area with "Bad" ecological status decreased by 32.1% compared to 2000. In contrast, the areas with a "Medium", "Good", or "Very Good" ecological status all showed a significant increase, with growth rates of 23.4%, 48.4%, and 87.8%, respectively. In 2000, the proportion of "Medium" and "Bad" ecological conditions was relatively similar, at about 40%, while areas with a "Very Bad", "Good", or "Very Good" ecological status accounted for about 4.71%, 14.12%, and 0.58%, respectively. In 2020, the ecological conditions of "Bad", "Medium", and "Good" accounted for 27.5%, 49.4%, and 20.9% respectively, indicating a clear change compared to 2000. Overall, in the past 20 years, about 20–25% of the areas categorized below the "Medium" ecological status have shown improvement. Initially, the northern regions were heavily impacted, classified as "Very Bad" and "Bad". Over the years, substantial efforts or natural recovery processes have led to a significant reduction in degraded areas, with a corresponding increase in regions classified as "Good" and "Very Good". By 2020, the majority of the study area demonstrated an improved ecological quality, indicating successful ecological restoration and management efforts. The changes in the RSEI values indicated that the eco-environment in resource-based cities of China's LP has improved over the past two decades.

Cities of different development types showed different change trends (Figure 3). Most cities of the recessionary type showed a bad eco-environment such as BY, WH, and SZS. Regenerative cities, including LY and BT, showed opposite ecological conditions, which were medium. Growing cities saw a big difference, but maintained a medium status. Grown-up cities, overall, showed a stable and good eco-environment. While the overall ecological status of resource-based cities in China's LP has improved, it is important to note that there are still areas that have poor ecological conditions. This could be attributed to human activities and environmental factors. Furthermore, the ecological circumstances in southern cities were generally better than those in northern cities, including BY, SZS, YL, and OFS, which are mainly characterized as recessionary and growing cities. In addition, no city with a very good degree of ecological quality appeared.

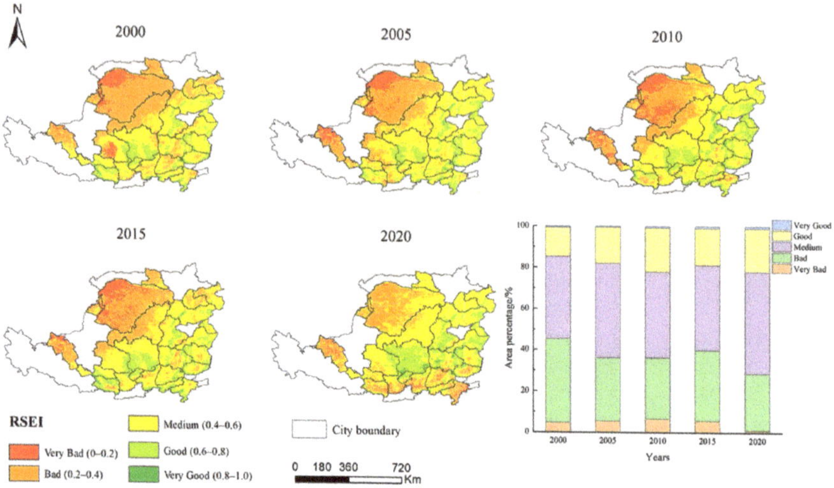

Figure 2. The RSEI map and area percentage in 2000, 2005, 2010, 2015, and 2020.

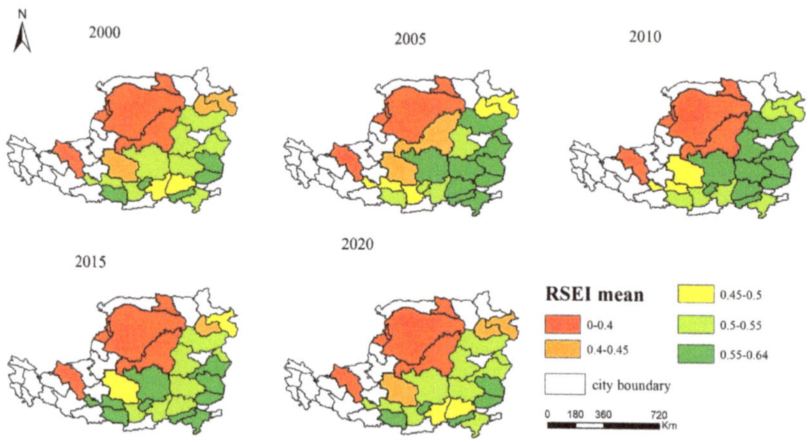

Figure 3. The RSEI mean value distribution between 2000 and 2020.

3.1.2. Changes in Comprehensive Urbanization Quality

Figure 4 depicts the nighttime light images of resource-based cities in China's Loess Plateau from 2000 to 2020, revealing a steady increase in both the lighting area and light pixels number. The MLI value, which was merely 0.033 in 2000, grew year by year and reached 0.065 in 2020, indicating a remarkable growth rate of about 96.6%. Moreover, the LAP increased from 23.5% in 2000 to 40.3% in 2020. Overall urbanization has been rapidly progressing since 2000, with southeastern cities showing better development than northwestern cities. While the urbanization rate in grown-up cities remained relatively stable, the other three city types underwent significant changes over the past 20 years. Despite this noticeable urbanization progress, the ecological conditions of resource-based cities in China's Loess Plateau still need improvement due to the impact of human activities on the environment. Additionally, southern cities generally exhibited better ecological conditions than northern cities, including BY, SZS, YL, and OFS, which mainly belong to the recessionary and growing city categories.

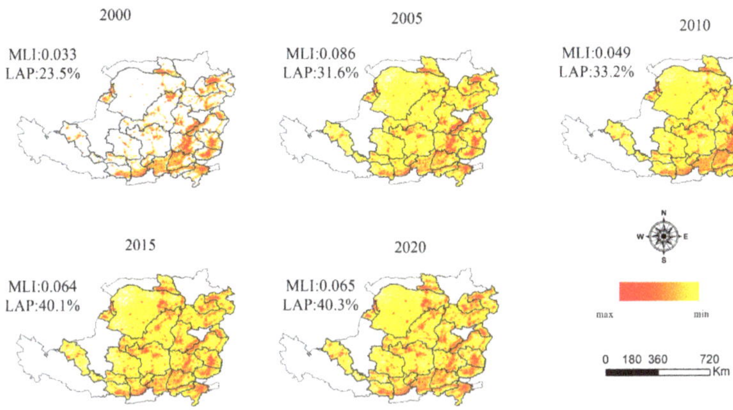

Figure 4. Resource-based cites in China's Loess Plateau light images and changes from 2000 to 2020.

The CNLI values and their changes were calculated for each city (Figure 5). The average CNLI values for all cities in the five periods were 1.5700, 4.4708, 4.6896, 5.8961, and 6.0864, with a change trend value of 4.5164. In 2020, the average CNLI value increased by 287.66% compared to 2000. The spatial distribution of CNLI was similar to the eco-environmental quality, and a high coupling coordination was mainly distributed in the southeastern region of the study area. The cities with the greatest improvement were WH, BJ, LY, BT, and SZ, which belong mainly to the growing and regenerative city categories, while LF, YZ, OFS, BY, and QY, which mainly belong to the growing city categories, saw less improvement. Despite the overall urbanization process of various resource-based cities having grown over the past 20 years, this development has been very uneven.

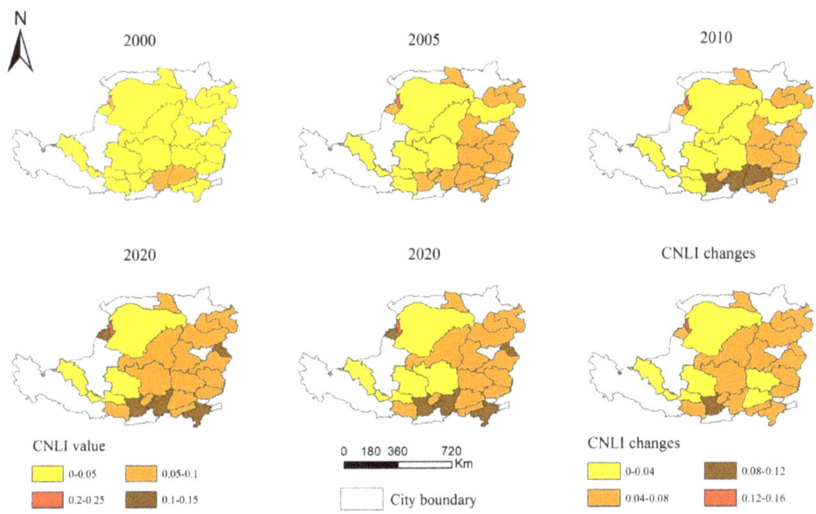

Figure 5. CNLI of resource-based cites in China's Loess Plateau and its changes.

3.2. The Spatio-Temporal Relationship Between Eco-Environment and Urbanization

The interaction between the eco-environment and urbanization was directly and effectively measured using MGWR in conjunction with mutual variables, taking into account spatial and temporal differences. Table 2 shows the remaining parameters, with urbanization as the dependent variable. The standard deviation of the coefficients reflects the degree of variation in the relationship between the eco-environment and urbanization, with larger values observed in 2000 compared to other years. In terms of the absolute mean value of the coefficient, 2000 had the greatest impact on the overall quality of the eco-environment, while the other years did not differ much. The year with the greatest overall impact on the ecosystem, in terms of the absolute value of the median, was also 2000, followed by 2015, 2020, 2005, and 2010. These results suggest that urbanization had a greater and more volatile impact on the eco-environment in 2000, indicating a strong spatial heterogeneity.

Table 2. Statistics of regression results of MGWR model.

Year	Mean	STD	Min	Median	Max
2000	0.042	0.725	−1.684	0.348	0.661
2005	−0.001	0.003	−0.005	−0.001	0.004
2010	−0.001	0.002	−0.004	−0.001	0.002
2015	−0.001	0.004	−0.008	−0.001	0.005
2020	−0.002	0.002	−0.006	−0.001	0.003

The feedback of the eco-environment on urbanization in different regions is presented from 2004 to 2018 (Figure 6). The MGWR coefficient map demonstrates a spatial pattern with positive estimates stretching from the northwest to the southeast. Specifically, in 2000, the areas with positive MGWR coefficient values were mainly located in the southeast of the study area, including cities such as LY, SMX, YC, WN, CZ, JC, and BJ. In 2005, great changes occurred, with positive MGWR coefficient value areas transferring to northeast regions, such as BT, DT, SZ, XZ, and YQ. The range of significant coefficients was between −1.684 and 0.6612, which showed large fluctuations. In addition, the overall value decreased from southeast to northwest. Since 2005, over the next 15 years, this showed that similar spatial distribution. The range of significant coefficients was between −0.0040 and −0.0010, which showed small fluctuations. The overall value decreased from northeast to southwest. With the development of time, cities in northeast areas showed a large correlation between the eco-environment and urbanization incidence.

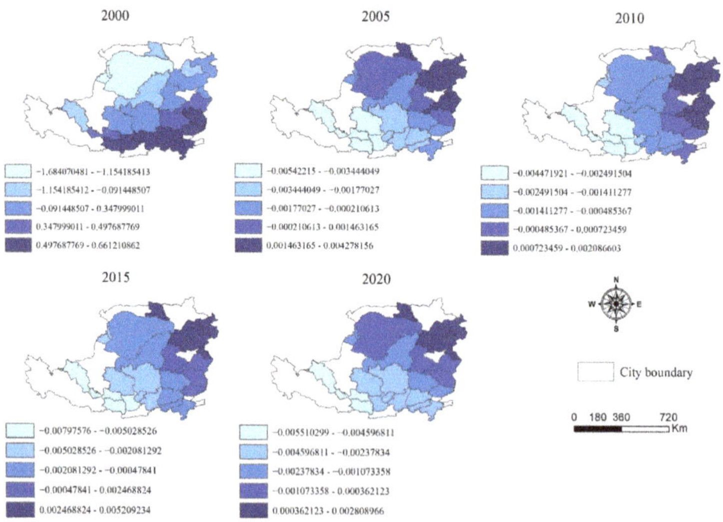

Figure 6. Spatial distribution of coefficients of multi-scale geographically weighted regression model.

3.3. Coupling Coordination Evaluation Results of Eco-Environment and Urbanization

The cluster maps (Figure 7) broadly illustrate the pattern of CCD spatial distribution. The results indicate that, in 2000, most resource-based cities in the LP were dominated by low and moderate coordination, and only a few cities such as Yuncheng and Weinan were at a moderate coordination level. All other cities had a low coupling coordination, of which Ordos was the lowest. In 2005 and 2010, the overall CCD of cities in the LP had the trend of a slow increase, a change compared with 2000, especially in cities in the southeast, where most were located in Shaanxi and Shanxi province. All cities had a low or medium coupling coordination. Additionally, more and more cities had an increasing coupling coordination degree at the same time. Many cities, about 72%, changed from a low coupling coordination to a medium coupling coordination, while only a few cities maintained a low coupling coordination. In 2015, the CCD continued to increase in most cities. In 2020, the CCD compared with 2010 changed to being better in general. Only six cities had a low coordination level, which were BY, OFS, YL, QY, PL, and BJ. There was no city with an extremely high coordination over the two decades.

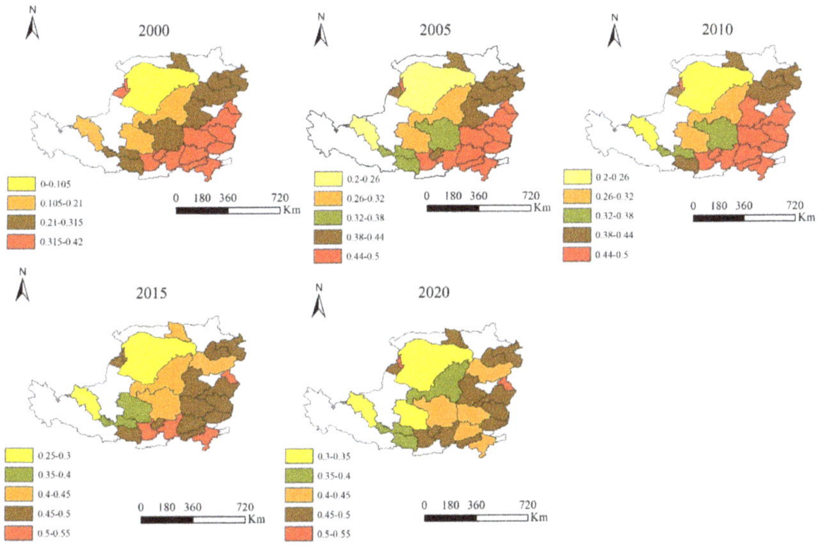

Figure 7. The coordination degree between eco-environment and urbanization in China's Loess Plateau.

From the perspective of temporal-scale changes (Figure 8), the CCD increased compared to 2000, regardless of the city type. The improvement of the coupling coordination degree can be attributed to many factors, including ecological restoration projects such as the Grain for Green Program, funding and policy protection, and active public participation in the LP. The cities saw CCD fluctuations of 36%. The CDD of the remaining cities increased persistently. The CCD of some cities continued to rise year by year, such as BY, Ordos, SZ, Yl, QY, and DT, which mostly belong to the recessionary and grown-up city categories. The CCD in the remaining cities remained stable or fluctuated after elevated development. In general, resource-based cities were more influenced by human activities, and economic development and ecological management have been deepening over the past 20 years.

Furthermore, the specific coupling coordination type assigned to each county highlighted the tension between urbanization and the state of the ecological environment (Table 3). The results showed that most resource-based cities on the LP experienced uncoordinated development, serious imbalances, and sluggish urbanization. In the later stages, some cities had transitioned to reluctant coordination, but continued to experience sluggish urbanization. Except for Wuhai, where the eco-environment and urbanization were in coordinated development in 2005, 2010, and 2015, other cities experienced sluggish urbanization in their comparative relationship between urbanization and the eco-environment. Most cities experienced a serious imbalance at the beginning of the 20th century and have since developed into a state of reluctant coordination. Additionally, cities like Changzhi, Yuncheng, and Weinan have remained in a state of reluctant coordination for the past 20 years. In a few cities, such as Shuozhou and Baoji, there have been repeated shifts in the coupling coordination between reluctant coordination and a serious imbalance.

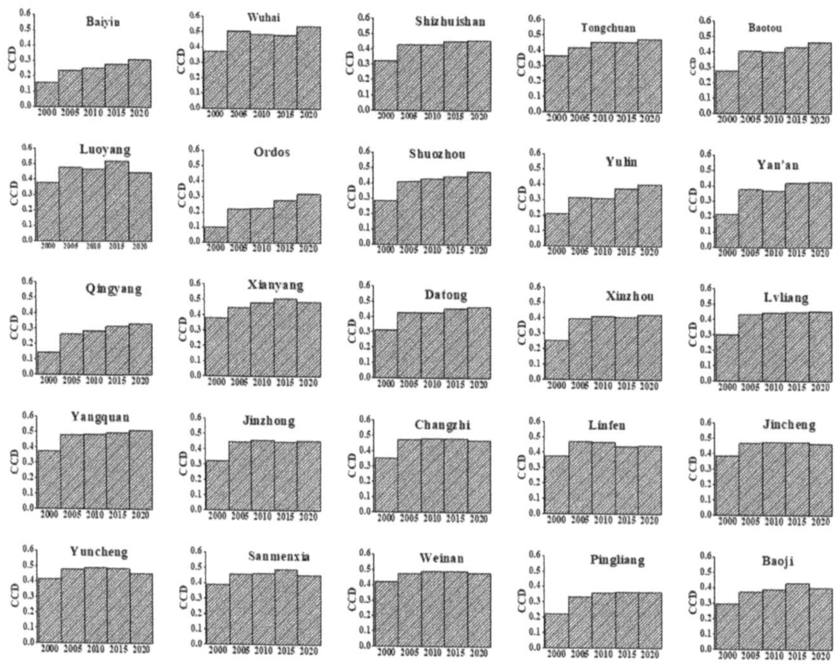

Figure 8. The temporal changes of the CCD between eco-environment and urbanization in China Loess Plateau about prefecture-level cities.

Table 3. Statistics of regression results of MGWR model.

City	Types	2000	2005	2010	2015	2020
Baiyin	Recessionary	I-1	I-1	I-1	I-1	I-1
Wuhai		I-1	II-3	II-3	II-3	II-1
Shizhuishan		I-1	II-1	II-1	II-1	II-1
Tongchuan		I-1	II-1	II-1	II-1	II-1
Baotou	Regenerative	I-1	II-1	I-1	II-1	II-1
Luoyang		I-1	II-1	II-1	II-1	II-1
Ordos	Growing	I-1	I-1	I-1	I-1	I-1
Shuozhou		I-1	II-1	II-1	II-1	II-1
Yulin		I-1	I-1	I-1	I-1	I-1
Yan'an		I-1	I-1	I-1	II-1	II-1
Qingyang		I-1	I-1	I-1	I-1	I-1
Xianyang		I-1	II-1	II-1	II-1	II-1
Datong	Grown-up	I-1	II-1	II-1	II-1	II-1
Xinzhou		I-1	I-1	II-1	II-1	II-1
Lvliang		I-1	II-1	II-1	II-1	II-1

Table 3. *Cont.*

City	Types	2000	2005	2010	2015	2020
Yangquan	Grown-up	I-1	II-1	II-1	II-1	II-1
Jinzhong		I-1	II-1	II-1	II-1	II-1
Changzhi		II-1	II-1	II-1	II-1	II-1
Linfen		I-1	II-1	II-1	II-1	II-1
Jincheng		I-1	II-1	II-1	II-1	II-1
Yuncheng		II-1	II-1	II-1	II-1	II-1
Sanmenxia		I-1	II-1	II-1	II-1	II-1
Weinan		II-1	II-1	II-1	II-1	II-1
Pingliang		I-1	I-1	I-1	I-1	I-1
Baoji		II-1	I-1	I-1	II-1	I-1

4. Discussion

4.1. Evolution and Relationship of Eco-Environment and Urbanization

Over the past two decades, urbanization has rapidly advanced in resource-based cities on the LP of China, resulting in significant changes in the eco-environment of these cities. Due to its natural background, the LP is predominantly covered with unique loess soil, leading to long-term soil erosion, low vegetation cover, and a poor ecological quality in most areas over the past century. Moreover, the local economic base of the region is weak, and the development of these cities relies on exploiting various resources to achieve rapid economic growth and significant socio-economic progress. This study utilized the Compound Night Light Index (CNLI) and Remote Sensing Ecological Index (RSEI) to estimate the urbanization and Eco-Environmental Quality (EEQ) of resource-based cities in the LP region to investigate the influence of urbanization on the eco-environment. The research findings revealed significant improvements in both the urbanization level and eco-environmental quality of these cities from 2000 to 2020.

Given the importance of resources for the success and sustainability of these cities, it is critical to address the prominent unsustainable problems they face. Conducting research is crucial to understanding the coupled and coordinated relationship between urbanization and the eco-environment in resource-based cities on the LP. Rapid urbanization in these cities has significantly impacted the eco-environment, but local ecological protection projects have been implemented to ensure stability. Recessionary resource-based cities, suffering from ecological degradation, declining urbanization rates, and limited economic diversification due to prolonged resource extraction, experience socio-economic stagnation. Priority should be given to economic diversification and environmental restoration, focusing on alternative industries like renewable energy and sustainable agriculture to rebuild the local economy. In contrast, regenerative cities are characterized by improving ecological conditions and moderate urbanization as restoration efforts progress. These cities should strengthen their green infrastructure and attract environmentally sustainable industries to reduce resource dependence, while advancing green technology. Growing cities, experiencing rapid urbanization, face challenges in balancing economic expansion with ecological preservation due to intensified environmental pressures. Lastly, grown-up cities represent a mature stage of urbanization with stable, diversified economies and a degree of ecological improvement. At the beginning of the 21st century, the coupling coordination in all cities of the study area was at a low to medium level. However, after implementing economic and ecological projects, the coupling coordination in most cities has continued to rise, although many cities still require improvement.

4.2. Implications and Suggestions for Resource-Based Cities Ecological Management and Sustainable Development

The findings of this study revealed that the coupling coordination degree between urbanization and the eco-environment in resource-based cities on the LP has been consistently low to medium over an extended period, with gradual improvement. These cities should aim to create sustainable, regenerative systems that support human and environmental health. Policy makers should prioritize investments in renewable energy, sustainable and green industry. In addition, it still is necessary to balance urban development with environmental conservation, protecting natural resources while supporting economic growth.

To address the challenges of slow urbanization, sluggish coordinated development, and delayed urbanization in resource-based cities, three targeted strategies are recommended. First, to enhance urbanization processes, policies should focus on diversifying the local economy beyond resource dependency by attracting new industries and supporting entrepreneurship, which can increase employment opportunities and reduce economic reliance on a single sector. Second, for improving coordinated development, there is a need to strengthen infrastructure, particularly in transportation and public services, to foster connectivity and integration with surrounding areas. This would facilitate the flow of goods, services, and labor, enhancing regional development and supporting balanced growth across sectors. Third, to address the slow urbanization specific to these cities, it is crucial to promote sustainable resource utilization and environmental rehabilitation efforts. This involves implementing stricter regulations on resource extraction, encouraging green technologies, and initiating ecological restoration projects to improve environmental conditions and make these cities more attractive for sustainable investment and migration. Together, these strategies aim to revitalize resource-based cities by fostering balanced, resilient, and sustainable urban development.

4.3. Implications, Limitations and Further Research

In this paper, we delved into an analysis of the evolutionary characteristics of the eco-environment and urbanization, with particular focus on their coupling coordination over the past two decades. Nonetheless, certain limitations need to be discussed regarding this research. With the results of principal component analysis, we confirmed that the data used in this study were relatively reliable and the results obtained are accurate and effective. However, the natural background to LP has long been extremely complex. In quantifying the intensity of ecological quality, only four directly perceived and remotely sensed easily inverted ecological factors were considered, which may lead to an inaccurate and complete representation of ecological quality. The application of a comprehensive system with multi-index evaluation and comparative analysis of subsystem transformations is valuable for future reference and analysis [45]. More accurate methods for extracting urbanization are needed, rather than relying solely on nighttime light data, such as incorporating multi-source data correction [46]. Furthermore, it should be noted that the situation such as energy use, land use, and soil erosion were not considered in this study. A comprehensive examination of the diverse driving factors contributing to the phenomenon is necessary, as the current quantitative explanations are insufficient. This oversight may hinder the establishment of a solid foundation for future ecosystem management and policy-making. For resource-based cities, the research scale should be expanded beyond the LP, with diverse future studies conducted across China and globally [47].

The direction of our future research will be concentrated on the following three areas: (1) selecting more targeted indicators to more accurately quantify the eco-environment of the LP in China, including about terrain, atmosphere, and biodiversity; (2) high-resolution data and multi-source data such as Synthetic Aperture Radar (SAR), unmanned aerial vehicle (UAV), and so on will be used to enhance quantization results about urbanization; and (3) more models should be employed to explore and analyze the coupling relationship about the different ecological factors, ecological elements, and ecological systems. Future local urbanization and ecological environment trends need to be quantitatively predicted based on historical data. The coupling and coordination of urbanization and ecological environment development are crucial steps toward achieving sustainable urban development [48]. Overall, our future research will focus on improving our understanding of the eco-environment of resource-based cities in China, with the ultimate aim of achieving sustainable development goals in different regions.

5. Conclusions

In this study, the eco-environment and urbanization over the period of 2000–2020 were evaluated by RSEI and CNLI, based on the MODIS data and nighttime light data, for 25 resource-based cities in the LP of China. This was based on the RSEI and CNLI, with data derived from MODIS and nighttime light sources. The main conclusions are as follows:

(1) The investigation of the relationship between the eco-environment and urbanization was conducted using a coupling coordination degree model and MGWR. Over the period from 2000 to 2020, the urbanization level and the eco-environmental quality of all cities in study area exhibited a steady increase. In addition, coupling coordination of all cities also rose at the same time. The development of different types of cities was very uneven, although there were continuous improvements in development over 20 years. Over those 20 years, CNLI values ranged from 1.57 to 6.08, while the RSEI values ranged from 0.46 to 0.48. The regression coefficient of economic urbanization in resource-based cities on the ecological environment across different regions was mainly negative, indicating that the ecological environment in these cities is considered to be seriously threatened. The CCD ranged from 0.30 to 0.43, indicating that all cities remained in uncoordinated development for the last 20 years.

(2) Based on an analysis of the four types of resource-based cities—recessionary, regenerative, growing, and grown-up—the coordination between urbanization and ecological environment status varied significantly. Recessionary cities showed a consistent trend of sluggish urbanization and environmental lag, indicating ongoing ecological challenges and limited urban expansion. In contrast, regenerative and growing cities demonstrated a moderate to high coordination, with a developing balance between urban development and ecological health.

(3) The grown-up cities exhibited signs of improved alignment between urbanization and ecological conditions. So, it is necessary to get rid of the industrial structure dependent on resources as soon as possible, continue to develop the economy and maintain the ecological balance, and constantly improve the degree of coupling and coordination. The findings of this study provide valuable insights into the internal mechanisms and fundamental conditions of the interaction and coupling between resource-based cities in the LP region. The findings of this study hold practical significance for promoting sustainable regional development and optimizing decision making related to eco-environmental issues. The profound interaction between spatial urbanization and the ecological environment suggests that limiting urbanization should not be viewed as the sole method of preventing ecological degradation. Therefore, with regard to the different coordination types of cities, there is still a requirement for specific suggestions and effective measures to be formulated in order to facilitate regional sustainable development.

Therefore, it is of paramount importance for policy makers and planners to acquire an in-depth comprehension of the coupled and coordinated development relationship between resource-based urban expansion and ecological environment. Such understanding serves as a scientific reference basis for making informed decisions. The present study generated spatially for understanding patterns of eco-environment and urbanization, along with their evolution trends and relationships. The findings of this study can raise public and governmental awareness of environmental protection and ecological security management. Moreover, the spatially varying impacts of urbanization can offer location-specific guidance for resource-based cities on urban landscape planning and effective ecological management in the future.

Author Contributions: Conceptualization, Yonghua Zhao; methodology, Shuaizhi Kang; validation, Ming Zhao; formal analysis, Xia Jia; investigation, Xia Jia; resources, Manya Luo; data curation, Huanyuan Wang; writing—original draft preparation, Shuaizhi Kang; writing—review and editing, Yonghua Zhao; visualization, Manya Luo; supervision, Xia Jia; project administration, Yonghua Zhao; funding acquisition, Yonghua Zhao. All authors have read and agreed to the published version of the manuscript.

Funding: This research was jointly supported by National Natural Science Foundation of China (U23A2061) and Innovation Capability Support Program of Shaanxi (2024RS-CXTD-55).

Data Availability Statement: The data used to support the findings of this study are available from the corresponding author upon request.

Conflicts of Interest: Author Huanyuan Wang was employed by the company Shaanxi Provincial Land Engineering Construction Group Co., Ltd. The remaining authors declare that the research was conducted in the absence of any commercial or financial relationships that could be construed as a potential conflict of interest.

References

1. Ji, J.; Tang, Z.; Zhang, W.; Liu, W.; Jin, B.; Xi, X.; Wang, F.; Zhang, R.; Guo, B.; Xu, Z.; et al. Spatiotemporal and Multiscale Analysis of the Coupling Coordination Degree between Economic Development Equality and Eco-Environmental Quality in China from 2001 to 2020. *Remote Sens.* **2022**, *14*, 737. [CrossRef]
2. Cai, J.; Li, X.; Liu, L.; Chen, Y.; Wang, X.; Lu, S. Coupling and Coordinated Development of New Urbanization and Agro-ecological Environment in China. *Sci. Total Environ.* **2021**, *776*, 145837. [CrossRef] [PubMed]
3. Sun, Y.; Liu, S.; Dong, Y.; An, Y.; Shi, F.; Dong, S.; Liu, G. Spatio-Temporal Evolution Scenarios and the Coupling Analysis of Ecosystem Services with Land Use Change in China. *Sci. Total Environ.* **2019**, *681*, 211–225. [CrossRef] [PubMed]
4. Guan, X.; Wei, H.; Lu, S.; Dai, Q.; Su, H. Assessment on the Urbanization Strategy in China: Achievements, Challenges and Reflections. *Habitat Int.* **2018**, *71*, 97–109. [CrossRef]
5. Wu, H.; Guo, B.; Fan, J.; Yang, F.; Han, B.; Wei, C.; Lu, Y.; Zang, W.; Zhen, X.; Meng, C. A Novel Remote Sensing Ecological Vulnerability Index on Large Scale: A Case Study of the China-Pakistan Economic Corridor Region. *Ecol. Indic.* **2021**, *129*, 107955. [CrossRef]
6. Fan, H.; Zhao, C.; Yang, Y. A Comprehensive Analysis of the Spatio-Temporal Variation of Urban Air Pollution in China during 2014–2018. *Atmosphere* **2020**, *220*, 117066. [CrossRef]
7. Guo, A.; Yang, J.; Xiao, X.; Xia, J.; Jin, C.; Li, X. Influences of Urban Spatial Form on Urban Heat Island Effects at the Community Level in China. *Sustain. Cities Soc.* **2020**, *53*, 101972. [CrossRef]
8. Zevenbergen, C.; Fu, D.; Pathirana, A. Transitioning to Sponge Cities: Challenges and Opportunities to Address Urban Water Problems in China. *Water* **2018**, *10*, 1230. [CrossRef]
9. Yu, H.; Liu, Y.; Shu, X.; Ma, L.; Pan, Y. Assessment of the Spatial Distribution of Organochlorine Pesticides (OCPs) and Polychlorinated Biphenyls (PCBs) in Urban Soil of China. *Chemosphere* **2020**, *243*, 125392. [CrossRef]
10. Han, Z.; Jiao, S.; Zhang, X.; Xie, F.; Ran, J.; Jin, R.; Xu, S. Seeking Sustainable Development Policies at the Municipal Level Based on the Triad of City, Economy, and Environment: Evidence from Hunan Province, China. *J. Environ. Manag.* **2021**, *290*, 112554. [CrossRef]
11. Yu, J.; Li, J.; Zhang, W. Identification and Classification of Resource-Based Cities in China. *J. Geogr. Sci.* **2019**, *29*, 1300–1314. [CrossRef]
12. Sachs, J.D.; Schmidt-Traub, G.; Mazzucato, M.; Messner, D.; Nakicenovic, N.; Rockström, J. Six Transformations to Achieve the Sustainable Development Goals. *Nat. Sustain.* **2019**, *2*, 805–814. [CrossRef]
13. Sun, Q.; Zhang, X.; Zhang, H.; Niu, H. Coordinated Development of a Coupled Social Economy and Resource Environment System: A Case Study in Henan Province, China. *Environ. Dev. Sustain.* **2018**, *20*, 1385–1404. [CrossRef]
14. Yang, C.; Zeng, W.; Yang, X. Coupling Coordination Evaluation and Sustainable Development Pattern of Geo-Ecological Environment and Urbanization in Chongqing Municipality, China. *Sustain. Cities Soc.* **2020**, *61*, 102271. [CrossRef]
15. Zou, C.; Zhu, J.; Lou, K.; Yang, L. Coupling Coordination and Spatiotemporal Heterogeneity between Urbanization and Ecological Environment in Shaanxi Province, China. *Ecol. Indic.* **2022**, *141*, 109152. [CrossRef]
16. Fu, S.; Zhuo, H.; Song, H.; Wang, J.; Ren, L. Examination of a Coupling Coordination Relationship between Urbanization and the Eco-Environment: A Case Study in Qingdao, China. *Environ. Sci. Pollut. Res.* **2020**, *27*, 23981–23993. [CrossRef]
17. Sui, Y.; Hu, J.; Zhang, N.; Ma, F. Exploring the dynamic equilibrium relationship between urbanization and ecological environment—A case study of Shandong Province, China. *Ecol. Indic.* **2024**, *158*, 111456. [CrossRef]
18. Xing, L.; Xue, M.; Hu, M. Dynamic Simulation and Assessment of the Coupling Coordination Degree of the Economy–Resource–Environment System: Case of Wuhan City in China. *J. Environ. Manag.* **2019**, *230*, 474–487. [CrossRef]
19. Yang, Y.; Wang, L.; Yang, F.; Hu, N.; Liang, L. Evaluation of the Coordination between Eco-Environment and Socioeconomy under the "Ecological County Strategy" in Western China: A Case Study of Meixian. *Ecol. Indic.* **2021**, *125*, 107585. [CrossRef]
20. Shi, F.; Liu, M.; Qiu, J.; Zhang, Y.; Su, H.; Mao, X.; Li, X.; Fan, J.; Chen, J.; Lv, Y.; et al. Assessing Land Cover and Ecological Quality Changes in the Forest-Steppe Ecotone of the Greater Khingan Mountains, Northeast China, from Landsat and MODIS Observations from 2000 to 2018. *Remote Sens.* **2022**, *14*, 725. [CrossRef]
21. Zhang, F.; Wang, Y.; Ma, X.; Wang, Y.; Yang, G.; Zhu, L. Evaluation of Resources and Environmental Carrying Capacity of 36 Large Cities in China Based on a Support-Pressure Coupling Mechanism. *Sci. Total Environ.* **2019**, *688*, 838–854. [CrossRef] [PubMed]
22. Yuan, Q.; Shen, H.; Li, T.; Li, Z.; Li, S.; Jiang, Y.; Xu, H.; Tan, W.; Yang, Q.; Wang, J.; et al. Deep Learning in Environmental Remote Sensing: Achievements and Challenges. *Remote Sens. Environ.* **2020**, *241*, 111716. [CrossRef]

23. Yu, J.; Li, X.; Guan, X.; Shen, H. A remote sensing assessment index for urban ecological livability and its application. *Geo-Spat. Inf. Sci.* **2024**, *27*, 289–310. [CrossRef]
24. Wang, D.; Shi, Y.; Wan, K. Integrated Evaluation of the Carrying Capacities of Mineral Resource-Based Cities Considering Synergy between Subsystems. *Ecol. Indic.* **2020**, *108*, 105701. [CrossRef]
25. Ruan, F.; Yan, L.; Wang, D. The Complexity for the Resource-Based Cities in China on Creating Sustainable Development. *Cities* **2020**, *97*, 102571. [CrossRef]
26. Fan, F.; Zhang, X. Transformation Effect of Resource-Based Cities Based on PSM-DID Model: An Empirical Analysis from China. *Environ. Impact Assess.* **2021**, *91*, 106648. [CrossRef]
27. Li, Q.; Liu, S.; Yang, M.; Xu, F. The Effects of China's Sustainable Development Policy for Resource-Based Cities on Local Industrial Transformation. *Resour. Policy* **2021**, *71*, 101940. [CrossRef]
28. Song, Y.; Xue, D.; Dai, L.; Wang, P.; Huang, X.; Xia, S. Land Cover Change and Eco-Environmental Quality Response of Different Geomorphic Units on the Chinese Loess Plateau. *J. Arid Land* **2020**, *12*, 29–43. [CrossRef]
29. Zhang, Y.; Chen, Y. Research Trends and Areas of Focus on the Chinese Loess Plateau: A Bibliometric Analysis during 1991–2018. *Catena* **2020**, *194*, 104798. [CrossRef]
30. Zhang, H.; Sun, X.; Ahmad, M.; Lu, Y.; Xue, C. A Step Towards a Green Future: Does Sustainable Development Policy Reduce Energy Consumption in Resource-Based Cities of China? *Front. Environ. Sci.* **2022**, *10*, 901721. [CrossRef]
31. Li, S.; Cao, X.; Zhao, C.; Jie, N.; Liu, L.; Chen, X.; Cui, X. Developing a pixel-scale corrected nighttime light dataset (PCNL, 1992–2021) combining DMSP-OLS and NPP-VIIRS. *Remote Sens.* **2023**, *15*, 3925. [CrossRef]
32. Xu, H.; Wang, Y.; Guan, H.; Shi, T.; Hu, X. Detecting Ecological Changes with a Remote Sensing Based Ecological Index (RSEI) Produced Time Series and Change Vector Analysis. *Remote Sens.* **2019**, *11*, 2345. [CrossRef]
33. Xu, H.; Li, C.; Shi, T. Is the z-score standardized RSEI suitable for time-series ecological change detection? Comment on Zheng et al. (2022). *Sci. Total Environ.* **2022**, *853*, 158582. [CrossRef]
34. Xu, Y.; Yang, Y.; Chen, X.; Liu, Y. Bibliometric Analysis of Global NDVI Research Trends from 1985 to 2021. *Remote Sens.* **2022**, *14*, 3967. [CrossRef]
35. Li, H.; Hong, L. Spatio-Temporal Land Use/Land Cover Dynamics and Its Driving Forces in the Mekong Basin Using Landsat Imageries from 1988 to 2017. *Geocarto Int.* **2022**, *37*, 14676–14698. [CrossRef]
36. Ermida, S.L.; Soares, P.; Mantas, V.; Göttsche, F.M.; Trigo, I.F. Google Earth Engine Open-Source Code for Land Surface Temperature Estimation from the Landsat Series. *Remote Sens.* **2020**, *12*, 1471. [CrossRef]
37. Tollerud, H.J.; Brown, J.F.; Loveland, T.R. Investigating the Effects of Land Use and Land Cover on the Relationship between Moisture and Reflectance Using Landsat Time Series. *Remote Sens.* **2020**, *12*, 1919. [CrossRef]
38. Gao, B.; Huang, Q.; He, C.; Ma, Q. Dynamics of Urbanization Levels in China from 1992 to 2012: Perspective from DMSP/OLS Nighttime Light Data. *Remote Sens.* **2015**, *7*, 1721–1735. [CrossRef]
39. Fu, H.; Shao, Z.; Fu, P.; Cheng, Q. The Dynamic Analysis between Urban Nighttime Economy and Urbanization Using the DMSP/OLS Nighttime Light Data in China from 1992 to 2012. *Remote Sens.* **2017**, *9*, 416. [CrossRef]
40. Liu, N.; Liu, C.; Xia, Y.; Da, B. Examining the coordination between urbanization and eco-environment using coupling and spatial analyses: A case study in China. *Ecol. Indic.* **2018**, *93*, 1163–1175. [CrossRef]
41. Li, P.; Liu, C.; Liu, L.; Wang, W. Dynamic analysis of supply and demand coupling of ecosystem services in Loess Hilly Region: A case study of Lanzhou, China. *Chin. Geog. Sci.* **2021**, *31*, 276–296. [CrossRef]
42. Pan, Z.; Gao, G.; Fu, B.; Liu, S.; Wang, J.; He, J.; Liu, D. Exploring the historical and future spatial interaction relationship between urbanization and ecosystem services in the Yangtze River Basin, China. *J. Clean. Prod.* **2023**, *428*, 139401. [CrossRef]
43. Oshan, T.M.; Li, Z.; Kang, W.; Wolf, L.J.; Fotheringham, A.S. mgwr: A Python Implementation of Multiscale Geographically Weighted Regression for Investigating Process Spatial Heterogeneity and Scale. *ISPRS Int. J. Geo-Inf.* **2019**, *8*, 269. [CrossRef]
44. Hu, J.; Zhang, J.; Li, Y. Exploring the Spatial and Temporal Driving Mechanisms of Landscape Patterns on Habitat Quality in a City Undergoing Rapid Urbanization Based on GTWR and MGWR: The Case of Nanjing, China. *Ecol. Indic.* **2022**, *143*, 109333. [CrossRef]
45. Feng, Y.; Li, G. Interaction between urbanization and eco-environment in the Tibetan Plateau. *J. Geogr. Sci.* **2021**, *31*, 298–324. [CrossRef]
46. Hou, Y.; Huang, Q.; Ren, Q.; Gu, T.; Zhou, Y.; Wu, P.; Fan, Y.; Zhu, G. Spatiotemporal dynamics of urban sprawl in China from 2000 to 2020. *GIScі. Remote Sens.* **2024**, *61*, 2351262. [CrossRef]
47. Liu, X.; Guo, P.; Yue, X.; Zhong, S.; Cao, X. Urban transition in China: Examining the coordination between urbanization and the eco-environment using a multi-model evaluation method. *Ecol. Indic.* **2021**, *130*, 108056. [CrossRef]
48. Zhang, X.; Fan, H.; Liu, F.; Lv, T.; Sun, L.; Li, Z.; Shang, W.; Xu, G. Coupling coordination between the ecological environment and urbanization in the middle reaches of the Yangtze River urban agglomeration. *Urban Clim.* **2023**, *52*, 101698. [CrossRef]

International Journal of
Geo-Information

MDPI

Article

Synergy of Remote Sensing and Geospatial Technologies to Advance Sustainable Development Goals for Future Coastal Urbanization and Environmental Challenges in a Riverine Megacity

Minza Mumtaz [1], Syed Humayoun Jahanzaib [2], Waqar Hussain [3], Sadia Khan [3], Youssef M. Youssef [4,*], Saleh Qaysi [5], Abdalla Abdelnabi [6], Nassir Alarifi [5] and Mahmoud E. Abd-Elmaboud [7]

1 Department of Civil Engineering, Ziauddin University, Karachi 74700, Pakistan; minza.mumtaz@zu.edu.pk
2 Department of Civil Engineering, NED University of Engineering and Technology, University Road, Karachi 75270, Pakistan; humayoun@jn.com.pk
3 Department of Environmental Engineering, NED University of Engineering and Technology, University Road, Karachi 75270, Pakistan; waqar.minhas@hawajpl.com (W.H.); sadiakhan@neduet.edu.pk (S.K.)
4 Geological and Geophysical Engineering Department, Faculty of Petroleum and Mining Engineering, Suez University, Suez 43518, Egypt
5 Department of Geology & Geophysics, College of Science, King Saud University, P.O. Box 2455, Riyadh 11451, Saudi Arabia; sqaysi@ksu.edu.sa (S.Q.); nalarifi@ksu.edu.sa (N.A.)
6 Department of Earth & Planetary Sciences, McGill University, Adams Building, 3450 University Street, Montreal, QC H3A 0E8, Canada; abdalla.abdelnabi@mail.mcgill.ca
7 Irrigation & Hydraulics Department, Faculty of Engineering, Mansoura University, Mansoura 35516, Egypt; mahmoud_elsayed@mans.edu.eg
* Correspondence: youssef.ibrahim@pme.suezuni.edu.eg

Abstract: Riverine coastal megacities, particularly in semi-arid South Asian regions, face escalating environmental challenges due to rapid urbanization and climate change. While previous studies have examined urban growth patterns or environmental impacts independently, there remains a critical gap in understanding the integrated impacts of land use/land cover (LULC) changes on both ecosystem vulnerability and sustainable development achievements. This study addresses this gap through an innovative integration of multitemporal Landsat imagery (5, 7, and 8), SRTM-DEM, historical land use maps, and population data using the MOLUSCE plugin with cellular automata–artificial neural networks (CA-ANN) modelling to monitor LULC changes over three decades (1990–2020) and project future changes for 2025, 2030, and 2035, supporting the Sustainable Development Goals (SDGs) in Karachi, southern Pakistan, one of the world's most populous megacities. The framework integrates LULC analysis with SDG metrics, achieving an overall accuracy greater than 97%, with user and producer accuracies above 77% and a Kappa coefficient approaching 1, demonstrating a high level of agreement. Results revealed significant urban expansion from 13.4% to 23.7% of the total area between 1990 and 2020, with concurrent reductions in vegetation cover, water bodies, and wetlands. Erosion along the riverbank has caused the Malir River's area to decrease from 17.19 to 5.07 km^2 by 2020, highlighting a key factor contributing to urban flooding during the monsoon season. Flood risk projections indicate that urbanized areas will be most affected, with 66.65% potentially inundated by 2035. This study's innovative contribution lies in quantifying SDG achievements, showing varied progress: 26% for SDG 9 (Industry, Innovation, and Infrastructure), 18% for SDG 11 (Sustainable Cities and Communities), 13% for SDG 13 (Climate Action), and 16% for SDG 8 (Decent Work and Economic Growth). However, declining vegetation cover and water bodies pose challenges for SDG 15 (Life on Land) and SDG 6 (Clean Water and Sanitation), with 16% and 11%, respectively. This integrated approach provides valuable insights for urban planners, offering a novel framework for adaptive urban planning strategies and advancing sustainable practices in similar stressed megacity regions.

Keywords: cellular automata–artificial neural networks; Landsat; SRTM-DEM; land use/land cover; urban flooding; coastal megacities; sustainable development goals (SDGs)

1. Introduction

Megacities, defined by the United Nations as urban agglomerations with populations exceeding 10 million inhabitants [1], encounter significant environmental challenges resulting from unprecedented rates of urbanization and climate change impacts [2]. These challenges are particularly intensified in coastal megacities due to their unique geographical settings and environmental vulnerabilities, including storm surge risks, sea-level rise impacts, tidal flooding, saltwater intrusion into groundwater aquifers, and accelerated coastal erosion [3,4]. The global distribution of megacities shows a significant presence along coastal zones, with a particularly high concentration in South Asia (Figure 1a). These urban agglomerations function as critical centers of industrial development and economic activity, offering substantial competitive advantages for developing nations through port facilities, international trade connections, and industrial infrastructure networks [5–7]. The socio-economic opportunities in these coastal zones continue to attract populations to already densely populated megacities, accelerating economic growth [8,9]. However, this rapid coastal urbanization has led to increased vulnerability to both anthropogenic pressures and climate-related environmental risks [10,11]. Understanding and addressing these compounded challenges is crucial for megacity communities worldwide, as the impacts of these environmental stressors significantly affect agricultural systems, hydrological patterns, and urban livelihoods—key factors in achieving sustainable development.

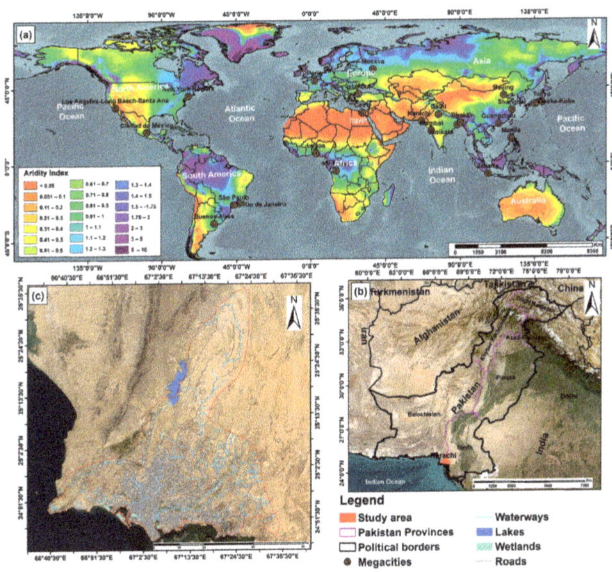

Figure 1. (**a**) The global aridity map categorizes Pakistan within the arid to semi-arid zone [4]. Panels (**b**,**c**) illustrate Karachi, the capital city of Sindh province, delineated by a red polygon in southern Pakistan using Landsat-8 imagery (RGB bands 7, 5, 2) highlighting the city's primary water resources and infrastructure.

Rapid urbanization significantly alters landscape patterns [12], resulting in adverse environmental challenges, including decreased wetland ecosystems, degradation of inland water resources [13], depletion of groundwater resources [14], accelerated soil erosion, and increased surface runoff [15]. Consequently, unplanned urban expansion adversely affects water resources and reduces agricultural land in riverine megacities, thereby threatening food security for growing populations [15]. The expansion of impervious surfaces during urbanization also reduces the land's natural capacity to absorb rainfall, resulting in higher runoff volumes and shorter response times [16]. Even gradual changes in land use/land cover (LULC) can contribute to these impacts, with major alterations in river boundaries further intensified by unpredictable climatic conditions [17–19]. Projections suggest that more frequent and intense precipitation events, along with extreme river flows [20,21], will further complicate flood management strategies in megacities. For example, Mumbai faced catastrophic flooding in 2005 when over 900 mm of rain fell within 24 h, worsened by high tides and insufficient drainage [22]. Jakarta contends with a combination of sea-level rise, land subsidence, and river flooding, with 40% of the city already below sea level [23]. Recent research has highlighted the increasing severity of these environmental challenges in Pakistan's urban centers. A comprehensive assessment of urbanization-led land cover changes across Pakistan from 1990–2020 revealed significant impacts on terrestrial carbon storage capacity, with major cities showing pronounced environmental degradation [24]. Additionally, studies examining land use/cover change associations with urban surface temperature have demonstrated concerning trends in Pakistani cities, where green-to-grey surface transitions have directly compromised urban livability through intensified heat island effects [25]. In New York City, Hurricane Sandy in 2012 exposed the vulnerability of critical infrastructure to storm surges and flooding [23]. These examples highlight the complex environmental conditions of coastal megacities, where risks from storm surges and heavy rainfall are often exacerbated by poor urban planning and aging urban infrastructure [10,13], particularly in south Asian developing nations [26]. The interaction of these factors generates complex environmental risks that challenge conventional urban planning and disaster management strategies.

To address these global challenges, the United Nations General Assembly established 17 Sustainable Development Goals (SDGs) in 2015 (UN Resolution A/RES/70/1), encompassing 169 targets that integrate environmental protection, human well-being enhancement, and economic development [26]. Given the crucial role of urbanization in achieving these goals, it is vital to evaluate urban growth in the context of the SDGs' various dimensions [27]. A thorough assessment of environmental, economic, and social objectives is necessary to determine sustainable urbanization levels [28]. However, forecasting spatiotemporal LULC changes in coastal megacities is complex, as future shifts are likely to escalate due to ongoing urbanization, population growth, climate change, and economic expansion [17,29]. Additionally, evaluating potential floodplain losses remains difficult due to data limitations at the watershed scale [30–32]. These constraints highlight the need for a structured approach to advance SDGs in developing countries, focusing on enhancing environmental quality and urban resilience [2,33,34].

Advancements in remote sensing (RS) sensors, combined with geographic information system (GIS) techniques, have greatly improved the ability to monitor LULC dynamics [8], enabling a deeper understanding and more effective management of urban environmental challenges [35]. Modern integrated approaches, incorporating artificial intelligence-based modeling, provide robust tools for tracking urban growth and environmental degradation [34]. These include high-resolution satellite imagery analysis, machine learning-driven land use prediction models, and advanced hydrological simulation systems [36,37]. Notably, cellular automata (CA) and artificial neural networks (ANN) have emerged as key technologies for urban growth modeling [38–40], along with advanced digital elevation modeling for precise hydrological analysis [9,11]. Such integrated frameworks facilitate a comprehensive understanding of complex urban–environmental interactions [35], equipping urban planners and policymakers with powerful, evidence-based tools to address challenges in rapidly expanding megacities [41].

While existing research has made significant advances in spatiotemporal monitoring of LULC dynamics using multispectral satellite imagery across diverse climatic regions [11,16,21,42,43], three critical gaps persist in current understanding. First, although studies have examined either urban growth patterns or flood hazards independently, limited research integrates both aspects within a unified analytical framework for coastal megacities. Second, while the importance of SDGs is widely recognized, quantitative methodologies linking LULC changes directly to SDG achievement metrics remain underdeveloped, particularly for economic indicators (SDGs 8 and 9), environmental measures (SDGs 13 and 15), and social benchmarks (SDGs 11 and 6). Third, existing predictive models often lack the integration of hydrological risk assessment with future urban development scenarios, limiting their practical application in sustainable urban planning.

This study addresses existing research gaps through an innovative methodological framework that integrates these geospatial approaches with advanced Modules for Land Use Change Evaluation (MOLUSCE) modeling techniques and hydrological simulations. The integration of these diverse analytical methods represents a substantial advancement in predicting environmental degradation and mitigating flood risks in urban environments. Our methodological framework uniquely quantifies the intricate relationships between urban expansion, environmental degradation, and Sustainable Development Goal (SDG) achievement potential in coastal megacities.

This research focuses on Karachi, Pakistan, a rapidly expanding coastal megacity facing significant environmental challenges due to unplanned urbanization and climate change impacts. Karachi serves as an ideal case study, exemplifying the complex interplay between urban growth and environmental degradation in stressed coastal regions.

The research aims to:

1. Monitor spatiotemporal LULC changes from 1990 to 2020 to identify primary drivers of land change and associated natural resource degradation.
2. Predict potential LULC alterations for 2025, 2030, and 2035.
3. Investigate shifts in stream channels for the Malir River Basin from 1990 to 2020 to inform hydrological stress prioritization strategies.
4. Simulate flood inundation scenarios for the 2020 torrential rain event using contemporary LULC data and projected maps for 2025–2035.
5. Explore the contributions of the integrated approach outputs to potential positive and negative impacts on SDGs.

This research plays a pivotal role in advancing land management practices in Karachi through systematic monitoring and prediction of LULC changes, encompassing historical analysis, future projections, and flood scenario simulations. Ultimately, this study aims to support informed decision-making processes and promote sustainable development by providing valuable insights for urban planners, policymakers, and environmental managers in Karachi and similar rapidly urbanizing regions, thus contributing to the development of more resilient and sustainable urban environments.

2. Materials and Methods

2.1. Study Area

Pakistan, the world's sixth most populous country, has experienced rapid, unplanned urbanization since 1947, driven largely by migration, particularly during the 1965 and

1971 wars and Afghan migration in the 1990s [44,45]. Pakistan's population distribution is markedly uneven, with Sindh, the second-most populous province, experiencing significant population clustering in Karachi city. Karachi, a megacity in southwestern Pakistan (Figure 1), is the world's third-largest city by population. With an estimated 21.2 million inhabitants occupying an area exceeding 3530 km^2, Karachi exemplifies the challenges of rapid urbanization [46,47].

The city has undergone significant urban sprawl, driven by rural–urban migration, as people seek better living conditions and economic opportunities, and the influx of immigrants fleeing natural calamities [28]. Between 2000 and 2013, Karachi's built-up area expanded at an annual rate of 2.2%, a growth trajectory closely linked to its burgeoning population. Projections suggest that the city's population, which stood at 14 million in 2010, could reach 25 million by 2030 [45]. Rapid urbanization in Karachi has disrupted its environmental dynamics, with LULC changes between 2009 and 2015 causing stream blockages due to urban expansion and human activities. These alterations in drainage patterns have increased the city's vulnerability to flash floods during heavy rainfall [46,47]. Karachi presents a clear case of urban stream syndrome, characterized by the extensive modification of natural waterways. The proliferation of paved surfaces and encroachment of informal settlements and infrastructure along natural water bodies have significantly altered the city's stream network, intensifying the runoff process.

The dynamics of surface runoff in Karachi are influenced by soil permeability, changes in LULC, and local topography [48]. In this context, monitoring and predicting LULC changes, particularly in and around waterways, is crucial for mitigating and preventing flash floods. Drastic LULC alterations have destabilized stream ecosystems in Karachi, making them more prone to "flashy" behavior. This phenomenon is characterized by rapid rises in the upper limb of the hydrograph, leading to an increased incidence of flash floods and overall watershed degradation. The city's uncontrolled and unplanned urban expansion has exacted a severe environmental toll, with natural resources being overexploited to meet the growing population's needs. Consequently, Karachi faces declining freshwater resources and biodiversity loss—challenges that conventional urban management methods have proven ill-equipped to address effectively. This complex interplay of rapid urbanization, environmental degradation, and increased flood risk underscores the need for innovative urban planning and water resource management approaches in Karachi.

Karachi serves as an ideal case study for investigating the interplay between urbanization and environmental and hydrological degradation in rapidly developing megacities. Studying LULC changes and their impact on natural resources and flood dynamics in this megacity offers valuable insights for developing more resilient and sustainable urban environments in rapidly growing coastal cities worldwide.

2.2. Research Methodology

The comprehensive methodological framework of this study, as illustrated in Figure 2, encompasses several key stages:

1. Implementation: This initial phase involves the identification of land change drivers to analyze the spatiotemporal dynamics of land use changes in Karachi. A series of satellite images spanning three decades (1990, 1995, 2000, 2010, 2015, and 2020) were processed and analyzed to establish historical land use patterns and trends.
2. Data Integration: Ancillary data, including demographic information, terrain characteristics (DEM, slope, and aspect), and accessibility factors (stream and road networks), were incorporated as spatial variables in the MOLUSCE (Modules for Land Use Change Evaluation) plugin. This integration enables a more robust prediction of future LULC changes.

3. Modeling: The MOLUSCE plugin was employed to forecast LULC changes for 2025 and 2035. This process involved the creation of transition probability matrices and change maps, as well as the application of artificial neural network (ANN) and cellular automata (CA) models to simulate future land use scenarios.
4. Validation: The model's accuracy was assessed through a validation process, ensuring the reliability of the predicted LULC maps for future years.
5. Hydrological Analysis: Using the calibrated LULC maps and precipitation data from 2020, hydrological processes were simulated for future scenarios (2025, 2030, and 2035) using HEC-RAS (Hydrologic Engineering Center's River Analysis System).
6. Impact Assessment: The final stage involved analyzing the land use types in inundated areas from 2025 to 2035, providing insights into the potential flood risks associated with projected land use changes.

This methodological approach enables a comprehensive examination of the interplay between land use changes, urban expansion, and flood risk in Karachi, offering valuable insights for urban planning and flood management strategies.

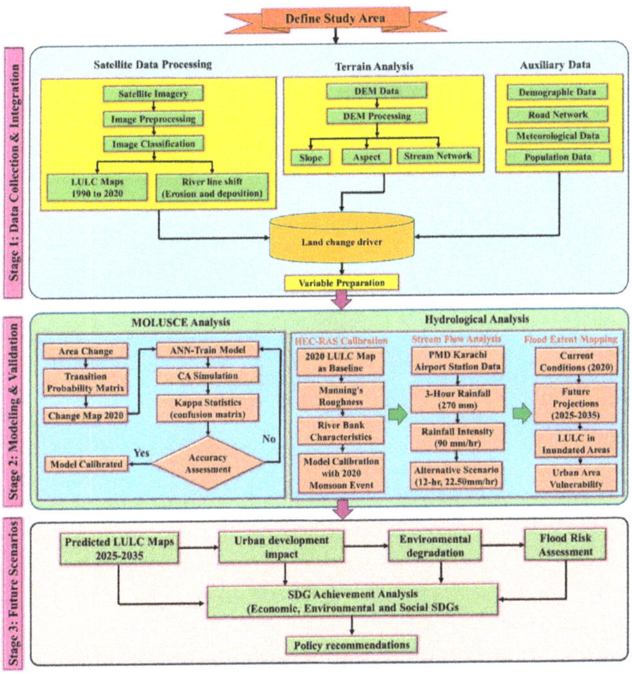

Figure 2. Workflow illustrating the methodology implemented in this study.

2.2.1. Data Sources and Preparation

This study integrated diverse datasets from multiple sources to thoroughly investigate the variations in LULC and hydrological processes in Karachi, as summarized in Table 1. Land use studies typically use intervals of five to ten years to detect significant LULC changes [35,49]. For this study, five-year intervals were selected (1990, 1995, 2000, 2010, 2015, and 2020) to capture land use dynamics over time and identify the driving forces behind these shifts [50]. Satellite imagery was the primary source of geospatial data, with LANDSAT-TM 5 images from 1990, 1995, and 2000 obtained from the United States Geological Survey (USGS). For the more recent years, LANDSAT-ETM 7 (2010) and LANDSAT-ETM 8 (2015, 2020) images from USGS were employed to create base maps and calibrate the model. The LANDSAT 7 data for 2005 were excluded due to scan line corrector failure, which resulted in data gaps. The topographic analysis relied on SRTM-DEM data from 2014, acquired from USGS, facilitating the extraction of slope information and stream networks.

Table 1. Source of extracted data used in the study.

Type	Year	Source	Description
LANDSAT-TM 4	1990, 1995, 2000	USGS	Extracted land use map using the image maximum likelihood classification method
LANDSAT-ETM 7	2010	USGS	Prepared base map for the year 2010
LANDSAT-ETM 8	2015, 2020	USGS	Prepared base map for the years 2015 and 2020 for calibration of the model
SRTM	2014	USGS	Used to extract the slope and stream network from DEM
Road network	2014	World Bank/ OSM	OSM was used to extract missing road networks from the downloaded tiff file and validate the map using the Google Earth time series engine.
Demographic data	2017	OCHA	Population data were extracted for Karachi city from the downloaded zip file containing census data of Pakistan.
Precipitation data	2020	PMD	Used as an input for hydrological analysis

A 2014 road network map supported the study's infrastructure component, developed using a combination of World Bank data and Open Street Map (OSM) information and subsequently verified using Google Earth's time series engine to ensure accuracy. The demographic analysis utilized population statistics from 2017, extracted from the United Nations Office for the Coordination of Humanitarian Affairs (OCHA) dataset, chosen due to the unavailability of more recent official census data for Pakistan. The hydrological analysis component was informed by precipitation data from the Pakistan Meteorological Department (PMD), focusing on the torrential rain event in 2020. This integration of satellite imagery, topographic information, infrastructure data, demographic statistics, and meteorological records enabled a robust examination of historical trends and current conditions, as well as future projections of land use changes and associated flood risks in Karachi's rapidly evolving urban landscape.

2.2.2. Preparation of Spatial Controlling Variables

To enhance the accuracy of land use change analysis, this study incorporated ancillary data to identify the most influential factors in land use dynamics. Three primary categories of land change drivers were selected: terrain, demographics, and accessibility data. These spatial controlling variables were prepared through a series of geospatial analyses and data processing techniques.

Terrain data derived from SRTM formed the foundation of the topographic analysis. The DEM was clipped to the study area boundary and processed to generate additional ancillary data. Figure 3a presents the DEM of Karachi city, illustrating the elevation variations across the urban landscape. Slope data depicting the elevation change rate were derived from the DEM using the Spatial Analyst Surface tool in ArcGIS. The resulting slope map, shown in Figure 3b, provides insights into the terrain's steepness and potential impact on urban development and water flow patterns. Aspect data, representing the direction and steepness of the slope, were similarly generated using the Spatial Analyst Surface tool, with the 2010 reference map as the basis. Figure 3c displays the aspect map, offering a comprehensive view of the terrain's directional characteristics.

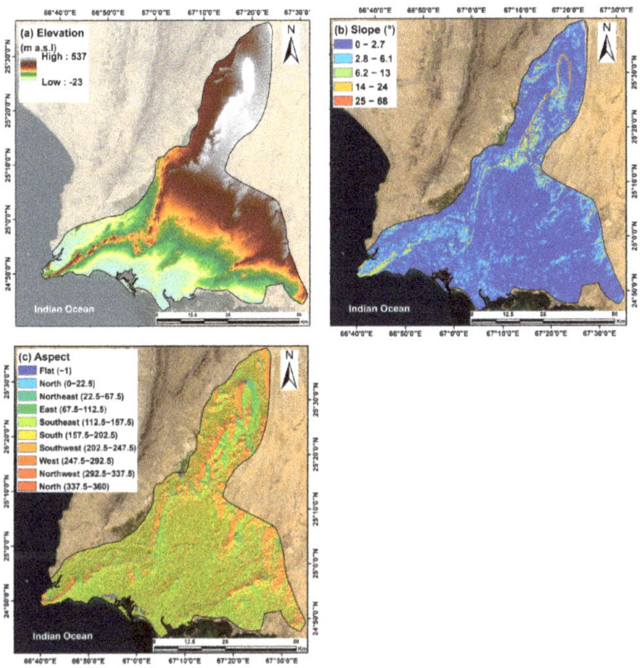

Figure 3. Topographical factors: (**a**) DEM, (**b**) slope, and (**c**) aspect.

Accessibility data, comprising stream and road networks, were prepared to analyze the influence of water bodies and transportation infrastructure on land use changes. The stream network data, illustrated in Figure 4a, were generated using the Euclidean Distance tool applied to a stream delineation model derived from the DEM using ArcGIS's hydrology tools. This process allowed for the quantification of proximity to water bodies, a crucial factor in urban development and flood risk assessment. Initially compiled from various sources as outlined in Table 1, the road network data underwent a similar Euclidean distance analysis. The resulting map, shown in Figure 4b, represents the distance from the road network, providing insights into the accessibility of different areas within the city. The raster data were reclassified into five distinct classes to facilitate analysis, ensuring a harmonized dataset for integration with other spatial variables.

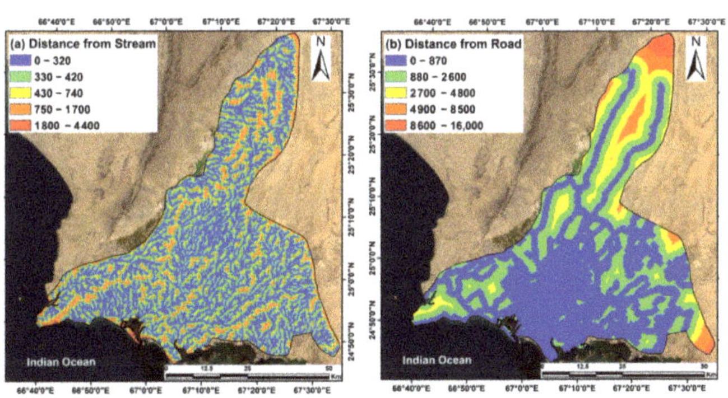

Figure 4. Spatial representation of accessibility data: (**a**) Distance from streams and (**b**) distance from roads.

Demographic data extracted from the most recent available census report were incorporated using the Euclidean Distance tool to represent the population distribution and density across the study area. This data layer adds a crucial socio-economic dimension to the analysis, allowing for the examination of how population patterns influence and are influenced by land use changes. Integrating these diverse spatial controlling variables, namely, terrain, accessibility, and demographics, provides a comprehensive framework for analyzing the complex interplay of factors driving land use dynamics in Karachi. This multifaceted approach enables a more nuanced understanding of the urban landscape's evolution and potential implications for flood risk and urban planning strategies.

2.2.3. Land Use/Land Cover Classification and Prediction

This study employed a comprehensive approach to LULC classification and prediction, utilizing multi-temporal satellite imagery and advanced geospatial modelling techniques. The methodology encompasses several key stages: data preparation, classification, transition probability analysis, model calibration, and future prediction.

Data Preparation and Classification

All satellite images were standardized to the WGS 1984 UTM Zone 42N projection to ensure spatial consistency. Land use/land cover (LULC) classifications were analyzed using supervised classification in ArcMap 10.8. The Maximum Likelihood Classifier (MLC) was applied to produce LULC maps for six time periods: 1990, 1995, 2000, 2010, 2015, and 2020, aligning with recent studies applying similar methodologies [51]. A random sample, representing 80% of the Regions of Interest (ROIs) database, was selected for training purposes. This dataset was derived from 2020 field observations, true-color Landsat imagery, and Google Earth images to train the LULC classification for each year. The accuracy of the classified images was validated using a confusion matrix (Table S1), comparing the remaining 20% of randomly selected ROIs with the corresponding classified pixels. The study area was categorized into five LULC classes: urbanized areas, vegetation, water bodies, wetlands, and barelands, as outlined in Table 2. The generated LULC maps for each year (Figure 5) illustrate the spatial and temporal dynamics of land use changes over the past three decades.

Table 2. Land use/land cover classification criteria based on Level I USGS classification.

Land Use Level I Classes	Description
Urbanized	Built-up areas, constructed or under-construction areas, commercial land, industrial land, service or utilities providers, urban or semiurban settlements
Vegetation	Agricultural land, shrubs, and other green areas
Water bodies	Streams, rivers, lakes, reservoirs
Wetlands	Afforested or non-forested wetlands
Barelands	Rocky areas, areas other than the above-mentioned classes

Transition Probability Analysis

A transition probability matrix was generated (Table 3) for the years 2010 and 2015 to simulate LULC changes for 2020. This matrix revealed significant land use dynamics. For instance, the probability of urbanized land converting to vegetation land was 6.6%, while there was an 18.18% chance of urbanized areas transforming into barelands. Notably, water bodies showed a 34.7% likelihood of transitioning to urbanized areas. Barelands demonstrated a 15.3% probability of shifting to urbanized areas, while other classes showed relatively minor transformations.

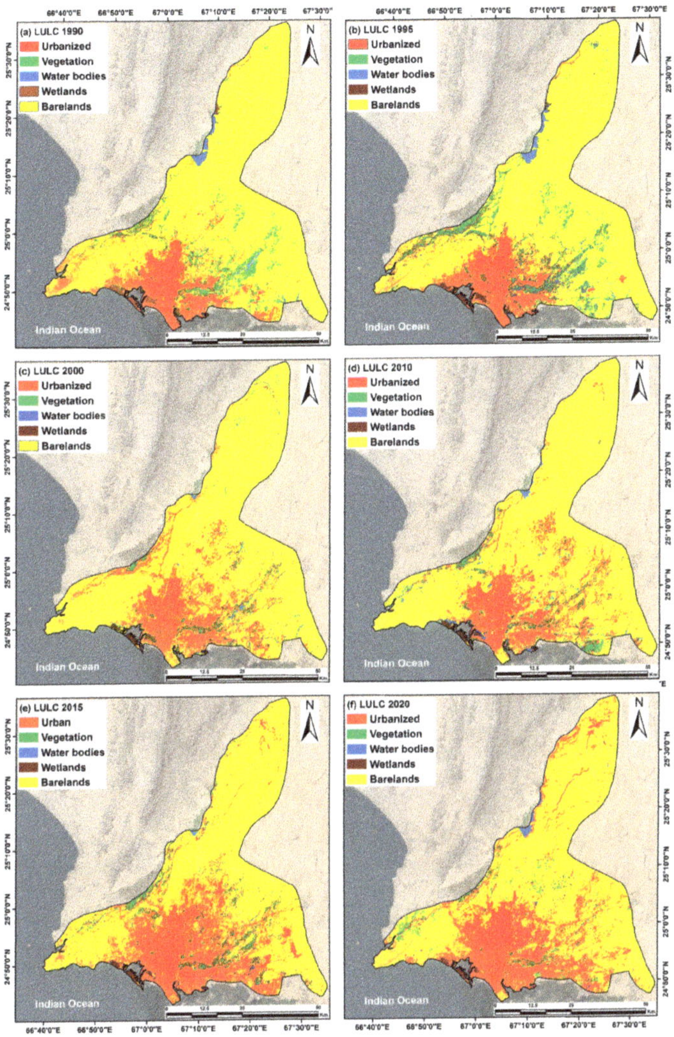

Figure 5. LULC map for Karachi city for the years (**a**) 1990, (**b**) 1995, (**c**) 2000, (**d**) 2010, (**e**) 2015, and (**f**) 2020.

Table 3. Transition probability matrix generated for 2010–2015.

		2015				
	Land use type	Urbanized	Vegetation	Water bodies	Wetlands	Barelands
2010	Urbanized	0.721	0.066	0.029	0.0004	0.182
	Vegetation	0.380	0.303	0.058	0.016	0.242
	Water bodies	0.347	0.089	0.074	0.038	0.452
	Wetlands	0.166	0.020	0.022	0.772	0.021
	Barelands	0.153	0.014	0.014	0.0001	0.818

Model Calibration and Validation

The Modules for Land Use Change Evaluation (MOLUSCE) plugin in QGIS was employed to train base maps from 2010 to 2015 using various combinations of spatial variables. The goal was to achieve the most accurate simulated map for 2020. Six iterations were executed using a trial-and-error method to obtain Kappa statistics greater than 0.5 between the simulated and referenced maps. Accuracy was evaluated using Cohen's Kappa, a widely recognized metric for assessing LULC [52], to test different spatial variable combinations (Table 4). The MOLUSCE plugin's built-in Kappa validation technique facilitated the assessment of multiple combinations [53], offering insights into their predictive significance for LULC changes (Table 5). All combinations displayed excellent alignment between simulated and reference maps, with Kappa values ranging from 0.90 to 0.97. The combination of DEM, slope, aspect, roads, streams, and population achieved the highest Kappa statistic (K = 0.97), indicating near-perfect agreement [54]. This optimal combination was subsequently employed to calibrate the model for predicting LULC maps for 2025 and 2035.

Table 4. Cohen's Kappa interpretation.

Cohen's Kappa	Accuracy Correspondence
>0.75	Excellent
0.4–0.75	Good
<0.4	Poor

Table 5. Kappa classification for accuracy assessment of different combinations of spatial variables between the classified and referenced LULC maps.

Arrangement of Spatial Variables	Kappa Achieved
DEM, slope, aspect, roads, stream, population	0.97
Slope, aspect, roads, stream, population	0.94
DEM, slope, roads, stream, population	0.93
Aspect, slope, roads, streams	0.96
Slope, roads, streams	0.90

Cellular Automata–Artificial Neural Network (CA-ANN) Implementation

The study employed the CA-ANN method, a hybrid approach that combines the spatial explicit modelling capabilities of cellular automata with the pattern recognition and learning abilities of artificial neural networks [55]. This integration allows for nuanced prediction of complex urban growth patterns, accounting for both spatial relationships and non-linear interactions between several factors influencing land use change.

Model Performance and Future Predictions

The performance of the CA-ANN model was evaluated by comparing the supervised classification-derived base map with the simulated map for 2020 (Table 6). The analysis revealed minor discrepancies: the urbanized class showed a 2.36% increase in the simulated map, vegetation cover decreased by 0.57%, water bodies and wetlands saw slight increases (0.29% and 0.09%, respectively), and barelands decreased by 2.18%. Despite these differences, the model achieved a high Kappa accuracy of 0.97, indicating strong agreement between the referenced and simulated maps. This robust performance validates the model's calibration and reliability for future LULC predictions. Consequently, the 2015 and 2020 maps now serve as a baseline for forecasting LULC changes in 2025 and 2035, ensuring accurate projections of land transformation trends. This comprehensive approach to LULC classification and prediction provides a solid foundation for understanding and anticipating urban growth patterns and their potential impacts on Karachi's environment and flood risk.

Table 6. Land use/land cover area in sq. km of referenced and simulated maps for the year 2020.

Classes	Referenced Map (sq. km)	Simulated Map (sq. km)	Δ%
Urbanized	850	934	2.36
Vegetation	111	91	−0.57
Water bodies	47	58	0.29
Wetlands	13	16	0.09
Barelands	2557	2479	−2.18

2.2.4. Simulating the Impact of Future LULC Scenarios on Flood Risk Extent

To assess the impact of land use dynamics on flood extent in Karachi, this study employed the HEC-RAS, a widely recognized open-source hydraulic model. The 2020 land use map, derived from the LULC classification process, served as the baseline for flood inundation analysis. A uniform Manning's roughness coefficient of 0.04 was applied to the riverbanks for model simplification while maintaining urban terrain characteristics. The model was calibrated using the 2020 monsoon rain event and corresponding flood extent, accurately representing the study area's hydraulic behavior under known conditions. Meteorological data, crucial for precise flood modelling, were obtained from the PMD Karachi, specifically utilizing a 3-hourly dataset from the Karachi Airport station. These data were then modified to reflect the extreme precipitation event of 27 August 2020, which saw an unprecedented 270 mm of rainfall within a 3 h period, resulting in an exceptionally high average rainfall intensity of approximately 90 mm/hr. An alternative scenario was also simulated to account for the uncertainty in the recurrence of such high-intensity rainfall events, distributing the same 270 mm of precipitation over a 12 h period, yielding a more moderate intensity of 22.50 mm/hr. This dual-scenario approach allows for a comprehensive flood risk assessment under extreme and more frequent rainfall patterns. Utilizing the calibrated HEC-RAS model in conjunction with the LULC predictions for 2025 and 2035, flood extent simulations were conducted for these future years under both rainfall scenarios. This methodology accounts for current conditions and anticipates future changes, providing valuable insights for long-term urban planning and flood mitigation strategies in Karachi's rapidly evolving urban environment.

3. Results

3.1. Spatiotemporal Analysis of LULC Changes in Karachi (1990–2020)

The spatiotemporal analysis of LULC changes in Karachi from 1990 to 2020 reveals a complex urban landscape transformation characterized by rapid urbanization and significant environmental alterations. Utilizing maximum likelihood supervised classification, the generated land use suitability maps (Figure 5) and corresponding quantitative data (Tables 7 and 8) provide a comprehensive overview of Karachi's urban evolution over three decades. The generated LULC maps, presented in Figure 5, were found to be reliable, achieving an overall accuracy exceeding 97%, with no user or producer accuracy for any class falling below 77%. The Kappa coefficient was close to 1, indicating a high level of agreement [25]. Table S1 provides the overall accuracy and Kappa coefficient for the classified maps corresponding to the years from 1990 to 2020. The most striking trend observed is the substantial expansion of urbanized areas, which increased from 13.4% of the total area in 1990 to 23.7% in 2020. This growth trajectory was particularly pronounced between 2010 and 2015, with urbanized areas surging from 15.3% to 25.2%, reflecting an intense period of urban development driven by population growth and infrastructure expansion. This pattern aligns with observations in other rapidly growing coastal megacities in developing countries [56]. The slight decrease in urbanized areas from 25.2% in 2015 to 23.7% in 2020 may indicate a stabilization phase or potential data classification variations, warranting further investigation.

Table 7. Percentage of change in magnitude for each year from 1990 to 2020 for the assigned LULC.

Classes	1990	1995	2000	2010	2015	2020
Urbanized	13.4	13.6	14.5	15.3	25.2	23.7
Vegetation	4.3	5.3	2.2	2.8	3.2	3.1
Water bodies	3.4	5.0	2.1	2.6	1.9	1.3
Wetlands	1.0	0.9	0.4	0.5	0.5	0.4
Barelands	77.9	75.2	80.7	78.8	69.2	71.4

Table 8. Land use type conversion between the years 1990 and 2020 in sq. km.

		2020						
		Urbanized	Vegetation	Water bodies	Wetlands	Barelands	Total	Loss
1990	Urbanized	348.72	10.13	6.61	0.40	112.92	478.78	130.06
	Vegetation	67.99	22.92	1.72	1.19	61.35	155.17	132.25
	Water bodies	30.35	3.99	13.45	0.04	73.08	120.91	107.46
	Wetlands	17.74	0.75	1.58	11.36	3.83	35.27	23.90
	Barelands	384.78	73.27	23.64	0.12	2304.96	2786.76	481.81
	Total	849.59	111.07	47.00	13.11	2556.13		
	Gain	500.87	88.14	33.55	1.75	251.17		
	Net change	370.81	−44.11	−73.91	−22.16	−230.64		

Conversely, other land cover categories experienced significant declines. Water bodies showed a marked reduction from 3.4% in 1990 to merely 1.3% in 2020, a trend that may be attributed to urban encroachment, climate change-induced drought, or land reclamation practices. Similarly, wetlands diminished from 1.0% to 0.4% over the same period, indicating ongoing degradation of these ecologically vital areas. These findings corroborate the global trend of coastal wetland loss in urban areas, as ref. [57] reported. The vegetation class exhibited a fluctuating pattern, initially increasing from 4.3% in 1990 to 5.3% in 1995, followed by a sharp decline to 2.2% in 2000, before stabilizing around 3.1% by 2020. This recovery in later years might be attributed to reforestation efforts or changes in agricultural practices, although further investigation is needed to confirm these hypotheses. Barelands, while remaining the dominant land cover type, showed a notable decrease from 77.9% in 1990 to 71.4% in 2020, with fluctuations in between. This decline primarily reflects the conversion of barren areas into urbanized zones, a common phenomenon in rapidly expanding cities [58].

Quantitative analysis of land transformation (Table 8) reveals that approximately 384.78 sq. km of barelands and 30.35 sq. km of water bodies were converted to urban areas between 1990 and 2020. The urbanized category experienced a net positive change of 370.81 sq. km, while all other land use categories showed net negative changes. This pattern underscores the dominant trend of urban expansion at the expense of different land cover types, particularly barelands, which saw a major negative change of 230.64 sq. km. The conversion matrix analysis [59] provides insights into the dynamics of land use changes. The diagonal values in Table 8 represent areas that remained unchanged, such as 13.45 sq. km of water bodies, highlighting the persistence of specific landscape features amidst widespread change. These findings underscore the profound environmental implications of Karachi's rapid urbanization, including the loss of water bodies, wetlands, and vegetation cover. Such changes can significantly impact local ecosystems, biodiversity, and the city's resilience to climate-related hazards like flooding [60]. The observed patterns of LULC change in Karachi align with global trends in coastal megacities, where urban expansion often occurs at the expense of natural landscapes, potentially exacerbating environmental vulnerabilities [56]. This comprehensive analysis provides crucial insights for urban planners and policymakers, emphasizing the need for sustainable urban development strategies that balance growth with environmental conservation.

3.2. Simulation and Projection of Land Use Dynamics (2020–2035)

The simulation of future land use dynamics in Karachi from 2020 to 2035 reveals a continuation of the urbanization trend observed in the historical analysis, with significant implications for the city's environmental and socio-economic landscape. Utilizing a robust modelling approach that incorporates multiple spatial variables, including DEM, slope, aspect, road networks, stream systems, and population data, we projected LULC changes for 2025, 2030, and 2035 (Figure 6 and Table 9). The modelling process employed an iterative approach, using the LULC maps of 2015 and 2020 as dependent variables to simulate 2025, then using 2020 and 2025 to project 2030, and finally, 2025 and 2030 to forecast 2035. This sequential method allows for incorporating cumulative changes and potential feedback loops in urban development patterns, enhancing the reliability of long-term projections [61].

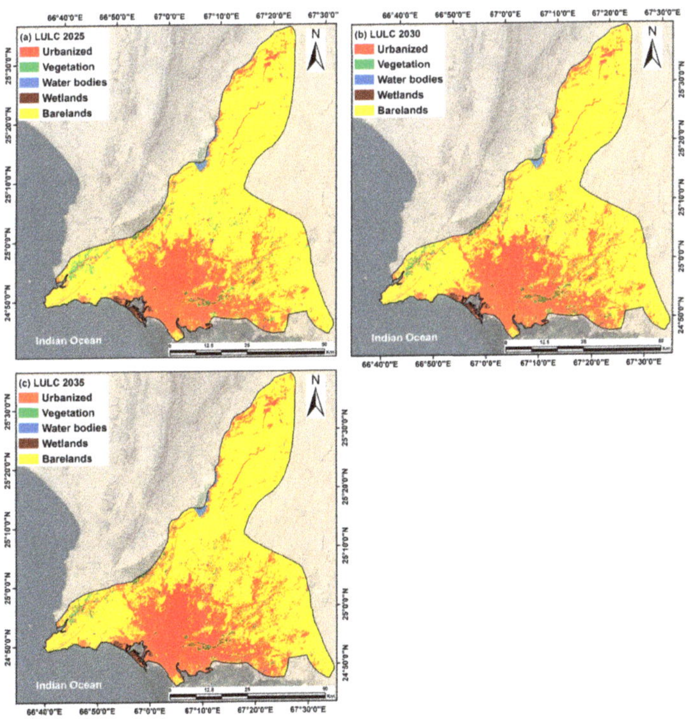

Figure 6. Predictive LULC maps for (**a**) 2025, (**b**) 2030, and (**c**) 2035 reveal a notable increase in urban class pixels, while those associated with water and vegetation classes show a decline, indicating conversion to urban land use.

Table 9. Statistics of land change dynamics from the year 2020 to 2035 in sq. km.

Year	Urbanized	Vegetation	Water Bodies	Wetlands	Barelands
2020	850	111	47	13	2557
2025	847	98	39	13	2582
2030	887	99	42	12	2539
2035	909	77	47	12	2534

As presented in Table 9, the simulation results indicate a consistent increase in urbanized areas from 850 sq. km in 2020 to a projected 909 sq. km by 2035, representing a net growth of 59 sq. km over 15 years. This expansion of urban land use aligns with global trends in rapidly growing megacities, particularly in developing countries. However, the rate of urban expansion appears to be non-linear, with a slight decrease between 2020 and 2025, followed by accelerated growth in subsequent periods. This pattern may reflect the complex interplay of factors influencing urban development, including policy interventions, economic cycles, and infrastructure projects. Concurrently, the model predicts fluctuations in other land use categories. Vegetation areas are projected to decrease from 111 sq. km in 2020 to 77 sq. km by 2035, indicating a significant loss of green spaces. This trend is concerning from an ecological perspective and may exacerbate urban heat island effects and reduce ecosystem services [62]. Water bodies show an interesting pattern, decreasing initially from 47 sq. km in 2020 to 39 sq. km in 2025 but then recovering to the 2020 level by 2035. This fluctuation may reflect the implementation of water body management strategies or the creation of artificial water bodies as part of urban development plans. Barelands, historically the dominant land cover type in Karachi, are projected to experience slight fluctuations, peaking at 2582 sq. km in 2025 before declining to 2534 sq. km by 2035. This pattern suggests that while urbanization continues to consume barren areas, the process is not uniform and may be influenced by factors such as land suitability and development policies.

The predictive land use maps for 2025 and 2035 (Figure 6) visually corroborate these quantitative findings. The maps clearly illustrate the intensification of urban pixels, particularly in areas adjacent to existing urban clusters, suggesting a pattern of urban sprawl and densification. Notably, the maps indicate that pixels associated with water bodies and vegetation classes are likely to be replaced by urban land use, highlighting the potential environmental trade-offs of continued urban expansion. These projections underscore the urgent need for sustainable urban planning strategies in Karachi. The continued loss of vegetation cover and potential pressures on water body resources, coupled with rapid urbanization, may exacerbate environmental challenges such as flooding, urban heat island effects, and biodiversity loss [60]. Moreover, the spatial patterns of urban growth suggested by these simulations may have implications for social equity and access to urban amenities, issues that are critical in the context of a megacity in a developing country [56].

While these simulations provide valuable insights into potential future land use patterns in Karachi, it is important to note that they represent one possible trajectory based on current trends and assumed relationships between variables. Many factors, including policy decisions, economic conditions, and climate change impacts, will influence future land use. Therefore, these projections should be viewed as a tool for informed decision-making rather than a definitive forecast, emphasizing the need for adaptive and responsive urban planning approaches.

3.3. *Land Use Dynamic Scenarios and Their Impact on the River Boundary*

The analysis of land use dynamics and their impact on the Malir River boundary in Karachi over three decades (1990–2020) reveals significant geomorphological changes characterized by pronounced erosional processes outweighing depositional ones. This trend, quantified in Table 10 and visually represented in Figure 7, underscores the complex interplay between urban development, land use changes, and fluvial geomorphology in rapidly expanding coastal megacities. The data presented in Table 10 illustrate a consistent pattern of net erosion along the Malir River, reaching across all three decades studied. From 1990 to 2000, erosion affected 4.54 sq. km of the reach, while depositing occurred in only 1.96 sq. km. This erosional trend intensified in the subsequent decade (2000–2010), with 9.20 sq. km eroded compared to a mere 0.76 sq. km of deposition. The most recent decade (2010–2020) showed a slight moderation in this trend, with 2.73 sq. km eroded and 1.63 sq. km deposited, yet still maintained a net erosional balance. This persistent dominance of erosion over deposition suggests a significant alteration in the river's natural equilibrium, likely influenced by anthropogenic factors associated with urban expansion. The predominance of erosion in urban river systems is a phenomenon observed globally in rapidly urbanizing areas [63]. This trend can be attributed to several factors, including increased surface runoff due to impervious urban surfaces, alterations in sediment supply, and direct human interventions in the river channel [64].

Table 10. Shift of the Malir River's river line over the decades from 1990 to 2020, highlighting the changes in the areas eroded and deposited around the streamline.

Year	Area in sq. km	
	Erosion	Deposition
1990–2000	−4.54	1.96
2000–2010	−9.20	0.76
2010–2020	−2.73	1.63

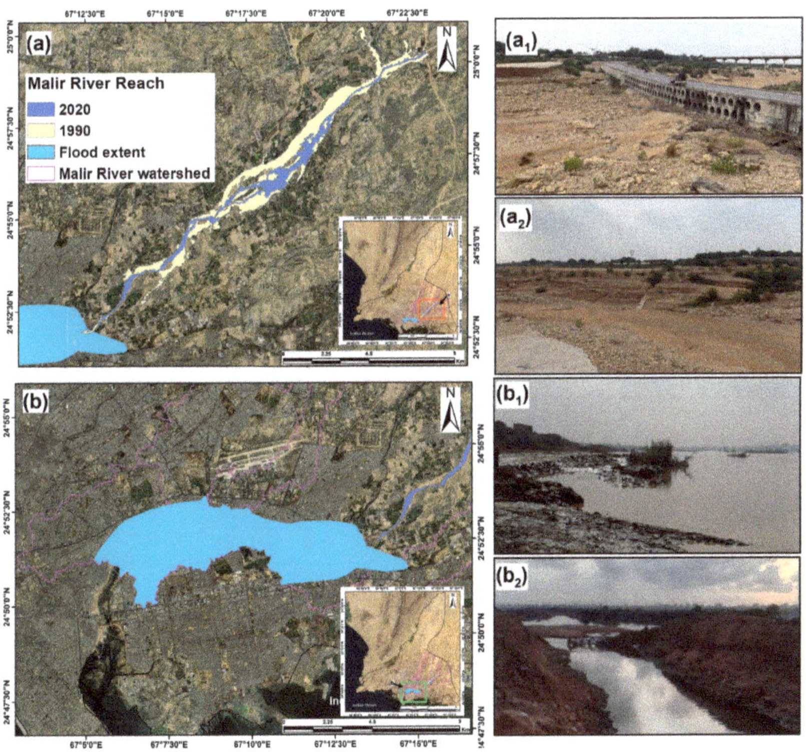

Figure 7. The location map of the Malir River watershed in Karachi shows (**a**) a 30-year shift in the river's course from 1990 to 2020, with the river occupying a larger area in 1990 (light yellow) compared to 2020 (blue), illustrating the effects of LULC changes, including significant erosion and conversion to barelands, as observed in the field (see sub-figures ($\mathbf{a_1}$,$\mathbf{a_2}$)). Panel (**b**) illustrates the effect of this shift on flood extent downstream near urban areas, as documented in field observations (see sub-figures ($\mathbf{b_1}$,$\mathbf{b_2}$)).

Figure 7a provides a striking visual representation of these changes, comparing the Malir River's course in 1990 (shown in yellow) to its 2020 configuration (in blue). The image clearly illustrates the river's reduced area in 2020, with significant portions of the 1990 riverbed being eroded or converted to other land use classes (Figure 7(a_1,a_2)). This visualization underscores the dynamic nature of the river–urban interface and highlights the potential for future conflicts between urban development and natural fluvial processes. The observed erosional dominance has significant implications for Karachi's flood risk management and urban planning. Increased erosion can lead to channel widening and deepening (Figure 7(b_1)), potentially altering flood conveyance capacities and undermining riverside infrastructure [65]. Moreover, the loss of riparian zones through erosion can diminish the river's natural flood mitigation capabilities and ecological functions (Figure 7(b_2)).

Notably, while the overall trend shows net erosion, the magnitude of erosion and deposition varied across the three decades. The period from 2000 to 2010 saw the most dramatic erosion, which could be linked to particularly intensive urban development or significant hydrological events during this time. The slight increase in deposition observed in the 2010–2020 period, while still outweighed by erosion, might indicate the onset of some stabilization processes or the implementation of river management strategies. These findings align with global observations of urban river channel changes in rapidly developing cities [66]. However, the magnitude of change observed in the Malir River, particularly the high erosion rates, suggests that Karachi may be experiencing these impacts more acutely than many other urban areas. This underscores the urgent need for integrated urban and river management strategies that consider the dynamic nature of the urban–river interface.

3.4. Forecasting Flood Risk Response to Land Use Changes and Rainfall Patterns

The analysis of projected flood risk in Karachi from 2020 to 2035, as influenced by land use changes and rainfall patterns, reveals a complex interplay between urban development and environmental vulnerability. Table 11 and Figures 7b and 8 provide comprehensive insights into flood-prone areas' spatial distribution and temporal evolution across different land use categories.

Table 11. Percentage of affected land use type versus flood (inundation) extent.

	2020	2025	2030	2035
Urbanized	64.77	64.49	64.22	66.65
Vegetation	15.04	14.06	17.23	14.27
Water bodies	2.29	1.60	1.41	2.11
Wetlands	0.005	0.005	0.002	0.002
Barelands	17.89	19.85	17.14	16.97

Urbanized areas consistently emerge as the most susceptible to flooding, with the affected percentage ranging from 64.22% to 66.65% over the study period. This trend culminates in 2035, where 66.65% of urbanized areas are projected to be at risk, underscoring the heightened vulnerability resulting from continued urban expansion. This finding aligns with global observations of increased flood risk in rapidly urbanizing areas, often attributed to the proliferation of impervious surfaces and alterations to natural drainage systems [67]. Vegetated land exhibits notable fluctuations in flood susceptibility, peaking at 17.23% in 2030. This non-linear trend may reflect the complex interactions between land cover changes, urban green space initiatives, and evolving precipitation patterns. The temporary increase in vegetation area vulnerability could be associated with the encroachment of urban development into flood-prone green spaces or changes in land management practices [68]. Water bodies and wetlands show minimal changes in flood impact over the projected period, with water body areas fluctuating between 1.41% and

2.29% and wetlands consistently below 0.005%. The relatively stable values for these categories suggest that existing water bodies and wetlands may have already adapted to or are inherently resilient to flood events. Barelands remain the second most affected category throughout the projection period, with values ranging from 16.97% to 19.85%. The fluctuations in this category might reflect the dynamic nature of urban fringe areas, where barren lands are often in transition between natural and developed states [69].

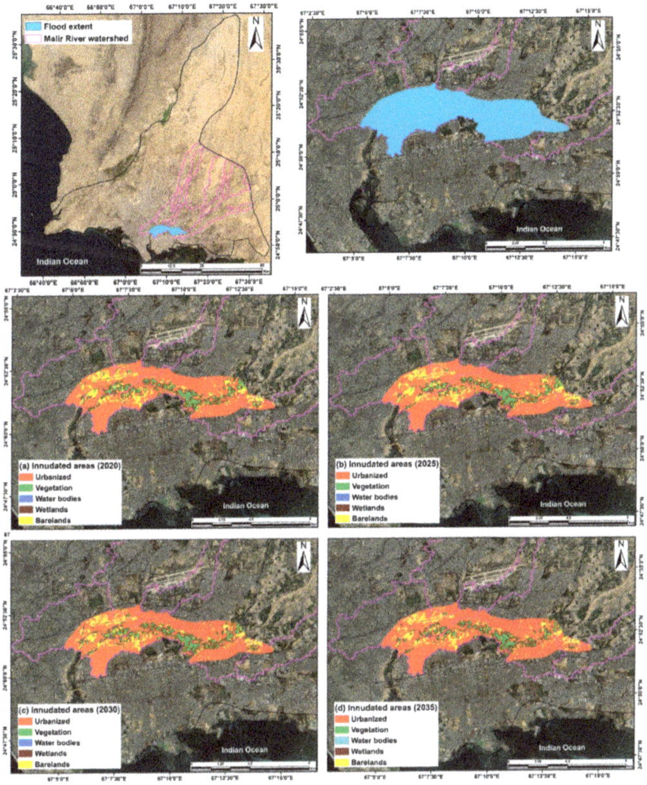

Figure 8. Flood inundation extent map for (**a**) 2020, (**b**) 2025, (**c**) 2030 and (**d**) 2035.

Figure 7b provides a spatial representation of the flood extent in relation to the Malir River reach, illustrating the expansion of flood-prone areas beyond the immediate river channel. This visualization highlights the potential for widespread urban inundation, particularly in areas where urban development has encroached upon natural floodplains. Figure 8 offers a temporal progression of flood inundation extents for 2020, 2025, 2030, and 2035. The predominance of blue areas, representing flooded urban zones, visually corroborates the quantitative data from Table 11. The spatial patterns depicted suggest a gradual expansion of flood-prone urban areas, likely due to the combined effects of urban sprawl and alterations to natural drainage patterns.

These findings underscore the critical need for integrated flood risk management strategies in Karachi's urban planning initiatives. The projected increase in flood vulnerability, particularly in urbanized areas, highlights the potential consequences of uncontrolled urban expansion into flood-prone zones. Moreover, the fluctuations observed in vegetation and barelands categories emphasize the dynamic nature of flood risk in rapidly evolving urban landscapes. The results align with global trends in urban flood risk, where the interaction between climate change, urbanization, and altered hydrological regimes contributes to increased flood vulnerability [70]. However, the specific patterns observed in Karachi, particularly the high percentage of affected urban areas, suggest that this megacity may face acute challenges in mitigating future flood risks.

4. Discussion

The comprehensive analysis of LULC changes in Karachi from 1990 to 2035 reveals significant urban expansion and its consequent environmental impacts, particularly on natural resources and drainage system degradation. This study addresses a critical gap in the literature, as limited research has been conducted on LULC dynamics in the Sindh province region [71].

4.1. Urban Expansion and Land Use Change in Karachi

The analysis of LULC changes in Karachi from 1990 to 2020 reveals a dramatic urban expansion, reflecting the city's rapid growth as a major economic center. The urbanized area increased from 13.4% in 1990 to 23.7% in 2020 (Table 7), a trend projected to continue, reaching 909 sq. km by 2035 (Table 9). This expansion aligns with global urbanization patterns in developing countries [69] but occurs at a particularly intense rate in Karachi. Concurrently, significant reductions were observed in other land cover types. Vegetation cover fluctuated but showed an overall decline from 4.3% in 1990 to 3.1% in 2020. Water bodies experienced a substantial decrease from 3.4% to 1.3% over the same period. These changes indicate a significant transformation of natural landscapes into urban areas, a pattern consistent with studies in other rapidly growing cities [72].

The scale and rate of LULC changes in Karachi appear more pronounced than in other Pakistani cities. For instance, studies in the Okara district revealed a decline in vegetation areas from 91.6% to 89.3% by 2020 [10], while research in Islamabad's water bodies demonstrated a shift from vegetation and water bodies to urban areas and bare soil [72]. The more dramatic changes in Karachi likely reflect its status as a megacity and prime economic hub, attracting rapid development and population growth. These findings underscore the urgent need for sustainable urban planning strategies in Karachi, particularly given the projected continuation of these trends. The balance between urban growth and the preservation of natural landscapes will be crucial for the city's long-term environmental and social sustainability.

4.2. Impact on Natural Drainage Systems

The rapid urbanization of Karachi has significantly impacted its natural drainage systems, particularly the Malir River. Our analysis reveals a pronounced river morphology shift from 1990 to 2020, characterized by a predominance of erosion over deposition (Table 10). Specifically, erosion affected 2.73 sq. km of the river reaches from 2010 to 2020, while deposition occurred in only 1.63 sq. km. This trend was consistent across the three decades studied, indicating a long-term alteration of the river's natural equilibrium. The observed changes in river morphology can be attributed to several factors associated with urban expansion. The conversion of natural and bare lands to impervious urban surfaces has likely altered the hydrological regime, increased surface runoff, and potentially exacerbated erosional processes.

Additionally, anthropogenic activities along the riverbank, such as encroachment and modification of the river channel, have contributed to these geomorphological changes. These findings align with global observations of urban river channel changes in rapidly developing cities [64,66]. However, the magnitude of change observed in the Malir River suggests that Karachi may be experiencing these impacts more acutely than many other urban areas. This underscores the critical need for river management strategies considering the dynamic nature of urban–river interactions. The alterations in river morphology have significant implications for flood risk management. Channel widening and deepening due to erosion can alter flood conveyance capacities, potentially increasing the vulnerability of adjacent urban areas to flooding. Moreover, the loss of natural floodplains and riparian zones through erosion and urban encroachment may diminish the river's natural flood mitigation capabilities.

4.3. Flood Risk Assessment and Projections

The current and projected flood risk analysis in Karachi reveals a concerning trend of increasing vulnerability, particularly in urbanized areas. Our findings indicate that urbanized areas are consistently the most affected by flooding, with the percentage of flood-prone urban land projected to increase from 64.77% in 2020 to 66.65% by 2035 (Table 11). This trend underscores the growing susceptibility of Karachi's urban infrastructure to flood-related hazards. The spatial distribution of flood risk, as illustrated in Figure 8, demonstrates a clear expansion of flood-prone areas beyond the immediate vicinity of the Malir River. This pattern suggests that the impacts of urbanization on flood risk extend well beyond the direct modification of river channels and floodplains. Conversion of natural surfaces to impervious urban areas likely plays a significant role in altering local hydrology, increasing surface runoff's volume and velocity.

Interestingly, our projections show some fluctuations in the flood susceptibility of different land use categories over time. For instance, vegetated land exhibits notable variations, peaking at 17.23% flood-prone areas in 2030. This non-linear trend may reflect the complex interactions between land cover changes, urban green space initiatives, and evolving precipitation patterns. It also highlights the potential for targeted land-use planning interventions to mitigate flood risks. The minimal changes observed in the flood impact on water bodies and wetlands (consistently below 2.29% and 0.005%, respectively) suggest that existing water body features may have already adapted to or are inherently resilient to flood events. However, this stability should not be taken for granted, as continued urban expansion and climate change could alter these dynamics in the future. Barelands remain the second most affected category throughout the projection period, with values ranging from 16.97% to 19.85%. The fluctuations in this category might reflect the dynamic nature of urban fringe areas, where barelands often transition between natural and developed states. These findings align with global trends in urban flood risk, where the interaction between climate change, urbanization, and altered hydrological regimes contributes to increased flood vulnerability [70]. However, the specific patterns observed in Karachi, particularly the high percentage of affected urban areas, suggest that this megacity may face particularly acute challenges in mitigating future flood risks.

4.4. Effectiveness of MOLUSCE and CA-ANN Modeling

The application of the MOLUSCE (Modules for Land Use Change Evaluation) plugin in QGIS, coupled with the cellular automata–artificial neural network (CA-ANN) modelling approach, has proven to be highly effective in predicting LULC changes in Karachi. Our model achieved a Kappa statistic of 0.97, indicating excellent agreement between simulated and referenced maps. This high accuracy underscores the robustness of our methodology in capturing complex urban growth patterns and land use transitions. The effectiveness of our approach aligns with and, in some cases, surpasses similar studies in other regions. For instance, research in Muscat, Oman using the MOLUSCE plugin achieved 88% and 92% accuracy with ANN and linear regression models, respectively [73].

Similarly, a study in Ulaanbaatar, Mongolia attained 94.2% accuracy in predicting urban expansion using MOLUSCE [38]. Our higher accuracy may be attributed to the comprehensive set of spatial variables incorporated in our model, including DEM, slope, aspect, roads, streams, and population data. Studies in other regions further corroborate the CA-ANN model's success in simulating land cover changes. Research in North Sumatra, Indonesia effectively used CA-ANN to analyze the impact of several factors on LULC changes from 1990 to 2000 [74].

4.5. Relevance to Sustainable Development Goals

The analysis of LULC changes and flood risk assessment in Karachi provides substantial support for the long-term governance and management of urban development in the region. This study's findings maintain several economic, social, and environmental SDGs (Figure 9). The achievement degree of different SDGs, considering the LULC changes and flood risk projections, can be summarized as follows:

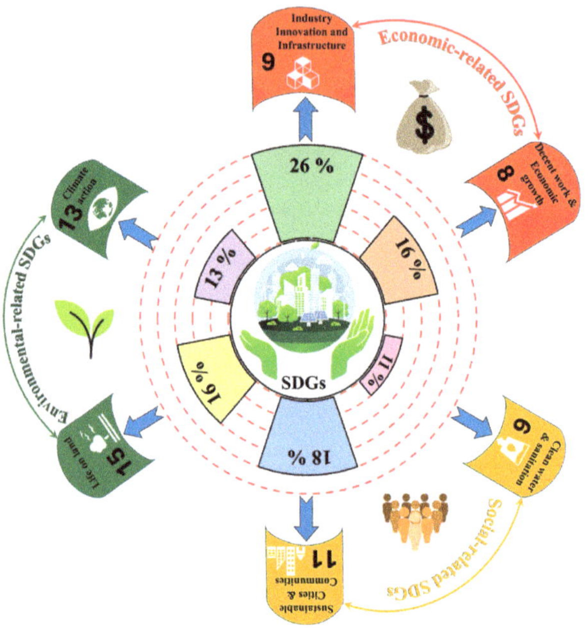

Figure 9. Quantitative contribution of land use change and flood risk analysis in Karachi to achieving Sustainable Development Goals (SDGs).

4.5.1. Economy-Related SDGs

SDG 8: Decent Work and Economic Growth

The contribution of this study to SDG 8 fulfilment is estimated at 16%. The projected urban expansion from 850 sq. km in 2020 to 909 sq. km by 2035 (Table 9) indicates potential for economic growth (Target 8.1) and job creation (Target 8.3). However, the associated flood risks highlight the need for sustainable economic development strategies that consider environmental vulnerabilities [75].

SDG 9: Industry, Innovation, and Infrastructure

This study shows a contribution of 26% to SDG 9 fulfilment. The application of advanced geospatial techniques and the high accuracy of predictions (97% Kappa statistic) demonstrate progress in innovative approaches to urban planning (Target 9.5). The findings provide a basis for developing resilient infrastructure (Target 9.1) that can withstand increasing flood risks [76].

4.5.2. Environment-Related SDGs

SDG 13: Climate Action

The study's contribution to SDG 13 fulfillment is estimated at 13%. The analysis of flood risk patterns, showing urban vulnerability increasing from 64.77% of urban areas in 2020 to 66.65% in 2035 (Table 11), is crucial for Target 13.1 (Strengthen resilience to climate-related hazards). These findings emphasize the urgent need for climate adaptation strategies in urban planning.

SDG 15: Life on Land

The contribution of this study to fulfilling SDG 15 is about 16%. The decrease in vegetation cover from 4.3% in 1990 to 3.1% in 2020 (Table 7) highlights challenges in achieving Target 15.1 (Sustainable use of terrestrial ecosystems) and Target 15.5 (Reducing degradation of natural habitats). The findings underscore the need to balance urban development with ecosystem preservation.

4.5.3. Social-Related SDGs

SDG 11: Sustainable Cities and Communities

This study significantly contributes to fulfilling SDG 11, with approximately 18%. The comprehensive analysis of urban growth and associated flood risks is particularly relevant to Target 11.3 (Enhance inclusive and sustainable urbanization) and Target 11.5 (Reduce the adverse effects of natural disasters). The findings provide a basis for developing more resilient urban planning strategies.

SDG 6: Clean Water Bodies and Sanitation

The contribution of this study to SDG 6 fulfilment is estimated at 11%. The analysis of water body reduction from 3.4% in 1990 to 1.3% in 2020 (Table 7) highlights challenges in achieving Target 6.6 (Protect water bodies-related ecosystems). These findings emphasize the need for integrated water body resource management in rapidly urbanizing areas.

This analysis underscores the interconnected nature of the SDGs and the need for integrated approaches to urban development. The estimated achievement percentages highlight the substantial work required to align Karachi's development trajectory with global sustainability goals. These findings provide a scientific basis for decision-makers to prioritize interventions and develop comprehensive strategies that address multiple SDGs simultaneously, potentially serving as a model for other rapidly growing cities in developing countries.

4.6. Research Limitations and Future Directions

This study faced several methodological and data-related limitations in analyzing land use/land cover changes and flood risk assessment in Karachi. While the model excels in capturing overall trends, it may not fully account for sudden policy changes, economic shocks, or extreme events that could dramatically alter land use patterns. These limitations arise from several factors. First, the temporal resolution of Landsat imagery (5, 7, and 8) between 1990 and 2020, with inconsistent observation intervals, likely missed critical transitional changes in urban development. The absence of 2005 data, resulting from the scan line corrector failure, further compounded this issue. Additionally, the reliance on 2017 demographic data and the assumption of a uniform Manning's roughness coefficient (0.04) for riverbanks in the HEC-RAS model introduced analytical constraints. Second, although the MOLUSCE plugin with CA-ANN modeling achieved a high Kappa statistic of 0.97, it exhibited varying sensitivity to input variables. While detailed uncertainty quantification was beyond the scope of this study, these findings lay the groundwork for refining sensitivity analyses in future research. The integration of World Bank road network data with OpenStreetMap (OSM) data introduced additional uncertainties due to discrepancies in temporal and spatial resolutions, despite validation efforts using Google Earth's time series engine. Moreover, differences in spatial resolutions between Landsat imagery (30 m) and OSM data introduced scaling uncertainties, particularly in rapidly changing urban fringe areas. These factors highlight the need for enhanced data harmonization and sensitivity analyses in future studies to improve the accuracy of LULC change and flood risk assessments. To maintain clarity and manageability, certain factors, such as the distance to business centers, were excluded. These factors are often indirectly captured by variables like proximity to roads and urbanized areas, which inherently reflect economic activity [77]. Lastly, a significant challenge in LULC change modeling is class imbalance, which can affect the reliability of predictions [78,79]. Addressing this issue is critical for improving the robustness and applicability of these models in sustainable land use planning.

Future research should address these limitations through several key approaches:

- Incorporating higher temporal resolution imagery and interpolation techniques (such as cubic spline interpolation or complementary historical datasets) to minimize data loss between sampling intervals.
- Implementing advanced machine learning techniques to handle class imbalances in LULC data [79].
- Developing more sophisticated hydrological models with variable surface roughness
- Enhancing validation metrics beyond the Kappa index, coupled with real-time monitoring systems, to improve LULC classification and flood risk prediction accuracy.

Specific priorities for future implementation include:

- Developing an integrated real-time monitoring system combining remote sensing with ground-based sensors for early flood warning, particularly in zones showing >64% flood susceptibility
- Creating detailed zoning proposals identifying priority areas for green infrastructure, especially in regions where vegetation cover has declined from 4.3% to 3.1%
- Establishing specific guidelines for waterway protection zones along the Malir River, informed by our erosion analysis showing 2.73 sq. km erosion in the last decade
- Implementing a GIS-based decision support system that combines flood risk projections with infrastructure planning, particularly in areas projected to face 66.65% flood risk by 2035

Key policy recommendations include:

- Revising building codes for enhanced flood resilience in high-risk zones
- Developing targeted conservation policies for the remaining water bodies and wetlands (declined to 1.3% and 0.4%, respectively)
- Establishing urban growth boundaries based on our CA-ANN model projections

Despite these challenges, this study establishes a robust framework for understanding urban growth dynamics and environmental change in coastal megacities. The model's ability to simulate future scenarios offers valuable insights for policymakers and urban planners, enabling them to anticipate challenges and devise proactive strategies for sustainable urban development.

5. Conclusions

Coastal megacities worldwide face increasing environmental challenges due to rapid urbanization and climate change impacts. This study addressed these critical challenges in Karachi, Pakistan, the world's third-largest city by population. An integrated approach combining remote sensing, geospatial analysis, and the MOLUSCE plugin with CA-ANN modelling was employed to analyze spatiotemporal LULC changes from 1990 to 2020 and project future scenarios for 2025, 2030, and 2035. The model was calibrated using a combination of spatial variables, including DEM, slope, aspect, roads, streams, and population data, achieving a high Kappa statistic of 0.97, indicating excellent agreement between simulated and referenced maps. This robust performance validates the model's reliability for future LULC predictions. The results revealed significant urban expansion from 13.4% to 23.7% of the total area between 1990 and 2020, with concurrent reductions in vegetation cover, water bodies, and wetlands. Notably, the Malir River's area decreased from 17.19 to 5.07 km^2 by 2020 due to erosion, highlighting a key factor in urban flooding vulnerability. Flood risk projections indicate that urbanized areas will be most affected, with 66.65% potentially inundated by 2035. These findings have significant implications for sustainable development, particularly in relation to SDGs 6, 8, 9, 11, 13, and 15. The study contributes substantially to SDG 11 (Sustainable Cities and Communities) and SDG 9 (Industry, Innovation and Infrastructure), with estimated achievement levels of 18% and 26%, respectively. The study also contributes to SDG 13 (Climate Action), with an estimated 13% achievement. However, the declining vegetation cover and water bodies present challenges for SDG 15 (Life on Land) and SDG 6 (Clean Water and Sanitation), with contributions of 16% and 11%, respectively. The study also shows a 16% contribution to SDG 8 (Decent Work and Economic Growth). These results underscore the urgent need for adaptive urban planning strategies that integrate flood risk management with sustainable development goals, addressing multiple SDGs simultaneously. These findings underscore the role of advanced geospatial techniques and predictive modelling in urban planning and environmental management, particularly in rapidly evolving urban environments like Karachi. The case of Karachi exemplifies the broader trend of rapid urbanization in Pakistan, particularly the growth of megacities. Future research should focus on developing nature-based solutions for flood mitigation, assessing the socio-economic impacts of projected changes, and exploring innovative policy frameworks to enhance urban resilience in the face of climate change and rapid urbanization while simultaneously addressing multiple SDGs.

Supplementary Materials: The following supporting information can be downloaded at: https://www.mdpi.com/article/10.3390/ijgi14010030/s1, Table S1: Accuracy assessments of the classified images of the years 1990, 1995, 2000, 2010, 2015, and 2020.

Author Contributions: Conceptualization, Minza Mumtaz, Youssef M. Youssef and Mahmoud E. Abd-Elmaboud; Data curation, Syed Humayoun Jahanzaib; Formal analysis, Waqar Hussain, Saleh Qaysi, Youssef M. Youssef and Mahmoud E. Abd-Elmaboud; Funding acquisition, Saleh Qaysi and Nassir Alarifi; Investigation, Waqar Hussain and Mahmoud E. Abd-Elmaboud; Methodology, Minza Mumtaz, Syed Humayoun Jahanzaib, Waqar Hussain, Sadia Khan and Youssef M. Youssef; Project administration, Sadia Khan, Saleh Qaysi and Nassir Alarifi; Resources, Minza Mumtaz, Abdalla Abdelnabi and Nassir Alarifi; Software, Minza Mumtaz, Waqar Hussain, Sadia Khan, Abdalla Abdelnabi and Mahmoud E. Abd-Elmaboud; Supervision, Sadia Khan; Validation, Minza Mumtaz, Sadia Khan and Youssef M. Youssef; Visualization, Sadia Khan, Abdalla Abdelnabi, Youssef M. Youssef and Mahmoud E. Abd-Elmaboud; Writing—original draft, Minza Mumtaz, Syed Humayoun Jahanzaib and Sadia Khan; Writing—review and editing, Saleh Qaysi, Abdalla Abdelnabi, Nassir Alarifi, Youssef M. Youssef and Mahmoud E. Abd-Elmaboud. All authors have read and agreed to the published version of the manuscript.

Funding: This research is supported by Researchers Supporting Project number (RSP2025R432), King Saud University, Riyadh, Saudi Arabia.

Data Availability Statement: The datasets generated during and/or analyzed during the current study are available from the corresponding author upon reasonable request.

Acknowledgments: The authors would like to express their gratitude to the Researchers Supporting Project number (RSP2025R432), King Saud University, Riyadh, Saudi Arabia.

Conflicts of Interest: The authors have no conflicts of interest to declare that are relevant to the content of this article.

References

1. Blackburn, S.; Pelling, M. *Megacities and the Coast: Risk, Resilience, and Transformation*; Earthscan from Routledge: Oxfordshire, UK, 2013.
2. van den Brandeler, F.; Gupta, J.; Hordijk, M. Megacities and rivers: Scalar mismatches between urban water management and river basin management. *J. Hydrol.* **2019**, *573*, 1067–1074. [CrossRef]
3. Arefin, R.; Rahman, A.T.M.S.; Das, J.; Jahan, C.S.; Mazumder, Q.H.; Gomaa, E.; Abd El Aal, A.K.; Radwan, A.E.; Youssef, Y.M. Megacity solid waste disposal suitability mapping in Dhaka, Bangladesh: An integrated approach using remote sensing, GIS and statistics. *Environ. Monit. Assess.* **2024**, *196*, 910. [CrossRef] [PubMed]
4. Kamerman, P. Using Raster Data to Calculate and Then Map Aridity Indices. *painblogR*, 15 December 2020.
5. Idowu, D.; Zhou, W. Global Megacities and Frequent Floods: Correlation Between Urban Expansion Patterns and Urban Flood Hazards. *Sustainability* **2023**, *15*, 2514. [CrossRef]
6. Barredo, J.I.; Demicheli, L. Urban sustainability in developing countries' megacities: Modelling and predicting future urban growth in Lagos. *Cities* **2003**, *20*, 297–310. [CrossRef]
7. Abdo, H.G.; Darwish, K.S.; Bindajam, A.A.; Niknam, A.; Youssef, Y.M.; Ahmed, M.F.M.; Mallick, J. Detection of flood vulnerable areas in urban basins using multi-criteria analysis and geospatial tools: A case study from eastern Mediterranean. *Environ. Earth Sci.* **2024**, *83*, 524. [CrossRef]
8. Arafa, N.A.; El-Said Salem, Z.; Abdeldayem, A.L.; Ghorab, M.A.; Moustafa, Y.M.; Soliman, S.A.; Farag, M.H.; Purohit, S.; Elhag, M.; Youssef, Y.M. Advancing Deltaic Aquifer Vulnerability Mapping to Seawater Intrusion and Human Impacts in Eastern Nile Delta: Insights from Machine Learning and Hydrochemical Perspective. *Earth Syst. Environ.* **2024**, 1–26. [CrossRef]
9. Akbar, H.; Gheewala, S.H. Impact of Climate and Land Use Changes on Flowrate in the Kunhar River Basin, Pakistan, for the Period (1992–2014). *Arab. J. Geosci.* **2021**, *14*, 707. [CrossRef]
10. Ahmad, W.; Iqbal, J.; Nasir, M.J.; Ahmad, B.; Khan, M.T.; Khan, S.N.; Adnan, S. Impact of land use/land cover changes on water quality and human health in district Peshawar Pakistan. *Sci. Rep.* **2021**, *11*, 16526. [CrossRef]
11. Salem, Z.E.-S.; Arafa, N.A.; Abdeldayem, A.L.; Youssef, Y.M. Machine Learning-Enhanced GALDIT Modeling for The Nile Delta Aquifer Vulnerability Assessment in the Mediterranean Region. *Groundw. Sustain. Dev.* **2025**, *28*, 101403. [CrossRef]
12. Siddika, S.; Sresto, M.A. Assessing urban resilience of Khulna City in response to environmental and socioeconomic challenges. *DYSONA-Appl. Sci.* **2025**, *6*, 134–144.
13. ul Haq, F.; Naeem, U.A.; Gabriel, H.F.; Khan, N.M.; Ahmad, I.; Ur Rehman, H.; Zafar, M.A. Impact of Urbanization on Groundwater Levels in Rawalpindi City, Pakistan. *Pure Appl. Geophys.* **2021**, *178*, 491–500. [CrossRef]
14. Saghafian, B.; Farazjoo, H.; Bozorgy, B.; Yazdandoost, F. Flood intensification due to changes in land use. *Water Resour. Manag.* **2008**, *22*, 1051–1067. [CrossRef]
15. Rigo, R.; Houet, T. Do Land Use and Land Cover Scenarios Support More Integrated Land Use Management? *Land* **2023**, *12*, 1414. [CrossRef]
16. Soomro, A.G.; Mahar, G.A.; Memon, A.H.; Gadeh, M.A.; Azam, M. Spatiotemporal Impact of Precipitation Trend on LULC Using Satellite Remote Sensing Technique in Khirthar National Park, Sindh Pakistan. *Int. J. Econ. Environ. Geol.* **2022**, *12*, 16–21. [CrossRef]
17. UNEP. *IPCC Report from Intergovernmental Panel on Climate Change*; UNEP: Nairobi, Kenya, 2007; pp. 1–19.
18. Huang, B.; Yang, Y.; Li, R.; Zheng, H.; Wang, X.; Wang, X.; Zhang, Y. Integrating Remotely Sensed Leaf Area Index with Biome-BGC to Quantify the Impact of Land Use/Land Cover Change on Water Retention in Beijing. *Remote Sens.* **2022**, *14*, 743. [CrossRef]
19. Raviteja, K.; Pujari, M.K. Effect of LULC changes on surface runoff in Brahmani river basin using Soil and Water Assessment Tool (SWAT). *Int. J. Health Sci.* **2022**, *6*, 1010–1028. [CrossRef]
20. Ranger, N.; Hallegatte, S.; Bhattacharya, S.; Bachu, M.; Priya, S.; Dhore, K.; Rafique, F.; Mathur, P.; Naville, N.; Henriet, F.; et al. An assessment of the potential impact of climate change on flood risk in Mumbai. *Clim. Change* **2011**, *104*, 139–167. [CrossRef]
21. Hallegatte, S.; Green, C.; Nicholls, R.J.; Corfee-Morlot, J. Future flood losses in major coastal cities. *Nat. Clim. Change* **2013**, *3*, 802–806. [CrossRef]
22. Abidin, H.Z.; Andreas, H.; Gumilar, I.; Fukuda, Y.; Pohan, Y.E.; Deguchi, T. Land subsidence of Jakarta (Indonesia) and its relation with urban development. *Nat. Hazards* **2011**, *59*, 1753–1771. [CrossRef]

23. Rosenzweig, C.; Solecki, W.D.; Blake, R.; Bowman, M.; Faris, C.; Gornitz, V.; Horton, R.; Jacob, K.; LeBlanc, A.; Leichenko, R.; et al. Developing coastal adaptation to climate change in the New York City infrastructure-shed: Process, approach, tools, and strategies. *Clim. Change* **2011**, *106*, 93–127. [CrossRef]
24. Waleed, M.; Sajjad, M.; Shazil, M.S. Urbanization-led land cover change impacts terrestrial carbon storage capacity: A high-resolution remote sensing-based nation-wide assessment in Pakistan (1990–2020). *Environ. Impact Assess. Rev.* **2024**, *105*, 107396. [CrossRef]
25. Mehmood, M.S.; Rehman, A.; Sajjad, M.; Song, J.; Zafar, Z.; Shiyan, Z.; Yaochen, Q. Evaluating land use/cover change associations with urban surface temperature via machine learning and spatial modeling: Past trends and future simulations in Dera Ghazi Khan, Pakistan. *Front. Ecol. Evol.* **2023**, *11*, 1115074. [CrossRef]
26. United Nations. *Transforming Our World: The 2030 Agenda for Sustainable Development*; United Nations, Department of Economic and Social Affairs: New York, NY, USA, 2015; Volume 1, p. 41.
27. Drosos, V.C.; Liampas, S.A.G.; Farmakis, D.E. Zoning Planning for Sustainable Development and Exploitation of the Mountainous Areas. *Int. J. Soc. Sci. Humanit.* **2016**, *6*, 451–455. [CrossRef]
28. Mangi, M.Y.; Yue, Z.; Kalwar, S.; Lashari, Z.A. Comparative Analysis of Urban Development Trends of Beijing and Karachi Metropolitan Areas. *Sustainability* **2020**, *12*, 451. [CrossRef]
29. Li, Y.; Gao, J.; Yin, J.; Liu, L.; Zhang, C.; Wu, S. Flood Risk Assessment of Areas under Urbanization in Chongqing, China, by Integrating Multi-Models. *Remote Sens.* **2024**, *16*, 219. [CrossRef]
30. Jongman, B.; Ward, P.J.; Aerts, J.C.J.H. Global exposure to river and coastal flooding: Long term trends and changes. *Glob. Environ. Chang.* **2012**, *22*, 823–835. [CrossRef]
31. Feng, B.; Zhang, Y.; Bourke, R. Urbanization impacts on flood risks based on urban growth data and coupled flood models. *Nat. Hazards* **2021**, *106*, 613–627. [CrossRef]
32. Masria, A.; Alshammari, T.O.; Ghareeb, M.; Seif, A.K.; Abd-Elmaboud, M.E.; Ammar, A.I. 2D and 3D Modeling of Resistivity and Chargeability to Identify the Type of Saturated Groundwater for Complex Sedimentary Facies. *Hydrology* **2024**, *11*, 120. [CrossRef]
33. Ibrahim, S.; Balzter, H. Evaluating Flood Damage to Paddy Rice Fields Using PlanetScope and Sentinel-1 Data in North-Western Nigeria: Towards Potential Climate Adaptation Strategies. *Remote Sens.* **2024**, *16*, 3657. [CrossRef]
34. Risks, C. *Climate Risks and Adaptation in Asian Coastal Megacities*; The World Bank: Washington, DC, USA, 2010.
35. Ahmadlou, M.; Karimi, M.; Al-Ansari, N. The use of maximum entropy and ecological niche factor analysis to decrease uncertainties in samples for urban gain models. *GIScience Remote Sens.* **2023**, *60*, 2222980. [CrossRef]
36. Demarquet, Q.; Rapinel, S.; Arvor, D.; Corgne, S.; Hubert-Moy, L. Satellite Long-Term Monitoring of Wetland Ecosystem Functioning in Ramsar Sites for Their Sustainable Management. *Sustainability* **2024**, *16*, 6301. [CrossRef]
37. Kamaraj, M.; Rangarajan, S. Predicting the future land use and land cover changes for Bhavani basin, Tamil Nadu, India, using QGIS MOLUSCE plugin. *Environ. Sci. Pollut. Res.* **2022**, *29*, 86337–86348. [CrossRef] [PubMed]
38. Gantumur, B.; Wu, F.; Vandansambuu, B.; Tsegmid, B.; Dalaibaatar, E.; Zhao, Y. Spatiotemporal dynamics of urban expansion and its simulation using CA-ANN model in Ulaanbaatar, Mongolia. *Geocarto Int.* **2022**, *37*, 494–509. [CrossRef]
39. Alam, N.; Saha, S.; Gupta, S.; Chakraborty, S. Prediction modelling of riverine landscape dynamics in the context of sustainable management of floodplain: A Geospatial approach. *Ann. GIS* **2021**, *27*, 299–314. [CrossRef]
40. Mabdeh, A.N.; Ajin, R.S.; Razavi-Termeh, S.V.; Ahmadlou, M.; Al-Fugara, A. Enhancing the Performance of Machine Learning and Deep Learning-Based Flood Susceptibility Models by Integrating Grey Wolf Optimizer (GWO) Algorithm. *Remote Sens.* **2024**, *16*, 2595. [CrossRef]
41. Rana, I.A.; Bhatti, S.S. Lahore, Pakistan—Urbanization challenges and opportunities. *Cities* **2018**, *72*, 348–355. [CrossRef]
42. The Intergovernmental Panel on Climate Change. *Climate Change 2007: The Physical Science Basis*; IPCC: Geneva, Switzerland, 2007; Volume 6, p. 333.
43. Fletcher, T.D.; Andrieu, H.; Hamel, P. Understanding, Management and Modelling of Urban Hydrology and Its Consequences for Receiving Waters: A State of the Art. *Adv. Water Resour.* **2013**, *51*, 261–279. [CrossRef]
44. Basit, M.; Sajjad, S.; Imran Khan, M.; Ali, A.; Khan Kurshid, S. Spatio-Temporal Trends of Urban Population in Pakistan. *Asian J. Multidiscip. Stud.* **2018**, *6*, 2348–7186.
45. Khan, M.; Khan, H. An Assessment of the Problems Faced by Karachi and Pakistan Due to the Rapid Population Growth of the City. *J. Hist. Soc. Sci.* **2016**, *7*, 1–30. [CrossRef]
46. Zafar, S.; Zaidi, A. Flash floods in Malir basin due to urbanization. In Proceedings of the 2016 IEEE International Geoscience and Remote Sensing Symposium (IGARSS), Beijing, China, 10–15 July 2016; pp. 4481–4484. [CrossRef]
47. Bano, S.; Shahnawaz. Concrete Expansion: Urban Growth Estimation Through Geo informatics, A Case Study of Karachi. *Int. J. Innov. Sci. Technol.* **2024**, *6*, 341–351.
48. Siddik, M.S.; Tulip, S.S.; Rahman, A.; Islam, M.N.; Haghighi, A.T.; Mustafa, S.M.T. The impact of land use and land cover change on groundwater recharge in northwestern Bangladesh. *J. Environ. Manag.* **2022**, *315*, 115130. [CrossRef] [PubMed]

49. Amgoth, A.; Rani, H.P.; Jayakumar, K.V. Exploring LULC changes in Pakhal Lake area, Telangana, India using QGIS MOLUSCE plugin. *Spat. Inf. Res.* **2023**, *31*, 429–438. [CrossRef]
50. Ayek, A.A.; Zerouali, B. Monitoring temporal changes of the Qattinah Lake surface area using Landsat data and Google Earth Engine. *DYSONA-Appl. Sci.* **2025**, *6*, 126–133.
51. Jasim, A.T. Assessing LULC dynamics in Kirkuk City, Iraq using Landsat imagery and maximum likelihood classification. *DYSONA-Appl. Sci.* **2025**, *6*, 113–119.
52. Congalton, R.G.; Green, K. *Assessing the Accuracy of Remotely Sensed Data: Principles and Practices*, 2nd ed.; CRC Press: Boca Raton, FL, USA, 2008; ISBN 9781420055139.
53. Muhammad, R.; Zhang, W.; Abbas, Z.; Guo, F.; Gwiazdzinski, L. Spatiotemporal Change Analysis and Prediction of Future Land Use and Land Cover Changes Using QGIS MOLUSCE Plugin and Remote Sensing Big Data: A Case Study of Linyi, China. *Land* **2022**, *11*, 419. [CrossRef]
54. Bol, T.T.; Randhir, T.O. Predicting Land Use and Land Cover Changes in the Chindwin River Watershed of Myanmar Using Multilayer Perceptron-Artificial Neural Networks. *Land* **2024**, *13*, 1160. [CrossRef]
55. Matlhodi, B.; Kenabatho, P.K.; Parida, B.P.; Maphanyane, J.G. Analysis of the Future Land Use Land Cover Changes in the Gaborone Dam Catchment Using CA-Markov Model: Implications on Water Resources. *Remote Sens.* **2021**, *13*, 2427. [CrossRef]
56. Seto, K.C.; Fragkias, M.; Güneralp, B.; Reilly, M.K. A Meta-Analysis of Global Urban Land Expansion. *PLoS ONE* **2011**, *6*, e23777. [CrossRef]
57. Davidson, N.C. How much wetland has the world lost? Long-term and recent trends in global wetland area. *Mar. Freshw. Res.* **2014**, *65*, 934–941. [CrossRef]
58. Dewan, A.M.; Yamaguchi, Y. Land use and land cover change in Greater Dhaka, Bangladesh: Using remote sensing to promote sustainable urbanization. *Appl. Geogr.* **2009**, *29*, 390–401. [CrossRef]
59. Zhang, B.; Zhang, Q.; Feng, C.; Feng, Q.; Zhang, S. Understanding Land Use and Land Cover Dynamics from 1976 to 2014 in Yellow River Delta. *Land* **2017**, *6*, 20. [CrossRef]
60. Grimm, N.B.; Faeth, S.H.; Golubiewski, N.E.; Redman, C.L.; Wu, J.; Bai, X.; Briggs, J.M. Global change and the ecology of cities. *Science* **2008**, *319*, 756–760. [CrossRef] [PubMed]
61. Verburg, P.H.; Alexander, P.; Evans, T.; Magliocca, N.R.; Malek, Z.; DA Rounsevell, M.; van Vliet, J. Beyond land cover change: Towards a new generation of land use models. *Curr. Opin. Environ. Sustain.* **2019**, *38*, 77–85. [CrossRef]
62. Wolch, J.R.; Byrne, J.; Newell, J.P. Urban green space, public health, and environmental justice: The challenge of making cities 'just green enough'. *Landsc. Urban Plan.* **2014**, *125*, 234–244. [CrossRef]
63. Gregory, K.J. The human role in changing river channels. *Geomorphology* **2006**, *79*, 172–191. [CrossRef]
64. Chin, A. Urban transformation of river landscapes in a global context. *Geomorphology* **2006**, *79*, 460–487. [CrossRef]
65. Gurnell, A.; Lee, M.; Souch, C. Urban Rivers: Hydrology, Geomorphology, Ecology and Opportunities for Change. *Geogr. Compass* **2007**, *1*, 1118–1137. [CrossRef]
66. Walsh, C.J.; Roy, A.H.; Feminella, J.W.; Cottingham, P.D.; Groffman, P.M.; Morgan, R.P. The urban stream syndrome: Current knowledge and the search for a cure. *J. N. Am. Benthol. Soc.* **2005**, *24*, 706–723. [CrossRef]
67. Jha, A.K.; Bloch, R.; Lamond, J. *Cities and Flooding: A Guide to Integrated Urban Flood Risk Management for the 21st Century*; World Bank Publications: Washington, DC, USA, 2012.
68. Thorne, C.R.; Lawson, E.C.; Ozawa, C.; Hamlin, S.L.; Smith, L.A. Overcoming uncertainty and barriers to adoption of Blue-Green Infrastructure for urban flood risk management. *J. Flood Risk Manag.* **2018**, *11*, S960–S972. [CrossRef]
69. Seto, K.C.; Güneralp, B.; Hutyra, L.R. Global forecasts of urban expansion to 2030 and direct impacts on biodiversity and carbon pools. *Proc. Natl. Acad. Sci. USA* **2012**, *109*, 16083–16088. [CrossRef]
70. Kundzewicz, Z.W.; Kanae, S.; Seneviratne, S.I.; Handmer, J.; Nicholls, N.; Peduzzi, P.; Mechler, R.; Bouwer, L.M.; Arnell, N.; Mach, K.; et al. Flood risk and climate change: Global and regional perspectives. *Hydrol. Sci. J.* **2014**, *59*, 1–28. [CrossRef]
71. Soomro, S.-E.; Hu, C.; Boota, M.W.; Soomro, M.H.A.A.; Jian, S.; Zafar, Z.; Li, X. Mapping flood extend and its impact on land use/land cover and settlements variations: A case study of Layyah District, Punjab, Pakistan. *Acta Geophys.* **2021**, *69*, 2291–2304. [CrossRef]
72. Butt, A.; Shabbir, R.; Ahmad, S.S.; Aziz, N. Land use change mapping and analysis using Remote Sensing and GIS: A case study of Simly watershed, Islamabad, Pakistan. *Egypt. J. Remote Sens. Space Sci.* **2015**, *18*, 251–259. [CrossRef]
73. Al-Rubkhi, A.N.M.; Talal, A.A.; Mohammed, A.B. Land Use Change Analysis and Modeling Using Open Source (QGis)-Case Study: Boasher Willayat. Bachelor's Thesis, Eastern Visayas State University, Tacloban City, Philippines, 2017.
74. Saputra, M.H.; Lee, H.S. Prediction of Land Use and Land Cover Changes for North Sumatra, Indonesia, Using an Artificial-Neural-Network-Based Cellular Automaton. *Sustainability* **2019**, *11*, 3024. [CrossRef]
75. Rehman, A.; Xue, L.; Islam, F.; Ahmed, N.; Qaysi, S.; Liu, S.; Alarifi, N.; Youssef, Y.M.; Abd-Elmaboud, M.E. Unveiling Groundwater Potential in Hangu District, Pakistan: A GIS-Driven Bivariate Modeling and Remote Sensing Approach for Achieving SDGs. *Water* **2024**, *16*, 3317. [CrossRef]

76. Saqr, A.M.; Nasr, M.; Fujii, M.; Yoshimura, C.; Ibrahim, M.G. Delineating suitable zones for solar-based groundwater exploitation using multi-criteria analysis: A techno-economic assessment for meeting sustainable development goals (SDGs). *Groundw. Sustain. Dev.* **2024**, *25*, 101087. [CrossRef]
77. Wan, J.; Ma, C.; Jiang, T.; Phillips, A.; Wu, X.; Wang, Y.; Wang, Z.; Cao, Y. A spatial econometric investigation into road traffic accessibility and economic growth: Insights from the Chengdu-Chongqing twin-city economic circle. *Humanit. Soc. Sci. Commun.* **2024**, *11*, 183. [CrossRef]
78. Hewitt, R.J.; Shadman Roodposhti, M.; Bryan, B.A. There's no best model! Addressing limitations of land-use scenario modelling through multi-model ensembles. *Int. J. Geogr. Inf. Sci.* **2022**, *36*, 2352–2385. [CrossRef]
79. Ahmadlou, M.; Karimi, M.; Sammen, S.S.; Alsafadi, K. Three novel cost-sensitive machine learning models for urban growth modelling. *Geocarto Int.* **2024**, *39*, 2353252. [CrossRef]

International Journal of
Geo-Information

MDPI

Article

The Impact of Built-Up Area Dispersion on the Cultural Heritage of the Region of the South Aegean, Greece

Efstratia Chatzi *, Evangelia-Theodora Derdemezi and Georgios Tsilimigkas

Department of Geography, University of the Aegean, 81100 Mytilene, Greece; geod17008@aegean.gr (E.-T.D.); gtsil@aegean.gr (G.T.)

* Correspondence: efi.chatzi@geo.aegean.gr

Abstract: Cultural heritage serves as a repository of a place's history and identity, integral to its landscape and central to all three sustainability goals. However, managing and protecting cultural heritage through regulatory planning often proves inadequate. This particularly applies to islands where infrastructure development pressures and urban sprawl lead to significant changes. This study quantifies the impact of uncontrolled built-up area dispersion on islands' cultural heritage, focusing on the Southern Aegean region. By identifying and delineating the boundaries of archaeological areas and historic sites, we assess pressures through the mapping and quantitative analysis of built-up areas derived from the Global Monitoring for Environment and Security (GMES) program. The results reveal spatial relations and potential conflicts, underscoring the insufficient protection of cultural heritage due to inadequate management and ineffective planning tools. Specifically, on islands like Mykonos and Naxos, over 80% of their archaeological areas are impacted by urban sprawl. Additionally, this study finds that six islands (Antiparos, Irakleia, Kea, Mykonos, Paros and Patmos) have over 60% of their built-up areas in exurban regions, exacerbating pressures on heritage sites. These findings highlight the need for an integrated spatial planning system that incorporates landscape and cultural heritage assets into strategic planning regulations, ensuring the preservation of these essential resources amidst ongoing development pressures.

Keywords: spatial planning; cultural heritage; built-up area; islands; Greece

1. Introduction

Islands, by their very nature, possess unique characteristics that set them apart from continental landmasses. They often feature distinct ecosystems, biodiversity and geological formations shaped by their isolation and specific climatic conditions [1,2]. Moreover, islands frequently exhibit a rich cultural diversity, influenced by centuries of maritime trade, colonization and indigenous traditions. The limited land area and restricted resources on islands have led to the development of resourceful and sustainable practices. These practices range from innovative agriculture and fishing techniques to efficient water management systems [3]. Additionally, island environments often foster tight-knit communities with strong social bonds and a deep connection to their surroundings [4]. As such, islands represent not only particular natural ecosystems but also significant cultural landscapes that need protection and management [5–7].

Islands possess a rich and unique cultural heritage, deeply rooted in their geography, history and interactions with neighboring cultures. The tangible dimension of this heritage endures and continues to emerge today, primarily expressed through archaeological sites and monuments. In addition to their historical significance, these sites also contribute to island development by attracting tourism and providing economic benefits [8–10].

Cultural heritage plays a pivotal role in shaping the spatial development patterns of built-up areas, influencing the distribution and expansion of urban landscapes. Communities, in their quest to preserve their cultural heritage, tend to focus development around historic sites, traditional settlements or areas of cultural significance. This phenomenon, known as cultural clustering, leads to the dispersion of built-up areas as development gravitates towards these culturally important locations [11,12]. The preservation of cultural heritage is becoming increasingly vital amidst the challenges posed by built-up area dispersion. Developmental pressures often threaten historic sites and traditional communities, placing invaluable cultural assets at risk of being lost [13–15].

In island territories, this issue becomes even more acute and complex due to escalating tourist development and the growing demand for second homes, which intensify residential pressures [16]. The location and expansion of built-up areas in the limited space of the islands are influenced by both natural and human-induced factors, such as topography, proximity to the sea, access to the road network and closeness to areas of cultural significance, such as traditional settlements and archaeological sites [17–19].

Hence, the situation on the Greek islands presents a complex interplay between urban sprawl and the management and preservation of cultural heritage [20–23]. The absence of a consistent and up-to-date spatial planning system has led to unregulated development, resulting in urban sprawl [24,25]. This unregulated development poses significant challenges to the preservation of the islands' rich cultural heritage, as historic sites, traditional settlements and areas of cultural significance become increasingly threatened. The rapid construction of small, individual properties, often outside the framework of spatial planning, exacerbates the problem by driving further urban sprawl in exurban areas—defined here as development that occurs beyond the institutional limits of the settlements and is not subject to the regulatory processes governing urban areas [26–30].

The balance between tourism development and the preservation of the islands' cultural heritage is delicate. On the one hand, cultural tourism is an important economic asset for the islands, also providing a strong incentive for the preservation of cultural heritage [31]. On the other hand, mass tourism creates intense pressures mainly through unregulated construction. An integrated spatial planning system is considered key for the proper protection and management of cultural heritage, combined with the promotion of sustainable tourism—tourism that supports economic growth while preserving cultural and natural resources for future generations [32–35].

In recent years, research on urban sprawl and cultural heritage preservation has largely focused on broader or more generalized case studies. This study, however, is the first to apply this specific methodology to the Southern Aegean archipelago. The region presents a unique combination of factors such as island size, population density and a high concentration of significant cultural monuments, making it an ideal subject for this analysis. The Cyclades and Dodecanese islands, in particular, feature complex patterns of urban expansion that directly intersect with areas of historical and archaeological importance. By examining built-up areas both within and outside official settlement boundaries, this study offers insights into the relation between urbanization and cultural heritage, filling a gap in the existing literature and providing a new perspective on the challenges of heritage preservation in these regions.

This paper aims to highlight this issue through the quantitative analysis of built-up areas located within cultural heritage sites on the islands of the South Aegean region. This phenomenon is driven by pressures to develop infrastructure for mass tourism and the inadequacies of spatial planning systems in preventing construction in these culturally sensitive areas. More specifically, the aim of this study is to examine the relation between deficiencies in spatial planning, urban sprawl and their effects on cultural heritage in island regions. The main hypothesis of this study is that the interplay of developmental pressures, leading to the dispersion of built-up areas, combined with shortcomings in spatial planning's capacity to protect cultural heritage areas, has contributed to significant spatial alterations on islands that threaten the cultural landscape.

The objectives of this research are as follows:

- To quantify built-up area dispersion: this involves assessing the extent of built-up areas within and outside settlement delineations to understand spatial development patterns and their intersection with cultural heritage sites.
- To evaluate the impact on cultural heritage: this study seeks to analyze the direct and indirect impacts of built-up area dispersion and infrastructure development on archaeological sites, traditional settlements and other culturally significant landscapes.
- To propose integrated management approaches: this study aspires to contribute to the discourse on spatial planning and heritage preservation by advocating for an integrated planning system that balances development with sustainability and cultural integrity.

Focusing on the South Aegean region in Greece, which comprises the Cyclades and Dodecanese archipelagos encompassing 38 islands, this research addresses two key research questions: (a) to assess the extent of the built-up area within and outside settlement delineations and (b) to analyze the impact of built-up dispersion on cultural heritage areas. These research questions will be answered through the quantitative analysis and mapping of pertinent variables, utilizing geoinformatics tools. The interpretation of the findings will be based on "Spatial Planning Principles".

The scientific significance of the South Aegean region lies in its rich cultural and historical identity, represented through its archaeological sites, traditional settlements and other culturally significant landscapes. As a region facing intense development pressures, it serves as an ideal case study for understanding how unregulated urban sprawl and deficiencies in spatial planning systems impact cultural heritage. The insights derived from this study are not only applicable to Greece but also offer valuable lessons for similar island territories globally, where the delicate balance between development and preservation is often at risk.

The structure of this paper is organized as follows: Section 2 explores spatial planning and cultural heritage protection, highlighting key policy considerations at various spatial levels. In Section 3, the data sources and methodology employed in this study are described, and the geographical context of the study region is also presented. Section 4 presents the empirical results, supported by tables and figures that clearly illustrate the findings. Finally, Section 5 provides a discussion of the results' implications and offers concluding insights along with recommendations for future research and policy development.

2. Policy Considerations for Cultural Heritage Protection at Various Spatial Levels

At the international and European levels, the importance of cultural heritage has been recognized for nearly a century. The First International Congress of Architects in Athens in 1931 initiated this effort with the formulation of the "Athens Charter", which highlighted the cultural elements of the broader urban environment as cultural assets [36,37]. In 1964, the adoption of the "Venice Charter" expanded the concept of a monument by recognizing that "The concept of a historic monument embraces not only the single architectural work but also the urban or rural setting in which the evidence of a particular civilization, a significant development or a historic event is identified." [38].

A milestone for the protection and management of world cultural heritage was the adoption of the 1972 Convention Concerning the Protection of the World Cultural and Natural Heritage [38]. In 1985, the Granada Convention for the Protection of the Architectural Heritage of Europe established the principles of the European cooperation on and coordination of architectural conservation policies [39]. In 1992, the Valetta Convention on the Protection of the Archaeological Heritage was adopted with the aim of protecting archaeological heritage from illegal excavations and major construction projects [40]. In 2003, a convention adopted at the international level highlighted the intangible attributes of cultural heritage [41,42]. The Convention for Safeguarding the Intangible Cultural Heritage introduced a new dimension to cultural heritage by including human traditions and ways of living [43–48].

The aforementioned conventions have been ratified by the Greek legislative system, further enriching the legislative context at the national level [49]. In Greece, the protection of cultural heritage is regulated, established and protected by the National Constitution, specifically Article 24. Antiquities legislation was governed by Law 5351/1932 "Amendments and Additions to Law 5356/1912 'Concerning Antiquities'" [50] and later by Law 1469/1950 "Protection of Buildings and Works of Art Post-1830" [51], which address their protection before and after 1930, respectively. However, these laws do not provide a comprehensive management framework for cultural heritage sites and the landscape. Law 3028/2002 "On the Protection of Antiquities and Cultural Heritage in General" [52] plays a central role in the legislative context, establishing the framework for the protection of the Greek cultural heritage from ancient times to the present and emphasizing the preservation of historical memory for present and future generations. Article 3 of this law emphasizes the importance of integrating cultural heritage protection and management into all levels of spatial planning across the country. Additionally, Law 4858/2021, titled "Ratification of the Code of Legislation for the Protection of Antiquities and Cultural Heritage in General" [53], addresses the protection of antiquities and cultural heritage. It introduces more detailed provisions for the spatial regulations surrounding the protection of antiquities and cultural heritage, focusing on how spatial planning must consider cultural heritage preservation as part of urban and land-use strategies.

The need to protect and manage cultural heritage through spatial planning is evident from the above brief reference to the national legislation on cultural heritage. In spatial planning laws and their reforms, although both natural and cultural heritage are acknowledged as invaluable and therefore highly prioritized assets in terms of protection and management at the national, regional and local spatial planning level, the implementation of this direction in spatial frameworks and plans is insufficient and fragmentary. Protection has mainly been attempted through regulatory measures at the local spatial planning level by means of General Urban Plans (GUPs) and Open City Spatial and Housing Organization Plans (OCSHOPs) [54,55]. However, these plans have so far been institutionalized for a very small percentage of the territory. The regulatory zones that have been used in Greece to address issues arising from the insufficient spatial planning context are the Urban Development Control Zones (UDCZs) [56]. These are typically exurban areas with strict terms and building conditions to prevent development outside the official settlement delineations. Although UDCZs were not designed to prioritize cultural heritage protection, they could have an indirect effect through the prevention of exurban built-up area expansion. However, this was not implemented, on the one hand, due to the small percentage of land of area they occupy and, on the other hand, due to the delay in their institutionalization.

Our study area is limited to only a few islands, including Syros, Mykonos, Paros, Naxos, Patmos, Andros, Santorini and parts of Kos and Rhodes. Despite their presence, the percentages of built-up areas outside settlement boundaries remain high. For example, the percentages of built-up areas outside settlements are 40% for Syros, 89% for Mykonos, 60% for Paros, 48% for Naxos, 86% for Patmos, 53% for Andros, 20% for Santorini, 52% for Kos and 46% for Rhodes. Of these, significant portions directly impact cultural heritage, with nearly 100% in Mykonos, 86% in Naxos, 80% in Santorini, 53% in Rhodes and lower percentages in Patmos, 2%, and Andros, 7%.

Cultural heritage is a spatial component that requires a cohesive and consistent approach to spatial planning for its preservation and management. UNESCO's concept of cultural landscapes emphasizes the connection between human activity and the natural environment, offering a key perspective for integrating both tangible and intangible heritage. This integration was introduced in spatial planning through the 2012 revision of the Regional Frameworks on Spatial Planning and Sustainable Development [53,57]. These landscapes represent the interaction between natural and cultural elements, highlighting the need to protect them from urbanization and infrastructure expansion.

Urban sprawl theories explain how built-up areas can spread unregulated beyond designated boundaries, posing risks to cultural heritage preservation [58,59]. Spatial planning theory stresses the importance of integrated frameworks to balance competing land uses and protect cultural heritage while attempting to address development pressures [28,60,61].

Though the 2012 revision of the regional frameworks is a positive step, more needs to be done. The landscape should be further incorporated into special spatial frameworks that focus on activities with a high spatial impact, along with more specific regulatory measures at the local planning level [62].

3. Materials and Methods

3.1. Data and Methodology

In this section, an overview of the spatial data and methodologies employed in this research is presented, with a focus on evaluating the impact of built-up areas on monuments and cultural heritage sites in the South Aegean region of Greece. This analysis follows the steps shown in Figure 1.

The research began by inventorying and mapping cultural assets across the study area, using Geographic Information Systems (GISs). Monument locations, sizes and settlement boundaries were systematically collected from various archival sources, including historical records, Official Government Gazette Sheets and existing geospatial datasets available in open databases.

Data on cultural heritage sites were primarily obtained from the Archaeological Cadaster managed by the Ministry of Culture and Sports [63,64], as well as the Thal-Chor Project [32,65]. This cross-border initiative aligned with Directive 2014/89/EU on Maritime Spatial Planning, providing valuable insights into both terrestrial and marine archaeological boundaries, particularly on islands where cultural assets extend into the sea.

The Official Government Gazette (OGG) [66] provided essential data for traditional settlement boundaries, downloaded from the government webpage as topographic charts and orthophoto maps. These were georeferenced using the Greek Geodetic Reference System 1987 (GGRS87) for spatial compatibility. Settlement boundaries for official settlements were sourced from open platforms such as e-poleodomia [67]. For traditional settlements without corresponding OGG records, their boundaries were digitized using information from OpenStreetMap. These steps ensured comprehensive mapping and analysis of settlement boundaries across the region.

Table 1. Comparison of traditional and non-traditional settlements on Cyclades and Dodecanese islands, Southern Aegean region, Greece. Source: own elaboration.

Islands of Southern Aegean				
Cyclades Settlements—Traditional Settlements				
Islands	Traditional	Surface Area km^2/Total Island Area km^2	Non-Traditional	Surface Area km^2/Total Island Area km^2
Amorgos	8	4.96	1	0.18
Anafi	-		1	3.97
Andros	7	4.02	34	15.61
Antiparos	-		1	1.037
Donousa	2	0.928	-	
Folegandros	2	1.131	3	2.666
Ios	3	7.58	-	
Irakleia	1	0.872	-	
Kea	5	3.22	-	
Kimolos	1	1.46	3	0.935
Kythnos	3	6.6	3	1.85
Koufonisi	-		1	5.201
Mykonos	2	12.33	5	7.22
Milos	5	3.69	5	6.35
Naxos	22	5.43	17	7.92
Paros	10	27.67	16	25.58
Schoinousa	-		1	1.926
Serifos	4	4.08	4	9.27
Sifnos	4	5.73	5	3.07
Sikinos	1	2.58	1	1.92
Syros	5	4.51	24	5.41
Santorini	7	2.598	14	8.285
Tinos	37	16.95	12	6.87
Dodecanese Settlements—Traditional Settlements				
Arkoi	-	-	1	1.662
Astipalaia	5	3.127	-	
Chalki	3	0.8572	-	
Kalimnos	4	1.066	8	8.16
Karpathos	3	0.65	13	18.71
Kasos	-		4	3.736
Kos	-		4	2.805
Leipsoi	-		1	0.9044
Leros	-		9	8.99
Nisyros	3	0.53	1	0.382
Patmos	1	2.42	-	
Rhodes	4	1.76	46	21.3
Simi	3	0.506	1	1.366
Tilos	3	1.334	-	-

The extensive network of traditional settlements in the Southern Aegean region reveals the architectural value of these settlements and highlights the role of urban planning in preserving and sustainably managing this value. Within the boundaries of traditional settlements, the legislation sets strict restrictions, as already described. However, integrated protection and management of the place that also takes into consideration the wider landscape of traditional settlements is significant and indispensable. One main cause of change in the islands' landscape is the continuous construction activity in exurban areas. In Table 2, a thorough presentation of the percentage of built-up areas on the Cyclades and Dodecanese islands is provided, distinguishing this between built-up areas inside and outside settlement boundaries and comparing them with the total area of each island. The percentage of built-up areas within settlement boundaries serves as a proxy for settlement density, while the percentage outside settlement boundaries acts as an indicator of urban sprawl. The intensity of urban sprawl varies across the islands studied, reflecting the unique geographic characteristics of each island. This variation highlights the need for spatial planning policies that address urban growth while preserving the islands' cultural heritage. The bar chart in Figure 4 illustrates these percentages, showcasing the differences in built-up area distribution across the case study example.

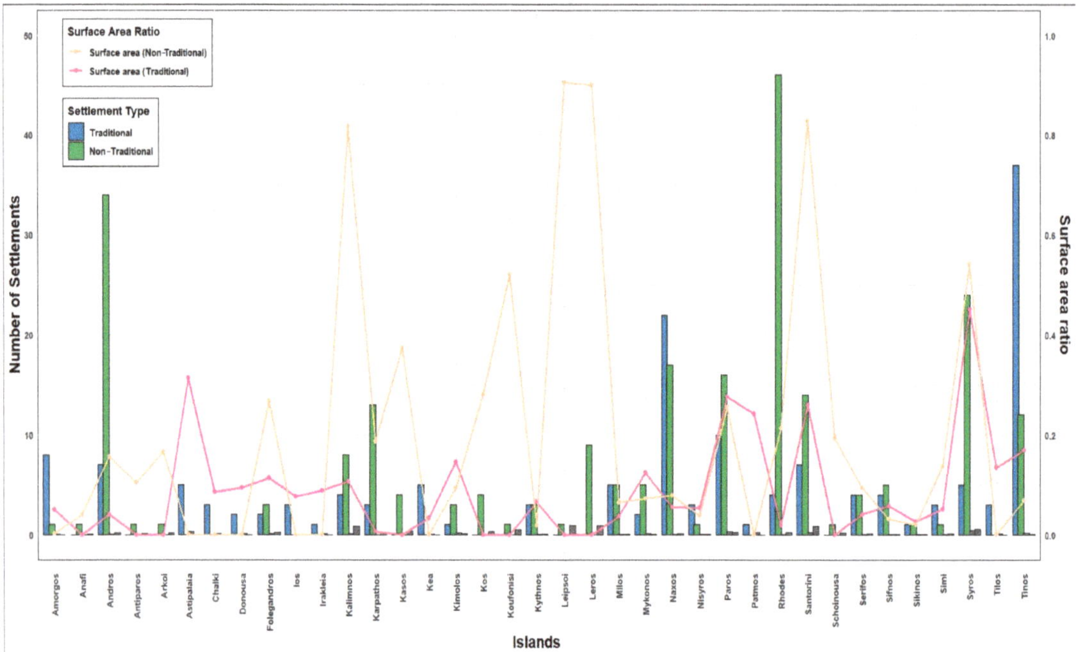

Figure 3. Traditional and non-traditional settlements on Cyclades and Dodecanese islands of the Southern Aegean region, Greece. Source: own elaboration.

For example, Amorgos in the Cyclades shows a pattern with 53.63% of its built-up area within settlements and 46.37% outside, indicating complex spatial development. Islands like Antiparos, Kea, Irakleia, Mykonos, Paros and Patmos have a high percentage of built-up area outside settlements, suggesting uncontrolled development, potentially driven by tourism. These outliers highlight the diverse dynamics of urbanization and land use, influenced by geographical, economic and social factors. In contrast, islands such as Kimolos and Simi show dense urban cores, with a significant share of built-up areas within the official settlement boundaries. Overall, the Dodecanese region generally has a higher percentage of built-up areas within the settlement boundaries, compared to the Cyclades region. The results reveal that six islands (Antiparos, Irakleia, Kea, Mykonos, Paros and Patmos) have substantial rates of built-up areas in exurban areas (over 60%), while several others show similar levels of built-up areas both within and outside settlements.

Although the islands of the South Aegean share several common characteristics related to built-up area dispersion and cultural heritage pressures, they also have important differences. These differences come from factors such as island size, tourism development, land availability and planning regulations. Due to the large number of islands included in this study, a detailed examination of each individual case is beyond this study's scope. Instead, our analysis highlights regional trends while recognizing that some islands may have unique patterns worth exploring in future research. This comprehensive examination underscores the multifaceted dynamics of urban sprawl across the Aegean islands.

Table 2. Built-up areas (%) inside and outside settlement boundaries and their relation to total island area and population information: Cyclades vs. Dodecanese. Source: own elaboration.

Islands	Population (Census 2021)	Built-Up Areas (%) Inside the Settlement Boundaries	Built-Up Areas (%) Inside the Settlement Boundaries/Total Area	Built-Up Areas (%) Outside the Settlement Boundaries	Built-Up Areas (%) Outside the Settlement Boundaries/Total Area
			Cyclades		
Amorgos	1961	53.63%	0.26%	46.37%	0.23%
Anafi	293	67.61%	0.18%	32.39%	0.09%
Andros	8826	46.47%	0.14%	53.53%	0.17%
Antiparos	1265	29.39%	0.6%	70.61%	1.44%
Donousa	213	54.45%	0.27%	45.55%	0.23%
Folegandros	719	75.68%	0.62%	24.32%	0.2%
Ios	2299	48.75%	0.26%	51.25%	0.28%
Iraklia	148	31.89%	0.09%	68.11%	0.18%
Kea	2335	20.03%	0.13%	79.97%	0.51%
Kimolos	810	86.81%	0.4%	13.19%	0.06%
Kythnos	1568	43.55%	0.3%	56.45%	0.39%
Koufonisi	391	56.57%	2.01%	43.43%	1.54%
Mykonos	10,704	10.88%	1.43%	89.12%	11.71%
Milos	5302	67.01%	0.55%	32.99%	0.27%
Naxos	14,708	51.71%	0.44%	48.29%	0.41%
Paros	14,520	39.6%	1.46%	60.4%	2.23%
Santorini	15,231	79.41%	4.94%	20.59%	3.71%
Schoinousa	229	58.43%	0.46%	41.57%	0.32%
Serifos	1241	64.01%	0.32%	35.99%	0.18%
Sifnos	2777	53.79%	0.71%	46.21%	0.61%
Sikinos	253	75.64%	0.25%	24.36%	0.08%
Syros	21,124	59.91%	5.35%	40.09%	3.58%
Thirasia	249	57.12%	0.7%	42.88%	0.17%
Tinos	8934	57.53%	1.12%	42.47%	0.83%
			Dodecanese		
Arkoi	44	78.79%	0.04%	21.21%	0.01%
Astipalaia	1376	73.59%	0.29%	26.41%	0.1%
Chalki	475	98.26%	0.33%	1.74%	0.01%
Kalimnos	17,752	93.17%	2.09%	6.83%	0.15%
Karpathos	6567	46.79%	0.25%	53.21%	0.28%
Kasos	1223	89.98%	0.36%	10.02%	0.04%
Kos	37,089	47.87%	0.77%	52.13%	0.84%
Leipsoi	778	74.27%	0.43%	25.73%	0.15%
Leros	7992	66.79%	1.88%	33.21%	0.94%
Nisyros	1048	78.31%	0.08%	21.69%	0.02%
Patmos	3283	13.42%	0.36%	86.58%	2.3%
Rhodes	125,113	53.33%	0.77%	46.67%	0.67%
Simi	2603	96.68%	0.79%	3.32%	0.03%
Tilos	746	87.61%	0.2%	12.39%	0.03%

For example, on islands such as Mykonos and Naxos, built-up areas outside settlement boundaries encroach upon archaeological sites and traditional landscapes, disrupting their integrity. This encroachment can lead to irreversible damage, including the loss of archaeological context and challenges in managing tourism flows to these sensitive areas. Additionally, urban sprawl often creates fragmented landscapes, reducing the coherence of traditional settlements and their cultural significance.

Figure 5 illustrates the spatial distribution of selected island examples of the study region, based on the findings presented in Table 2.

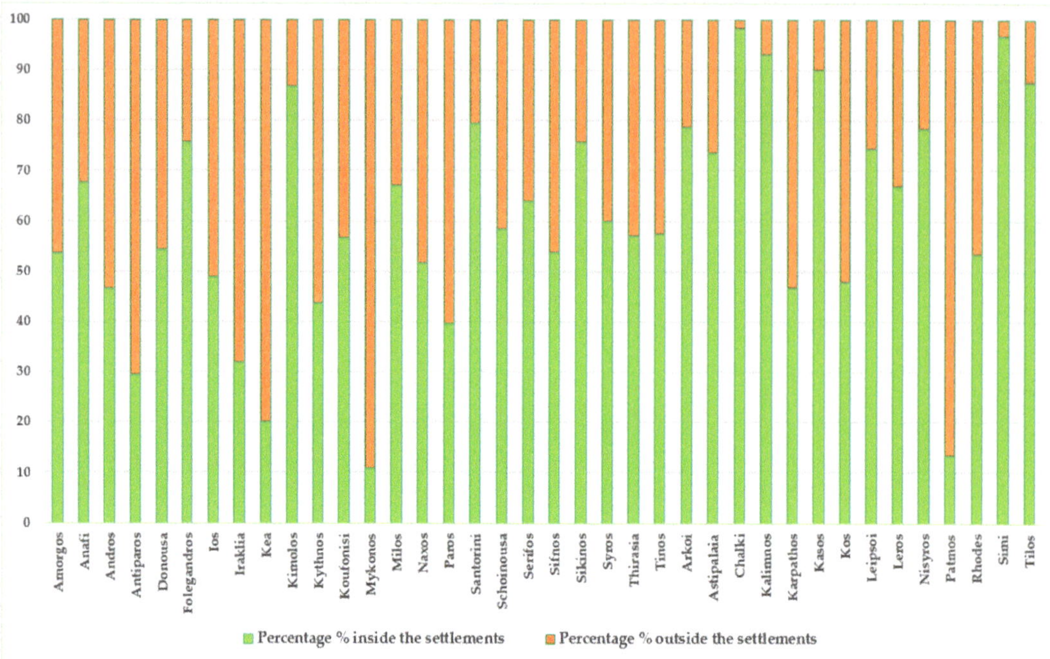

Figure 4. Built-up areas (%) falling inside and outside the settlement boundaries: Cyclades vs. Dodecanese. Source: own elaboration.

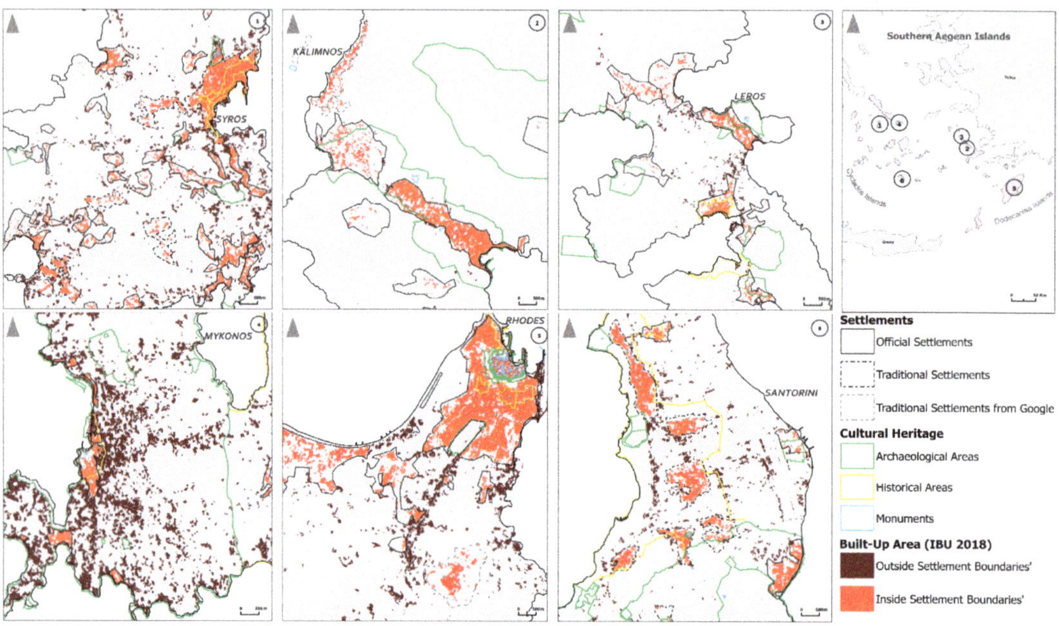

Figure 5. Indicative examples of the spatial distribution of built-up areas inside and outside the settlement boundaries on Cyclades and Dodecanese islands and their proximity to cultural heritage assets. Source: own elaboration.

Table 3 and Figure 6 demonstrate the impact of built-up areas lying outside official boundary delineations on cultural heritage across islands in the Southern Aegean region.

Table 3. Impact of built-up areas on cultural heritage sites on the islands of the Southern Aegean, including percentage and number of affected sites. Source: own elaboration.

Effect of Built-Up Areas on the Cultural Heritage Sites on Islands of Southern Aegean Region						
Cyclades	Historical %	Historical No.	Archaeological %	Archaeological No.	Monuments %	Monuments No.
Amorgos	-	-	39.34%	23	1.02%	62
Anafi	-	-	23.26%	1	-	4
Andros *	-	-	7.68%	18	1.26%	260
Antiparos	-	-	4.39%	4	1.27%	4
Donousa	-	-	-	3	-	-
Folegandros	-	-	-	-	16.28%	17
Ios	-	-	7.83%	7	1.87%	21
Irakleia	-	-	100%	2	-	1
Kea *	9.82%	1	38.20%	17	0.67%	13
Kimolos *	-	-	24.89%	4	5.33%	10
Kythnos *	-	-	-	5	0.60%	22
Koufonisi	-	1	-	1	-	-
Mykonos	88%	2	99.71%	7	0.03%	34
Milos	-	1	72.49%	12	1.07%	29
Naxos *	60.17%	39	85.66%	13	2.34%	214
Paros	31.25%	2	43.04%	13	0.51%	77
Schoinousa *	-	-	-	1	-	-
Serifos	-	-	87.20%	2	2.77%	18
Sifnos *	14.63%	1	15.16%	6	1.13%	67
Sikinos	-	-	-	-	39.82%	15
Syros	46.70%	3	2.23%	8	0.26%	26
Santorini	49.92%	17	80.68%	4	0.56%	35
Tinos	7.14%	3	8.01%	12	1.32%	45
Dodecanese						
Arkoi	-	-	-	2	-	-
Astipalaia	74.77%	1	14.10%	17	-	9
Chalki	-	-	-	5	-	25
Kalimnos *	-	-	4.53%	6	0.59%	29
Karpathos	-	-	48.39%	6	0.19%	34
Kasos	-	-	24.95%	3	8.27%	16
Kos	1.17%	4	39.10%	10	1.17%	76
Leipsoi	-	-	14.23%	4	-	1
Leros *	76.44%	2	57.09%	11	0.34%	15
Nisyros	13.59%	4	15.20%	3	-	19
Patmos	-	-	-	1	2.28%	41
Rhodes	10.53%	7	52.88%	39	6.20%	247
Simi	2%	2	99.80%	6	24.05%	28
Tilos	-	1	7.89%	3	1.16%	59

* Where certain archaeological sites are located in marine areas.

The cultural heritage areas analyzed in this study are strictly those officially designated by the Archaeological Cadaster. This means that our findings pertain only to formally recognized sites and do not consider the broader cultural or socioeconomic aspects of heritage preservation. Consequently, the analysis focuses on spatial relations rather than interpretative or historical dimensions of cultural heritage management.

Cultural heritage is classified into three categories: historical, archaeological and monuments. Each cell in Table 3 represents the percentage of built-up areas affecting each category on each island. For example, on the island of Mykonos, Cyclades island complex region, built-up areas affect 88% of the historical sites, 99.71% of the archaeological sites and 0.03% of monuments.

The percentage of archaeological areas might be slightly lower than the actual figures due to the fact that certain archaeological areas extend into coastal regions, as indicated by an asterisk (*) in Table 3. This applies to islands such as Andros, Kea, Kimolos, Kythnos, Naxos, Schoinousa, Sifnos, Kalimnos and Leros, where certain archaeological sites are located in marine areas.

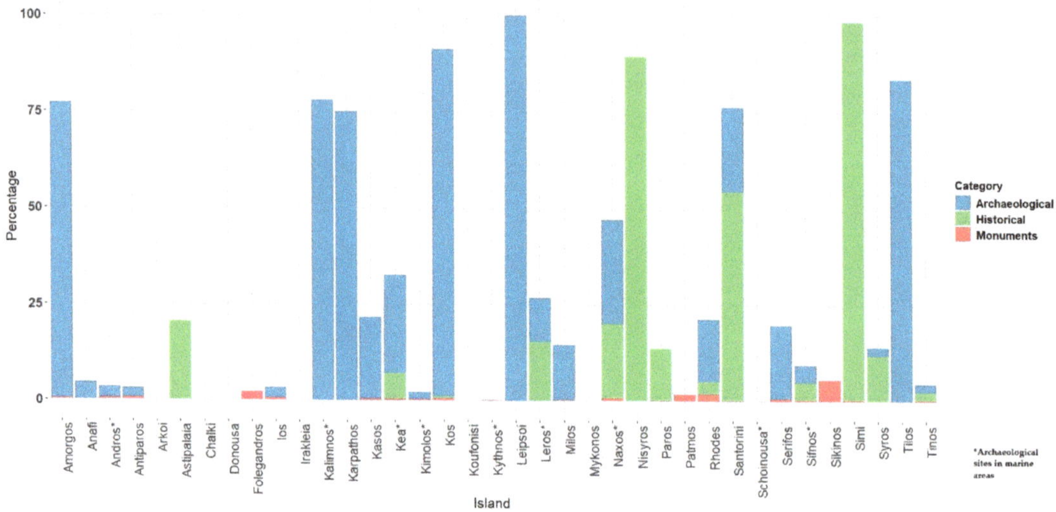

Figure 6. Cultural heritage impact assessment: built-up areas on the islands of the Southern Aegean region. Source: own elaboration.

Table 3 enables a comparison of the impact of built-up areas on cultural heritage across different islands, providing insights into the preservation efforts needed and developmental patterns. Higher percentages indicate a greater impact and potential areas for undertaking cultural heritage protection initiatives, while lower percentages suggest better preservation of cultural heritage.

The results highlight diverse needs for cultural heritage preservation and varying degrees of integration of cultural heritage into the urban context. Unlike prior studies that broadly discuss the conflict between development and heritage conservation, this research quantifies the specific impacts of unregulated development on cultural assets across islands, providing actionable insights for regional planning. Notably, islands like Serifos, Simi and Mykonos have high percentages of historical and archaeological sites, reflecting their rich cultural significance. Additionally, islands such as Santorini and Leros show a remarkable concentration of cultural sites, suggesting the need for careful spatial planning that is capable of balancing developmental and heritage conservation objectives. The bar chart, with green for historical sites, blue for archaeological sites and red for monuments, illustrates these varying impacts and highlights the need for careful urban planning to protect the islands' cultural heritage.

As mentioned in the previous sections, in this study, the correlation between built-up areas and cultural heritage across islands is explored. This study not only identifies significant correlations between urban growth and cultural heritage impacts but also highlights critical areas where preservation efforts must be prioritized. These findings offer a detailed framework for addressing the global challenge of balancing heritage conservation with development in culturally rich but vulnerable regions. Understanding the relation between these two factors is crucial for spatial planning, conservation efforts and sustainable development.

In Table 4, correlations calculated using the Pearson correlation coefficient are presented, expressing the linear relation between two variables. In the Pearson equation, the variables Xi and Yi represent the individual data points for "Built-up Area" and "Cultural Heritage" for each specific island. The variables $\overline{X}$ and $\overline{Y}$ are the means of "Built-up Area" and "Cultural Heritage", respectively. The correlation coefficient indicates both the strength and direction of the relation between "Built-up Area" and "Cultural Heritage" for each island. Values closer to 1 or −1 indicate a stronger correlation, while values closer to 0 indicate a weaker correlation.

Table 4. Correlation coefficients between built-up areas and cultural heritage across various islands. Source: own elaboration.

Islands	Amorgos	Anafi	Andros	Antiparos	Arkoi
Correlation of cultural heritage with built-up area	0.721	−0.179	0.629	−0.004	NA
Islands	Astipalaia	Chalki	Donousa	Folegandros	Ios
Correlation of cultural heritage with built-up area	0.654	NA	0.751	0.981	0.507
Islands	Irakleia	Kalimnos	Karpathos	Kasos	Kea
Correlation of cultural heritage with built-up area	0.659	−0.332	0.753	−0.125	0.576
Islands	Kimolos	Kythnos	Kos	Koufonisi	Leipsoi
Correlation of cultural heritage with built-up area	NA	0.932	0.844	0.626	−0.195
Islands	Leros	Mykonos	Milos	Naxos	Nisyros
Correlation of cultural heritage with built-up area	0.757	0.805	0.815	0.819	−0.137
Islands	Paros	Patmos	Rhodes	Santorini	Schoinousa
Correlation of cultural heritage with built-up area	0.968	0.826	0.874	0.710	−0.030
Islands	Serifos	Sifnos	Sikinos	Simi	Syros
Correlation cultural heritage with built-up area	−0.273	0.830	0.189	0.757	−0.602
Islands	Tilos	Tinos			
Correlation of cultural heritage with built-up area	−0.273	0.894			
Strong positive correlation $r > 0.7$			Weak negative correlation $-0.3 < r \leq 0$		
Moderate positive correlation $0.3 < r \leq 0.7$			Moderate negative correlation $-0.7 < r \leq -0.3$		
Weak positive correlation $0 < r \leq 0.3$			Strong negative correlation $r < -0.7$		

NA = Not Applicable.

Using the Pearson correlation coefficient to evaluate the entire Cyclades and Dodecanese regions showed a correlation coefficient of 0.774, indicating a positive relation. This suggests that as built-up areas expand without restrictions, they often encroach upon cultural heritage sites located within protected areas. Consequently, this expansion poses a threat to heritage sites, emphasizing the urgent need for policy initiatives targeting the protection of these invaluable areas from unregulated development. Analyzing the correlation coefficients for each island in the Cyclades and Dodecanese regions reveals distinct relations between cultural heritage and urban growth, more specifically the following:

- Islands such as Folegandros, Tinos and Kythnos exhibit highly positive correlations, indicating a robust connection between cultural significance and built-up areas. This implies that the built-up areas significantly impact the cultural heritage sites, occupying substantial portions of these areas. The unregulated expansion of built space, often beyond official settlement boundaries, threatens these valuable cultural sites.
- Islands like Syros, Kalimnos and Serifos stand out with negative correlations, implying a weak negative relation between cultural heritage areas and built-up areas. This suggests that the built-up areas do not negatively affect the cultural heritage areas on these islands.

Table 4 underscores the importance of tailoring policies to island-specific conditions. For islands with strong positive correlations, strict spatial planning regulations and enhanced monitoring of built-up area dispersion (urban sprawl) are essential. In contrast, for islands with weak or negative correlations, efforts should focus on maintaining existing protections and monitoring for potential future risks.

5. Conclusions

Landscapes are inherently dynamic, evolving over time in response to societal changes. The main factors of this change include natural processes and human activities, such as changes in production techniques, built-up area expansion, tourism infrastructures, transport networks, etc. However, when these changes occur rapidly and without strategic planning, they can lead to significant and often irreversible alterations in landscapes, raising concerns about the preservation of the unique natural and cultural characteristics of a region.

In island regions, this phenomenon is particularly evident due to their fragile socio-spatial systems and the immense pressures exerted by mass tourism. Our findings indicate that uncontrolled urban sprawl poses significant threats to the cultural heritage of the South Aegean islands. Current spatial planning tools are inadequate for protecting archaeological areas and historic sites, primarily due to ongoing pressures from expanding construction activities. Tourism development further intensifies these challenges, leading to built-space expansion that disrupts the cultural landscape.

Tourism, the primary economic driver for the islands, generates demand for infrastructure such hotels and second homes, particularly in areas with high cultural value. This unregulated development often compromises the historical integrity of heritage sites. Seasonal population increases due to tourism exacerbate land-use pressures, driving construction in peripheral and exurban areas.

Urban sprawl is particularly concentrated in coastal zones and areas near cultural landmarks, intensifying development pressures. While this study focuses on spatial analysis, tourism-driven urbanization, second-home expansion and coastalization also play a key role in shaping these trends. However, due to this study's methodological focus on geospatial datasets, these socioeconomic factors were not quantitatively analyzed. Future research should integrate socioeconomic data to provide a deeper understanding of urban expansion and its impact on cultural heritage areas.

Development often occurs outside settlement boundaries due to several factors. Land outside these boundaries is typically cheaper, lacks urban infrastructure and offers more availability and environmental qualities. Moreover, it is not subject to the institutional regulations that govern land within designated settlements, making it an attractive option for unregulated construction.

The patterns of urban sprawl tend to favor coastal zones and areas near cultural landmarks, fragmenting traditional settlements and encroaching upon heritage sites. Inconsistent spatial planning exacerbates these pressures, as it struggles to balance development with preservation.

The key threats include encroachment on cultural zones, landscape degradation and the loss of traditional settlement character, all of which undermine heritage preservation. Addressing these issues requires a clear understanding of the drivers of urban sprawl and their impact on cultural landscapes.

This study highlights the urgent need for an integrated approach to spatial planning that prioritizes the preservation of cultural heritage alongside sustainable development. While this study highlights common trends in built-up area dispersion and its impact on cultural heritage across the South Aegean, it is important to acknowledge that individual islands may experience these pressures differently. Factors such as tourism intensity, land availability and local planning policies contribute to these variations. Future research could explore island-specific case studies to provide a more detailed understanding of the unique challenges faced by different locations.

Establishing a comprehensive spatial planning framework is essential, incorporating clear guidelines for landscape management and regulations to protect cultural heritage.

To improve the effectiveness of spatial planning and heritage protection policies, it is crucial to strengthen policy implementation and enforcement. Social and economic driving forces often override planning's ability to regulate space. The General Urban Plans (GUPs and OCSHOPs) appear insufficient to mitigate the local dynamics being created. Improving the spatial planning system should involve action at multiple levels: (a) at the local level, through the implementation of general and special urban plans; (b) at the regional level, by completing and revising regional frameworks; and (c) at the national level, through the completion and revision of special spatial frameworks.

Ensuring the effectiveness of spatial planning requires the establishment and operation of a spatial planning observatory. This observatory would play a critical role in monitoring, evaluating and improving spatial planning practices across all levels of governance.

A significant challenge is the inconsistent application of spatial planning laws, which allows development in protected zones. Policymakers need to improve coordination between various government agencies, local authorities and stakeholders to ensure that regulations are properly enforced and that there is a shared commitment to cultural heritage preservation.

Advanced geoinformatics tools can play a pivotal role in assisting policymakers by enhancing their understanding of spatial dynamics and their implications for cultural heritage. In Greece, ongoing revisions to spatial planning and the development of new frameworks offer a valuable opportunity to address these issues. Future research should focus on the implementation of these tools and frameworks, exploring innovative methods for landscape management that effectively integrate cultural preservation with development goals.

Policymakers and stakeholders must recognize the delicate balance between preserving cultural heritage and fostering tourism development in island contexts. The pressures of mass tourism and unregulated growth can alter landscapes and threaten the integrity of cultural resources, endangering sustainability objectives and the viability of the tourism sector. By prioritizing strategic planning and thoughtful management, it is possible to safeguard the rich cultural heritage of these islands while promoting sustainable development objectives. Implementing a balanced approach will ensure that both cultural heritage and economic growth can coexist harmoniously, to the benefit of both current and future generations.

Importantly, this study introduces a quantitative framework for heritage protection, offering an innovative approach that can be applied globally. While this research is based on the specific context of the Southern Aegean islands, the methodology and findings are relevant to any region facing the pressures of urbanization and cultural heritage conservation. The framework developed in this study provides a valuable tool for local, regional and international policymakers, enabling the creation of more effective spatial planning strategies that balance cultural heritage preservation with necessary development. This work sets a model for the sustainable management of urban expansion in culturally sensitive areas worldwide, offering a foundation for future research and policy efforts in regions facing similar challenges.

Author Contributions: Conceptualization, Efstratia Chatzi and Georgios Tsilimigkas; methodology, Efstratia Chatzi and Georgios Tsilimigkas; software, Efstratia Chatzi and Evangelia-Theodora Derdemezi; formal analysis, Efstratia Chatzi; resources, Efstratia Chatzi and Evangelia-Theodora Derdemezi; data curation, Efstratia Chatzi; writing—original draft preparation, Efstratia Chatzi; writing—review and editing, Efstratia Chatzi, Evangelia-Theodora Derdemezi and Georgios Tsilimigkas; visualization, Efstratia Chatzi; supervision, Georgios Tsilimigkas and Evangelia-Theodora Derdemezi. All authors have read and agreed to the published version of the manuscript.

Funding: This research received no external funding.

Data Availability Statement: The Impervious Built-Up 2018 (raster 10 m) data can be accessed through Copernicus Land at https://land.copernicus.eu/en/products/high-resolution-layer-impervious-built-up/impervious-built-up-2018 (accessed on 19 April 2024), The archaeological, historical and monument data are available from the Archaeological Cadaster at https://www.arxaiologikoktimatologio.gov.gr/ (accessed on 19 April 2024), The settlement boundaries can be found through the GIS of the Greek Ministry at http://gis.epoleodomia.gov.gr/v10/index.html (accessed on 19 April 2024), while the traditional settlement boundaries were digitized from the Official Government Gazettes for the Cyclades and Dodecanese islands.

Conflicts of Interest: The authors declare no conflicts of interest.

Abbreviations

GMS	Global Monitoring for Environment and Security
GUPs	General Urban Plans
IBU	Impervious Built-Up
MSP	Marine Spatial Planning
NUTS	Nomenclature of territorial units for statistics
OCSHOPs	Open City Spatial and Housing Organization Plans
OGGs	Official Government Gazettes
UDCZ	Urban Development Control Zone

References

1. García-Rodríguez, J.-L.; García-Rodríguez, F.J. Human heritage and sustainable development on arid islands: The case of the Eastern Canary Islands. *Isl. Stud. J.* **2016**, *11*, 113–130. [CrossRef]
2. García-Hernández, M.; De la Calle-Vaquero, M.; Yubero, C. Cultural Heritage and Urban Tourism: Historic City Centers under Pressure. *Sustainability* **2017**, *9*, 1346. [CrossRef]
3. Tweed, C.; Sutherland, M. Built cultural heritage and sustainable urban development. *Landsc. Urban Plan.* **2007**, *83*, 62–69. [CrossRef]
4. Hadjimitsis, D.; Agapiou, A.; Alexakis, D.; Sarris, A. Exploring Natural and Anthropogenic Risk for Cultural Heritage in Cyprus Using Remote Sensing and GIS. *Int. J. Digit. Earth* **2013**, *6*, 115–142. [CrossRef]
5. Savo, V.; Caneva, G.; McClatchey, W.; Reedy, D.; Salvati, L. Combining Environmental Factors and Agriculturalists' Observations of Environmental Changes in the Traditional Terrace System of the Amalfi Coast (Southern Italy). *Ambio* **2014**, *43*, 297–310. [CrossRef]
6. Derdemezi, E.-T.; Tsilimigkas, G.; Kizos, T. Mining activity and island landscape issues: Evidence from Cyclades islands, Greece. *Eur. Plan. Stud.* **2021**, *30*, 384–404. [CrossRef]
7. Swensen, G.; Jerpåsen, G.B. Cultural Heritage in Suburban Landscape Planning: A Case Study in Southern Norway. *Landsc. Urban Plan.* **2008**, *87*, 289–300. [CrossRef]
8. Kizos, T.; Koulouri, M. Agricultural landscape dynamics in the Mediterranean: Lesvos (Greece) case study using evidence from the last three centuries. *Environ. Sci. Policy* **2006**, *9*, 293–306. [CrossRef]
9. Vecco, M. A Definition of Cultural Heritage: From the Tangible to the Intangible. *J. Cult. Herit.* **2010**, *11*, 321–324. [CrossRef]
10. Theodora, Y. Aegean Sea—Challenges and dilemmas in management and planning for local development in fragmented insular regions. *Heritage* **2019**, *2*, 108. [CrossRef]
11. Gupta, V.; Hanges, P.J.; Dorfman, P. Cultural clusters: Methodology and findings. *J. World Bus.* **2002**, *37*, 11–15. [CrossRef]
12. Knippschild, R.; Zöllter, C. Urban Regeneration between Cultural Heritage Preservation and Revitalization: Experiences with a Decision Support Tool in Eastern Germany. *Land* **2021**, *10*, 547. [CrossRef]
13. Guzman, P.; Pereira Roders, A.R.; Colenbrander, B. Impacts of Common Urban Development Factors on Cultural Conservation in World Heritage Cities: An Indicators-Based Analysis. *Sustainability* **2018**, *10*, 853. [CrossRef]
14. Lane, P.J. Future Urban Growth and Archaeological Heritage Management: Some Implications for Research Activity in Africa. *Conserv. Manag. Archaeol. Sites* **2011**, *13*, 134–159. [CrossRef]
15. Guzmán, P.C.; Pereira Roders, A.R.; Colenbrander, B.J.F. Measuring links between Cultural Heritage management and sustainable urban development: An overview of global monitoring tools. *Cities* **2017**, *60*, 192–201. [CrossRef]
16. Kizos, T.; Tsilimigkas, G.; Karampela, S. What drives built-up area expansion on islands? Using soil sealing indicators to estimate built-up area patterns on Aegean islands, Greece. *Tijdschr. Econ. Soc. Geogr.* **2017**, *108*, 35–52. [CrossRef]
17. Tsilimigkas, G.; Kizos, T.; Derdemezi, E.-T. Urbanization and land cover change in Aegean islands: Assessing the socio-economic impact of tourism development. *Land* **2021**, *10*, 545. [CrossRef]
18. Tsilimigkas, G.; Derdemezi, E.-T. Unregulated built-up area expansion on Santorini Island, Greece. *Eur. Plan. Stud.* **2020**, *28*, 1790–1811. [CrossRef]
19. Leka, A.; Lagarias, A.; Panagiotopoulou, M.; Stratigea, A. Development of a Tourism Carrying Capacity Index (TCCI) for Sustainable Management of Coastal Areas in Mediterranean Islands—Case Study Naxos, Greece. *Ocean Coast. Manag.* **2022**, *216*, 105978. [CrossRef]
20. Lagarias, A.; Stratigea, A. Coastalization Patterns in the Mediterranean: A Spatiotemporal Analysis of Coastal Urban Sprawl in Tourism Destination Areas. *GeoJournal* **2023**, *88*, 2529–2552. [CrossRef]

21. Williams, M. Managing the Cultural Heritage of the Mediterranean Islands: A Case Study of Cyprus. *J. Mediterr. Archaeol.* **2020**, *33*, 155–178.
22. Lekakis, S. Distancing and Approaching: Local Communities and Monuments in the Aegean Sea—A Case Study from the Island of Naxos. *Conserv. Manag. Archaeol. Sites* **2013**, *15*, 76–93. [CrossRef]
23. Lekakis, S. Managing Archaeological Sites in the Aegean Sea: The Context and a Critique on Recent Management Trends in Sites Around the Mediterranean Sea. Master's Thesis, Institute of Archaeology, University College London, London, UK, 2005.
24. Lagarias, A.; Stratigea, A. High-Resolution Spatial Data Analysis for Monitoring Urban Sprawl in Coastal Zones: A Case Study in Crete Island. *Lect. Notes Comput. Sci.* **2021**, *12792*, 55–69. [CrossRef]
25. Christofilopoulos, D.G. *Cultural Environment—Spatial Planning and Sustainable Development*; Sakkoulas: Athens, Greece, 2002; ISBN 960-420-165-4.
26. Kristy, G. The Impact of Urban Sprawl on Cultural Heritage in Herat, Afghanistan: A GIS Analysis. *Digit. Appl. Archaeol. Cult. Herit.* **2018**, *11*, e00086. [CrossRef]
27. Cori, B. Spatial Dynamics of Mediterranean Coastal Regions. *J. Coast. Conserv.* **1999**, *5*, 105–112. [CrossRef]
28. Muñoz, F. Lock Living: Urban Sprawl in Mediterranean. *Cities* **2003**, *20*, 381–385. [CrossRef]
29. Zervos, G. The evolution of traditional settlements and modern planning in the institutional texts of protection of the immovable cultural and natural heritage of Europe. In Proceedings of the Conference "Protection of Traditional Settlements & Contemporary Architectural Design", Athens, Greece, 23 January 2012.
30. Tsilimigkas, G.; Gourgiotis, A. Unregulated urban sprawl and spatial distribution of fire events: Evidence from Greece. *Environ. Hazards* **2023**, *17*, 436–455. [CrossRef]
31. Stratigea, A.; Katsoni, V. A strategic policy scenario analysis framework for the sustainable tourist development of peripheral small island areas—The case of Lefkada-Greece Island. *Eur. J. Futures Res.* **2015**, *3*, 5. [CrossRef]
32. Pavlogeorgatos, G.; Aliouris, K.; Ioannou, N.; Chatzimitsis, D. Maritime Spatial Planning in Greece and Cyprus. 2023. ISBNs 9786185705077/9786185705084. Available online: https://www.propobos.gr/wp-content/uploads/2023/01/Thalassios_xwrotaxikos_sxediasmos_LowRes.pdf (accessed on 5 May 2024).
33. Agapiou, A.; Alexakis, D.D.; Lysandrou, V.; Sarris, A.; Cuca, B.; Themistocleous, K.; Hadjimitsis, D.G. Impact of urban sprawl on Cultural Heritage monuments: The case study of Paphos area in Cyprus. *J. Cult. Herit.* **2015**, *16*, 671–680. [CrossRef]
34. De Noronha Vaz, E.; Cabral, P.; Caetano, M.; Nijkamp, P.; Painho, M. Urban Heritage Endangerment at the Interface of Future Cities and Past Heritage: A Spatial Vulnerability Assessment. *Habitat Int.* **2012**, *36*, 287–294. [CrossRef]
35. He, Q.S.; He, W.S.; Song, Y.; Wu, J.Y.; Yin, C.H.; Mou, Y.C. The impact of urban growth patterns on urban vitality in newly built-up areas based on an association rules analysis using geographical 'big data'. *Land Use Policy* **2018**, *75*, 689–696. [CrossRef]
36. Athens Charter for the Restoration of Historic Monuments. Adopted at the First International Congress of Architects and Technicians of Historic Monuments. 1931. Available online: http://www.icomos.org/en/charters-and-texts/179-articles-en-francais/ressources/charters-and-standards/167-the-athens-charter-for-the-restoration-of-historic-monuments (accessed on 5 May 2024).
37. Antoniou, T. The development of the national cultural identity in the light of the current Constitution and Community law. In *The Cultural Heritage and the Law—Proceedings of the Conference, Athens Greece, 3–4 June 2003*; Sakkoula Publications: Athens/Thessaloniki, Greece, 2004; p. 561. ISBN 978-960-301-864-3.
38. ICOMOS. *Charter for the Conservation and Restoration of Monuments and Sites (The Venice Charter)*; ICOMOS: Paris, France, 1964. Available online: https://www.icomos.org/en/participer/179-articles-en-francais/ressources/charters-and-standards/157-thevenice-charter (accessed on 5 May 2024).
39. Council of Europe: Council of Europe. *European Convention on the Protection of the Architectural Heritage of Europe*; Council of Europe: Strasbourg, France, 1987.
40. Council of Europe: Council of Europe. *European Convention on the Protection of the Archaeological Heritage (Revised)*; Council of Europe: Strasbourg, France, 1995.
41. Blake, J. UNESCO's 2003 Convention on Intangible Cultural Heritage: The Implications of Community Involvement in 'Safeguarding'. In *Intangible Heritage*, 1st ed.; eBook; Routledge: Oxfordshire, UK, 2008; p. 29. ISBN 9780203884973.
42. Kurin, R. Safeguarding Intangible Cultural Heritage in the 2003 UNESCO Convention: A Critical Appraisal. *Int. J. Cult. Prop.* **2004**, *56*, 221–222, ISSN 1350-0775. [CrossRef]
43. UNESCO. Convention for the Safeguarding of the Intangible Cultural Heritage. Paris. Available online: https://ich.unesco.org/doc/src/01852-EN.pdf (accessed on 8 May 2024).
44. UNESCO. *Convention for Safeguarding the Intangible Cultural Heritage*; United Nations Educational, Scientific and Cultural Organization: Paris, France, 2003.

45. UNESCO. Convention Concerning the Protection of the World Cultural and Natural Heritage. World Heritage Convention, UNESCO. 1972. Available online: https://whc.unesco.org/en/conventiontext (accessed on 8 May 2024).
46. Application of UNESCO Guidelines. Available online: https://whc.unesco.org/en/guidelines/ (accessed on 8 May 2024).
47. Sevieri, G.; Galasso, C.; D'Ayala, D.; De Jesus, R.; Oreta, A.; Grio, M.E.D.A.; Ibabao, R. A multi-hazard risk prioritization framework for Cultural Heritage assets. *Nat. Hazards Earth Syst. Sci.* **2020**, *20*, 1391–1406. [CrossRef]
48. ICOMOS. *Charter for the Conservation of Historic Towns and Urban Areas*; ICOMOS: Paris, France, 1987; Available online: https://www.icomos.org/images/DOCUMENTS/Charters/towns_e.pdf (accessed on 5 May 2024).
49. OGG (Official Government Gazette). *On Urban, Villages and Settlements Plans of the State and these Construction*; Greek Official Gazette 228/A/16.08.1923; Government Printing Office: Athens, Greece, 1923.
50. OGG (Official Government Gazette). *On Archaeological Legislation*; Greek Official Gazette 275/A/24.08.1932; Government Printing Office: Athens, Greece, 1932; pp. 50, 52.
51. OGG (Official Government Gazette). *On the Protection of Antiquities*; Greek Official Gazette 169/A/07.08.1950; Government Printing Office: Athens, Greece, 1950; pp. 1–6.
52. OGG (Official Government Gazette). *Law 3028/2002 "For the Protection of Antiquities and Cultural Heritage in General"*; Official Gazette 153/28.6.2002 A; Government Printing Office: Athens, Greece, 2002.
53. OGG (Official Government Gazette). *Law 4858 Sanction Code of Legislation for the Protection of Antiquities and Cultural Heritage in General*; Official Gazette A 220/19.11.2021; Government Printing Office: Athens, Greece, 2021.
54. OGG (Official Government Gazette). *Extension of Urban Plans, Residential Development and Related Regulations*; Official Government Gazette 33/A/1983; Government Printing Office: Athens, Greece, 1983.
55. OGG (Official Government Gazette). *Sustainable Urban Development of Cities and Settlements in the Country and Other Provisions*; Official Government Gazette 124/A/13.06.1997; Government Printing Office: Athens, Greece, 1997.
56. OGG (Official Government Gazette). *Modernization of Regional and Urban Planning Legislation and Other Provisions*; Official Government Gazette A 245/09.12.2020; Government Printing Office: Athens, Greece, 2020.
57. OGG (Official Government Gazette). *Special Framework for Spatial Planning and Sustainable Development for Tourism and Strategic Environmental Impact Assessment*; Official Government Gazette 1138 B/11.06.2009; Government Printing Office: Athens, Greece, 2009.
58. Wu, W.; Zhao, S.; Zhu, C.; Jiang, J. A Comparative Study of Urban Expansion in Beijing, Tianjin, and Shijiazhuang Over the Past Three Decades. *Landsc. Urban Plan.* **2015**, *134*, 93–106. [CrossRef]
59. Hanzl, M. Urban Sprawl in Europe: Landscapes, Land-Use Change, and Policy. *Plan. Pract. Res.* **2010**, *25*, 273–274. [CrossRef]
60. Catalán, B.; Saurí, D.; Serra, P. Urban Sprawl in the Mediterranean?: Patterns of Growth and Change in the Barcelona Metropolitan Region 1993–2000. *Landsc. Urban Plan.* **2008**, *85*, 174–184. [CrossRef]
61. Koutsi, D.; Stratigea, A. Unburying hidden land and maritime cultural potential of small islands in the Mediterranean for tracking heritage-led local development paths: Case study Leros-Greece. *Heritage* **2019**, *2*, 938–966. [CrossRef]
62. Athanasiou, A. Dynamics and Design of Centers in the City of the Coming Decades—Towards Centralized or Decentralized Models? *AeiXoros* 2002. [CrossRef]
63. NCMA (National Cadastre and Mapping Agency). *Viewing Orthophotos*; Greek Cadaster, Greek Government: Athens, Greece, 2023. Available online: https://www.ktimatologio.gr/ (accessed on 10 May 2024).
64. Archaeological Cadastre Greece. Available online: https://www.arxaiologikoktimatologio.gov.gr/ (accessed on 19 April 2024).
65. Thal-Chor. Available online: https://thalchor-2.ypen.gov.gr/ (accessed on 10 May 2024).
66. Official Government Gazette. Available online: https://www.gov.gr/en/ipiresies/dikaiosune/nomothesia/phulla-ephemeridas-tes-kuberneses-phek (accessed on 5 May 2024).
67. E-Poleodomia. Available online: https://gis.epoleodomia.gov.gr/v11/ (accessed on 10 May 2024).
68. EEA (European Environment Agency). *Impervious Built—Up 2018*, Copernicus, Land Monitoring Service: Copenhagen, Denmark, 2018. Available online: https://land.copernicus.eu/en/products/high-resolution-layer-imperviousness/impervious-built-up-2018 (accessed on 19 April 2024).
69. Waldo, T. Resolution, resampling and all that. In *Building Data Bases for Global Science*; Mounsey, H., Tomlinson, R., Eds.; Taylor and Francis: London, UK, 1988; pp. 129–137.
70. Tobler, W. Measuring Spatial Resolution. In Proceedings of the Land Resources Information Systems Conference, Beijing, China, 25–28 May 1987; pp. 12–16. Available online: https://www.researchgate.net/publication/291877360 (accessed on 19 April 2024).
71. Damaskos, D.; Plantzos, D. *A Singular Antiquity: Archaeology and Hellenic Identity in Twentieth-Century Greece*; Benaki Museum: Athens, Greece, 2008; p. 424. ISBN 978-960-8347-96-0.
72. Lekakis, S. To the Splendour of Our Byzantine Heritage: Managing Orthodox Churches in the Island of Naxos. In *Naxos and the Byzantine Aegean: Insular Responses to Regional Change*; Crow, J., Hill, D., Eds.; Norwegian Institute at Athens: Athens, Greece, 2018; pp. 371–390.

73. Lekakis, S. Here Be Dragons: Historical and Contemporary Archaeology and Heritage in the Aegean Sea. *Heritage* **2023**, *6*, 1391–1412. [CrossRef]
74. EL-STAT (Hellenic Statistical Authority). 2020. Available online: http://www.statistics.gr/en/home (accessed on 10 May 2024).
75. NUTS—Nomenclature of Territorial Units for Statistics. Available online: https://ec.europa.eu/eurostat/web/nuts (accessed on 10 May 2024).

International Journal of
Geo-Information

MDPI

Article

A Methodological Framework for Assessing Overtourism in Insular Territories—Case Study of Santorini Island, Greece

Akrivi Leka [1,*], Apostolos Lagarias [2], Anastasia Stratigea [1] and Panayiotis Prekas [1]

[1] Department of Geography and Regional Planning, School of Rural, Surveying and Geoinformatics Engineering, National Technical University of Athens, Athens 15780, Greece; stratige@central.ntua.gr (A.S.); panagiotisprk@gmail.com (P.P.)

[2] Department of Planning and Regional Development, School of Engineering, University of Thessaly, Volos 38334, Greece; lagarias@iacm.forth.gr

* Correspondence: akrleka@central.ntua.gr; Tel.: +30-210-772-4031

Abstract: This paper aims at addressing sustainability concerns in vulnerable insular territories. Such concerns are due to the rising overtourism phenomenon that affects islands at a rapidly escalating pace and renders sustainable local development of these outstanding areas—from a natural and cultural viewpoint—at stake. Towards this end, this work capitalizes on current literature and attempts to structure a methodological framework and a respective set of indicators' groups that are capable of assessing dimensions of overtourism in each single tourism destination, thus providing evidence-based and more robust guidelines for articulating policy decisions that can remedy incidents of overtourism. The proposed methodological framework follows a place-based approach and combines tourism demand and supply data with environmental, social, economic and spatial data and respective indicators for assessing the tourism density and intensity of each destination's tourism pattern and related multi-dimensional impacts. Validation of both the proposed framework and indicators' groups is conducted in Santorini Island, Greece, i.e., an island that lies at top positions of many lists of destinations, marked as suffering by overtourism. Results show that Santorini Island is confronted with severe overtourism impacts, which are highly affecting its identity, productive model and spatial pattern, while endangering its natural and cultural wealth.

Keywords: sustainable insular territories; overtourism; indicators; geospatial data analysis; Santorini Island

1. Introduction

Globally, the tourism industry is one of the fastest growing economic sectors, trending to reach 1.8 billion in 2030, i.e., 54% increase since 2015 [1], and a major driving force of destinations' economic growth. Concurrently, however, severe concerns have been expressed as to the detrimental impacts of the rapidly expanding travel and tourism industry on destinations' environment and host communities [2]. These have given rise to the articulation of the sustainable tourism concept, expressed by the UNWTO [3] as the tourism that takes full account of its current and future economic, social and environmental impacts, addressing the needs of visitors, the industry, the environment and the host communities, while this concept permeates also the Sustainable Development Goals of the UN Agenda 2030 (SDGs 8, 12, and 14—targets 8.9, 12b, and 14.7) [4,5]. Nevertheless, the sustainable tourism concept is marked by ambiguity. In fact, a precise definition of this multifaceted term in each particular destination [6] and the diverse perspectives and stakes inherent in its interpretation within a certain destination are lacking [7].

Among the globally, highly-rated tourism destinations are coastal and particularly *insular areas* in the Mediterranean, where the 3S (Sun-Sea-Sand) mass tourism model prevails. The prevalence of this model is particularly affecting the stability and sustainability of these areas by means of land use changes in the vulnerable coastal parts (construction of large tourist complexes, large infrastructure facilities, spread of built constructions outside the settlements' boundaries, etc.); pollution as a result of the irrational management of waste and water resources; CO_2 emissions due to intense transportation, etc. [8]. Additionally, insular regions are particularly vulnerable to problems relating to unsustainable levels and forms of tourism that can lead to overtourism [9]. Moreover, these regions are more dependent on tourism than their counterparts in the mainland, while, concurrently, they are confronted with a number of insularity-related challenges and vulnerabilities, e.g., limited resources and access to the global supply chain, high import costs, cultural and environmental fragility, climate change vulnerability [10], to name a few. Furthermore, islandness has historically resulted in employment and income barriers. Such barriers, in older times were confronted by emigration. Currently, however, tourism is often featured as a potential source of livelihood, a developmental lever that can provide jobs, income, and a subsistence level of existence [11], and a catalyst for social and territorial cohesion as well as local economic growth on an island scale [12]. However, although tourism constitutes a key driver of islands' economic growth, it is also a source of significant pressures on multiple levels (social, spatial, cultural, environmental, etc.). In addition, under conditions of climate change, largely affecting insular territories, as well as other recent global crises (economic, energy, health, geopolitical tensions, etc.), the rapidly evolving monocultural model of tourism in island regions endangers sustainability objectives in the long run, while, in many occasions, gives prominence to the overtourism phenomenon, an issue that is currently at the heart of the interest of the research and policy making community, as well as host communities in island areas.

In fact, overcrowded destinations, insular ones as well, are not a new phenomenon. This was already identified at the end of the mid-60s, where the roots of overtourism are traced [13]. However, the term overtourism per se is a rather new one, firstly appearing in tourism-related literature in 2016 [14]. Overtourism aims to express the impacts of destinations' overcrowding or, stated differently, the pressure exerted by an excessive volume and type of tourist flows in a certain destination [14]. Thus, the term intends to summarize the various negative impacts that are related to tourism mainly in city contexts as well as in nature-based destinations, such as national parks, coastal areas or islands. In fact, although overtourism is primarily noticed in urban areas, such as Venice, Barcelona, Amsterdam, etc. [13,15,16], the most vulnerable destinations are not necessarily cities, but rather coastal, insular, and rural heritage sites [17].

But what exactly does the *overtourism* term imply? The answer to this question is not a clear-cut one, with overtourism representing another vague term in the policy and planning realm. Indeed, a commonly-agreed definition of overtourism—occasionally referred to as overcrowding or visitor pressure [14]—does not yet exist, while viewpoints from different disciplines (e.g., tourism, planning) seem to be contradictory or even conflicting. Furthermore, the overtourism concept raises particular concerns in the scientific community with respect to even its very existence [9]. That said, Capocchi et al. [18] wonder whether overtourism is really anything more than over-development under a new name. Chung and Koo [19] and Huertas [20] consider overtourism as a media-generated term, experiencing a rush of media popularity, while the UNWTO [21] expresses the view that overtourism can eventually be simply a type of management problem. Additionally, Dredge [22] argues whether overtourism is simply a common phenomenon which, despite the fact that it has already been experienced for a long time, it is just now receiving widespread attention.

Further to the above arguments, a literature search unveils a number of overtourism definitions, expressing different viewpoints on the topic. Generally speaking, overtourism is commonly grasped as unsustainable tourism [23]; a state of "*too many tourists at a place and in an unsustainable way*" [24] (p. 48); a threat to quality of life [25], heritage [26], natural and socio-cultural resources [25], and place attachment [27]; and a phenomenon mainly associated with the volume, type, and time frame of tourist flows, coupled with the carrying capacity and fragility of the destination. In addition, literature-based definitions stress the physical, cultural, economic, social, psychological, political, spatial, etc., dimensions of overtourism. In this respect, Peeters et al. [17] (p. 22) define overtourism as "*... a state that is tightly linked to a situation in which tourism impacts, at certain times and in certain locations, exceed physical, ecological, social, economic, psychological, and/or political capacity thresholds*", and a state that is mainly associated with the volume of visitors. Mihalic [23] highlights the impacts of overtourism on the natural and cultural resources, and perceives overtourism as the rapid escalation and growth of tourism supply and demand, which result in the excessive use of destinations' natural resources and the destruction of their cultural attractions, while also affecting negatively the local socioeconomic environments in the long run. Stressing the social dimension, the UNWTO [21] perceives overtourism on the ground of the impacts of tourism on a destination, or parts thereof that are excessively influencing the perceived quality of life of citizens and/or the quality of visitors' experiences in a negative way. Accordingly, Vourdoubas [28] presents overtourism as a phenomenon occurring mainly in popular tourism destinations and being mostly expressed by the large number of tourists—a "tourism storm" or "tourism invasion"—that overwhelms the destination and results in negative impacts on the local communities, the environment, and the visitors' experience. The definition of overtourism by Milano et al. [29] and García-Buades et al. [30] lies in the same direction, perceiving overtourism as a state of an excessive growth of visitors, which leads to overcrowding in specific areas, while, in turn, renders residents the "victims" of the impacts of temporary and seasonal tourism peaks that cause permanent changes in their lifestyles, limit their access to amenities, and damage, in general, their well-being. The social dimension is also stressed by Saveriades [31], claiming that overtourism is associated with the maximum level of use (in terms of numbers and activities) that can be absorbed by a destination so that an acceptable decline in the quality of experience of visitors and an acceptable adverse impact on the host community can be ensured. Placing emphasis on the psychological dimension of overtourism, Goessling et al. [27] conceptualize overtourism as a psychological reaction to tourist pressure, which is the result of the damage of "place-residents" interrelationships and causes the shift of residents' attitudes towards tourism. Capocchi et al. [18] stress the subjective nature of overtourism, where locals, tourists or both realize a destination as having that number of tourists, which can change its character, affect its authenticity, and lead to irritation and annoyance. Milano et al. [32] claim that overtourism can be grasped as a multifaceted and self-perpetuating system, placing priority to economic growth through mass tourism and gradually mobilizing a "visitor economy" that leads to economic restructuring at the destination level. In this respect, questionable unfair practices are unleashed, e.g., rent expropriation, labor exploitation, displacement and expulsion of indigenous community, and natural resource extraction.

In seeking to achieve sustainable tourism objectives at the destination level, in alignment with the 2030 SDGs, the issue of overtourism is currently at the heart of the scientific research discussions, and is a central matter at the community and decision-making level in overcrowded destinations [18]. Despite the recent, exponentially expanding literature that has been produced on this topic [33], displaying relative concerns, the overtourism context still remains obscure and lacks a clear-cut, widely accepted definition, as well as a well-documented and commonly agreed methodological approach (steps, criteria, and related indicators) in order for this to be identified and assessed. Thus, no general guidelines for determining a destination's tourism capacity are identified as each tourism destination environment can display multiple relevant thresholds [15]. Furthermore, since overtourism is subject to various interpretations, emanating from different viewpoints and disciplines [34,35], a certain kind of disagreement with regard to policy reaction is also noticed.

That said, the present paper attempts to add to the growing literature by indulging into the overtourism concept and elaborating on methodological concerns and relevant indicators for assessing, managing, and monitoring overtourism or the risk of overtourism as the means for relevant decision-making to be facilitated. The contribution of this work lies in: systematizing current methodological endeavors into a coherent, integrated, of universal use, sound and planning-oriented methodological framework; and identifying relevant indicators' groups that can support the implementation of this framework. This, apart from the main two pillars of such an exploration, i.e., tourist demand and supply, which is mostly the ground of relevant studies in the literature, also attempts to highlight and quantify spatial, environmental, land use, economic, etc., impacts of overtourism that can potentially lead to the degradation of the tourist destination per se and the loss of those critical attributes upon which a destination's attractiveness is based [36,37]. In addition, considering the limited number of overtourism assessment studies that attempt to place the impacts of the tourism sector's growth (demand and supply) in the spatial context of the highly vulnerable insular territories by means of related spatial repercussions, this work attempts to partially fill this gap and enrich the current literature. Towards this end, it illuminates the methodological ground and produces empirical evidence from its application in a specific case study, Santorini Island, namely, a highly-rated tourist destination, located in the Southern Aegean Region, Greece that lies at the top positions of destinations characterized by overtourism [38–41]. The methodological framework and data elaboration procedures used for the indicators' analysis are designed in a way to permit replication to other case study areas, thus adding to the universal applicability of this research work.

The paper is structured as follows: in Section 2, a literature review is carried out, aiming to identify/classify sets of indicators that can be of relevance for exploring the overtourism condition; in Section 3, the methodological framework developed in this work, the selected sets of indicators, as well as the data sources are shortly outlined; in Section 4, the dynamics of the tourism sector in the study region are delineated; in Section 5, an assessment of overtourism of the selected case study—Santorini Island—is demonstrated; in Section 6, a discussion of the results is carried out; and finally in Section 7, the key conclusions are drawn.

2. Literature Review on Indicators for Assessing Overtourism

As noticed by Butler and Dodds [9], the vulnerability of islands to overtourism is largely grounded in their geographical size and location, size and cohesion of local population, accessibility and links to tourist markets, as well as relative power or control over their own development. That said, an assessment of overtourism per se is not a straightforward task [24,33]. Moreover, a certain gap is noticed in terms of both a common approach for such an assessment and a commonly-defined and widely accepted set of indicators that can be used in this respect. This, in turn, raises obstacles to conducting, on a common ground, comparative studies among various destinations and documenting potential thresholds, beyond which the tourism pattern in a specific destination can be featured as "overtourism" [33,36,42].

In seeking to feature types of indicators that are relevant when exploring the overtourism state, a literature search on the topic is conducted, highlighting methodological pathways and related indicators on the one hand, as well as case study examples and the pool of specific indicators that are applied, on the other hand. Following this search, a range of indicators' groups can be identified that are pertinent to the exploration of the overtourism state or risk (Figure 1). Further to these groups, it has to be stressed that the overtourism assessment has to follow a place-based stream that takes into consideration the particular attributes of each single destination, e.g., natural and cultural resources, value system, local economic structure, attributes of host population at hand, etc. The locally-induced political power for controlling future pathways is also critical in this respect [9]. In addition to the aforementioned, it should be kept in mind that the complexity of assessment of the overtourism state per se as well as its high reliance on each single destination's peculiarities introduces uncertainties as to the establishment of an upper threshold of tourists, beyond which a destination's state is at risk of overtourism [36].

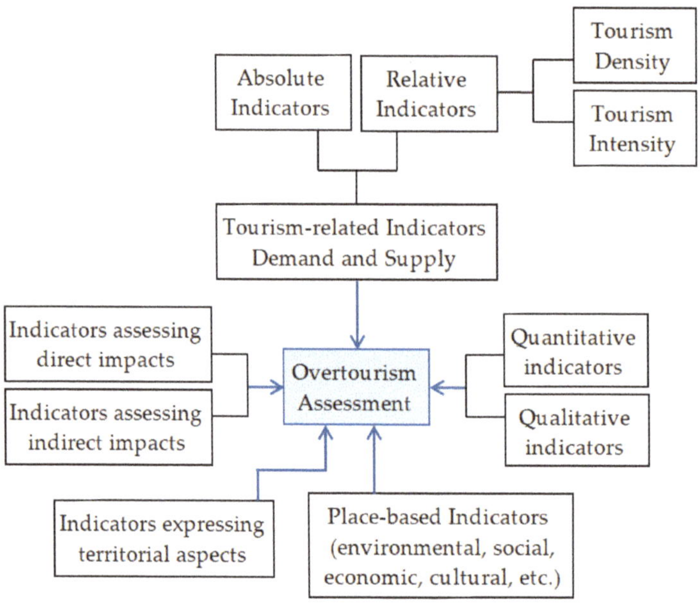

Figure 1. Types/groups of indicators for assessing the overtourism phenomenon. Source: Own elaboration.

Following this classification of indicators, overtourism assessments can follow both *quantitative and qualitative* paths, where the former is established on a set of quantitative indicators, e.g., tourism accommodation in the destinations in terms of the number of hotels or beds [17] or land surface consumed due to urban sprawl in a destination [36], while the latter is realized through more qualitative data, e.g., attitudes of host community of the destination against tourism/visitors or tourists' perceptions about the destination [30] as the means for grasping the number of visitors a destination can bear [43].

A literature review also demonstrates one additional group of indicators for assessing overtourism—actually a quite important one, as demonstrated by the various literature-based case studies—that is purely linked to tourism activity per se in a destination, including both the demand and the supply side. These indicators are further distinguished to those marked as "absolute indicators", e.g., the number of beds of a certain destination, which can be used for, e.g., assessing tourism dynamics in this destination, and those that relate tourism variables to other general aspects in the destination, attributed as "relative indicators" (Figure 1) [33], which is mainly used for overtourism assessments. The group of relative indicators is further discriminated into two broad categories, namely, a first one that expresses tourism intensity and a second associated with tourism density. Tourism intensity indicators are those emanating from absolute indicators divided by the resident population of the geographical area concerned, and display the pressure of tourism on the population. Those demonstrating tourism density emerge from absolute indicators divided by the size of the geographical area concerned, and are used to ascertain the pressure of tourism on the territory. In addition, indicators identified in the literature can be further distinguished to those that assess vulnerability to overtourism, e.g., the rate of GDP reliance on tourism or territorial aspects of overtourism, e.g., urban sprawl due to tourism accommodation and infrastructure deployment, loss of agricultural land in a destination. Finally, indicators for overtourism assessment are classified into those expressing direct impacts of overtourism, e.g., ecological damage, land use changes due to tourism infrastructure deployment, loss of agricultural land, etc. [7], and the ones highlighting indirect impacts of overtourism, e.g., degradation or loss of cultural heritage [44], displacement of communities from certain core tourism settlements due to rising costs or congestion [7], loss of social cohesion of the host population [5,36].

3. Methodology and Data

3.1. Methodology

The issue of overtourism is currently highly rated in the agendas of policy makers, planners and local administration/communities. That said, within the scope of this paper, the development and implementation of an integrated, planning-oriented, of universal use, methodological framework (Figure 2) is presented, which can act as a guide for exploring overtourism signs or state in various tourism destinations.

The proposed framework (Figure 2) is grounded in:

- The adaptation of and capitalization on relevant literature works [14,17,33], in the effort to establish a coherent and conceptually sound *methodological framework* for overtourism exploration and identification.
- The very essence of the tourism phenomenon and its main constituents, i.e., supply, demand, and destination's tourism and other infrastructure as well as their impacts on the particular natural, cultural and human context of each single destination.
- The *PSR—Pressure/State/Response model* as a structured, well-proven and robust framework [45] for better understanding actions and activities that are affecting the state of a system, and are feeding an appropriate policy response for addressing them. This builds upon a range of properly selected indicators and can orient the evidence- and place-based policy reaction of, among others, the overtourism phenomenon.

The starting point of this framework is associated with the destination itself and more specifically with the *driving forces*, upon which its attractiveness is relied, reflecting thus the place-based rationale of such a framework. As driving forces of a destination are considered the: (i) natural and cultural assets, which actually constitute the main motive for tourism attraction; (ii) unimpeded accessibility, implying options available for smoothly reaching the destination; (iii) tourism accommodation infrastructure and related entrepreneurship (food, facilities, etc.)—the tourism supply—for serving the needs of tourist flows; and (iv) public infrastructure per se (transport networks, health infrastructure, water and energy infrastructure, etc.), which is also part of the quality of tourism experience in a destination. A positive assessment of those driving forces by travelers are largely defining the level of tourism demand while, in turn, developments in the destination (e.g., further deployment of public infrastructure or accessibility networks, expansion of accommodation infrastructure), coupled with the rising volume of tourism flows (rising tourism demand) are the main sources of impacts occurring on the natural, cultural and social scenery of the destination. In addition, based on temporal data related to tourist supply and demand, the dynamics of the tourism sector in the destination can be featured.

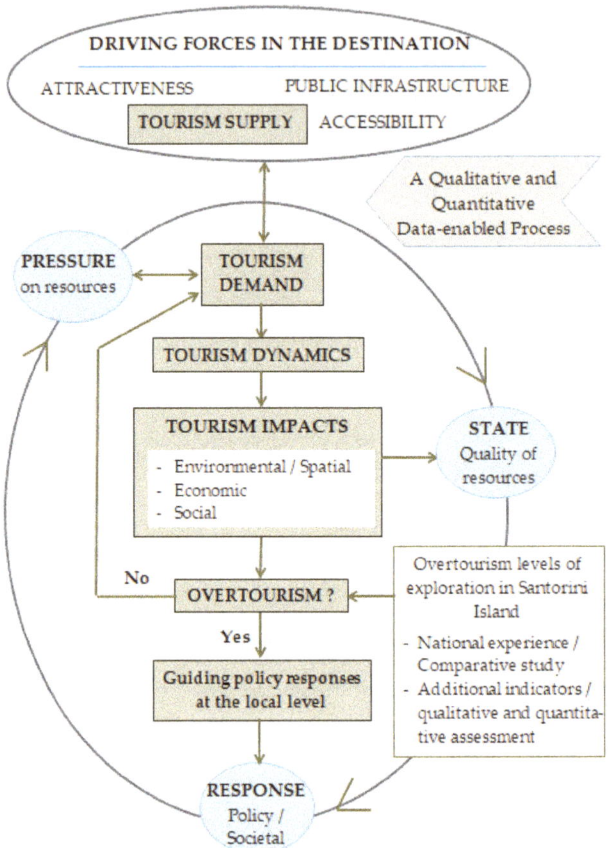

Figure 2. Steps of the methodological framework for assessing and monitoring overtourism. Source: Own elaboration.

The next step of the proposed framework aims at identifying—both quantitatively and qualitatively—the impacts of the aforementioned components on the destination. These refer mainly to impacts on the valuable natural and cultural assets (degradation of ecosystems, landscape, cultural heritage, etc.), the spatial pattern (land use changes), as well as the economic context (changes in the economic profile and prevalence of the tertiary sector) and social structure (loss of local identity, social cohesion, etc.) of the destination. In order for such an assessment to be accomplished, a relevant set of indicators, drawn upon the international literature/experience is used, where indicators are selected and properly adjusted to reflect the specific peculiarities of the case study region.

Based on the outcome of the above-mentioned step, the response to the question on whether the destination under examination is suffering from overtourism is based on a two-stage approach, grounded in the following: (a) national experience through comparative study and related outcomes, and (b) combined interpretation of qualitative and quantitative results, emanating from the selected indicators. In the case of a negative response as to whether an overtourism state is noticed, a steady monitoring of the destination is following in order for early overtourism signs to be identified and be properly addressed in the policy realm. In the opposite case, i.e., the destination suffers from overtourism, the proposed framework is capable of explicitly addressing the critical *overtourism dimensions,* and thus properly guiding the policy (or community) reaction to effectively cope with them.

The implementation of the aforementioned framework in this work uses as a case study the island of Santorini, South Aegean Region, Greece. Taking into account that this specific island is rating high in the lists of destinations around the globe that suffer from overtourism [38–41], Santorini Island consists of a fruitful ground for validating the usefulness and effectiveness of the proposed framework and documenting Santorini's overtourism context.

3.2. Data and Indicators

Drawn upon literature review, the list of indicators used in this work for assessing the main components of the proposed framework, i.e., tourist supply, demand and their impacts on the environment, economy, society as well as space (Figure 2) are demonstrated in Tables 1–3. The majority of indicators of these tables are falling into two categories, namely, tourism intensity and tourism density (Tables 1 and 2—11 indicators). This indicators' set is further expanded to incorporate additional indicators (Table 3), which are grasped as useful for exploring environmental/spatial (six indicators), economic (four indicators), and social (three indicators) impacts of tourism in Santorini Island. A geospatial analysis is conducted in this respect in order for the spatial implications of tourism to be identified and quantified. It should be noted here that the permanent population data, used in this work, refer to the year 2021 (National Census Data). However, for the purpose of calculations in Tables 1 and 2, conducted for the years 2023 and 2022, respectively, the population data in 2021 are used, assuming that the permanent population of the study area remains unchanged.

Table 1. Tourism supply (TS_i) indicators.

a/a	Indicator	Description of Indicator (Calculations in the Year 2023)
1	TS1—Tourism Intensity Index	[Total Tourism Accomodation (Hotel and Airbnb Beds)/Permanent Population (Residents)] TS1 features the pressure on population that is due to tourism accommodation infrastructure.
2	TS2—Airbnb Intensity Index	[Total Airbnb Beds/Permanent Population (Residents)] TS2 highlights the pressure on population due to the Airbnb tourism supply.
3	TS3—Tourism Infrastructure Indensity Index	[No. of Hotels and Airbnb Accommodation/Permanent Population (Residents)] TS3 assesses the pressure on population that is due to tourism infrastructure deployment.
4	TS4—Tourism Index	[Total Tourist Accomodation (Hotel and Airbnb Beds) * 100]/[Permanent Population (Residents) * Total Area in km^2] TS4 sketches the tourism saturation level.

Table 1. *Cont.*

a/a	Indicator	Description of Indicator (Calculations in the Year 2023)
5	TS5—Tourism Infrastructure Index	[(No. of Hotels and Airbnb) * 100]/[Permanent Population (Residents) * Total Area in km^2] TS5 demonstrates the pressure of tourism infrastructure on the destination (local population and area).
6	TS6—Tourism Infrastructure Density Index	[No. of Hotels and Airbnb/Total Area in km^2] TS6 reflects the spatial pressure on the destination that is due to the deployment of accommodation infrastructure.

* implies "multiplied by". Source: Own elaboration.

Table 2. Tourism demand (TD_i) indicators.

a/a	Indicator	Description of Indicator (Calculations in the Year 2022)
1	TD1—Tourism Intensity Index	[Total Hotel Arrivals * 100]/[Permanent Population] TD1 features the pressure of tourist flows on the population.
2	TD2—Attracti- veness Index	[Foreign Arrivals (Hotels)/Domestic Arrivals (Hotels)] TD2 expresses the attractiveness of a destination to foreigners compared to domestic tourist flows.
3	TD3—Tourism Penetration Index	[Overnights of Foreigners (Hotels) * 100]/[Permanent Population * 360] TD3 determines the penetration of foreign visitors in a place in relation to the local resident population.
4	TD4—Tourism Concentration Index (density)	[Total Overnight Stays (Hotels)/Total Area in km^2] TD4 demonstrates visitors' willingness to stay in a specific destination, and thus the pressure exerted on its local resources.
5	TD5—Tourism Overnight Stays Index (intensity)	[Total Overnight Stays (Hotels)/Permanent Population] TD5 demonstrates the potential pressure on the local population.

* implies "multiplied by". Source: Own elaboration.

Table 3. Indicators for assessing sustainability implications of tourism in a destination.

Indicator		Description of Indicator
a/a	Environmental/Spatial	
1	EN1—Environmental Vulnerability	[Percentage of the area falling into the NATURA 2000 network] EN1 displays areas with high natural value, manifesting vulnerability/risk of environmental degradation due to tourism activities.
2	EN2—Share of Built-up Space	[Percentage of built-up space in 2018] EN2 demonstrates the environmental pressure due to urban sprawl, i.e., the evolving built-up areas pattern due to mainly tourism accommodation deployment.
3	EN3—Change in Built-up Space	[Mean annual rate of built-up area change (2006–2018) multiplied by 1000] EN3 displays the evolving land use change patterns that are mainly due to developments in the tourism sector.
4	EN4—Vegetation Change Index	[Percentage of change (2017–2024)] EN4 features the loss of green space with repercussions in the microclimate and degradation of the landscape.
5	EN5—Built-up Areas within Protected Zones (Natura 2000, archaeological zones, etc.)	[Percentage of built-up areas within protected zones in 2018] EN5 highlights the penetrating nature of built-up areas due to mainly tourism demand, expanding into protected zones.
6	EN6—Built-up growth within Protected Zones (Natura 2000, archaeological zones, etc.)	[Mean annual rate of built-up areas within protected zones in 2006–2018 multiplied by 1000] EN6 displays the mean annual rate of the aforementioned expansion.

Table 3. *Cont.*

Indicator		Description of Indicator
a/a	Economic	
1	EC1—Available Agricultural Land	[Percentage of agricultural zones (2019)]
2	EC2—Agricultural Cultivation	Percentage of (wider) agricultural zones still cultivated
3	EC3—Loss of Agricultural Land	[Percentage of agricultural land change (1990–2018)] EC3 features the pressure of mainly tourism on the agricultural sector, placing at stake traditional production patterns, especially in insular destinations.
4	EC4—Employment Pattern	[Share of employment percentage (primary, secondary and tertiary sector, rate of change 2001–2011, 2011–2021)] EC4 features the impacts of tourism on the employment pattern of a destination.
a/a	Social	
1	S1—Large-scale Tourism Infrastructure	[Percentage of built-up space in 2018 that is attributed to large-scale tourism infrastructure] S1 features the social/psychological pressure due to large-scale tourism development and relevant land-take.
2	S2a—Loss of Housing Affordability	[Ratio 1: Ratio between land rent for housing in the destination and mean value at the territorial level] S2a features the pressure on permanent residents that are not able to find affordable housing due to tourism pressure on land values.
3	S2b—Loss of Housing Affordability	[Ratio 2: Ratio between land sale value in the destination and mean value at the territorial level] S2b features the pressure on permanent residents that are not able to acquire their own house due to tourism pressure on land values.

Source: Own elaboration.

Data with regard to the indicators used in this study are gathered from the sources displayed in Table 4. It should be noted here that all data are georeferenced to the LAEA-Europe projection system (ETRS89-extended—EPSG:3035), while the territory of Santorini Island is split into six local zones, based on aggregations that are drawn upon the official administrative boundaries (https://www.statistics.gr/digital-cartographical-data) (accessed on 15 November 2024) [46].

Table 4. Data sources.

Type/Source of Data
Imperviousness Classified Change Layer (raster 20 m) covering the period 2006–2018, 3-yearly provides, at the pan-European level, the degree of imperviousness change in the categories of sealing change (unchanged no sealing, new cover, loss of cover, unchanged sealed, increased sealing, decreased sealing) between each of the two Imperviousness Density status layers [47].
Archaeological zones: obtained from the geospatial platform of the Archaeological Cadastre of Greece, recording and documenting Greece's immovable monuments, archaeological sites, historical sites, and their protection zones [48].
Natura 2000 areas and Wildlife Repositories of Greece: obtained from the Geodata portal of Greece [49]. The "Kaldera" volcanic zone of Santorini Island is simultaneously part of both networks, and is also recently at the exploration stage as a candidate to the UNESCO World Geological Sites.

Table 4. *Cont.*

Type/Source of Data
Agricultural land change: obtained for a large time span, based on Corine Land Cover database availability, DOI (vector 2018) [50], and DOI (vector 1990) [51]. Codes 221, 242, and 243 are applicable in Santorini Island and the changes are identified by comparing all relevant classes. To more accurately estimate the remaining agricultural zones in 2019, a fine resolution global dataset is used, namely the COPERNICUS/Landcover/100m/Proba-V-C3/Global/2019 (km). To identify agricultural land still cultivated (vineyards and arable land) and their relationship to artificial areas, the Coastal Zones Land Use/Landcover Database [52] and part of the European Union's Copernicus Land Monitoring Service are used.
NDVI (Normalized Difference Vegetation Index): obtained by both Landsat 8 satellite images, at 30 m resolution and Harmonized Sentinel-2 MSI, Multispectral images, at 10 m resolution. Data are processed through Google Earth Engine, and the difference of NDVI between 2016 and 2024 (Landsat 8) and 2017 and 2024 is obtained at the pixel level. NDVI values are obtained at each case as the maximum pixel value during the summer period of each reference year.
Population: data on the permanent population are derived from the 2021 National Census of the Hellenic Statistical Authority [53].
Tourism arrivals: derived from the annual report of the Institute of the Association of Greek Tourism Enterprises (INSETE) [54]. The Annual Report of each Region/District includes Demographic data, Macroeconomic data, Tourism infrastructure, Arrivals-Night stays and Revenues, key indicators (Average Per Capita Expenditure, Expenditure per Night and Average Length of Stay), Employment, etc. It should be noted here the lack of data on Airbnb arrivals.
Airbnb accommodation: gathered through the open data source platform [55].
Hotel accommodation: emanates from the Hellenic Chamber of Hotels [56].
Land values: estimated through the commercial real estate platform Spiti24 [57], by defining a specific zone and obtaining mean values for rental and sale.

4. Featuring the Evolving Tourism Dynamics in Santorini Island

Santorini Island (or Thera) is part of the Cyclades complex, Southern Aegean Region, Greece (Figure 3). Administratively, the island belongs to the Municipality of Thera, which, apart from Santorini Island—the main tourist destination of the municipality—it also incorporates the smaller island of Thirasia and the uninhabited islets of Palaia Kameni, Nea Kameni, Aspronisi, Christiani, Askania, and Esxati. The area of the municipality is 90.69 km^2 and its population amounts to 15,480 inhabitants, according to the 2021 Census. It is worth noticing the remarkable increase in the municipality's population in the time span 2001–2021 (+12.79%) which surpasses the rising population trends of the Region of Southern Aegean (+9.84%), while in contrast to the generally declining population trend of the Greek country as a whole (−4.13%).

The island of Santorini (or Thera) is currently a highly-reputed tourist destination on a global scale, featured by remarkable natural and cultural assets, and a special geomorphology, witnessing its volcanic origin. The whole area of the island is marked as a "Landscape of Outstanding Natural Beauty" (Code AT5011063), while the same holds for the island of Thirasia and the islets of Aspronisi, Palaia Kameni and Nea Kameni (Code AT5010091). In addition, Santorini Island, as well as the whole area of Thera Municipality, have a rich cultural heritage, incorporating historical/archaeological sites and traditional settlements. Moreover, the agricultural production has a long tradition in the island, which is so far well known for its Protected Designation of Origin (PDO) products, such as wine, small tomatoes and fava bean that have, among others, strengthened the island's extroversion.

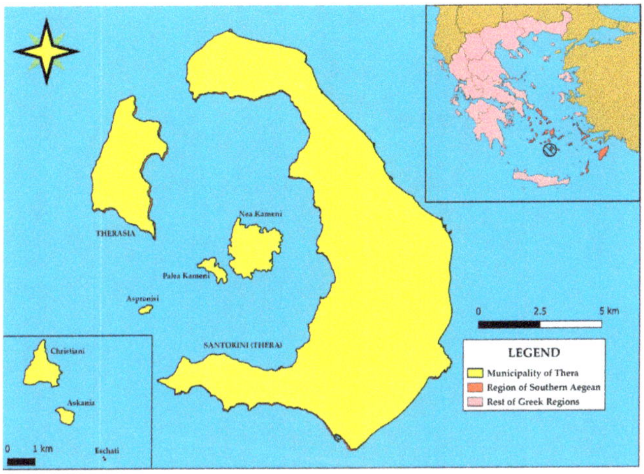

Figure 3. Location of Thera Municipality in the Greek territory. Source: Own elaboration.

Taking into consideration that overtourism is the outcome of a rapidly evolving tourism demand and supply in a destination [33], the dynamics of the tourism sector—demand and supply—in Santorini Island are explored as a first step in this work, seeking to grasp their evolution over time. Towards this end, a range of absolute indicators as to the tourism demand and supply are investigated, demonstrating the growth rates of the tourism sector in the study region as well as their spatial distribution. More specifically, the evolution of tourism supply and demand is delineated by the use of the absolute indicators demonstrated in Table 5.

Table 5. Indicators used for exploring tourism dynamics in Santorini Island.

Tourism Supply	Tourism Demand
No. of hotels 2012–2023	Arrivals by air 2010–2022
No. of hotel rooms 2012–2023	Arrivals in hotels 2010–2022
No. of hotel beds 2012–2023	Arrivals by cruise ships 2013–2022
No. of Airbnb beds 2023	No. of cruise ships 2013–2022

In Figures 4–6, the dynamics of tourism supply in the time span 2012–2023, its spatial pattern in 2023, and the built-up space in Santorini Island in 2018, respectively are delineated. More specifically, the number of hotel units in the Municipality of Thera seems to follow a strongly increasing course in the time span 2012–2023, as shown in Figure 4a, where the majority of hotel accommodations of the Municipality of Thera are located in Santorini (Thera) Island. The same holds for hotel rooms and beds (Figure 4b,c). Actually in the period 2012–2023, a noticeable increase of 16.7%, 37.0%, and 43.13% in hotel units, rooms and beds, respectively has occurred, with a remarkable prevalence of the 5* hotel category. The tourism supply of the Municipality of Thera in general and Santorini Island in particular is also further enriched by the rapid increase in Airbnb accommodation. In 2023, the number of Airbnb accommodations rises to 4783 units and 10,750 beds [57], which seems to be quite a large number, taking into consideration that the Athens metropolitan capital disposes 11,314 Airbnb beds.

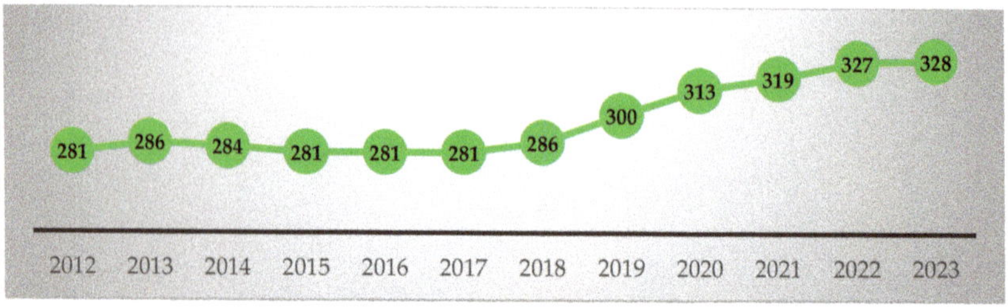

(**a**) No. of hotels

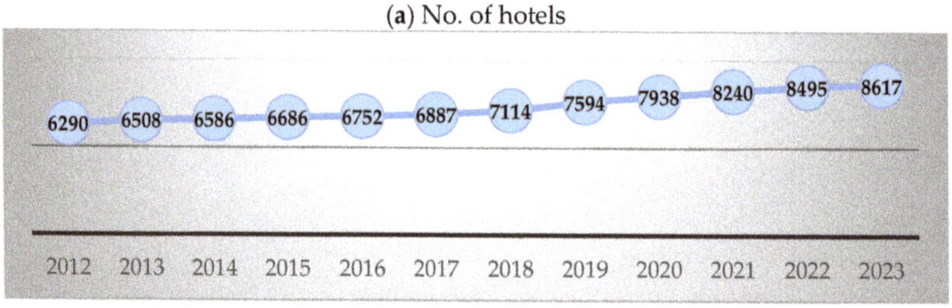

(**b**) No. of hotel rooms

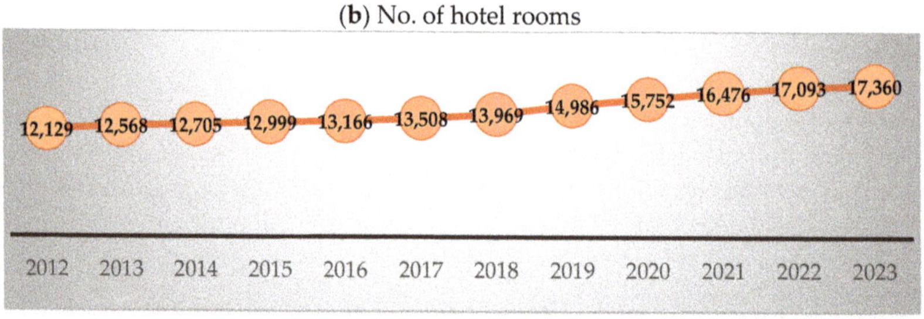

(**c**) No. of hotel beds

Figure 4. Dynamics of tourism supply in the time span 2012–2023. Source: Own elaboration based on data from the Institute of the Association of Greek Tourism Enterprises [54] and the Hellenic Chamber of Hotels [56].

In fact, it is also worth noticing that Santorini Island owns the largest number of Airbnb accommodations (4783 units), compared to any other island of the Cyclades complex and the Southern Aegean Region as a whole, and is actually followed by the islands of Mykonos (3418 Airbnb accommodations) and Paros (3134 Airbnb accommodations), where these three islands—Santorini, Mykonos and Paros—are marked as the flagship of Airbnb development in the Southern Aegean Region. Furthermore, the locational choice of the bulk of hotel and Airbnb accommodations in Santorini Island displays a similar spatial pattern (Figure 5a,b, where the Kernel density estimation calculates the density for the whole hotels and Airbnb dataset by applying a quartic Kernel shape and a radius of 1000 m so that the best visual representation of these data can be attained). The same holds for the pattern of built-up space in 2018 (Figure 6) that is in line with the escalating pressure for tourism accommodation infrastructure, serving the rising tourist demand in Santorini Island. In addition, in Figure 6, the pressure exerted by tourism accommodation on the arable land and vineyards in Santorini Island is evident, as well as the deployment of tourism accommodation in proximity or inside protected natural areas and archaeological sites.

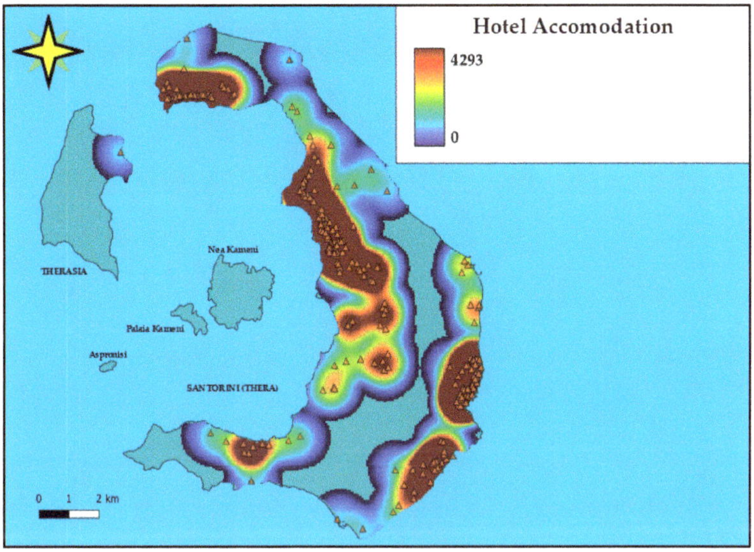

(a)

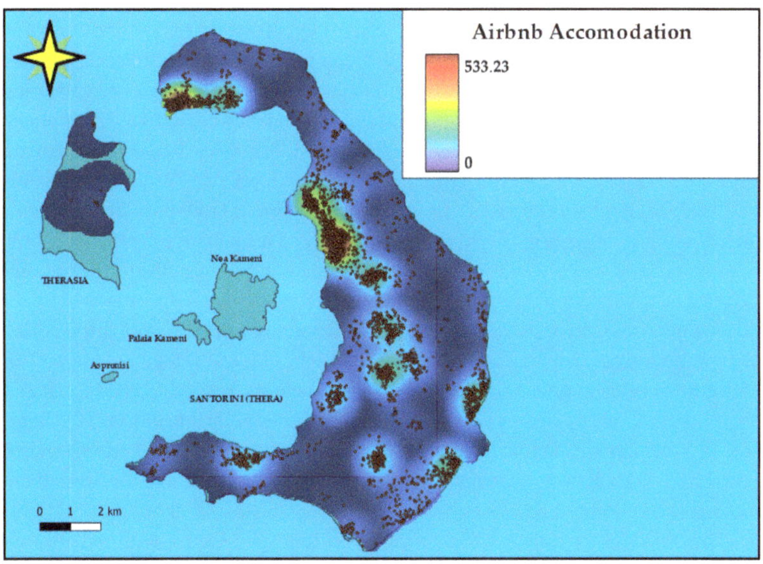

(b)

Figure 5. Spatial pattern of tourism accommodation. (**a**) Kernel density estimation of hotel accommodation (radius 1000 m) in 2023. Source: Own elaboration based on data from the Hellenic Chamber of Hotels [56]. (**b**) Kernel density estimation of Airbnb accommodation (radius 1000 m) in 2023. Source: Own elaboration based on data from the Airbnb Platform [55].

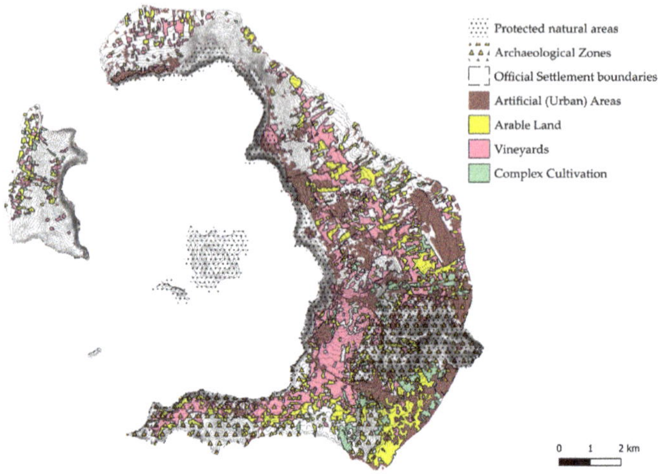

Figure 6. Pattern of built-up space in 2018. Source: Own elaboration based on Coastal Zones Land Cover/Land Use 2018 [52].

The aforementioned built-up pattern of Santorini Island can be also partly justified by the spatial planning system in Greece. More specifically, spatial planning in Greece, as evolved during the '90s, follows a top-down hierarchical scheme. The containment of urban sprawl is set as a major priority target under the National Framework of Spatial Planning and Sustainable Development that is accompanied by a set of Special Spatial Planning Frameworks. However, the absence of a coherent regional and local planning system throughout most of the 20th century has led to the tendency of built-up development "outside of the city plan areas". In the effort to control and/or confine this tendency, urban sprawl has been regulated by a Presidential Decree dating back to 1978. This basically permits urban development in almost all private as well as public land, provided that some requirements are met regarding the size of the plots. The minimum plots' size is set to 4000 m^2, with specified exceptions for smaller plots, depending on, e.g., the formation date of the property, combined with direct accessibility to the road network. As estimated by the Technical Chambers of Greece, 30% of illegal construction declarations during the period 2011–2020 concerns areas that are still not embedded in the city and settlement plans. This percentage corresponds to over 1 million illegal constructions, related to residential and second house formations as well as various tourism and commercial facilities. The illegal development particularly expands in areas where infrastructure is mostly inadequate, and in close proximity with forests, stream banks, and the coastline. Since the '80s, illegally or legally-built constructions are known to largely penetrate the insular landscape of highly touristic islands, like Santorini and Mykonos. In addition, in coping with the absence of established General Urban Plans in many territories of Greece, zones of Settlement Development Control have been implemented since the '90s, aiming at articulating land use constraints in ex-urban space. These zones have been used in priority for areas with severe environmental degradation problems including, among others, coastal areas in a number of islands (Crete, Mykonos, Kos, Santorini, Samos, Andros, etc.).

In Figure 7, the dynamics of tourism demand in Santorini Island are roughly sketched. More specifically, arrivals by air in the time span 2016–2023 (Figure 7a) show a strong upward trend, demonstrating an increase of 62.61% in this period, with a short break in the COVID-19 period. The same holds for arrivals in hotel accommodation (Figure 7b), which in the time span 2010–2022 demonstrate a remarkable increase of 487.84%. On the contrary, arrivals by cruise ships in Santorini Island in the time span 2013–2022 (Figure 7c) display a more stable pattern, rising in this period by 6.24%, while a certain escalating trend is noticed in the number of cruise ships arrivals in the same time span (Figure 7d), reaching an increase of 17.86%.

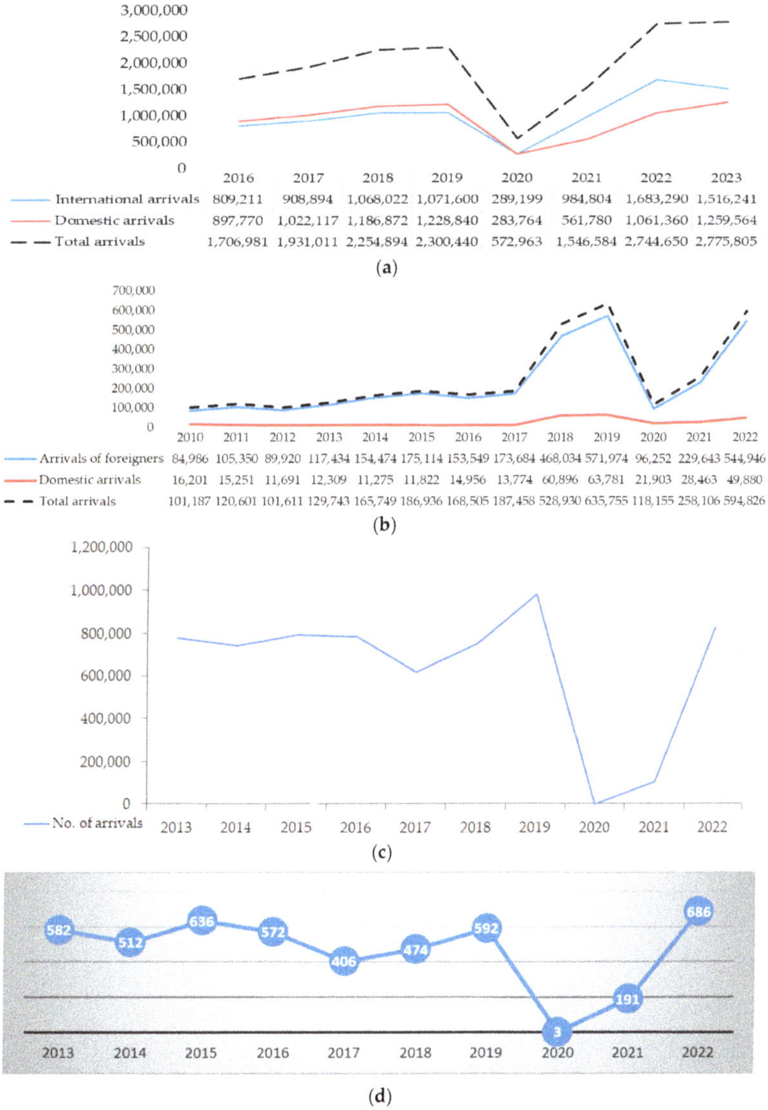

Figure 7. The dynamics of tourism demand in the Municipality of Thera. (**a**) Arrivals in Santorini airport in the time span 2016–2023. Source: Own elaboration by the use of data from JTR-Airport.gr [58]. (**b**) Arrivals in hotel accommodations in the Municipality of Thera 2010–2022. Source: Own elaboration by the use of data from the Hellenic Statistical Authority (ELSTAT) [53]. (**c**) No. of arrivals by cruise ships in Santorini Island in the time span 2013–2022. Source: Own elaboration by the use of data from the Institute of the Association of Greek Tourism Enterprises (INSETE) [54] and the Hellenic Ports Association [59]. (**d**) No. of cruise ships in the time span 2013–2022. Source: Own elaboration by the use of data from the Institute of the Association of Greek Tourism Enterprises (INSETE) [54] and the Hellenic Ports Association [59].

The tourism dynamics in the island of Santorini, as previously depicted by means of absolute indicators, clearly demonstrate the escalating trends of the tourism sector—both in terms of supply and demand—in this island. In order to explore overtourism signs, however, absolute values of these indicators have to be scaled by the island's surface and size of population, thus placing them in relation to the destination's capacity. In addition, they need to be combined or co-interpreted with other indicators, a task that is accomplished in the following section.

5. Assessing Overtourism Signs in Santorini Island—Empirical Results

The impacts of the tourism sector in the island of Santorini are exhibited in this section by calculating values of the indicators presented in Tables 1–3. The results of such calculations are demonstrated in Tables 6–9 below.

Table 6. Assessment of tourism supply values (TSi indicators) in the Cyclades island complex.

Indicator / Island	TS1	TS2	TS3	TS4	TS5	TS6
Santorini/Thera	1.83	0.69	0.33	2.00	0.36	56.40
Mykonos	2.70	1.16	0.34	2.57	0.32	34.63
Folegandros	1.82	0.43	0.22	5.68	0.69	4.97
Ios	1.45	0.47	0.18	1.33	0.17	3.88
Paros	1.20	0.70	0.23	0.61	0.11	16.68
Naxos	0.76	0.41	0.16	0.15	0.03	6.51
Amorgos	0.67	0.38	0.17	0.53	0.13	2.57
Sifnos	1.07	0.68	0.24	1.44	0.32	8.86
Antiparos	1.01	0.68	0.19	2.24	0.43	5.38
Milos	0.74	0.40	0.20	0.46	0.12	6.60
Sikinos	0.49	0.34	0.12	1.14	0.28	0.71
Serifos	0.94	0.92	0.31	1.25	0.41	5.04
Anafi	0.45	0.37	0.16	1.12	0.40	1.16
Kea	1.14	0.89	0.22	0.76	0.15	3.47
Andros	0.47	0.32	0.11	0.12	0.03	2.54
Syros	0.25	0.13	0.05	0.24	0.05	10.23

Source: Own elaboration.

Table 7. Assessment of Tourism demand values (TD_i indicators) in the Cyclades island complex.

Indicator / Island	TD1	TD2	TD3	TD4	TD5
Santorini/Thera	3837.16	10.93	28.50	19,137.23	112.03
Mykonos	5501.23	11.22	49.01	19,369.64	190.34
Folegandros	3102.19	7.65	22.63	2114.82	94.76
Ios	2372.99	3.86	17.04	1669.13	79.15
Paros	976.69	3.08	8.17	2835.09	38.33
Naxos	616.25	2.87	5.37	1026.74	24.74
Amorgos	571.85	2.64	5.12	375.60	24.20
Sifnos	703.10	2.76	4.13	755.06	20.10
Antiparos	503.64	1.61	3.73	571.44	20.41
Milos	524.52	1.98	2.86	492.45	14.87
Sikinos	326.20	7.68	2.38	59.30	9.96
Serifos	485.01	1.42	2.24	219.01	13.27
Anafi	275.18	10.91	2.11	60.32	8.31
Kea	342.01	0.48	1.10	153.55	9.79
Andros	300.95	0.52	1.08	227.17	9.78
Syros	275.18	0.57	1.04	1819.11	8.78

Source: Own elaboration.

Table 8. Employment structure per productive sector in Santorini Island (indicator EC4).

Sector / Share (Rate of Change)	Share % in 2001	Share % in 2011 (% Change 2001–2011)	Share % in 2021 (% Change 2011–2021)
Primary	5.52	4.18 (−1.34)	3.67 (−0.51)
Secondary	25.72	17.63 (−8.09)	5.60 (−12.03)
Tertiary	68.77	78.18 (+9.41)	90.66 (+12.48)

Source: INSETE [60].

Table 9. Assessment of sustainability implications of overtourism in Santorini Island.

a/a	Indicator / Santorini Island	Total Area	Zone 1	Zone 2	Zone 3	Zone 4	Zone 5	Zone 6
EN1	Environmental vulnerability	23.0	19.8	29.1	21.4	33.5	66.8	**
EN2	Share of built-up space	9.2	5.7	8.0	15.9	14.1	4.9	1.0
EN3	Change in built-up space	3.1	2.9	7.2	2.1	1.7	0.3	0.0
EN4	Vegetation change index	−20.7	−15.3	−18.4	−28.1	−25.6	−23.6	−14.6
EN5	Built-up within protected zones	6.4	8.5	3.7	12.7	6.3	4.0	**
EN6	Built-up growth within protected zones	0.4	0.2	4.4	0.0	−0.3	0.0	0.0
EC1	Available agricultural land	53.1	57.3	47.0	55.4	57.7	60.8	40.1
EC2	Agricultural cultivation	56.4	23.8	46.7	62.5	76.4	69.7	38.5
EC3	Loss of agricultural land	24.3	40.7	15.0	20.4	15.0	23.8	62.0
S1	Large-scale tourism infrastructure	11.9	13.5	17.3	7.8	6.5	34.5	0.0
S2a	Loss of housing—Afford.1	1.56 *	-	-	-	-	-	-
S2b	Loss of housing—Afford.2	2.00 *	-	-	-	-	-	-

* Estimated only at the island level due to data availability constraints. Source: Own elaboration. ** This index was not estimated for Zone 6 due to the absence of protected zones in this part.

Based on these results, a two-stage assessment is conducted as follows (see Figure 2):

- Firstly, a comparative assessment is carried out, based on indicators' calculation—tourist supply TS_i and demand TD_i indicators—for all islands of the Cyclades complex in order to position Santorini Island in the immediate insular context, namely, the Cyclades Regional Entity (part of the Southern Aegean Region), which is a highly touristic insular entity of the Greek territory (Tables 6 and 7).
- Secondly, an integrated assessment of the state of Santorini Island is conducted, taking into account the values of indicators associated with the tourism supply and demand; sustainability indicators (environmental, economic and social); and the spatial pattern of tourism development and related implications (Tables 8 and 9).

More specifically, at the *first stage* of assessment, a comparative analysis is accomplished. In this respect, values of the proposed indicators TS_i and TD_i for the 18 Islands/Municipalities of the Regional Entity of Cyclades complex—Santorini is part of this complex—are calculated and compared (Tables 6 and 7). This analysis aims at highlighting the position of Santorini Island in terms of tourism density and intensity in this Regional Entity. An inspection of the results demonstrates that Santorini Island gains the highest or the second highest score (ranking second only after Mykonos) in the majority of TS_i and TD_i indicators (Tables 6 and 7). It is worth mentioning the value of TS6 (Table 6), which displays the pressure of tourism infrastructure (hotel and Airbnb) on the island's area. Actually, in TS6, Santorini Island is placed in the first position, at a large distance from the rest of the islands of the Cyclades complex, including Mykonos Island. Additionally, despite the pressure exerted on local resources and the resulting unsustainable state of the island, Santorini Island still remains a highly attractive destination, marked also by a high level of tourist service, as values of TD4 and TD5 demonstrate. Additionally, a high extroversion, as shown by the indicators TD2 and TD3, features the share of foreign tourists as the prevalent one in the island. Tourism flows, however, seem to exert high pressure on the local population (TD1). Furthermore, the high tourism intensity of Santorini Island—interpreted as pressure to the local population—compared to the rest of the islands of the Cyclades complex is clearly depicted by the value of TS1 (second position after Mykonos Island). Finally, Santorini Island falls into the group with a high tourism saturation level in terms of hotel and Airbnb beds with respect to the population and the surface of the island (TS4), and the group with high tourism infrastructure density (TS5). It should be noted here that if a comparison takes place only between the large islands of Cyclades complex in terms of population and surface, then Santorini Island in TS5 ranks first (smaller islands in terms of population and surface are definitely altering the calculation results in this respect). Here, it should also be noticed that the volcanic origin of Santorini Island renders potential areas for tourism infrastructure deployment rather limited, thus exacerbating the density of such infrastructure and related pressure on specific parts of the island.

In conclusion, the comparative analysis witnesses that Santorini Island follows a definitely intense tourism growth pattern, both in terms of supply and demand. The sustainability implications of this pattern need to be evaluated in order for a complete insight, interpretation, and assessment of overtourism state or risk to overtourism of Santorini Island to be sketched.

The calculations related to the *second stage*, targeting an integrated assessment of the overtourism state of Santorini Island, are presented in Tables 8 and 9. The results in Table 8 (EC4 indicator) demonstrate the prevalence of the tourism sector that is witnessed by the employment share of the tertiary sector. This rates quite high and displays an escalating trend to the detriment of the primary and secondary sectors, which demonstrate downturn signs through time.

In addition, based on the results of geospatial analysis of the indicators EN1-EN6, EC1-EC3, and S1 (see Table 9) for the six local zones that correspond to a spatial differentiation based on administrative boundaries [48], the following remarks can be drawn (indicative results are presented in Figure 8):

- The central and southern part of the island (Zones 3 and 4) are the most heavily impacted by construction, followed by the northern part (Zone 2 mainly, and next Zone 1).
- All zones, apart from Zone 6, are characterized by significant environmental vulnerability (maximized in Zone 5 that includes the archaeological zone of Akrotiri), while building within protected areas can reach up to 12.7% of land (Zone 4).
- Vegetation health and coverage (as captured by NDVI) is rapidly decreasing in the past years, with these attributes displaying a more rapid pace in Zones 3, 4, and 5.
- Agricultural land (as depicted by Corine Land Cover) is rapidly shrinking over the past three decades, rising to 25% on the island level and above 40% in the northern part of Santorini, as well as the Therasia land. A significant percentage of the remaining wider agricultural zones is now left uncultivated, especially in the central-southern part (Zones 3, 4, and 5).
- A large-scale tourism infrastructure that includes 4* and 5* hotels of significant bed capacity, pools and other auxiliary spaces, covering large plots, is estimated to a significant 12–18% coverage of total artificial land, located in the northern and central part, while such infrastructure dominates (35%) in the Akrotiri territory (Zone 5), where no significant local settlement is established.

- Santorini Island at an overall level presents a significant loss of housing affordability as the mean rental price is 56% higher than the average for Cyclades and Attika Region, while the mean price for sale is 100% higher. This is clearly manifested in the extreme difficulties reported especially for deputy school teachers to find a residence when they arrive to the island, but certainly it also affects various local population groups.

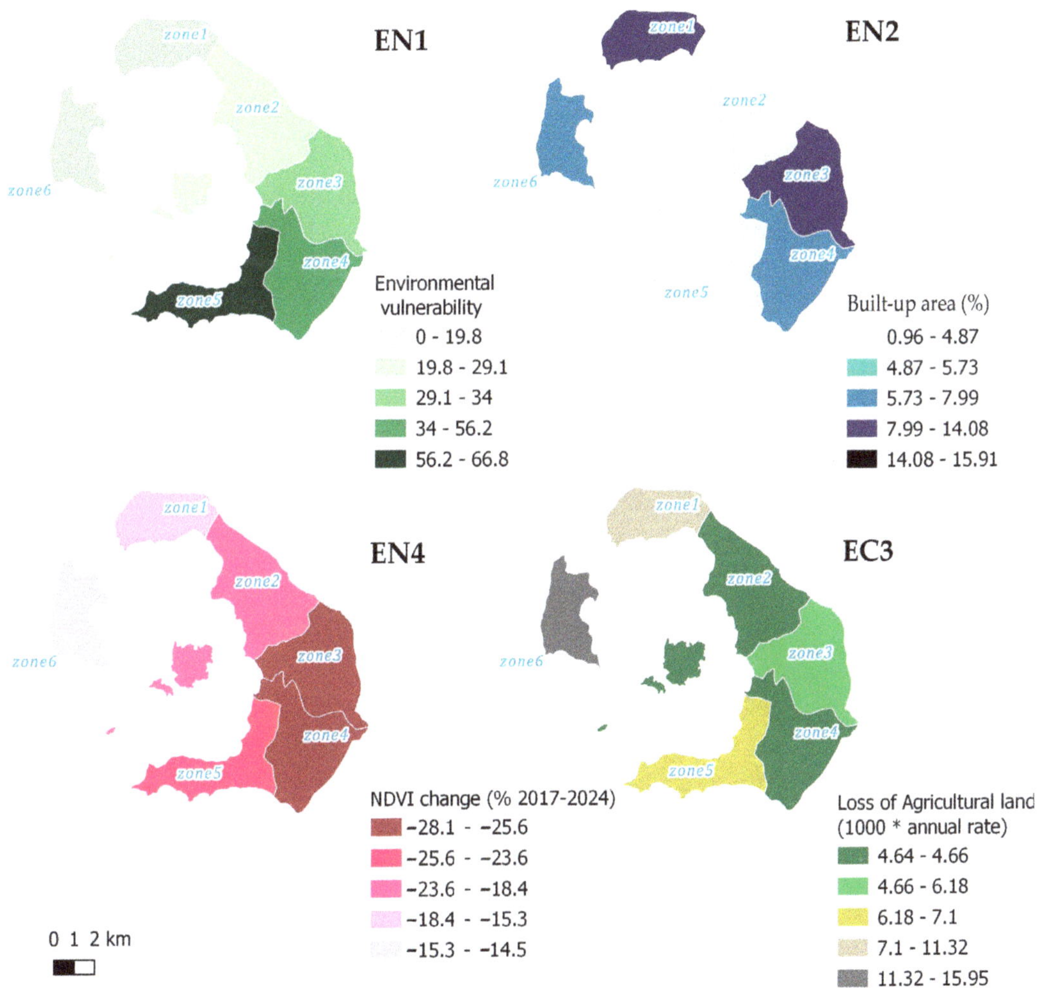

Figure 8. Mapping of the geospatial indicators EN1, EN2, EN4, and EC3 (as defined in Table 3) in the six administrative zones of Santorini Island (* in EC3 implies multiplication). Source: Own elaboration.

As shown by the EN3 indicator, built-up areas continue to expand, usually taking the form of large-scale (or in certain cases medium-scale) tourism infrastructure deployment in ex-urban areas. Moreover, a close inspection of the EN4 indicator (NDVI loss) shows that vegetation degradation and loss are considerably affecting the vineyards (Figure 9), i.e., a very important resource for the famous (PDO product) local wine of Santorini Island and a milestone of agricultural tradition. This degradation can be a combined effect of climate change and drought, as well as the sprawling of new constructions (or relevant preparatory works) taking place since 2017 in such fields (Figure 10). According to most indicators, the central part is very heavily affected by built-up areas, while significant pressure is currently identified in the northern part (Oia settlement) and towards the outstanding archaeological space of Akrotiri in the southern part of Santorini Island.

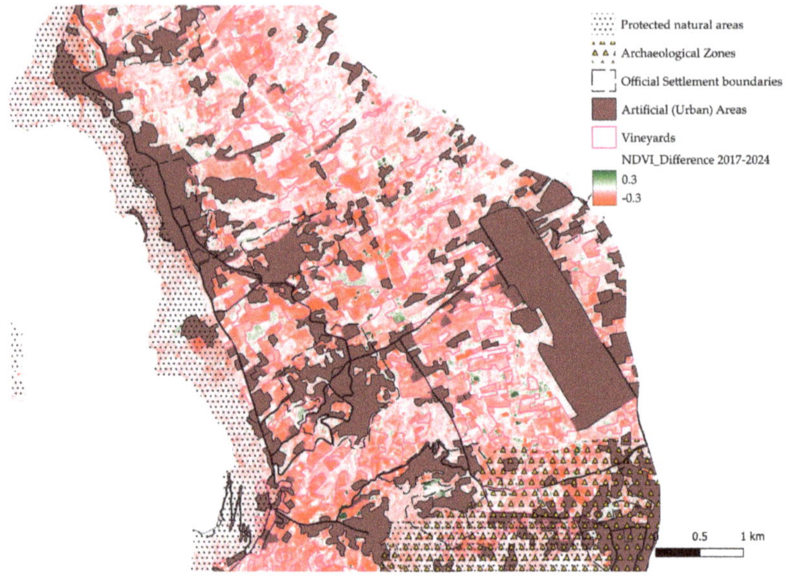

Figure 9. NDVI loss in comparison to vineyards in Santorini Island. Source: Own elaboration from Coastal Zones Land Use/Land Cover database [52] and Sentinel-2 images.

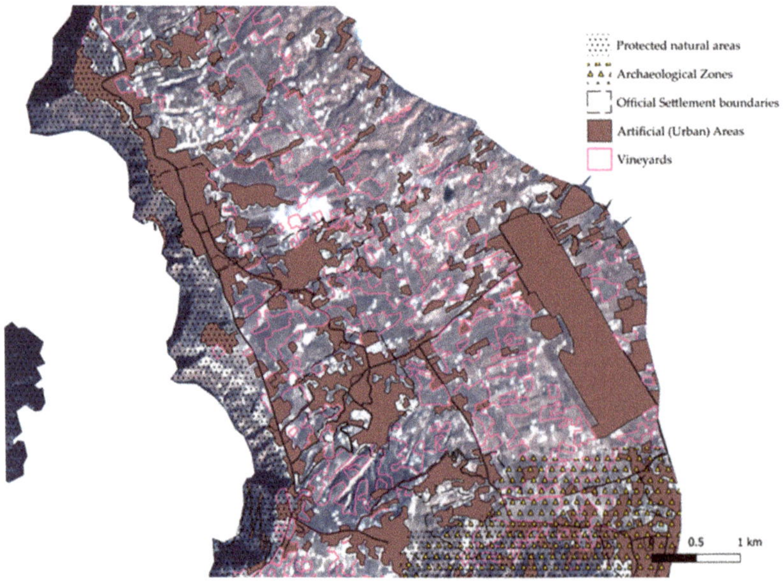

Figure 10. Overlay of the newest available Sentinel-2 image (September 2024) with the vineyards, showing some new built-up developments within the fields. Source: Own elaboration from Coastal Zones Land Use/Land Cover database [52] and Sentinel-2 images.

6. Discussion

In recent years, the integration of sustainability objectives into tourism planning and management has been advocated by the scientific and policy community, bringing forward a paradigm shift that is in alignment with the 2030 UN Agenda and the SDGs, and aims to achieve a balance among economic growth, environmental stewardship and social responsibility [61]. Such a shift stresses the importance attached to the adverse impacts—both quantitative and qualitative—of tourism overload in destinations, especially in those that are suffering from overtourism, and takes the form of a transition from "tourism growth"—in terms of the number of visitors, economic output, etc.—to "sustainable tourism" that delves into the interaction of tourism with other sectors, hosting communities, natural and cultural capital, as well as other policy directions at the destination level. Concurrently, the rising awareness as well as the experience of the repercussions of overcrowded destinations and related impacts on the quality of life of population has given rise to social movements at the community level, as a means to protest against the escalating tourism load in destination territories [14]. Remarkable examples of such movements have taken place especially in urban environments, such as Barcelona [61] or Venice [62], to name a few. However, as clearly stated by a range of research works, coastal, insular, and rural heritage sites seem to be more vulnerable to the overtourism phenomenon [17].

That said, the focus of the present work is on insular territories and particularly those lying in the Mediterranean Region, including Greek ones, as the world's leading sun, sea and sand (3S) mass tourism destinations and—as evidence shows—distinct examples for studying the overtourism condition. Indeed, insular territories as geographically constrained land parcels, where the insularity state is largely conditioning their developmental trajectory [5,8,36], are quite vulnerable to overtourism. These territories, in contrast to their insularity-related structural weaknesses, are endowed by a highly-valued natural and cultural capital that renders them: (i) unique and highly attractive places for tourists; (ii) highly competitive destinations in the global tourism scenery; and (iii) challenging places for both global and local real estate interests, tourism-oriented entrepreneurship and establishment of a "visitor" economy. As a result, tourism development in insular regions has been perceived as the cornerstone for coping with developmental barriers, and a way out to the flourishing of local economies and communities by the wealth of employment opportunities tourism provides as a labor-intensive sector. The so far continuously escalating tourism influx in Mediterranean insular territories, however, has through time unveiled its detrimental consequences, raising today severe sustainability and resilience concerns. This holds even more true in the context of today's quite challenging external decision environment, featured by climate change threats [63], water scarcity [64], rapid urbanization of coastal space and urban sprawl [28], to name a few. Within such a scenery, the concept of overtourism in insular territories gains much attention, and is presented by many as "an old wine in a new bottle" [22] or a new term to describe an already known, although at present highly exaggerating, tourism overcrowding trend. Concurrently, it constitutes today a new task in the scientific and policy agenda by means of assessing such a state and its implications in the highly sensitive and fragile insular territories. Such an assessment aims at highlighting the key dimensions of overtourism in each single destination—e.g., environmental deterioration, water stress, urban sprawl, loss of agricultural land, energy deficit, structural changes in the local economic profile, threatening of natural/cultural heritage, loss of authenticity and local identity, rising of living costs and real estate, damage to land- and sea-scape, economic inequalities and social exclusion—in order for a relevant "win-win-win" (local community, visitors and natural/cultural environment) policy reaction to be articulated for coping with the multiple repercussions of overtourism and remedying sustainability inefficiencies, developmental lag, and identity spoiling of insular territories.

In such an effort, a literature research unveils certain critical issues that seem to be the source of overtourism in insular territories. A first and very critical one relates to the developments of the sharing economy and more specifically to the peer-to-peer accommodation providers, such as the case of Airbnb, Booking.com, and Uber. As Hall et al. [65] claim, the globally-dominant platform-based intermediaries can severely affect the rising of tourist flows and their spatial implications in destinations. In fact, Airbnb is often blamed for some of the largest impacts in the accommodation sector and destinations [33,65], mainly due to the fact that it remains, in most of the cases, neither regulated nor monitored. In addition, the role of Airbnb is further intensified by the use of the Information and Communication Technologies (ICTs), playing a crucial role as an accelerator of tourism growth as well as temporal and geographical concentration of tourist flows and volumes in certain locations [17]. Starting as a private accommodation development on a small individual scale, Airbnb has currently been transformed into a massive accommodation structure, resulting in destinations' overtourism [9]. Such problems have so far been noticed in Mallorca [30], as well as Greek islands [66]. In fact, developments associated with Airbnb, second houses, and time share developments have led plans, addressing sustainable tourism objectives, to failure and have allowed for the rising of overtourism [67]. The issue of Airbnb's influence seems to be proven in Santorini Island, as shown in this work, where the Airbnb as well as hotel accommodation infrastructure are a major source of built-up expansion and urban sprawl, remaining also largely uncontrolled and expanding in vulnerable parts of Santorini Island, e.g., natural and cultural sites of remarkable value or the Caldera rim.

A second critical issue, leading to islands' overtourism, relates to their overdependence on offshore tour operators to bring tourism in and promote an island, since in most cases local agencies do not dispose adequate resources, power or capacity to achieve this [9]. The image promoted by such operators may lead to overtourism, where the small-scale accommodation businesses of the islands are not able to oppose against this narrative and keep track with local identity and traditions, considering the risk of losing the powerful customers' suppliers [68]. Santorini Island is a typical example of overdependence on offshore operators that are largely affecting overtourism in the island, e.g., cruise companies that constitute one of the major factors for the noticeable overcrowding phenomenon in the island. In addition, the dependence on offshore tour operators is evident by indicators demonstrating the share of foreign tourists in the island (TD2 and TD3 indicators in Table 7).

Taking into consideration that problems of overtourism are primarily and usually experienced at the local level, the handling of such overdependence on external tour operators is tightly interwoven to a third critical issue, the one of political structure of islands as well as their power and independency to control the rate and nature of tourism development. This is also tightly linked to the options they have in order for alternative sources of employment and income for local population to be ensured when rejecting or strictly limiting tourism development to avoid overtourism [9]. Speaking of the island regions, both the power of political structures—local administrations as parts placed at the lowest level of the policy-making pyramid, with considerably confined flexibility and freedom to define the own desirable future of its territory—and the options available in such geographically constrained and resource inefficient regions are rather limited. This holds true for the study area of Santorini Island as well, which as part of broader administrative entities—Regional Entity of Cyclades Complex and the Region of Southern Aegean, as well as the Greek state as a whole in terms of its jurisdiction for setting up strategic policy directions for tourism development, which is defined at the central political scene—lacks, to a great extent, political control over the island's own destiny in general, including developments in the tourist sector.

Finally, a fourth critical issue is associated with the role of the local community in planning the tourist development of their island, an issue that highlights the residents' perspective, which is often given insufficient attention or, in many cases, is largely ignored. In literature, this issue is stressed, featuring the little attention that is paid to the residents' perception and the social implications of overtourism [9,14,33]. The lack of local community engagement in tourism planning seems to be the case in the majority of Greek islands, as relevant studies prove [5].

Having these critical issues in mind, this paper attempts to establish a methodological framework for overtourism assessment that is of universal use and is accompanied by a relevant set of indicators that can prove useful for targeted assessments in both mature destinations, as the case of Santorini Island, but also emerging ones, offering thus the chance for either reactive or proactive policy-making, respectively. The methodological framework follows the rationale of a dynamic, systemic, integrated and feedback-based process and is grounded in the robust PSR model. It attempts to link driving forces of each single destination (see Figure 2) with tourism demand and related repercussions, the intensity of which can guide inferences as to the overtourism state or risk to overtourism of the destination at hand, while highlighting specific, spatially-defined dimensions for policy action undertaking.

The indicators' set used for feeding this framework in the present study is in alignment with the current theoretical ground and empirical studies with regard to the tourism supply and demand. In fact, a set of quantitative indicators is selected, aiming at highlighting the tourism density and intensity in Santorini Island. These, combined with the study of the tourism dynamics can, in a way, guide inferences as to the evolution of the tourism sector in the study region. Further to the tourism supply/demand indicators and the dynamics of the tourist sector, an additional indicators' set is proposed, which is both of quantitative and qualitative nature and aims at featuring the environmental, economic, social and spatial implications of tourism that can define overtourism or risk to overtourism conditions. A combined assessment/interpretation of all indicators' sets abovementioned is used for responding to the question of whether the specific destination—Santorini Island—suffers from overtourism or not. The selected set of indicators has proven successful in sketching the overtourism state of Santorini Island, unveiling, among others, critical overtourism hotspots, sectoral conflicts in terms of land use, environmental repercussions as well as economic implications for local population (cost of renting or buying a house). More specifically, the results of the implementation of this framework in Santorini Island demonstrate that tourism so far has contributed to the transition of Santorini Island from an agrarian society to a thriving service-oriented economy, mainly a "visitor economy". Both quantitative and qualitative indicators demonstrate an alarming island's state, implying that certain capacity thresholds have already been reached or even overcome. Quite alarming is also the impact on the agricultural sector, placing at stake well-established, traditional production patterns and PDO products, steering for many decades the island's extroversion and being part of the local identity, while the same holds with the pattern of land use change and the ongoing process of tourist accommodation expansion even within Natura 2000 and Archaeological Zones or in close proximity to them.

The proposed methodological framework can form the ground for further empirical work with respect to the overtourism phenomenon in insular as well as non-insular destinations (e.g., cities, rural areas, coastal regions of mainland). For the successful implementation of this framework, of critical importance is the choice of the most relevant indicators' set. Provided that each single destination owns a distinct socio-economic, environmental, cultural, etc., profile that is largely defining its vulnerability and fragility to tourism impacts, it rests on the relevant research groups to make the most wise choice of indicators that can better serve the scope of their study, reflect the specific peculiarities of the region at hand, ensure the availability of trustworthy data with respect to these indicators, and highlight the critical dimensions of overtourism that need to be dealt with by policy makers.

Advantages/Limitations of the Study

Currently, a commonly-agreed definition as well as a unified set of indicators for overtourism identification does not exist. In fact, when seeking to assess or identify the overtourism phenomenon, a large set of indicators is proposed by a number of research endeavors [9,14,17,28,30,42]. The majority of them rely on indicators reflecting the tourism pressure in terms of supply and demand, both lying at the heart of the current discussion on overtourism, while a number of studies expands the discussion on overtourism by encompassing additional indicators, related to the spatial and temporal concentration of tourist flows, rate of growth, added value of tourism in terms of its interaction with other activities, etc. Taking into consideration that there is no "one size fits all" approach, the decision on the indicators' set needs to adopt an integrated and place-based stand. In fact, various place-based peculiarities can dictate specific indicators' choices, e.g., in the case study of Santorini, its volcanic origin and related island's morphology calls for highlighting the spatial dimension, taking into consideration the relative risks related to this origin. This fact is today highly underlined by the immense seismic activity occurring in the island, and the scepticism as to the pattern of the built-up space (e.g., in the Caldera rim) and the risks rising out of this.

In addition, it should also be noted that the choice of indicators is largely driven by data availability, which can evidently guide or affect each specific endeavor and lead to an incomplete overtourism assessment or certain approximation to the detriment of its criticality for establishing more informed overtourism remedy policies in a destination. Generally speaking, lacking are spatially-disaggregated data on tourism demand and supply; socio-economic data that can accurately capture a destination's profile; societal responses to tourism; high-resolution environmental assessment data; and data on climate-related threats. An additional obstacle with respect to data availability relates to data that are not monitored nor are available for the requested territory, or are just not up-to-date. Furthermore, since data are often not available, new forms of relevant sources need to be discovered in order for overtourism assessments to be properly accomplished. Moreover, it is worth mentioning the comparative credibility of data, since apart from land use and environmental data, obtained through global or pan-European geospatial databases like Copernicus Land Monitoring Service, most tourism-related data come from local/national databases and are occasionally incomplete or simply not comparable.

In order to cope with the aforementioned obstacles and establish a methodological framework that can be of universal use, considerable effort has been made in this research to make use of databases that are available at a global or at least pan-European level, in order for the proposed framework to be replicable in other areas and respective case studies for overtourism assessment. This is to the benefit of the proposed approach and is particularly achieved in most of the spatial databases regarding the environmental/spatial impact, and the economic impact that was based on land use transition data, focused on cultivated areas and agricultural land. Data stemming from the Copernicus Land Cover database and Google Earth Engine satellite databases are considered to provide a common and directly comparable ground for future case studies, while official Census data (as provided by the Greek authorities) could also potentially be comparable, since in most cases the indexes and classifications applied among Census Bureau Offices are similar and easy to adapt. Airbnb data could also be considered as universal and can be gathered from various local sources or directly through Airbnb and/or other platforms. Local sources have only to be used in order to evaluate land rent and house prices. In our view, and in the light of the rapidly expanding global-scale databases, limitations regarding data availability are efficiently handled in the proposed framework.

Another obstacle for making inferences as to the overtourism state of a region is the lack of threshold values. In fact, there are no commonly accepted thresholds, beyond which tourism development of a destination is demonstrated as overtourism. That said, it is clear that such thresholds are rather quite difficult to be defined, since they are largely grounded in the destination's carrying capacity, which is perceived as the amalgam of specific or place-related factors, including tourism type; tourist flows through time and space; stage of development; environmental, economic and social peculiarities; fragility of the destination; cultural aspects; geomorphological characteristics, etc. In coping with this difficulty, the integrated approach adopted in Santorini Island and the multiple dimensions explored, as these are reflected by the choice of the indicators' set, coupled with the spatial representation of those indicators, has proven successful in revealing those parts of the island that manifest the most acute signs of overtourism load.

Finally, of critical importance is the way the issue of overtourism in insular territories is perceived or assessed by various groups at the destination level. As previously noticed, a certain literature deficit is identified in this respect, where studies related to the diversified and even conflicting views of the various stakeholders' groups, the local community and the administrative bodies, as well as the visitors of the destination are limited. Such research efforts attempt to identify the attitudes of the host community of a destination against tourism/visitors or the tourists' perceptions about the destination, and are currently identified mostly in cases of urban overtourism studies, giving birth to social movements presented under the neologism term of "tourismphobia". Regarding the Santorini Island case study, an effort to embed social (e.g., social cohesion, identity) and/or psychological aspects of overtourism or the way overtourism affects the quality of life of both the local population and visitors did not take place in this study. However, an attempt to identify the implications of overtourism in local life is conducted through studying the rising costs for the local community that are due to overtourism. This is accomplished by embedding in the indicators' set the indicators S2a and S2b, representing the land rent and land sale value, respectively. This dimension is definitely of critical importance for local residents and, as shown in the Santorini Island example, is largely affected by overtourism trends.

7. Conclusions

The pressure of tourism growth in environmental, social, cultural, territorial, and economic terms is expected to intensify in the years to come in order for the needs of the globally rapid escalating tourism flows to be met. Such a growth in the Mediterranean region in general and Greece in particular, especially in the geographically fragmented insular territories, is associated with the major wealth of employment and income it can offer at the local level, as well as its contribution to the national economy. However, this growth is also a source of considerable pressure, resource depletion and degradation, spoiling of insular identity, and eventually of unbearable and/or irreversible damages in tourism-prominent areas. Overtourism and its related impacts, in this respect, are considered a state that needs to be seriously grasped by relevant destinations, and has to be dealt with the proper coordination of spatial planning and related tourism policies as the essential control mechanisms, especially in highly-rated tourism destinations, such as the insular ones. The articulation of policies has to definitely be drawn upon a well-defined, data-enabled and evidence-based ground, assessing the overtourism state and its key dimensions in each single destination in order for a dedicated policy reaction to be articulated, especially when it refers to fragile socio-spatial systems as in the case of the islands. However, as various researchers claim [14,17], it is necessary to realize that overtourism is a new concept in the planning and policy arena, and thus is currently at the early stages of the policy cycle. This is justified by the fact that overtourism, in its present exaggerating form, is a phenomenon of approximately the last few decades. Thus, it lacks extensive empirical research and comparative analysis as well as those data that are necessary for its documentation and assessment of its impacts on territories. This, in turn, implies that there is still a long way to go before effective policies can be formulated to adequately cope with it. In order to proceed to the next step, i.e., the one of policy analysis and formulation, of critical importance is its exploration by the use of multi-level spatially-defined quantitative and qualitative data at requested spatial levels that are capable of highlighting tourism density and intensity as well as their implications in each specific environmental, economic, and socio-cultural context. Of critical importance is also the establishment of a pool of indicators that are fed by available, spatially-disaggregated and local-level data and can form the ground for place-based integrated assessments of the overtourism phenomenon. In any case, the response to the question "How much tourism is too much tourism?" articulated by Weber et al. [14] does not have a 'one-size-fits-all' response, but has to be answered by properly prioritizing the well-being of locals in a destination, i.e., protect and preserve the natural and cultural capital, local identity and values, social cohesion, etc., and using spatial and developmental planning as a means to promote "win-win-win" solutions (local community—tourists—natural/cultural environment) in order for a fruitful future developmental trajectory that aligns with sustainability and resilience objectives to be attained.

Author Contributions: Writing—review and editing; methodology; data curation; writing—original draft (Section 3, Section 4, Section 5, and Section 6), Akrivi Leka; writing—review and editing; methodology; writing—original draft; data curation (Section 3, Section 4, Section 5, and Section 6), Apostolos Lagarias; conceptualization, writing—review and editing; methodology; writing—original draft (Section 1, Section 2, Section 3, Section 4, Section 6, and Section 7), Anastasia Stratigea; data curation (Section 4), Panayiotis Prekas. All authors have read and agreed to the published version of the manuscript.

Funding: This research received no external funding.

Data Availability Statement: Data used are contained within the article.

Conflicts of Interest: The authors declare no conflicts of interest. On behalf of all authors, the corresponding author states that there is no conflict of interest.

References

1. Misrahi, T. What Will Travel Look Like in 2030? World Economic Forum. 2016. Available online: https://www.weforum.org/agenda/2016/09/what-will-travel-look-like-in-2030/ (accessed on 22 October 2024).
2. Zolfani, S.H.; Sedaghat, M.; Maknoon, R.; Zavadskas, E.K. Sustainable tourism: A comprehensive literature review on frameworks and applications. *Econ. Res.* **2015**, *28*, 1–30. [CrossRef]
3. UNWTO. Available online: https://www.unwto.org/sustainable-development (accessed on 16 October 2024).
4. OECD. Rethinking tourism success for sustainable growth. In *Tourism Trends and Policies 2020*; OECD, Ed.; OECD Publishing: Paris, France, 2020; ISSN 20767773. Available online: https://www.oecd-ilibrary.org/docserver/6b47b985-en.pdf?expires=1731843712&id=id&accname=guest&checksum=9306352C569EF690D59EECBEC18DA594 (accessed on 16 September 2024). [CrossRef]
5. Koutsi, D.; Lagarias, A.; Stratigea, A. Evidence-based exploration as the ground for heritage-led pathways in insular territories: Case study Greek Islands. *Heritage* **2022**, *5*, 2746–2772. [CrossRef]
6. Butler, R.W. Sustainable tourism: A state-of-the-art review. *Tour. Geogr.* **1999**, *1*, 7–25. [CrossRef]

7. Jurgens, S.S.; Mijts, E.; Van Rompaey, A. Are there limits to growth of tourism on the Caribbean islands? Case-study Aruba. *Front. Sustain. Tour.* **2024**, *3*, 1292383. [CrossRef]
8. Leka, A.; Lagarias, A.; Panagiotopoulou, M.; Stratigea, A. Development of a Tourism Carrying Capacity Index (TCCI) for sustainable management of coastal areas in Mediterranean Islands–Case Study Naxos, Greece. *Ocean. Coast. Manag.* **2022**, *216*, 105978. [CrossRef]
9. Butler, R.; Dodds, R. Island tourism: Vulnerable or resistant to overtourism? *Highlights Sustain.* **2022**, *1*, 54–64. [CrossRef]
10. UN Tourism. Tourism in Small Island Developing States (SIDS). Available online: https://www.unwto.org/sustainable-development/small-islands-developing-states (accessed on 9 November 2024).
11. Graci, S.; Dodds, R. *Sustainable Tourism in Island Destinations*; Routledge: Oxford, UK, 2010; ISBN 9781844077809.
12. Coppin, F.; Guichard, V. Tourism as a tool for social and territorial cohesion-Exploring the innovative solutions developed by INCULTUM pilots. In *Innovative Cultural Tourism in European Peripheries*; Borowiecki, K.J., Fresa, A., Civantos, J.M.M., Eds.; Routledge: London, UK, 2025; pp. 7–27. ISBN 9781032728995.
13. Koens, K.; Postma, A.; Papp, B. Is overtourism overused? Understanding the impact of tourism in a city context. *Sustainability* **2018**, *10*, 4384. [CrossRef]
14. Weber, F.; Stettler, J.; Crameri, U.; Eggli, F. *Measuring Overtourism. Indicators for Overtourism: Challenges and Opportunities*; Lucerne University of Applied Sciences and Arts, Hochschule Luzern, Institut für Tourismus Wirtschaft: Lucerne, Switzerland, 2020; ISBN 978-3-033-07773-7. Available online: https://www.researchgate.net/publication/344282239_Measuring_Overtourism_Indicators_for_overtourism_Challenges_and_opportunities (accessed on 12 October 2024). [CrossRef]
15. Wall, G. From carrying capacity to overtourism: A perspective article. *Tour. Rev.* **2020**, *75*, 212–215. [CrossRef]
16. Zemla, M. Overtourism and local political debate: Is the over-touristification of cities being observed through a broken lens? *Entrep.–Educ.* **2024**, *20*, 215–225. [CrossRef]
17. Peeters, P.; Gössling, S.; Klijs, J.; Milano, C.; Novelli, M.; Dijkmans, C.; Eijgelaar, E.; Hartman, S.; Heslinga, J.; Isaac, R.; et al. *Research for TRAN Committee-Overtourism: Impact and Possible Policy Responses*; European Parliament, Policy Department for Structural and Cohesion Policies: Brussels, Belgium, 2018.
18. Capocchi, A.; Vallone, C.; Amaduzzi, A.; Pierotti, M. Is 'overtourism' a new issue in tourism development or just a new term for an already known phenomenon? *Curr. Issues Tour.* **2020**, *23*, 2235–2239. [CrossRef]
19. Chung, N.; Koo, C. The use of social media in travel information search. *Telemat. Inform.* **2015**, *32*, 215–229. [CrossRef]
20. Huertas, A. How live videos and stories in social media influence tourist opinions and behaviour. *Inf. Technol. Tour.* **2018**, *19*, 1–28. [CrossRef]
21. UNWTO. *"Overtourism"? Understanding and Managing Urban Tourism Growth Beyond Perceptions*; United Nations World Tourism Organization: Madrid, Spain, 2018.
22. Dredge, D. "Overtourism"-Old Wine in New Bottles? 2017. Available online: https://www.linkedin.com/pulse/overtourism-old-wine-new-bottles-dianne-dredge/ (accessed on 14 November 2024).
23. Mihalic, T. Conceptualizing overtourism: A sustainability approach. *Ann. Tour. Res.* **2020**, *84*, 103025. [CrossRef]
24. Gowreesunkar, V.G.; Thanh, T.V. Between overtourism and under- tourism: Impacts, implications, and probable solutions. In *Overtourism: Causes, Implications and Solutions*; Séraphin, H., Gladkikh, T., Vo Thanh, T., Eds.; Palgrave Macmillan: Cham, Switzerland, 2020; pp. 45–68. [CrossRef]
25. Mihalic, T.; Kuščer, K. Can overtourism be managed? Destination management factors affecting residents' irritation and quality of life. *Tour. Rev.* **2021**, *77*, 16–34. [CrossRef]
26. Adie, B.A.; Falk, M. Residents' perception of cultural heritage in terms of job creation and overtourism in Europe. *Tour. Econ.* **2021**, *27*, 1185–1201. [CrossRef]
27. Goessling, S.; McCabe, S.; Chen, N. A socio-psychological conceptualisation of overtourism. *Ann. Tour. Res.* **2020**, *84*, 102976. [CrossRef] [PubMed]
28. Vourdoubas, J. Evaluation of overtourism in the island of Crete, Greece. *Eur. J. Appl. Sci. Eng. Technol.* **2024**, *2*, 21–32. [CrossRef]
29. Milano, C.; Cheer, J.M.; Novelli, M. Overtourism: An evolving phenomenon. In *Overtourism: Excesses, Discontents and Measures in Travel and Tourism*; Milano, C., Cheer, J.M., Novelli, M., Eds.; CAB International: Oxfordshire, UK, 2019; pp. 1–17.
30. García-Buades, M.E.; García-Sastre, M.A.; Alemany-Hormaeche, M. Effects of overtourism, local government, and tourist behavior on residents' perceptions in Alcúdia (Majorca, Spain). *J. Outdoor Recreat. Tour.* **2022**, *39*, 100499. [CrossRef]
31. Saveriades, A. Establishing the social tourism carrying capacity for the tourist resorts of the east coast of the Republic of Cyprus. *Tour. Manag.* **2000**, *21*, 147–156. [CrossRef]
32. Milano, C.; Novelli, M.; Russo, A.P. Anti-tourism activism and the inconvenient truths about mass tourism, touristification and overtourism. *Tour. Geogr.* **2024**, *26*, 1313–1337. [CrossRef]
33. Buitrago, E.M.; Yñiguez, R. Measuring overtourism: A necessary tool for landscape planning. *Land* **2021**, *10*, 889. [CrossRef]
34. Dodds, R.; Butler, R.W. Introduction. In *Overtourism: Issues, Realities and Solutions*; Dodds, R., Butler, R.W., Eds.; Walter de Gruyter GmbH: Berlin, Germany; Boston, MA, USA, 2019; pp. 1–21.

35. Gretzel, U. The role of social media in creating and addressing overtourism. In *Overtourism: Issues, Realities and Solutions*; Dodds, R., Butler, R.W., Eds.; Walter de Gruyter GmbH: Berlin, Germany; Boston, MA, USA, 2019; pp. 263–276.
36. Lagarias, A.; Stratigea, A.; Theodora, Y. Overtourism as an emerging threat for sustainable island communities–Exploring indicative examples from the South Aegean Region, Greece. In Proceedings of the Computational Science and Its Applications–ICCSA 2023 Workshops, Athens, Greece, 3–6 July 2023; Lecture Notes in Computer Science. Gervasi, O., Murgante, B., Rocha, A.M.A.C., Garau, C., Scorza, F., Karaca, Y., Torre, C.M., Eds.; Springer Nature: Cham, Switzerland, 2023. Part VII. Volume 14110, pp. 404–421. [CrossRef]
37. Constantoglou, M.; Klothaki, T. How much tourism is too much? Stakeholder's perceptions on overtourism, sustainable destination management during the pandemic of COVID 19 era in Santorini Island Greece. *J. Tour. Hosp. Manag.* **2021**, *9*, 288–313. [CrossRef] [PubMed]
38. Stanchev, R. The Most Affected European Destinations by Over-Tourism. Universitat de les Illes Balears, Facultat de Turisme, Grau de Turisme, Any acadèmic 2017–18. Available online: https://dspace.uib.es/xmlui/bitstream/handle/11201/148140/Stanchev_Rostislav148140.pdf?sequence=3&isAllowed=y (accessed on 18 October 2024).
39. Wilson, N. 30 Destinations Cracking Down on Overtourism, from Venice to Bhutan. *Independent*. 2024. Available online: https://www.independent.co.uk/travel/sustainable-travel/destinations-cracking-down-overtourism-venice-b2587510.html (accessed on 18 October 2024).
40. Bateman, J. Santorini Under Pressure: The Threat of Overtourism. 2019. Available online: https://www.greece-is.com/santorini-pressure-threat-overtourism/ (accessed on 18 October 2024).
41. O'Toole, L. Beautiful European Island under Pressure Where Even Tourists Complain About the Tourists. *Daily Express*. 2024. Available online: https://www.express.co.uk/news/world/1910789/santorini-greece-overtourism-tourists-news (accessed on 18 October 2024).
42. Simancas Cruz, M.; Pilar Peñarrubia Zaragoza, M. Analysis of the accommodation density in coastal tourism areas of insular destinations from the perspective of overtourism. *Sustainability* **2019**, *11*, 3031. [CrossRef]
43. Veríssimo, M.; Moraes, M.; Breda, Z.; Guizi, A.; Costa, C. Overtourism and tourismphobia. *Tour. Int. Interdiscip. J.* **2020**, *68*, 156–169. [CrossRef]
44. Bambrick, H. Resource extractivism, health and climate change in small islands. *Int. J. Clim. Change Strateg. Manag.* **2017**, *10*, 272–288. [CrossRef]
45. OECD. *Environment at a Glance 2013: OECD Indicators*; OECD Publishing: Paris, France, 2013. [CrossRef]
46. Official Administrative Boundaries. Available online: https://www.statistics.gr/digital-cartographical-data (accessed on 15 November 2024).
47. Imperviousness Classified Change Layer. Available online: https://sdi.eea.europa.eu/catalogue/copernicus/api/records/73f0e3eb-ea21-4083-9b49-9c345c766775?language=all (accessed on 18 February 2025).
48. Archaeological Zones. Available online: https://www.arxaiologikoktimatologio.gov.gr/en (accessed on 18 February 2025).
49. Natura 2000 Areas and Wildlife Repositories of Greece. Available online: https://geodata.gr/en/ (accessed on 3 November 2024).
50. Agricultural Land Change. (Vector 2018). Available online: https://sdi.eea.europa.eu/catalogue/copernicus/api/records/71c95a07-e296-44fc-b22b-415f42acfdf0?language=all (accessed on 6 November 2024).
51. Agricultural Land Change. (Vector 1990). Available online: https://sdi.eea.europa.eu/catalogue/copernicus/api/records/5c1f2e03-fcba-47b1-afeb-bc05a47bada0?language=all (accessed on 6 November 2024).
52. Coastal Zones Land Cover/Land Use 2018 (Vector), Europe, 6-yearly, Feb. 2021. Available online: https://sdi.eea.europa.eu/catalogue/copernicus/api/records/205e2db2-4e35-4b1b-bf84-271c4a82248c?language=all (accessed on 18 February 2025).
53. Hellenic Statistical Authority (ELSTAT). Available online: https://www.statistics.gr/en/home/ (accessed on 15 November 2024).
54. Institute of the Association of Greek Tourism Enterprises (INSETE). Available online: https://insete.gr/districts/?lang=en (accessed on 8 October 2024).
55. Airbnb Accommodation. Available online: https://bfortune.opendatasoft.com/explore/dataset/Airbnb-listings/export/?disjunctive.neighbourhood&disjunctive.column_10&disjunctive.city&q=santorini (accessed on 8 October 2024).
56. Hellenic Chamber of Hotels. Available online: https://www.grhotels.gr/en/tourist-guide/search-hotels-and-camping/ (accessed on 8 October 2024).
57. Spiti24. Available online: https://www.spiti24.gr/ (accessed on 18 October 2024).
58. JTR-Airport (Santorini Airport). Available online: https://www.jtr-airport.gr/en (accessed on 15 June 2024).
59. Hellenic Ports Association. Available online: https://www.aivp.org/en/aivp/our-members/directory/hellenic-ports-association/ (accessed on 12 October 2024).
60. Institute of the Association of Greek Tourism Enterprises (INSETE). The Contribution of Tourism in the Greek Economy in 2023. 2024. Available online: https://insete.gr/wp-content/uploads/2024/04/23_04_Tourism_and_Greek_Economy_2019-2023.pdf (accessed on 18 February 2025).

61. Donaire, A.J.; Galí, N.; Coromina, L. Evaluating tourism scenarios within the limit of acceptable change framework in Barcelona. *Ann. Tour. Res. Empir. Insights* **2024**, *5*, 100145. [CrossRef]
62. Seraphin, H.; Sheeran, P.; Pilato, M. Over-tourism and the fall of Venice as a destination. *J. Destin. Mark. Manag.* **2018**, *9*, 374–376. [CrossRef]
63. Mycoo, M.; Wairiu, M.; Campbell, D.; Duvat, V.; Golbuu, Y.; Maharaj, S.; Nalau, J.; Nunn, P.; Pinnegar, J.; Warrick, O. Small Islands. In *Climate Change 2022: Impacts, Adaptation and Vulnerability. Contribution of Working Group II to the Sixth Assessment Report of the Intergovernmental Panel on Climate Change*; Pörtner, H.-O., Roberts, D.C., Tignor, M., Poloczanska, E.S., Mintenbeck, K., Alegría, A., Craig, M., Langsdorf, S., Löschke, S., Möller, V., et al., Eds.; Cambridge University Press: Cambridge, UK; New York, NY, USA, 2022; pp. 2043–2121. [CrossRef]
64. Grofelnik, H.; Maradin, M. Challenges of the water supply of Croatian islands in conditions of climate change and tourism growth. *Geoadria* **2023**, *28*, 2–35. [CrossRef]
65. Hall, C.M.; Prayag, G.; Safonov, A.; Coles, T.; Gössling, S.; Naderi Koupaei, S. Airbnb and the sharing economy. *Curr. Issues Tour.* **2022**, *25*, 3057–3067. [CrossRef]
66. Panousi, S.; Petrakos, G. Overtourism and tourism carrying capacity-A regional perspective for Greece. In *Culture and Tourism in a Smart, Globalized, and Sustainable World*; Katsoni, V., Ciná van Zyl, C., Eds.; Springer: Cham, Switzerland, 2021; pp. 215–229. [CrossRef]
67. Butler, R.W. Sustainable Tourism in Sensitive Environments: A Wolf in Sheep's Clothing? *Sustainability* **2018**, *10*, 1789. [CrossRef]
68. Bastakis, C.; Buhalis, D.; Butler, R. The perception of small and medium sized tourism accommodation providers on the impacts of tour operators' power in Eastern Mediterranean. *Tour. Manag.* **2004**, *25*, 151–170. [CrossRef]

International Journal of
Geo-Information

MDPI

Article

Spatial-Temporal Changes in Ecosystem Service Value and Its Overlap with Coal Mining Intensity in the Yellow River Basin, China, During 2000–2030

Yongjun Yang [1], Renjie Gong [1], Qinyu Wu [2,3,*] and Fu Chen [4]

1 School of Environment and Spatial Informatics, China University of Mining and Technology, Xuzhou 221116, China; y.yang@cumt.edu.cn (Y.Y.); rj.gong@cumt.edu.cn (R.G.)
2 School of Social & Environmental Sustainability, University of Glasgow, Dumfries DG1 4ZL, UK
3 School of Public Policy & Management, China University of Mining and Technology, Xuzhou 221116, China
4 School of Public Administration, Hohai University, Nanjing 211100, China; chenfu@hhu.edu.cn
* Correspondence: qinyu.wu@glasgow.ac.uk

Abstract: Understanding the ecosystem services and their interaction with coal resource development is crucial for formulating sustainable development policies. In this study, we focused on the Yellow River Basin, characterized by both rich coal resources and ecological fragility. The key findings are that (1) the ecosystem service value (ESV) in the Yellow River Basin exhibited significant spatial heterogeneity during 2000–2030, decreasing from the southeast to northwest, and decreasing the most notably in the southern part of the upper reaches of the river basin; (2) the high-high clustering area of the ESV shifted from the upper-middle reaches in 2000 to the middle-lower reaches in 2020, while the low-low clustering area remained within Inner Mongolia. By 2030, the high-high clustering area is expected to stabilize in southern Shaanxi Province, and the low-low area will potentially spread eastward; (3) the overall ESV is low, and it experienced a significant decline from 2000 to 2020, with water supply emerging as a major limiting factor, although some policy-supported counties had better ecological service values and trends. (4) From 2000 to 2020, the coal mining intensity (CMI) was concentrated in the upper and middle reaches, particularly at the junctions of Shanxi, Shaanxi, and Inner Mongolia, and the pattern remained stable, but local areas experienced increased mining intensity; (5) the overlap of the CMI and ESV primarily exhibited a low-high clustering pattern in the middle and upper reaches of the Yellow River Basin and eastern Ordos City, and a high-high clustering pattern in the middle reaches of the basin in Shanxi Province, which remained stable and slightly expanded from 2000 to 2030; (6) the trade-off between the ecosystem services in the overlap area intensified, especially between the provisioning and support services, and was significantly impacted by the coal mining activities. The findings indicate that the area that overlaps with the coal mining area in the Yellow River Basin has expanded and has had an increasing negative impact on the ESV. It is also essential to address the trade-offs between the provisioning and support services and to implement ecological restoration measures to mitigate the risk of ESV loss. Future efforts should focus on the regions where the CMI and ESV overlap and have poor coordination and the adverse effects of resource extraction on ecosystem services are becoming more pronounced. The results of this study demonstrate that spatial overlap analysis is effective in identifying the hotspots and provides a foundation for developing sustainable and high-quality policies for ecologically fragile basins.

Keywords: ecosystem service value; ecological restoration; mining; trade-offs and synergies; Yellow River Basin

1. Introduction

Ecosystem services could contribute to human survival, health, livelihood, and well-being. The human welfare generated by the ecosystem services is economically treated as the ecosystem service value (ESV) [1]. The significance of these services has generated interest in safeguarding them. Since the groundbreaking study conducted by Costanza et al. [2], there has been a significant increase in the global research focused on estimating, measuring, and mapping ecosystem services. This growth has been particularly evident since the millennium ecosystem assessment was conducted [3], an international study involving more than 1300 scientists. The millennium ecosystem assessment provided crucial evidence that about 60% of the world's ecosystems have degraded over the past fifty years [4]. Given the range of the consequences of the human exploitation of ecosystems, it has become vitally important to prevent ecosystem degradation and to sustainably manage ecosystems and landscapes sustainably to ensure future ecosystem supply and meet the growing demands for ecosystem services [5]. This requires an understanding of the spatiotemporal interactions between human activities and ecosystems, enabling decision makers and policymakers to adopt sustainable management strategies and nature-based solutions, thereby enhancing the resilience and adaptability of ecosystems. Such an understanding can be derived from assessing the past land use/land cover dynamics [6], but it also necessitates forecasting the future situation, predicting the potential impacts of current decisions and trends, and revealing the trade-offs [7].

The Yellow River Basin is located in northern China and characterized by a wide range of natural features and socio-economic conditions [8]. It serves as a significant freshwater source for the northern region of China. The basin is commonly referred to as the "energy basin" because of its abundant reserves of coal, oil, natural gas, and non-ferrous metals. This makes it a crucial hub for China's energy industry, chemical industry, raw materials, and basic industries [9]. Due to the challenging ecological conditions and economic growth pressures, the natural and semi-natural ecosystems in the Yellow River Basin have become more vulnerable [10,11]. The depletion of water resources and contraction of wetlands in the upper Yellow River pose a threat to the river's resource supply to the upper Yangtze River [12]. The deep loess layer, arid climate, and intricate topography have resulted in significant soil erosion [13]. The ecological fragility of the region has worsened due to the increased resource destruction, environmental pollution, and geological disasters resulting from large-scale mineral resource development. To address these issues, a set of ecological restoration initiatives have been implemented in the Yellow River Basin, aiming to facilitate the rejuvenation of forests and grasslands. However, the effort to strengthen the ecosystem's ability to withstand disruptions has been rather restricted. The conflict between the exploitation of resources and the preservation of the ecological environment continues to be a significant concern [14]. Hence, examining the correlation between the alterations of the ecosystem and the exploitation of mineral resources is of great importance for enhancing the ecological environment in the Yellow River Basin and achieving superior regional development.

Most previous studies have focused on the coupling and coordination relationship between resource exploitation and ecosystems [15–17], often forming a singular relationship at a global scale and neglecting the cluster patterns and spatial dependency of ecosystems and resource exploitation at the local scale. This has limited policymakers in formulating differentiated policies for resource exploitation and ecosystem protection, thereby reducing the efficiency of decision making. Moreover, most research on the impact of mineral resource development on ecosystem services has been based on static analyses of historical periods or single time points [15,18,19]. Assessing the trends of ecosystem services and predicting future ecosystem services are crucial for accurately grasping the future macro trends and formulating sustainable regional development policies.

Research conducted on continuous timescales is of great significant as it not only effectively identifies spatial and temporal changes but also demonstrates resilience in intricate landscapes, thereby closely correlating the results with the spatiotemporal dependencies [20]. An important problem is that time series image data have rarely been utilized to calculate the ecosystem service value despite the continuous land use/land cover changes over time [16,21,22]. This is primarily due to the time-consuming and labor-intensive nature of acquiring and processing time series remote sensing data, which deter most researchers interested in studying the ESV. The Google Earth Engine has been made available to the scientific community, providing an efficient resolution of this issue. The Google Earth Engine has compiled an extensive assortment of commonly utilized spatial datasets that can be analyzed through online tools. In addition, the Google Earth Engine has created a web-based interactive development environment for users. The Google Earth Engine application programming interface (API) and interactive development environment facilitate the swift development of applications and visualization of outcomes [23]. The use of the Google Earth Engine has enabled the analysis of time series images, allowing researchers to obtain a more profound comprehension of the ecosystem services. Hence, the Google Earth Engine provides researchers with a wide range of analytical techniques and resources, facilitating comprehensive investigations of spatial data. This research will foster the sustainable advancement of ecosystems and the preservation and administration of ecosystems.

In summary, the correlation between the alterations of the ecosystem services and the exploitation of mineral resources at the scale of river basins has not been examined. Considering the knowledge gaps and the strategic importance of the Yellow River Basin, in this study, we assessed the evolution of the ESV under the mining activities in the Yellow River Basin and its interactive mechanisms. The specific objectives of this study were (1) to develop a dynamic ecosystem service assessment module based on the Google Earth Engine platform and spatial analysis technique; (2) to analyze the spatiotemporal characteristics of the ecosystem services in the Yellow River Basin from 2000 to 2020; (3) to predict the ESV in the Yellow River Basin in 2030; (4) to analyze the evolution of the trade-offs and synergies between the ecosystem services; and (5) to reveal the overlap between the ESV and coal mining intensity (CMI) at the global and regional scales. The overall goal was to assist policymakers in understanding the spatiotemporal interactions between mineral resource development and the ecosystem services to promote the formulation of refined and sustainable management strategies, thereby providing scientific support for balancing economic development and ecological restoration in large-scale basins.

2. Materials and Methods

2.1. Study Area

The Yellow River Basin (31°31′–43°31′ N, 89°19′–119°39′ E) is the sixth-largest basin in the world, and it has a significant impact on both China's ecology and economy. It spans nine provinces and regions, covering an area of almost 1.86 million km^2 at elevations ranging from −64 m to 6819 m [14] (Figure 1). The basin contains 30.3% of China's population and contributes 26.5% of the national gross domestic product (GDP) [24]. The climate ranges from semi-humid in the southeast to semi-arid and arid in the northwest, with corresponding decreases in temperature and precipitation. The basin functions as a crucial ecological corridor that links the Tibetan Plateau, the Loess Plateau, and the North China Plain. It also serves as an important ecological barrier in Northern China [9,12]. Nevertheless, the ongoing exploitation of resources has resulted in a substantial decrease in water flow, and the basin's capacity to support resources and the ecosystem is approaching its critical threshold.

As China's energy basin, the Yellow River Basin has a staggering 1563 coal mines, accounting for almost one-third of the nation's coal mines. China's total coal production capacity is roughly 2.97 billion tons per year, and the coal production of the study area accounts for nearly two-thirds of the country's overall capacity. In 2017, Shanxi, Shaanxi, Inner Mongolia, Ningxia, and Gansu provinces, situated in the middle and upper parts of the basin, collectively generated 2.46 billion tons of coal, accounting for almost 70% of the entire coal production in the country [9]. This region is widely regarded as the most economically valuable and has the greatest potential for coal production and distribution in China. Coal mining in the Yellow River Basin and the rest of the country has contributed to

China's socio-economic growth, but it has also caused severe disturbances to the ecological environment [9,14,16,18,19].

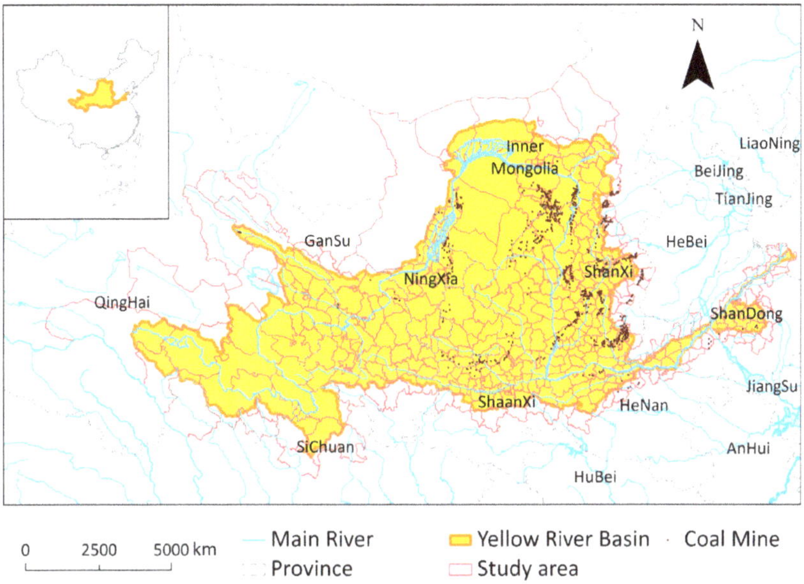

Figure 1. Location of the study area and the counties. The mainstream of the Yellow River and its tributaries pass through nine provinces and regions in China. The spatial distribution of the coal mining areas is presented as points on the map.

2.2. Research Framework

The primary sequence of activities in this study was as follows. (1) ESV calculation: Utilizing the Google Earth Engine platform, the ESV in the Yellow River Basin in China during 2000–2020 was calculated. Considering the regional heterogeneity, spatial heterogeneity indices such as the net primary productivity and standardized precipitation evapotranspiration index were used to adjust the equivalent weights annually. Furthermore, to illustrate the influence of socio-economic progress on the ecosystem, the ESV was dynamically adjusted using the GDP and Engel's coefficient. (2) Trend analysis: the Sen+Mann-Kendall test was employed to examine the change trend of the ESV. (3) ESV prediction: the statistical spatiotemporal kriging model was used to predict the spatial distribution of the ESV in the Yellow River Basin in 2030. (4) Autocorrelation analysis: The local indicators of the spatial association (LISA) model were applied to analyze the autocorrelation of the ecosystem services. The bi-variate LISA (BILISA) model was employed to examine the correlation between the ESV and the extent of mining development. (5) Trade-off and synergy evolution analysis: correlation analysis was conducted to explore the evolution of the trade-offs and synergies between the ecosystem services in the overlap area. The research framework is illustrated in Figure 2.

2.3. Data

The land use classification data were obtained from the moderate resolution imaging spectroradiometer (MODIS) Land Cover Type product (MCD12Q1 v061). The net primary productivity (NPP) data were obtained from the MOD17A3HGF v6.1 datasets, which are accessible in Google Earth Engine's MODIS collection. The standardized precipitation evapotranspiration index (SPEI) data were obtained from the SPEIbase v2.9 datasets, which are hosted on the Google Earth Engine platform. The data for further elements of the ESV, such as the GDP and annual GDP index, were obtained from the statistical yearbooks published by the National Bureau of Statistics. The production data for mineral resources were obtained from the Global Coal Mine Tracker database. Considering that the earliest available MODIS data began in January 2000, and the latest data are updated to early 2021 due to the lag of remote sensing products and statistical yearbooks, we chose a 20-year range (2000–2020) for the research. The data utilized in this study are described in Table 1.

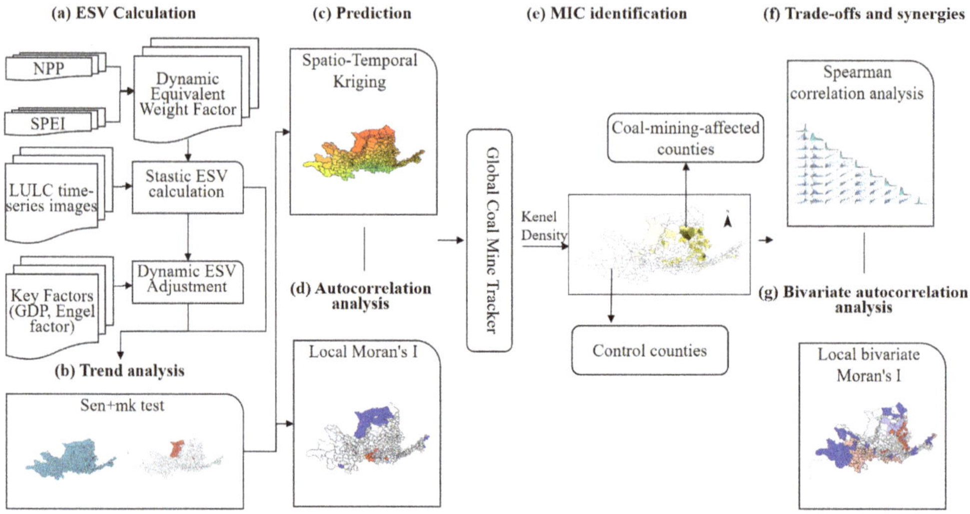

Figure 2. Framework for the analysis of the spatial-temporal changes of the ecosystem service value and its overlap with the coal mining intensity in the study area. (**a**) Calculation of ESV; (**b**) trend analysis of ESV from 2000 to 2020; (**c**) prediction of ESV in 2030; (**d**) autocorrelation analysis of ESV; (**e**) identification of spatial distribution and mining intensity of coal mines; (**f**) trade-off and synergies of ESV under mining activities; and (**g**) bi-variate autocorrelation analysis of ESV and CMI.

Table 1. Variables utilized in the computation of the ecosystem service value.

Data Application	Unit	Source
Delineation of Land Systems	Class Index (0–6)	MCD12Q1 v061, 2000–2020 https://doi.org/10.5067/MODIS/MCD12Q1.061 (accessed on 10 September 2024)
Gross Domestic Product (GDP)	Chinese Yuan (CNY)/km^2	National Bureau of Statistics, 2000–2020
Per Capita Net Income (PCNI)	CNY	(https://www.stats.gov.cn/) (accessed on 23 September 2024)
Net Primary Productivity (NPP)	kg·C/m^2	MOD17A3HGF v6.1, 2000–2020 (https://lpdaac.usgs.gov/resources/data-action/aster-ultimate-2018-winter-olympics-observer/) (accessed on 10 September 2024)
Standardized Precipitation Evapotranspiration Index (SPEI)	mm/month	The Global SPEI database (SPEIbase) v2.9, 2000–2020 (doi:10.20350/digitalCSIC/15470) (accessed on 10 September 2024)
Coal Mining Intensity (CMI)	Continuous Index (0–1)	Global Coal Mine Tracker, Global Energy Monitor, April 2024 release (https://globalenergymonitor.org/projects/global-coal-mine-tracker/tracker-map/) (accessed on 10 September 2024)

2.4. Assessment of the ESV

In this study, we used the millennium ecosystem assessment methodology to categorize the ecosystem services into four main types: provisioning services (raw materials, food, and water), regulating services (climate and gas regulation and waste treatment), support services (biodiversity protection and soil formation), and cultural services (recreation and culture). An ecosystem service accounting model based on the Google Earth Engine platform [25] was applied to calculate the ESV. This process involved determining the land use and ecological function parameters followed by static and dynamic calculations incorporating social development variables.

The land use types were classified based on the system developed by the National Committee of the State Ministry of Agriculture (1984), including woodland, grassland, cropland, wetlands, water bodies, unused land, and built-up land. The calculation was based on the equivalence weight factors proposed by Xie et al. [26]. Xie et al. built on the methodology of Costanza et al. [2] by conducting surveys among 200 ecological experts in China, and developing a value equivalent factor table for terrestrial ecosystems. This table, updated in 2015 using the literature data, expert knowledge, and biomass methods, defines one equivalent factor as 1/7 of the market value of grain produced per hectare annually. In this study, we used data from the National Statistical Yearbook, and we determined that the average grain yield was 4405 kg/ha with a price of 3.05 CNY/kg in 2000, resulting in an ESV per hectare of CNY 1919.32.

Given the heterogeneity of the ecosystems in China, directly adopting the national average equivalence coefficient could lead to inaccurate estimates. Therefore, we recalculated these coefficients according to the specific conditions in the study area. Additionally, spatial and temporal adjustments were made by incorporating dynamic factors such as the NPP and SPEI to account for the evolving nature of the ecosystem services over time and across regions. This was achieved using the following formula:

$$F_{nij} = \begin{Bmatrix} P_{ij} \times F_1 \\ R_{ij} \times F_2 \end{Bmatrix} \tag{1}$$

where F_{nij} is the equivalent value per unit area of an ecosystem in the nth ecosystem service category in year j in study area i; P_{ij} is the spatiotemporal adjustment factor of the NPP for this type of ecosystem in year j in study area i; and R_{ij} is the spatiotemporal adjustment factor of the SPEI for this type of ecosystem in year j in study area i. F_1 includes gas regulation, raw materials, soil formation and retention, climate regulation, waste treatment, biodiversity protection, and recreation and culture. F_2 includes water supply and food production.

To assure the comparability of all the ESVs at different points in time, it is essential to compute the corresponding economic value, taking into consideration the impacts of the price level, inflation, and other relevant aspects. The studied basin relies on the development of coal mining resources as a key industry, which has a significant influence on the GDP. The equivalent economic value is calculated using the following equation, with 2000 as the constant and the base year for determining the yearly average ESV:

$$ESV_C = \frac{ESV_s}{E_{avg}} \times E_{an} \tag{2}$$

$$E_{an} = \frac{E_m}{\prod_{i=n}^{m} \varnothing_i} \tag{3}$$

where ESV_c is the equivalent ESV. $ESVs$ is *the static ESV*. E_{avg}, E_{an}, and E_m are economic values associated with a single weight factor. E_{avg} is the average value, and E_{an} is derived based on the E_m value for the current year m within the research period. The constant year (start year) is denoted by n. $\varnothing_i$ is the yearly GDP index in year i, which is a measure of the annual percentage change in the GDP.

As civilization progresses and the quality of life improves, individuals will become more concerned about ecosystem services, resulting in a greater inclination to financially support these services. Hence, it is necessary to modify the fixed ESV by incorporating the socio-economic coefficient. The S-shaped development curve of Pearl can be employed to depict the attributes of the ecological value [27].

$$ESV = ESV_c \times A_c \tag{4}$$

$$A_c = \frac{1}{1 + exp^{-t}} \tag{5}$$

$$t = \frac{1}{E_n} - 3 \tag{6}$$

where A_c is the dynamic adjustment coefficient; t is the stage of social development; and E_n is Engel's coefficient.

To implement the methodology described above, we adapted the Google Earth Engine code developed by Liang et al. [25] to compute the ecosystem services and the ESV in the study area. This was accomplished utilizing the API and the code editor on the Google Earth Engine platform.

To implement the methodology described above, we adapted the Google Earth Engine code developed by Liang et al. [25] to compute the ecosystem services and the ESV in the study area. This was accomplished utilizing the API and the code editor on the Google Earth Engine platform. Google Earth Engine is a cloud computing tool developed by Google, which is characterized by the ability to process remote sensing and geographic information data of large areas online. Compared with standalone processing software such as ENVI and ArcGIS (Environmental Systems Research Institute, USA), it can process data online without downloading it to a local computer. And these data and their processing code are shared in the form of links. The calculation model included the following steps:

(1) equivalent weight coefficient adjustment (https://code.earthengine.google.com/5d8a6b6287dddb24181111e7ae06965e) (accessed on 10 September 2024);
(2) ESV calculation (https://code.earthengine.google.com/91cbd359658c52bf747cd0298cc0af95) (accessed on 10 September 2024);
(3) dynamic ESV adjustment (https://code.earthengine.google.com/91cbd359658c52bf747cd0298cc0af95) (accessed on 10 September 2024).

2.5. *Spatial-Temporal Change Analysis of the ESV*

2.5.1. Trend of the ESV

The Sen+Mann-Kendall (MK) test can measure the trend of the ESV. The Sen slope estimator can be utilized to compute the trend value of the changes in the ESV. Typically, it has been employed in conjunction with the MK non-parametric test. This involves calculating the Sen trend slope and then using the MK test to assess the significance of the trend [19,28].

Theil–Sen median trend analysis, also known as Sen trend analysis, is a reliable non-parametric statistical technique used to calculate trends. Sen trend analysis has an advantage over linear regression trend analysis as it can mitigate the impact of missing data and the shape of the data distribution in the time series. Additionally, it can reduce the interference caused by outliers in the time series. The computation formula for the Sen trend degree is as follows [29].

$$\beta_{ESV} = median\left(\frac{ESV_j - ESV_i}{j - i}\right), \forall j > i \tag{7}$$

where ESV_j and ESV_i are the ESV time series data. $\beta_{ESV} > 0$ indicates an increasing trend, while $\beta_{ESV} < 0$ indicates a decreasing trend.

The Mann-Kendall test is commonly employed in conjunction with Sen trend analysis [29]. This method is a non-parametric statistical test that is robust to missing values and outliers, and it does not require any specific distribution for the sample data. The statistical test approach can be expressed as follows:

$$Z = \begin{cases} \frac{S-1}{\sqrt{Var(S)}} & (S > 0) \\ 0 & (S = 0) \\ \frac{S-1}{\sqrt{Var(S)}} & (S < 0) \end{cases} \tag{8}$$

$$S = \sum_{j=1}^{n-1} \sum_{i=j+1}^{n} sign(ESV_j - ESV_i) \tag{9}$$

$$Var(S) = \frac{n(n-1)(2n+5)}{18} \tag{10}$$

$$sign(\theta) = \begin{cases} 1 & (\theta > 0) \\ 0 & (\theta = 0) \\ -1 & (\theta < 0) \end{cases} \tag{11}$$

where ESV_j and ESV_i are the ESV time series data; the sign function is denoted by the *sign* symbol; S is the test statistic; and Z is the standardized test statistic, which is employed to assess the presence of a significant trend in the time series data. A higher Z-value indicates a more pronounced trend [30,31]. n is the number of data points. For significance level α, $|Z| > Z_{1-\alpha/2}$ indicates that there is a significant change trend. In this study, the significance level α was set to 0.05 to determine the significance of the trend shift in the ESV time series.

2.5.2. Prediction of the ESV

Kriging is an optimal spatiotemporal prediction method, and it considers the statistical relationships of the process in both space and time. This technique was developed by Matheron [32], who built upon the research conducted by Krige [33], a mining engineer hailing from South Africa. Kriging was subsequently extended to spatiotemporal applications of the ESV by utilizing the spatiotemporal statistical modeling and interpolation features in the R package gstat.

2.5.3. Spatial Hotspot Analysis of the ESV

Spatial autocorrelation analysis was used to express the spatial correlation and spatial patterns of the ESV across the studied basin. The spatial hotspot analysis was conducted using the local Moran's I statistic, which has been commonly used for detecting local spatial clustering patterns in data. The local Moran's I measures the degree of spatial autocorrelation between a given location and the surrounding areas. The formula for the local Moran's I is as follows:

$$I_i = Z_i \sum_{j=1}^{n} w_{ij} Z_j \tag{12}$$

where I_i is the local bi-variate Moran's I in spatial unit i. Z_i is the standardized value of the ESV at location i. The standardization is achieved by subtracting the mean and dividing by the standard deviation. Z_j is the standardized value of the ESV at neighboring location j. w_{ij} is the spatial weight matrix, which defines the spatial relationship between locations i and j. In this study, a distance-based spatial weight matrix was employed, in which adjacent areas were assigned a weight of 1 and non-adjacent areas were assigned a weight of 0.

2.5.4. Correlation Analysis

In this study, we utilized Pearson's correlation analysis to reveal the trade-offs and synergies among various ecosystem services. Pearson's correlation analysis is generally considered to yield more accurate and consistent results compared to other correlation analysis methods, such as Spearman correlation analysis and Kendall correlation analysis, when examining the relationships between continuous data variables [34]. Therefore, it has been commonly employed to reduce the interactions between ecosystem services [17,35].

2.6. Assessment of the CMI

The mining intensity refers to the degree to which mining activities disturb a specific geographic area. In this study, we calculated the mining intensity of coal mines by considering a single administrative county as the unit of computation. The calculation was based on three indicators: the production of coal mines, the proven coal reserves, and the number of mining areas within a single county. These indicators were weighted to determine the overall mining intensity. Initially, we employed the kernel density method [32] to analyze and describe the geographic arrangement of the coal mine production and proven coal reserves. Furthermore, we applied Equation (13) to normalize the three indicators and subsequently combined them into a comprehensive index, namely the CMI, using the equal-weight summation method, which is expressed by Equation (14).

$$U'_{ij} = \frac{U_{ij} - U_{j,min}}{U_{j,max} - U_{j,min}} \tag{13}$$

where U'_{ij} is the normalized value of mining intensity indicator j in study unit i, U_{ij} is the initial value of mining intensity indicator j, and $U_{j,max}$ and $U_{j,min}$ are the maximum and minimum values of mining intensity indicator j, respectively.

$$CMI_i = \sum_{j=1}^{3} U'_{ij} \times w_j \tag{14}$$

where CMI_i is the composite level index of the mining intensity, w_j is the weight of mining intensity indicator j, and its value is one-third in this study.

2.7. Overlap Analysis of the ESV and CMI

Local bivariate spatial autocorrelation analysis, which is expressed by Equation (15), examines the spatial correlation of the attributes inside various geographic units using LISA cluster maps [36]. The exact equation is as follows:

$$I_i^{xy} = Z_i^x \sum_{j=1}^{n} w_{ij} Z_j^y \tag{15}$$

where I_i is the local bi-variate Moran's I in spatial unit i, and Z_i^x and Z_j^y are the normalized values of the variance for the observed ESV and CMI values in spatial units i and j, respectively. Four possible cluster patterns can be generated. (i) High-high aggregation refers to the combination of a high ESV and high CMI values. (ii) Low-low aggregation refers to the combination of a low ESV and low CMI values. (iii) Low-high aggregation refers to the combination of a low ESV and high CMI values. (iv) High-low aggregation refers to the combination of a high ESV and low CMI values.

3. Results

3.1. Spatial Distribution of the ESV from 2000 to 2030

Figure 3 illustrates the spatiotemporal patterns of the ESV from 2000 to 2030. Overall, the ESV exhibits significant heterogeneity, decreasing gradually from the southeastern regions (Shandong Province, Henan Province, the downstream areas of Shanxi Province, and southern Shaanxi Province) to the northwestern regions where the ecological conditions are poorer. Additionally, the ESVs are relatively low in the downstream areas of Henan Province and Shandong Province, as well as the Weihe River Basin in the middle reaches. From the spatiotemporal perspective, the ESV generally exhibits a diffusion-like decline originating from the northwest between 2000 and 2030, and the most pronounced decrease occurred in the southern areas of the upper reaches of the Yellow River where the rate of change is mostly less than −75.71. In these regions, the fragile ecological environment, severe water scarcity, serious soil erosion, strong wind erosion, and expanding desertification and soil salinization are the primary contributing factors. However, the rate of decline has slowed in the southern upper reaches, Weihe River Basin, and Qinling Mountains. As a major beneficiary of the Returning Farmland to Forest project, the increase in forested land in Shaanxi Province has contributed to improvements in both the ecological environment and ecosystem. In contrast, the downstream regions with widespread urban land use have experienced the smallest reduction in the ESV.

3.2. The Hotspots of the ESV

As shown in Figure 4, the LISA cluster and significance analysis revealed the spatial clustering characteristics from 2000 to 2030. The clustering of the ESV in most areas of the Yellow River Basin was not spatially significant, with significant areas primarily characterized by high-high and low-low ESV clusters. In 2000, the southernmost part of the study area was a high-value ESV cluster, mainly including northern Sichuan Province, southern Shaanxi Province, southern Gansu Province, and northern Henan Province. The central-western region of the Inner Mongolia section of the studied basin was primarily a low-value cluster. By 2010, the high-high cluster in northern Sichuan Province had diminished, a high-value cluster began to appear in the central-southern part of Shaanxi Province, and there were no significant changes in the low-value clusters. In 2020, the high-value cluster in Sichuan Province disappeared, and a high-high cluster evolved in the border area between Shanxi and Henan provinces. The low-value cluster shifted toward the northeast. However, according to the predicted ESV's LISA clustering results for 2030, the high-high clustering pattern will be distributed in the southern part of Shaanxi Province, indicating that only this region's high ESV will exhibit temporal and spatial stability. The low-low cluster will be distributed in the northwestern part of the study area, suggesting a long-term trend of spreading eastward.

3.3. The Trend of the ESV

According to Figure 5, there are significant differences among the ecosystem services across the studied basin, and each service exhibits a certain degree of spatial autocorrelation. Biodiversity protection, recreation and culture, climate regulation, and soil formation and retention are more prominent in the southeastern and southwestern parts of the study area, and they exhibit relatively strong performances in the central region. However, in the western and northwestern areas, such as the Inner Mongolia Plateau and the Qinghai-Tibet Plateau, these services exhibit low values and an extremely significant decreasing trend. Similar patterns can be observed in the food production and raw material services. Notably, although the values are relatively low in the upper reaches of the Yellow River, specifically in the Alxa League and Bayannur City, they remain stable. The water supply is the most significant limiting factor across the studied basin, exhibiting negative values in most regions. The waste treatment services are primarily concentrated in the upper reaches of the Yellow River and the Qinling Mountains, exhibiting a significant increasing trend, while most of the other regions exhibit a notable decrease. The value of the gas regulation service is generally low and exhibits an extremely significant decreasing trend during the study period, and it only performs well in certain areas near the Taihang Mountains in Shanxi Province. Overall, the spatial heterogeneity of the average ESV and its change trend across the studied basin is not pronounced. The overall value is low and experiences a sharp decline between 2000 and 2020. However, certain counties, such as Qingjian County in Shanxi Province, a national ecotourism county, Wudi County in Shanxi Province, a model county for the Returning Farmland to Forest project, and the Shiguai District in Baotou City, Inner Mongolia, a model autonomous county, exhibit better average ESVs and trends. These exceptional areas clearly benefit from the targeted policy support and incentives.

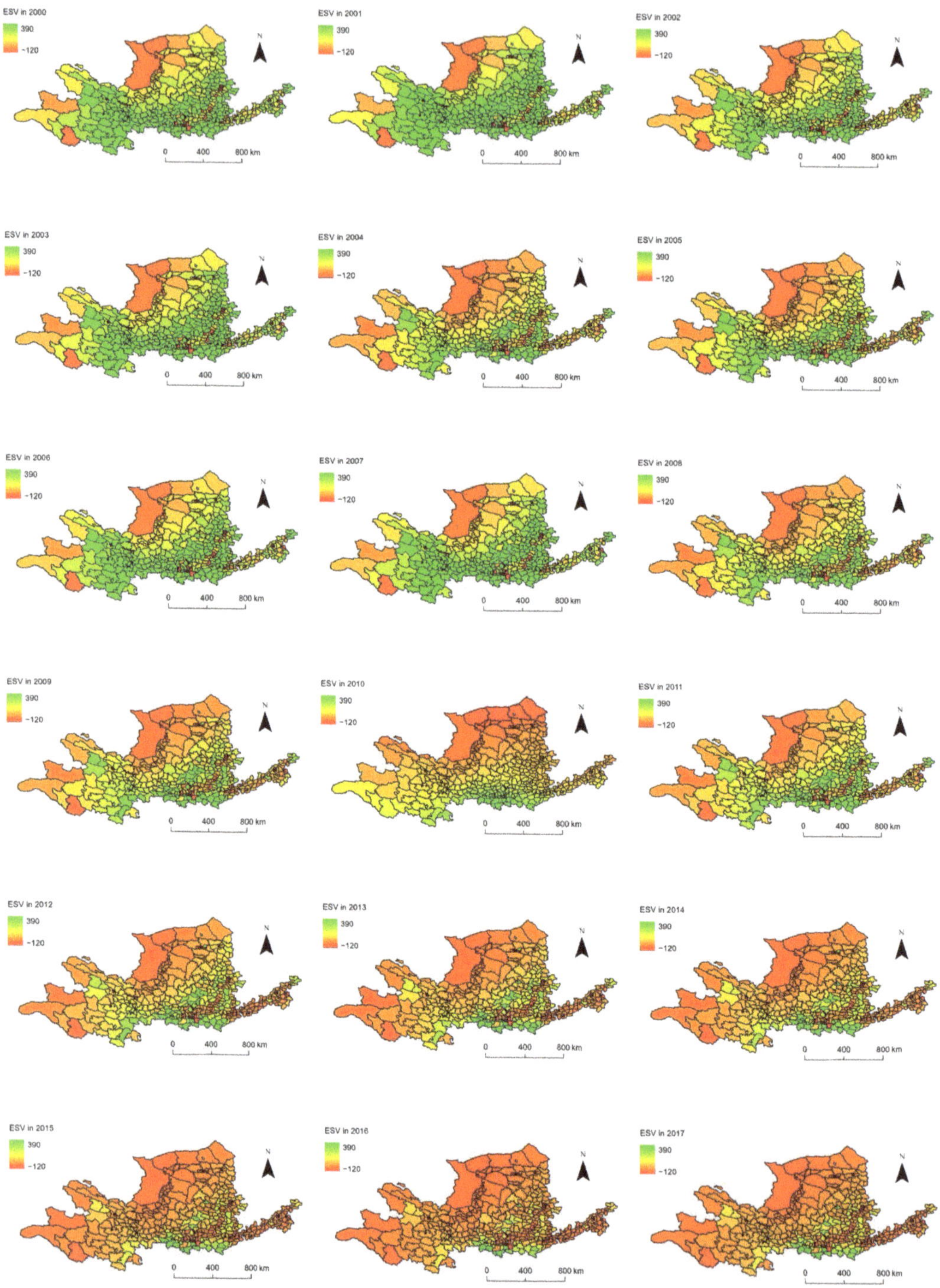

Figure 3. *Cont.*

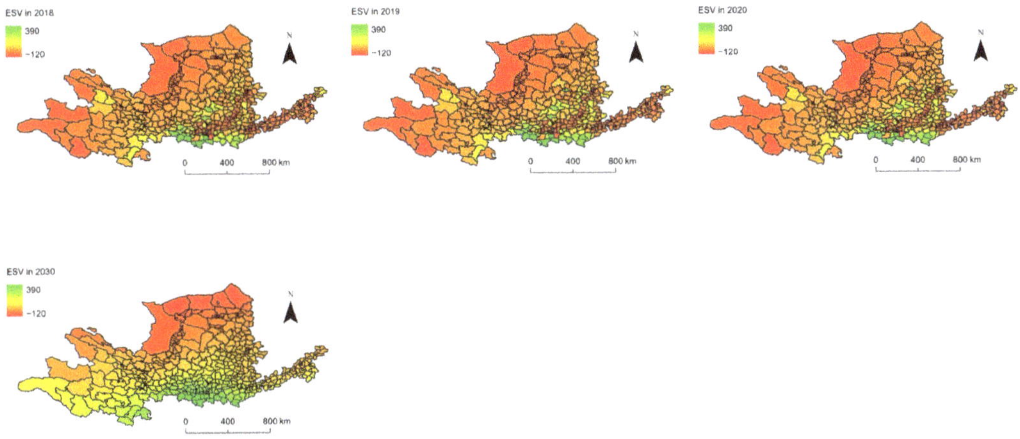

Figure 3. Spatial distribution of ESV in the Yellow River Basin from 2000 to 2030 (units: 10^4 CNY/km^2).

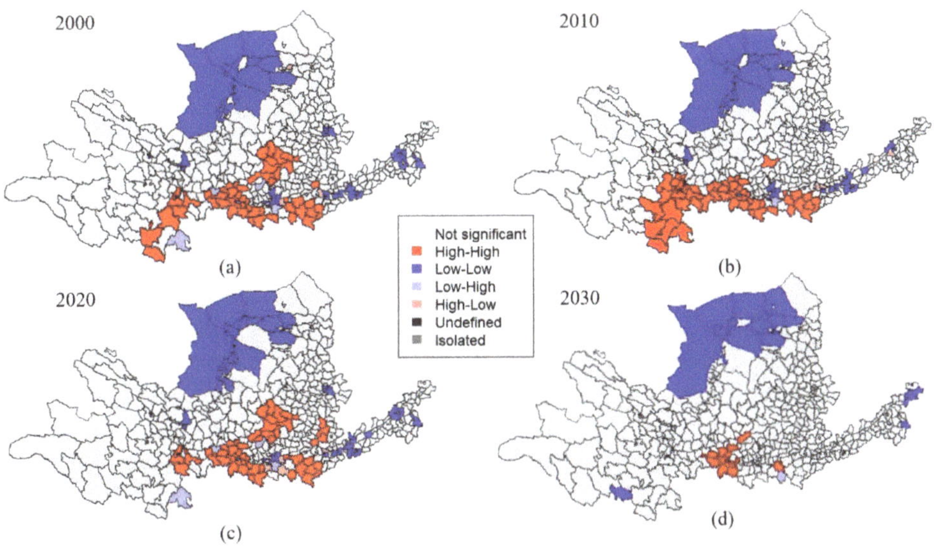

Figure 4. LISA clusters of ESV in the study area. (**a–d**) LISA clusters of ESV in the study area in 2000, 2010, 2020, and 2030, respectively.

3.4. CMI During 2000–2030

As shown in Figure 6, from 2000 to 2020, the spatiotemporal distribution of the CMI index was primarily high in the upper and middle reaches of the studied river basin, particularly in Shanxi, Shaanxi, Inner Mongolia, Ningxia, and Gansu provinces. The CMI was notably high in the junction areas between Shanxi, Shaanxi, and Inner Mongolia. During the two decades, the overall spatiotemporal pattern of the CMI remained relatively stable, although the intensity shifted in certain areas. For example, the CMI continued to increase in eastern Ordos City, Inner Mongolia, and Shenmu City, Shaanxi Province. Additionally, the scale and intensity of the coal mining expanded in the southern parts of Shaanxi and Gansu provinces, located in the middle and upper reaches of the Yellow River Basin. In contrast, the CMI gradually weakened in northern Gansu Province, located in the upper reaches.

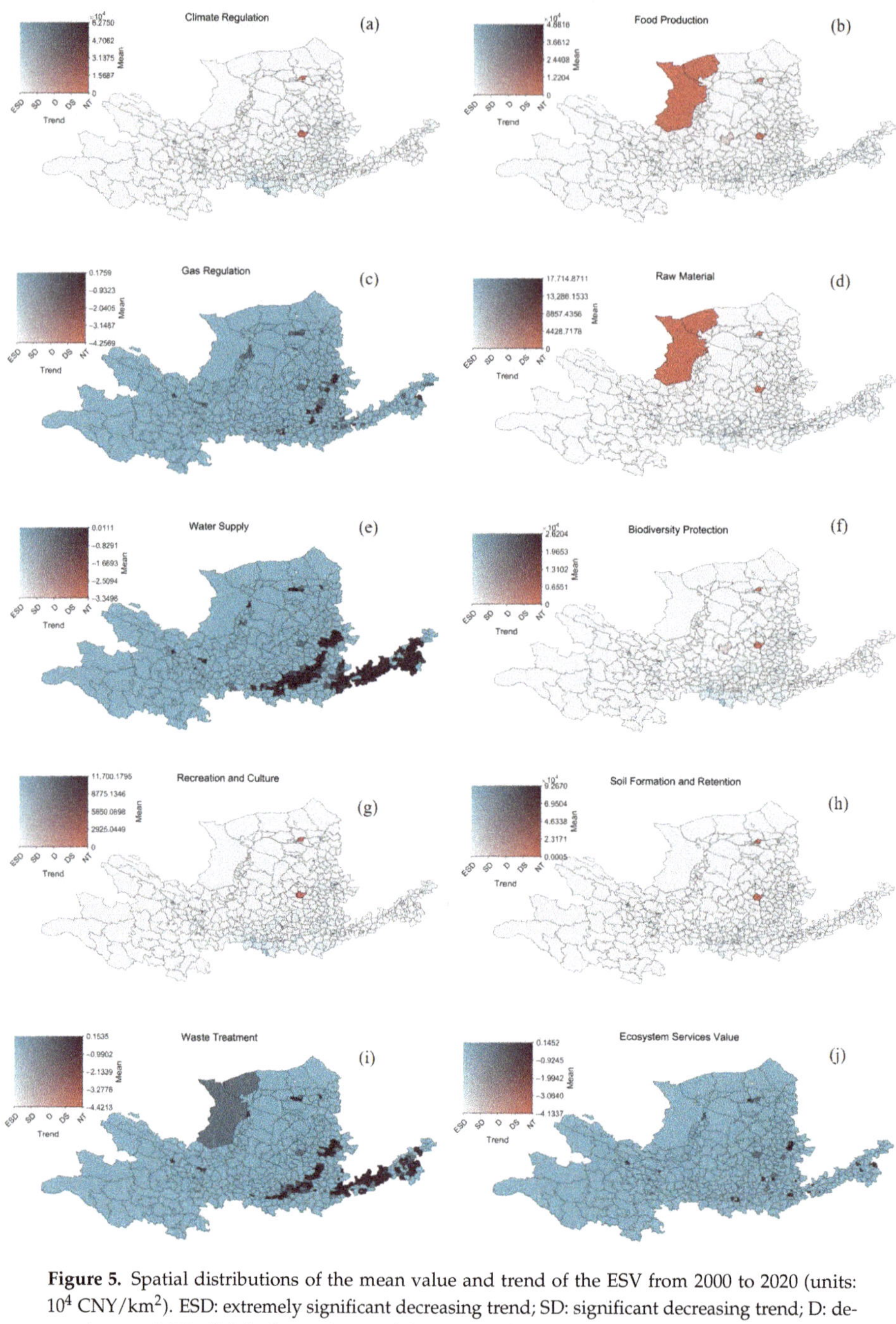

Figure 5. Spatial distributions of the mean value and trend of the ESV from 2000 to 2020 (units: 10^4 CNY/km^2). ESD: extremely significant decreasing trend; SD: significant decreasing trend; D: decreasing trend; DS: slightly decreasing trend; NT: no trend; IS: slightly increasing trend; I: increasing; ESI: extremely significant increasing trend. (**a**), Climate Regulation; (**b**), Food Production; (**c**), Gas regulation; (**d**), Raw material; (**e**), Water Supply; (**f**), Biodiversity Protection; (**g**), Recreation and culture; (**h**), Soil Formation and Retention; (**i**), Waste Treatment; (**j**). ESV(Ecosystem Services Value).

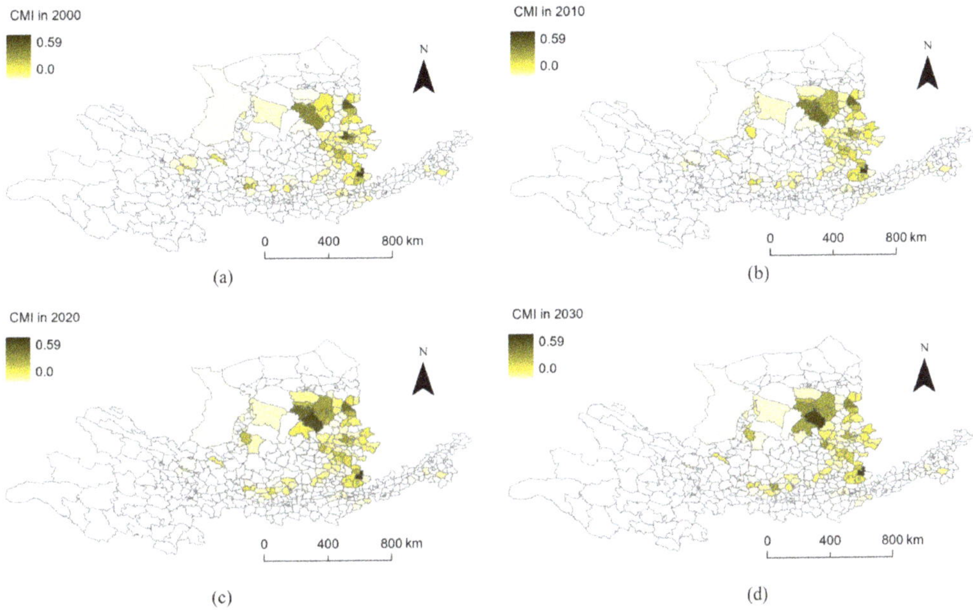

Figure 6. Spatial distributions of CMI in the Yellow River Basin from 2000 to 2030. (**a–d**) Coal mining intensity (CMI) in 2000, 2010, 2020, and 2030, respectively.

Coal mining, a key industry in the study area, produced 2.463 billion tons of coal in 2017, accounting for over 69.7% of the national coal production, and this trend is expected to continue. Due to the inability of domestic oil and gas production to meet the demands of national economic development, the focus of coal resource development has increasingly shifted to the upper and middle reaches of the studied river basin. However, these regions are located in arid and semi-arid areas with fragile ecosystems. The inevitable ecological damage caused by coal mining further exacerbates the problems of soil erosion, loss of arable land, and vegetation degradation [37].

3.5. Overlap of the ESV and CMI

As shown in Figure 7, the bivariate LISA clustering results for the ESV and CMI reveal a significant low-high clustering pattern, and the clusters are particularly concentrated in the upper and middle reaches of the studied river basin, especially in the eastern region of Ordos City. This area has hilly gullies and the Mu Us Desert, which experience severe wind erosion. Resource exploitation has further damaged the surface vegetation, intensifying land desertification, dust storms, soil erosion, grassland degradation, and resource depletion, thus making the ecosystem highly fragile [38]. However, the clustering of the ESV difference and CMI in this region exhibits a significant high-high clustering pattern, indicating that policymakers have effectively considered the dual pressures of ecological vulnerability and resource exploitation in their planning and have achieved positive outcomes.

The high-high clustering of the ESV and CMI is mainly distributed in Shanxi Province, in the middle reaches of the study area, and is interspersed with the low-high clustering areas. The region's ecosystem services are strongly influenced by the topography. The high-high clusters are concentrated in the Taihang and Lüliang Mountains, while the low-high clusters are in the Taiyuan and Xinzhou basins. This is primarily due to the difficulty of developing mountainous areas, which experience less human interference. The different topographic conditions directly impact the land use patterns, creating a locked-in effect on the regional ecosystem service capacity. The mountain areas have a stronger ecosystem service supply capacity than the plains and basins, in which the ecosystem service performance often falls below the resource extraction intensity, leading to nearly imbalanced coupling coordination. However, the high-high clustering pattern of the ESV difference and CMI in this region has significantly contracted, while the low-high clustering pattern has notably strengthened, especially between 2010 and 2020. This change suggests that with increasing resource extraction intensity, particularly in flat and ecologically fragile basin areas, the negative impacts of resource exploitation on the ESV are becoming more pronounced. This trend requires increased attention to prevent the further degradation of the ecosystem and to implement effective measures to alleviate the conflict between resource utilization and ecological protection.

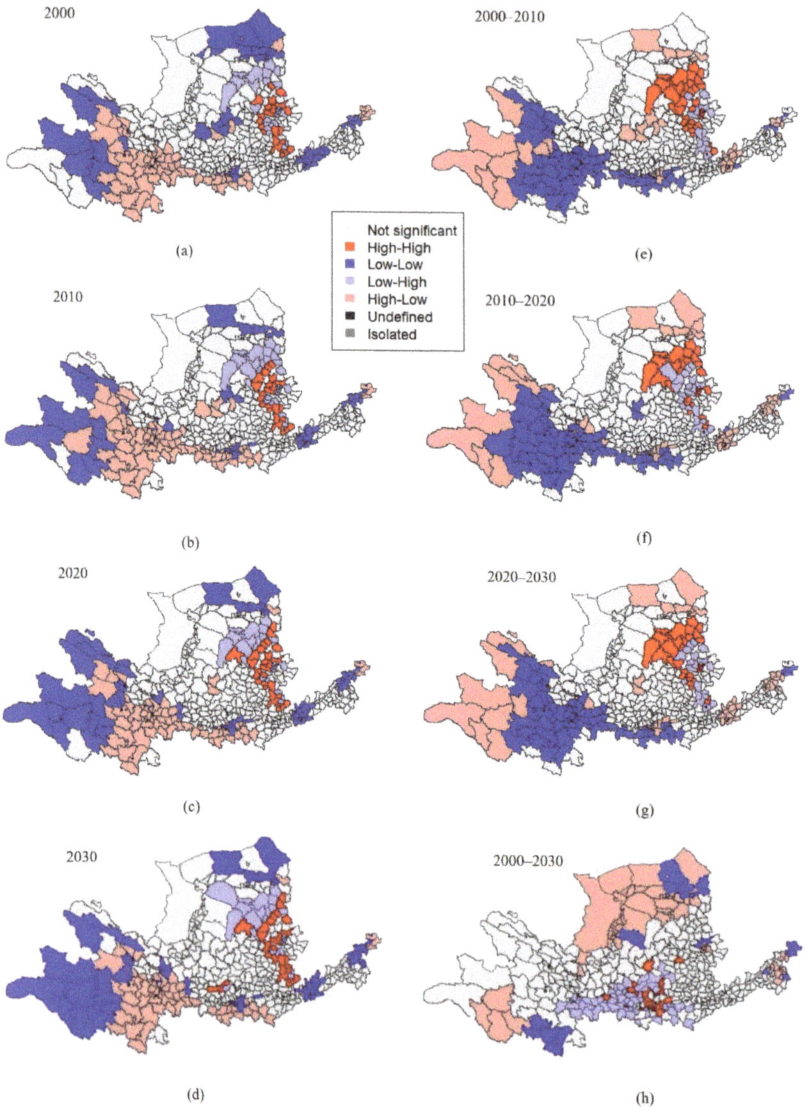

Figure 7. Bivariate LISA cluster map between ESV and CMI. (**a–d**) bivariate LISA cluster between ESV and CMI from 2000 to 2030; (**e–h**) bivariate LISA cluster between ESV change and CMI.

From 2000 to 2030, the spatial patterns of the high-high and high-low clusters of the ESV and CMI are remaining relatively stable, with a slight expansion. The high-high clustering pattern, bordered by the Lüliang Mountains, is expanding from east to west, while the low-low clustering pattern, centered in northeastern Inner Mongolia, is expanding outward. In addition, the high-high clustering pattern of the ESV change and CMI during 2000–2020 is in the Shanxi–Shaanxi–Inner Mongolia adjacent region, where the ESV decreases significantly and the CMI is high. This indicates that the resource exploitation activities in these regions have intensified, potentially placing greater pressure on the already fragile ecosystems. During 2020–2030, the central area of Shannxi Province will exhibit a high-high clustering pattern of the ESV change and CMI. Therefore, future attention should focus on enhancing the ecosystem protection and restoration in the Shanxi–Shaanxi–Inner Mongolia adjacent region and central Shaanxi Province to prevent further ecological degradation.

4. Discussion

4.1. Trade-Offs of the ESV in the Overlap Areas

In 2000, among the 36 pairs of ecosystem services, significant correlations were found between 34 pairs ($p \leq 0.05$), and 30 pairs occurred in the overlap areas ($p \leq 0.05$), with 30 pairs observed in overlap areas (Figure 8). In the Yellow River Basin, 31 pairs exhibited positive correlations, and 3 pairs exhibited negative correlations. The overlap areas contained 28 pairs with positive correlations and 2 pairs with negative correlations. Of these pairs, seventeen pairs were highly correlated (Pearson coefficient $r \geq 0.7$), five pairs were moderately correlated ($0.5 \leq r \leq 0.7$), and eight pairs were weakly correlated ($r < 0.7$). The synergy between the support services and cultural services was the strongest in the Yellow River Basin. This was mainly because the cultural services are primarily related to economic variables and do not require extensive land resources. For instance, agricultural and forest landscapes can also become leisure spaces that attract tourists. However, the synergy between the support services and provisioning services was weaker, and it even exhibited a trade-off with the food production services. A trade-off existed between the climate regulation and provisioning services. This was likely because the areas with high soil retention had extensive forest vegetation, and there was land use competition between the forested and cultivated lands. Therefore, in this study, the trade-off may have been due to the spatial incompatibility caused by the dependency of the provisioning and regulating services on the specific land use types.

A comparison of the correlation coefficients of the ecosystem services (Figures 8–10) revealed that the synergy between the raw materials and services such as water supply, waste treatment, climate regulation, and recreation and culture services decreased significantly in the overlap areas compared to the control counties. Specifically, the difference coefficients for the raw materials and waste treatment were 0.361 (2000) and 0.168 (2010), indicating that coal mining-related ecological damage hindered the promotion of other services by the raw materials [18]. Additionally, the trade-offs between food production and regulating services (e.g., climate regulation) strengthened, suggesting an increase in the agricultural burden due to coal mining [39]. The synergy between the soil formation and retention services and gas regulation services declined by 0.510 (2000) and 0.121 (2010), indicating that the degradation of the soil quality affected the gas regulation capacity [40]. Interestingly, the overlap areas had a stronger synergy between the water supply services and the supporting and regulating services compared to the control counties. This could be because the underground water was influenced by the mining activities. By 2020, the overlap areas shifted from a balanced relationship between food production and regulating services to a synergistic relationship. The waste treatment services also improved, possibly due to the implementation of sustainable management practices [41].

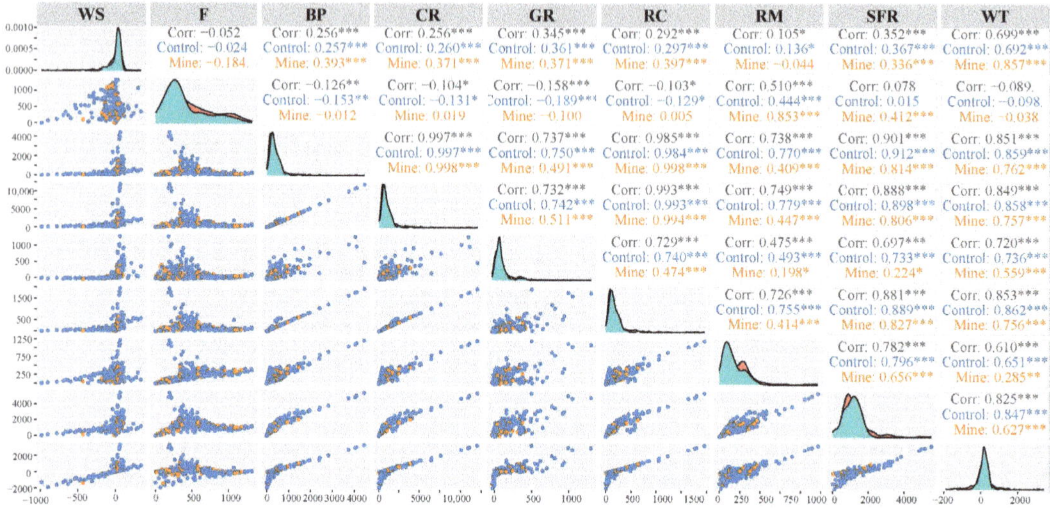

Figure 8. Trade-offs and synergies among ecosystem services in Yellow River Basin, overlap areas (Mine), and control areas (Control) in 2000. WS: water supply; F: food; BP: biodiversity protection; CR: climate regulation; GR: gas regulation; RC: recreation and culture; RM: raw material; SFR: soil formation and retention; WT: waste treatment.

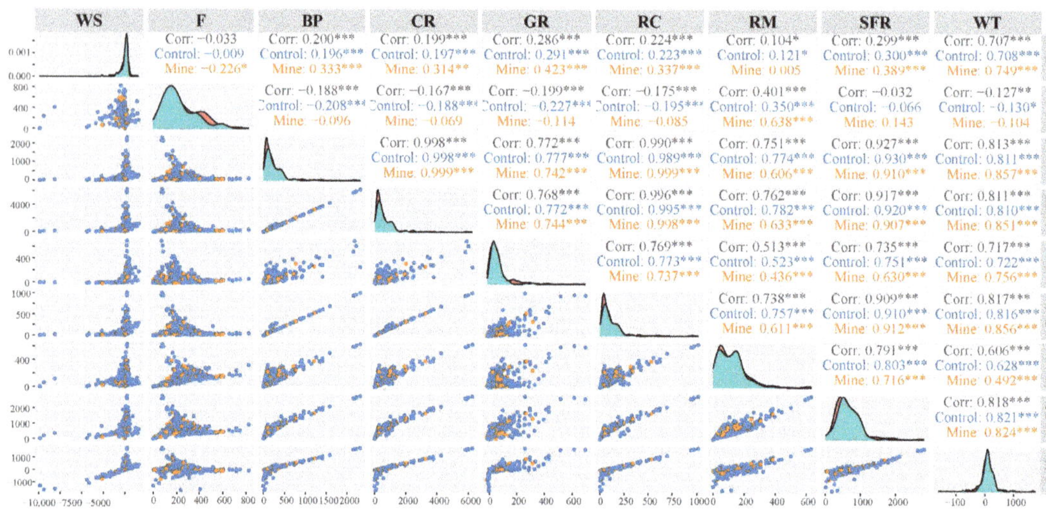

Figure 9. Trade-offs and synergies among ecosystem services in Yellow River Basin, overlap areas (Mine), and control areas (Control) in 2010. WS: water supply; F: food; BP: biodiversity protection; CR: climate regulation; GR: gas regulation; RC: recreation and culture; RM: raw material; SFR: soil formation and retention; WT: waste treatment.

A comparison of the changes in the trade-off and synergy coefficients of the overlap areas throughout the study period revealed that the synergy coefficients between the water supply services and the recreation, waste treatment, and climate regulation services remained consistent. However, the correlation coefficient between the water supply and food production services decreased from −0.184 to −0.226 (by 0.043). The synergy coefficient between the food production and raw materials decreased by 0.215 and that between the food production and soil formation and retention services decreased by 0.269. Furthermore, the trade-offs between the food production services and the waste treatment, climate regulation, and gas regulation services strengthened. In particular, the gas regulation services exhibited a stronger synergy with the other services, especially the soil formation and retention services. This may be because China's new energy practices have contributed to the reduction in greenhouse gas emissions and thus have improved the soil quality [42]. The synergy between the raw material sand food production continued to decrease, particularly between 2010 and 2020, with a decrease of 0.502. The shift from a synergistic relationship to a trade-off between the raw materials and water supply, as indicated by the change in the correlation coefficient from 0.005 to −0.050 between 2010 and 2020, likely resulted from the insufficient water supply and environmental pressures caused by the mining activities [19]. In conclusion, the supportive role of the water supply in the food production and raw materials in the study area decreased over time. The coal mining activities have affected the equilibrium between the raw materials and supporting, regulating, and cultural services and have impeded the promotion of the regulating services by the soil formation and retention services. The changes in the environmental services within the study area reflect the conflict between economic development and environmental protection [18,43].

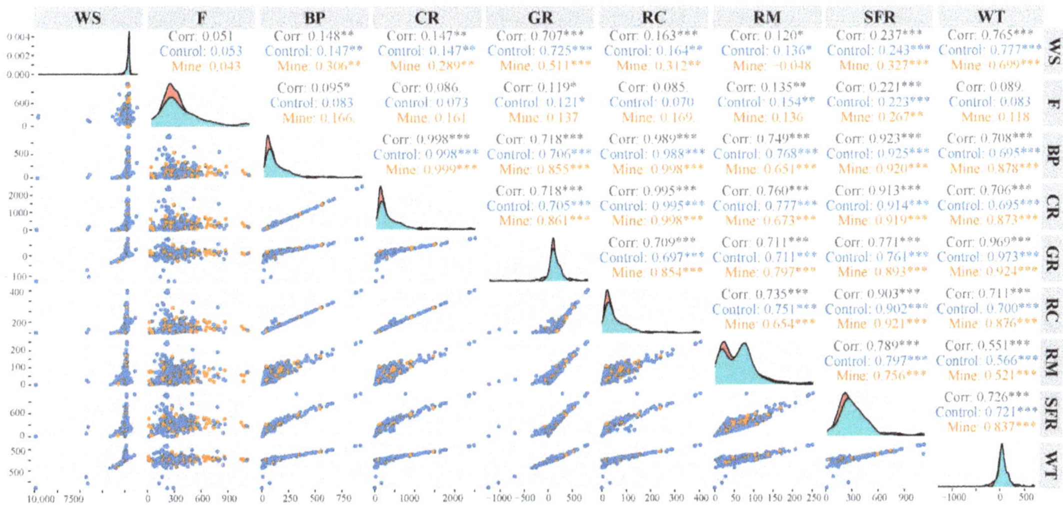

Figure 10. Trade-offs and synergies among ecosystem services in Yellow River Basin, overlap areas (Mine), and control areas (Control) in 2020. WS: water supply; F: food; BP: biodiversity protection; CR: climate regulation; GR: gas regulation; RC: recreation and culture; RM: raw material; SFR: soil formation and retention; WT: waste treatment.

4.2. Implications, Limitations, and Future Work

Starting in the twentieth century, the Chinese government has undertaken a range of ecological and environmental conservation initiatives in the study area [44]. As shown in Figure 3, these have somewhat improved the regional ecological environment. However, as shown in Figure 6, the synergies between the ecosystem services in coal mining-affected counties, specifically between raw materials and provisioning services and between gas regulation and supporting services, have continuously decreased. Moreover, the gap with the control countries has widened. Simple re-vegetation appears not to have formed a well-functioning ecosystem with multiple services and functions working in synergy. This may be due to coal mining disrupting the surface and causing subsidence and fissures, which directly affect the soil structure and hydrological conditions, causing soil nutrient loss and subsequently impacting the vegetation growth and soil nutrient cycling [41,45]. Additionally, the re-planted vegetation struggles to replicate the structure of the native communities, and its recovery may require a long period of time [46]. This suggests that the ecological restoration should focus on the synergistic effects of soil formation and conservation, diversity protection, and gas regulation services to enhance the ecosystem stability, thereby improving the recovery efficiency and capacity of the ecosystem.

Ecological restoration should be tailored to the local conditions. Figure 7 shows that the issue of uneven regional development is particularly pronounced in the coal mining provinces. Taking Shanxi Province as an example, according to the announcement by the Shanxi Provincial Energy Bureau regarding the production capacity of coal mines in Shanxi Province, there were a total of 668 coal mines in Shanxi at the end of December 2020 [47]. Long-term intensive coal mining has left the ecological environment fragile and the ecological carrying capacity low. The analysis of the spatiotemporal evolution of the ESVs in counties in Shanxi Province revealed that the ecosystem service capacity of the study area is mainly influenced by the topography, and the areas with better ecosystem service capacities are concentrated in the Taihang and Lüliang mountain regions. However, the ecosystem service capacity in the Lüliang Mountain area is significantly lower than that in the Taihang Mountain area. This is mainly because the Lüliang Mountains are located in a typical loess hilly and gully region, where severe soil erosion has occurred. The water conservation service per unit area is only 320.9 m^3/ha, 20% lower than the provincial average, and the mountainous region is adjacent to the area where mineral resources have been found. Human activities are regularly conducted in this area. In 2021, the Lüliang Mountain area had a total of 91 coal mines, accounting for 13.62% of the province's total number of coal mines. The extensive operations and insufficient management measures have amplified the strain on the ecosystem services and diminished the quality of the regional ecosystem services in this area [48]. The northwestern part of the study area in Inner Mongolia has been identified as a location with limited value in terms of ecosystem services. This is mostly due to the unfavorable natural climate conditions and frequent occurrence of sandstorm catastrophes in this area. This area has a mid-temperate continental semi-arid monsoon climate and experiences frequent strong winds and little rainfall, resulting in a low upper limit of its ecosystem service capacity [49–51]. Therefore, it is strongly recommended that policymakers and decision makers consider these factors during the implementation of ecological restoration, develop more targeted restoration policies, and consult experts during the restoration process to enhance the effectiveness of the restoration measures. New technology of coal mining such as water-preserved mining, and backfill mining, should be widely applied to protect the ecosystem services such as water supply.

In this study, we used the administration cells, namely the counties within the Yellow River Basin, to assess the changes in the ESV and its relationship with the CMI. It is beneficial to assess the official statistical data and make management decisions for each administration cell. Nevertheless, the spatial division of the towns and cities is relatively imprecise [52], which hinders the comprehension of the cause-and-effect relationships between the ecosystem services [53]. Through a survey, we found that the mean size of the towns and cities was 27,979 km^2, ranging from 1656 to 253,015 km^2. To mitigate the impact of the changes in the geographic unit area, the indicators of ecosystem services and socio-ecological factors pertaining to the towns and cities were measured by normalizing them with respect to the area. Nevertheless, the spatial unit areas analyzed in this study were significantly larger compared to those used in previous studies that employed the same ecosystem service valuation approach [54–56]. This technique was unsuccessful in capturing the detailed ecological and human-related factors within these huge spatial units, which are especially crucial for understanding and predicting the outcomes regarding the ecosystem services [56]. Additionally, the dynamic ESV adjustment variables used in this study, such as the GDP and Engel's coefficient, have a national-scale precision, which may lead to the overestimation of the ESV in some of the western provinces due to the uneven economic development in these areas [57]. Therefore, future research should assess and analyze the ESV across the entire study area at multiple spatial scales such as the geographic grid, individual coal mine, and watershed scales.

5. Conclusions

In this study, we utilized abundant data from the Google Earth Engine platform, comprehensively assessed the ecosystem services in the Yellow River Basin from 2000 to 2020, analyzed the evolution patterns and trends of the ecosystem services, predicted the future ESV, and determined the interactive stress mechanisms between the ESV and coal mining. The findings of this study are summarized below:

1. From 2000 to 2020, the overall spatial distribution of the ESV increased from east to west; however, from 2000 and 2030, the area with a decreasing ESV trend is expanding from the initial position in the northwest. The most significant decreases occurred in the southern upstream region of the Yellow River Basin where most of the rates of change were less than −75.71.
2. From 2000 to 2030, the high-high clustering areas of the ESV in the Yellow River Basin are shifting from Sichuan and southern Shaanxi in 2000 to the central-southern part of Shaanxi and the border area between Shanxi and Henan, while the low-low clustering areas are remaining stable in the central-western part of Inner Mongolia, with a trend of spreading eastward. By 2030, the high-high areas are expected to be stable and to be in the southern part of Shaanxi, while the low-low areas may continue to spread eastward.
3. The ecosystem services in the study area have exhibited significant spatial heterogeneity. The gas regulation, biodiversity protection, and recreation and culture services are stronger in the southeastern and southwestern regions and weaker in the northwestern regions such as Inner Mongolia and the Tibetan Plateau. The provision of the ecosystem services has decreased over the past 20 years, but certain services along the Yellow River, such as water supply and waste treatment services, have exhibited increasing trends.
4. From 2000 to 2030, the CMI in the study area is mainly concentrated in the upstream and midstream areas, especially at the junction of Shanxi, Shaanxi, and Inner Mongolia, where the overall pattern is stable but the intensity of the mining continues to increase in localized areas such as Ordos and Shenmu cities. From 2000 to 2020, the ESVs of these areas decreased significantly and the CMI intensified.
5. The development of the coal resources in the Yellow River Basin must take into account the value of the ecosystem services. It is necessary to avoid setting up coal mines in highly sensitive areas during the planning phase. During the mining phase, methods such as water-preserved mining should be adopted to reduce the impact on the hydrological system, while ecological restoration measures should be taken to restore the ecosystem functions, thereby reducing the trade-off among ecosystem services and enabling more ecosystem services to be preserved during mining.

Author Contributions: Yongjun Yang: conceptualization, supervision, methodology, and writing—review and editing. Renjie Gong: methodology and writing—review and editing. Qinyu Wu: conceptualization, data curation, methodology, software, visualization, writing—original draft, and writing—review and editing. Fu Chen: methodology and writing—review and editing. All authors have read and agreed to the published version of the manuscript.

Funding: This research is supported by the National Science Foundation of China (Grant No. 52474197 and No. 52394193).

Data Availability Statement: The data presented in this study are available on request from the corresponding author.

Conflicts of Interest: The authors declare no conflict of interest.

References

1. Semeraro, T.; Scarano, A.; Santino, A.; Emmanuel, R.; Lenucci, M. An innovative approach to combine solar photovoltaic gardens with agricultural production and ecosystem services. *Ecosyst. Serv.* **2022**, *56*, 101450. [CrossRef]
2. Costanza, R.; Limburg, K.; Naeem, S.; O'Neill, R.V.; Paruelo, J.; Raskin, R.G.; Sutton, P. The value of the world's ecosystem services and natural capital. *Nature* **1997**, *387*, 253–260. [CrossRef]
3. De Groot, R.; Brander, L.; Van Der Ploeg, S.; Costanza, R.; Bernard, F.; Braat, L.; Christie, M.; Crossman, N.; Ghermandi, A.; Hein, L.; et al. Global estimates of the value of ecosystems and their services in monetary units. *Ecosyst. Serv.* **2012**, *1*, 50–61. [CrossRef]
4. Millennium Ecosystem Assessment. Available online: https://www.millenniumassessment.org/en/index.html (accessed on 12 May 2024).
5. Xu, J.; Renaud, F.G.; Barrett, B. Modelling land system evolution and dynamics of terrestrial carbon stocks in the Luanhe River Basin, China: A scenario analysis of trade-offs and synergies between sustainable development goals. *Sustain. Sci.* **2022**, *17*, 1323–1345. [CrossRef]
6. Kizos, T.; Verburg, P.H.; Bürgi, M.; Gounaridis, D.; Plieninger, T.; Bieling, C.; Balatsos, T. From concepts to practice: Combining different approaches to understand drivers of landscape change. *Ecol. Soc.* **2018**, *23*, 25. [CrossRef]
7. Malek, Z.; Boerboom, L.G.J. Participatory scenario development to address potential impacts of land use change: An example from the Italian Alps. *Mt. Res. Dev.* **2015**, *35*, 126–138. [CrossRef]
8. Chen, Y.; Fu, B.; Zhao, Y.; Wang, K.; Zhao, M.M.; Ma, J.; Wu, J.-H.; Xu, C.; Liu, W.; Wang, H. Sustainable development in the Yellow River Basin: Issues and strategies. *J. Clean. Prod.* **2020**, *263*, 121223. [CrossRef]
9. Ma, L.; Tian, H.; Kang, L. Eco-environmental impact and spatial control of mineral resources exploitation in the Yellow River Basin. *Resour. Sci.* **2020**, *42*, 137–149. [CrossRef]
10. Fu, G.; Chen, S.; Liu, C.; Shepard, D. Hydro-climatic trends of the Yellow River basin for the last 50 years. *Clim. Chang.* **2004**, *65*, 149–178. [CrossRef]
11. Jiang, W.; Yuan, L.; Wang, W.; Cao, R.; Zhang, Y.; Shen, W. Spatio-temporal analysis of vegetation variation in the Yellow River Basin. *Ecol. Indic.* **2015**, *51*, 117–126. [CrossRef]
12. Yin, D.; Yu, H.; Shi, Y.; Zhao, M.; Zhang, J.; Li, X. Matching supply and demand for ecosystem services in the Yellow River Basin, China: A perspective of the water-energy-food nexus. *J. Clean. Prod.* **2023**, *384*, 135469. [CrossRef]
13. Wohlfart, C.; Kuenzer, C.; Chen, C.; Liu, G. Social–ecological challenges in the Yellow River basin (China): A review. *Environ. Earth Sci.* **2016**, *75*, 1066. [CrossRef]
14. Bian, Z.; Yu, H.; Lei, S.; Yin, D.; Zhu, G.; Mu, S.; Yang, D. A strategic study on coal resources development and ecological restoration strategy in the Yellow River Basin. *J. China Coal Soc.* **2021**, *46*, 1378–1391.
15. Boldy, R.; Annandale, M.; Erskine, P.D.; Sonter, L.J. Assessing impacts of mining on provisioning ecosystem services in a culturally diverse landscape of Western Cape York, Australia. *Landsc. Ecol.* **2023**, *38*, 4467–4481. [CrossRef]
16. Hu, F.; Xu, L.; Guo, Y.; Zhang, C. Spatio-temporal evolution and trend prediction of urban ecosystem service value based on CLUE-S and GM (1,1) compound model. *Environ. Monit. Assess.* **2023**, *195*, 1282.
17. Sun, B.; Tang, J.; Yu, D.; Song, Z. Coupling coordination relationship between ecosystem services and water-land resources for the Daguhe River Basin, China. *PLoS ONE* **2021**, *16*, e0257123. [CrossRef]
18. Bian, Z.; Lu, Q. Ecological effects analysis of land use change in coal mining area based on ecosystem service valuing: A case study in Jiawang. *Environ. Earth Sci.* **2013**, *68*, 1619–1630. [CrossRef]
19. Li, S.; Zhao, Y.; Xiao, W.; Yellishetty, M.; Yang, D. Identifying ecosystem service bundles and the spatiotemporal characteristics of trade-offs and synergies in coal mining areas with a high groundwater table. *Sci. Total Environ.* **2022**, *807*, 151036. [CrossRef]
20. Guo, D.; Song, X.; Hu, R.; Cai, S.; Zhu, X.; Hao, Y. Grassland type-dependent spatiotemporal characteristics of productivity in Inner Mongolia and its response to climate factors. *Sci. Total Environ.* **2021**, *775*, 145644. [CrossRef]
21. Chen, S.; Feng, Y.; Tong, X.; Liu, S.; Xie, H.; Gao, C.; Lei, Z. Modeling ESV losses caused by urban expansion using cellular automata and geographically weighted regression. *Sci. Total Environ.* **2020**, *712*, 136509. [CrossRef]
22. Fang, L.; Wang, L.; Chen, W.; Sun, J.; Cao, Q.; Wang, S.; Wang, L. Identifying the impacts of natural and human factors on ecosystem service in the Yangtze and Yellow River Basins. *J. Clean. Prod.* **2021**, *314*, 127995. [CrossRef]
23. Gorelick, N.; Hancher, M.; Dixon, M.; Ilyushchenko, S.; Thau, D.; Moore, R. Google Earth Engine: Planetary-scale geospatial analysis for everyone. *Remote Sens. Environ.* **2017**, *202*, 18–27. [CrossRef]
24. Xiao, F.; Liu, Q.; Xu, Y. Estimation of terrestrial net primary productivity in the Yellow River Basin of China using light use efficiency model. *Sustainability* **2022**, *14*, 7399. [CrossRef]
25. Liang, J.; Xie, Y.; Sha, Z.; Zhou, A. Modeling urban growth sustainability in the cloud by augmenting Google Earth Engine (GEE). *Comput. Environ. Urban Syst.* **2020**, *84*, 101542. [CrossRef]
26. Xie, G.; Zhang, C.; Zhen, L.; Zhang, L. Dynamic changes in the value of China's ecosystem services. *Ecosyst. Serv.* **2017**, *26*, 146–154. [CrossRef]
27. Li, B.; Yang, Z.; Cai, Y.; Xie, Y.; Guo, H.; Wang, Y.; Zhang, P.; Li, B.; Jia, Q.; Huang, Y.; et al. Prediction and valuation of ecosystem service based on land use/land cover change: A case study of the Pearl River Delta. *Ecol. Eng.* **2022**, *179*, 106612. [CrossRef]

28. Xiao, W.; Ren, H.; Sui, T.; Zhang, H.; Zhao, Y.; Hu, Z. A drone- and field-based investigation of the land degradation and soil erosion at an opencast coal mine dump after 5 years' evolution of natural processes. *Int. J. Coal. Sci. Technol.* **2022**, *9*, 42. [CrossRef]
29. Sen, P.K. Estimates of the Regression Coefficient Based on Kendall's Tau. *J. Am. Stat. Assoc.* **1968**, *63*, 1379–1389. [CrossRef]
30. Gao, H.; Liu, S.; Feng, S.; Peng, X.; Ning, Y.; Shi, Y.; Wang, Z.; Wei, B.; Yan, W. Nonlinear imprints of forest coverage on the relationships between gross primary production (GPP) and landscape patterns. *Ecol. Indic.* **2023**, *146*, 109783. [CrossRef]
31. Peng, T.; Tian, H.; Singh, V.P.; Chen, M.; Liu, J.; Ma, H.; Wang, J. Quantitative assessment of drivers of sediment load reduction in the Yangtze River basin, China. *J. Hydrol.* **2020**, *580*, 124242. [CrossRef]
32. Matheron, G. Principles of geostatistics. *Econ. Geol.* **1963**, *58*, 1246–1266. [CrossRef]
33. Krige, D.G. A statistical approach to some basic mine valuation problems on the Witwatersrand. *J. South. Afr. Inst. Min. Metall.* **1951**, *52*, 119–139.
34. Niu, L.; Shao, Q.; Ning, J.; Huang, H. Ecological changes and the tradeoff and synergy of ecosystem services in western China. *J. Geogr. Sci.* **2022**, *32*, 1059–1075. [CrossRef]
35. Kim, I.; Kwon, H.; Kim, S.; Jun, B. Identification of landscape multifunctionality along urban-rural gradient of coastal cities in South Korea. *Urban Ecosyst.* **2020**, *23*, 1153–1163. [CrossRef]
36. Amaral, P.V.; Anselin, L. Finite sample properties of Moran's I test for spatial autocorrelation in tobit models *. *Pap. Reg. Sci.* **2014**, *93*, 773–782. [CrossRef]
37. Peng, S.; Bi, Y. Strategic consideration and core technology about environmental ecological restoration in coal mine areas in the Yellow River basin of China. *J. China Coal Soc.* **2020**, *45*, 1211–1221.
38. Han, J.; Hu, Z.; Mao, Z.; Li, G.; Liu, S.; Yuan, D.; Guo, J. How to account for changes in carbon storage from coal mining and reclamation in Eastern China? Taking Yanzhou Coalfield as an example to simulate and estimate. *Remote Sens.* **2022**, *14*, 2014. [CrossRef]
39. Dorn, F.M.; Hafner, R.; Plank, C. Towards a climate change consensus: How mining and agriculture legitimize green extractivism in Argentina. *Extr. Ind. Soc.* **2022**, *11*, 101130. [CrossRef]
40. Yan, M.; Cui, F.; Liu, Y.; Zhang, Z.; Zhang, J.; Ren, H.; Li, Z. Vegetation type and plant diversity affected soil carbon accumulation in a postmining area in Shanxi Province, China. *Land Degrad. Dev.* **2020**, *31*, 181–189. [CrossRef]
41. Du, H.; Xie, S.; Bi, Y.; Liu, Y.; Sun, H.; Liu, Y. Influence of mining ground fissures development on seedling bank and seedling establishment factors in semi-arid mining area. *Coal Sci. Technol.* **2024**, *52*, 350–362.
42. Agus, C.; Putra, P.B.; Faridah, E.; Wulandari, D.; Napitupulu, R.R.P. Organic carbon stock and their dynamics in rehabilitation ecosystem areas of post open coal mining at tropical region. *Procedia Eng.* **2016**, *159*, 329–337. [CrossRef]
43. Bag, S.; Rahman, M.S.; Srivastava, G.; Chan, H.-L.; Bryde, D.J. The role of big data and predictive analytics in developing a resilient supply chain network in the South African mining industry against extreme weather events. *Int. J. Prod. Econ.* **2022**, *251*, 108541. [CrossRef]
44. Yang, Y.; Li, H.; Qian, C. Analysis of the implementation effects of ecological restoration projects based on carbon storage and eco-environmental quality: A case study of the Yellow River Delta, China. *J. Environ. Manag.* **2023**, *340*, 117929. [CrossRef]
45. Zhao, Y.; Sun, B.; Liu, S.; Zhang, C.; He, X.; Xu, D.; Tang, W. Identification of mining induced ground fissures using UAV and infrared thermal imager: Temperature variation and fissure evolution. *ISPRS J. Photogramm. Remote Sens.* **2021**, *180*, 45–64. [CrossRef]
46. Yang, Y.; Tang, J.; Zhang, Y.; Zhang, S.; Zhou, Y.; Hou, H.; Liu, R. Reforestation improves vegetation coverage and biomass, but not spatial structure, on semi-arid mine dumps. *Ecol. Eng.* **2022**, *175*, 106508. [CrossRef]
47. Li, J.; Li, H.; Zhang, Q.; Shao, H.; Gao, C.; Zhang, X. Effects of fertilization and straw return methods on the soil carbon pool and CO_2 emission in a reclaimed mine spoil in Shanxi Province, China. *Soil Tillage Res.* **2019**, *195*, 104361. [CrossRef]
48. Cui, Y.; Fan, L.; Liu, S.; Sun, T. Evaluation of forest ecosystem services value in Shanxi Province. *Acta Ecol. Sin.* **2019**, *39*, 4732–4740.
49. Dou, H.; Li, X.; Li, S.; Dang, D.; Li, X.; Lyu, X.; Li, M.; Liu, S. Mapping ecosystem services bundles for analyzing spatial trade-offs in inner Mongolia, China. *J. Clean. Prod.* **2020**, *256*, 120444. [CrossRef]
50. Fan, F.; Liang, C.; Tang, Y.; Harker-Schuch, I.; Porter, J.R. Effects and relationships of grazing intensity on multiple ecosystem services in the Inner Mongolian steppe. *Sci. Total Environ.* **2019**, *675*, 642–650. [CrossRef]
51. Liu, L.; Ma, Q.; Shang, C.; Wu, J. How does the temporal relationship between ecosystem services and human wellbeing change in space and time? Evidence from Inner Mongolian drylands. *J. Environ. Manag.* **2023**, *339*, 117930. [CrossRef]
52. Raudsepp-Hearne, C.; Peterson, G.D.; Bennett, E.M. Ecosystem service bundles for analyzing tradeoffs in diverse landscapes. *Proc. Natl. Acad. Sci. USA* **2010**, *107*, 5242–5247. [CrossRef] [PubMed]
53. Cord, A.F.; Bartkowski, B.; Beckmann, M.; Dittrich, A.; Hermans-Neumann, K.; Kaim, A.; Lienhoop, N.; Locher-Krause, K.; Priess, J.; Schröter-Schlaack, C.; et al. Towards systematic analyses of ecosystem service trade-offs and synergies: Main concepts, methods and the road ahead. *Ecosyst. Serv.* **2017**, *28*, 264–272. [CrossRef]
54. Gómez-Creutzberg, C.; Lagisz, M.; Nakagawa, S.; Brockerhoff, E.G.; Tylianakis, J.M. Consistent trade-offs in ecosystem services between land covers with different production intensities. *Biol. Rev.* **2021**, *96*, 1989–2008. [CrossRef] [PubMed]
55. Jaligot, R.; Chenal, J.; Bosch, M. Assessing spatial temporal patterns of ecosystem services in Switzerland. *Landsc. Ecol.* **2019**, *34*, 1379–1394. [CrossRef]

56. Spake, R.; Lasseur, R.; Crouzat, E.; Bullock, J.M.; Lavorel, S.; Parks, K.E.; Schaafsma, M.; Bennett, E.M.; Maes, J.; Mulligan, M.; et al. Unpacking ecosystem service bundles: Towards predictive mapping of synergies and trade-offs between ecosystem services. *Glob. Environ. Chang.* **2017**, *47*, 37–50. [CrossRef]
57. Ou, D.; Wu, N.; Li, Y.; Ma, Q.; Zheng, S.; Li, S.; Yu, D.; Tang, H.; Gao, X. Delimiting ecological space and simulating spatial-temporal changes in its ecosystem service functions based on a dynamic perspective: A case study on Qionglai City of Sichuan Province, China. *J. Resour. Ecol.* **2022**, *13*, 1128–1142.

International Journal of *Geo-Information*

MDPI

Article

Optimizing Site Selection for Construction: Integrating GIS Modeling, Geophysical, Geotechnical, and Geomorphological Data Using the Analytic Hierarchy Process

Doaa Wahba [1], Awad A. Omran [2], Ashraf Adly [3], Ahmed Gad [4,5,*], Hasan Arman [4] and Heba El-Bagoury [6]

1 Geology Department, Faculty of Science, New Valley University, Kharga 72511, Egypt
2 Geology Department, Faculty of Science, Assiut University, Assiut 71526, Egypt
3 Seismology Department, National Research Institute of Astronomy and Geophysics (NRIAG), Cairo 11421, Egypt
4 Geosciences Department, College of Science, United Arab Emirates University, Al Ain 15551, United Arab Emirates
5 Geology Department, Faculty of Science, Ain Shams University, Cairo 11566, Egypt
6 Geography Department, Faculty of Arts, Port Said University, Port Said 42511, Egypt
* Correspondence: ahmed_g@uaeu.ac.ae

Abstract: Identifying suitable sites for urban, industrial, and tourist development is important, especially in areas with increasing population and limited land availability. Kharga Oasis, Egypt, stands out as a promising area for such development, which can help reduce overcrowding in the Nile Valley and Delta. However, soil and various environmental factors can affect the suitability of civil engineering projects. This study used Geographic Information Systems (GISs) and a multi-criteria decision-making approach to assess the suitability of Kharga Oasis for construction activities. Geotechnical parameters were obtained from seismic velocity data, including Poisson's ratio, stress ratio, concentration index, material index, N-value, and foundation-bearing capacity. A comprehensive analysis of in situ and laboratory-based geological and geotechnical data from 24 boreholes examined soil plasticity, water content, unconfined compressive strength, and consolidation parameters. By integrating geotechnical, geomorphological, geological, environmental, and field data, a detailed site suitability map was created using the analytic hierarchy process to develop a weighted GIS model that accounts for the numerous elements influencing civil project design and construction. The results highlight suitable sites within the study area, with high and very high suitability classes covering 56.87% of the land, moderate areas representing 27.61%, and unsuitable areas covering 15.53%. It should be noted that many settlements exist in highly vulnerable areas, emphasizing the importance of this study. This model identifies areas vulnerable to geotechnical and geoenvironmental hazards, allowing for early decision-making at the beginning of the planning process and reducing the waste of effort. The applied model does not only highlight suitable sites in the Kharga Oasis, Egypt, but, additionally, it provides a reproducible method for efficiently assessing land use suitability in other regions with similar geological and environmental conditions around the world.

Keywords: geotechnical parameters; seismic data; site suitability; AHP; GIS model; Kharga Oasis; Egypt

1. Introduction

Rapid urbanization and industrialization around the world demand careful sustainable urban planning and exploration of potential sites. Sophisticated tools and techniques are needed to identify suitable locations for construction, particularly in areas with complex geographic and environmental conditions [1,2]. Within the framework of Geographic Information System (GIS) modeling, various analytical methods can be employed to estimate different site hazard and suitability scenarios. According to Marinoni [3], one of the most promising multiple criteria decision-making (MCDM) techniques is the Analytical Hierarchy Process (AHP). AHP, pioneered by Saaty [4,5], is a structured, mathematical decision-making method that evaluates multiple criteria. It incorporates both quantitative data and qualitative factors to assess and rank different scenarios, helping to identify the most suitable option. GISs and AHP have been employed to provide a structured framework for integrating and analyzing mixed datasets [2,6].

The integration of GISs with AHP for a single site selection has been widely explored in different contexts. For instance, the optimization of photovoltaic power and solar power plant site selection [7,8]. Similarly, GIS-AHP models have been used for the construction of waste landfills [9], fire stations in cities [10], and wastewater and sewage treatment plants [11]. Additionally, rainwater harvesting site selection for drought-prone areas, flood hazard mapping, and seismic vulnerability assessments have benefited from this approach [12–14]. On a larger scale, some researchers have investigated the integration of GISs and AHP for land use suitability assessments and sustainable urban development. Pramanik [15] conducted a site suitability analysis for agricultural land use in Darjeeling District, India, using AHP and GIS techniques to emphasize the importance of determining adequate land for agriculture in hilly areas. Suthar et al. [16] utilized nine key factors, including slope, elevation, land use/land cover, distance from faults, rivers, roads, urban amenities, soil erosion, and landslide density, to categorize land into five suitability classes for urban development in Himalayan districts using GISs and AHP. Their study emphasized the challenges of urbanization in a tectonically active mountain region. Al-Ghorayeb et al. [1] recommended considering socio-economic, environmental, and geophysical factors when identifying suitable areas for sustainable urban expansion, utilizing a GIS-AHP model to achieve sustainability. Ramya and Devadas [17] identified suitable locations for establishing agro-based industries in the Tehri Garhwal District of Uttarakhand State, India. They demonstrated the efficiency of GIS, AHP, and TOPSIS modeling for evaluating land use suitability for industrial development. Although GIS-AHP models have been widely utilized for site suitability analysis, the combination of subsurface geotechnical parameters derived from seismic data with surface parameters is not investigated to accurately assess site suitability. This integration is necessary for regions with complex geological and environmental settings.

The assessment of subsurface conditions and the accurate characterization of geotechnical properties are keys to determining the feasibility and safety of construction projects. Geotechnical parameters are an important factor in the design and implementation of geotechnical engineering projects [18–20]. The engineering and physical properties of soil are more variable than those of any other construction material, and these properties can vary significantly both horizontally and vertically [21]. Traditionally, obtaining geotechnical measurements has been a lengthy and invasive process. However, recent technological advancements have made it possible to determine these parameters from non-destructive geophysical measurements such as seismic wave velocity data [22,23]. The shallow seismic method has become a valuable tool for determining geotechnical parameters, providing better insight into the engineering conditions of soils and subsurface rocks by measuring P- and S-wave velocities, which are governed by the fundamental parameters of elastic strength and density [24,25]. Parameters such as Poisson's ratio, stress ratio, concentration index, material index, N-value, and foundation-bearing capacity derived from seismic velocity data provide basic insight into the geotechnical characteristics of a construction site [18,23]. In addition, a comprehensive analysis of the soil mechanical properties of a construction site, such as soil plasticity, water content, unconfined compressive strength, and consolidation, is necessary to assess site suitability [26,27].

The case of development projects in Egypt illustrates the importance of this comprehensive approach. Rapid population growth has led to the creation of new settlements, industrial centers, and tourist destinations in desert areas [28]. However, many of these projects lack thorough investigation before development, resulting in long-term sustainability and safety concerns. For example, inadequate assessment of rock suitability has resulted in infrastructure being constructed on unsuitable foundations, which could pose a risk of construction failure, along with insufficient risk mapping for flooding and unpredictable seismic activity, threatening lives and infrastructure [29–31]. Future development projects can be designed and implemented with greater confidence regarding their long-term environmental sustainability. Integrating different concepts into land suitability assessment can further enhance sustainable urban planning and development in Egypt. However, these concerns are widespread worldwide, as urbanization encroaches on various land settings.

This study aims to advance sustainable urban development by integrating geophysical, geotechnical, and surface parameters into a GIS-AHP-based site suitability analysis. Using the Kharga Oasis in Egypt as a case study, it demonstrates how geotechnical parameters derived from seismic data enhance land use planning in complex geological and environmental settings. This approach involves evaluating geotechnical parameters through seismic analysis, multivariate factor analysis of borehole data, and surface factors such as geology, slope, and proximity to highways and streams. The weighted linear combination method generates hazard/suitability maps that provide a replicable model for informed decision-making and sustainable urban planning in similar regions.

2. Materials and Methods

2.1. The Study Area

Kharga Oasis, the largest of Egypt's oases, is a north–south-oriented depression approximately 160 km long, located in the hyper-arid north–central Western Desert. It lies about 550 km south of Cairo and 200 km west of the Nile River (Figure 1a). It is situated in deeply deflated depressions below steep, limestone-capped escarpments that form part of the Egyptian Limestone Plateau [32]. The plain is open to the south and southwest, while a continuous escarpment, 300–500 m high, borders its northern, northwestern, and eastern edges. This region, enclosed by steep plateau borders, functions as a single, unified geophysical unit [33]. Geomorphologically, the oasis is mostly covered by various types of dunes, sand sheets, and sand drifts. The topography reflects the differential deflation of exposed rock units, including Cretaceous to Tertiary bedrock and younger sediments [34]. Tectonically, Kharga Oasis belongs to the stable shelf province of Egypt, with nearly horizontal strata, dipping between 1° and 2° to the west–southwest and east–northeast. Faults and large-scale gentle folds are present on the oasis's surface. The southern part of Kharga has a thin sedimentary cover and is dominated by east–west trending faults, while the northern part is characterized by north–south trending normal faults. The sedimentary sequence in the study area ranges from the Early Cretaceous to the Quaternary, unconformably overlying Precambrian basement rocks (Figure 1b). This sequence comprises the following formations, from oldest to youngest: the Maghrabi Formation (Cenomanian), consisting of coastal mud-plain deposits and channel sandstone, transitioning to deltaic and marginal marine sand and mudstone paleosols [35]. Quseir Formation (Campanian) consists of varicolored shale, siltstone, and flaggy sandstone, containing freshwater gastropods, plants, and vertebrate remains [35]. Dakhla Formation (Maastrichtian–Danian) is characterized by yellowish carbonate-bound siltstone and sandstone intercalated with clay, representing submarine fan deposits [36]. Tarawan Formation (Upper Paleocene) begins with marly intercalations of shale and clay, grading upward into white chalky limestone and hard siliceous limestone, as observed in the Abu Minqar area, north of Kharga Oasis [37]. Esna Shale Formation (Landanian and Ypresian) consists of thick marl and greenish-grey open marine shales with allochthonous carbonate intercalations [38]. El Rufuf Formation (Ypresian) comprises a well-bedded lagoonal platform limestone with brown chert bands [34]. Quaternary deposits are represented by sabkha deposits (lacustrine mud and evaporites) and sand dunes composed of coarse, rounded sands that cover vast areas, forming sandy plains northeast and northwest of Kharga town [39].

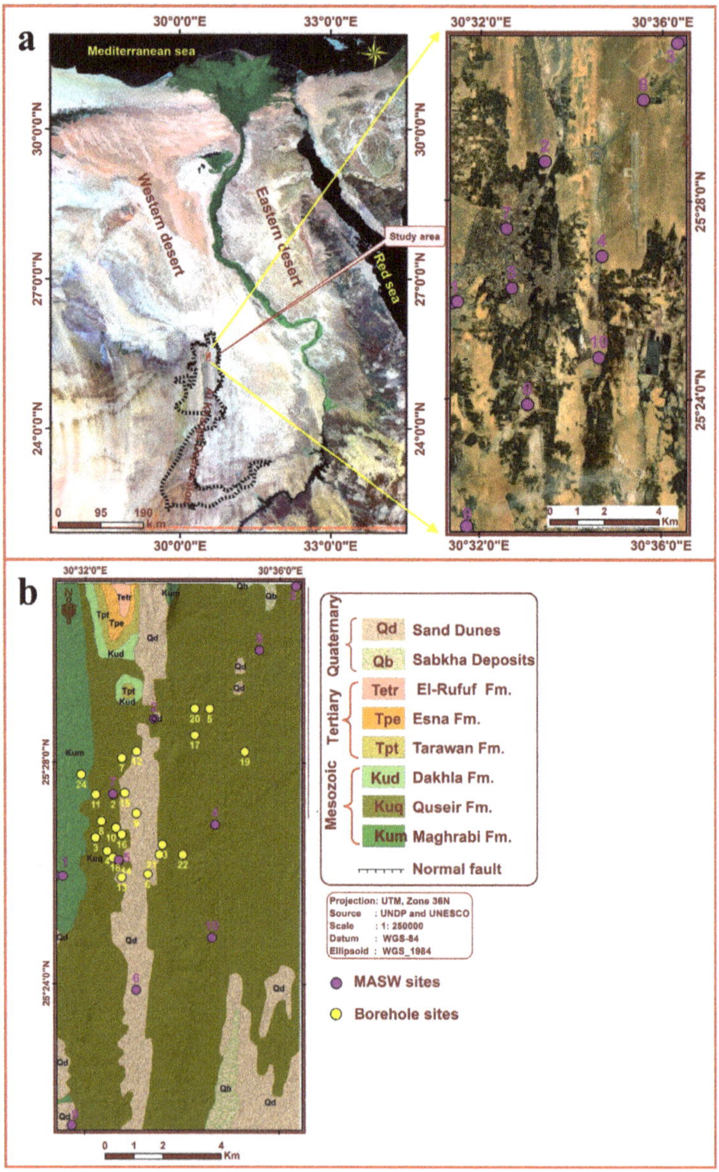

Figure 1. (**a**) Location and topographic provinces; (**b**) geologic setting in the study area.

2.2. MASW and Ambient Vibration Data

Multichannel analysis of surface waves (MASW) is a widely used geophysical seismic technique for near-surface geotechnical characterization that provides shear wave velocity profiles important for subsurface exploration. Its efficiency and accuracy are often compared to other earlier technical methods such as electrical resistivity tomography (ERT) and induced polarization imaging (IPI). Its ease of use, non-destructiveness, and cost-effectiveness make it a valuable tool in subsurface exploration. It provides important geotechnical parameters such as soil material stiffness and is highly resistant to field conditions due to the high signal-to-noise ratio [25,40,41].

In this study, MASW data were collected using 48 evenly spaced vertical geophones (4.5 Hz) arranged in a straight line on the ground surface, with a 10 kg sledgehammer serving as the energy source to generate waves. A metallic plate was used to enhance signal excitation, and wave recordings were initiated. The distance between the geophones and the offset between the source and the first geophone was 1 m, resulting in a total spread length of 47 m. To account for heterogeneity at shallow depths, measurements were repeated seven times, with the deployment shifted by 5 m after each measurement [40]. The MASW measurement locations (10 sites) provided good coverage of the study area, including recognized locations such as New Valley University, historical sites, Kharga Stadium, Kharga Airport, and several schools (Figure 1b). The beamforming approach was applied to derive the frequency–velocity spectrum, which was computed in the frequency range of 5–100 Hz and the velocity range of 100–1000 m/s. Ambient vibrations were recorded using a Taurus digital recorder with a sampling interval of 0.01 s, and recordings at each site lasted for more than an hour. Horizontal-to-vertical (H/V) Fourier spectral ratios were used to analyze the ambient vibration wavefield. Dispersion curves from MASW measurements and the ellipticity function from the H/V ratios of ambient vibrations were combined to derive the subsurface velocity structure using a geophysical inversion method.

2.3. Geotechnical Parameters from Seismic Measurements

Traditional geotechnical survey techniques, such as borehole drilling and laboratory testing, are accurate but costly, expensive, and limited in space. In contrast, seismic methods allow for the estimation of subsurface elastic properties over a larger area with greater efficiency. Seismic wave velocity (P-wave and S-wave) is directly related to basic soil and rock properties such as density and elastic strength and provides a reliable basis for deriving geotechnical parameters. This approach not only reduces the time and cost of data collection but also complements borehole data by filling gaps in spatial areas [22,24,41].

2.3.1. Kinetic Elastic Moduli

Seismic methods are often used to determine the in situ modulus of rock. The dynamic Young's modulus can be estimated by combining the compressional wave velocity with the rock's mass density. Similarly, the dynamic rigidity modulus can be estimated using the shear wave velocity.

Poisson's Ratio (σ)

Poisson's ratio (σ) is the ratio of lateral deformation (perpendicular to applied stress) to longitudinal deformation (parallel to the applied stress). It is defined as the negative ratio of lateral strain to axial strain, with the negative sign accounting for the contraction in the lateral direction when the material stretches in the axial direction. Poisson's ratio can be calculated using Equation (1) [42].

$$\sigma = \frac{1}{2}\frac{\left(\frac{V_p}{V_s}\right)^2 - 2}{\left(\frac{V_p}{V_s}\right)^2 - 1} = \frac{3K - 2G}{2(3k + G)} = \frac{\lambda}{2(\lambda + \mu)} \tag{1}$$

where V_p and V_s are the compressional and shear wave velocities in m/s, respectively, K is the bulk modulus, G = μ is the shear modulus, and λ is Lamé constant.

Rigidity Modulus (μ)

The modulus of rigidity, or shear modulus (μ), measures a solid material's resistance to shear deformation caused by parallel forces acting in opposite directions. The shear modulus is the ratio of shear stress (force per unit area) to shear strain (deformation). It can be calculated using Equation (2) [43].

$$\mu = \rho V_s^2 = \frac{E}{2(1 + \sigma)} \tag{2}$$

where ρ, E, and σ are average density, Young's modulus, and Poisson's ratio, respectively.

Young's Modulus (E)

Young's modulus (E) measures how much a material stretches or compresses when stressed. It is the ratio of stress to strain on the loading plane in the direction of the applied force. A higher value indicates a stiffer material, which means it is more resistant to stretching or compressing. In engineering design, it is used to calculate the strength and stiffness of structures. In geology, it helps determine the mechanical properties of rocks, which are often subjected to stresses such as those caused by earthquakes or tectonic forces. Since a material under uniaxial tension must elongate, Young's modulus (E) is always non-negative. The seismic P- and S-wave velocities in rocks are proportional to the square root of this modulus, meaning that stiffer rocks have faster seismic wave velocities. Young's modulus can also be related to the shear modulus (μ) and Poisson's ratio (σ), as described in Equations (3) and (4) [44,45].

$$E = \rho V_s^2 \frac{3\left(\frac{V_p}{V_s}\right)^2 - 4}{\left(\frac{V_p}{V_s}\right)^2 - 1} = \frac{9KG}{3K + G} \tag{3}$$

$$E = 2(1 + \sigma)\mu = \frac{\mu(3\lambda + 2\mu)}{\lambda + \mu} \tag{4}$$

Bulk Modulus (K)

The bulk modulus (K) measures a material's resistance to changes in volume. It is defined as the ratio of the pressure applied to a material to the resulting volumetric strain, which represents the fractional change in the material's volume. The bulk modulus can be calculated using Equations (5) and (6) [42].

$$K = \rho\left(V_p^2 - \frac{4V_s^2}{3}\right) = \frac{E}{3(1 - 2\sigma)} \tag{5}$$

$$K = \frac{2\mu(1 + \sigma)}{3(1 - 2\sigma)} = \frac{3\lambda + 2\mu}{3} \tag{6}$$

2.3.2. Material Competence Scales

N-Value

The liquefaction resistance of soil is typically evaluated using the in situ Standard Penetration Test (SPT) or the Cone Penetration Test (CPT). However, when both compressional P- and S-wave travel times are measured, SPT N-values and the elastic properties of the material can also be determined. For known rock types, P-wave data can be used to estimate the ease of excavation or rippability, while S-wave data can be used to estimate SPT N-values, which, in turn, can be used to calculate the soil's bearing capacity using Equation (7) [46].

$$N = \left(\frac{V_s}{76.55}\right)^{2.24719} \tag{7}$$

The American Society for Testing and Materials (ASTM) [47] classifies soil into five categories based on Standard Penetration Test (SPT) N-values: very loose sediments (0 to 4), loose materials (5 to 10), medium dense (11 to 30), dense (31 to 50), and very dense or rock (51 and above). These classifications are used by engineers and geotechnical professionals to assess soil properties and design foundations and other structures.

Competence (Concentration) Index (C_i)

The status of soil compaction serves as an indicator of its suitability for foundations and civil engineering applications. C_i is a material-dependent engineering parameter that quantifies the degree of compaction in a material. It depends on both the soil's elastic moduli and the pressure distribution at depth. A higher C_i indicates greater material compaction, enhancing the material's suitability for foundations and other civil engineering applications [48]. C_i can be calculated using Equation (8) [49].

$$C_i = \frac{1+\sigma}{\sigma} \tag{8}$$

Abd El-Rahman [50] defined C_i as a function of the velocity squared ratio, as presented in Equation (9).

$$C_i = \frac{\left[3-4\left(\frac{V_s^2}{V_P^2}\right)\right]}{\left[1-2\left(\frac{V_s^2}{V_P^2}\right)\right]} \tag{9}$$

C_i ranges were categorized into five classes to correspond to the degree of soil competence: very loose (3.5–4.0), loose (4.0–4.5), medium dense (4.5–5.0), dense (5.0–5.5), and very dense (5.5–6.0) [50].

Material Index (M_i)

M_i is a quantitative measure of a material's compactness and resistance to deformation. It is influenced by several factors, including the composition of the material, solidification, presence of cracks, liquid in porous spaces, and environmental conditions. A higher M_i indicates a more compact and stable material, making it suitable for foundation and civil engineering applications. M_i is expressed by the elastic modulus in Equation (10) [51].

$$M_i = 1-4\sigma = \frac{\mu-\lambda}{\mu+\lambda} \tag{10}$$

Soil properties for construction purposes can be categorized into four main groups based on their σ and M_i: very soft sediments (<−0.6), soft materials (−0.6 to −0.2), fairly compacted (−0.2 to 0.2), moderately compacted (0.2 to 0.6), and compacted (0.6 to 1.0) [52,53].

Stress Ratio (S_i)

S_i is a dimensionless parameter used in soil mechanics to quantify the relationship between effective stress and pore pressure in soil. Generally, formations under high fluid pressure exhibit low differential pressure and abnormally low seismic velocities. This is because high fluid pressure in the formation reduces effective stress, which, in turn, decreases the shear strength of the soil. The propagation velocity of seismic waves is proportional to the shear wave velocity, which measures the shear strength of the soil. Cordier [54] describes formations where vertical stresses at a specific depth are lower than horizontal stresses due to pore pressure as sub-compacted or over-pressure zones. These zones are commonly found in young, un-consolidated sedimentary layers. Studies have shown a relationship between σ of normally consolidated soils and S_i. Equation (11) has been used to calculate this value [55].

$$S_i = \frac{\sigma}{1-\sigma} \tag{11}$$

Also, the stress ratio can be expressed in terms of the speed-squared ratio, as presented in Equation (12) [51].

$$S_i = 1 - 2\left(\frac{V_s^2}{V_p^2}\right) \tag{12}$$

S_i can be used to classify soils into four categories: very soft (0.7–0.61), soft (0.61–0.52), fairly compacted (0.52–0.43), moderately compacted (0.43–0.34), and compacted (0.34–0.25) [51].

Reaction Modulus (R_m)

The reaction modulus (R_m), a crucial parameter in geotechnical engineering, characterizes the relationship between applied stress and the resulting deformation in a soil or rock mass. It is typically determined through plate load tests (PLTs), in which a circular plate, ranging from 15 cm to 75 cm in diameter, is subjected to increasing loads, and the corresponding displacements are measured [56]. Several factors influence R_m, including soil properties, stress levels, depth, and foundation geometry. The accurate estimation of R_m is essential for designing foundations and other geotechnical structures to ensure their stability and performance. According to Khalil and Hanafy [48], R_m can also be calculated from V_s using Equation (13).

$$R_m = \left(\frac{V_s}{80.5}\right)^{1/0.357} \tag{13}$$

R_m values and their corresponding rock quality were categorized into five classes to represent the degree of soil competence: very loose (0–1200), loose (1200–2400), medium dense (2400–3600), dense (3600–4800), and very dense (4800–6000) [52,53].

2.3.3. Material Bearing Capacity (Q)

Natural and artificial cyclic dynamic loading can induce additional loads on the building, potentially leading to liquefaction and/or shear failure if the total load exceeds the ultimate bearing capacity of the material. Therefore, it is essential to assess the bearing capacity of foundation materials, including the ultimate bearing capacity (Q_{ult}) and allowable bearing capacity (Q_{all}), for areas underlain by recent deposits.

Ultimate Bearing Capacity (Q_{ult})

Q_{ult} of a foundation material is the maximum load that can be applied before the soil fails due to shear or sand liquefaction. ASTM [57] defines Q_{ult} as the average load per unit area required to cause the rupture of a supporting soil or rock mass. This definition applies to structures founded directly on rock or shallow foundations on rock that are less than four times the foundation width in depth. The final foundation design is typically determined by factors such as settlement, sliding stability, and overturning, rather than just bearing capacity. Q_{ult} of cohesionless soils can be estimated using the SPT N-value, as presented in Equation (14) [58].

$$Q_{ult} = 30N = 10^{2.932\ (\log V_s - 1.45)} \tag{14}$$

Allowable Bearing Capacity (Q_{all})

Q_{all} is the maximum pressure that can be applied to foundation soil (or rock mass) without causing shear failure or liquefaction. It is determined by considering various factors, such as soil type, foundation depth, foundation size and shape, and acceptable levels of settlement. Q_{all} is calculated using Equation (15) [59].

$$Q_{all} = \frac{Q_{ult}}{F_s} \tag{15}$$

The factor of safety (F_s) for cohesionless soil is 2, while that for cohesive soil is 3.

2.4. Soil Geotechnical Properties from Borehole Data

Geological and geotechnical data from 24 boreholes were used in this study to determine the depth of underlying layers, as well as the physical and chemical characteristics of various subsurface formations. The borehole depths range from 10 to 15 m. Five of the boreholes are located at the same sites as the MASW measurements, while the rest are situated nearby (Figure 1b). The boreholes were mechanically drilled using a rotary drilling technique, utilizing bentonite slurry as the drilling fluid. During drilling in rock, Core Recovery (CR) and Rock Quality Designation (RQD) were determined for each core sample. On the other hand, Standard Penetration Tests (SPTs) were performed in sandy soil. Figure S1 shows representative drilled borehole lithological log.

The rock cuttings and cores from these boreholes were used for various laboratory measurements. These measurements included soil chemistry, such as total dissolved solids (TDS), chloride (Cl^-), and sulfate ($SO_4{}^{-2}$); soil consistency, including the natural water content (NWC%), liquid limit (LL%), plastic limit (PL%), plasticity index (PI%), and consistency index (CI); soil composition, particularly grain size distribution; unconfined compressive strength (UCS); dry density (DD); and swelling potential and moisture content. The samples were tested in accordance with relevant standards [60–65].

The inclusion of swelling properties in routine geotechnical investigations in Egypt remains limited. This limited use stems from the lack of application of such data in engineering design and construction. Consequently, estimates based on readily available index parameters, such as the liquid limit (LL), the plasticity index (PI), and density, often serve as alternative sources of information [66]. LL and PI are widely used for evaluating soil swelling potential due to their availability and ease of measurement [67]. In this study, the swell index (C_s) was determined based on Equation (16) [68].

$$C_s = 0.00194\ (PI - 4.6) \tag{16}$$

2.5. Statistical Analysis

Multivariate statistical analysis aims to reveal patterns, trends, and relationships among variables in a dataset, providing deeper and more comprehensive insights into the variables under investigation. Key techniques for this analysis include correlation and clustering methods. Correlation Analysis (CA) evaluates the relationships between two or more variables to identify those with the strongest links and to understand their mutual influence. Clustering methods examine similarities and differences in a dataset, which facilitates the identification of variable correlations. Principal component analysis (PCA) focuses on describing the variation in a dataset, while Hierarchical Aggregate Clustering (HAC) groups samples and reveals patterns by combining related groups of variables [45,69,70]. The measured geotechnical parameters, soil water chemistry, and swelling properties were analyzed using a combined statistical approach for data interpretation. Four statistical methods were applied using IBM SPSS Statistics (IBM Corp., Version 25, 2017): coefficient of variation (COV), CA, PCA, and HAC.

2.6. GIS-AHP Method for Suitability Assessment

2.6.1. Data Used and Factor Selection

In this study, Landsat-8 satellite imagery was used to process and enhance various algorithms and methods. The Landsat-8 data were downloaded from the USGS website [71]. The VNIR and SWIR bands, with a spatial resolution of 30 m, were layer-stacked and merged with the Panchromatic band, which has a 15 m spatial resolution, to create a high-resolution image at 15 m. Additionally, topographic maps at a 1:50,000 scale and a geological map [72] were utilized. The satellite data were geo-referenced using ground control points collected via GPS and enhanced for the delineation and interpretation of various thematic information using Erdas IMAGINE (Erdas IMAGINE; version 14.0; 2014). A digital elevation model (DEM) was prepared from the ALOS PALSAR DEM with a resolution of 12.5 [73], and both slope and elevation maps were generated from the developed digital elevation model. Land use was extracted through unsupervised classification from the Sentinel 2-A database [74]. Figure 2 displays a flowchart of the data collection and integration in the GIS-AHP modeling.

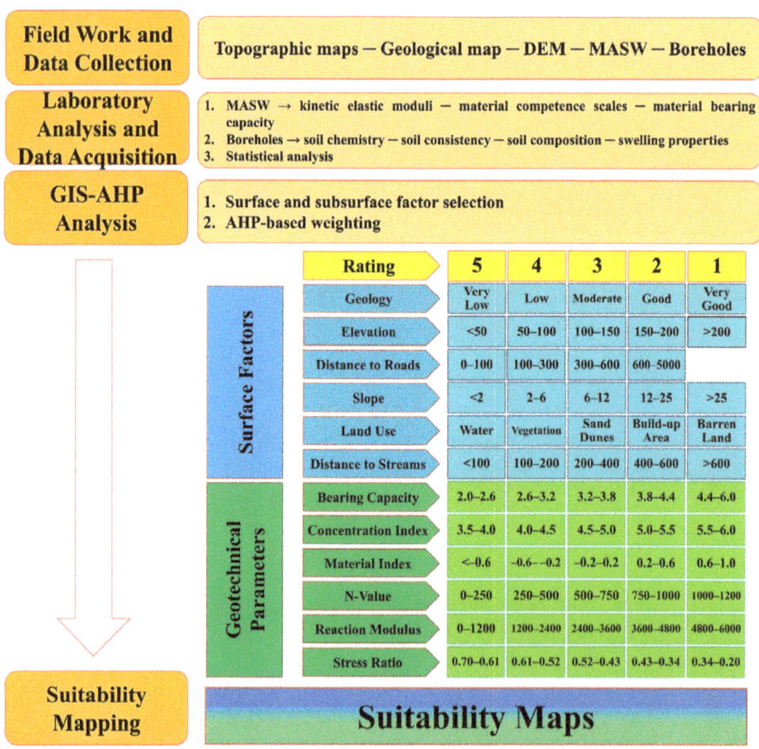

Figure 2. Flowchart of data integration and rating of surface and subsurface factors.

The selection of factors, the number of classes, and their boundary values were determined based on a review of relevant prior studies [1,75,76]. Eleven geotechnical, geomorphological, geological, and environmental factors were identified as the most important in assessing the suitability of the study area. A spatial database was constructed using ArcGIS (ArcGIS; version 10.8.4.1; 2020), and the thematic layers of the parameters were converted from vector to raster format with a 30 m cell size to facilitate spatial analysis. Distribution and suitability zonation maps for various geotechnical parameters were created using GIS spatial analysis tools. The factors were then divided into two categories:

Surface Factors

Surface factors include distances to major streams and roads, slope angles, elevation, land use, and geology. Major streams were interpreted and mapped from Landsat-8 and topographic maps (1:50,000), while distances to roads and streams were calculated using ArcGIS's Euclidean distance function. Proximity to roads affects bedrock stability and can impact construction suitability, while closeness to streams increases flood risks, thereby reducing site suitability. This study classified slopes into five categories, assigning the highest rating to steeper slopes (>25°) as the least vulnerable to water accumulation. Elevation, which influences urban expansion due to infrastructure limitations, was also categorized into five ranges: (1) <50 m, (2) 50 m, (3) 100 m, (4) 150 m, and (5) >200 m above sea level. A lithological map was created from the geological map of the study area. Land use was classified into five categories: water, vegetation, sand dune, build-up area, and barren land.

Subsurface Factors

A geotechnical investigation focused on the second subsurface layer, due to the first layer's looseness and minimal thickness. Subsurface parameters include the stress ratio (soil stiffness), material index (soil properties), concentration index (pore fluid influence), N-value (soil density and strength), and bearing capacity (weight limit). Seismic velocities, densities, and thicknesses of the soil layers were measured to calculate these parameters. Additionally, soil water chemistry, consistency, and swelling potential were assessed. ArcGIS and Surfer (Golden Software Surfer; version 25.1.229; 2023) were used to visualize the spatial distribution of these factors across the study area, informing foundation decisions for the construction project.

2.6.2. Rating of the Categories of the Parameters

Based on statistical analysis, the Egyptian civil engineering codes [77], and field observations, class values were assigned to surface factors, following the framework adopted by Aly et al. [28], which considers both terrain suitability and engineering safety. Terrain suitability was assessed using geotechnical data (fracture density, slope), while engineering safety assessed structural requirements in line with national codes. Proximity to streams was also considered to mitigate risks from floods and slope instability. This study used a five-grade scale [78,79], where 1 indicates the least favorable and 5 indicates the most favorable conditions (Figure S2).

For subsurface factors, different classification ranges were applied. The material index classification followed Birch [52] and Bowles [49], while the stress ratio analysis used Thomsen's [80] framework, and density gradients were based on Cordier's [54] approach. Each factor's rating was rescaled to a uniform suitability scale to ensure consistency. A positive correlation between assigned values and suitability was maintained, using integer values from 1 to 5 to represent five hazard or suitability classes: very low, low, medium, high, and very high (Figure 2).

2.6.3. Weighting of the Parameters

The AHP method assesses land suitability for urban and industrial development by using pairwise comparisons to determine the relative importance of factors (Table S1). By constructing a pairwise comparison matrix and calculating the principal eigenvector, AHP assigns weights to each factor, which are then used to create a site suitability map, considering local geological, geotechnical, and environmental conditions. Saaty [4] introduced a procedure for determining the consistency index, known as the consistency ratio (CR). The CR serves as a numerical measure to verify the consistency of judgments within a comparison matrix. It helps mitigate the potential for arbitrary choices, leading to more reliable decision outcomes. It is expressed in Equations (17) and (18) as:

$$CR = \frac{CI}{RI} \tag{17}$$

$$CI = \frac{\lambda_{max} - n}{n - 1} \tag{18}$$

where RI is the random index, CI is the consistency index, λ_{max} is the largest eigenvalue of the matrix, and n is the matrix order. To ensure logical coherence in the weighting process, the consistency ratio (CR) was computed through the pairwise comparison. All the CR values in this study were less than 0.1, indicating that the ratings were consistent.

AHP was used to determine the relative weights of various factors in the assessment. A 6 × 6 matrix was created with surface and subsurface factors on the axes, and multiple matrices were linearly combined to prepare final matrices, which were used to generate a hazard/suitability map. All pair-wise comparisons, eigenvectors, weights, and the CR were calculated using the AHP Priority Calculator (Tables 1 and 2).

Table 1. Weight factors using the AHP method of surface factors.

Surface Factors	Geology	DEM	Distance to Stream	Slope	Distance to Road	Land Use	Weight
Geology	1	0.5	2	2	1	2	0.199553
DEM	2	1	2	2	2	2	0.282211
Distance to Stream	0.5	0.5	1	1	1	1	0.125711
Slope	0.5	0.5	1	1	1	1	0.125710
Distance to Road	1	0.5	1	1	1	1	0.141105
Land use	0.5	0.5	1	1	1	1	0.125710

Table 2. Weight factors using the AHP method of subsurface factors.

Subsurface Factors	Q	C_i	M_i	N-Value	R_m	S_i	Weight
Q	1	1	1	2	4	4	0.252975
C_i	1	1	2	2	4	3	0.272779
M_i	1	0.50	1	2	2	2	0.185225
N-value	0.50	0.50	0.50	1	2	4	0.146404
R_m	0.25	0.25	0.50	0.50	1	2	0.080642
S_i	0.25	0.33	0.50	0.25	0.50	1	0.061975

2.6.4. The Overall Suitability Indices

To calculate an overall suitability index for each factor, thematic layers of parameters, along with their corresponding AHP weights, were combined using the weighted linear combination method [3,81]. This method assigns weights to factors based on their impact on suitability, which are then summed up to produce a detailed suitability map using ArcGIS-Raster Calculator. The method can be mathematically expressed using Equation (19) [82].

$$S = \sum_{i=1}^{n} W_i \, X_i \quad (19)$$

where S is the suitability degree, n is the number of factors, W_i is the weight of factor i, and X_i is the rating of factor i.

The weight assignments were validated by comparing the GIS-AHP-derived suitability map to field observations and borehole geotechnical data. For example, sites classified as high suitability on the map correlated with sites that had good soil and rock geotechnical attributes.

3. Results and Discussion

3.1. Geotechnical Parameters Derived from Seismic Measurements

In this study, the P- and S-wave velocities of the surface and second layers are shown in Figure 3. The depth to the top of the bedrock ranges between 1 and 8 m from the ground surface. The elastic moduli for the surface and second layers are listed in Tables 3 and 4. Figures 4 and 5 present the spatial distribution of the stress ratio (S_i), material index (M_i), concentration index (C_i), N-value, and bearing capacity for both layers. The spatial distribution maps of the geotechnical parameters indicate that the northwestern corner of the study area is characterized by subsurface rocks of lower quality compared to the rest of the area. According to the ranges of engineering parameters and the corresponding rock/soil type [50–53,55] (Table S2), the values of the minimum and maximum geotechnical parameters in the study area fall within the moderate to high rock quality range.

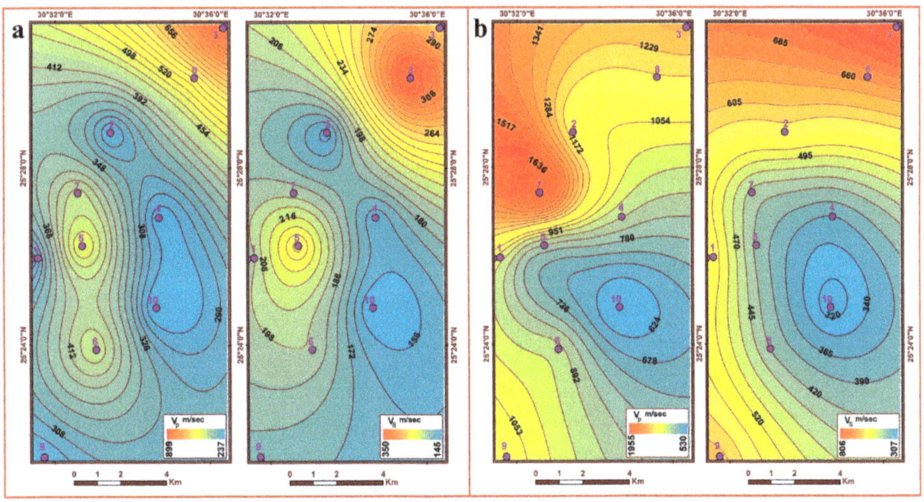

Figure 3. Spatial distribution of the P- and S-wave velocities for the (**a**) surface and (**b**) second layers.

Table 3. Geotechnical parameters of the surface layer.

Sites	Thickness (m)	V_p m/s	V_s m/s	ρ gm/cm³	σ	M Dyn/cm²	E Dyn/cm²	K Dyn/cm²	M_i	C_i	S_i	N-Value	R_m (KPa/m)	Q_{Ult} Kg/cm²	Q_{Ult} KPa
S 1	2	290	173	1.28	0.22	3.83×10^8	9.37×10^8	5.65×10^8	0.10	5.47	0.29	6.82	8.52	0.20	0.02
S 2	1	263	151	1.25	0.25	2.85×10^8	7.14×10^8	4.83×10^8	−0.02	4.93	0.34	4.58	5.82	0.14	0.01
S 3	1.5	899	259	1.70	0.45	1.14×10^9	3.31×10^9	1.22×10^{10}	−0.82	3.20	0.83	22.27	26.40	0.67	0.07
S 4	3.2	253	157	1.24	0.19	3.05×10^8	7.23×10^8	3.85×10^8	0.25	6.35	0.23	5.13	6.50	0.15	0.02
S 5	8.5	516	251	1.48	0.35	9.31×10^8	2.50×10^9	2.69×10^9	−0.38	3.90	0.53	20.31	24.18	0.61	0.06
S 6	0.5	467	197	1.44	0.39	5.59×10^8	1.56×10^9	2.39×10^9	−0.57	3.55	0.64	9.98	12.27	0.30	0.03
S 7	8.4	450	194	1.43	0.39	5.37×10^8	1.49×10^9	2.17×10^9	−0.54	3.59	0.63	9.54	11.75	0.29	0.03
S 8	4	567	350	1.51	0.19	1.85×10^8	4.42×10^9	2.39×10^9	0.23	6.20	0.24	53.84	61.36	1.61	0.16
S 9	2.9	265	175	1.25	0.11	3.83×10^8	8.53×10^8	3.67×10^8	0.55	9.82	0.13	7.05	8.80	0.21	0.02
S 10	2.5	237	145	1.22	0.20	2.56×10^8	6.14×10^8	3.42×10^8	0.20	5.98	0.25	4.06	5.20	0.12	0.01

Table 4. Geotechnical parameters of the second layer.

Sites	Thickness (m)	V_p m/s	V_s m/s	ρ gm/cm^3	σ	M Dyn/cm^2	E Dyn/cm^2	K Dyn/cm^2	M_i	C_i	S_i	N-Value	R_m (KPa/m)	Q_{Ult} Kg/cm^2	Q_{Ult} KPa
S 1	>28	1004	567	1.74	0.27	5.61×10^9	1.42×10^{10}	1.01×10^{10}	−0.06	4.76	0.36	221.58	237.00	6.64	0.65
S 2	>29	1023	568	1.75	0.28	5.66×10^9	1.44×10^{10}	1.08×10^{10}	−0.11	4.61	0.38	222.72	238.18	6.67	0.65
S 3	>28.5	1312	806	1.87	0.20	1.21×10^{10}	2.90×10^{10}	1.59×10^{10}	0.21	6.08	0.25	621.55	634.79	18.62	1.83
S 4	>26.5	897	354	1.70	0.41	2.13×10^9	5.99×10^9	1.08×10^{10}	−0.63	3.45	0.69	55.67	63.34	1.67	0.16
S 5	>21	599	409	1.53	0.06	2.57×10^9	5.46×10^9	2.08×10^9	0.75	16.80	0.07	85.02	94.93	2.55	0.25
S 6	11	940	418	1.72	0.38	3.00×10^9	8.26×10^9	1.12×10^{10}	−0.51	3.65	0.60	90.62	100.89	2.72	0.27
S 7	>21	1955	427	2.06	0.47	3.76×10^9	1.11×10^{10}	7.37×10^{10}	−0.90	3.11	0.90	96.46	107.10	2.89	0.28
S 8	>26	1165	667	1.81	0.26	8.06×10^9	2.02×10^{10}	1.38×10^{10}	−0.02	4.90	0.34	356.77	373.55	10.69	1.05
S 9	>27	1097	620	1.78	0.27	6.86×10^9	1.74×10^{10}	1.23×10^{10}	−0.06	4.77	0.36	287.96	304.41	8.63	0.85
S 10	11	530	307	1.49	0.25	1.40×10^9	3.50×10^9	2.31×10^9	0.01	5.04	0.33	36.66	42.50	1.10	0.11

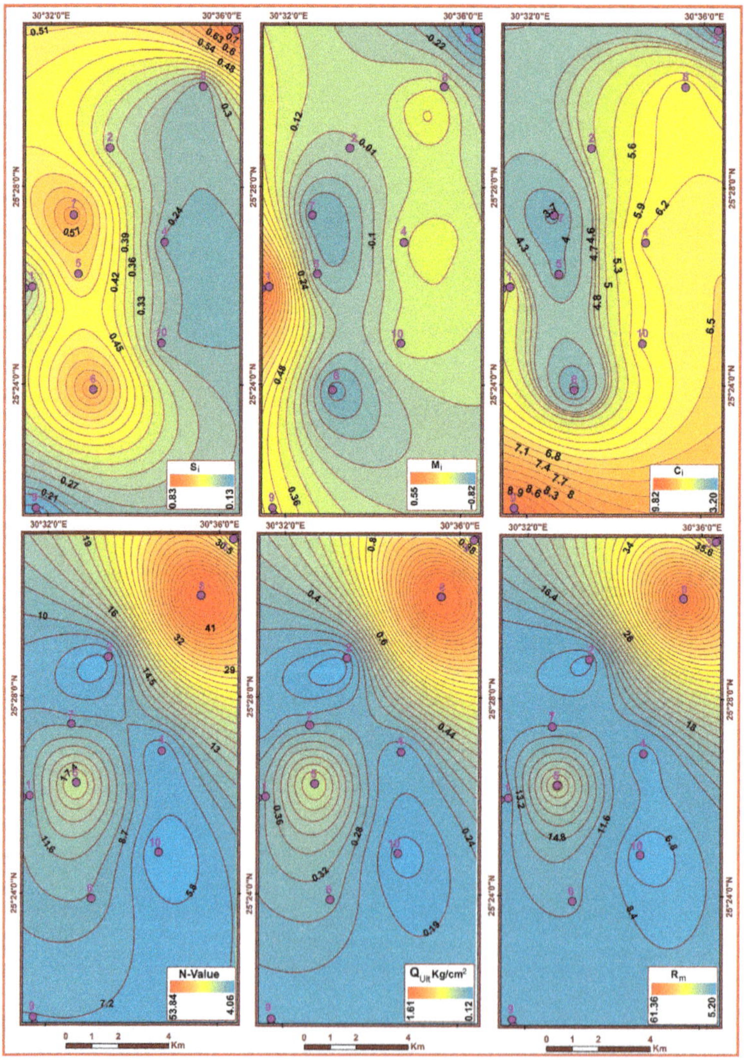

Figure 4. Spatial distribution of S_i, M_i, C_i, N-value, Q_{ult}, and R_m for the surface layer.

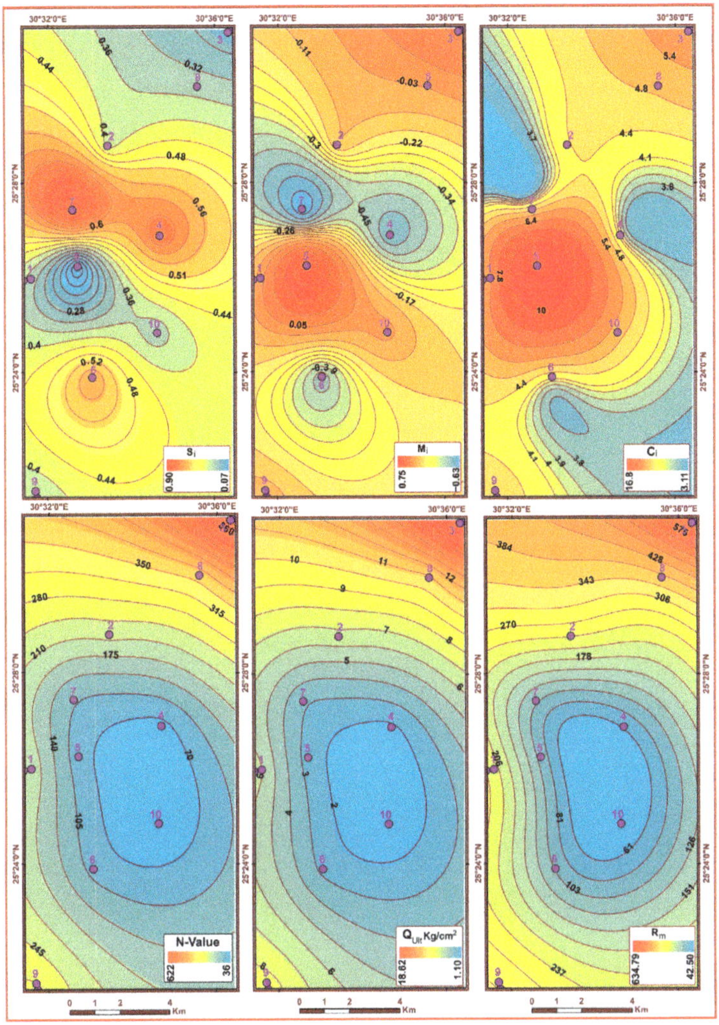

Figure 5. Spatial distribution of S_i, M_i, C_i, N-value, Q_{ult}, and R_m for the second layer.

For S_i values, the obtained ratios range from 0.13 to 0.83 for the surface layer and from 0.07 to 0.90 for the second layer. Figure 4 shows that the smallest values are present in the eastern and southwestern parts of the area, while the largest values prevail in the central zone, trending northwest. This indicates that the central zone contains less competent materials compared to other parts of the area. Figure 5 reveals that the highest values for the second layer are located in the central and southern parts of the area.

M_i values range from −0.82 to 0.55 for the first layer and from −0.899 to 0.747 for the second layer. The surface layer exhibits a gradual decrease in material index values toward the southeastern zone (Figure 4), while the second layer shows lower values in the upper central part (Figure 5). Lower material index values indicate less competent materials, and vice versa.

C_i values for the surface layer range from 3.20 to 9.82 and from 3.11 to 16.80 for the second layer. The surface layer shows a general decrease in C_i values toward the western zone, suggesting less competent materials occupy this area (Figure 4). For the second layer, the smallest concentration index values are found in the southeastern and western parts of the area (Figure 5). Again, smaller concentration index values correspond to less competent materials.

The surface layer (Figure 4) has an N-value ranging from 4.06 to 53.84. The smallest values are found in the central and southern zones, representing loose material according to ASTM [47], although some sites show medium-dense soil. The second layer has N-values ranging from 36.66 to 621.55. N-values above 50 indicate the hard rock nature of the second layer (Figure 5).

R_m values for the surface layer range from 5.20 to 61.36 and from 42.50 to 634.79 for the second layer. In the surface layer, there is a general decrease in R_m values toward the southern and western zones, indicating that less competent materials occupy these areas (Figure 4). In contrast, the second layer shows the smallest R_m values in the central and eastern parts of the area (Figure 5). Lower Rm values correspond to less competent materials.

Q_{Ult} values vary from 0.12 to 1.61 kg/cm^2 for the surface layer (Figure 4) and from 1.099 to 18.620 kg/cm^2 for the second layer (Figure 5). The central and southern zones of the surface layer exhibit smaller Q_{Ult} values, while the central part of the second layer shows lower values. Smaller bearing capacity values correspond to materials with lower geotechnical competency [83].

3.2. Geotechnical Parameters from Borehole Data

The descriptive statistics of 20 commonly used geotechnical parameters and measurements at target sites across Kharga Oasis are summarized in Table 5.

Table 5. Descriptive statistics of the studied geotechnical parameters in Kharga Oasis, Egypt.

Geotechnical Parameter	Min	Max	Mean	Median	St. Dev.	CV%
TDS (ppm)	1100	8200	3220	2630	1976	61.4
Cl^- (ppm)	340	5560	1430	1048	1252	87.5
SO_4^{-2} (ppm)	100	3560	809	550	799	98.8
Natural water content (NWC) (%)	11.88	36.00	21.51	21.04	5.19	24.1
Liquid limit (LL) (%)	50.00	98.00	74.69	72.00	12.39	16.6
Plastic limit (PL) (%)	20.60	33.50	28.42	30.00	3.83	13.5
Plasticity index (PI) (%)	24.70	68.00	47.92	47.85	9.98	20.8
Consistency index (CI)	0.89	1.33	1.13	1.13	0.09	8.2
Swelling pressure (SP) (kg/cm^2)	2.10	4.00	3.38	3.63	0.59	17.3
Unconfined compression strength (UCS) (kg/cm^2)	0.88	4.00	2.29	2.29	0.68	29.5
Dry density (DD) (g/cm^3)	1.46	1.89	1.60	1.60	0.09	5.8
Moisture content (MC) (%)	15.69	27.73	21.29	21.34	3.52	16.5
Depth to water (DW) (m)	0.40	3.50	1.94	2.00	0.69	35.3
Swelling potential (C_s)	0.53	1.23	0.86	0.84	0.17	19.6
Foundation layer thickness (FLT) (m)	2.00	15.00	8.76	8.50	3.53	40.3
Initial void ratio (IVR)	0.44	1.20	0.78	0.82	0.14	18.4
Compression index (C_c)	0.01	0.20	0.13	0.13	0.04	34.4
Pre-consolidation pressure (PP) (kg/cm^2)	1.90	3.90	2.91	3.27	0.67	23.2
Effective overburden pressure (EOP) (kg/cm^2)	0.37	2.00	0.72	0.53	0.42	58.8
Over-consolidation ratio (OCR)	1.00	7.09	5.27	5.83	2.15	40.9

The plasticity chart (Figure 6) shows that the analyzed soils are predominantly inorganic cohesive clays with high plasticity, indicating a significant potential for volume changes due to moisture fluctuations [84]. This is evidenced by the fact that most samples plot above the A-line, with liquid limits (LLs) exceeding 50% and plasticity indices (PIs) over 30%. Only one sample displays moderate plasticity. Notably, Figure 6 illustrates a strong positive correlation between PI and LL, indicating that PI consistently increases with higher LL values [85].

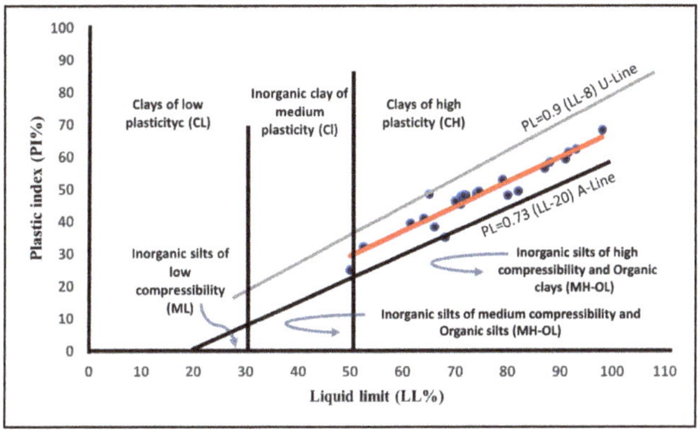

Figure 6. Plasticity chart for the cohesive soils of Kharga Oasis.

The spatial distribution of key geotechnical properties, including soil salinity (Cl^- and SO_4^{-2}), swelling potential (C_s), and unconfined compressive strength (UCS), is shown in Figure 7. Arid regions with shallow water tables and extensive drainage networks experience increased soil and groundwater salinity due to intense evaporation and evapotranspiration, resulting in secondary salinization [86]. El May et al. [87] found that shallow groundwater in Tunisian alluvial regions reduced construction suitability. An elevated water table intensifies cyclic seasonal evapoconcentration, increasing soil salinity and accelerating corrosion, thereby significantly reducing site suitability. Figure 7 shows that the area primarily exhibits moderate salinity severity with SO_4^{-2} and Cl^-, while severe zones are concentrated in the central part of Kharga Oasis.

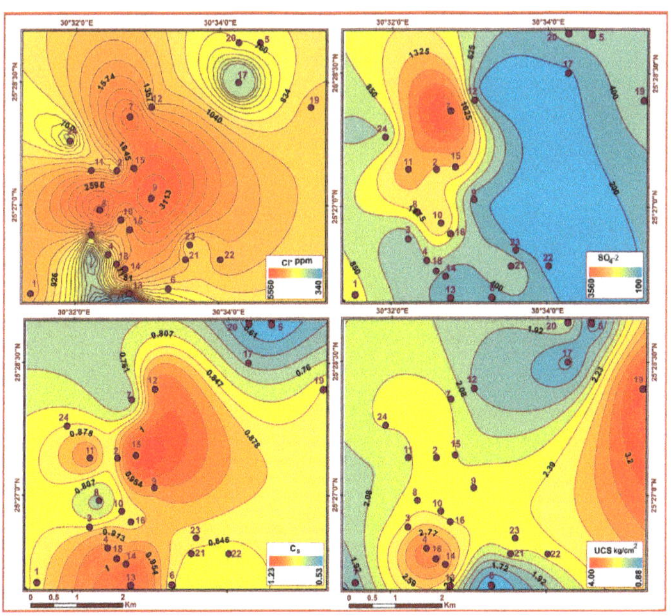

Figure 7. Spatial distribution maps showing Cl^-, SO_4^{-2}, C_s, and UCS.

According to American Concrete Institute Committee (ACI) [88] the severity limits for concrete exposed to solutions containing sulfates and chlorides are outlined in Table S3. Both chemicals, when exceeding safe limits (2000 ppm for chlorides and 1500 ppm for sulfates), can compromise the structural integrity of concrete. High concentrations of these chemicals significantly weaken and deteriorate the material. Chloride exposure accelerates the corrosion of embedded steel reinforcements by disrupting the protective oxide layer. Additionally, sulfates cause expansive reactions within the concrete matrix, resulting in softening and cracking.

Areas with high swelling potential typically exhibit low compressive strength. This study identifies the central and southwestern regions as most susceptible to swelling, while the eastern areas of Kharga exhibit the highest compressive strength. As expected, areas with high swelling potential generally exhibit lower compressive strength (Figure 7). Geographically, the swelling potential (C_s) is highest in the central and southern portions of the district, while the eastern parts of Kharga Oasis boast the highest compressive strength.

The analyzed soils exhibit an over-consolidated nature, with over-consolidation ratios (OCRs) ranging from 1.00 to 7.09. Compression index (C_c) values vary between 0.01 and 0.20, indicating different levels of compressibility. Spatial analysis of these parameters identified distinct zones characterized by the initial void ratio (IVR), OCR, and C_c, as shown in Figure 8. Notably, the lowest values for all three parameters are concentrated in the central region, extending southwestward of Kharga Oasis.

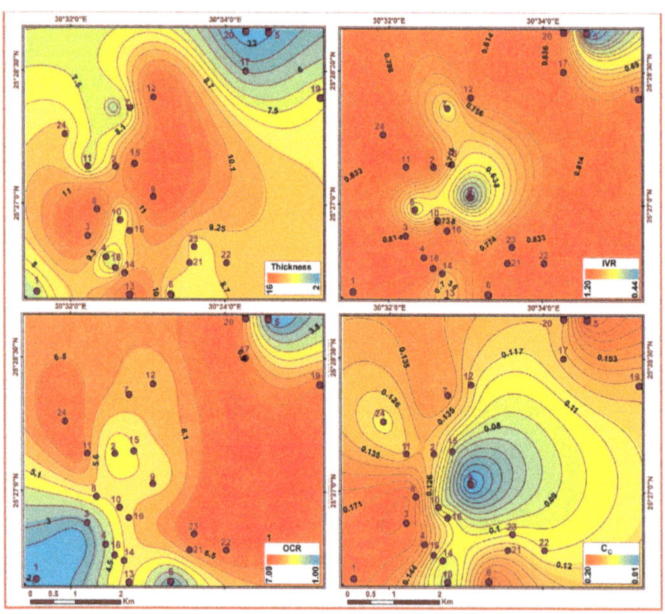

Figure 8. Spatial distribution maps showing thickness, IVR, OCR, and Cc.

3.3. *Variability in the Measured Parameters in Sites*

This study applied the coefficient of variation (CV) to 20 commonly used geotechnical parameters across Kharga Oasis, revealing variability ranging from low to high (Table 5). The consistency index (CI) and dry density (DD) showed low variability (CV < 10%), with CI negatively correlated with natural water content (NWC) and moisture content (MC) and positively correlated with DD (Table S4). DD also exhibited a negative correlation with OCR.

Moderate variability (CV 10–20%) was observed for parameters like LL, PL, PI, SP, MC, and C_s, with significant correlations among them. LL showed a positive correlation with PL, PI, C_s, and OCR, but a negative correlation with EOP. PL was positively correlated with NWC, PI, and C_s. PI exhibited a perfect positive correlation with C_s.

High variability (CV > 20%) was found in parameters such as TDS, Cl^-, SO_4^{-2}, UCS, NWC, FLT, IVR, C_c, PP, EOP, OCR, and depth to water (DW). TDS was positively correlated with Cl^- and SO_4^{-2} and showed a negative correlation with DW and IVR. TDS, Cl^-, and SO_4^{-2} were linked to hydrogeological processes and varied with depth. Greater soil profile thickness has been found to inversely correlate with infiltration capacity [89], thereby enhancing water-holding capacity and accelerating salt accumulation due to seasonal evaporation cycles [90]. The strong correlations between TDS and Cl^-, along with their negative relationship with DW, suggest that both TDS and Cl^- originate from similar sources and decrease moderately with greater depth. The moderate correlation between TDS and SO_4^{-2}, as well as foundation layer thickness, coupled with the weak, insignificant correlation between SO_4^{-2} and DW, indicates that TDS may dissolve from multiple sources, while SO_4^{-2} concentrations are only weakly influenced by increasing depth. The weak correlation between Cl^- and SO_4^{-2} further supports the hypothesis that their sources differ. This study emphasizes the need for multiple measurements due to spatial variability and the complexity of geotechnical and hydrogeological interactions in the region.

3.4. Variability in the Geotechnical Factors

The analysis identified five factors with eigenvalues greater than 1, explaining over 79.2% of the total variance in the dataset (Table S5). The first three factors accounted for 60.61% of the variance, with the most common factors valued above 0.8, indicating a significant influence on geotechnical variability (Figure 9).

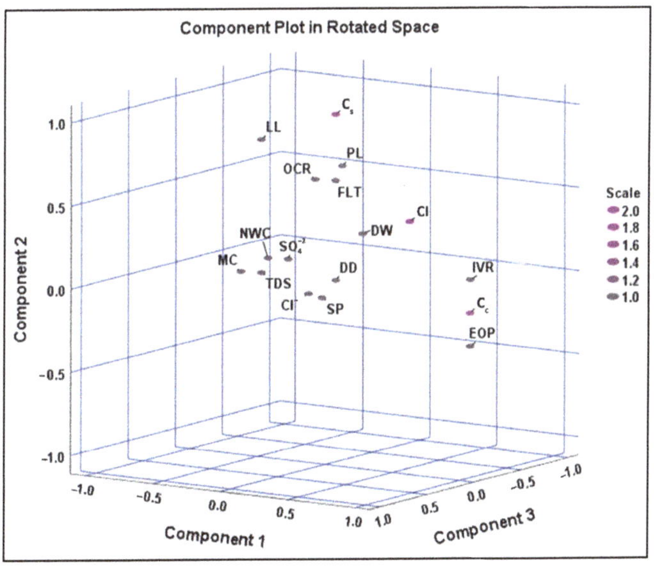

Figure 9. Principal component analysis by component plot in rotated space.

Factor 1 (26.6%) was mainly influenced by consolidation parameters (PP, C_c, EOP, IVR), with a moderate influence from plasticity (SP, PL). This grouping into one factor is consistent with the strong positive correlations among these parameters (Table S4). Factor 2 (17.5%) was dominated by plasticity parameters (C_s, PI, LL, PL) and foundation layer thickness (FLT), reflecting a relationship between plasticity and infiltration. The inclusion of the foundation layer thickness is likely due to its significant correlation with PI and

C_s, where an increase in the layer's thickness reduces the infiltration of dissolved salts or water, thereby lowering the swelling rate. Factor 3 (16.5%) was associated with water content parameters (NWC, MC) and had inverse effects on strength (UCS, CI). These four parameters are believed to be indirectly controlled by the water content at the sites. Factor 4 (11.1%) was driven by soil water salinity (TDS, Cl^-, SO_4^{-2}) and associated depth to water (DW) and OCR. Factor 5 (7.5%) highlighted dry density with minor contributions from PL, CI, and OCR.

3.5. Cluster Analysis

This study applied HCA to all geotechnical parameter data from 24 sampling points (118 observations) in Kharga Oasis, dividing them into four classes based on spatial similarities. Two main groups were identified: Class IV as one group and a larger group split into Class III and a combination of Classes I and II (Figure S3). Each class displayed different levels of variability in geotechnical parameters: Class I displayed high variability in most parameters, with low UCS, PP, and OCR values, and moderate Cl^- and SO_4^{-2} corrosion hazards. Class II displayed the highest values for TDS, Cl^-, and SO_4^{-2}, with low values for NWC, plasticity, and consolidation parameters, making it the least suitable for construction. Class III displayed high values for water content, plasticity parameters, and shear parameters but low C_c values. Class IV displayed the most favorable for construction, with high values for UCS, SP, IVR, and OCR. Overall, Class IV soils offer the best conditions for construction, followed by Classes III and I, with Class II being the least suitable. The spatial distribution of these classes in Kharga Oasis is presented in Figure 10.

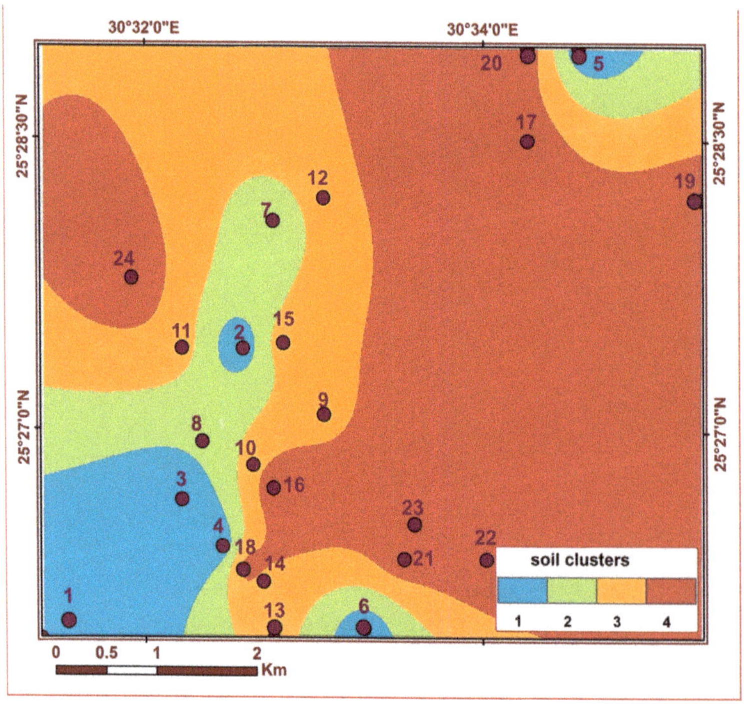

Figure 10. Spatial distribution of geotechnical clusters.

3.6. Suitability Maps

A spatial database was created using ArcGIS software, and the thematic layers of the parameters were converted from vector to raster format with a 30 m cell size. The elevation map was derived from the DEM and classified into five categories, as shown in Figure 11a. A slope map (Figure 11b), generated from the processed digital elevation model (DEM) data, was integral to this study. Slopes were categorized into five distinct classes: (1) <2°, (2) 2–5°, (3) 5–8°, (4) 8–10°, and (5) >10°. Slope is a key criterion in site suitability analysis, as steep slopes increase construction costs, erosion risks, and landslide susceptibility [91,92]. The slope distribution also affects flood and landslide risks, with gentler slopes attracting floodwater [93].

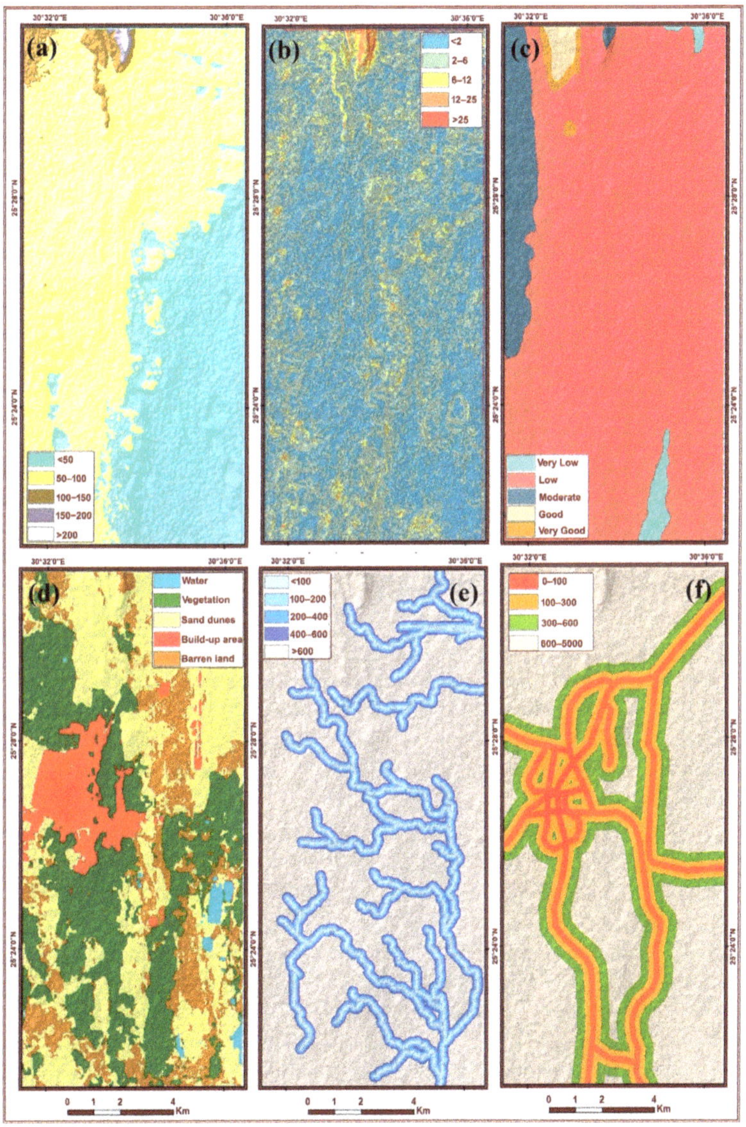

Figure 11. The thematic layers of the parameters involved in the analysis: (**a**) elevation; (**b**) slope map; (**c**) lithology; (**d**) land use; (**e**) distance from streams; (**f**) distance from road network.

Lithology significantly impacts urban development. Solid rocks offer excellent foundations but often require costly excavations. The geological formations (Figure 11c) in the study area were classified into five categories based on their geotechnical engineering behavior. These include soil (e.g., loess deposits, gravels, and sands) and hard formations, primarily found in the northern and western parts of the study area, where shale, sandstone, and siltstone predominate (Figure 11c). The hard rock category was assigned to have the highest weight, as it affects infiltration rates and supports the occurrence of floods.

Land use plays a crucial role in landslide susceptibility and flood occurrence due to the inverse relationship between vegetation density and rainfall [94]. Additionally, land use and land cover significantly affect rainfall interception, infiltration, evapotranspiration, and groundwater retention capacity in certain areas [95,96]. Land use also plays a crucial role in water management in this study. Consequently, barren land received the highest score (indicating the highest potential hazard), followed by build-up areas, sand dunes, and cultivated land. Water bodies received the lowest score due to their inherent hydrological function (Figure 11d).

Proximity to streams (Figure 11e) is a primary factor in analyzing flood hazards. The closer an area is to a stream, the more vulnerable it is to flooding, particularly during flash floods or flow reversals [76]. Recognizing this critical relationship, the drainage network was integrated into the suitability analysis. Sixth- to eighth-order streams, identified using the Strahler classification [97], were extracted from the DEM and used to create a dedicated spatial layer. Buffer zones of 100, 200, 300, and 400 m were then established around these streams to highlight areas potentially affected by flood events. To define the appropriate criteria for constructing the suitability maps, geographical factors, particularly the main national road network (Figure 11f), were analyzed. Based on this analysis, buffer zones of 100, 300, 600, and 5000 m were established around the road network, reflecting the varying influence of the roads on nearby areas.

This study included three suitability models: the surface model (Figure 12a), the subsurface model (Figure 12b), and the final overarching model that integrates surface and subsurface data (Figure 12c). The comprehensive model, based on the combined assessment of potential hazards and suitability, provides a critical basis for informed decision-making regarding future development in the area. It provides a crucial basis for informed decision-making regarding future development projects in the area, balancing both surface conditions and the underlying subsurface parameters that may impact construction. Among the models, site risk/suitability was categorized into four categories: low, medium, high, and very high suitability. This classification was developed to identify areas that are suitable for development and those with high risk. In Figure 12, the fitted parameters exhibit considerable spatial variation in the study area, with some regions emerging as ideal for urban expansion, while others exhibit more constrained development potential. This variability is directly linked to several critical factors. The final model is dominated by subsurface geotechnical properties such as the ultimate bearing capacity (Q_{ult}), bearing capacity, concentration index (C_i), material index (M_i), N-value, and stress ratio (S_i). In contrast to these, surface factors such as elevation, geology, proximity to existing roads, proximity to streams, slope, and land use also contribute to suitability determination. This study concludes that approximately 24.5% of the land is classified as highly suitable for urban development, indicating areas that combine good surface and subsurface conditions. Additionally, 32.37% of the land is considered highly suitable, providing development capabilities with minimal precautions. Moderately suitable areas, comprising 27.61% of the land area, require extensive geotechnical site investigations to ensure stability and safety. Finally, 15.53% of the study area is classified as having low suitability, where development would likely face significant challenges due to adverse geotechnical conditions or environmental hazards.

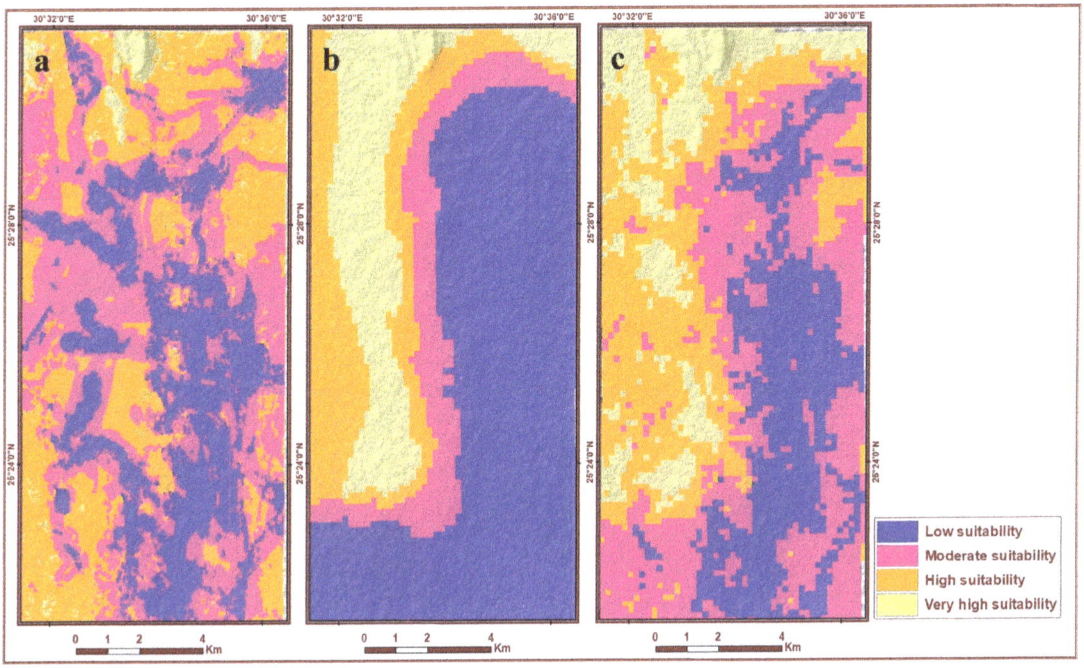

Figure 12. (**a**) The final suitability map for surface factors. (**b**) The final suitability map for subsurface factors. (**c**) The overall hazard/suitability map for a combination of the surface and subsurface models.

By comparing the land use map (Figure 11d) to the final suitability map (Figure 12c), we can see that some of the current facilities (build-up) are located in low-suitability regions. These results provide a robust framework for sustainable urban planning, identify optimal site selections for development, and highlight areas where geotechnical improvements may be necessary to mitigate the risk. Furthermore, this model can be a valuable tool for stakeholders in prioritizing land use decisions and monitoring urban development while minimizing potential risks.

3.7. Advantages, Limitations, and Future Research

The selection of suitable construction sites is an invaluable decision-making process that examines various factors. Recent advancements in technology and methodology contribute to improved methods for site suitability assessment. Boreholes (invasive) and seismic surveys (non-invasive) are considered prominent methods for examining the subsurface characteristics for design purposes [98–100]. Seismic data boost borehole and laboratory data by providing continuous subsurface data along survey lines. This integration is especially significant in areas with limited borehole data, allowing for better subsurface exploration and characterization [25,100].

This study's findings provide valuable insights into the geotechnical and geophysical aspects of Kharga Oasis, with a focus on integrating MASW seismic data with GIS-AHP models to assess land suitability for urban and industrial development. The integration of seismic data for non-invasive geotechnical analysis, as well as borehole data and environmental and geological parameters, is a significant improvement in site suitability modeling.

The accuracy of seismic velocity data (P- and S-wave) in assessing geotechnical parameters, as well as their integration with surface variables, is superior to the model used when compared to commonly used methods that use costly and time-consuming geotechnical laboratory measurements or that apply surface factors modeling only (e.g., slope, geology, land use) in suitability assessment.

The GIS-AHP model presented in this study is not limited to the Kharga Oasis; it is a flexible model that could be applied to similar regions around the world, especially for urban planning, industrial site selection in tectonically active areas, and tourism development in areas with changing environmental conditions. By adapting geotechnical and environmental factors to local conditions, this model could generate a sustainable development plan across diverse geographic and environmental contexts.

Although we ensured consistency in the AHP method's pairwise comparisons ($CR < 0.1$), minor differences in weight assignments may affect the final suitability assessment. Using advanced machine learning algorithms for parameter weighting may improve the objectivity and dependability of GIS-AHP models. Future research can improve our model by including locally adapted geotechnical, hydrological, and environmental parameters to optimize suitability models even more.

4. Conclusions

For the first time, the potential suitability of the Kharga area for urban growth and industrial development was assessed, and respective maps were produced by integrating diverse data (remote sensing, geological, environmental, geomorphological, and geotechnical) within a GIS-AHP approach. This study presented spatial distribution maps of elastic moduli and geotechnical parameters derived from compressional velocity (V_p), shear velocity (V_s), and density measurements. The results showed that areas located from the northwest to southeast, northeast, and south of the study area are characterized by more competent soil deposits, while other parts, with moderately to weakly compacted sediments, should be given special consideration during construction operations.

Moreover, the geotechnical hazards of the soil in Kharga Oasis were illustrated and confirmed using detailed geological and geotechnical information from 24 boreholes. Advanced spatial mapping techniques identified zones at high risk for soil and water hazards, including concrete corrosion and fluctuating groundwater levels, which can impact soil plasticity and other geotechnical properties. By combining these spatial distributions with cluster analysis, the area was zoned according to its suitability for construction, accounting for potential hazards.

This study demonstrates how integrating diverse data, including both surface and subsurface information, through GIS-AHP enables informed land use planning that safeguards against natural hazards. This research provides decision-makers, including city planners, geotechnical professionals, and civil engineers, with valuable tools for proactive planning and sustainable development. The resulting geotechnical cluster and suitability maps can be used to (1) evaluate and plan urban areas by considering both engineering conditions and potential geohazards; (2) optimize the scope of engineering geological investigations in specific construction zones, saving time and resources; (3) select appropriate foundation types and construction designs, leading to safer and more cost-effective infrastructure; and (4) predict changes in engineering geological conditions and potential hazardous events, enabling proactive mitigation strategies.

Supplementary Materials: The following supporting information can be downloaded at https://www.mdpi.com/article/10.3390/ijgi14010003/s1: Table S1: Pair-wise comparison matrix. Table S2: The range of engineering parameters and the corresponding rock/soil type. Table S3: Degree of severity on concrete exposed to sulfate- and chloride-containing solutions. Table S4: Pearson correlation matrix of geotechnical parameters in Kharga Oasis, Egypt. Table S5: Varimax rotated principal component analysis for target geotechnical parameters determined in the study area. Figure S1; Representative lithological log. Figure S2: AHP Priority Calculator. Figure S3: Dendrogram illustrating the hierarchical clusters of the analyzed borehole sites.

Author Contributions: Conceptualization, Doaa Wahba, Awad A. Omran, Ashraf Adly, Ahmed Gad, and Heba El-Bagoury; methodology, Doaa Wahba, Awad A. Omran, Ashraf Adly, and Heba El-Bagoury; formal analysis Doaa Wahba, Awad A. Omran, Ashraf Adly, and Heba El-Bagoury; software, Doaa Wahba, and Heba El-Bagoury; validation, Doaa Wahba, Ahmed Gad, Hasan Arman, and Heba El-Bagoury; investigation, Doaa Wahba, Awad A. Omran, Ashraf Adly, and Heba El-Bagoury; data curation, Doaa Wahba, Ahmed Gad, and Heba El-Bagoury; visualization, Doaa Wahba, Ahmed Gad, and Heba El-Bagoury; writing—original draft preparation, Doaa Wahba, Ahmed Gad, and Heba El-Bagoury; writing—review and editing, Doaa Wahba, Awad A. Omran, Ahmed Gad, Hasan Arman, and Heba El-Bagoury. All authors have read and agreed to the published version of the manuscript.

Funding: This research received no external funding.

Data Availability Statement: The data are contained within this article or the Supplementary Materials. The data presented in this study are available in the Supplementary Materials.

Acknowledgments: The authors would like to acknowledge the laboratory facilities offered by the Geology Department, Faculty of Science, New Valley University, Kharga, Egypt. Sincere thanks to Abelbaset Abudeif (Applied Geophysics at the Geology Department, Faculty of Science, Sohag University) for his guidance and for providing technical assistance during the present study.

Conflicts of Interest: The authors declare no conflicts of interest.

References

1. Al-Ghorayeb, A.; Al-Shaar, W.; Elkordi, A.; Faour, G.; Al-Shaar, M.; Attalah, Y. Land Suitability Analysis for Sustainable Urban Development: A Case of Nabatiyeh Region in Lebanon. *J* **2023**, *6*, 267–285. [CrossRef]
2. Rane, N.L.; Choudhary, S.P.; Saha, A.; Srivastava, A.; Pande, C.B.; Alshehri, F.; Roy, R.; Katipoğluf, O.M.; Abdo, H.G. Delineation of environmentally sustainable urban settlement using GIS-based MIF and AHP techniques. *Geocarto Int.* **2024**, *39*, 2335249. [CrossRef]
3. Marinoni, O. Implementation of the analytical hierarchy process with VBA in ArcGIS. *Comput. Geosci.* **2004**, *30*, 637–646. [CrossRef]
4. Saaty, T.L. *Theory and Applications of the Analytic Hierarchy Process*; RWS Publ.: Pittsburgh, PA, USA, 1977.
5. Saaty, T.L. *The Analytic Hierarchy Process*; McGraw–Hill: New York, NY, USA, 1980.
6. Yu, D.; Fang, C. Urban Remote Sensing with Spatial Big Data: A Review and Renewed Perspective of Urban Studies in Recent Decades. *Remote Sens.* **2023**, *15*, 1307. [CrossRef]
7. Gacu, J.G.; Garcia, J.D.; Fetalvero, E.G.; Catajay-Mani, M.P.; Monjardin, C.E.F. Suitability Analysis Using GIS-Based Analytic Hierarchy Process (AHP) for Solar Power Exploration. *Energies* **2023**, *16*, 6724. [CrossRef]
8. Islam, M.R.; Aziz, M.T.; Alauddin, M.; Kader, Z.; Islam, M.R. Site suitability assessment for solar power plants in Bangladesh: A GIS-based Analytical Hierarchy Process (AHP) and Multi-Criteria Decision Analysis (MCDA) approach. *Renew. Energy* **2024**, *220*, 119595. [CrossRef]
9. Abdelouhed, F.; Ahmed, A.; Abdellah, A.; Yassine, B.; Mohammed, I. GIS and remote sensing coupled with analytical hierarchy process (AHP) for the selection of appropriate sites for landfills: A case study in the province of Ouarzazate, Morocco. *J. Eng. Appl. Sci.* **2022**, *69*, 19. [CrossRef]
10. Nyimbili, P.H.; Erden, T. GIS-Based Fuzzy Multi-Criteria Approach for Optimal Site Selection of Fire Stations in Istanbul, Turkey. *Socio-Econ. Plan. Sci.* **2020**, *71*, 100860. [CrossRef]
11. Al Nasiri, N.; Al Fazari, A.; Ali, W.; Agyekum, B.; Ramadan, E. Multi-Criteria Decision-Making Approach for Siting Sewer Treatment Plants in Muscat, Oman. *Urban Sci.* **2023**, *7*, 82. [CrossRef]
12. Bojer, A.K.; Bekalo, D.J.; Debelee, T.G.; Nadarajah, S.; Al-Quraishi, A.M.F. Rainwater Harvesting Site Selection for Drought-Prone Areas in Somali and Borena Zones, Oromia Regional State, Ethiopia: A Geospatial and Multi-Criteria Decision Analysis. *Water* **2024**, *16*, 1789. [CrossRef]
13. Vashist, K.; Singh, K.K. Flood hazard mapping using GIS-based AHP approach for Krishna River basin. *Hydrol. Process.* **2024**, *38*, e15212. [CrossRef]

14. Alizadeh, M.; Hashim, M.; Alizadeh, E.; Shahabi, H.; Karami, M.R.; Beiranvand Pour, A.; Pradhan, B.; Zabihi, H. Multi-Criteria Decision Making (MCDM) Model for Seismic Vulnerability Assessment (SVA) of Urban Residential Buildings. *ISPRS Int. J. Geo-Inf.* **2018**, *7*, 444. [CrossRef]
15. Pramanik, M.K. Site suitability analysis for agricultural land use of Darjeeling district using AHP and GIS techniques. *Model. Earth Syst. Environ.* **2016**, *2*, 56. [CrossRef]
16. Suthar, N.; Das, D.; Mallik, J. Land-use suitability assessment for urban development using Multi-Criteria Decision-Making Analysis in the Himalayan districts of Shimla, Nainital, and Darjeeling, India. *Discov. Environ.* **2024**, *2*, 90. [CrossRef]
17. Ramya, S.; Devadas, V. Integration of GIS, AHP and TOPSIS in evaluating suitable locations for industrial development: A case of Tehri Garhwal district, Uttarakhand, India. *J. Clean. Prod.* **2019**, *238*, 117872. [CrossRef]
18. Khan, M.S.; Park, J.; Seo, J. Geotechnical Property Modeling and Construction Safety Zoning Based on GIS and BIM Integration. *Appl. Sci.* **2021**, *11*, 4004. [CrossRef]
19. Arman, H.; Gad, A.; Abdelghany, O.; Mahmoud, M.; Aldahan, A.; Paramban, S.; Abu Saima, M. Geotechnical data compilation for evaporitic rocks in Abu Dhabi, UAE: A resource for engineers. *Data Brief* **2024**, *54*, 110322. [CrossRef]
20. Arman, H.; Abdelghany, O.; Mahmoud, B.; Aldahan, A.; Paramban, S.; Gad, A.; Abu Saima, M. Effects of groundwater and distilled water on the durability of evaporitic rocks. *Sci. Rep.* **2023**, *13*, 5667. [CrossRef]
21. Chu, C.; Wang, Y.; Song, H.; Xu, L.; Hou, X.; Sheng, S.; Liu, L. Evaluating the Characteristics of Spatial Variability of Soil in Vertical Direction Highly Heterogeneous Region Based on Cone Penetration Test. *Lithosphere* **2024**, *2024*, 184. [CrossRef]
22. Othman, A.A. Construed geotechnical characteristics of foundation beds by seismic measurements. *J. Geophys. Eng.* **2005**, *2*, 126–138. [CrossRef]
23. Qaher, M.; Eldosouky, A.M.; Saada, S.A.; Basheer, A.A. Integration of ERT and shallow seismic refraction for geotechnical investigation on El-Alamein Hotel Building Area, El-Alamein new city, Egypt. *Geomech. Geophys. Geo-Energ. Geo-Resour.* **2023**, *9*, 115. [CrossRef]
24. Khalil, M.H.; Hanafy, S.M. Engineering applications of seismic refraction method: A field example at Wadi Wardan, Northeast Gulf of Suez, Sinai, Egypt. *J. Appl. Geophys.* **2008**, *65*, 132–141. [CrossRef]
25. Alexopoulos, J.D.; Dilalos, S.; Voulgaris, N.; Gkosios, V.; Giannopoulos, I.-K.; Kapetanidis, V.; Kaviris, G. The Contribution of Near-Surface Geophysics for the Site Characterization of Seismological Stations. *Appl. Sci.* **2023**, *13*, 4932. [CrossRef]
26. Ma, S.; Yao, Y.; Bao, P.; Guo, C. Effects of moisture content on strength and compression properties of foundation soils of cultural relics in areas flooded by the Yellow River. *Front. Mater.* **2023**, *10*, 1186750. [CrossRef]
27. Umar, I.H.; Lin, H. Marble Powder as a Soil Stabilizer: An Experimental Investigation of the Geotechnical Properties and Unconfined Compressive Strength Analysis. *Materials* **2024**, *17*, 1208. [CrossRef]
28. Aly, M.H.; Giardino, J.R.; Klein, A.G. Suitability assessment for New Minia City Egypt: A GIS approach to engineering geology. *Environ. Eng. Geosci.* **2005**, *11*, 259–269. [CrossRef]
29. Abdelalim, A.M. Risks Affecting the Delivery of Construction Projects in Egypt: Identifying, Assessing and Response. In *Project Management and BIM for Sustainable Modern Cities*; GeoMEast 2018. Sustainable Civil Infrastructures; Shehata, M., Rodrigues, F., Eds.; Springer: Cham, Switzerland, 2019. [CrossRef]
30. Yousri, E.; Sayed, A.E.B.; Farag, M.A.M.; Abdelalim, A.M. Risk Identification of Building Construction Projects in Egypt. *Buildings* **2023**, *13*, 1084. [CrossRef]
31. El-Bagoury, H.; Gad, A. Integrated Hydrological Modeling for Watershed Analysis, Flood Prediction, and Mitigation Using Meteorological and Morphometric Data, SCS-CN, HEC-HMS/RAS, and QGIS. *Water* **2024**, *16*, 356. [CrossRef]
32. Bazeen, Y.S.; Hewaidy, A.G.A.; Samir, A.; El Sayed, M.M.; Abd El-Moghny, M.W.; Ayyad, H.M. Sequence stratigraphy of the Paleocene succession in the Kharga Oasis, Western Desert, Egypt: Insights from microplankton biostratigraphy and benthic foraminifer paleoenvironments. *Palaeoworld* **2024**, *33*, 188–204. [CrossRef]
33. Embabi, N.S. *Landscapes and Landforms of Egypt*; Springer: Cham, Switzerland, 2018.
34. Hermina, M. The surroundings of Kharga, Dakhla and Farafra oases. In *The Geology of Egypt*; Routledge: London, UK, 2017; pp. 259–292.
35. Barthel, K.W.; Herrmann–Gegen, W. Late Cretaceous and early Tertiary stratigraphy in the great sand sea and SE margins (Farafra and Dakhla oasis), SW Desert, Egypt. *Mitt. Bayer. Staats Paleontol. Hist. Geol.* **1981**, *21*, 141–182.
36. Tantawy, A.A.; Keller, G.; Adatte, T.; Stinnesbeck, W.; Kassab, A.; Schulte, P. Maastrichtian to Paleocene depositional environment of the Dakhla Formation, Western Desert, Egypt: Sedimentology, mineralogy, and integrated micro-and macrofossil biostratigraphies. *Cretac. Res.* **2001**, *22*, 795–827. [CrossRef]
37. Issawi, B. Review of Upper Cretaceous–Lower Tertiary Stratigraphy in Central and Southern Egypt. *AAPG Bull.* **1972**, *56*, 1448–1463. [CrossRef]
38. Said, R. Planktonic foraminifera from the Thebes Formation, Luxor. *Micropalaeontol.* **1960**, *6*, 277–286. [CrossRef]
39. Bubenzer, O.; Embabi, N.S.; Ashour, M.M. Sand Seas and Dune Fields of Egypt. *Geosciences* **2020**, *10*, 101. [CrossRef]

40. Adly, A.; Hassoup, A.; Farrag, A.; Wahba, D. Characterization of geotechnical parameters from measurements of surface waves at the Kharga Oasis, Egypt. *J. Appl. Geophys.* **2020**, *182*, 104–159. [CrossRef]
41. Abudeif, A.M.; Abdel Aal, G.Z.; Abdelbaky, N.F.; Abdel Gowad, A.M.; Mohammed, M.A. Evaluation of Engineering Site and Subsurface Structures Using Seismic Refraction Tomography: A Case Study of Abydos Site, Sohag Governorate, Egypt. *Appl. Sci.* **2023**, *13*, 2745. [CrossRef]
42. Telford, W.M.; Geldast, L.P.; Sheriff, R.E.; Keys, D.A. *Applied Geophysics*; Cambridge Univ. Press: Cambridge, UK, 1976.
43. Zainorabidin, A.; Said, M.J. Determination of shear wave velocity using multi-channel analysis of surface wave method and shear modulus estimation of peat soil at Western Johore. *Procedia Eng.* **2015**, *125*, 345–350. [CrossRef]
44. Adams, L.H. *Elastic Properties of Materials of the Earth's Crust*; Dover: New York, NY, USA, 1951.
45. Akinyemi, O.D. Multivariate interrelatedness of geotechnical and petrophysical properties towards developing near-surface lithology clusters in a sedimentary terrain. *Bull. Eng. Geol. Environ.* **2022**, *81*, 258. [CrossRef]
46. Stümpel, H.; Kähler, S.; Meissner, R.; Milkereit, B. The use of seismic shear waves and compressional waves for lithological problems of shallow sediments. *Geophys. Prospect* **1984**, *32*, 662–675. [CrossRef]
47. *ASTM D1586-84*; (American Society for Testing and Materials). Standard Test Method for Standard Penetration Test (SPT) and Split-Barrel Sampling of Soils. ASTM International: West Conshohocken, PA, USA, 2011.
48. Khalil, M.H.; Hanafy, S.M. Geotechnical parameters from seismic measurements: Two field examples from Egypt and Saudi Arabia. *J. Environ. Eng. Geophys.* **2016**, *21*, 13–28. [CrossRef]
49. Bowles, J.E. *Physical and Geotechnical Properties of Soils*; McGraw–Hill. Inc.: New York, NY, USA, 1984.
50. Abd El-Rahman, M. Evaluation of the kinetic elastic moduli of the surface materials and application to engineering geologic maps at Maba–Risabah area (Dhamar Province) Northern Yemen. *Egypt. J. Geol.* **1989**, *33*, 229–250.
51. Abd El-Rahman, M. The potential of absorption coefficient and seismic quality factor in delineating less sound foundation materials in Jabal Shib Az Sahara area, Northwest of Sanaa, Yemen Arab Republic. *Egypt MERC Earth Sci.* **1991**, *5*, 181–187.
52. Birch, F. Handbook of physical contents. *Geol. Soc. Am. Mem.* **1966**, *97*, 97–173.
53. Sheriff, R.E.; Geldart, L.P. *Exploration Seismology*; Cambridge University Press: Cambridge, UK, 1986.
54. Cordier, J.P. Velocities in Reflection Seismology. D. Reidel Publishing Company: Dordrecht, The Netherlands, 1985.
55. Tatham, R.H.; Krug, E.H. Vp/Vs interpretation. *Dev. Geophys. Explor. Methods* **1985**, *6*, 139–188.
56. Terzaghi, K. Evalution of conefficients of subgrade reaction. *Geotechnique* **1955**, *5*, 297–326. [CrossRef]
57. *ASTM D1194-94*; (American Society for Testing and Materials). Standard Test Method for Bearing Capacity of Soil for Static Load and Spread Footings. ASTM International: West Conshohocken, PA, USA, 2003.
58. Parry, R.H. Estimating bearing capacity in sand from SPT values. *J. Geotech. Eng. Div.* **1977**, *103*, 1014–1019. [CrossRef]
59. Georgiadis, M.; Butterfield, R. Displacements of footings on sand under eccentric and inclined loads. *Can. Geotech. J.* **1988**, *25*, 199–212. [CrossRef]
60. *ASTM D4318-17e1*; (American Society for Testing and Materials). Standard Test Methods for Liquid Limit, Plastic Limit, and Plasticity Index of Soils. ASTM International: West Conshohocken, PA, USA, 2018.
61. *ASTM D422-63R07*; (American Society for Testing and Materials). Standard Test Method for Particle-Size Analysis of Soils. ASTM International: West Conshohocken, PA, USA, 2007.
62. *ASTM D2166-06*; (American Society for Testing and Materials). Standard Test Method for Unconfined Compressive Strength of Cohesive Soil. ASTM International: West Conshohocken, PA, USA, 2006.
63. *ASTM D4546-21*; (American Society for Testing and Materials). Standard Test Methods for One-Dimensional Swell or Collapse of Soils. ASTM International: West Conshohocken, PA, USA, 2021.
64. *ASTM D2216-19*; (American Society for Testing and Materials). Standard Test Methods for Laboratory Determination of Water (Moisture) Content of Soil and Rock by Mass. ASTM International: West Conshohocken, PA, USA, 2019.
65. *ASTM D7263-21*; (American Society for Testing and Materials). Standard Test Methods for Laboratory Determination of Density and Unit Weight of Soil Specimens. ASTM International: West Conshohocken, PA, USA, 2021.
66. Alptekin, A.; Taga, H. Prediction of Compression and Swelling Index Parameters of Quaternary Sediments from Index Tests at Mersin District. *Open Geosci.* **2019**, *11*, 482–491. [CrossRef]
67. Thomas, K.L.; Baker, J.C.; Zelanzy, L.W. An Expensive Soil Index for predicting shrink–swell potential. *Soil Sci. Soc. Am.* **2000**, *64*, 268–274. [CrossRef]
68. Nakase, A.; Kamei, T.; Kusakabe, O. Constitutive parameters estimated by plasticity index. *J. Geotech. Eng.* **1988**, *114*, 844–858. [CrossRef]
69. García-Ros, G.; Villalva-León, D.X.; Castro, E.; Sánchez-Pérez, J.F.; Valenzuela, J.; Conesa, M. Multivariate Statistical and Correlation Analysis between Acoustic and Geotechnical Variables in Soil Compression Tests Monitored by the Acoustic Emission Technique. *Mathematics* **2023**, *11*, 4085. [CrossRef]
70. Gad, A.; Abdelghany, O.; Arman, H.; Mahmoud, B.; Aldahan, A.; Paramban, S.; Saima, M. Geochemistry and Sr, S, and O stable isotopes of Miocene Abu Dhabi evaporites, United Arab Emirates. *Heliyon* **2023**, *9*, e16033. [CrossRef] [PubMed]

71. USGS Earth Explorer. Esri, i-cubed, USDA, USGS, AEX, GeoEye, Getmapping, Aerogrid, IGN, IGP, UPR-EGP, and the GIS User Community, ESRI. Available online: https://earthexplorer.usgs.gov/ (accessed on 10 December 2023).
72. Conoco, C. *Stratigraphic Lexicon and Explanatory Notes to the Geological Map of Egypt, 1: 500,000*; Institute Fur Angewandte geodasie: Berlin, Germany, 1989.
73. Earthdata. 2024. Available online: https://search.earthdata.nasa.gov/search (accessed on 11 December 2023).
74. Dataspace. Copernicus Browser. 2023. Available online: https://browser.dataspace.copernicus.eu (accessed on 12 December 2023).
75. Abd El Karim, A.; Alogayell, H.M.; Alkadi, I.I.; Youssef, I. Mapping of GIS-Land Use Suitability in the Rural–Urban Continuum between Ar Riyadh and Al Kharj Cities, KSA Based on the Integrating GIS Multi Criteria Decision Analysis and Analytic Hierarchy Process. *Environments* **2020**, *7*, 75. [CrossRef]
76. Chelariu, O.E.; Minea, I.; Ia̧tu, C. Geo-hazards assessment and land suitability estimation for spatial planning using multi-criteria analysis. *Heliyon* **2023**, *9*, e18159. [CrossRef] [PubMed]
77. *ECP-203-2007*; Egyptian code for Design and Construction the Reinforced Concrete Building. National Housing and Building Research Center: Cairo, Egypt, 2007.
78. Dai, F.C.; Lee, C.F.; Zhang, X.H. GIS–based geo–environmental evaluation for urban land–use planning: A case study. *Eng. Geol.* **2001**, *61*, 257–271. [CrossRef]
79. Wang, X.D.; Zhong, X.H.; Liu, S.Z.; Liu, J.G.; Wang, Z.Y.; Li, M.H. Regional assessment of environmental vulnerability in the Tibetan Plateau: Development and application of a new method. *J. Arid Environ.* **2008**, *72*, 1929–1939. [CrossRef]
80. Thomsen, L. Weak elastic anisotropy. *Geophysics* **1986**, *51*, 1954–1966. [CrossRef]
81. Malczewski, J. *GIS and Multicriteria Decision Analysis*; John Wiley & Sons: Hoboken, NJ, USA, 1999.
82. Zhang, J.; Su, Y.; Wu, J.; Liang, H. GIS based land suitability assessment for tobacco production using AHP and fuzzy set in Shandong province of China. *Comput. Electron. Agric.* **2015**, *114*, 202–211. [CrossRef]
83. Almalki, H.; El-Werr, A.K.; Abdel-Rahman, K. Estimation of near-surface geotechnical parameters using seismic measurements at the proposed KACST expansion site, Riyadh, KSA. *Arab. J. Geosci.* **2011**, *4*, 1131–1150. [CrossRef]
84. Masoud, A.A. Geotechnical site suitability mapping for urban land management in Tanta District, Egypt. *Arab. J. Geosci.* **2016**, *9*, 340. [CrossRef]
85. Karakan, E.; Demir, S. Observations and findings on mechanical and plasticity behavior of sand-clay mixtures. *Arab. J. Geosci.* **2020**, *13*, 983. [CrossRef]
86. Masoud, A.A. Geotechnical Evaluation of the Alluvial Soils for Urban Land Management Zonation in Gharbiya Governorate, Egypt. *J. Afr. Earth Sci.* **2015**, *101*, 360–374. [CrossRef]
87. El May, M.; Dlala, M.; Chenini, I. Urban Geological Mapping: Geotechnical Data Analysis for Rational Development Planning. *Eng. Geol.* **2010**, *116*, 129–138. [CrossRef]
88. American Concrete Institute (ACI) Committee 318. *Building Code Requirements for Structural Concrete (ACI 318–08) and Commentary (ACI 318R–08)*; ACI Farmington Hills: Farmington Hills, MI, USA, 2008.
89. Fu, Z.; Li, Z.; Cai, C.; Shi, Z.; Xu, Q.; Wang, X. Soil thickness effect on hydrological and erosion characteristics under sloping lands: A hydropedological perspective. *Geoderma* **2011**, *167*, 41–53. [CrossRef]
90. Ansal, A.; Iyisan, R.; Yıldırım, H. The cyclic behaviour of soils and effects of geotechnical factors in microzonation. *Soil Dyn. Earthq. Eng.* **2001**, *21*, 445–452. [CrossRef]
91. Rao, K.D. Multi–criteria spatial decision analysis for forecasting urban water requirements: A case study of Dehradun city India. *Landsc. Urban Plan.* **2005**, *71*, 163–174. [CrossRef]
92. Bathrellos, G.D.; Kalivas, D.P.; Skilodimou, H.D. GIS–based landslide susceptibility mapping models applied to natural and urban planning in Trikala Central Greece. *Estud. Geol.* **2009**, *65*, 49–65. [CrossRef]
93. Ullah, K.; Wang, Y.; Fang, Z.; Wang, L.; Rahman, M. Multi–hazard susceptibility mapping based on Convolutional Neural Networks. *Geosci. Front.* **2022**, *13*, 101425. [CrossRef]
94. Tehrany, M.S.; Pradhan, B.; Mansor, S.; Ahmad, N. Flood susceptibility assessment using GIS–based support vector machine model with different kernel types. *Catena* **2015**, *125*, 91–101. [CrossRef]
95. Aksha, S.K.; Resler, L.M.; Juran, L.; Carstensen, L.W. A geospatial analysis of multi–hazard risk in Dharan Nepal. *Geomat. Nat. Hazard Risk* **2020**, *11*, 88–111. [CrossRef]
96. Ursu, A.; Stoleriu, C.C.; Ion, C.; Jitariu, V.; Enea, A. Romanian Natura 2000 Network: Evaluation of the Threats and Pressures through the Corine Land Cover Dataset. *Remote Sens.* **2020**, *12*, 2075. [CrossRef]
97. Strahler, A. Quantitative Geomorphology of Drainage Basins and Channel Networks. In *Handbook of Applied Hydrology*; Chow, V., Ed.; McGraw Hill: New York, NY, USA, 1964; pp. 439–476.
98. Garofalo, F.; Foti, S.; Hollender, F.; Bard, P.-Y.; Cornou, C.; Cox, B.R.; Dechamp, A.; Ohrnberger, M.; Perron, V.; Sicilia, D.; et al. InterPACIFIC project: Comparison of invasive and non-invasive methods for seismic site characterization. Part II: Inter-comparison between surface-wave and borehole methods. *Soil. Dyn. Earthq. Eng.* **2016**, *82*, 241–254. [CrossRef]

99. Anbazhagan, P. Subsurface Investigation—Integrated and Modern Approach. In *Geotechnics for Natural and Engineered Sustainable Technologies*; Developments in Geotechnical Engineering; Krishna, A., Dey, A., Sreedeep, S., Eds.; Springer: Singapore, 2018. [CrossRef]
100. Bačić, M.; Librić, L.; Kaćunić, D.J.; Kovačević, M.S. The Usefulness of Seismic Surveys for Geotechnical Engineering in Karst: Some Practical Examples. *Geosciences* **2020**, *10*, 406. [CrossRef]

International Journal of
Geo-Information

MDPI

Article

Exploring the Equality and Determinants of Basic Educational Public Services from a Spatial Variation Perspective Using POI Data

Hejie Wei [1], Wenfeng Ji [1], Ling Li [1], Yi Yang [2,*] and Mengxue Liu [3]

[1] College of Resources and Environmental Sciences, Henan Agricultural University, Zhengzhou 450046, China; hjwei@henau.edu.cn (H.W.); wfji@stu.henau.edu.cn (W.J.); liling@henau.edu.cn (L.L.)
[2] College of Forestry, Henan Agricultural University, Zhengzhou 450046, China
[3] Lhasa Plateau Ecosystem Research Station, Key Laboratory of Ecosystem Network Observation and Modeling, Institute of Geographic Sciences and Natural Resources Research, Chinese Academy of Sciences, Beijing 100101, China; mengxueliu@mail.bnu.edu.cn
* Correspondence: yangyi023@henau.edu.cn

Abstract: The equitable distribution of basic educational services is crucial for attaining educational fairness and promoting balanced demographic and economic growth. This research leverages point-of-interest (POI) data to analyze the spatial arrangement of basic educational service facilities in the Yellow River Basin of China. Employing kernel density analysis and spatial autocorrelation with a geographic information system tool, this study examines the spatial distribution of these facilities. It also applies geographically weighted regression to identify the primary factors influencing their spatial layout. This study reveals a pronounced disparity between the four downstream and five upstream provinces of the Yellow River Basin in terms of basic educational facility availability. In the downstream provinces, facilities constitute 82.45% of the total, markedly surpassing the level of 17.55% in the upstream provinces. The kernel density analysis shows that areas with a high concentration of educational facilities often align with provincial capitals, including Taiyuan in Shanxi Province, Xi'an in Shaanxi Province, Zhengzhou in Henan Province, and regions around Shandong Province. Significant regional differences exist within the Yellow River Basin. Preprimary, primary, and secondary education facilities exhibit strong spatial clustering, with Moran's I indices of 0.26, 0.19, and 0.09, respectively. High–high clusters of preprimary education are predominantly found in the western region of the basin, whereas low–low clusters appear in some eastern and northern areas. Primary and secondary educational facilities show high–high clustering in the north. The spatial distribution of these educational facilities is chiefly influenced by the permanent population and the proportion of the tertiary industry. Per capita gross domestic product (GDP) and educational fiscal expenditure play secondary roles in influencing the spatial layout. The results have important practical significance for promoting the equalization of basic education public services and equal educational opportunities for the school-age population in the Yellow River Basin.

Keywords: basic education; point-of-interest data; spatial pattern; influencing factors; Yellow River Basin

1. Introduction

Basic public services include essential social conditions necessary for sustaining national economic and social stability, safeguarding individuals' fundamental rights to survival and development, and facilitating comprehensive human development [1]. These services are determined based on specific societal concepts in alignment with the stage and overall level of a nation's socioeconomic development. Scholars have explored various research directions concerning basic public services [2]. Some studies primarily focus on the spatial distribution of basic public services [3,4] and residents' satisfaction [5,6]. Others emphasize government functions and supply-side structural reforms [7–9]. For example, drawing from the theory of market failure, Samuelson [10] proposed a model wherein public services should be supplied by the government as a "single center". This model examines the spatial layout and supply efficiency of urban public services and facilities. Other scholars have investigated the equalization of urban public services and facilities [11–13]. Lan et al. [14] and Song et al. [15] analyzed the influence factors and effects of spatial differentiation in public services. These scholars' research, characterized by distinctive perspectives, provides an in-depth analysis of the spatial distribution and efficiency of public services.

Basic educational public services constitute a crucial component of basic public services, and they include preprimary, primary, and secondary education. The high quality of these services is directly linked to the overall construction of a modern socialist nation, serving as the foundation for the nation's steady progress, a basic guarantee for the people's right to education, and the cornerstone for building a high-quality education system [16]. Basic education plays a pivotal role in nurturing talents of various levels and types [17], enhancing the overall quality of the nation, and promoting the construction of socialist modernization. However, current disparities in China's basic educational public services across regions, urban and rural areas, schools, and different educational groups are pronounced, necessitating urgent solutions to achieve regional equalization in basic educational public services [18–20]. The spatial distribution of basic educational resources constitutes a critical aspect of the geographical imbalances in public services [21], with scholars focusing on several key areas in their research on basic educational public services.

For quantitative analyses of the level of basic educational public services, researchers have employed a variety of methods and techniques for measurement [22,23]. For instance, Wen et al. [24] utilized principal component analysis, the Gini coefficient, and factor analysis to quantitatively measure the level of basic education in China. Additionally, some studies have adopted geospatial querying and spatial interpolation methods for calculating and modeling spatial indicators [25–27]. Michael employed the gravity model to assess the accessibility of secondary school facilities [28]. Concurrently, Malczewski and Jackson [29] applied GIS technology to outline educational resource regions, providing a spatial analysis perspective for the study of the level of basic educational public services. The application of these diverse methods and technologies enhances the range of quantitative research on the level of basic educational public services.

The spatial distribution characteristics of basic educational resources constitute a significant area of focus. For instance, Wang et al. [30] utilized point-of-interest (POI) data of educational facilities to investigate the spatial distribution characteristics of educational resources in the primary urban area of Xi'an. Additionally, Yeates [31] and Maybee [32] analyzed the site selection, scale, and facility layout of high schools and middle schools, delineating reasonable boundaries. Researchers such as Zhao et al. [33] observed a declining trend in basic educational resources in rural China from the northeast to the southwest. Furthermore, some studies have conducted satisfaction measurements concerning the spatial equity of basic educational resources [34–36]. These investigations offer valuable insights into comprehending the distribution of basic educational resources.

Exploring the factors influencing the spatial distribution of basic educational resources includes the equalization of basic educational facilities and their determinants [37,38].

Lü and Liu [39] employing fixed-effect models and multiple linear regression methods and scrutinized the factors influencing the equalization of basic educational facilities in China. Zhao et al. [33] identified that the spatial arrangement of basic educational resources in rural China is positively correlated with rural economic development levels and the administrative capacity of local governments. Wang et al. [16] uncovered that regions with lower value in terms of basic educational facilities in China are predominantly situated west of the Hu Huanyong Line, with this disparity being driven by factors such as population size, industrial structure, and urban area size. Xu et al. [40] detected a significant "core–periphery" spatial distribution pattern of basic educational resources concerning residential areas, underscoring the relationship between basic education and population concentration. Cai et al. [41] contended that the "urban–rural dual disparity" constitutes a primary factor affecting the equality of educational opportunities between urban and rural areas, highlighting the issue of urban–rural educational inequality. Zhang et al. [42] explored the relationship between construction land expansion and basic education schools in Shanghai based on POI data. Jiang et al. [43] found that the population distribution dominated the spatial distribution of primary school facilities in Chengdu. Huang et al. [44] suggested that the spatial distribution and accessibility of educational service facilities in Lincang County, Yunnan Province, an underdeveloped region in China, were positively correlated with population density, aging, and income levels. Furthermore, studies have explored the connection between educational policies and the spatial layout of basic educational resources, furnishing crucial insights for a more profound understanding of the distribution and influencing factors of basic educational resources [45–47].

Moreover, some research has focused on the spatial allocation and optimization of basic educational resources. These research domains include aspects such as spatial optimization paths [48,49] and policies [50–52] for achieving the equalization of basic educational resources. Additionally, the examination of the equalization of basic educational resources with respect to societal development attracts substantial interest among researchers [53–55]. Studies have shown that the equalization of basic educational resources contributes to fostering balanced population distribution and development [56] and enhancing the overall human resource pool [57].

Although significant strides have been made in the study of basic educational public services [58,59], most of these inquiries have lacked a geographical spatial perspective. Furthermore, their focus has predominantly revolved around compulsory education, with insufficient attention being given to preprimary education. As society progresses, research primarily centered on primary and secondary education may no longer adequately address evolving societal needs. The integration of big data into research methodologies has become increasingly prevalent, with big POI data offering distinct advantages in the study of spatial patterns of phenomena. POI data, characterized by large volumes, cost-effectiveness, and frequent updates [60], have the potential to enhance the efficiency and precision of research in the domain of basic public services, including basic educational facilities.

The Yellow River Basin stands as a crucial economic region in China, playing a pivotal role in the country's socioeconomic development, particularly in poverty alleviation (e.g., implementing education subsidies, building rural schools, improving teacher quality, and increasing employment opportunities) endeavors. Originating in the Bayan Har Mountains in Qinghai Province, the Yellow River traverses nine provinces: Qinghai, Sichuan, Gansu, Ningxia, Inner Mongolia, Shaanxi, Shanxi, Henan, and Shandong. Research within the Yellow River Basin has primarily concentrated on topics such as ecological protection and high-quality development [61–71]. Nonetheless, issues regarding the adequacy of the overall supply of educational resources and regional disparities remain pressing concerns within China's basic educational infrastructure and are focal points of academic research [72–74]. Specific research on the spatial pattern of basic education and related aspects in the Yellow River Basin or its sub-regions has been limited [25,30]. Investigating the spatial distribution of basic education bears practical significance. This study, which takes the Yellow River Basin as its subject for researching basic educational services, serves as a representative example.

Based on the literature review presented above, this study addresses several knowledge gaps in the field of public service research related to basic education. These gaps include insufficient attention to preschool education and a lack of watershed and geospatial research perspectives. The study is founded on a big data analysis model concerning the layout of public educational service facilities. It employs kernel density analysis and spatial autocorrelation methods in ArcGIS to research and analyze the spatial pattern of basic educational facilities within the Yellow River Basin. Additionally, by collecting socioeconomic data and utilizing geographically weighted regression (GWR) and other methodologies, this study explores the factors influencing the spatial layout. This quantitative analysis of regional disparities and the equalization of the spatial pattern of basic educational facilities in the Yellow River Basin, combined with statistical data, aims to investigate the factors affecting the spatial pattern. The objective is to provide a theoretical foundation and scientific reference for the equalization of basic educational facilities and the equitable allocation of basic educational resources within the Yellow River Basin.

2. Materials and Methods

2.1. Study Area

The Yellow River Basin includes the geographical and ecological region influenced by the Yellow River, extending from its source to its estuary (Figure 1). The topography of the Yellow River Basin exhibits significant variations, characterized by elevated terrain in the western regions and lower elevations in the east, resulting in substantial geomorphological diversity across different areas. This basin holds significant importance as an energy and heavy industry hub in China, spanning the eastern, central, and western regions of the country. Its origins can be traced back to Qinghai, and it stretches all the way to Shandong, covering nine provinces and regions: Qinghai, Sichuan, Gansu, Ningxia, Shaanxi, Shanxi, Inner Mongolia, Henan, and Shandong. Within this basin, a diverse and complex ecosystem prevails, including forests, grasslands, deserts, and wetlands. These complex natural resource conditions have profoundly shaped the fundamental characteristics of socioeconomic development within the Yellow River Basin. Presently, the socioeconomic development of the Yellow River Basin is complexly linked with the ecological environment, facing a multitude of challenges. These include increasingly stringent resource limitations, overall inefficiencies in economic development, disparities in development between the upper, middle, and lower reaches, and the absence of a fully functional regional collaborative development mechanism. The gross domestic product (GDP) of the nine provinces and regions within the basin constitutes approximately 26% of the national total. The population residing in this basin accounts for approximately 30% of the national total. According to the statistical yearbooks of the provinces in the Yellow River Basin, the permanent population residing in the Yellow River Basin by the end of 2020 totaled 219.7076 million, and the region boasted a total of 65,909 basic educational facilities. Furthermore, the proportion of annual financial expenditure on education relative to the general public budget expenditure has demonstrated a consistent upward trajectory.

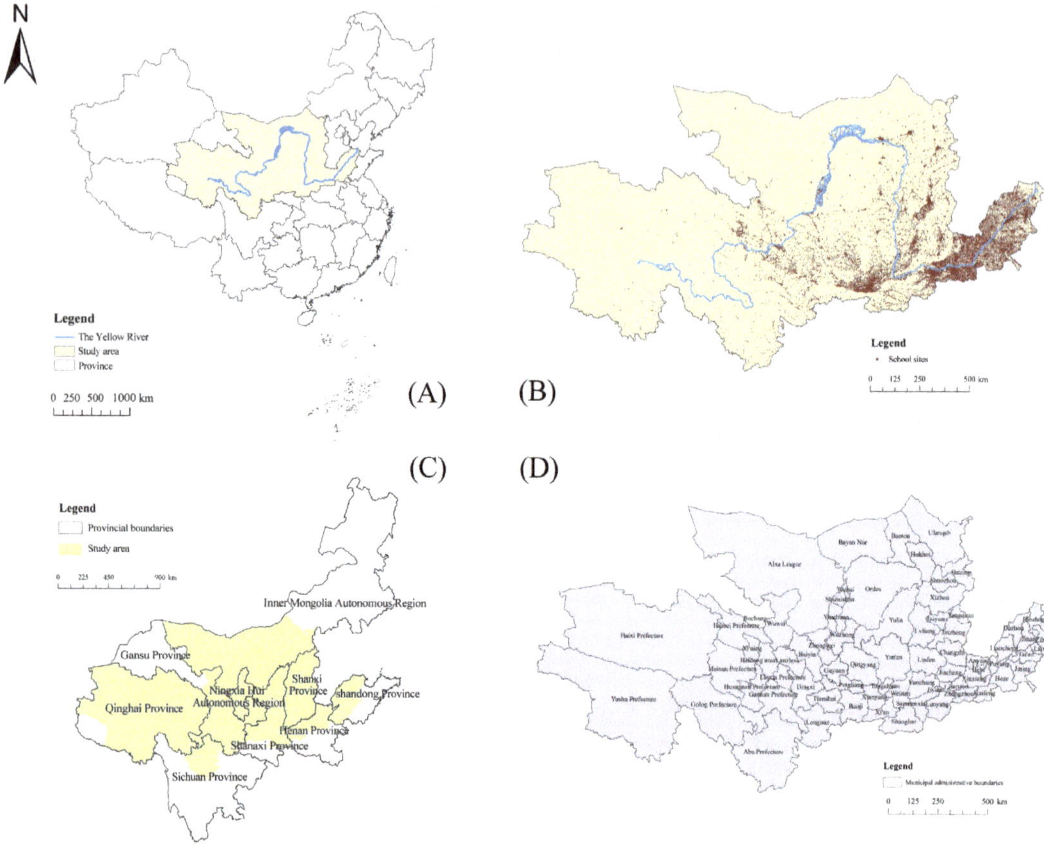

Figure 1. The location of the Yellow River Basin (**A**), the distribution of school sites (**B**), the provinces through which the Yellow River flows (**C**), and administrative divisions at the prefecture level in the Yellow River Basin (**D**).

2.2. Data Sources

POI data from 2020 related to basic education (preprimary education, primary education, and secondary education) and fundamental map data, including vector maps of cities within the Yellow River Basin and the Yellow River water system, were obtained from the Resource and Environment Science and Data Center of the Chinese Academy of Sciences (https://www.resdc.cn/, accessed on 4 January 2022). The original POI data were cleaned up by removing duplicate records, correcting incorrect locations, and processing missing data. The study used an algorithm based on unique coordinates to detect and delete duplicate records. Incorrect location data were identified and corrected by verifying the matching degree between the geographic coordinates of the POIs and the online map service. A comprehensive collection of socioeconomic data, which were crucial for analyzing the influencing factors, was obtained from the 2021 statistical yearbooks of each province. This compilation includes relevant socioeconomic information pertaining to the cities involved in this study.

2.3. Methods

2.3.1. Kernel Density Analysis Method

The study of the spatial pattern of basic educational facilities in the Yellow River Basin primarily focuses on the distribution of their quantity and the degree of concentration. In order to examine the concentration of basic educational facilities and visually depict their spatial distribution pattern, the kernel density analysis method is employed. The precise calculation formula is as follows [75]:

$$f_n(x) = \frac{1}{nh}\sum_{i=1}^{n} k\left(\frac{x-X_i}{h}\right) \tag{1}$$

In Formula (1), $k\left(\frac{x-X_i}{h}\right)$ represents the kernel function; $h > 0$ is the bandwidth; n is the number of POIs of basic education in the study area; $x - X_i$ represents the distance from the estimation point to the sample point X_i.

2.3.2. Spatial Autocorrelation Analysis

The global Moran's I statistic is employed to assess the spatial association and level of disparity in basic educational facilities among cities within the Yellow River Basin. If Moran's I > 0, this indicates a positive spatial correlation among all spatial units of basic educational facilities, indicating that a higher I value corresponds to a greater degree of clustering. If Moran's I = 0, this indicates a random distribution of basic educational facilities, devoid of spatial correlation. If Moran's I < 0, this suggests a negative spatial correlation in basic educational facilities within the Yellow River Basin, with a smaller *I* value indicating a higher degree of dispersion [76].

$$I = n\sum_{i=1}^{n}\sum_{j=1}^{n} W_{ij}(x_i - \overline{x})(x_j - \overline{x}) / \sum_{i=1}^{n}\sum_{j=1}^{n} W_{ij}\sum_{i=1}^{n}(x_i - \overline{x})^2 \tag{2}$$

Here, n is the sample size, i.e., the number of spatial locations, x_i and x_i are the observed values at spatial locations i and j, and W_{ij} represents the spatial relationship between locations i and j. When i and j are adjacent spatial locations, $W_{ij} = 1$; otherwise, $W_{ij} = 0$.

2.3.3. Geographically Weighted Regression Analysis

GWR is employed to analyze the influencing factors affecting the spatial pattern distribution of basic educational public facilities within the Yellow River Basin. Ordinary linear regression models seek to minimize the sum of squares of the distances from the parameters to the regression line. However, when dealing with spatial data, particularly data characterized by spatial non-stationarity, ordinary linear regression models prove inadequate for comprehensive modeling and analysis. In contrast, the GWR model incorporates the spatial structure into the linear regression framework, performing local regressions on spatial data. This approach captures the variations in the relationship between independent and dependent variables that occur with geographical spatial changes [77,78]. Consequently, it is well suited for handling data featuring spatial non-stationarity. The calculation formula is as follows.

$$y_i = \beta_{0(u_i,v_i)} + \sum_k \beta_{k(u_i,v_i)} x_{jk} + \varepsilon_i \tag{3}$$

In Formula (3), y_i is the dependent variable, $\beta_{0(u_i,v_i)}$ is the constant term, (u_i, v_i) represents the spatial location of the i sampling point; $\beta_{k(u_i,v_i)}$ is regression parameter k at sample point i; ε_i is the random error at sample point i.

The selection of influencing factors of the spatial distribution of basic education service facilities is explained as follows. The development of these facilities is intrinsically linked to economic support. Education, as a people-centered endeavor, relies on educational infrastructure to serve the population. The construction and maintenance of educational infrastructure fall within the domain of tertiary industry. Given the influence of the household registration system, a significant portion of the floating population leaves their children in their original household registration areas for education, thereby exerting a minimal effect on the spatial distribution of basic educational facilities within the region. Consequently, this study conducts a regression analysis by employing the per capita resource quantity of basic educational facilities as the dependent variable and the per capita GDP, permanent population, proportion of tertiary industry, and educational fiscal expenditure as explanatory variables.

Prior to conducting a GWR analysis on the influencing factors, it is essential to first employ the ordinary least squares (OLS) method to assess the presence of multicollinearity among all variables and verify it using the variance inflation factor (VIF). The results reveal that the VIF for each variable is below 5, signifying the absence of multicollinearity among them. Consequently, all four selected factors can be utilized as explanatory variables. The influencing factors of basic educational facilities at the prefectural level of administrative units exhibit spatial instability. In comparison with conventional regression analysis models, GWR can effectively account for this spatial instability. To illustrate the advantages of GWR, both OLS and GWR models were employed to analyze and compare the data, as presented in Table 1. The coefficient of determination (R^2) derived from the GWR analysis, amounting to 0.939, surpasses the R^2 value of 0.914 obtained from the OLS analysis. Furthermore, the corrected value of the Akaike information criterion for GWR is lower than that of the OLS analysis, indicating that the GWR model is more suitable for examining the influencing factors of the spatial pattern of basic educational facilities compared with OLS.

Table 1. Results of comparison between OLS and GWR.

	AICc	R^2	Adjusted R^2
OLS	1006.98	0.914	0.909
GWR	998.37	0.939	0.925

3. Results

3.1. Distribution Characteristics of Basic Educational Facilities

3.1.1. Spatial Distribution of Absolute Numbers

The spatial distribution of basic educational facilities within the Yellow River Basin was uneven, and the absolute numbers of preprimary, primary, and secondary educational facilities were 33,558, 21,254, and 11,097, respectively (Figure 1). By examining the distribution of basic educational facilities across various cities within the basin, the proportion of basic educational resources in each city was determined. Regarding the number of preprimary basic educational resources, the proportions in Shaanxi Province, Shanxi Province, Henan Province, and Shandong Province were, respectively, 14.8%, 14.5%, 29.7%, and 25.7%. The proportions in Qinghai Province, Sichuan Province, Gansu Province, Inner Mongolia Autonomous Region, and Ningxia Hui Autonomous Region were less than 5.0%. It is evident that the upstream provinces of Qinghai, Sichuan, Inner Mongolia, and Ningxia Hui Autonomous Region face distinct disadvantages in terms of preprimary education resources compared with the downstream provinces. Henan and Shandong provinces, situated downstream, enjoy a clear advantage, collectively accounting for over half of the total basic educational resources, with each province contributing more than 25.0% of the total. While there may be slight variations in the proportions of primary and secondary educational resources among the provinces, the overall trend remains consistent, with Henan and Shandong provinces maintaining their advantageous positions and the five upstream provinces continuing to face disadvantages. In summary, when considering the statistics of basic educational resources, Shandong and Henan provinces exhibit a substantial advantage, followed by Shanxi and Shaanxi provinces, while the other five provinces contend with a disadvantageous position.

3.1.2. Density Distribution Characteristics of Absolute Numbers

Using the POI data for preprimary education, primary education, and secondary education in the Yellow River Basin as the data source, kernel density analysis was employed to analyze the density distribution of basic educational facilities and to perform a visual analysis (Figure 2).

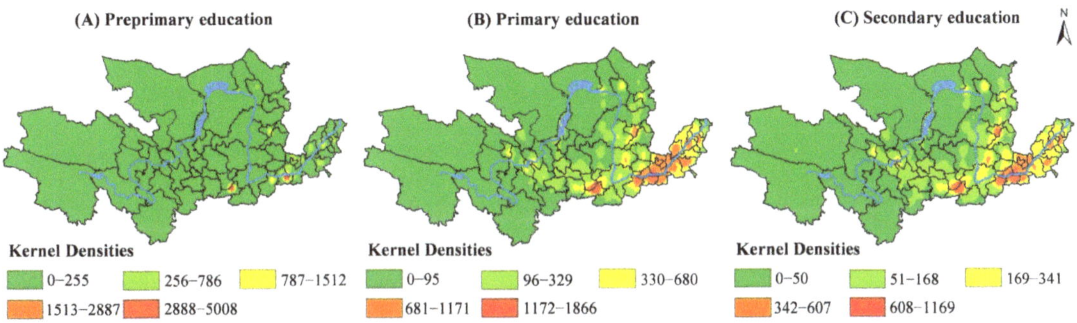

Figure 2. Kernel density analysis of basic educational facilities in the Yellow River Basin, including preprimary education (**A**), primary education (**B**), and secondary education (**C**).

The analysis findings indicate that preprimary education exhibits a higher density of distribution in the eastern region, with some areas in the central region displaying moderate density, while the western region experiences a more dispersed distribution. Notably, the distribution of high-value areas corresponds to the geographical concentration of certain central cities. These central cities, such as the regions centered around Taiyuan in Shanxi Province, Xi'an in Shaanxi Province, Zhengzhou in Henan Province, and the area centered around Shandong Province, constitute concentrated high-value areas. Conversely, most of the upstream regions within the Yellow River Basin exhibit a lower-density distribution. The density of the distribution of primary and secondary education closely mirrors the pattern observed in preprimary education, with variations in the regions where the median values are situated. The visual representation of the comprehensive POI data pertaining to basic education highlights that the high-value areas of basic educational facilities are prominently situated within the high-density zones of the eastern part of the Yellow River Basin. These zones include contiguous high-value areas in the capitals of the eastern provinces. In contrast, the educational resources in the western region exhibit a clear dispersion. This observation underscores the eastern region's advantageous position concerning basic educational facilities while also highlighting the significant regional disparities in the availability of basic educational infrastructure.

3.2. Distribution Characteristics of Per Capita Basic Educational Facilities

3.2.1. Spatial Distribution Characteristics

The per capita quantity of basic educational facilities at the city level within the Yellow River Basin has been categorized into five distinct groups, as illustrated in Figure 3. High-value cluster areas for per capita preprimary public services are predominantly situated in the eastern region, with some northern regions such as Baotou in the Inner Mongolia Autonomous Region and Liaocheng in Shandong Province also forming part of this high-value cluster. In contrast, low-value cluster areas are primarily found in the western region, including areas such as Yushu Tibetan Autonomous Prefecture and Golog Tibetan Autonomous Prefecture in Qinghai Province. The per capita primary education public services exhibit a substantial high-value cluster area in select southern regions of Qinghai, Gansu, and Sichuan, with additional high-value clusters evident in certain eastern areas, notably Luoyang in Henan Province. Conversely, low-value clusters are primarily concentrated in the northern areas, such as Alxa, Bayannur, and Baotou in the Inner Mongolia Autonomous Region. Regarding per capita secondary education public services, high-value areas are concentrated in a few eastern regions, including Yuncheng in Shanxi Province and Xuchang in Henan Province. Conversely, low-value areas are predominantly located in the northern regions, such as Bayannur and Ordos in the Inner Mongolia Autonomous Region. The pattern of equalization for per capita preprimary public services closely aligns with the pattern observed for the absolute-number equalization, displaying significant disparities between the eastern and western regions. However, the issue of non-equalization in per capita primary and secondary education public services is not as pronounced as that observed in absolute numbers.

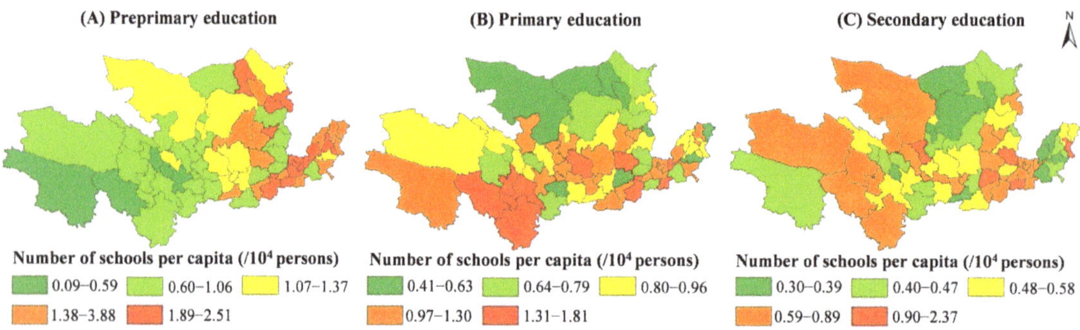

Figure 3. A distribution map of per capita basic educational facilities in the Yellow River Basin, including preprimary education (**A**), primary education (**B**), and secondary education (**C**).

3.2.2. Aggregation Characteristics

The global Moran's I values for preprimary, primary, and secondary education in the Yellow River Basin are 0.26, 0.19, and 0.09, respectively (Table 2). A Moran's I value greater than 0 indicates a positive spatial correlation among all study units of basic educational facilities, with a higher I value corresponding to a greater degree of spatial clustering. The global Moran's I index provides insight into the overall spatial aggregation pattern of basic educational facilities. Specifically, the concentration of preprimary educational facilities is relatively high, followed by primary education and secondary education.

Table 2. The global Moran's I index of per capita basic educational facilities in the Yellow River Basin.

Value	Preprimary Education	Primary Education	Secondary Education
Moran's I	0.258	0.190	0.085
Expected index	−0.014	−0.014	−0.014
Variance	0.002	0.006	0.006
Z score	6.391	2.706	1.334
p value	0.000	0.007	0.182

To gain a more nuanced understanding of the internal variations and localized spatial clustering of basic educational facilities within the Yellow River Basin, it is essential to assess the distribution pattern and level of the local Moran's I value (Figure 4). H-H indicates high–high clustering, signifying that the region possesses a high quantity of per capita educational resources, and neighboring areas also exhibit high quantities. H-L indicates high–low clustering, where the region itself has high quantities, but neighboring areas have low quantities. L-H indicates low–high clustering, indicating that the region itself has low quantities, but neighboring areas exhibit high quantities. L-L indicates low–low clustering, where both the region and its neighboring areas have low quantities. The L-L cluster areas of per capita preprimary educational resources are predominantly located in the western regions, including a substantial area. Examples include Haixi Mongol and Tibetan Autonomous Prefecture, Haibei Tibetan Autonomous Prefecture, Hainan Tibetan Autonomous Prefecture in Qinghai Province, and Gannan Tibetan Autonomous Prefecture, Longnan, and Dingxi in Gansu Province. On the other hand, H-H cluster areas are primarily concentrated in the eastern regions, including cities such as Sanmenxia, Luoyang, Zhengzhou, and Kaifeng in Henan Province and Jinan, Binzhou, Zibo, and Liaocheng in Shandong Province. For per capita primary educational resources, the L-L cluster areas are situated in Baotou, Ulanqab, and Hohhot in the Inner Mongolia Autonomous Region, while H-H cluster areas are primarily found in Sanmenxia, Luoyang, Jiaozuo, and Xinxiang in the southern part of the region. In the case of per capita secondary educational resources, the L-L cluster areas are located in Bayannur, Baotou, and Ordos in the Inner Mongolia Autonomous Region, with H-H cluster areas mainly being centered around cities such as Luoyang, Jiaozuo, and Linfen. Compared with primary and secondary education, the distribution of preschool education resources reveals significant disparities between the eastern and western regions, particularly in terms of aggregation characteristics. Unlike primary and secondary education, preschool education is heavily influenced by market dynamics, resulting in the swift concentration of superior resources in the east, which is considered a high-return area. Consequently, addressing the issue of unequal public services in preschool education has become increasingly urgent. Therefore, the government should assume greater responsibility in promoting the equitable distribution of public services in preschool education.

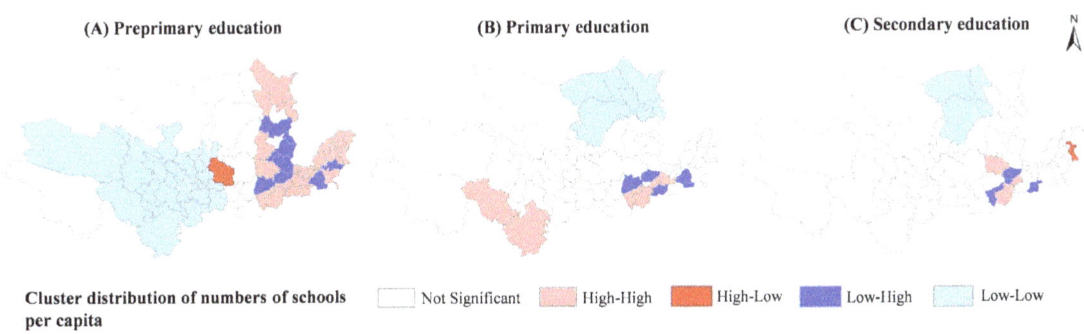

Figure 4. Cluster distribution map of per capita basic educational facilities in the Yellow River Basin, including preprimary education (**A**), primary education (**B**), and secondary education (**C**).

3.3. *Influencing Factors of the Spatial Pattern of Basic Educational Facilities*

In the GWR model, each research factor corresponds to a specific regression coefficient. The statistical analysis of the model's regression results is summarized in Table 3. The analysis reveals that the regression coefficients associated with the proportion of the tertiary industry and permanent population exhibit the highest maximum values, with the coefficient for the tertiary industry displaying the most significant fluctuations. This indicates that these two factors exert a considerable influence on the spatial distribution of basic educational facilities. The regression coefficients related to the permanent population consistently demonstrate positive values, implying a positive effect of the permanent population on basic educational facilities. Conversely, the regression coefficients for per capita GDP, proportion of tertiary industry, and educational fiscal expenditure exhibit both positive and negative values. This suggests that these three factors exert both positive and constraining effects on basic educational facilities in various regions.

Table 3. Statistics and VIF of GWR model's regression coefficients.

Variables	Minimum Value	Median Value	Maximum Value	Average Value	Standard Deviation	VIF
Per Capita GDP	−0.136	−0.0678	0.000321	−0.00182	0.0003388	1.005001
Permanent Population	2.527	3.038	3.549	2.888	0.1998	1.922066
Proportion of Tertiary Industry	−2.664	5.9535	14.571	5.538	3.8875	1.327498
Educational Fiscal Expenditure	−1.459	0.392	2.243	0.445	1.3697	1.327498

As illustrated in Figure 5, the regression coefficients associated with the permanent population are consistently positive, with the highest coefficient being observed in Baotou City and Hohhot City at more than 3.25. Coefficients with low values were distributed in the lower reaches of the Yellow River Basin. It is crucial to emphasize that addressing the ratio of the permanent population to basic educational facilities in the downstream region of the basin is pivotal for promoting balanced development in terms of population and education. Regarding the proportion of tertiary industry, the regression coefficients display both positive and negative values, demonstrating an increasing trend from the upstream to the downstream areas of the Yellow River Basin. The low-value areas for the proportion of the tertiary industry are predominantly situated in the upstream region, while downstream areas exhibit a higher proportion. This indicates a pronounced synergistic effect between tertiary industry and education, with an increase in the proportion of tertiary industry exerting a clear positive effect on the quantity of basic educational facility resources in the region.

In contrast, the regression coefficients linked to per capita GDP and education expenditure are comparatively smaller, and they are categorized as secondary influencing factors for basic educational facilities. The coefficients for per capita GDP include both positive and negative values, with the highest coefficient being recorded in Baoji City at 0.000321, while the lowest coefficient is −0.136 in Jining City. The regions characterized by lower per capita GDP values are predominantly situated in the downstream region of the Yellow River Basin. In these areas, the impact of an increase in per capita GDP on the expansion of basic educational facilities is relatively modest and may even manifest a negative effect in certain areas. This phenomenon can be attributed to an upgraded focus on educational quality over quantity in these downstream regions as a result of increased per capita GDP. Consequently, an elevation in per capita GDP might lead to a deviation from an increase in the quantity of basic educational facilities in these areas. Moreover, the regression coefficients associated with educational fiscal expenditure also include both positive and negative values, with a concentration of high-value areas in the central region of the Yellow River Basin. An increase in educational fiscal expenditure and its proportion significantly influences the increase in the quantity of basic educational facilities in the region. In summary, the primary influencing factors shaping the spatial pattern of basic educational facilities in the Yellow River Basin include the permanent population and the proportion of tertiary industry. The secondary factors include the per capita GDP and education expenditure.

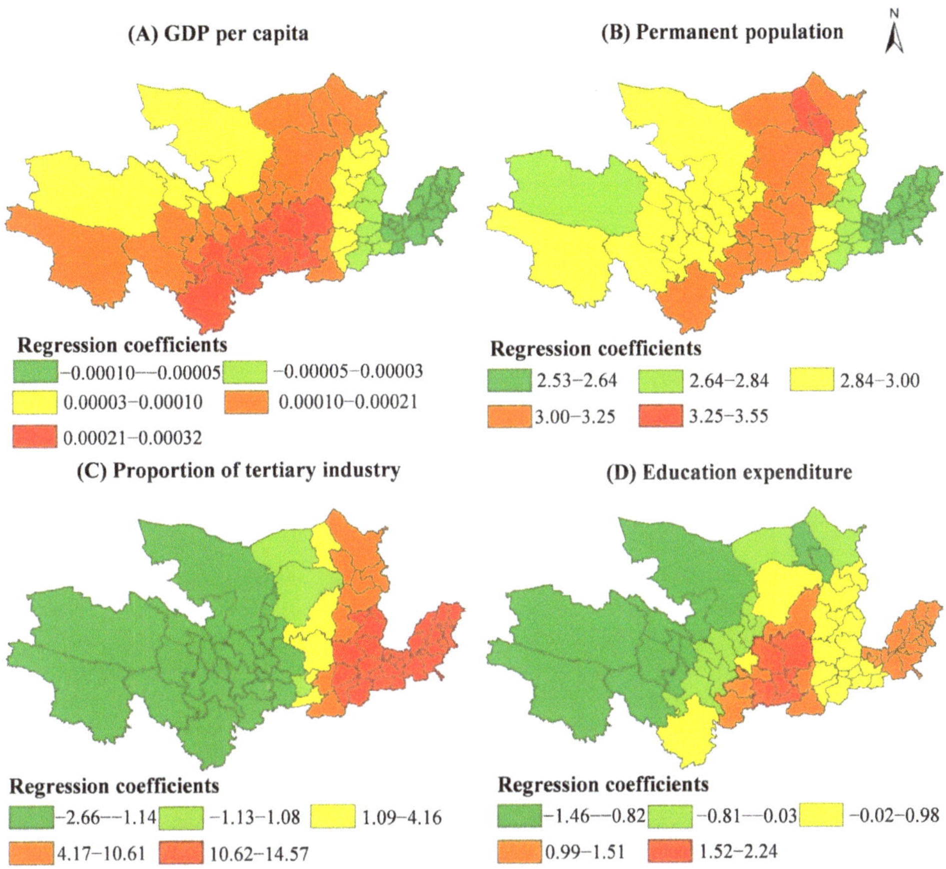

Figure 5. The distribution of the GWR model's regression coefficients, including GDP per capita (**A**), permanent population (**B**), proportion of tertiary industry (**C**), and education expenditure (**D**).

4. Discussion

4.1. Influencing Factors of Regional Disparities and Policy Recommendations

The findings highlight the key role of the permanent population and the proportion of tertiary industry in shaping the spatial pattern of basic educational facilities in the Yellow River Basin, with per capita GDP and educational fiscal expenditure playing a secondary role. The density of the population is crucially important for the distribution of basic educational public service facilities [16,40]. High-population-density areas typically necessitate more basic educational resources to meet the growing educational demands. In the downstream regions of the Yellow River Basin, the influence of the permanent population on the demand for basic educational public service facilities is relatively lower. Consequently, it is crucial to fine-tune the ratio of the permanent population to basic educational public service facilities, thereby achieving a harmonious development of both the population and educational resources. The proportion of tertiary industry significantly shapes the distribution of basic educational public service facilities, often mirroring the level of economic development and the extent of the service industry within a region [37,38]. The research findings underscore that the downstream areas of the Yellow River Basin exhibit a relatively higher proportion of tertiary industry, while this is lower in the upstream areas. This suggests that elevating the proportion of tertiary industry can exert a positive effect on the quantity of basic educational public service resources in the region. This phenomenon may be attributed to the fact that the development of tertiary industry drives economic growth and regional prosperity, thereby generating additional resources and funding for basic educational public service facilities, which, in turn, can expand their reach. The high-value areas of basic educational public service facilities in the Yellow River Basin tend to align with provincial capital cities, which serve as regional centers. These cities attract more economic, educational, and resource investments, consequently boasting a richer array of basic educational public service facilities. This disparity is indicative of the unevenness in regional economic development, resource distribution, and investment patterns.

The research further reveals that per capita GDP and educational fiscal expenditure assume a secondary role in influencing basic educational public service facilities. An elevation in per capita GDP, signifying improved economic conditions, may stimulate the distribution and provision of basic educational public service resources [79,80]. The regression coefficients for educational fiscal expenditure exhibit both positive and negative values, overall remaining relatively low. The high-value areas of basic educational public service facilities are primarily concentrated in the middle and lower reaches of the Yellow River Basin. A moderate increase in educational fiscal expenditure contributes to the enhancement of the supply and quality of basic educational public service facilities. Consequently, as secondary influencing factors, per capita GDP and educational fiscal inputs exert a certain degree of influence on the quantity and provision of basic educational public service facilities.

The spatial distribution of fundamental educational public service facilities in diverse regions is influenced by various factors. Consequently, it is crucial to attain rational resource utilization and allocation strategies tailored to local circumstances. The optimization of the comprehensive structure of fundamental educational public service resources and their proportion in the total permanent resident population holds primary significance. Through efficacious resource allocation and judicious planning, educational resources can be equitably disseminated across regions, ensuring an equitable provision of fundamental educational public service facilities. This endeavor will contribute to narrowing the educational resource gap among different regions, thereby enhancing the universality and equity of fundamental educational public service facilities. The enhancement of the tertiary sector's share and per capita GDP levels is important. The development of the tertiary sector and the increase in per capita GDP will lead to higher financial support, thereby strengthening government investment in the field of education [24,39,50]. This, in turn, will bolster the construction and enhancement of educational infrastructure, elevate teacher compensation and professional development, and foster the more effective provisioning of quality educational resources, thereby further enhancing the quality of fundamental educational public service facilities. Throughout this process, it is crucial to moderately increase the allocation of government public financial budget expenditure on education. Augmenting government financial support for education will bolster the coverage and quality of fundamental educational public service facilities, thereby ameliorating educational facility conditions and fortifying the teaching staff. This collective endeavor is poised to contribute significantly to the ongoing optimization and enhancement of the spatial arrangement of fundamental educational public service facilities.

4.2. Advantages and Limitations

This research harnessed the potential of big POI data, including a myriad of geographical location information, such as schools, hospitals, and communities, among others. These data sources provide a comprehensive and complex repository of geographic information, rendering them invaluable for research purposes [81–83]. Moreover, their accessibility and cost-effectiveness make them an attractive choice for data collection. Presented in point form, POI data represent diverse geographical location information within urban areas, enhancing research efficiency and accuracy [84,85]. Furthermore, they facilitate visual assessments of the clustering and dispersion of basic educational public service facilities. In this study, a range of established spatial analysis methods, including kernel density analysis, spatial autocorrelation, and GWR, were judiciously applied [75,76,86]. These methodologies are widely recognized and employed in the field of spatial analysis, ensuring the robustness of the research framework. In particular, the utilization of the GWR model enables an in-depth exploration of the spatial connections between basic educational public service facilities and their influencing factors. By conducting localized regressions on spatial data, this approach reveals the spatial nuances and fluctuations of basic educational public services. Consequently, it furnishes more targeted recommendations for policy formulation and strategic planning.

In this study, the per capita resource quantity of basic educational public service facilities in the Yellow River Basin served as the dependent variable, while the permanent population, per capita GDP, proportion of tertiary industry, and educational fiscal expenditure functioned as explanatory variables. One limitation of this study pertains to the relatively limited number of explanatory variables, and future research should contemplate the inclusion of additional variables, such as transportation networks, school-age children, and population mobility. These factors also exert an influence on the spatial configuration of basic educational public service facilities, and their integration into the analysis can enhance the comprehensiveness and precision of the study, enabling a more profound understanding of the influencing factors of spatial patterns and facilitating more tailored resource allocation. The current analysis predominantly emphasized the relationship between the quantity of basic educational public service facilities and their influencing factors, with relatively less attention being dedicated to the assessment of the quality of basic educational public service resources. Aspects such as school size, infrastructure, software, and teaching facilities constitute vital components of educational quality. Future research endeavors should aim to include more quality-related factors, including school size, teacher strength, and the completeness of teaching facilities, to holistically assess the quality of basic educational public service resources. Such comprehensive research would significantly contribute to a more profound understanding of the actual provision of basic educational public service facilities, offering comprehensive recommendations for enhancing education quality and equity.

5. Conclusions

Based on the dataset of basic education POIs, this study employed kernel density analysis and spatial autocorrelation methods within ArcGIS software 10.8 to scrutinize the density and spatial clustering of the basic educational public service pattern within the Yellow River Basin. Furthermore, GWR was utilized to investigate and elucidate the factors contributing to its spatial configuration. The findings can be summarized as follows:

(1) Regarding the quantity of basic educational facilities in the Yellow River Basin, it is evident that the four downstream provinces exhibit a substantially higher concentration of resources compared with the five upstream provinces. The downstream provinces account for a substantial 82.45% of the basic educational resources. A kernel density analysis of basic education within the Yellow River Basin revealed that areas with high-density basic educational facilities coincide geographically with certain provincial capitals. Notable examples include regions centered around Taiyuan in

Shanxi Province, Xi'an in Shaanxi Province, Zhengzhou in Henan Province, and areas surrounding Shandong Province.

(2) Concerning the spatial clustering of basic educational facilities in the Yellow River Basin, H-H is observed in the western region for preprimary education, while L-L is prevalent in certain eastern and northern areas. Primary and secondary education exhibit H-H in the north, and all three levels manifest spatial associations.

(3) The primary determinants influencing the spatial distribution of basic educational facilities in the Yellow River Basin are the proportion of tertiary industry and the permanent population. The secondary influencing factors include educational fiscal expenditure and per capita GDP, exerting a relatively lesser influence.

According to the spatial pattern and determinants of basic educational public services in the Yellow River Basin, a series of policy measures should be implemented to coordinate and optimize the structure of basic education service resources. These measures include increasing the proportion of the tertiary sector, moderately raising government public fiscal budget expenditures on education to optimize the spatial distribution of basic education public service facilities in the Yellow River Basin, and promoting balanced development in basic education.

In this study, due to data limitations, the assessment results and the selected explanatory variables with their relation to regional disparities in basic educational public service facilities were limited, and they cannot fully represent the quality and the influencing factors of basic educational public services. With the improvement in basic educational public service evaluation systems and the utilization of multi-source data, future research could integrate a diverse array of factors to fully comprehend the spatial layout and influencing determinants of basic educational public service facilities, thereby facilitating more detailed and exhaustive investigations.

Author Contributions: Conceptualization, Hejie Wei; funding acquisition, Hejie Wei; methodology, Mengxue Liu; visualization, Wenfeng Ji; writing—original draft, Hejie Wei; writing—review and editing, Ling Li and Yi Yang. All authors have read and agreed to the published version of the manuscript.

Funding: This research was funded by the Second Tibetan Plateau Scientific Expedition and Research Program (2019QZKK0608) and Henan Agricultural University Teaching Reform Project 2023 (2023XJGLX111).

Data Availability Statement: The data presented in this study are available from the corresponding author upon request.

Conflicts of Interest: The authors declare no conflicts of interest.

References

1. Anand, S.; Ravallion, M. Human Development in Poor Countries: On the Role of Private Incomes and Public Services. *J. Econ. Perspect.* **1993**, *7*, 133–150. [CrossRef]
2. Dick-Sagoe, C. Decentralization for improving the provision of public services in developing countries: A critical review. *Cogent Econ. Financ.* **2020**, *8*, 1804036. [CrossRef]
3. Li, T.; Zhao, Y.; Kong, X. Spatio-Temporal Characteristics and Influencing Factors of Basic Public Service Levels in the Yangtze River Delta Region, China. *Land* **2022**, *11*, 1477. [CrossRef]
4. Wang, Y.; Chi, Y. Measuring spatiotemporal changes of rural basic public service in poverty-stricken area of China. *Int. Reg. Sci. Rev.* **2018**, *41*, 510–539. [CrossRef]
5. Song, X.Q.; Deng, W.; Liu, Y.; Zhao, C. Residents' satisfaction with public services in mountainous areas: An empirical study of southwestern Sichuan province, China. *Chin. Geogr. Sci.* **2017**, *27*, 311–324. [CrossRef]
6. Dong, W.; Zhang, Q.; Yan, C.; Fu, W.; Xu, L. Residents' satisfaction with primary medical and health services in Western China. *BMC Health Serv. Res.* **2017**, *17*, 298. [CrossRef]

7. Ferlie, E.; Freeman, G.; McDonnell, J.; Petsoulas, C.; Rundle-Smith, S. Introducing choice in the public services: Some supply side issues. *Publ. Money Manag.* **2006**, *26*, 62–72. [CrossRef]
8. Song, X.; Zhao, C.; Han, J.; Zhang, Q.; Liu, J.; Chi, Y. Measurement and Influencing Factors Research of the Energy and Power Efficiency in China: Based on the Supply-Side Structural Reform Perspective. *Sustainability* **2020**, *12*, 3879. [CrossRef]
9. Zeng, S.; Xia, F.; Yang, P. The Equalization Gap of Basic Public Services Supply in China: An Empirical Study on Education, Health Care and Social Security. *China Financ. Econ. Rev.* **2018**, *7*, 24–51.
10. Samuelson, P.A. The Pure Theory of Public Expenditure. *Rev. Econ. Stat.* **1954**, *36*, 387–389. [CrossRef]
11. Li, Z.; He, S.; Su, S.; Li, G.; Chen, F. Public services equalization in urbanizing China: Indicators, spatiotemporal dynamics and implications on regional economic disparities. *Soc. Indic. Res.* **2020**, *152*, 1–65. [CrossRef]
12. Xu, J.; Yan, Z.; Hu, S.; Pu, C. The Spatial Distribution and Optimization of Medical and Health Land from the Perspective of Public Service Equalization: A Case Study of Urumqi City. *Sustainability* **2022**, *14*, 7565. [CrossRef]
13. Zhang, S.; Xu, S.; Wan, D.; Liu, H.; Zhao, L.; Guo, L.; Ren, J. Equalization Measurement and Optimization of the Public Cultural Facilities Distribution in Tianjin Central Area. *Sustainability* **2023**, *15*, 4856. [CrossRef]
14. Lan, F.; Wu, Q.; Zhou, T.; Da, H. Spatial Effects of Public Service Facilities Accessibility on Housing Prices: A Case Study of Xi'an, China. *Sustainability* **2018**, *10*, 4503. [CrossRef]
15. Song, X.; Deng, W.; Liu, Y. Spatial spillover and the factors influencing public service supply in Sichuan province, China. *J. Mt. Sci.* **2014**, *11*, 1356–1371. [CrossRef]
16. Wang, F.; Bai, Y.P.; Zhou, L.; Ji, X.P.; Xu, Z.B.; Qiao, F.W. Spatial pattern and influencing factors of the equalization of basic education public service in China. *Geogr. Res.* **2019**, *38*, 285–296.
17. Glaser, R. Education and thinking: The role of knowledge. *Am. Psychol.* **1984**, *39*, 93. [CrossRef]
18. Zhang, D.; Li, X.; Xue, J. Education inequality between rural and urban areas of the People's Republic of China, migrants' children education, and some implications. *Asian Dev. Rev.* **2015**, *32*, 196–224. [CrossRef]
19. Zhang, X.; Kanbur, R. Spatial Inequality in Education and Health Care in China. In *Regional Inequality in China*; Routledge: London, UK, 2009; pp. 92–110.
20. Wang, L. Social exclusion and inequality in higher education in China: A capability perspective. *Int. J. Educ. Dev.* **2011**, *31*, 277–286. [CrossRef]
21. Liu, Q.; Song, J. Spatial distribution pattern of basic education facilities at county level in Tibet Autonomous Region. *J. Arid Land Resour. Environ.* **2022**, *36*, 84–92.
22. Hajnal, G. Diversity and convergence: A quantitative analysis of European public administration education programs. *J. PAfE* **2003**, *9*, 245–258. [CrossRef]
23. Huang, Q.; Cui, X.; Ma, L. The Equity of Basic Educational Facilities from the Perspective of Space. *Sustainability* **2023**, *15*, 12031. [CrossRef]
24. Wen, J.; Jiang, H. Changes in China's inter-provincial equalization of basic education service. *Financ. Trade Econ.* **2014**, *34*, 44–53.
25. Ni, Q.; Wu, X.; Cui, P. Research on the Equity of Educational Facilities in Counties of the Loess Plateau Gully Area: Chengcheng County, Shaanxi Province as an Example. *Sustainability* **2022**, *14*, 13106. [CrossRef]
26. Cobb, C. Geospatial Analysis: A New Window into Educational Equity, Access, and Opportunity. *Rev. Res. Educ.* **2020**, *44*, 97–129. [CrossRef]
27. He, G.; Huang, Q. Geospatial Analysis and Research on Social and Spatial Inequality of Compulsory Education: A Case Study of Hangzhou, China. *Complexity* **2021**, *2021*, 6265751. [CrossRef]
28. Pacione, M. Access to urban services—The case of secondary schools in Glasgow. *Scott. Geogr. Mag.* **1989**, *105*, 12–18. [CrossRef]
29. Malczewski, J.; Jackson, M. Multicriteria spatial allocation of educational resources: An overview. *Socio-Econ. Plan. Sci.* **2000**, *34*, 219–235. [CrossRef]
30. Wang, X.; Bai, Y.P.; Wang, F.; Zhang, M.; Che, L. Characteristics of spatial distribution of basic education resources in Xi'an City on street scale. *Arid Land Geogr.* **2019**, *42*, 1470–1477.
31. Yeates, M. Hinterland delimitation: A distance minimizing approach. *Prof. Geogr.* **1963**, *15*, 7–10. [CrossRef]
32. Maybee, G.D. What Do We Believe About Grades to be Included, Desirable Size, Appropriate Locations, and the Facilities for Junior High Schools? *NASSP Bull.* **1962**, *46*, 5–7. [CrossRef]
33. Zhao, L.; Wu, D.T.; Wang, Z.H.; Qu, L.P.; Yu, W. Spatio-temporal pattern and influencing factors of the allocation of rural basic education resources in China. *Econ. Geogr.* **2018**, *38*, 39–49. (In Chinese)
34. Pizzol, B.; Giannotti, M.; Tomasiello, D.B. Qualifying accessibility to education to investigate spatial equity. *J. Transp. Geogr.* **2021**, *96*, 103199. [CrossRef]
35. Lu, W.; Li, Y.; Zhao, R.; He, B.; Qian, Z. Spatial Pattern and Fairness Measurement of Educational Resources in Primary and Middle Schools: A Case Study of Chengdu–Chongqing Economic Circle. *Int. J. Environ. Res. Public Health* **2022**, *19*, 10840. [CrossRef]

36. Li, T.; Fang, X.; Zhu, J.; Peng, Q.; Zhao, W.; Fu, X. Horizontal and Vertical Spatial Equity Analysis Based on Accessibility to Living Service Amenities: A Case Study of Xi'an, China. *Land* **2024**, *13*, 1113. [CrossRef]
37. Jiang, L.; Chen, J.; Tian, Y.; Luo, J. Spatial Pattern and Influencing Factors of Basic Education Resources in Rural Areas around Metropolises—A Case Study of Wuhan City's New Urban Districts. *ISPRS Int. J. Geo-Inf.* **2022**, *11*, 576. [CrossRef]
38. Chen, B.; Zhang, H.; Wong, C.U.I.; Chen, X.; Li, F.; Wei, X.; Shen, J. Research on the Spatial Distribution Characteristics and Influencing Factors of Educational Facilities Based on POI Data: A Case Study of the Guangdong–Hong Kong–Macao Greater Bay Area. *ISPRS Int. J. Geo-Inf.* **2024**, *13*, 225. [CrossRef]
39. Lv, W.; Liu, G. Research on some influencing factors of China's educational equalization. *J. Quant. Tech. Econ.* **2010**, *5*, 20–33.
40. Xu, H.; Zhou, Z.; Kong, D.; Zhang, S.; Zhang, W.; Hao, Q. Spatial pattern of basic education resources in Wuhan based on spatial coupling algorithm. *Econ. Geogr.* **2019**, *39*, 88–94.
41. Cai, H.; Wang, S.; Yu, K. Preliminary study of basic education facilities configuration in rural area under the perspective of fairness and beneficial result: A case study of Jingyang, Shaanxi. *Mod. Urban Res.* **2016**, *31*, 83–91.
42. Zhang, Z.; Luan, W.; Tian, C.; Su, M.; Li, Z. Spatial Distribution Equilibrium and Relationship between Construction Land Expansion and Basic Education Schools in Shanghai Based on POI Data. *Land* **2021**, *10*, 1059. [CrossRef]
43. Jiang, J.; Wang, Z.; Yong, Z.; He, J.; Yang, Y.; Zhang, Y. Spatial Distribution and Accessibility Analysis of Primary School Facilities in Mega Cities: A Case Study of Chengdu. *Sustainability* **2024**, *16*, 723. [CrossRef]
44. Huang, C.; Feng, Y.; Wei, Y.; Sun, D.; Li, X.; Zhong, F. Assessing Regional Public Service Facility Accessibility Using Multisource Geospatial Data: A Case Study of Underdeveloped Areas in China. *Remote Sens.* **2024**, *16*, 409. [CrossRef]
45. Wang, W.; Zhou, Z.; Chen, J.; Cheng, W.; Chen, J. Analysis of Location Selection of Public Service Facilities Based on Urban Land Accessibility. *Int. J. Environ. Res. Public Health* **2021**, *18*, 516. [CrossRef]
46. Lange, W.; Alves, F. Spatial index of educational opportunities: Rio de Janeiro and Belo Horizonte. *Procedia Soc. Behav. Sci.* **2011**, *21*, 287–293. [CrossRef]
47. Xu, Y.; Song, W.; Liu, C. Social-Spatial Accessibility to Urban Educational Resources under the School District System: A Case Study of Public Primary Schools in Nanjing, China. *Sustainability* **2018**, *10*, 2305. [CrossRef]
48. Xie, Z.; Liu, X. New urbanization in rural basic education resources optimization configuration: Based on an empirical analysis of the basic education in Hebei Province. *J. Taiyuan Urban Voc. Coll.* **2014**, *43*, 92–94.
49. Ning, Z. *Study on Allocation of Resources of Rural Basic Education Under the Fairview: Taking Shandong Province as an Example*; Liaoning University: Shenyang, China, 2016.
50. Bunar, N. Choosing for quality or inequality: Current perspectives on the implementation of school choice policy in Sweden. *J. Educ. Policy* **2010**, *25*, 1–18. [CrossRef]
51. Law, W.; Pan, S. Legislation and Equality in Basic Education for All in China. *Interchange* **2009**, *40*, 337–372. [CrossRef]
52. Levin, H. Equal Educational Opportunity and the Distribution of Educational Expenditures. *Educ. Urban Soc.* **1973**, *5*, 149–176. [CrossRef]
53. Zhang, K.; Zhang, L.; Zhu, D. Research on the influence of basic educational resources on housing price: A case study of Haidian district in Beijing. *Educ. Econ.* **2018**, *32*, 27–34.
54. Zhang, L.; Guo, W. Basic education resources of the influence of the spatial layout of new urbanization. *Hous. Real Est.* **2016**, *42*, 259.
55. Meng, Z.; Pan, X. Basic education resource configuration space evolution of the population in Shanghai. *Urban Probl.* **2018**, *32*, 34–35.
56. Qian, F.; Yi, S. Research on Equalization of Public Services Based on Changes of Urban and Rural Population Structure. In Proceedings of the Eleventh International Conference on Management Science and Engineering Management 11, Kanazawa, Japan, 28–31 July 2018; Springer: Cham, Switzerland, 2018.
57. Glewwe, P.W.; Hanushek, E.A.; Humpage, S.D.; Ravina, R. *School Resources and Educational Outcomes in Developing Countries: A Review of the Literature from 1990 to 2011*; Working Paper No. 17554; National Bureau of Economic Research: Cambridge, MA, USA, 2011.
58. Yu, Y.; Appiah, D.; Zulu, B.; Adu-Poku, K.A. Integrating Rural Development, Education, and Management: Challenges and Strategies. *Sustainability* **2024**, *16*, 6474. [CrossRef]
59. Yang, G.; Xue, Y.; Ma, Y. Social Organization Participation, Government Governance and the Equalization of Basic Public Services: Evidence from China. *Int. J. Environ. Res. Public Health* **2019**, *16*, 2996. [CrossRef]
60. Xue, B.; Li, J.; Xiao, X.; Xie, X.; Lu, C.; Rren, W.; Jiang, L. Overview of man-land relationship research based on POI Data: Theory, method and application. *Geogr. Geo-Inf. Sci.* **2019**, *35*, 51–60.
61. Liu, L.; Liang, L.; Gao, P.; Fan, C.; Wang, H. Coupling relationship and interactive response between ecological protection and high-quality development in the Yellow River Basin. *J. Nat. Resour.* **2021**, *36*, 176–195. [CrossRef]
62. An, S.; Zhang, S.; Hou, H.; Zhang, Y.; Xu, H.; Liang, J. Coupling Coordination Analysis of the Ecology and Economy in the Yellow River Basin under the Background of High-Quality Development. *Land* **2022**, *11*, 1235. [CrossRef]

63. Zhang, J.; Liu, Y.; Liu, C.; Guo, S.; Cui, J. Study on the Spatial and Temporal Evolution of High-Quality Development in Nine Provinces of the Yellow River Basin. *Sustainability* **2023**, *15*, 6975. [CrossRef]
64. Zhang, K.; Dong, Z.; Guo, L.; Boyer, E.W.; Liu, J.; Chen, J.; Fan, B. Coupled coordination spatiotemporal analyses inform sustainable development and environmental protection for the Yellow River Basin of China. *Ecol. Indic.* **2023**, *151*, 110283. [CrossRef]
65. Wang, A.; Wang, S.; Liang, S.; Yang, R.; Yang, M.; Yang, J. Research on Ecological Protection and High-Quality Development of the Lower Yellow River Based on System Dynamics. *Water* **2023**, *15*, 3046. [CrossRef]
66. Feng, D.; Han, J.; Jia, H.; Chang, X.; Guo, J.; Huang, P. Regional Economic Growth and Environmental Protection in China: The Yellow River Basin Economic Zone as an Example. *Sustainability* **2023**, *15*, 10790. [CrossRef]
67. Du, H.; Wang, Z.; Li, H.; Zhang, C. Analysis of Spatial and Temporal Pattern Evolution and Decoupling Relationships of Land Use Functions Based on Ecological Protection and High-Quality Development: A Case Study of the Yellow River Basin, China. *Land* **2024**, *13*, 862. [CrossRef]
68. Kong, X.; Han, M.; Li, Y.; Kong, F.; Sun, J.; Zhu, W.; Wei, F. Spatial Differentiation and Formation Mechanism of Ecological Sensitivity in Large River Basins: A Case Study of the Yellow River Basin, China. *Ecol. Indic.* **2024**, *158*, 111571. [CrossRef]
69. Zhou, M.; Li, Z.; Gao, M.; Zhu, W.; Zhang, S.; Ma, J.; Ta, L.; Yang, G. Revealing the Eco-Environmental Quality of the Yellow River Basin: Trends and Drivers. *Remote Sens.* **2024**, *16*, 2018. [CrossRef]
70. Liu, Z.; Chang, Y.; Pan, S.; Zhang, P.; Tian, L.; Chen, Z. Unfolding the spatial spillover effect of urbanization on composite ecosystem services: A case study in cities of Yellow River Basin. *Ecol. Indic.* **2024**, *158*, 111521. [CrossRef]
71. Wei, H.; Xue, D.; Huang, J.; Liu, M.; Li, L. Identification of Coupling Relationship between Ecosystem Services and Urbanization for Supporting Ecological Management: A Case Study on Areas along the Yellow River of Henan Province. *Remote Sens.* **2022**, *14*, 2277. [CrossRef]
72. Liu, W.; Li, J.; Zhao, R. The effects of rural education on poverty in China: A spatial econometric perspective. *J. Asia Pac. Econ.* **2023**, *28*, 176–198. [CrossRef]
73. Rao, N.; Yang, Y.; Su, Y.; Cohrssen, C. Promoting Equity in Access to Quality Early Childhood Education in China. *Children* **2023**, *10*, 1674. [CrossRef] [PubMed]
74. Guo, L.; Huang, J.; Zhang, Y. Education Development in China: Education Return, Quality, and Equity. *Sustainability* **2019**, *11*, 3750. [CrossRef]
75. West, M. Kernel Density Estimation and Marginalization Consistency. *Biometrika* **1991**, *78*, 421–425. [CrossRef]
76. Getis, A. Spatial autocorrelation. In *Handbook of Applied Spatial Analysis*; Springer: Berlin/Heidelberg, Germany, 2010; pp. 255–278.
77. Brunsdon, C.; Fotheringham, A.S.; Charlton, M.E. Geographically Weighted Regression: A Method for Exploring Spatial Nonstationarity. *Int. Encycl. Hum. Geogr.* **1996**, *28*, 281–298. [CrossRef]
78. Yang, W.; Li, T.; Cao, X. The spatial pattern of Community Travel Low Carbon Index (CTLCI) and spatial heterogeneity of the relationship between CTLCI and influencing factors in Guangzhou. *Geogr. Res.* **2015**, *34*, 1471–1480.
79. Villela, R.; Paredes, J.J. Empirical Analysis on Public Expenditure for Education, Human Capital and Economic Growth: Evidence from Honduras. *Economies* **2022**, *10*, 241. [CrossRef]
80. Efthalitsidou, K.; Zafeiriou, E.; Spinthiropoulos, K.; Betsas, I.; Sariannidis, N. GDP and Public Expenditure in Education, Health, and Defense. Empirical Research for Greece. *Mathematics* **2021**, *9*, 2319. [CrossRef]
81. Yeow, L.W.; Low, R.; Tan, Y.X.; Cheah, L. Point-of-Interest (POI) Data Validation Methods: An Urban Case Study. *ISPRS Int. J. Geo-Inf.* **2021**, *10*, 735. [CrossRef]
82. Wei, H.; Yang, Y.; Han, Q.; Li, L.; Huang, J.; Liu, M.; Chen, W. Evaluation and System Coupling of Beautiful Qinghai–Tibet Plateau Construction Based on Point of Interest Data. *Systems* **2022**, *10*, 149. [CrossRef]
83. Qu, X.; Xu, G.; Qi, J.; Bao, H. Identifying the Spatial Patterns and Influencing Factors of Leisure and Tourism in Xi'an Based on Point of Interest (POI) Data. *Land* **2023**, *12*, 1805. [CrossRef]
84. Zhang, S.; Xia, Y.; Li, Z.; Li, X.; Wu, Y.; Liu, P.; Du, S. An Assessment of Urban Residential Environment Quality Based on Multi-Source Geospatial Data: A Case Study of Beijing, China. *Land* **2024**, *13*, 823. [CrossRef]
85. He, H.; Wang, W.; Wang, Z.; Li, S.; Chen, J. Enhancing Seismic Landslide Susceptibility Analysis for Sustainable Disaster Risk Management through Machine Learning. *Sustainability* **2024**, *16*, 3828. [CrossRef]
86. Comber, A.; Brunsdon, C.; Charlton, M.; Dong, G.; Harris, R.; Lu, B.; Lü, Y.; Murakami, D.; Nakaya, T.; Wang, Y.; et al. The GWR route map: A guide to the informed application of Geographically Weighted Regression. *arXiv* **2020**, arXiv:2004.06070.

International Journal of
Geo-Information

MDPI

Article

Enhancing Precision Beekeeping by the Macro-Level Environmental Analysis of Crowdsourced Spatial Data

Daniels Kotovs [1,*], Agnese Krievina [2] and Aleksejs Zacepins [1]

[1] Institute of Computer Systems and Data Science, Faculty of Engineering and Information Technologies, Latvia University of Life Sciences and Technologies, Liela iela 2, LV-3001 Jelgava, Latvia; iitfdek@lbtu.lv

[2] Department of Bioeconomics, Institute of Agricultural Resources and Economics, Struktoru iela 14, LV-1039 Riga, Latvia; arei@arei.lv

* Correspondence: daniels.kotovs@lbtu.lv

Abstract: Precision beekeeping focuses on ICT approaches to collect data through various IoT solutions and systems, providing detailed information about individual bee colonies and apiaries at a local scale. Since the flight radius of honeybees is equal to several kilometers, it is essential to explore the specific conditions of the selected area. To address this, the aim of this study was to explore the potential of using crowdsourced data combined with geographic information system (GIS) solutions to support beekeepers' decision-making on a larger scale. This study investigated possible methods for processing open geospatial data from the OpenStreetMap (OSM) database for the environmental analysis and assessment of the suitability of selected areas. The research included developing methods for obtaining, classifying, and analyzing OSM data. As a result, the structure of OSM data and data retrieval methods were studied. Subsequently, an experimental spatial data classifier was developed and applied to evaluate the suitability of territories for beekeeping. For demonstration purposes, an experimental prototype of a web-based GIS application was developed to showcase the results and illustrate the general concept of this solution. In conclusion, the main goals for further research development were identified, along with potential scenarios for applying this approach in real-world conditions.

Keywords: precision beekeeping; GIS; open data; OpenStreetMap; environmental analysis

1. Introduction

In recent years, beekeeping activities have gained increasing recognition for their importance in biodiversity, ecosystems, agriculture, and human health [1]. Besides being a productive branch of agriculture focused on breeding and maintaining bees for the harvest of various bee products [2], which have well-documented human health benefits [3], beekeeping plays a vital role in broader ecological and agricultural contexts [4]. The pollination services provided by bees help maintain ecosystem balance and are essential for the productivity of agricultural crops [5]. Pollination is essential for the production of many fruits, vegetables, and field crops, including oilseed production [6,7]. The presence of managed beehives in agricultural landscapes can increase pollination rates, improve crop quality and yield, and contribute to the overall productivity and profitability of agricultural systems [8], as well as to the development of rural areas [9]. Considering the decline of insects on both local and global scales, the EU has committed to reversing the decline in pollinators, as outlined in the EU Biodiversity Strategy for 2030 [10]. One possible short-term strategy to safeguard biodiversity and food security worldwide is to increase the number of farmed pollinators, particularly honeybees, relying on beekeepers' ability to preserve them [11].

The effectiveness of beekeeping largely depends on the competence of the beekeepers, who perform various activities based on decision-making processes [12]. Beekeeping activities involve a wide range of tasks and responsibilities aimed at maintaining healthy and productive bee colonies [13,14]. These activities primarily include bee colony monitoring, hive inspections, feeding and nutrition management, as well as pest and disease control [15,16]. Beyond these core tasks, a beekeeper's responsibilities also involve assessing local conditions to ensure optimal apiary placement, such as evaluating the availability of sufficient food resources [17]. Beekeeping also intersects with economic considerations, particularly in commercial operations [18,19]. In such cases, foresight in apiary development, strategic planning, and productivity forecasting are essential. All these responsibilities and tasks undoubtedly demand significant effort and resources from the beekeeper [20,21].

Modern beekeeping has evolved significantly with advancements in technology, science, and sustainable practices, transforming it into precision beekeeping [22]. By using modern technologies and data-driven approaches, precision beekeeping enables beekeepers to monitor beehive health, track environmental variables, and optimize management practices with unprecedented precision and efficiency [23]. Precision beekeeping has emerged as a response to the need for optimal bee colony management, integrating technology and statistical methods to manage apiaries efficiently and minimize the risk of bee population losses [2,24].

The term 'Precision Beekeeping' or 'Precision Apiculture' was first defined in the study 'Application of Information Technologies in Precision Apiculture' [25], where the authors, as computer science engineers, highlighted the potential for applying modern approaches from precision agriculture to beekeeping. Similar to precision agriculture, precision beekeeping is based on a three-phase cycle consisting of data collection, data analysis, and application [26]. While these phases involve a certain level of automation, they still require the beekeeper's involvement, particularly in decision-making and the application process. Precision beekeeping, in its original sense, is described as an apiary management strategy focused on monitoring individual bee colonies to minimize resource consumption and maximize bee productivity [25]. Thus, it can be stated that precision beekeeping, in its original concept, primarily targets the local scale, focusing on individual hives (bee colonies), and is confined to the territory of the apiary and its immediate surroundings. This statement is further supported by the most commonly used solutions in precision beekeeping, which are based on the use of various sensors and Internet of Things devices that focus on the detailed monitoring of individual colonies and apiary parameters [27]. The most common monitoring data include hive parameters such as temperature, hive weight, humidity levels, sounds, gases, and bee activity. Particular attention is also given to external factors affecting the apiary, such as weather conditions, which have a significant impact on the health of the bee colony [11]. Thus, it is possible to track bee activity, colony health and local conditions of apiaries in real-time [28]. However, given that the flight radius of a bee around the hive is approximately 3 km [29], it is very important to take into account the conditions of the surrounding environment on a larger scale. This statement is supported by the fact that, in addition to precision beekeeping's emphasis on the local scale of the apiary, the authors of the concept of precision beekeeping also mention the regional level as a specific geographic area encompassing neighboring bee farms as one of the levels of data collection and exchange [30]. Subsequent studies have expanded on this direction, with one of the key approaches involving the use of spatial data and geographic information systems (GISs) [27,31,32].

There are various environmental factors that can have a negative impact on bees that were also mentioned in several studies related to beekeeping [33–35]. These bring various challenges for beekeepers that threaten the health and productivity of bee colonies worldwide. Urbanization, pollution, and agricultural expansion reduce the availability of suitable areas for beekeeping, as they impact foraging and the persistence of bee colonies [36]. One of the most serious concerns is caused by intensive farming and the resulting presence of pesticides on farmland, as they can be toxic to bees [35]. This may be especially important in the case of migratory beekeepers who change the location of their apiaries. In fact, due to their high sensitivity to chemical and phytosanitary products, as well as their ability to collect samples from the air, vegetation, water, and soil, honeybees serve as bioindicators of disturbances in ecosystems and the environment [5,37,38].

Addressing these challenges requires a multifaceted approach, combining macro-level environmental data analysis and a suitability assessment of interested areas. A possible solution in this case could be using geographic information in combination with analysis and classification methods to perform the pre-exploration of the territory of interest. When using geographic information for beekeeping needs, much of the research has focused on analyzing a specific territory to perform suitability assessment or obtain important information for decision-making support. In many cases, the outcomes of such studies were static maps based on data from a specific period and relevant at the time of development [31,39–43]. The next stage of development of GIS-based solutions can be considered interactive GIS solutions that allow interaction with spatial data and use mechanisms for updating data over time. One of the clear examples of such solutions is the web-based solution described in [44], called Beescape. It allows one to interactively explore surrounding environment, assessing the quality of the landscapes for supporting bees and other pollinators [45]. The conducted research and proposed solutions demonstrate the relevance of using GIS in beekeeping and also contribute to the development of new ideas and the search for innovative approaches.

In this study, the authors focused on investigating the feasibility of spatial crowdsourcing to explore the potential application of this approach for real-time environmental analysis to support decision-making in the field of beekeeping. The primary objective was to demonstrate the feasibility of using a crowdsourced spatial database in combination with WEB and GIS technologies, highlighting potential scenarios for the technical implementation of the solution. The solution presented in this research paper should be regarded solely as an experimental prototype, demonstrating the concept of the approach from a technical perspective to support further studies in developing similar solutions.

2. Materials and Methods

To conduct this study, an attempt was made to go beyond region-specific data and extend the scope to an international scale. The main goal of this approach was to explore the possibility of developing a universal solution that is not restricted to a specific country or region. For this purpose, the term 'large-scale' data was introduced within the framework of this article. The term 'large-scale' refers to a territorial scope that is significantly larger than an apiary, where most common solutions in precision beekeeping, such as hive or apiary monitoring, are typically applied. Specifically, 'large-scale' encompasses relatively extensive areas, starting from the radius of bees' flight (several kilometers) and extending to the exchange of information at the national or even international level, such as in the case of migratory beekeepers.

Since it is necessary to perform a spatial analysis over areas with a radius of several kilometers at any location worldwide, obtaining data of this scale is challenging due to their variety and volume. Geodatabases from individual countries or regions provide environmental information at various levels of detail. Examples include Field Register Data from the Rural Support Service of Latvia [46] as well as open data provided by The geoPortal of Energy and Geology (Portugal) [47]. However, there are limitations. Solutions developed from such sources are limited to specific countries or regions. At the same time, the use of several databases is complicated by the unification of data. Moreover, it is important to consider data availability and the limitations of their use, including licenses and restrictions.

To overcome these barriers and obtain data on an international scale, open geographic data can be used. One of the crucial approaches for the exchange of open data is data crowdsourcing, which is especially valuable in cases when traditional data collection methods are time-consuming or impractical. There is a large amount of open geographic data sources available worldwide, examples of which are published on Free GIS Data by Robin Wilson [48].

To demonstrate the concept of the approach and the implementation of the solution, the authors of this study used OpenStreetMap (OSM) [49], one of the most well-known open geospatial databases, as a source of spatial data. It is a crowdsourced and open mapping project that includes geographic information about the entire world and relies on contributions from volunteers and stakeholders worldwide who collect, edit, and update spatial data [50,51]. OSM data have gained significant attention for their potential applications in various domains, including navigation, logistics services [52,53], and urban and regional planning [54–56]. They are also widely used to produce land use maps through data reclassification [57–59], which is the primary data processing method employed in this study. Due to its versatility and popularity, the OpenStreetMap database is one of the most widely used spatial data sources for research purposes, offering a valuable alternative to commercial and government datasets.

OSM data represents real-world objects on the ground (buildings, roads, areas, etc.) using tags attached to the corresponding data structures (nodes, ways, and relations), with each tag describing a geographic attribute of the spatial feature. A tag consists of a key and a value. The key is used to define a feature group or type (e.g., highway, building, etc.), while the value provides specific information for the feature belonging to a particular group (key) (e.g., motorway, shop, etc.) [60].

In addition to viewing the map, OSM allows users to obtain spatial data directly from the database. The data from OSM can be obtained in multiple ways, including through the Overpass API (https://wiki.openstreetmap.org/wiki/Overpass_API, accessed on 17 January 2025), which enables the retrieval of OSM spatial data using queries. An example of OSM data obtained using Overpass API query through Overpass turbo (https://overpass-turbo.eu, accessed on 17 January 2025) can be seen in Figure 1.

To better understand the concept and idea of the approach of this research, a high-level architecture was developed (see Figure 2) that illustrates the main components of the solution.

A central component of the approach is an open geographic database (e.g., OpenStreetMap) that allows data to be freely accessed and edited, ensuring that the data are current and rich. The idea is to provide data to the database from interested parties. These data can later be used by beekeepers to analyze the area of interest (by the data processing phase) and obtain valuable information. Such an approach could be beneficial for agriculture in general, as it would also enable the exchange of information between beekeepers and other farmers.

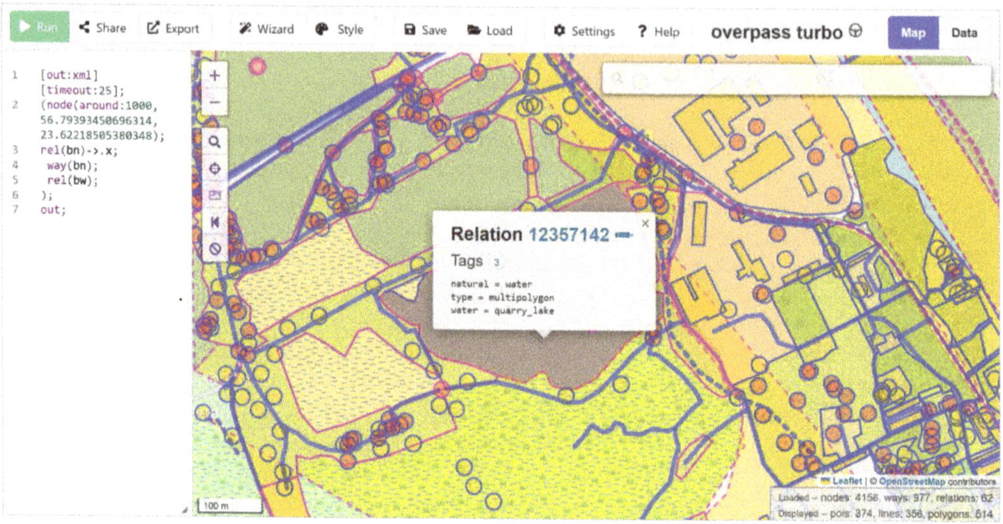

Figure 1. Example of OSM data obtained using the Overpass API query via the Overpass turbo.

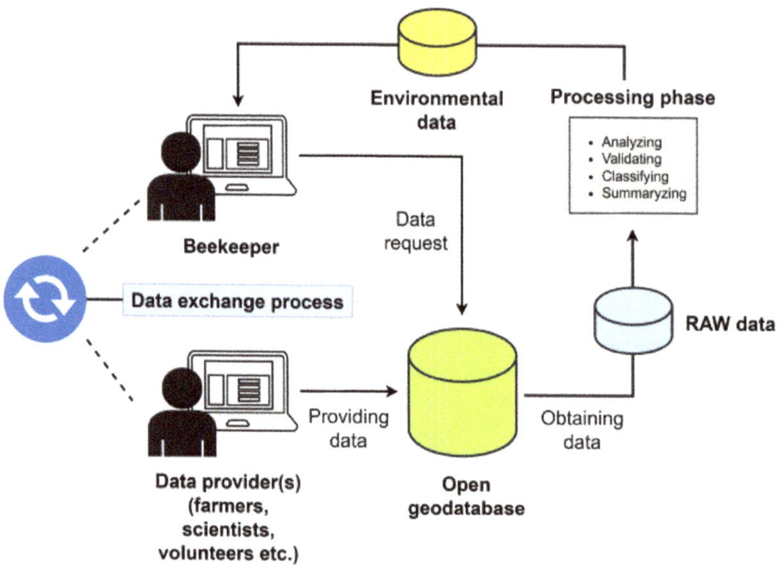

Figure 2. High-level architecture of the proposed approach.

At the core of the solution from a technical perspective is the Data Processing Module (DPM), which includes all the main functionalities related to data analysis, such as data acquisition, processing, and result generalization. The architecture of the DPM is shown in Figure 3.

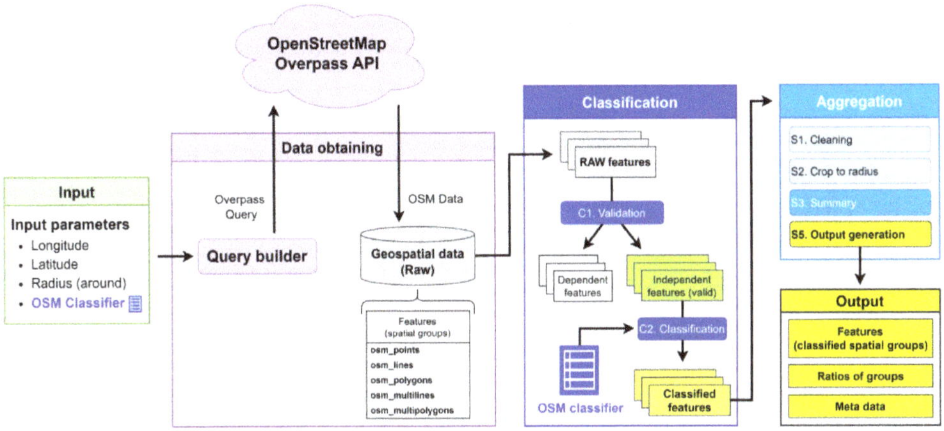

Figure 3. Architecture of the Data Processing Module (DPM).

The Data Processing Module consists of five stages:

1. Input stage—defining input data, which include
    a. The coordinates of the point of interest (latitude, longitude),
    b. The radius around the point of interest,
    c. A classifier that represents a dictionary of all OSM objects and their belonging to certain classification groups.
2. Data obtaining stage—creating a query for Overpass API and obtaining data from OSM.
3. Classification stage—performing crucial manipulation operations with OSM data, including the validation and classification of each OSM geospatial feature.
4. Aggregation stage—performing secondary-level manipulation operations with OSM data, including cleaning/cropping, summarizing, and the generation of output dataset.
5. Output stage—returning the output data set, which includes
    a. Classified OSM spatial features (as spatial groups),
    b. A summary, including ratios of classification groups,
    c. Metadata.

The solution was implemented using R 4.3.3, a programming language and environment for statistical computing [61]. Also, various R libraries were used during the implementation. The following are worth noting as the most significant:

- 'osmdata'—a library for downloading and using data from OpenStreetMap (OSM);
- 'dplyr'—a library for data manipulation;
- 'sf'—a library that provides simple features access for R;
- 'leaflet'—an open-source library (based on JavaScript) that is used to develop interactive maps;
- 'plotly'—a graphing library to develop interactive, publication-quality graphs;
- 'shiny'—a free and open-source R library for the development of web applications.

Also, during development, RStudio IDE (an integrated development environment for R) and Microsoft Excel were used.

3. Results and Discussion

OpenStreetMap spatial data were obtained using the coordinates of the point of interest (latitude and longitude) and a radius around it (with a maximum of 3 km set as the general bee flight radius around an apiary or beehive). The OSM data, retrieved using R and the 'osmdata' library, were represented as a List object, consisting primarily of metadata and five OSM spatial groups (as simple feature data frames):

- Points (separate objects, like bus stops, street lamps, etc.),
- Lines and multilines (roads, watercourses, etc.),
- Polygons and multipolygons (areas, fields, etc.).

It is important to note that the OSM data obtained at this point were raw, meaning they required further processing. An example of the data retrieved from OSM can be seen in Figure 4.

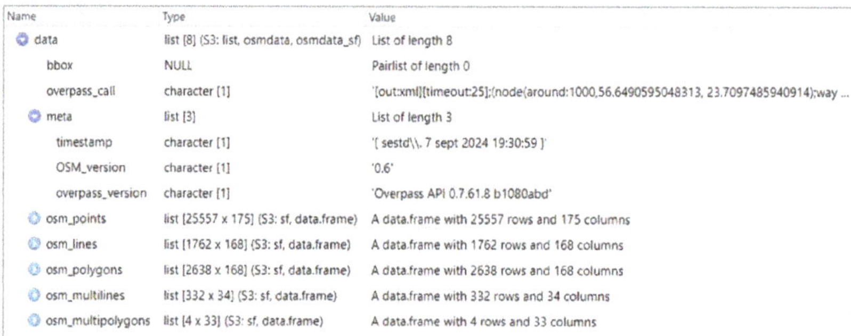

Name	Type	Value
data	list [8] (S3: list, osmdata, osmdata_sf)	List of length 8
bbox	NULL	Pairlist of length 0
overpass_call	character [1]	'[out:xml][timeout:25];(node(around:1000,56.6490595048313, 23.7097485940914);way ...
meta	list [3]	List of length 3
timestamp	character [1]	'[sestd\\. 7 sept 2024 19:30:59]'
OSM_version	character [1]	'0.6'
overpass_version	character [1]	'Overpass API 0.7.61.8 b1080abd'
osm_points	list [25557 x 175] (S3: sf, data.frame)	A data.frame with 25557 rows and 175 columns
osm_lines	list [1762 x 168] (S3: sf, data.frame)	A data.frame with 1762 rows and 168 columns
osm_polygons	list [2638 x 168] (S3: sf, data.frame)	A data.frame with 2638 rows and 168 columns
osm_multilines	list [332 x 34] (S3: sf, data.frame)	A data.frame with 332 rows and 34 columns
osm_multipolygons	list [4 x 33] (S3: sf, data.frame)	A data.frame with 4 rows and 33 columns

Figure 4. View of OSM data obtained (screenshot of RStudio IDE).

The next step after obtaining the OSM data was processing. Functionally, the processing stage can be divided into the validation, classification, and aggregation phases.

3.1. Validation

Before classification can be performed, it is important to validate whether each spatial object is suitable for classification. Each OSM spatial object is characterized by tags that include a unique identifier (osm_id), data about the represented object, and geometry. Since the obtained data are raw, they may contain spatial objects that can be considered as dependent objects (parts of other objects). For example, a point may be a geometric component of another object, such as a line segment. In this case, the point does not represent any real object and therefore lacks valuable data. The example below illustrates such a case of a dependent spatial object (see Figure 5).

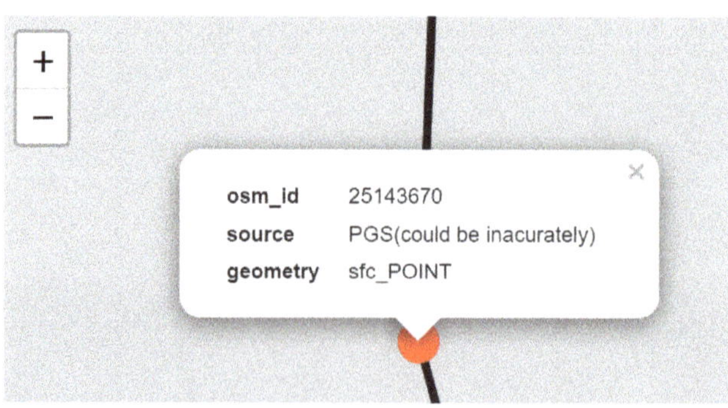

Figure 5. The example of a dependent (invalid) spatial object (point).

In another case, a point may represent a real object, for example, a bus station. In this case, it will have relevant data describing the specific bus station. Therefore, it is crucial to remove dependent objects from the spatial groups and keep independent spatial features that provide valuable information about specific objects. The example below illustrates such a case of an independent (valid) spatial object (see Figure 6).

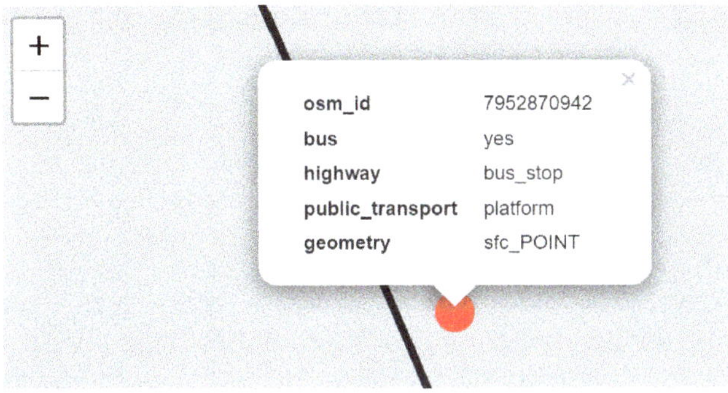

Figure 6. The example of an independent (valid) spatial object (point).

3.2. Classification

The purpose of the classification phase is to assign each spatial object to a specific class. It is crucial to note that the development of the classification stage was based on studying the technical side of the approach. The accuracy of the classification in this case was not verified; therefore, classification results can only be used as an experimental source of information. The classification process is based on using a classifier that serves as a dictionary for all existing OSM spatial features (tags). The idea was to use a list of all existing (known) OSM features and to indicate their belonging to certain classes. At the moment of a development of the classifier (autumn 2023), 1325 unique spatial features of OSM were identified. All of them were analyzed, and, as a result, 1130 unique features were determined as the primary ones, which will be used in the process of spatial object classification.

Classifier has the following classes:

- Zone: Natural/Rural (zNR), Industrial/Urban (zIU), Both (Neutral) (zB);
- Origin: Natural (oN), Anthropogenic (oA), Both (Neutral) (oB);
- Important: truc/falsc (1/0);
- Danger: true/false (1/0).

Classifier itself represents a table, containing existing OSM tags (keys and values) and their classification according to their belonging to certain classes. An example of the OSM classifier table can be seen below (see Table 1).

Table 1. An example of the OSM classifier table.

Key	Value	Zone	Origin	Primary	Important	Danger
abutters	commercial	zIU	oA	1	0	0
abutters	industrial	zIU	oA	1	0	0
abutters	mixed	zB	oA	1	0	0
abutters	residential	zB	oA	1	0	0
abutters	retail	zIU	oA	1	0	0
access	agricultural	zNR	oB	1	1	0
landuse	allotments	zNR	oA	1	1	0
leisure	fitness_centre	zIU	oA	1	0	0
leisure	fitness_station	zIU	oA	1	0	0

The classification of objects in the classifier was carried out on the basis of their description and information about their most probable position and role in relation to the surrounding geographic area. For example, 'landuse = allotments' in most cases describes allotment gardens, made by humans, used for gardening (growing vegetables, fruits and flowers) that in most cases are not used for permanent residential purposes. Such an object will probably be located outside cities and urban areas, in rural areas (zones). Moreover, this type of object can be noted as important and may be of particular interest as the flowering plants growing there can serve as a good source of food resources for bees. In another case, the object 'leisure = fitness_centre', describing a place with exercise machines and other fitness equipment, most likely will be located in a populated area (urban zone) to ensure the proper attendance of active people. Objects of this type can be useful in determining the level of urbanization of a particular region, which is also important for the exploration of the surrounding environment. Objects that may be potentially dangerous for bees belong to the 'Danger' class and may serve as an important factor for avoiding a specific area, such as waste landfills, which may pollute surrounding areas. It is important to note that bees are highly sensitive to the environment. During the classification process, all obtained and valid OSM spatial objects from all spatial groups were classified according to the classifier and supplemented with the corresponding classes.

3.3. Aggregation

The aggregation phase is aimed to perform secondary-level processing of OSM data in order to clean and summarize classified data as well as generate the output dataset. One of the most important operations of the aggregation stage is 'Crop to radius', the purpose of which is to limit OSM data to the initially selected area of interest (radius). An example of the 'Crop to radius' operation can be seen on Figure 7.

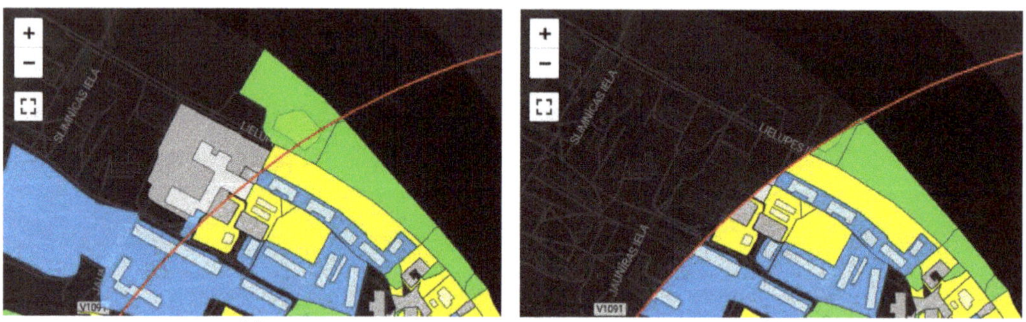

(a) (b)

Figure 7. 'Crop to radius' operation: (**a**) OSM data before cropping; (**b**) OSM data after cropping.

The operation 'Crop to radius' is crucial as the OSM spatial objects obtained may be located outside the defined area of the query. There are several reasons for this. First, OSM objects may have a geometry, which goes beyond the boundaries of a defined area. Second, OSM objects may have related objects that are located outside the defined area.

Another important operation to highlight is the summary preparation. This operation is necessary to prepare a generalized view of the statistics of the results. First, it generates common metadata about the total count and total area of OSM feature groups (see Table 2). Second, it prepares a summary dataset of performed classification that includes the count and areas of each classification group, as well as the ratios (in percentage) between them (see Table 3). Both these datasets are used in the visualization process of the results.

Table 2. Metadata (as a table) of the performed classification.

Object Group	Meta Parameter
Points	40
Lines	109
Multilines	3
Polygons	33
Multipolygons	3
Objects, total	188
Polygons, area (ha)	164
Multipolygons, area (ha)	2416
Total, area (ha)	2580

Table 3. Summary dataset (as a table) of the performed classification.

Class	Count (Objects)	Count Ratios, %	Area (Polygons), ha	Area Ratios, %
zNR	54	28.72	2364	91.63
zIU	5	2.66	1	0.03
zB	129	68.62	215	8.33
oN	8	4.26	2231	86.47
oA	143	76.06	17	0.65
oB	37	19.68	332	12.88
Important	14	-	122	-
Danger	0	-	0	-

The final operation of the aggregation phase is output generation, which aggregates all resulting data as an output dataset. It contains classified OSM features (five spatial groups), including a summary of each group, as well as general summary and metadata datasets (as previously described). The output dataset itself is a list object, which is further used by a prototype of a web-based GIS application as the main data source.

3.4. Web-Based GIS Application

An experimental prototype of a web-based GIS application named BeeLand Macro uses the Data Processing Module as a core and clearly demonstrates all its functional capabilities. The application was developed with a focus on a user-friendly experience, allowing users to freely explore areas of interest by performing real-time spatial analysis. The main component of the BeeLand Macro is an interactive map that allows selecting a specific area (as a circle) worldwide by clicking on the map or by adjusting the location settings available in the 'Location' tab on the sidebar (see Figure 8).

When the location and radius of the area are selected, the user needs to click on the 'Get data' button to execute the DPM. The user will be informed about the progress of the execution via the 'Please wait' window. When the DPM finishes execution, the application presents the results by providing access to map filters of classified spatial features, as well as providing a classification summary in the form of interactive pie charts and information boxes (see Figure 9).

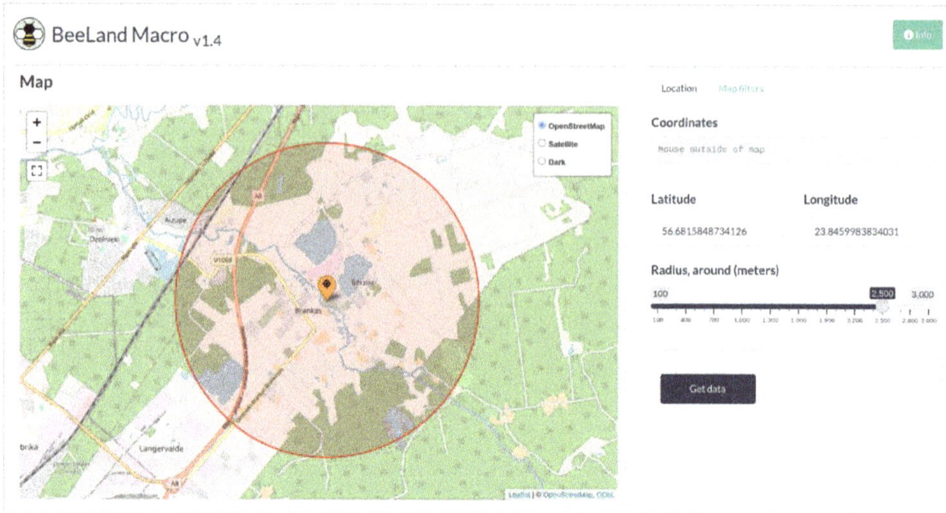

Figure 8. Interactive map of BeeLand Macro application.

Figure 9. General view of the BeeLand Macro application.

By using map filters, it is possible to manage the visibility of OSM spatial features (as layers), encoded by different colors depending on a class (described on the legend of the map). Each spatial feature can be explored individually and interactively by inspecting its properties, which are displayed in a pop-up window that appears by clicking on the specific feature (see Figure 10).

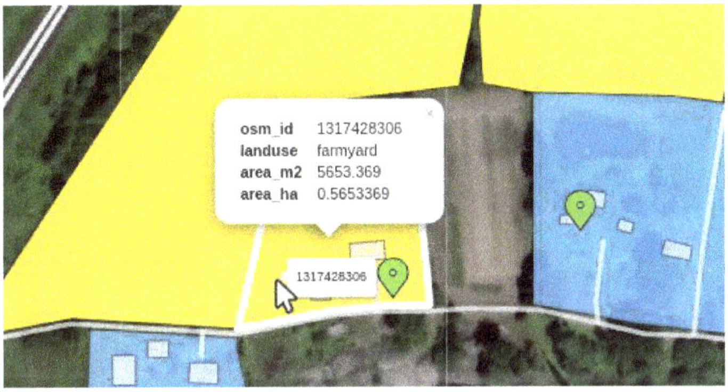

Figure 10. Properties of spatial features (polygon).

The classification summary represents an interactive visualization of DPM results, developed with an emphasis on the ease of perception. Classification results are grouped by the object count and object area and allow assessing the situation regarding various environmental characteristics, such as the approximate level of urbanization, anthropogenesis, and the greenness of territories. Individual information blocks highlight specific information regarding individual groups of objects that require special attention. These include important and dangerous objects in particular. Using summary pie charts makes it easy to evaluate the percentage ratio of different zones and the structure of the origin of spatial objects. As a result, the user could have a general understanding of the environment in a particular location, which could be helpful in the decision-making process as well as in planning further activities.

3.5. Crowdsourced Data

Since this study focused on exploring the potential of using crowdsourced data, it is important to highlight the key characteristics of such data, namely their quality and completeness. OSM data, often regarded as one of the best examples of crowdsourced data and Volunteered Geographic Information (VGI), have quality and reliability as some of their most widely discussed attributes [62–66]. Various approaches and methods have been proposed to assess and improve the quality of OSM data, including comparisons with reference datasets [67,68], the use of machine learning techniques [69,70], and intrinsic analysis [71,72]. Studies have examined different aspects of OSM data quality, such as completeness, logical and topological consistency, as well as positional and thematic accuracy [66,73,74].

Thus, due to the diverse aspects and characteristics of crowdsourced data, opinions on their quality and applicability vary. Some studies emphasize the positive aspects of crowdsourced data, such as accessibility, relevance, and practicality, while others focus on their quality issues, highlighting potential shortcomings. The authors of this study believe that these differing opinions among researchers are well founded. Crowdsourced data often exhibit significant heterogeneity in their characteristics due to the nature of their origin [75]. One of the most important factors contributing to this heterogeneity is the motivation of participants in the crowdsourcing process, specifically the presence of perceived benefits from their activity. This is reflected in the levels of data completeness and accuracy, which often vary by region. For example, more populated areas tend to achieve higher positioning accuracy and data completeness compared to sparsely populated regions [76]. Such outcomes are to be expected, as urban areas with developed infrastructure are likely to be of greater importance to a larger number of people than remote rural areas [77].

In beekeeping, it is also important to recognize the value of a specific territory for participants and their commitment to fostering symbiosis with other agricultural sectors. For instance, the territory where beekeepers and other farmers operate holds significance for both parties. Their collaboration can provide mutual benefits, as pollination is essential for agriculture [78,79], while also supporting the production of bee products. A crowdsourced data solution could serve as an information exchange platform to facilitate this cooperation. For example, crop farmers could update information about their activities, such as pesticide use, changes in crops and field structures, or the timing of upcoming plant flowering. This information would be highly valuable for beekeepers planning their activities, particularly migratory beekeepers seeking suitable areas to support their apiaries [80].

At the same time, since bees are highly sensitive to environmental factors, it is important for beekeepers to monitor the current situation in a specific region. Timely updates on issues such as pollution, new development projects, or the creation of a blooming garden could provide valuable information about changes in the environment. However, given the experimental nature of this study, the assumptions mentioned above are hypothetical and would require more detailed research if the concept were to be implemented for practical use.

3.6. Accuracy of Results

Although the main objective of this study was to investigate and demonstrate the technical approach, the accuracy of the results remains an important consideration. Since the solution presented in this study is a demonstration prototype, it is not intended to produce reliable data in real-world conditions at its current stage of development. To apply such a solution in real-world scenarios and obtain meaningful results, it must be thoroughly adapted to meet specific needs and scale. This adaptation should include a more detailed development of the data processing phase, particularly the classification and validation of crowdsourced data, in collaboration with relevant experts.

A particularly important step in developing a real product based on the presented prototype is attracting the target audience—potential users such as beekeepers, farmers, and other interested parties. This is a crucial point, as such a solution will primarily serve as an information exchange platform, where the main role will be played by its users and stakeholders.

The accuracy of the results in such a solution is highly dependent on the quality of the input data. Therefore, in the case of OSM and similar crowdsourced databases, the approach described in this study can be applied in real-world settings if interested parties and contributors in specific regions take collective responsibility for providing accurate and complete data for the common good. However, it is important to note that crowdsourcing, as an approach, inherently carries the risk of inaccuracies, which is an inevitable limitation despite the significant resource opportunities it provides.

4. Conclusions

The use of modern technologies in precision beekeeping helps to overcome various types of difficulties, making it sustainable and relevant to the nowadays realities. Research aimed at developing this branch of agriculture has led to the emergence of various effective

25. Zacepins, A.; Stalidzans, E.; Meitalovs, J. Application of information technologies in precision apiculture. In Proceedings of the 13th International Conference on Precision Agriculture (ICPA 2012), Indianapolis, IN, USA, 15–18 July 2012; Volume 7.
26. Brase, T.A. *Precision Agriculture*; Delmar Learning: Thomson, IL, USA, 2006.
27. Alleri, M.; Amoroso, S.; Catania, P.; Verde, G.L.; Orlando, S.; Ragusa, E.; Sinacori, M.; Vallone, M.; Vella, A. Recent developments on precision beekeeping: A systematic literature review. *J. Agric. Food Res.* **2023**, *14*, 100726. [CrossRef]
28. Bumanis, N. Data fusion challenges in precision beekeeping: A review. *Res. Rural. Dev.* **2020**, *35*, 252–259. [CrossRef]
29. Zacepins, A.; Ozols, N.; Kviesis, A.; Gailis, J.; Komasilovs, V.; Komasilova, O.; Zagorska, V. Evaluation of the honey bee colonies weight gain during the intensive foraging period. *Agron. Res.* **2022**, *20*, 457–468. [CrossRef]
30. Zacepins, A.; Stalidzans, E. Information processing for remote recognition of the state of bee colonies and apiaries in precision beekeeping (apiculture). *Biosyst. Inf. Technol.* **2013**, *2*, 6–10. [CrossRef]
31. Kotovs, D.; Zacepins, A. Importance of GIS solutions for beekeepers: A review. *Agron. Res.* **2023**, *21*, 799–813. [CrossRef]
32. Abou-Shaara, H.F. Geographical information system for beekeeping development. *J. Apic. Sci.* **2019**, *63*, 5–16. [CrossRef]
33. Albacete, S.; Sancho, G.; Azpiazu, C.; Rodrigo, A.; Molowny-Horas, R.; Sgolastra, F.; Bosch, J. Bees exposed to climate change are more sensitive to pesticides. *Glob. Chang. Biol.* **2023**, *29*, 6248–6260. [CrossRef]
34. Lima MA, P.; Cutler, G.C.; Mazzeo, G.; Hrncir, M. Editorial: The decline of wild bees: Causes and consequences. *Front. Ecol. Evol.* **2022**, *10*, 1027169. [CrossRef]
35. Tosi, S.; Sfeir, C.; Carnesecchi, E.; vanEngelsdorp, D.; Chauzat, M.P. Lethal, sublethal, and combined effects of pes-ticides on bees: A meta-analysis and new risk assessment tools. *Sci. Total Environ.* **2022**, *844*, 156857. [CrossRef] [PubMed]
36. LeBuhn, G.; Vargas Luna, J. Pollinator decline: What do we know about the drivers of solitary bee declines? *Curr. Opin. Insect Sci.* **2021**, *46*, 106–111. [CrossRef] [PubMed]
37. Margaoan, R.; Papa, G.; Nicolescu, A.; Cornea-Cipcigan, M.; Kösoğlu, M.; Topal, E.; Negri, I. Environmental pollution effect on honey bees and their derived products: A comprehensive analysis. *Environ. Sci. Pollut. Res.* **2024**, 1–22. [CrossRef] [PubMed]
38. Ferrante, F.; Pasquini, E.; Cappa, F.; Bellocchio, L.; Baracchi, D. Unravelling the microplastic menace: Different polymers additively increase bee vulnerability. *Environ. Pollut.* **2024**, *352*, 124087. [CrossRef] [PubMed]
39. Estoque, R.C.; Murayama, Y. Suitability analysis for beekeeping sites in La Union, Philippines, using GIS and multi-criteria evaluation techniques. *Res. J. Appl. Sci.* **2010**, *5*, 242–253. [CrossRef]
40. Marnasidis, S.; Kantartzis, A.; Malesios, C.; Hatjina, F.; Arabatzis, G.; Verikouki, E. Mapping priority areas for apiculture development with the use of geographical information systems. *Agriculture* **2021**, *11*, 182. [CrossRef]
41. Al-Khalaf, A.A. Modeling the potential distribution of the predator of honey bees, Palarus latifrons, in the Arabian deserts using Maxent and GIS. *Saudi J. Biol. Sci.* **2021**, *28*, 5667–5673. [CrossRef]
42. Cappa, V.; Cerioli, M.P.; Scaburri, A.; Tironi, M.; Farioli, M.; Nassuato, C.; Bellini, S. Analysis of Bee Population Decline in Lombardy during the Period 2014–2016 and Identification of High-Risk Areas. *Pathogens* **2021**, *10*, 1004. [CrossRef]
43. Triantomo, V.; Widiatmaka, W.; Fuah, A.M. Land use planning for beekeeping using geographic information system in Sukabumi Regency, West Java. *J. Pengelolaan Sumberd. Alam Dan Lingkung. (J. Nat. Resour. Environ. Manag.)* **2016**, *6*, 168. [CrossRef]
44. Prestby, T.J.; Robinson, A.C.; McLaughlin, D.; Dudas, P.M.; Grozinger, C.M. Characterizing user needs for Beescape: A spatial decision support tool focused on pollinator health. *J. Environ. Manag.* **2023**, *325*, 116416. [CrossRef]
45. The Center for Pollinator Research. Beescape. Available online: https://pollinators.psu.edu/bee-health/beescape (accessed on 15 January 2025).
46. The Rural Support Service of Latvia (RSS), Field Register Data (in Latvian). Available online: https://www.lad.gov.lv/lv/lauku-registra-dati (accessed on 17 January 2025).
47. The National Laboratory of Energy and Geology (LNEG), The geoPortal of Energy and Geology, Open Data. Available online: https://geoportal.lneg.pt/en/opendata (accessed on 17 January 2025).
48. Dr Robin Wilson, Free GIS Data. Available online: https://freegisdata.rtwilson.com (accessed on 17 January 2025).
49. OpenStreetMap Contributors, OpenStreetMap. Available online: https://www.openstreetmap.org (accessed on 17 January 2025).
50. Bennett, J. *OpenStreetMap*; Packt Publishing Ltd.: Birmingham, UK, 2010.
51. Jokar Arsanjani, J.; Zipf, A.; Mooney, P.; Helbich, M. An introduction to OpenStreetMap in Geographic Information Science: Experiences, research, and applications. *OpenStreetMap GIScience Exp. Res. Appl.* **2015**, 1–15. Available online: https://link.springer.com/book/10.1007/978-3-319-14280-7 (accessed on 20 January 2025).
52. Amirian, P.; Basiri, A.; Gales, G.; Winstanley, A.; McDonald, J. The next generation of navigational services using OpenStreetMap data: The integration of augmented reality and graph databases. *OpenStreetmap GIScience Exp. Res. Appl.* **2015**, 211–228. Available online: https://link.springer.com/chapter/10.1007/978-3-319-14280-7_11 (accessed on 20 January 2025).
53. Rahmig, C.; Kluge, A. Digital maps for railway applications based on OpenStreetMap data. In Proceedings of the 16th International IEEE Conference on Intelligent Transportation Systems (ITSC 2013), The Hague, The Netherlands, 6–9 October 2013; IEEE: Piscataway, NJ, USA, 2013; pp. 1322–1327.

54. Al Furjani, A.; Younsi, Z.; Abdulalli, A.; Elsaeh, M.; Almahdi, A.; Jouil, K.; Lashihar, S.B. Enabling the city information modeling CIM for urban planning with OpenStreetMap, O.S.M. In Proceedings of the 4th International Conference on Geospatial Technologies–Libya GeoTec, Tripoli, Libya, 3–5 March 2020; Volume 4.
55. Basiouka, S.; Potsiou, C.; Bakogiannis, E. OpenStreetMap for cadastral purposes: An application using VGI for official processes in urban areas. *Surv. Rev.* **2015**, *47*, 333–341. [CrossRef]
56. Yang, D.; Fu, C.S.; Smith, A.C.; Yu, Q. Open land-use map: A regional land-use mapping strategy for incorporating OpenStreetMap with earth observations. *Geo-Spat. Inf. Sci.* **2017**, *20*, 269–281. [CrossRef]
57. Fonte, C.C.; Patriarca, J.A.; Minghini, M.; Antoniou, V.; See, L.; Brovelli, M.A. Using openstreetmap to create land use and land cover maps: Development of an application. In *Volunteered Geographic Information and the Future of Geospatial Data*; IGI Global: Hershey, PA, USA, 2017; pp. 113–137.
58. Grippa, T.; Georganos, S.; Zarougui, S.; Bognounou, P.; Diboulo, E.; Forget, Y.; Lennert, M.; Vanhuysse, S.; Mboga, N.; Wolff, E. Mapping urban land use at street block level using openstreetmap, remote sensing data, and spatial metrics. *ISPRS Int. J. Geo-Inf.* **2018**, *7*, 246. [CrossRef]
59. Schultz, M.; Voss, J.; Auer, M.; Carter, S.; Zipf, A. Open land cover from OpenStreetMap and remote sensing. *Int. J. Appl. Earth Obs. Geoinf.* **2017**, *63*, 206–213. [CrossRef]
60. Mooney, P.; Minghini, M. A review of OpenStreetMap data. *Mapp. Citiz. Sens.* **2017**, 37–59.
61. R Core Team. *R: A Language and Environment for Statistical Computing*; R Foundation for Statistical Computing: Vienna, Austria, 2024; Available online: https://www.R-project.org/ (accessed on 21 January 2025).
62. Al-Bakri, M. Ten years of OpenStreetMap project: Have we addressed data quality appropriately?–Review Paper. *J. Eng.* **2015**, *21*, 158–175. [CrossRef]
63. Biljecki, F.; Chow, Y.S.; Lee, K. Quality of crowdsourced geospatial building information: A global assessment of OpenStreetMap attributes. *Build. Environ.* **2023**, *237*, 110295. [CrossRef]
64. Antoniou, V.; Skopeliti, A. The impact of the contribution microenvironment on data quality: The case of OSM. *Mapp. Citiz. Sens.* **2017**, 165–196.
65. Mooney, P.; Corcoran, P.; Winstanley, A.C. Towards quality metrics for OpenStreetMap. In Proceedings of the 18th SIGSPATIAL International Conference on Advances in Geographic Information Systems, San Jose, CA, USA, 2–5 November 2010; pp. 514–517.
66. Kaur, J.; Singh, J.; Sehra, S.S.; Rai, H.S. Systematic literature review of data quality within OpenStreetMap. In Proceedings of the 2017 International Conference on Next Generation Computing and Information Systems (ICNGCIS), Jammu, India, 11–12 December 2017; IEEE: Piscataway, NJ, USA, 2017; pp. 177–182.
67. Forghani, M.; Delavar, M.R. A quality study of the OpenStreetMap dataset for Tehran. *ISPRS Int. J. Geo-Inf.* **2014**, *3*, 750–763. [CrossRef]
68. Zhang, Y.; Zhou, Q.; Brovelli, M.A.; Li, W. Assessing OSM building completeness using population data. *Int. J. Geogr. Inf. Sci.* **2022**, *36*, 1443–1466. [CrossRef]
69. Jacobs, K.T.; Mitchell, S.W. OpenStreetMap quality assessment using unsupervised machine learning methods. *Trans. GIS* **2020**, *24*, 1280–1298. [CrossRef]
70. Jilani, M.; Bertolotto, M.; Corcoran, P.; Alghanim, A. Traditional vs. machine-learning techniques for OSM quality assessment. In *Geospatial Intelligence: Concepts, Methodologies, Tools, and Applications*; IGI Global: Hershey, PA, USA, 2019; pp. 469–487.
71. Barron, C.; Neis, P.; Zipf, A. A comprehensive framework for intrinsic OpenStreetMap quality analysis. *Trans. GIS* **2014**, *18*, 877–895. [CrossRef]
72. Minghini, M.; Frassinelli, F. OpenStreetMap history for intrinsic quality assessment: Is OSM up-to-date? *Open Geospat. Data Softw. Stand.* **2019**, *4*, 9. [CrossRef]
73. Kounadi, O. Assessing the Quality of OpenStreetMap Data. Msc Geographical Information Science. Master's Thesis, University College of London Department of Civil, Environmental and Geomatic Engineering, London, UK, 2009. Volume 19.
74. Borkowska, S.; Pokonieczny, K. Analysis of OpenStreetMap data quality for selected counties in Poland in terms of sustainable development. *Sustainability* **2022**, *14*, 3728. [CrossRef]
75. Yang, S.; Shen, J.; Konečný, M.; Wang, Y.; Štampach, R. Study on the spatial heterogeneity of the POI quality in OpenStreetMap. In Proceedings of the 7th International Conference on Cartography and GIS, Sozopol, Bulgaria, 18–23 June 2018; pp. 18–23.
76. Vandecasteele, A.; Devillers, R. Improving volunteered geographic information quality using a tag recommender system: The case of OpenStreetMap. *OpenStreetMap GIScience Exp. Res. Appl.* **2015**, 59–80.
77. Tian, Y.; Zhou, Q.; Fu, X. An analysis of the evolution, completeness and spatial patterns of OpenStreetMap building data in China. *ISPRS Int. J. Geo-Inf.* **2019**, *8*, 35. [CrossRef]
78. Klatt, B.K.; Holzschuh, A.; Westphal, C.; Clough, Y.; Smit, I.; Pawelzik, E.; Tscharntke, T. Bee pollination improves crop quality, shelf life and commercial value. *Proc. R. Soc. B Biol. Sci.* **2014**, *281*, 20132440. [CrossRef]

79. Puteri, G.; Herwina, H. The Potential of Integration of Stingless Bee Keeping and Agricultural Holdings. *J. Environ. Sci. Sustain. Soc.* **2022**, *11*, PP03_p9–PP03_p11. [CrossRef]
80. Pilati, L.; Prestamburgo, M. Sequential relationship between profitability and sustainability: The Case of Migratory Beekeeping. *Sustainability* **2016**, *8*, 94. [CrossRef]

International Journal of *Geo-Information*

MDPI

Article

Strategies for Glacier Retreat Communication with 3D Geovisualization and Open Data Sharing

Federica Gaspari [1,*], Federico Barbieri [1], Rebecca Fascia [1], Francesco Ioli [1,2], Livio Pinto [1] and Federica Migliaccio [1]

1 Department of Civil and Environment Engineering, Politecnico di Milano, 20133 Milan, Italy; federico2.barbieri@polimi.it (F.B.); rebecca.fascia@polimi.it (R.F.); francesco.ioli@polimi.it or francesco.ioli@geo.uzh.ch (F.I.); livio.pinto@polimi.it (L.P.); federica.migliaccio@polimi.it (F.M.)

2 Department of Geography, University of Zurich, 8057 Zurich, Switzerland

* Correspondence: federica.gaspari@polimi.it

Abstract: Images of melting ice have become powerful symbols of climate change, attracting both public attention and scientific interest. This research uses web technologies to document and communicate the ongoing retreat of the Belvedere Glacier in the Italian Alps. By combining historical and contemporary 2D and 3D geospatial data, the paper presents a comprehensive digital platform that allows visualization of long-term changes of the Belvedere Glacier. To increase public understanding and engagement, we develop a user-friendly web platform that provides interactive tools for exploring glacier data. By fostering a deeper understanding of the complex processes involved in glacier retreat by different audiences (students, general public, and technical experts), this work aims to inspire further research and cooperation, also thanks to the reproducibility of the open-source code.

Keywords: glacier monitoring; 3D documentation; geovisualization; web application; spatial database

1. Introduction

Images of melting ice in various forms have dominated the climate change discourse and communication since the early 2000s, sparking the interest of researchers in multidisciplinary studies dedicated to investigating their impact on public perception, awareness, and engagement [1]. Initial attempts to communicate the tangible impacts of global warming through images have led to the widespread use of polar ice images, often perceived by print media audiences far removed in time and space [2]. However, in the last decade, media coverage and attention has gradually shifted to mountain glaciers as disappearing natural objects and endangered heritage deeply connected to local communities, conveying place-based narratives while linking them to global challenges and threats [3,4]. Indeed, the retreat of glaciers and the disappearance of ice in mountain areas combine various aspects of the relationship between human society and natural landscapes [5–7].

The literature has highlighted and documented how climate change impacts on glaciers affect the economy and sustainability of local high mountain communities [8], tourism [9,10], energy production and supply [11,12], as well as the preservation of collective memories of the past [13,14], with both historical and social implications for cultural identities [15,16]. However, most attempts at visual communication rely mainly on a qualitative approach, using feelings of environmental loss under a predominantly aesthetic perspective [17]. Indeed, due to the complex nature of the phenomenon combining science and society, information and communication on climate change requires not only multidisciplinary approaches, but also multimedia solutions capable of integrating both the qualitative and quantitative aspects of the transformation of the natural environment [18]. In particular, there is a need to implement locally meaningful visualization, not limited to images, capable of shifting from a visual icon to a technical tool able to express both spatial and temporal dimensions, while efficiently engaging different audiences in understanding the challenges and developing solutions [19,20].

Traditionally, multitemporal photography has been widely used to evaluate changes in glacier environment, assess the impact of climate change on mountain environments, and raise climate change awareness [21,22]. This involves acquiring repeated images at different epochs, replicating identical viewpoints to detect changes over time. The need to understand and quantify long-term glacier changes dating back to the Little Ice Age has also led researchers to apply this method to historical documents, including high-quality illustrations or paintings, some of them dating back to the 17th century, and compare them with contemporary images taken from similar viewpoints [23–26]. Various georeferencing approaches, such as monoplotting [27], have been used to derive metric information on glacier changes from single oblique images by establishing a correspondence between the pixels of the image and real-world coordinates by means of existing georeferenced map and Digital Surface Models (DSMs).

Since the mid-20th century, photogrammetry has been used to extract 3D information from overlapping images captured by high-quality metric cameras onboard aircraft for mapping purposes [28,29]. Over the past 20 years, Structure-from-Motion (SfM) [30] and Multi-View Stereo (MVS) [31] have become standard techniques for generating high-resolution 3D point clouds, orthophotos, and DSMs from digital images. These techniques have enabled precise glacier mapping, ice volume variation estimation, and mass balance computation (e.g., [32–35]). SfM estimates a camera's exterior (position and orientation angles) and interior parameters (geometry and lens distortions), producing a sparse point cloud. MVS then refines this into a dense 3D reconstruction, generating dense point clouds, orthophotos, DSMs, and triangulated mesh models. The combined approach, SfM-MVS, is widely used in geospatial analysis [36,37].

Photogrammetry has been applied to glacier monitoring across various scales and platforms, from satellite stereo imagery [33–35,38–41] and terrestrial cameras [24,42–44]. The increasing availability of satellite data and drones has enhanced traditional terrestrial methods, enabling cross-validation and more frequent data collection. These approaches produce high-quality orthophotos and digital surface models, supporting detailed glacier evolution analysis. Additionally, low-cost single and stereo camera systems now provide higher-frequency imaging, enriching glacier monitoring timelines and enabling near real-time case study assessments.

However, despite the wide range of technologies available for monitoring purposes, there remains a notable gap in effective tools and strategies for communicating both qualitative and quantitative aspects of ice loss. While the literature highlights numerous climate change awareness initiatives and tools [45,46], only a few incorporate 3D geovisualization, and these are often tailored to technical or academic audiences [47,48]. This happens despite more than a decade of evidence demonstrating the effectiveness of 3D representations for geointerpretation and awareness raising, particularly in high mountain environments [49,50]. The problem is particularly pronounced for smaller glaciers, which are often overlooked in large-scale studies [51]. For this reason, it is necessary to explore the potential of web-based solutions that are accessible to different types of users, broadening engagement to include professionals, students, tourists, as well as local people [52]. The combined use of geospatial data management and visualization tools—especially when they are open source—can respond to the need to develop interactive, user-friendly, and flexible outreach platforms [53,54], effectively contributing to the communication and dissemination of research results while allowing their easy re-use in different contexts.

In this context, this paper aims to describe a digital geospatial framework designed for the management, visualization, and dissemination of research results, implementing communication strategies in a web platform for different types of users: students, researchers, and the general public. The Belvedere Glacier in the Italian Alps is the supersite test field for the application of the proposed methodology. Indeed, in the last two decades, this site has become a common ground for different groups of researchers in the field of geomatics, geology, and hydrology, collecting and generating large amounts of data about the glacier conditions [55–57].

The paper is divided into the following sections: Section 2 presents the geographical and research context of the Belvedere Glacier monitoring activities; Section 3 illustrates how the research communication strategies for different audiences are defined and then implemented in a web platform, describing its modular components and the technologies involved; current results and the impact of the platform on different end-user targets are then discussed in Section 4. Final conclusions and future work are included in Section 5.

2. The Belvedere Glacier

The Belvedere Glacier, located in the Anzasca Valley on the eastern side of the Monte Rosa Massif in Italy (Figure 1), is an alpine glacier of particular scientific interest due to its rapid evolution [55]. Covering an area of approximately 1.8 km^2 and extending from 2250 m to 1800 m a.s.l., the glacier's lower section is a temperate, debris-covered glacier fed by icefalls and avalanches from the Monte Rosa East Face [56]. It has historically been characterized by hazardous events, such as floods and slope instabilities, which pose significant risks to nearby areas, including the village of Macugnaga and the Zamboni Zappa Hut (2070 m a.s.l.) [57].

Figure 1. Map of the Belvedere Glacier area, with RGB orthophotomosaic from a 2023 Unmanned Aerial Vehicle (UAV) survey [58] showing the extent of glacier coverage and the surrounding topography from SWISS Topo. The inset highlights the glacier's location in the Italian Alps. The maps are framed in the global reference system RDN2008/UTM zone 32N (EPSG: 7791).

Already in the early 21st century, the glacier was the subject of a wide variety of studies because it experienced surge-type dynamics, with surface velocities increasing from an average of 30–45 m/y to over 200 m/y during 2001 [59–61]. This surge led to a significant thickening of the ice by more than 20 m and to the formation of a temporary super-glacial lake, Lago Effimero [62]. Emergency measures were taken due to the risk of flooding, and, although the lake was reformed in 2003, it eventually drained without causing major damage.

Despite the protective debris cover, the Belvedere Glacier has been retreating rapidly in recent years due to global warming (Figure 2), with an average thinning rate of ~2 m/y and a reduction in volume of ~2 $\times$ 10^6 m^3/y [34]. More recently, since 2015, the Politecnico di Milano has carried out annual surveys using UAV photogrammetry and Global Navigation Satellite System (GNSS) measurements of ground control points (GCPs) along the glacier [34]. In addition, in order to quantify surface and volume variations over a longer time period, historical aerial photographs taken for regional cartographic purposes have been used to generate georeferenced point clouds, extending the Belvedere digital archive back to 1977 [39].

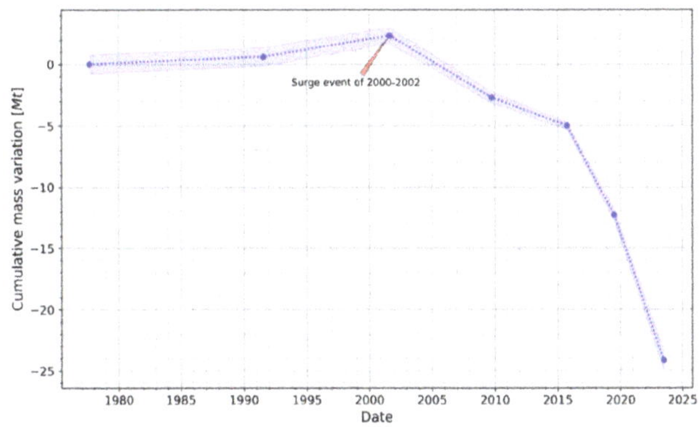

Figure 2. Trend of cumulative mass variation—expressed in Megaton—of the Belvedere Glacier (1980–2025), highlighting the 2000–2002 surge event [63].

Currently, the yearly documentation of the glacier's condition is ensured by the organization of an annual summer school that involves Bachelor and Master of Science students from the Politecnico di Milano in data collection and processing. In this way, the ongoing project aims to monitor the glacier dynamics and long-term evolution. Indeed, from the data collected, which are also made openly available, the researchers obtain time series of target displacements, velocities and accelerations, and reconstruct precise 3D models of the glacier together with orthophotos DSMs.

However, the sheer volume and variety of the data collected over the past year require effective approaches not only for archiving and managing the information, but, above all, for communicating what is happening to the glacier in both quantitative and qualitative terms, and for promoting awareness and participation among a wider community equally affected by the consequences of climate change.

3. Project Design

The project presented in this paper is a digital platform (https://thebelvedereglacier.it/ (accessed on 2 February 2025) that aims to engage educational, technical, and scientific audiences in the long-term research project focused on monitoring the Belvedere Glacier in the Italian Alps.

To ensure that the platform meets the needs of a diverse audience, a requirement analysis was conducted to identify three main user groups for crafting distinct experiences: students, the general public, and technical experts (including researchers and practitioners in geomatics and cartography). This analysis informed the design of three interconnected modules (Figure 3), each tailored to a specific target audience with the goal of fostering educational enrichment, technical dissemination, and public engagement.

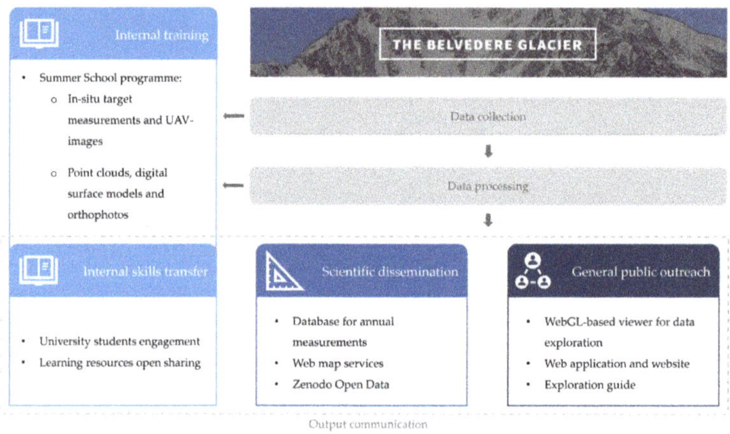

Figure 3. Components of the Belvedere Glacier project engaging different target audiences in the research communication of collected and processed products.

The platform's three components include:

- Internal Training and Skill Transfer component: designed for students, not only for educational purposes but also to offer hands-on experience of data collection and processing techniques. During a dedicated summer school, students engage in practical exercises, such as in situ target measurements and UAV image acquisition, which feed into the data processing pipeline. This integration ensures that students gain real-world skills in data acquisition and analysis, forming a bridge between theoretical learning and applied geomatics skills. This dual focus explains the division within the internal training and skill transfer components in Figure 3. The internal training aspect emphasizes capacity building through structured programs like the summer school, while the skill transfer highlights the dissemination of these skills within the academic environment, ensuring that university students can engage beyond their initial training.
- Scientific Dissemination component: geared toward a technical audience already familiar with geomatics and cartography, this module facilitates access to 2D and 3D geo-data collected and processed during annual in situ surveys. By making this high-quality digital documentation of the glacier openly available, the platform aims to support multidisciplinary research and technical studies on a rapidly changing natural heritage site.
- General Public Outreach component: targeted at the general public, this module centers around the main website, which provides an accessible entry point to the project. Through a user-friendly interface, citizens can navigate georeferenced 2D and 3D data, visualize quantitative information on glacier evolution, and access a range of multimedia resources that illustrate changes over time.

3.1. Internal Training and Skill Transfer

The Belvedere Glacier project heavily relies on the interface between research and higher education. Indeed, since 2016, the Department of Civil and Environmental Engineering of the Politecnico di Milano has been organizing the summer school "Design and Implementation of Topographic Surveys for Territorial Monitoring in Mountain Environments", an innovative teaching program aimed at undergraduate, graduate, and doctoral students in civil and environmental engineering, geoinformatics, and architecture [64]. The annual program aims to immerse participants in a hands-on experience of topographic land surveying in mountainous regions, while actively contributing to an ongoing long-term research project. This area of study is often covered only in lectures, with few opportunities for in situ practice, which is often confined to controlled learning environments. Over the course of a full week, students, staying in a traditional mountain hut, have the opportunity to turn knowledge into skills by first attending preparatory theory classes and seminars, and then by engaging in practical activities in photogrammetry, GNSS positioning, and Geographic Information System (GIS) processing and analysis with the guidance of a Professor as well as young researchers from the organizing team (Figure 4).

Figure 4. Images of a student group with tutors during the 2024 Belvedere Glacier summer school activities.

The summer school learning resources are divided into six modules, starting with a short introduction to the history of monitoring activities on the Belvedere Glacier during the last decades. The different modules are the starting point for the knowledge transfer of essential theoretical concepts of photogrammetry and drone surveying, GNSS positioning, GIS and database design, stereo camera processing, and 3D web geovisualization (Figure 5). Through focused explanations and guided tutorials, students can understand the theory and then apply it using commonly used software such as Agisoft Metashape 2.2.0, CloudCompare 2.13.1, QGIS 3.34.4, PostgreSQL 17, PostGIS 3.5.2, and Potree 1.8.

The material is published online on a website created using MkDocs and GitHub pages based on open-source code and is openly accessible via a URL (https://tars4815.github.io/belvedere-summer-school/ (accessed on 2 February 2025)). Such a flexible solution not only allows resources to be easily and continuously adapted and updated, but also unfolds opportunities for online collaboration with other educators interested in developing similar experiences [53,65]. The content of the website is shared with the students prior to the fieldwork and is one of the core elements of the preparatory step, along with the 3D viewer described in Section 3.3.1, which allows for virtual interactive exploration of the site in advance. In this way, prospective participants become more familiar with both the theoretical and practical framework that they will experience during the days of the summer schools, ensuring greater awareness and responsibility.

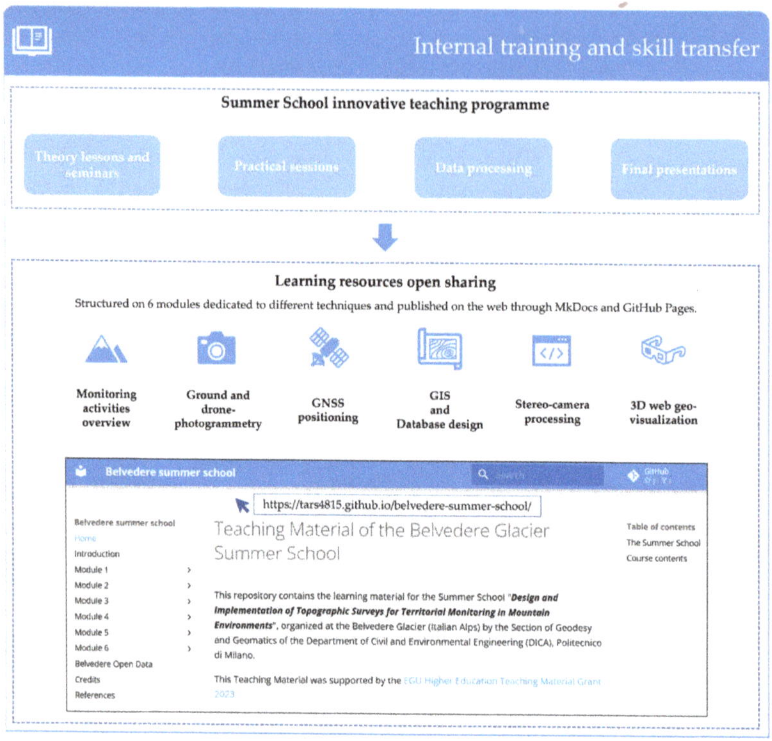

Figure 5. Overview of the Internal Training and Skill Transfer component, showcasing the innovative summer school teaching programme and the structured open sharing of learning resources across six technical modules, accessible through MkDocs and GitHub Pages.

This experience provides participants with the skills and familiarity necessary to use state-of-the-art equipment and software to carry out fieldwork activities, as well as developing critical thinking skills that allow them to understand the potential and limitations of geomatics techniques in a challenging environment. Such an approach is essential, as commercial drones and surveying equipment have become increasingly accessible in recent years, but the availability of resources to teach the correct use of these technologies is still limited outside of academia [66].

Furthermore, staying in a traditional alpine hut dating from the 1920s allows both students and tutors to be fully immersed in the mountain environment and gain an insight into the culture and daily life of the local community. In this way, the summer school can become an effective way of rediscovering both the territory and its community, building links not only between the participants but also with local citizens, while understanding the challenges and threats of an ever-changing natural heritage site in the era of climate change [67,68].

At the end of the summer school, the students, divided into groups, give presentations in front of an audience of researchers from the Department of Civil and Environmental Engineering of Politecnico di Milano. In such an occasion, which reinforces their communication and public speaking skills, they present the results of the processing of the data collected in the field, highlighting the evolution of glacier displacements and volume variations, with quantitative outputs coming from the analysis of point clouds and digital elevation models. These outputs, after a validation process conducted by the tutors of the summer school, are then shared with a wider audience in an updated version of the Belvedere glacier Open Data on Zenodo described in Section 3.2.2.

3.2. Scientific Dissemination

The Scientific Dissemination component provides a structured repository of data and tools for technical users, enabling access to detailed measurements and geospatial outputs from the Belvedere Glacier monitoring campaigns. This component is made of three main resources (Figure 6):

- A database for annual measurements: a PostGIS database which stores GNSS measurements, survey dates, instrument specifications, and displacement and velocity data. This centralized database allows researchers to track changes in the glacier over time with high precision.
- Open raster and 3D data: each campaign's results, including photogrammetric point clouds, orthophotos, and digital surface models, that are openly shared via a Zenodo repository. This open access approach supports the transparency and reusability of the data for future studies and interdisciplinary research.
- Web map services: hosted on GeoServer, the platform provides Web Map Service (WMS) links for orthophotomosaics and digital surface models, allowing technical users to integrate these layers into their own GIS projects.

Additionally, an interactive web 2D map built with Leaflet overlays key measurement points from the database, giving users a visual overview of GNSS measurements and changes across the glacier surface.

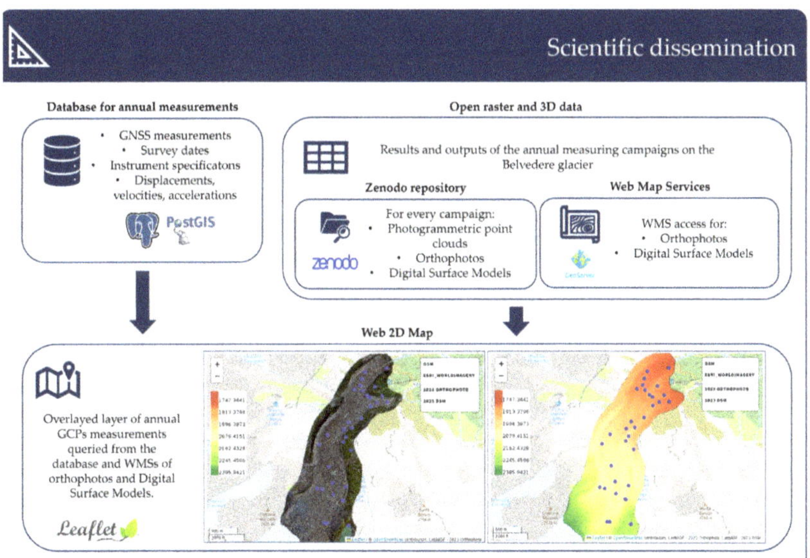

Figure 6. Design component of web resources for accessing and visualizing annual glacier monitoring data from the Belvedere Glacier, including GNSS measurements, photogrammetry, and derived products for a scientific audience.

3.2.1. Data for Annual Measurements

Access to monitoring products was previously restricted to researchers already involved in the data collection and processing, or to scientists outside the project team who explicitly requested the data. In addition, most of the information on the measured target analytics was stored in an unstructured format that was difficult to edit and query in an effective and flexible way. For all these reasons, a relational database was designed to become the primary repository for the wide variety of data and information coming from the annual in situ campaigns as well as from additional research activities connected to the Belvedere Glacier project. Its main goal is to facilitate efficient storage, accessibility, and sharing of the data among users with different roles. PostgresSQL 17 [69] was chosen for its flexible open-source license and for the possibility of installing PostGIS 3.5.2 [70], an extension that allows the management of geospatial data and formats. By customizing user permissions with the definition of specific roles, such a software infrastructure also enables direct links to GIS environments for both routine operations and remote public access. The first case of use is described in the scientific/technical outreach section, while the connection to the database for read-only purposes is discussed in the 3D viewer interface implementation.

In accordance with relational database principles, data are organized across multiple tables, with access enabled through established relationships using primary and foreign keys. The database Entity Relationship Diagram (ERD), shown in Figure 7, illustrates the main entities and their relationships.

The core entities of the schema are represented by: the *Surveys* table, which contains relevant information about in situ activities such as dates and year of organization; the Points table, which represents a register of all points on the glacier that are periodically surveyed using GNSS; and the Measurements table, which contains the measured coordinates for each point surveyed as part of an annual monitoring campaign. The relationship between these tables, shown in the ERD, allows survey and point information to be retrieved for each measurement record. This structure, therefore, allows the execution of a query in Structured Query Language (SQL), such as the extraction of all measurements from a given survey year or the time series of measurements for a selected point. Additionally, more information associated with each survey can be retrieved thanks to the relations with the entities Instrument (e.g., employed equipment specifics), Flight (e.g., camera mounted on the drone, number of images captured, average ground sampling distance), and Product (which contains all the information about the outputs of the photogrammetric processes like DSMs, orthophotos, point clouds, and meshes). This last entity is solved with a generalization with a downward collapse, taking into account the peculiarities of the 2D and 3D visualizations, including their specific attributes.

The schema shown in Figure 7 represents the core of the database, accessible in editing mode only to the team members responsible for updating and managing the records. To facilitate external access to the quantitative data collected and processed in the project, a combination of useful routine queries was generated to create read-only views in a separate public schema. Such a solution simplifies access to information such as the measured points for each year. They also identify active points (i.e., targets that remain on the glacier and have not been lost) by providing their most recent measurements and automatically compute time series data for displacements, velocities, and accelerations for each point by differentiating successive measurements. Other functionalities include the automatic conversion of GNSS coordinates between the WGS84 geographic reference system and the projected RDN2008/UTM Zone 32N reference system. In addition, the database allows the calculation of orthometric heights for each measurement using bilinear interpolation based on the Italian geoid model ITALGEO05 [71].

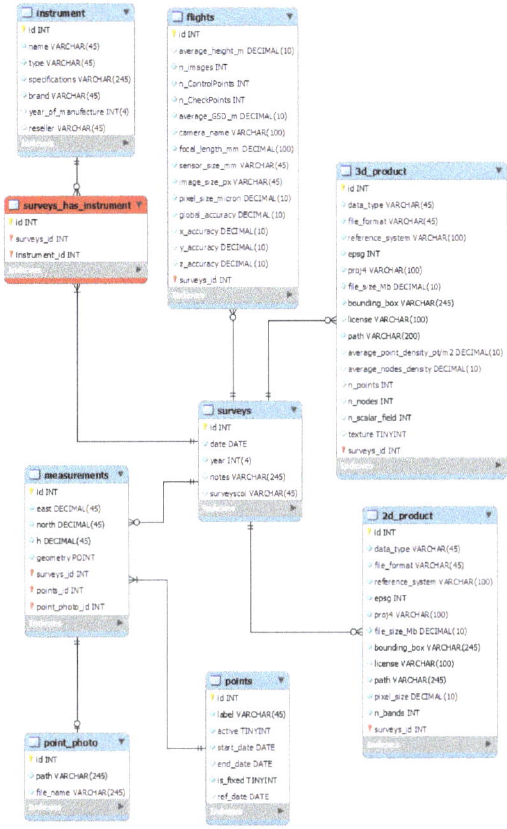

Figure 7. Relational database model for the Belvedere Glacier monitoring project, organizing data on instruments, surveys, points, measurements, and derived products.

3.2.2. Open Raster and 3D Data

To promote collaboration and open science, all point clouds, DSMs, and orthophotos covering the glacier are shared publicly under the Creative Commons Attribution 4.0 International on a Zenodo repository [58]. This dataset includes a structured JSON file with essential metadata, enabling researchers to use the data effectively and fostering interdisciplinary exploration. These resources support the replication of analyses conducted as part of the glacier monitoring project—such as glacier velocity and volume change estimates—and provide a foundation for studying other geomorphological processes, like moraine collapse, or gaining new insights into glacier dynamics.

The data are organized by year in individually downloadable zip folders named belvedere_YYYY.zip. Each folder contains all data for that specific year (including photogrammetric point clouds, orthophotos, and DEMs) along with the corresponding metadata. Metadata are provided in a JSON file, detailing essential information for data use. Point clouds are saved in a compressed LAZ format. Orthophotos and DEMs are georeferenced TIFF images. For large point clouds, the data are divided into regular tiles, numbered sequentially from the bottom-left corner of the bounding box, progressing row by row.

Additionally, the DSMs and orthophotos from the glacier monitoring campaign are available as WMS through GeoServer [72], an open-source server specifically designed for sharing geospatial data on the web. GeoServer supports multiple data formats and standards, including Open Geospatial Consortium (OGC) services like WMS and Web

Feature Service (WFS), which ensures broad compatibility with GIS software. Hosting the data on GeoServer allows users to visualize the rasters directly in GIS applications without needing to download large files, allowing for efficient remote access and interactive exploration. This approach enhances accessibility and supports interdisciplinary research by enabling the instant integration of glacier monitoring data into various analytical workflows.

3.2.3. Web 2D Map

The 2D web map represents the general access point for external users to the yearly target measurements archived in the database. Based on Leaflet.js, it provides users with an interactive platform to visualize and access measurement data for a specific campaign, yearly conducted along the glacier. Each data point is clickable, showing precise coordinates and allowing for easy data retrieval. The map also includes overlays of various WMS, such as DSM and orthophotos from different years, enhancing spatial analysis by offering visual context of the glacier terrain. Users can select a year from the dropdown menu to filter the available data for a specific period, then fetch the data for on-screen display or download them in Comma Separated Values (CSV) format for further analysis. The map combines base layers from sources like OpenStreetMap and Esri World Imagery, allowing users to explore changes over time and conduct comparative studies with the most recent datasets, such as the 2023 DSM and orthophoto layers (Figure 8).

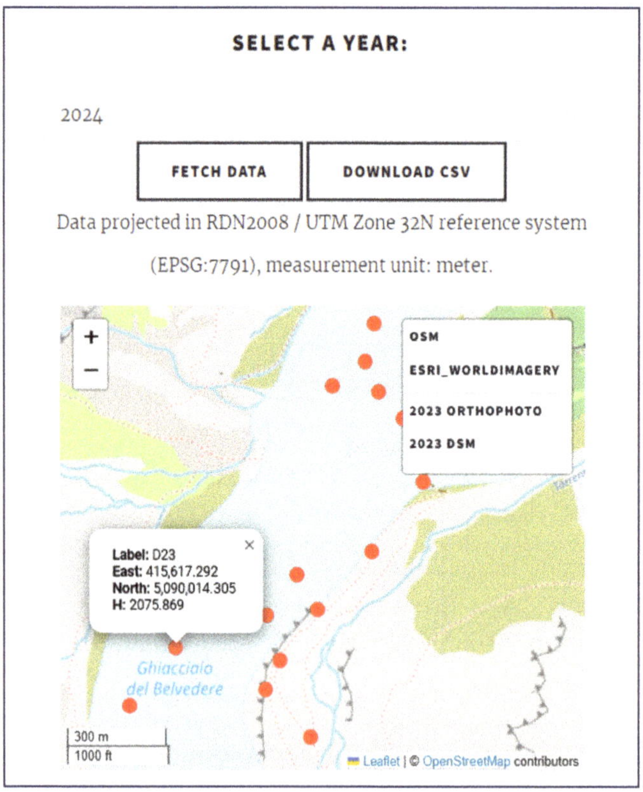

Figure 8. User-friendly web interface for visualizing and downloading annual glacier monitoring data on the Belvedere Glacier, allowing users to select years and download data in CSV format.

3.3. General Public Outreach

A decade of in situ surveys and monitoring activities on the Belvedere Glacier has produced a rich archive of data documenting the glacier evolution. These data, including point clouds, digital surface models, and GNSS measurements of target points, capture changes in displacement, velocity, and acceleration measured annually. Given the urgency to document a vanishing natural heritage site and increase awareness about climate change direct impacts, we developed an accessible web platform, The Belvedere Glacier (thebelvedereglacier.it (accessed on 2 February 2025)). This platform enables the public to stay informed about the latest activities on the glacier and explore the 3D products generated through ongoing monitoring which were previously available only to a limited audience.

To achieve these outreach goals, the web application (Figure 9) was designed with the following modules:

- Data Storage component: this module, detailed in Section 3.2.1, provides an organized and regularly updated repository of the collected data, enabling accurate archiving and management within the Belvedere Glacier project.
- 3D Viewer User Interface: the viewer offers a space for interactive, digital navigation of the site, where users can qualitatively and quantitatively explore the results of monitoring campaigns.
- User Exploration Guide: created as a digital handbook, this module helps users navigate the web application, explains data characteristics, and provides guidance for exploring the platform capabilities.
- Open-Source Code: to support transparency and reproducibility, the entire website codebase is published on GitHub, promoting collaboration and enabling users to replicate or contribute to the project. The repository includes a Docker configuration, streamlining deployment and scalability by defining services such as the Apache webserver with environmental variables for database connectivity. The application is hosted online through DigitalOcean, ensuring reliable access and performance. This setup simplifies deployment for developers and encourages seamless contributions to the project evolution.

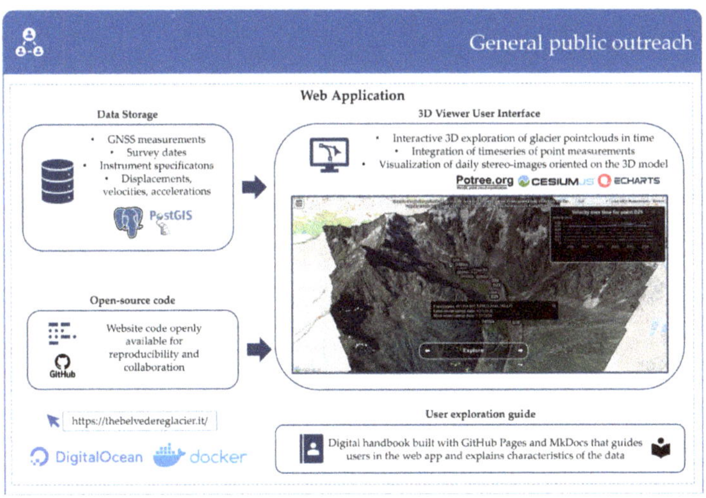

Figure 9. Design of the web application for public outreach and interactive exploration of the Belvedere Glacier monitoring project. Users can visualize 3D point clouds, time series data, and daily stereo images oriented on the 3D model. The application is built using open-source technologies and is hosted on DigitalOcean.

3.3.1. 3D Viewer User Interface

The public can access the products of the annual glacier survey campaigns through a dedicated web-based 3D viewer. It is based on a combination of the Potree [73] and Cesium JS libraries, where Potree enables the efficient streaming and rendering of massive point cloud datasets directly in a browser, while Cesium JS provides high-performance 3D geospatial visualization on the web [74–76]. Together, these libraries enable seamless, real-time exploration of complex 3D products over the web without the need for local software installation—only a stable internet connection is required for a smooth interaction with large datasets. In particular, the viewer includes point clouds of the glacier from different time periods, overlaid on a dynamic OpenStreetMap base map with background terrain rendering based on the Maptiler provider.

In addition to the native functionalities of the Potree library, custom features are implemented, such as the ability to navigate point clouds from different epochs in sequence and the connection to the database described in Section 3.2.1. In particular, the latter allows the database to be queried year by year via a simple drop-down menu (Figure 10). When a year is selected, the targets measured in the selected campaign are located in the 3D scene along the glacier. By clicking on one of them, it is then possible, thanks to the integration of the Echarts.js library, to visualize the corresponding time series of velocities, calculated over all the measurements of that point stored in the database.

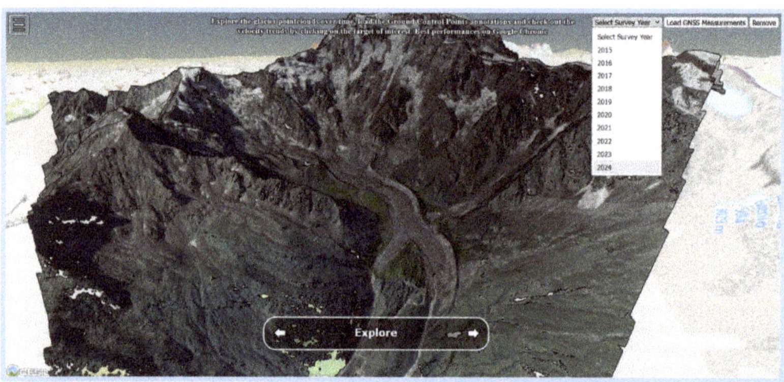

Figure 10. Potree-based interactive 3D viewer for exploring glacier monitoring data on the Belvedere Glacier.

3.3.2. User Exploration Guide

To further support the general public's experience on the platform, a practical guide is made available online (https://labmgf-polimi.github.io/thebelvedereglacier-handbook/ (accessed on 2 February 2025)) as a constantly updated platform. The Belvedere Glacier handbook, based on the same technologies used for the open publication of the summer school learning resources, is a digital platform where end users can find tips and tutorials to enhance their interactive experience with the different tools developed as part of the monitoring project, with a special focus on the online 3D viewer. For example, it provides tutorials on how to navigate through the 3D scene, rotate, and zoom on different objects, as well as how to measure distances and extract cross-sections from different point clouds and compare them qualitatively or quantitatively. Special attention is given to the custom features of the Belvedere viewer, such as the synchronized connection to the targets measurement database and the temporal exploration of the point clouds of glacier surfaces in different epochs. To guide the user through the different application functionalities, visual instructions and practical exploration examples are given.

3.3.3. Open-Source Code

To promote transparency and encourage further development within the community, the source code of the 3D viewer used in this project is openly available on GitHub (https://github.com/Tars4815/thebelvedereglacier (accessed on 2 February 2025)). This approach supports reproducibility and allows developers and researchers to adapt or extend the viewer functionalities according to their needs. By providing access to the underlying code, we aim to facilitate collaborative improvements, particularly around custom features such as temporal point cloud comparison and integration with the targets measurement database. This commitment to open-source principles reinforces project alignment with collaborative and accessible science.

4. Results and Discussion

Glacier monitoring, using a variety of techniques, has a long tradition and plays a crucial role in providing successive generations of researchers and geoscientists with the data needed to address the challenges of climate change. In this context, an efficient system for data management and visualization—capable of integrating both morphological and spatial dimensions of information—can not only deepen the understanding of glacier dynamics and foster collaboration, but also facilitate the development of dynamic tools for effective communication and dissemination of research results.

To meet these diverse needs, the digital platform developed for the Belvedere Glacier project has been designed to address three main audiences: students, researchers and technicians, and the public. Each audience benefits from tailored tools and functionalities that balance accessibility, usability, and depth of information.

Table 1 summarizes the characteristics of these solutions, highlighting the differences in terms of:

- Accessibility: level of effort or requirements to access the corresponding tools (e.g., logins, hardware, software, licenses, etc.).
- User friendliness: how intuitive and simple the tools and their GUIs are.
- Data access level: whether the audience can access raw data, pre-processed products, or only visualizations.
- Required prior knowledge: the level of expertise needed to use the tools effectively.
- User role: whether users are simply passive recipients (readers) or active participants (producers) of the research output through the developed platform.

Table 1. Comparison of the different experiences of the target users according to the characteristics of the platform. The qualitative evaluation presented is based on the authors' estimate.

Audience Component	Educational	Scientific	General Public
Accessibility	Moderate	Low	High
User friendliness	Moderate	Low	High
Data access level	High	Moderate	Low
Required prior knowledge	Moderate	High	Low
User role	Producer and Reader	Reader	Reader

4.1. *Internal Training and Skill Transfer*

The educational component of the Belvedere Glacier provides an innovative learning framework that combines theoretical knowledge with hands-on experience in glacier monitoring and data analysis. This approach directly aligns with the Sustainable Development Goal (SDG) target 4.7, which promotes education for sustainable development and empowers students as global citizens equipped to address environmental challenges, and SDG target 13.3, which focuses on building knowledge and capacity for climate action [77].

As shown in Table 1, the educational component is moderately accessible as the project includes openly available learning materials and tutorials published on a MkDocs website, making the core resources widely available to any interested student. This approach ensures a long-term impact beyond the fieldwork experience, providing resources and materials for the development of specific exercises and examples from real-world challenges and ongoing research to practical sessions and seminars within the university curriculum.

However, access to the full potential of the experience is more limited, as it requires participation in structured programs such as the summer school. While the summer school provides a comprehensive and immersive learning environment, logistical and practical constraints mean that only a limited number of students can participate each year. In particular, constraints such as limited accommodation capacity in the mountain hut, restrictions that may affect students with disabilities, and the current registration policy, which is primarily limited to internal students, present significant barriers. These limitations highlight the need for potential adaptations to improve inclusivity and extend the reach of the educational experience without compromising the quality of the practical training provided during the field sessions.

Students engage with both raw and processed datasets, such as GNSS measurements, photogrammetric point clouds, and digital surface models, gaining full control and access to data through hands-on exercises that provide guidance from acquisition to analysis. This approach not only enhances their ability to manage and interpret complex datasets, but also addresses the need for basic prior knowledge in fields like geomatics, GIS, or photogrammetry through introductory sessions and structured modules that build foundational skills.

By moving from passive learners to active contributors, students actively produce and analyze data, fostering a sense of ownership and deeper involvement in real-world glacier monitoring efforts. The impact of this aspect also extends beyond the summer school experience, as, over the past decade, many students, both undergraduate and graduate, have continued to work on different aspects of the project as part of their thesis work, also developing specific plugins for processing and extracting additional information from the data.

4.2. Scientific Dissemination

The technical component is primarily aimed at researchers and technicians, providing high accessibility to data and tools tailored for advanced analysis and monitoring. This component includes openly available pre-processed datasets, such as photogrammetric point clouds, orthophotos, and digital surface models, made available through WMS services and a dedicated Zenodo repository. The open format in which the data are released also allows easy access through free and open-source geospatial software (FOSS4G) such as QGIS and CloudCompare. As a result, these resources allow users to explore the datasets independently and perform advanced analysis, even in interdisciplinary contexts. In this way, even if the target users act primarily as simple readers of pre-processed data, by interpreting existing datasets, they indirectly become contributors to the wider impact of the research, developing their own research questions and implementing solutions with their custom tools. Since the release of the first version in April 2023, the historical long-term Zenodo dataset has been viewed more than 1500 times, with almost 500 downloads as of December 2024.

This technical component aligns strongly with target 11.4 of the SDGs which emphasises the importance of protecting natural heritage. By enabling detailed documentation and analysis of glacier dynamics, the platform not only helps to monitor environmental change, but also contributes to preserving the cultural and natural value of the Belvedere Glacier as a vital part of the Alpine landscape. This dual role reinforces the importance of long-term data management, highlighting the need for open science and collaborative efforts to address future challenges.

Another important aspect of this component is the possibility not only to replicate and enrich the analysis of glacier dynamics using the open data, but also to adapt the whole digital framework to other case studies. In fact, as the entire code of the platform is available as open source, all the tools of the different components can be reused for other research sites. Small glaciers monitored by independent research groups can easily benefit from this low-cost solution which can be adapted to their specific needs.

4.3. General Public Outreach

The public component focuses on promoting accessibility and awareness by providing tools that are highly accessible and intuitive for non-specialist users. The web-based 3D viewer allows anyone with an internet connection to interactively explore 3D glacier reconstructions without the need for locally installed specialized software or technical expertise. This open access approach ensures that a wide audience, regardless of technical background, can engage with and understand the effects of climate change on the glacier.

The usability of the platform is enhanced by its intuitive interface, simplified navigation, and additional user guides, making it suitable for people with no prior knowledge of geomatics or geospatial analysis. The design encourages exploration, providing a visual and interactive experience that transforms complex scientific data into an accessible graphical format for non-experts. In particular, point clouds, usually considered complex objects to understand, are made accessible and familiar to users thanks to the simple measurement tool provided to extract distances and cross-sections. (Figure 11). The platform also allows users to download the extracted point clouds in various formats for further analysis in other pieces of software (Figure 12).

Figure 11. Example of a procedure for cross-sections extraction from the interactive 3D web viewer used for public outreach, with the section command available on the lateral sidebar and its interactive definition in the 3D scene.

In this framework, users primarily take on the role of readers, as they consume and interact with preprocessed data and visualizations of graphs (Figure 13). While they do not contribute to data collection or processing, their interaction with the platform fosters greater engagement in the dissemination of glacier monitoring efforts and climate research. By bridging the gap between science and the public, the component supports broader educational and awareness-raising goals.

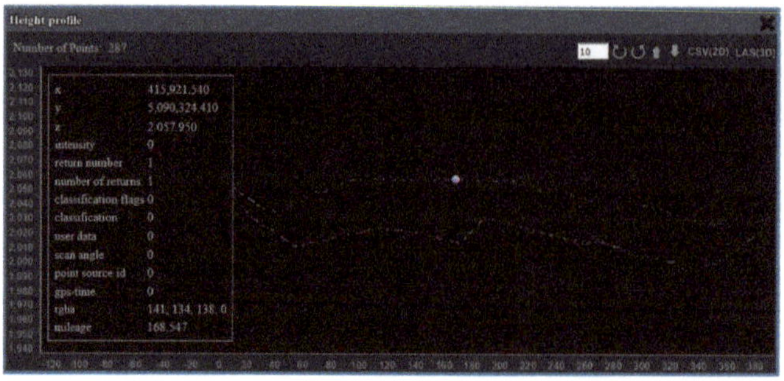

Figure 12. Dedicated window for the defined cross-section. By activating multiple point clouds, it is possible to quantitatively compare the changes between distinct epochs.

Figure 13. Example of a graphical visualization of the velocity time series of a user-selected target measured along the glacier. The real-time connection to the geodatabase allows the query of targets measured on a given survey. The pop-up windows show additional details on each point as well as their calculated velocity trends.

5. Conclusions and Future Perspectives

This paper presents the design workflow for the implementation of web-based geospatial solutions to communicate to different target audiences the evolution in time and space of the Belvedere Glacier, in the Italian Alps. After identifying the end users in the educational, scientific, and public components, this study demonstrates the power of geovisualization and data management to facilitate the dissemination of a comprehensive, long-term monitoring project. By using open-source technologies and platforms, the final web platform can also be easily adapted to different contexts, making other research focused on small glaciers accessible to a wider audience.

A fundamental advancement of this project is the development of a comprehensive, digital, and accessible 3D documentation of the glacier. This contrasts with traditional 2D imaging methods and provides a more immersive and informative representation of this rapidly changing natural environment, thanks to the development of dedicated solutions that bridge the gap between users familiar with geospatial data and the general public. This digital platform will serve as an invaluable resource for researchers, educators, and the wider community, enabling a deeper understanding of the impacts of climate change and fostering a greater appreciation of the importance of glacial ecosystems.

However, building an open access web platform for geospatial visualization and communication presents several challenges. Supporting efficient processing and rendering of large datasets while maintaining accessibility for the different targets of users requires an evaluation of the balance between performance and usability. Also, interoperability with evolving geospatial standards and data formats for web dissemination and publication is another key consideration to be considered for ensuring long-term sustainability of the project. In addition, maintaining an open access platform requires continuous updates to keep pace with technological advances, as well as addressing potential issues related to data licensing, privacy, and infrastructure costs. Despite these challenges, the open-source nature of this approach promotes transparency, reproducibility, and collaboration within the scientific community. By making data openly available, we aim to contribute to the growing body of evidence supporting the need for urgent action on climate change and foster a virtuous cycle of knowledge sharing and awareness, both for local communities and for society at large.

Future work will include addressing the current limitations of the components, while keeping them up-to-date and interoperable with current and future trends in geovisualization technologies. Indeed, the rapid development of new tools and the obsolescence of previous versions pose a challenge both in terms of feasibility and in terms of the educational component, which requires access to the latest versions of software and code libraries. Regarding the educational component, the implementation of virtual or mixed-reality solutions is under study: this approach would make the summer school accessible to all students, despite the constraints connected to the limited number of accommodations in the hut or to the physical accessibility of the study area.

Author Contributions: Conceptualization, Federica Gaspari and Francesco Ioli; data curation, Federico Barbieri and Francesco Ioli; funding acquisition, Livio Pinto; investigation, Federica Gaspari, Federico Barbieri, Rebecca Fascia and Francesco Ioli; methodology, Federica Gaspari, Rebecca Fascia and Francesco Ioli; project administration, Livio Pinto; software, Federica Gaspari and Francesco Ioli; supervision, Livio Pinto and Federica Migliaccio; validation, Federico Barbieri and Francesco Ioli; visualization, Federica Gaspari; writing—original draft, Federica Gaspari; writing—review and editing, Federico Barbieri, Rebecca Fascia, Livio Pinto and Federica Migliaccio. All authors have read and agreed to the published version of the manuscript.

Funding: The development of the open access learning resources for the summer school was supported by the EGU Higher Education Teaching Material Grant 2023. The web hosting of the Belvedere Glacier web platform is partially funded by the Open Knowledge Foundation through the 2023 Open Data Day mini grant.

Data Availability Statement: The source code of the platform is openly accessible in the GitHub repository: https://github.com/Tars4815/thebelvedereglacier (accessed on 17 November 2024). Point clouds, digital surface models, and orthophotos of the glacier from 1977 to 2023 are openly available in a Zenodo repository: https://zenodo.org/records/10817029 (accessed on 17 November 2024).

Acknowledgments: The authors would like to thank the Department of Civil and Environmental Engineering (DICA) of the Politecnico di Milano, in particular Carlo De Michele and Alberto Bianchi, and the Department of Environmental, Land and Infrastructure Engineering (DIIATI) of the Politecnico di Torino, in particular Alberto Cina, Paolo Maschio, and Marco Piras, for their contributions to the Belvedere Glacier long-term monitoring project. Special thanks to the team members of the DREAM (Drone tEchnology for wAter resources and hydrologic hazard Monitoring) and Kuoleva Jäätikko projects of the Alta Scuola Politecnica. Finally, this work would not have been possible without the valuable contributions of the students who have made the whole summer school experience so special in recent years.

Conflicts of Interest: The authors declare no conflicts of interest.

References

1. O'Neill, S.; Smith, N. Climate change and visual imagery. *Wiley Interdiscip. Rev. Clim. Chang.* **2014**, 5, 73–87. [CrossRef]
2. O'Neill, S. More than meets the eye: A longitudinal analysis of climate change imagery in the print media. *Clim. Chang.* **2020**, 163, 9–26. [CrossRef]
3. Gagné, K.; Rasmussen, M.B.; Orlove, B. Glaciers and society: Attributions, perceptions, and valuations. *Wiley Interdiscip. Rev. Clim. Chang.* **2014**, 5, 793–808. [CrossRef]
4. Gaudio, D.; Gobbi, M. Glaciers in the Anthropocene: A Biocultural View. *Nat. Cult.* **2022**, 17, 243–261. [CrossRef]
5. Jackson, M. Glaciers and climate change: Narratives of ruined futures. *Wiley Interdiscip. Rev. Clim. Chang.* **2015**, 6, 479–492. [CrossRef]
6. Salim, E.; Ravanel, L.; Deline, P. Does witnessing the effects of climate change on glacial landscapes increase pro-environmental behaviour intentions? An empirical study of a last-chance destination. *Curr. Issues Tour.* **2023**, 26, 922–940. [CrossRef]
7. Wang, S. Opportunities and threats of cryosphere change to the achievement of UN 2030 SDGs. *Humanit. Soc. Sci. Commun.* **2024**, 11, 44. [CrossRef]
8. Palomo, I. Climate Change Impacts on Ecosystem Services in High Mountain Areas: A Literature Review. *Mt. Res. Dev.* **2017**, 37, 179–187. [CrossRef]
9. Nesur, K.; Salim, E.; Girault, C.; Ravanel, L. Sharing scientific knowledge on glaciers to the general public: The role of glacier interpretation centres in mountain tourism diversification strategies. *J. Alp. Res. Rev. De Géographie Alp.* **2022**, 110, 1. [CrossRef]
10. Salim, E.; Ravanel, L.; Bourdeau, P.; Deline, P. Glacier tourism and climate change: Effects, adaptations, and perspectives in the Alps. *Reg. Environ. Chang.* **2021**, 21, 120. [CrossRef] [PubMed]
11. Schaefli, B.; Manso, P.; Fischer, M.; Huss, M.; Farinotti, D. The role of glacier retreat for Swiss hydropower production. *Renew. Energy* **2019**, 132, 615–627. [CrossRef]
12. Day, J.C.; Heron, S.F.; Markham, A. Assessing the climate vulnerability of the world's natural and cultural heritage. In *Parks Stewardship Forum*; UC Berkeley: Berkeley, CA, USA, 2020; p. 36.
13. Carey, M. The history of ice: How glaciers became an endangered species. *Environ. Hist.* **2007**, 12, 497–527. [CrossRef]
14. Dannevig, H.; Rusdal, T. Caring for melting glaciers. *Tour. Geogr.* **2023**, 25, 1679–1695. [CrossRef]
15. Carey, M.; Molden, O.C.; Rasmussen, M.B.; Jackson, M.; Nolin, A.W.; Mark, B.G. Impacts of glacier recession and declining meltwater on mountain societies. *Ann. Am. Assoc. Geogr.* **2017**, 107, 350–359. [CrossRef]
16. Ramírez, N.; Melfo, A.; Resler, L.M.; Llambí, L.D. The end of the eternal snows: Integrative mapping of 100 years of glacier retreat in the Venezuelan Andes. *Arct. Antarct. Alp. Res.* **2020**, 52, 563–581. [CrossRef]
17. Buhr, L. Picturing finitude: Photography of mountain glaciers as a multiple practice of dealing with environmental loss. *Environ. Values* **2024**, 33, 550–570. [CrossRef]
18. Schroth, O.; Angel, J.; Sheppard, S.; Dulic, A. Visual climate change communication: From iconography to locally framed 3D visualization. *Environ. Commun.* **2014**, 8, 413–432. [CrossRef]
19. Amrhein, C.; Balaban, D.C. The effectiveness of 360° videos for environmental communication. Exploring the impact of different types of storytelling. *Comput. Hum. Behav.* **2024**, 13, 100368. [CrossRef]
20. Li, W.; Zhu, J.; Pirasteh, S.; Zhu, Q.; Guo, Y.; Luo, L.; Dehbi, Y. A 3D virtual geographic environment for flood representation towards risk communication. *Int. J. Appl. Earth Obs. Geoinf.* **2024**, 128, 103757. [CrossRef]
21. On the Trail of the Glacier. Available online: https://onthetrailoftheglaciers.com/ (accessed on 2 February 2025).
22. Nüsser, M.; Schmidt, S. Glacier changes on the Nanga Parbat 1856–2020: A multi-source retrospective analysis. *Sci. Total Environ.* **2021**, 785, 147321. [CrossRef] [PubMed]
23. Kaufmann, V. The evolution of rock glacier monitoring using terrestrial photogrammetry: The example of Äusseres Hochebenkar rock glacier (Austria). *Austrian J. Earth Sci.* **2012**, 105, 63–77.
24. Kropáček, J.; Nüsser, M.; Schmidt, S. Exploring the potential of historical images for the investigation of glacier changes: The case of Belvedere Glacier, Italian Alps. *Acta Univ. Carol. Geogr.* **2024**, 59, 240–254. [CrossRef]
25. Nussbaumer, S.U.; Nesje, A.; Zumbühl, H.J. Historical glacier fluctuations of Jostedalsbreen and Folgefonna (southern Norway) reassessed by new pictorial and written evidence. *Holocene* **2011**, 21, 455–471. [CrossRef]
26. Zumbühl, H.J.; Nussbaumer, S.U. Little Ice Age glacier history of the Central and Western Alps from pictorial documents. *Cuad. De Investig. Geográfica* **2018**, 44, 115–136. [CrossRef]
27. Wiesmann, S.; Steiner, L.; Pozzi, M.; Bozzini, C.; Bauder, A.; Hurni, L. Reconstructing historic glacier states based on terrestrial oblique photographs. In Proceedings of the AutoCarto International Symposium on Automated Cartography, Columbus, OH, USA, 16–18 September 2012; Volume 1618.
28. Kääb, A.; Haeberli, W.; Gudmundsson, G.H. Analysing the creep of mountain permafrost using high precision aerial photogrammetry: 25 years of monitoring Gruben rock glacier, Swiss Alps. *Permafr. Periglac. Process.* **1997**, 8, 409–426. [CrossRef]
29. Paul, F.; Kääb, A.; Maisch, M.; Kellenberger, T.; Haeberli, W. The new remote-sensing-derived swiss glacier inventory: I. *Methods. Ann. Glaciol.* **2002**, 34, 355–361. [CrossRef]

30. Westoby, M.; Brasington, J.; Glasser, N.; Hambrey, M.; Reynolds, J. 'Structure-from-motion' photogrammetry: A low-cost, effective tool for geoscience applications. *Geomorphology* **2012**, *179*, 300–314. [CrossRef]
31. Seitz, S.M.; Curless, B.; Diebel, J.; Scharstein, D.; Szeliski, R. A comparison and evaluation of multi-view stereo reconstruction algorithms. In Proceedings of the IEEE Computer Society Conference on Computer Vision and Pattern Recognition, New York, NY, USA, 17–22 June 2006; Volume 1, pp. 519–526.
32. Bamber, J.L.; Rivera, A. A review of remote sensing methods for glacier mass balance determination. *Glob. Planet. Chang.* **2007**, *59*, 138–148. [CrossRef]
33. Berthier, E.; Cabot, V.; Vincent, C.; Six, D. Decadal region-wide and glacier-wide mass balances derived from multi-temporal raster satellite digital elevation models. Validation over the Mont-Blanc area. *Front. Earth Sci.* **2016**, *4*, 63. [CrossRef]
34. Ioli, F.; Bianchi, A.; Cina, A.; De Michele, C.; Maschio, P.; Passoni, D.; Pinto, L. Mid-term monitoring of glacier's variations with UAVs: The example of the belvedere glacier. *Remote Sens.* **2022**, *14*, 28. [CrossRef]
35. Di Rita, M.; Fugazza, D.; Belloni, V.; Diolaiuti, G.; Scaioni, M.; Crespi, M. Glacier volume change monitoring from UAV observations: Issues and potential of state-of art techniques. *Int. Arch. Photogramm. Remote Sens. Spat. Inf. Sci.* **2020**, *43*, 1041–1048. [CrossRef]
36. James, M.R.; Robson, S. Straightforward reconstruction of 3D surfaces and topography with a camera: Accuracy and geoscience application. *J. Geophys. Res. Earth Surf.* **2012**, 117. [CrossRef]
37. Eltner, A.; Kaiser, A.; Castillo, C.; Rock, G.; Neugirg, F.; Abellán, A. Image-based surface reconstruction in geomorphometry—Merits, limits and developments. *Earth Surf. Dyn.* **2016**, *4*, 359–389. [CrossRef]
38. Giulio Tonolo, F.; Cina, A.; Manzino, A.; Fronteddu, M. 3D glacier mapping by means of satellite stereo images: The Belvedere glacier case study in the Italian Alps. *Int. Arch. Photogramm. Remote Sens. Spat. Inf. Sci.* **2020**, *43*, 1073–1079. [CrossRef]
39. De Gaetani, C.I.; Ioli, F.; Pinto, L. Aerial and UAV images for photogrammetric analysis of Belvedere glacier evolution in the period 1977–2019. *Remote Sens.* **2021**, *13*, 3787. [CrossRef]
40. Immerzeel, W.W.; Kraaijenbrink, P.D.; Shea, J.; Shrestha, A.B.; Pellicciotti, F.; Bierkens, M.F.; de Jong, S.M. High-resolution monitoring of himalayan glacier dynamics using unmanned aerial vehicles. *Remote Sens. Environ.* **2014**, *150*, 93–103. [CrossRef]
41. Belloni, V.; Fugazza, D.; Di Rita, M. UAV-based glacier monitoring: GNSS kinematic track post-processing and direct georeferencing for accurate reconstructions in challenging environments. *Int. Arch. Photogramm. Remote Sens. Spat. Inf. Sci.* **2022**, *43*, 367–373. [CrossRef]
42. Mannerfelt, E.S.; Dehecq, A.; Hugonnet, R.; Hodel, E.; Huss, M.; Bauder, A.; Farinotti, D. Halving of Swiss glacier volume since 1931 observed from terrestrial image photogrammetry. *Cryosphere* **2022**, *16*, 3249–3268. [CrossRef]
43. Dematteis, N.; Troilo, F.; Scotti, R.; Colombarolli, D.; Giordan, D.; Maggi, V. The use of terrestrial monoscopic time-lapse cameras for surveying glacier flow velocity. *Cold Reg. Sci. Technol.* **2024**, *222*, 104185. [CrossRef]
44. Ioli, F.; Dematteis, N.; Giordan, D.; Nex, F.; Pinto, L. Deep Learning Low-cost Photogrammetry for 4D Short-term Glacier Dynamics Monitoring. *PFG–J. Photogramm. Remote Sens. Geoinf. Sci.* **2024**, *92*, 657–678. [CrossRef]
45. Schaffland, A.; Vornberger, O.; Heidemann, G. An interactive web application for the creation, organization, and visualization of repeat photographs. In Proceedings of the 1st Workshop on Structuring and Understanding of Multimedia Heritage Contents, Nice, France, 21 October 2019; pp. 47–54.
46. Clason, C.; Rangecroft, S.; Kallis, G.; Lewin, S.; Mullier, T. GlacierMap: A virtual opportunity to explore the Andes' vanishing glaciers. *Geography* **2021**, *106*, 148–153. [CrossRef]
47. Rastner, P.; Joerg, P.C.; Huss, M.; Zemp, M. Historical analysis and visualization of the retreat of Findelengletscher, Switzerland, 1859–2010. *Glob. Planet. Chang.* **2016**, *145*, 67–77. [CrossRef]
48. Geissler, J.; Mayer, C.; Jubanski, J.; Münzer, U.; Siegert, F. The potentials of high-resolution photogrammetry for analyzing glacier retreat in the Ötztal Alps, Austria. *Cryosphere* **2020**, *1*, 1–22.
49. Martin, S. Interactive visual media for geomorphological heritage interpretation. Theoretical approach and examples. *Geoheritage* **2014**, *6*, 149–157. [CrossRef]
50. Santos, I.; Henriques, R.; Mariano, G.; Pereira, D.I. Methodologies to represent and promote the geoheritage using unmanned aerial vehicles, multimedia technologies, and augmented reality. *Geoheritage* **2018**, *10*, 143–155. [CrossRef]
51. Triglav Čekada, M.; Radovan, D.; Lipuš, B.; Mongus, D. Very small glaciers as geoheritage: Combining a spatio-temporal visualisation of their development and related effects of climate change. *Geoheritage* **2020**, *12*, 85. [CrossRef]
52. Aldighieri, B.; Testa, B.; Bertini, A. 3D exploration of the San Lucano Valley: Virtual geo-routes for everyone who would like to understand the landscape of the Dolomites. *Geoheritage* **2016**, *8*, 77–90. [CrossRef]
53. Senger, K.; Betlem, P.; Birchall, T.; Buckley, S.J.; Coakley, B.; Eide, C.H.; Flaig, P.P.; Forien, M.; Galland, O.; Gonzaga, L.; et al. Using digital outcrops to make the high Arctic more accessible through the Svalbox database. *J. Geosci. Educ.* **2021**, *69*, 123–137. [CrossRef]
54. Pasquaré Mariotto, F.; Corti, N.; Drymoni, K. Advanced technologies for Geosite visualization and valorization: A review. *Appl. Sci.* **2023**, *13*, 5598. [CrossRef]

55. Macelloni, M.M.; Corte, E.; Ajmar, A.; Cina, A.; Giulio Tonolo, F.; Maschio, P.F.; Pisoni, I.N. Multi-platform, Multi-scale and Multi-temporal 4D Glacier Monitoring. The Rutor Glacier Case Study. In *Italian Conference on Geomatics and Geospatial Technologies;* Springer International Publishing: Cham, Switzerland, 2022; pp. 392–404.
56. Haeberli, W.; Kääb, A.; Paul, F.; Chiarle, M.; Mortara, G.; Mazza, A.; Deline, P.; Richardson, S. A surge-type movement at Ghiacciaio del Belvedere and a developing slope instability in the east face of Monte Rosa, Macugnaga, Italian Alps. *Nor. Geogr. Tidsskr. Nor. J. Geogr.* **2002**, *56*, 104–111. [CrossRef]
57. Diolaiuti, G.; D'Agata, C.; Smiraglia, C. Belvedere Glacier, Monte Rosa, Italian Alps: Tongue thickness and volume variations in the second half of the 20th century. *Arct. Antarct. Alp. Res.* **2003**, *35*, 255–263. [CrossRef]
58. Ioli, F.; De Gaetani, C.; Barbieri, F.; Gaspari, F.; Pinto, L.; Rossi, L. *Belvedere Glacier Long-Term Monitoring Open Data (2.0) [Data set];* Zenodo: Geneva, Switzerland, 2024. [CrossRef]
59. Kääb, A.; Huggel, C.; Barbero, S.; Chiarle, M.; Cordola, M.; Epifani, F.; Haeberli, W. Glacier Hazards At Belvedere Glacier and the Monte Rosa East Face, Italian Alps: Processes and Mitigation. *Inter. Natl. Symp. Interpraevent* **2004**, *1*, 67–78.
60. VAW-ETH. Studi sul comportamento del Ghiacciaio del Belvedere, Macugnaga, Italia. *Relazione* **1985**, *97*, 76–109.
61. Kääb, A.; Huggel, C.; Fischer, L.; Guex, S.; Paul, F.; Roer, I.; Salzmann, N.; Schlaefli, S.; Schmutz, K.; Schneider, D.; et al. Remote sensing of glacier-and permafrost-related hazards in high mountains: An overview. *Nat. Hazards Earth Syst. Sci.* **2005**, *5*, 527–554. [CrossRef]
62. Mortara, G.; Tamburini, A.; Ercole, G.; Semino, P.; Chiarle, M.; Godone, F.; Guiliano, M.; Haeberli, W.; Kääb, A.; Huggel, C.; et al. *Il Ghiacciaio del Belvedere e L'emergenza del Lago Effimero; Regione Piemonte;* Societa' Meteorologica Subalpina: Moncalieri, TO, Italy, 2009.
63. Ioli, F. Multi-Scale and Multi-Temporal Photogrammetry for Alpine Glacier Monitoring. Ph.D. Thesis, Politecnico di Milano, Milan, Italy, 2024. Available online: https://hdl.handle.net/10589/224092 (accessed on 2 February 2025).
64. Gaspari, F.; Ioli, F.; Barbieri, F.; Bonora, S.; Fascia, R.; Pinto, L.; Migliaccio, F. Bridging geomatics theory to real-world applications in alpine surveys through an innovative summer school teaching program. *Int. Arch. Photogramm. Remote Sens. Spat. Inf. Sci.* **2024**, *XLVIII-4/W12-2024*, 59–66. [CrossRef]
65. Potůčková, M.; Albrechtová, J.; Anders, K.; Červená, L.; Dvořák, J.; Gryguc, K.; Höfle, B.; Hunt, L.; Lhotáková, Z.; Marcinkowska-Ochtyra, A.; et al. E-TRAINEE: Open e-learning course on time series analysis in remote sensing. *Int. Arch. Photogramm. Remote Sens. Spat. Inf. Sci.* **2023**, *48*, 989–996.
66. Balletti, C.; Capra, A.; Calantropio, A.; Chiabrando, F.; Colucci, E.; Furfaro, G.; Guastella, A.; Guerra, F.; Lingua, A.; Matrone, F.; et al. The SUNRISE summer school: An innovative learning-by-doing experience for the documentation of archaeological heritage. *Int. Arch. Photogramm. Remote Sens. Spat. Inf. Sci.* **2023**, *48*, 147–154. [CrossRef]
67. Achille, C.; Fiorillo, F. Teaching and learning of cultural heritage: Engaging education, professional training, and experimental activities. *Heritage* **2022**, *5*, 2565–2593. [CrossRef]
68. Achille, C.; Fassi, F.; Marquardt, K.; Cesprini, M. Learning geomatics for restoration: ICOMOS summer school in Ossola Valley. *Int. Arch. Photogramm. Remote Sens. Spat. Inf. Sci.* **2017**, *42*, 631–637. [CrossRef]
69. Stonebraker, M.; Rowe, L.A. The design of Postgres. *ACM Sigmod Rec.* **1986**, *15*, 340–355. [CrossRef]
70. McKenna, J. *PostGIS Team. PostGIS 3.2.0, The Olivier Courtin Release (3.2.0);* Zenodo: Geneva, Switzerland, 2021. [CrossRef]
71. Barzaghi, R.; Borghi, A.; Carrion, D.; Sona, G. Refining the estimate of the italian quasi-geoid. *Boll. Di Geod. E Sci. Affin.* **2007**, *3*, 145–160.
72. Aime, A.; Fabiani, A.; Hards, B.; Bühner, N.; Romagnoli, D.; Tajariol, E.; Miño, F.; Roldan, G.; Turton, I.; Hughes, J.; et al. *GeoCat BV, Open Source Geospatial Foundation. GeoServer (2.26.1);* Zenodo: Geneva, Switzerland, 2024. [CrossRef]
73. Schütz, M. Potree: Rendering Large Point Clouds in Web Browsers. Ph.D. Dissertation, Technische Universität Wien, Vienna, Austria, 2015.
74. Campiani, A.; McAvoy, S.; Lercari, N.; Stuardo, R.L.; Delgado, G.J.; Mejía, J.L.; Rissolo, D.; Kuester, F. Developing an interoperable cloud-based visualization workflow for 3D archaeological heritage data: The Palenque 3D Archaeological Atlas. *Digit. Appl. Archaeol. Cult. Herit.* **2023**, *31*, e00293. [CrossRef]
75. Gaspari, F.; Barbieri, F.; Fascia, R.; Ioli, F.; Pinto, L. An Open-Source Web Platform for 3D Documentation and Storytelling of Hidden Cultural Heritage. *Heritage* **2024**, *7*, 517–536. [CrossRef]
76. Fascia, R.; Barbieri, F.; Gaspari, F.; Ioli, F.; Pinto, L. From 3D Survey to Digital Reality of a Complex Architecture: A Digital Workflow for Cultural Heritage Promotion. *Int. Arch. Photogramm. Remote Sens. Spat. Inf. Sci.* **2024**, *48*, 205–212. [CrossRef]
77. Biermann, F.; Hickmann, T.; Sénit, C.A.; Beisheim, M.; Bernstein, S.; Chasek, P.; Grob, L.; Kim, R.E.; Kotzé, L.J.; Nilsson, M.; et al. Scientific evidence on the political impact of the Sustainable Development Goals. *Nat. Sustain.* **2022**, *5*, 795–800. [CrossRef]

International Journal of
Geo-Information

MDPI

Article

Climate Change Maps for the Atlas of Switzerland

Luca Gaia *, Andreas Neumann and Lorenz Hurni

Institute of Cartography and Geoinformation, Swiss Federal Institute of Technology Zurich (ETH), 8093 Zurich, Switzerland; andreas.neumann@ethz.ch (A.N.); lhurni@ethz.ch (L.H.)
* Correspondence: gaial@ethz.ch

Abstract: Climate change has global consequences, and Switzerland is no exception. The communication of climate change poses various challenges, and maps are often part of this process. This work presents three maps illustrating the impacts of climate change, developed for the Atlas of Switzerland (AoS), an interactive digital national atlas. The aim is to make climate change impacts understandable and visible. Three different indicators of climate change were visualized: the rise of the zero degree line, the evolution of glacial lakes, and changes in the flowering dates of plants. Various approaches were employed that leverage the strengths of the AoS, including temporal navigation, interactivity, 3D data visualizations, and map combinations. The feasibility of these visualizations are demonstrated through the presented maps and analysis of key considerations for their creation. We believe these map types should be included in national atlases and can contribute to the achievement of Sustainable Development Goal 13: "Climate Action". Further research is needed to assess the effectiveness and user understandability of the proposed maps.

Keywords: climate change; atlas; cartography; digital maps; 3D maps; sustainability

1. Introduction

1.1. Climate Change

Global surface temperature in the period 2011–2020 has risen by approximately 1.1 °C compared to the pre-industrial period of 1850–1900, driven primarily by greenhouse gas emissions from human activities, and has shown a notable acceleration since 1970, as was reported by the IPCC [1]. Climate change impacts natural systems and human societies, as evidenced, for example, by a rise in global mean sea level, an increased frequency of extreme events such as heatwaves and droughts, or adverse effects on food security. In the future, a continued increase in greenhouse gas emissions will drive further global warming. In accordance with IPCC projections based on different emission scenarios, compared to the pre-industrial period, global temperatures are expected to increase by 1.4 °C until the period 2081–2100 under an optimistic very low greenhouse gas emissions scenario and could rise as much as 4.4 °C under a scenario with very high emissions. Urgent action is required to limit the impacts of climate change.

Switzerland is experiencing significant impacts from climate change, including a 1.5 °C increase in average temperature between the periods 1864–1900 and 1981–2010, a 60% loss in glacier volume relative to the 1850s, a 50% reduction in snowfall days at elevations below 800 m since the 1970s, and an extension of the vegetation period by two to four weeks since the 1960s [2].

1.2. Maps and Climate Change

The communication of climate change presents several significant challenges. These include the extreme complexity of the topic itself, the invisibility of the primary cause of climate change—greenhouse gas emissions—the temporal and geographical distance of the consequences of climate change, and the difficulty in perceiving the benefits of taking action to mitigate climate change [3]. For a layperson, climate change can be particularly challenging to comprehend, making it essential to use simple and clear communication [3]. Visual communication plays an important role in this process [4], graphics are also used [5], and maps are considered a key communication tool [6,7]. Climate change has a geographical component, which makes this topic particularly well-suited to be represented on maps [6,8]. In the climate change discussion, maps are employed to enhance awareness among the general public, communicate research findings by the scientific community, and support policymakers in making decisions on climate action [7].

Various aspects of the use of maps in climate change communication have already been investigated. Given the importance of the topic and the need for effective communication, the quality of climate change maps has been assessed based, i.e., on a map from the 2007 IPCC report [7]. Considering the maps in media, the content of maps used in American media during the climate change debate was investigated by Fish, focusing on cartographic design decisions and the elements that make a vivid map [6,9]. She also highlighted the importance of making climate change more tangible through maps [6]. Fish and Kreitzberg [10] analyzed the role of maps in conservative media in the United States, with a focus on the misuse of maps. They give guidelines for map creation to reduce the risk of such misuse and demonstrate that conservative media utilize maps significantly less than other media. Tsang [11] examined the impact of cartographic representation on the perception of media bias in climate change communication: he compared the effects of animated versus static maps and single versus double maps. Becsi et al. [12] examine the design principles of climate change impact maps for policymakers in Austria, drawing on participatory studies. Their findings highlight the importance of avoiding an excessive amount of information in map design and of visualizing uncertainties, despite the fact that the audience may not always understand the information correctly. Continuing with the topic of uncertainties, Kaye et al. [13] provide an overview of various techniques for visualizing climate variables and their associated uncertainties on maps. They place particular emphasis on the use of color and provide general guidelines. Maps related to climate change can also be found on social media. Liu et al. [14] conducted a user study to investigate the role of maps in social media posts, demonstrating that the inclusion of maps enhances the credibility of the posts. Considering 3D visualizations more specifically, there are already examples of 3D landscape visualizations showing the effects of climate change [15]. One such example is a visualization that illustrates areas above and below an altitude with reliable snow coverage in the future and the ski slopes in the region of Entlebuch (Switzerland) [16]. Furthermore, other forms of climate change data visualization are available online. For example, in the IPCC's interactive data viewer [17], users can visualize past values of various climate variables and projected future values based on different models and scenarios, or the Climate Atlas of Canada [18], which employs an interactive map and multimedia elements to illustrate the impacts of climate change across Canada.

1.3. Atlases, Sustainable Development Goals, and Climate Change

The Atlas of Switzerland (AoS), the official national atlas of Switzerland, was initiated in 1961 by a resolution of the Federal Council. The Institute of Cartography and Geoinformation at ETH Zurich is responsible for the publication. The printed version of the atlas was released in 13 separate deliveries between 1965 and 1997, comprising more than 600 maps [19]. In 2000, the first version of a digital and interactive atlas was launched, offering 2D and 3D visualizations. Over the following years, new versions were developed. The current version is a desktop application [20] based on an interactive virtual 3D globe, allowing for the visualization of interactive 2D and 3D maps. Currently, the atlas includes over 400 maps covering various topics. The Atlas is intended for a broad audience, and the topics should also be easily understandable by non-experts.

The focus of national atlases, which was initially to highlight the strengths and splendor of a nation, has now shifted to also visualizing various socio-cultural phenomena [21] by focusing on the needs of humans rather than those of the country [19]. In the printed version of the Atlas of Switzerland, the subject of climate was represented with four pages and 18 visualizations. These visualizations focused solely on describing various characteristics of Switzerland's climate, with no connection to climate change or its impacts.

A parallel publication, the printed Climatological Atlas of Switzerland [22] (not directly related to the AoS), contains maps and diagrams describing the climate in Switzerland. This atlas explains the limited number of climate maps in the printed AoS. The Climatological Atlas presents factual data, with some comparisons (not in map form) of past climate conditions. However, it does not include any future projected values. This atlas was published between 1982 and 2000 [23].

Nowadays, with climate change becoming an increasingly important topic, maps addressing this issue are also featured in national atlases. For example, the National Atlas of Hungary [24], a print publication, includes visualizations related to climate change, such as projections of expected temperature and precipitation changes. These maps incorporate historical data and future projections, aiming to raise awareness of the impacts of climate change within the country.

In 2015, the United Nations established 17 Sustainable Development Goals (SDGs) to enhance global well-being by 2030, with goal 13 (*Climate Action*) specifically focused on addressing climate change [25]. As highlighted by Kent et al. [26], cartography can contribute to achieving these goals by using maps to raise awareness of the issues and effectively communicate spatial information. The important role of cartography in advancing the SDGs is also highlighted by a joint publication [8] by the International Cartography Association (ICA) and the United Nations , which outlines key considerations for the maps creation that support sustainable development goals.

1.4. Goal

The goal of this work is to present three maps developed for the Atlas of Switzerland, aiming to make the effects of climate change more comprehensible and visible. We visualized three different indicators of climate change (zero degree line, glacial lakes, and flowering dates) using various approaches, leveraging the strengths of the Atlas of Switzerland, such as temporal navigation, interactivity, 3D data visualizations, and map combinations.

The focus of this paper is mainly on the map of the zero degree line; for this map, the methodology, the results, and the discussion are particularly in-depth. The remaining two maps, glacial lakes and plant phenology, merely scratch the surface of the subject matter, offering only a few general considerations to illustrate alternative approaches. In the discussion, key aspects are presented that should be considered when working with maps that show the effects of climate change.

This article is a revised and expanded version of a paper entitled *Cartographic visualization of the effects of climate change: a practical application for the Atlas of Switzerland* [27], which was presented at the EuroCarto conference in Vienna in September 2024. The final maps can be found in the current version of the AoS [20].

2. Materials and Methods

2.1. Zero Degree Line Map

This map is based on two primary datasets: elevation data for Switzerland and data on the altitude of the zero degree line at different time stamps. The height data are represented by a Digital Height Model (DHM) provided by Swisstopo [28] with a spatial resolution of 25 m, whereas MeteoSwiss (the Swiss meteorological and climatological national service) provides zero degree line data. The zero degree line is an isotherm connecting points with a temperature of 0°C and can also be interpreted as the elevation above sea level where this temperature occurs [29]. The zero degree line data include values based on historical observations and future climate projections. Each value consists of an elevation (e.g., 1525 m above sea level) representing the altitude at which the average temperature is expected to be 0°C in winter (considering the months December, January, and February) or in summer (considering the months June, July, and August).

The historical values of the zero degree line provided for this work were calculated based on past temperature values derived from observations taken at measurement stations. Based on these values, 30-year average values of the zero degree line from 1864 to 2017, considering the entire of Switzerland, were computed [30] and provided for this work by MeteoSwiss.

The projected future values of the zero degree line are based on simulations considering the period from 1981 to 2099, for the details, see [30]. These projections have been calculated using climate models that simulate temperature changes under two distinct greenhouse gas emissions scenarios [30]. The *low emissions scenario* assumes that substantial mitigation efforts will be made to limit greenhouse gas emissions. In contrast, the *high emissions scenario* assumes that there will be continued increases in greenhouse gas emissions and no mitigation efforts [2].

For each scenario, a median value is calculated, representing the central estimate, as well as an uncertainty range defined by an upper and lower boundary [2,30]. The upper boundary is defined as the second highest value of the models' simulations, and the lower boundary as the second lowest value of the models' simulations [2,30]. Based on the projected temperature values, the elevation of the zero degree line is calculated as described in [30]. The resulting data are averages over 30-year periods across the entire country of Switzerland for the median value, the upper boundary, and the lower boundary of the elevation of the zero degree line. MeteoSwiss provided these data. The boundaries give a range within which the zero degree line may vary, reflecting model uncertainty in future temperature trends. For the sake of simplicity, the upper boundary will be referred to as the *maximum* throughout this work, and the lower boundary as the *minimum*. Table 1 presents an example of the data provided for this work.

The data processing consists of basics GIS operations conducted using Python (version 3.9.13), GDAL (version 3.0.2) [31] and Mapshaper (version 0.6.59) [32]. Vector files are generated for each period of interest, containing polygons representing areas below or above specified elevation boundaries.

The process begins with the DHM in raster format, whereby regions that meet the defined elevation constraints are vectorized into polygons. In the case of observed data, the value of the zero degree line serves as the elevation boundary, resulting in the generation of one vector file for each time period. In the case of projected data, three vector files are created for each time period. One with the minimum value as a height boundary, one with the median value as a boundary, and one with the maximum value as a boundary. The process is repeated independently for each period of interest, for summer and winter, and (for the projections) for each scenario. The results are vector files comprising two polygon classes: those below the elevation boundary and those above it. In order to enhance map readability, areas below a defined size threshold are assigned to the other class to eliminate regions too small for effective visualization. Furthermore, Mapshaper [32] is applied to simplify the polygons. This tool enables the simplification of geometries by preventing polygons from disappearing. The Visvalingam algorithm [32,33] is used and 30% of the removable vertices are retained. See steps a, b and c in Figure 1.

Table 1. The table presents the average altitude of the zero degree line across Switzerland, calculated as a 30-year average for both summer and winter. The data presented here are derived from observations (1864–2017) and simulations (1981–2099). The simulations are based on two climate scenarios: one with low greenhouse gas emissions and another with high greenhouse gas emissions. The table also includes the range of uncertainties for each scenario, showing an upper and a lower boundary for the estimated values. The table presents data for selected periods as an example.

Time Period	Saison	Type of Data	Height (Observed)	Height Min. (Projected)	Height Median (Projected)	Height Max. (Projected)
1864–1893	Winter	Observations	438 m	-	-	-
1864–1893	Summer	Observations	3253 m	-	-	-
...	...	...	...	...	...	...
2070–2099	Winter	Scenario Low Emissions	-	989 m	1073 m	1208 m
2070–2099	Winter	Scenario High Emissions	-	1523 m	1734 m	1891 m
2070–2099	Summer	Scenario Low Emissions	-	3865 m	4044 m	4325 m
2070–2099	Summer	Scenario High Emissions	-	4464 m	4892 m	5628 m

For the projections, the individual vector files are aggregated into a single vector file containing three distinct areas: regions that fall below the minimum height value, regions that fall between the minimum and maximum height values, and regions that fall above the maximum value. This aggregation is conducted separately for each period, for both scenarios, and for the winter and summer datasets. See steps d and e in Figure 1.

To create specific visualizations (the so-called *Comparisons*), the individual vector files of the projections with the median height as the height boundary and considering the two different scenarios within the same period are combined in a single file. This approach is applied for the periods 2040–2069 and 2070–2099, separately for winter and summer. See steps d and e in Figure 1.

Once the steps mentioned above have been completed, vector files comprising polygons that represent areas situated above or below the specified elevation boundaries are prepared for visualization in the Atlas of Switzerland. This is done using the APS-Editor plugin [34], a Python QGIS plugin that serves as an editorial system. It enables the conversion of common geodata files into formats compatible with the AoS, as well as the definition of map and layer structures, associated styles, and additional parameters such as tiling settings for each layer. The plugin was developed specifically for the AoS and is not publicly available. See steps f in Figure 1.

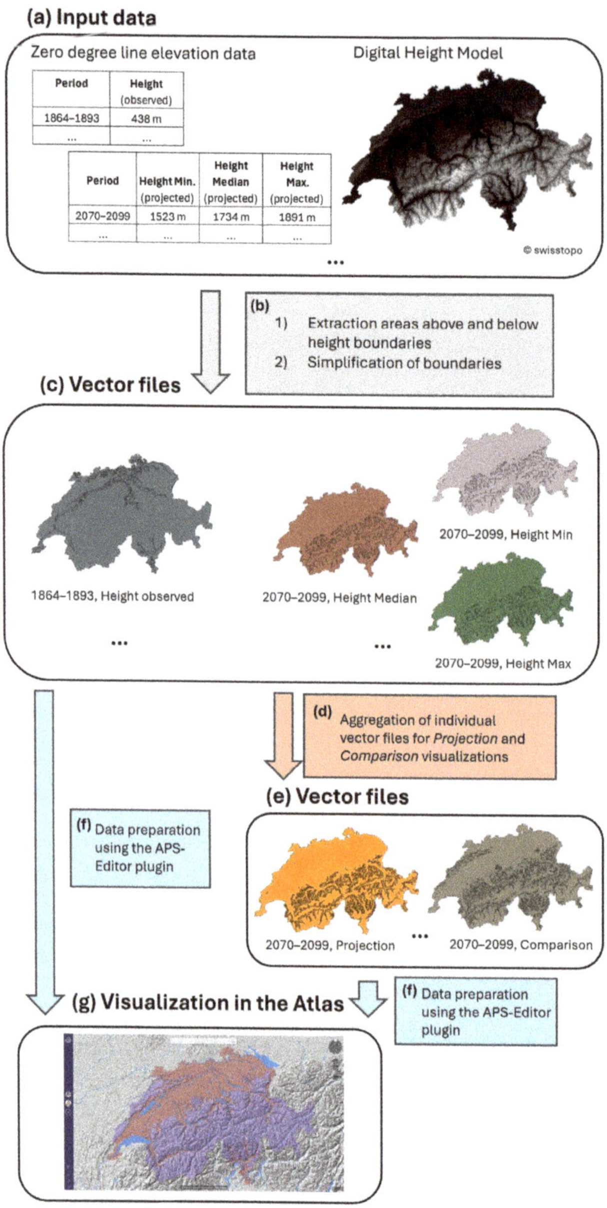

Figure 1. Workflow for the zero degree line map preparation. (**a**) Starting with zero degree line elevation data and a digital height model as input data. (**b**) Regions meeting the defined elevation constraints are vectorized into polygons, and their boundaries are simplified. (**c**) This results in vector files with two polygon classes: areas below and above the elevation boundary. (**d**) For projection and comparison visualizations, individual vector files are aggregated—considering minimum and maximum heights for projections and median heights for comparisons. (**e**) This process generates two types of vector files. (**f**) The vector files are then prepared for visualization using the APS-Editor plugin. (**g**) Finally, they are displayed in the Atlas of Switzerland. The process is repeated for different time periods, two scenarios, and both summer and winter data.

The map is composed of 28 layers. For observed data from the past, the layers display areas below and above the zero degree line for both winter and summer across the following periods: 1864–1893, 1885–1914, 1910–1939, 1935–1964, 1960–1989, and 1985–2014 (defined as *Observations*). For projected future values, the layers illustrate areas below the minimal projected height, areas between the minimal and maximal projected heights (which is an uncertainty range), and areas above the maximal projected height for both summer and winter. The periods 2010–2039, 2040–2069, and 2070–2099 are covered, and the two distinct scenarios are considered separately (these data are defined as *Projections*).

For the comparison of future consequences of the two different scenarios, the layer illustrates areas below the median value based on a high emission scenario, areas between this median value and the median value for a low emission scenario, and areas above the latter value. These comparisons are made for the periods 2040–2069 and 2070–2099, considering both summer and winter (these data are defined as *Comparisons*).

The layers are organized into six thematic groups: observations, projections, and comparisons for both summer and winter. The areas situated above the zero degree line, which indicate temperatures below 0 °C, are colored in a bluish color, which is associated with colder temperatures. Conversely, the areas situated below the zero degree line, where temperatures are above 0 °C, are colored reddish, which corresponds to warmer temperatures. More neutral yellows are chosen for other aspects, like the areas showing uncertainties. An explanatory text accompanies the map, facilitating comprehension of the topic and map structure.

The temporal periods were selected to provide sufficient data to observe trends. In addition, the periods chosen should ideally cover the range from the beginning of the measurements to the last years of projected values, and the periods should be approximately equally distributed. Six periods are chosen for the observations. Three periods are selected for the projections, and two of the future periods are used for comparison. The period 2010–2039 is excluded from the comparison because, in this period, the median zero degree line elevation is higher in the high emissions scenario than in the low emissions scenario.

2.2. Plant Phenology Map

Phenology deals with recurring stages in the life cycles of plants and animals, influenced by environmental factors [35,36]. Plant phenology focuses on tracking specific growth stages of plants and documenting when these occur each year [37]. Phenology is affected by climate change [38]. For example, Meier et al. [39] indicated an advance in the date of the leaf unfolding for trees in Switzerland quantifiable in up to 3 days earlier per decade since 1985.

MeteoSwiss also provides data [40] for the map about plant phenology. These data consist of information about the measurement stations (i.e., altitude and coordinates) and phenological observations, namely the day of the year (DOY) when a phenological phase (i.e., the start of flowering) begins for different plant species. The observations cover the period 1951–2023. The following approach is applied to prepare the data for visualization: The dandelion is chosen because it occurs at all altitudes. The phenological observation used is the DOY for flowering, defined as the time when 50% of the flowers have bloomed. The selection of measurement stations is based on the following criteria: stations must have a sufficient number of observations for the periods 1951–1970 and 2004–2023, they should be well distributed over Switzerland, and stations that are too close to each other should be avoided, as this could create issues with the symbolization. The mean DOY for flowering in the two periods is computed for each selected station. The resulting dataset is prepared using the APS-Editor. We want to visualize changes in flowering dates between the two time periods at the same locations, working with a timeline representation.

A custom-designed timeline was created since the APS-Editor has limited visualization capabilities and does not provide a default timeline option. This custom symbol resulted in a 3D representation of a timeline with the mean DOY for the two periods marked. On the map, the timelines are placed above the location of the measurement stations. The same procedure is also applied to the flower field daisy. The resulting map consists of two layers, allowing the user to choose whether to view the DOYs of the dandelion or the field daisy.

2.3. Glacial Lakes Map

Glaciers are also strongly affected by climate change. Glacial lakes, which are formed by melting glaciers, are used as an indicator of climate change. Agarwal et al. [41] reported an increase in the number and area of glacial lakes in the Himalayas between 1975 and 2017, while Zhang et al. [42] documented this trend on a global scale between 1990 and 2020. To illustrate the changes in Switzerland, the data used for the glacial lake map are based on the work of Mölg et al. [43] and Steffen et al. [44]. The first study provides data on the past development of glacial lakes at six different points in time (1900, 1946, 1973, 1979–1985, 2006, and 2016), while the second study provides data on possible future glacial lakes, which may appear if all glaciers in the Swiss Alps completely melt. For the visualization on the map, the shapes of the lakes are displayed on the terrain model. As many lakes are small and can be difficult to see from a distance, a 3D sphere is added above each lake to improve visibility, in addition to the shape on the terrain. The map consists of seven layers: six representing past periods and one showing the future situation. The user can select the desired layer using the temporal slider bar. To facilitate better comparison and understanding of the lake dimensions, the lakes are divided into five categories. The spheres are colored and scaled according to their category. A yellow-blue color palette is used for past observations, while an orange-red color palette is used for future data. In addition, the glaciers that existed in the first decade of the 2000s are also visualized on the map.

3. Results

3.1. Zero Degree Map

3.1.1. General Overview

Figure 2 shows one layer of the resulting map, presented as a view from above (2D map) of the whole of Switzerland. The map divides the country into two types of polygons: regions below the altitude of the zero degree line (colored red) and regions above the zero degree line (colored blue). The transition in color marks the altitude corresponding to the zero degree line. The displayed data are based on observations from winter months during the period 1985–2014. The map clearly shows that the Swiss plateau (Mitteland), valley floors in the Alps (e.g., in Valais and Grisons), and large parts of Ticino are exposed to a winter temperature above 0°C, in contrast to alpine regions with a temperature below 0°C. This visualization does not rely on 3D symbols but uses the ability of the AoS to display the terrain in a 3D perspective when the globe is tilted. In addition to the map itself, Figure 2 shows the graphical user interface elements of the AoS. The digital map is interactive, enabling users to explore it in various ways, such as navigating, tilting, or zooming using the keyboard, mouse, or on-screen buttons. Users can select other data for visualization variants, such as projected future values, via a dropdown menu or change the time period by interacting with the temporal selection bar. Base maps and their elements can also be customized interactively. Explanatory text, accompanied by images, is available in the upper-left part of the interface. In addition, clicking on specific points on the map opens a pop-up window with additional information, such as the value of the altitude of the zero degree line. The interactive map allows users to freely explore the content. There are 28 different layers available for exploration. While the map in Figure 2 provides a static snapshot of the situation at a particular time period, one of the key strengths of the AoS is its ability to incorporate temporal dimensions.

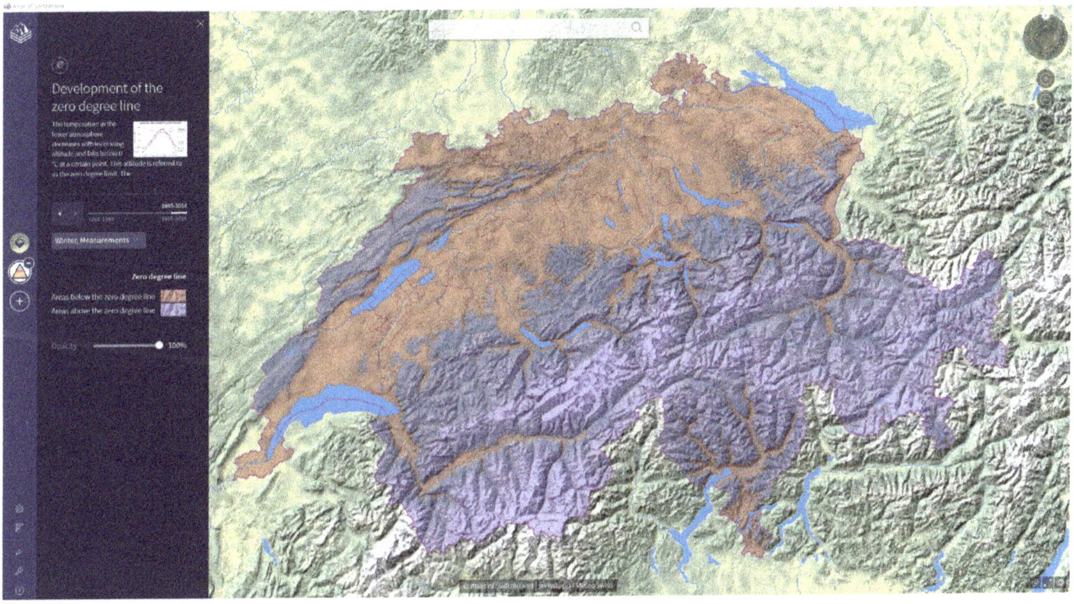

Figure 2. Visualization of observed winter data for the period 1985–2014: red areas lie below the zero degree line (temperatures above 0 °C), blue regions lie above the zero degree line (temperatures below 0 °C). The figure also showcases additional elements from the Atlas of Switzerland, i.e., navigation buttons, legend, or accompanying text.

3.1.2. Temporal Control

Figure 3 illustrates the zero degree line in winter, based on observations, across six distinct time periods in the region, including Zurich, Zug, and Einsiedeln. Blue areas represent regions above the zero degree line, red areas denote regions below it, and lakes and rivers are also included for spatial context. City labels are manually added for better orientation. This visualization enables a clear understanding of how the altitude of the zero degree line has changed over time. The figure highlights a clear trend from 1864–1893 to 1985–2014: the altitude of the zero degree line has risen. Consequently, regions situated below the zero degree line (red areas) have expanded, while areas situated above the zero degree line (blue areas) have diminished. Notably, the zero degree line shows no change between the periods 1864–1893 and 1885–1914. Furthermore, for the period 1935–1964, a temporary drop of the zero degree line altitude is observed. However, the upward trend in the zero degree line is evident when examining the entire timeline. This type of visualization is particularly effective for comparing the altitude of the zero degree line across different time periods, providing insights into climatic changes in the region.

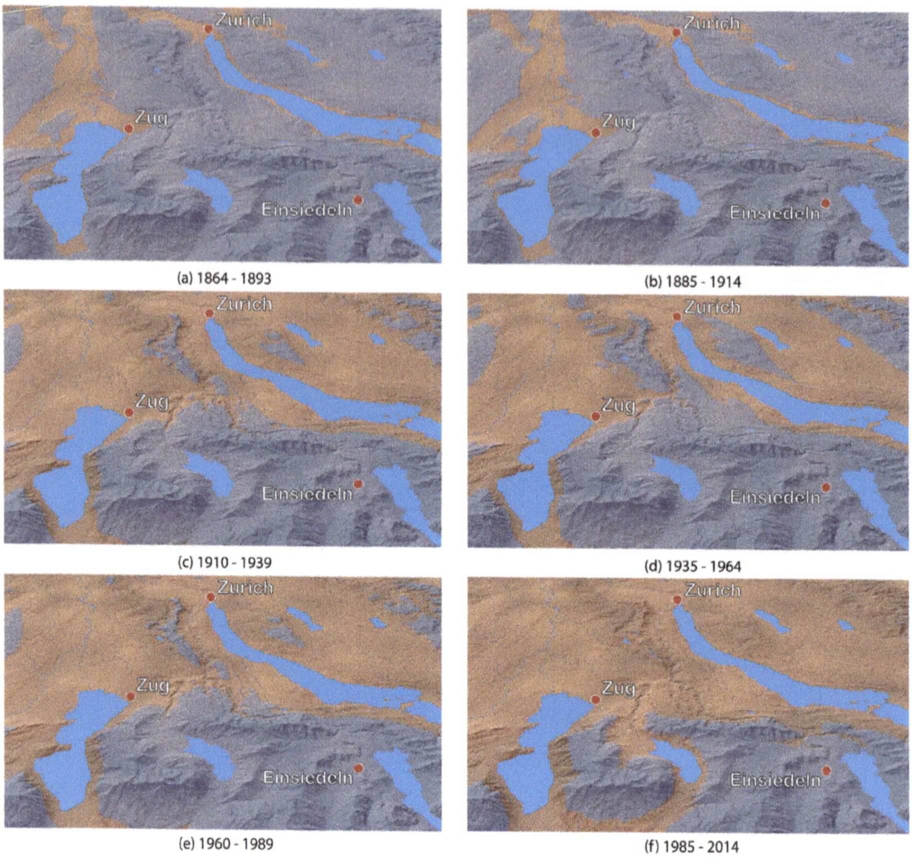

Figure 3. Details of the map showing the zero degree line in winter based on observations. The blue areas represent regions situated above the zero degree line, while the red areas denote regions situated below. The figure illustrates the situation at six different periods: (**a**) 1864–1893, (**b**) 1885–1914, (**c**) 1910–1939, (**d**) 1935–1964, (**e**) 1960–1989, and (**f**) 1985–2014, revealing a clear trend of increasing elevation of the zero degree line from the late 19th century to the early 21st century.

3.1.3. Future Scenario and Uncertainties

Figure 4 illustrates the projected future altitudes of the zero degree line in winter under a high emissions scenario across three periods: 2010–2039, 2040–2069, and 2070–2099. The red areas indicate regions that were consistently simulated to be below the zero degree line by the majority of the models. The boundary of this area is defined by the minimum value of the elevation of the zero degree line projections. The blue areas represent regions that were simulated to be above the zero degree line by the majority of the models. The boundary of this area is defined by the maximum value of the elevation of the zero degree line projections. The yellow areas indicate areas of uncertainty. The exact altitude of the future zero degree line remains uncertain. Based on the model projections, this line is expected to fall within the range defined by the maximum and minimum values of the projections, with the yellow areas on the map highlighting this uncertainty. These zones represent the regions where the future zero degree line is most likely to occur, reflecting the range of variability across the different model projections (see [2]). When the three time periods are considered, it is evident that the red areas (below the zero degree line) are increasing in extent, while the blue areas (above the zero degree line) are decreasing. This reflects an overall climb of the altitude of the zero degree line. Furthermore, a comparison of the first period (2010–2039) and the last period (2070–2099) periods reveals that the uncertainty range (yellow areas) increases over time, indicating a growing variability in future projections. Analogous visualizations have been produced for a low-emission scenario, although these are not displayed in the results presented here.

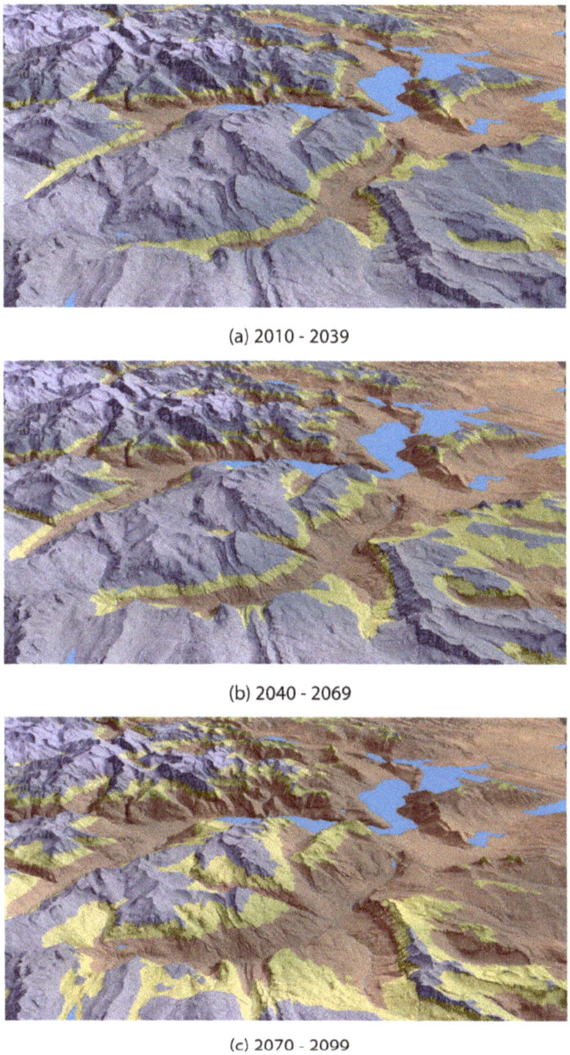

Figure 4. Projected future values of the zero degree line in winter considered in the context of a high-emission scenario for three time periods: (**a**) 2010–2039, (**b**) 2040–2069, and (**c**) 2070–2099. The red areas represent regions that, according to the climate models, will be below the zero degree line, whereas the blue areas denote regions above it. The yellow areas indicate the range of uncertainty, highlighting where the zero degree line is expected to occur in the future.

3.1.4. Comparison of the Effects of the Two Emission Scenarios

It can be challenging to compare the effects of the two different emissions scenarios on the zero degree line when the information is presented across separate layers [12]. To facilitate an easier comparison, additional layers have been created that combine the results of both scenarios into a single layer. Figure 5 illustrates an example of this combined result. It shows a comparison of the altitude of the future zero degree line during the 2070–2099 period for the winter months, based on projections considering two distinct scenarios. The red areas represent regions situated below the median altitude of the zero degree line under the scenario with climate change mitigation efforts (low emissions). In contrast, the blue areas indicate regions situated above the median altitude of the zero degree line under the scenario with high emissions. The gray areas indicate regions that may be preserved through climate change mitigation efforts. In the absence of sufficient climate protection measures, the high emissions scenario will be reached, and these gray areas would belong to the red areas, indicating they would fall below the zero degree line, with temperatures above 0°C. Conversely, if emissions are reduced and the low emissions scenario is achieved, these gray areas would lie above the zero degree line, with temperatures below 0°C, and will belong to the blue areas. This visualization demonstrates how the altitude of the zero degree line differs between the two scenarios and, importantly, which areas could be preserved through climate change mitigation efforts. It effectively highlights the positive impact of climate protection measures, as limiting greenhouse gas emissions would also limit the surge of the altitude of the zero degree line, as visible on the map. It is important to note that this visualization shows only the median value of the zero degree line, excluding the range of uncertainties.

Figure 5. A comparison of the altitude of the zero degree line for the time period 2070–2099 is presented, combining the projections considering two different scenarios in a single layer. The red areas represent regions situated below the median altitude of the zero degree line under the low emissions scenario, while the blue areas indicate regions situated above the median altitude under the high emissions scenario. The gray areas indicate regions that could potentially be preserved through climate change mitigation efforts.

3.1.5. Combination with Other Topics

Figure 6 illustrates one of the strengths of the AoS: the possibility to combine different maps. In this case, the map of the zero degree line is combined with the map of the ski resorts and ski slopes. The resulting visualization shows the projected values of the height of the zero degree line in winter, considering a high emissions scenario for 2070–2090. Red areas represent regions below the zero degree line, blue areas represent regions above it, and yellow areas indicate uncertain regions where it is unclear whether they will be above or below the zero degree line. Additionally, cable cars are represented by yellow lines, and colored lines show the tracks of the different ski slopes. The resulting visualization reveals that a large portion of the ski resort will be situated below the zero degree line in the future, which could present considerable challenges for maintaining adequate snow coverage for skiing. This type of visualization effectively links an indicator of climate change, the altitude of the zero degree line, with another topic, in this case, skiing, by clearly illustrating the potential impacts of climate change on this activity. While uncertainty remains regarding the future height of the zero degree line (shown in yellow), it is evident that a substantial part of the ski resort will be below the zero degree line. This raises concerns about the possibility economic and ecologic viability of skiing in these areas.

Figure 6. A combination of two maps: the map showing the zero degree line and the map showing the ski resorts. For the zero degree line, the projected values during the winter months in the period 2070–2090 are shown considering a scenario with high gas emissions. Red areas represent regions below the zero degree line, blue areas represent regions above it, and yellow areas indicate uncertain regions where the exact classification (above or below the zero degree line) is unclear. The yellow lines indicate the locations of cable cars, while the colored lines show the different ski slopes.

3.2. Plant Phenology

Figure 7 shows a detail of the plant phenology map showing data for dandelion. A 3D timeline is shown above the location of each station. To aid interpretation, the timeline is divided into six parts, each representing 15 days. The timeline starts on 17 March and ends on 14 June. The coloration of the bars visually indicates temporal changes. Two markers are placed on each timeline: a blue marker indicates the mean DOY for 1951–1970, while a pink marker represents the mean DOY for 2004–2023. This allows for an immediate comparison of how the mean DOY has shifted between the two periods. Considering the general trend, it is possible to see that the DOY has shifted earlier, even though there are some exceptions in the data.

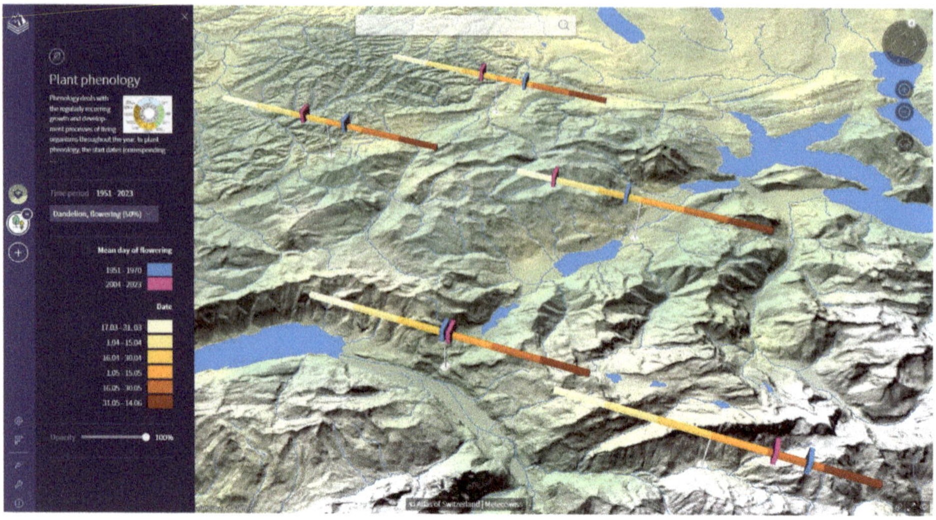

Figure 7. The figure displays a detail of the plant phenology map. A 3D timeline from mid-March to mid-June is shown above each measurement station. The mean flowering DOY for the dandelion for 1951–1970 is marked in blue, while the mean flowering day for 2004–2023 is marked in pink.

3.3. Glacial Lakes

Figure 8 displays the map of the glacial lakes, showing the glacial lakes in 1973 and the glacier extent in the first decade of the 2000s. The shape of the glacial lakes is visible on the terrain, colored in dark blue. The lakes are subdivided into five categories based on surface area. A sphere is displayed above each lake, varying in color and size according to the category. The user interface also includes a temporal bar that allows the user to select different time periods to visualize the data.

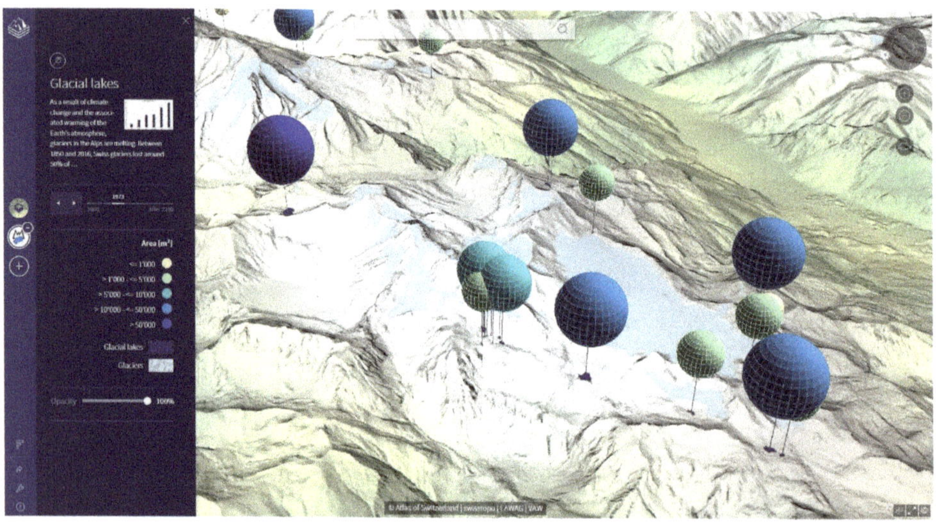

Figure 8. The figure displays the map of glacial lakes, showing the glacial lakes in 1973 and the glacier extent in the first decade of the 2000s. The shape of the glacial lakes is visible, colored dark blue. Above each lake, a sphere is visualized, with its color and size corresponding to the lake's category.

4. Discussion

This section discusses important aspects to consider when dealing with maps that show the effects of climate change. The schema in Figure 9 summarizes the key aspects of the discussion in three categories: the first covers map design choices, addressing the question of how to visualize the topic, the second focuses on the map content, including what should be shown on the maps, and the third category discusses how the maps can be used, interpreted, and perceived. For each key point, the corresponding subsection is indicated.

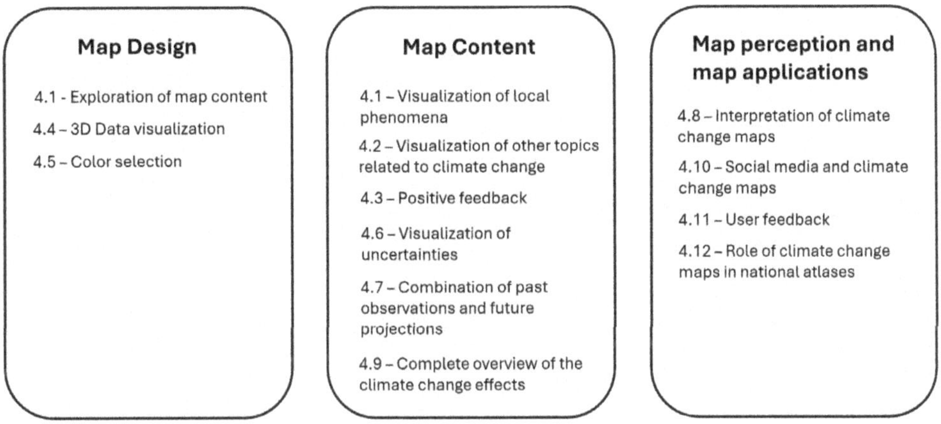

Figure 9. Schema summarizing the key points to consider when dealing with maps that show the effects of climate change: individual points are presented as subsections of the discussion.

4.1. Exploration of Map Content and Visualization of Local Phenomena

As illustrated in Figure 2, users can interact with the zero degree line map in various ways, exploring its content thematically, spatially, and temporally. They can select specific themes, navigate in the map, or choose a period of interest. Regarding the maps about the glacial lakes and plant phenology two of the three exploration options are available. Providing users with these navigation options gives them the means to gain a comprehensive overview of the topic.

Climate change is inherently tied to the spatial [6,45] and temporal dimensions [45]. Showing changes over time is particularly important in climate change maps [9].

Furthermore, allowing users to explore regions they are familiar with, through spatial navigation and by observing a phenomenon that can directly affect these areas can also help raise awareness of the issue [4,46]. Indeed, previous research about climate change communication has highlighted the importance of not communicating the topic too distantly. For example, Sheppard [4] affirms the importance of localizing the topic at the local community level to make it more concrete and better understandable. Based on a user study, Johannsen [46] also confirmed the importance of presenting local and familiar phenomena—those that can be directly observed in the user's region—rather than distant or remote events to effectively raise concerns about climate change.

4.2. Visualization of Other Topics Related to Climate Change

Lorenzoni et al. [45] define engagement with climate change as encompassing three essential aspects: cognitive, affective and behavioral. It means that the individual must be concerned about the topic, must feel motivated and must be able to act [45].

It is assumed that combining the maps about climate change with other maps of concern to the user available in the AoS could help to increase the motivational aspect of user engagement [45]. Figure 6 shows a combined visualization of the zero degree and ski resort maps. Linking the impacts of climate change to a topic that users may feel more emotionally connected to or interested in—such as skiing—can help raise awareness of the issue and foster greater user engagement [15,45].

This assumption is supported by the example of a successful application of this type of visualization described in the work of Schroth et al. [16]. In a participatory planning process in the Entlebuch region, a 3D landscape visualization, including mountains and ski slopes, was used to depict areas expected to have good snow conditions in 50 years. The resulting visualization was similar to our result in Figure 6. Most participants in the planning process were highly impressed by the visualization. Therefore, we believe that having such combined visualizations in the AoS would impact the users [15].

Sheppard [4] also illustrated the importance of making the maps personal and connecting the issue of climate change to individuals, locations and context.

4.3. Positive Feedback

As stated by Chess and Johnson [47], it is important in climate change communication to highlight the effects of climate change mitigation efforts. Moser [3] also notes that one of the challenges in climate change communication is the difficulty people have in perceiving the positive effects of mitigation efforts on the climate.

Our approach of comparing the altitude of the zero degree line under two different scenarios on the same layer (see Figure 5) addresses this limitation. By presenting the regions that are expected to remain below the zero degree line under a low-emission scenario and contrasting them with the much larger regions that will fall below the zero degree line under a high-emission scenario, we highlight the potential impact of climate change mitigation efforts. This approach emphasizes the areas that can be preserved through effective climate action.

4.4. 3D Data Visualization

The important role of 3D visualizations in the context of climate change communication is discussed by Sheppard [4]. In another work, he also investigated the use of landscape visualizations for improving climate change communication, highlighting both their potential and their limitations or risks [15]. Schroth et al. [48] conducted a small user test to investigate the effects of visualizing the local impacts of climate change using an interactive 3D game environment. Their findings indicate that this approach has a positive effect on climate change communication.

The elevation of the zero degree line is inherently linked to altitude. The AoS, based on a globe and digital terrain models, allows users to have a 2D view from above for an overview (see Figure 2) or to tilt the view to reveal 3D topography (Figures 3 and 4). Also based on the literature mentioned above, it can be supposed that this AoS feature is particularly valuable for visualizing the zero degree line in Switzerland, as the 3D terrain could allow for a more effective visualization of the phenomenon, especially in a mountainous country like Switzerland.

Additionally, the map of the plant phenology utilizes three-dimensional symbols to illustrate the timeline, while the map of the glacial lakes also incorporates three-dimensional symbols to enhance the visibility of the lakes' locations and dimensions.

However, it is important always to consider the limitations of 3D visualization. Compared to 2D maps, 3D maps present additional challenges, including the difficulty of comparing objects at different distances due to variations of the scale across a 3D scene or the risk of some objects being occluded and, consequently, not visible from the user's viewpoint [49]. Navigation in 3D visualization can also create problems for users, i.e., when the mouse is used both for navigation and additional tasks such as object selection [50]. According to Niedomysl et al. [51], the advantages and disadvantages of 2D and 3D maps depend on their purpose. Therefore, it is important to always consider whether the use of 3D could provide an advantage in the communication process [52]. 3D is not always the most effective choice for data visualization [49], for instance, a study by Niedomysl et al. [51] in another context has shown that a 3D map is not always more helpful or effective in improving learning outcomes than a 2D map. However, 3D visualizations are better suited for creating a more immersive experience in a scene [52]. Additionally, since 3D maps are less common than traditional 2D maps, they still have a fascinating effect on users [53]. These factors can serve as further motivations for using 3D visualizations to illustrate the effects of climate change.

4.5. Color Selection

To facilitate the interpretation of the zero degree line map, colors commonly associated with temperature values were employed on the map of the zero degree line. Areas with temperatures below 0°C (above the zero degree line) were represented by a bluish tone, while a reddish tone was used to indicate areas with temperatures above 0°C (below the zero degree line). This decision is consistent with Fish's considerations [9] regarding the use of color in vivid maps. She emphasizes that colors should, among other factors, be associated with common experiences. In addition, yellow is used for uncertain areas, while a gray tone is used for the areas that could be preserved. Green was specifically avoided for these areas because the map already has a reddish color. This decision was made to prevent difficulties for individuals with color vision deficiencies, as the combination of red and green can be problematic.

4.6. Visualization of Uncertainties

For some visualizations, such as those shown in Figures 3 and 6, an uncertainty range is visualized in relation to the projected future values. This range is shown as a yellow-shaded area, representing the zone between the maximum and the minimum projected value of the zero degree line. Aitken et al. [54] investigated different approaches to visualizing uncertainty in projected climate change-induced flood hazards in New Zealand, including a the use of a fog layer in a virtual reality environments with 3D visualizations. Becsi et al. [12], emphasized the importance of showing uncertainties when communicating climate change through maps; however, the audience does not always correctly understand such information. For the visualization in the map about the zero degree line, we worked with a simple visualization of the uncertainty. For the other layers, i.e., the layer enabling the comparison of the effects of the two scenarios (see Figure 5), and for the other two maps, we did not work with the uncertainties. However, showing uncertainties increases the reliability of the map [12], so it would be important to take uncertainties into account when possible.

4.7. Combination of Past Observations and Future Projections

As shown in the results, the maps about the zero degree line and glacial lakes include two types of data: values based on past observations and values based on future projections. It is useful to visualize both, providing users with comprehensive information about past conditions and possible future outcomes, when possible, based on different scenarios [4,9]. This approach enables users to understand how the situation has evolved and what might happen in the future. Therefore, we emphasize the importance of not limiting the analysis to historical data alone, but of incorporating future projections whenever possible to provide a more complete and insightful perspective. The importance of visualizing changes over time and the consideration of using projections for future values are thematized in Fish's work [9].

4.8. Interpretation of Climate Change Maps

When interpreting the maps, it is important to remember that the visualized data should be regarded as approximations. Observations inherently have uncertainties; the methods used to combine past observations introduce additional uncertainties, and projections of future values carry uncertainties. Even the vector data preparation process can introduce some additional approximations. In some cases (see Figures 4 and 6), the uncertainty of the projected zero degree line is explicitly shown on the map. However, no uncertainty is shown for the other layers and maps.

Considering the map about the zero degree line, the values presented are averages calculated over a period of three months (either summer or winter) across 30 years. Furthermore, it is important to note that Switzerland has diverse regions, but the data used in this work represent nationwide averages. As a consequence, the regional variations, such as those highlighted by Scherrer et al. [29], are not accounted for in the visualization. For these reasons, it is essential to consider the visualized data on the final map as approximations. It is not correct to extract specific values at precise locations; instead, general trends should be considered. Also if the user can zoom to the extent that their own residence is visible (a concept frequently referred to as "endless zoom" [7]), the visualized data always represents an average value. The level of detail of community proposed by Sheppard [4] is, at this time, not feasible within our own visualizations.

Similar considerations apply when analyzing maps of plant phenology and glacial lakes, as these also have limitations. For example, the glacial lake map displays all lakes that could potentially form in the future, even though not all of them will appear simultaneously, and almost half will disappear [44]. Even for these maps, it is particularly important to focus on the general trends when interpreting them.

As stated, for example, by McKendry and Machlis [7], the potential of misuse exists when dealing with maps. Therefore, it is of particular importance to inform the user also about the map's limitations. Indeed, transparency is important in communicating climate change [4]. In the case of the AoS, the limitations are provided in the accompanying text of the maps. Furthermore, when preparing these types of visualizations, it is important to consider the risks and ethical points raised by Sheppard [15], despite some of them being more specifically applicable to landscape visualizations. The objective is to convey the information as correctly as possible [4].

However, also considering the above-mentioned limitations, we think that these types of climate change maps can still contribute to better understanding the problematic and the negative impacts of climate change.

4.9. Complete Overview of the Climate Change Effects

As illustrated in Figure 3, an examination of the general trends reveals a discernible increase in the altitude of the zero degree line over approximately 150 years. That shows the importance of considering the general trend when interpreting climate data. A closer analysis of the maps reveals that there were no observable changes in the altitude of the zero degree line between the periods 1864–1893 and 1885–1914. Furthermore, a decrease in the altitude of the zero degree line is evident during the period 1935–1964. These observations might seem counterintuitive but reflect the complex nature of climate patterns, which are not always linear and often exhibit oscillations. However, the overall trend across the entire measurement period remains clear. Additionally, it is important to present results even when they do not align with expected outcomes [4]. Those who deny the existence of climate change frequently concentrate on specific years with anomalies [10], such as unusually low temperatures, to dispute broader climate change trends. By providing a comprehensive overview and a complete picture, we aim to avoid this problem.

Similar considerations can also be made for the map of plant phenology. In this map, for some stations, the DOY of dandelion flowering during the period 1950–1970 is earlier than in the period 2004–2023, which is not what it is expected because of climate change [38]. However, when considering the overall trend across all stations, the DOY of flowering in the 2004–2023 period is, in general, earlier than in the 1951–1970 period. This is in line with expectations based on the effects of climate change [38].

4.10. Social Media and Climate Change Maps

The map of the zero degree line was employed in external contexts, namely in a blog article [55] and an Instagram post [56]. The latter comprised eight maps of Switzerland, showcasing regions situated above and below the zero degree line across past time periods and projections for a future time period under two emission scenarios, accompanied by brief explanatory text. By scrolling, users could follow the story of the zero degree line's evolution. The map in the post exhibited certain limitations compared to the version in the AoS, and a few adjustments were necessary. The post was successful, reaching over 45,000 accounts and generating more than 2100 interactions [56]. This result suggests that these types of maps engage the public and can be effectively used outside of a national atlas to reach a broader audience, which might be more challenging to engage through traditional atlas formats alone. This experience highlights the important role of maps on social media, a point also observed in the work of Liu et al. [14].

4.11. User Feedback

A user study was not conducted with the three maps, but a user survey would help assess whether these visualization types are comprehensible and beneficial for the broader public of the AoS, and if they meet their intended objectives. This is an aspect that should be considered in future work. The creation of effective maps about climate change requires the active involvement of users in a co-design process and the implementation of empirical testing, as emphasized i.e., by Lorenz et al. [57] and Becsi et al. [12]. For the map about the zero degree line, the only feedback we received was in the form of comments on Instagram, where users could comment on the post. It is important to note that, on Instagram, users do not have access to all the information provided in the full map available in the AoS. Furthermore, it is likely that users spent less time engaging with the social media post than they would have done with the map in the AoS. Nevertheless, we believe that the comments can still provide a preliminary idea of potential issues. One critical point that emerged from the comments was the use of the terms *areas below the zero degree line* and *areas above the zero degree line*, which some users found confusing. The first term refers to areas with temperatures above 0°C, while the second refers to areas below 0°C. To address this issue in future map versions, it might be more effective to directly label areas using the temperature. This approach could make the visualization more straightforward to understand, particularly for audiences lacking familiarity with concepts such as the zero degree line.

4.12. Role of Climate Change Maps in National Atlases

National atlases were initially developed as tools to showcase and describe a country's strengths, however over time, their focus shifted towards visualizing more socio-economic indicators [21]. The digital version of the AoS now contains six maps that are directly related to climate change. We believe that maps addressing climate change should be an integral part of a national atlas, given the importance and timeliness of this issue and its concrete, local impacts. It is important that the public be made aware of these issues and encouraged to develop an informed opinion [6]. Furthermore, as future projections become increasingly accessible and the quality of such data improves, it is important to integrate this type of information into a national atlas. Maps should also illustrate future conditions, rather than limiting their scope to describing the current or historical situation.

5. Conclusions

This paper presents three maps for a national atlas, which illustrate the impacts of climate change. These maps have been developed for an interactive atlas that supports 3D visualizations. The focus is on Switzerland, and the three key climate change indicators that are shown are the rising elevation of the zero degree line, the evolution of glacial lakes in the Alps, and changes in the flowering dates of plants.

The main goal of these maps is to make the effects of climate change more visible and concrete. This paper demonstrates that it is feasible to create these types of visualizations for a national atlas. Three distinct visualization approaches were presented: a surface visualization for the map depicting changes in the zero degree line, a 3D timeline symbolization for the plant phenology map, and a combination of surface visualization and 3D symbols for the glacial lakes map. All these visualizations leverage the capabilities of 3D representation within the AoS. Our maps use several key features to visualize the effects of climate change. These include the use of 3D visualizations, the possibility for the user to explore the map content both temporally and spatially, the focus on regions familiar to the user, the emphasis on the positive impacts of climate change mitigation, the integration of past observations and future projections, the consideration of uncertainties and the possibility to combine the maps with other topics of personal relevance and interest to the user. We apply these important principles to our maps for effective visualization of the effects of climate change in an understandable way. We recognize that not all these features can be incorporated into every type of map or topic. However, we encourage cartographers to consider integrating at least some of these elements when designing maps related to climate change.

When interpreting these maps, it is important to consider their limitations. Climate change data are derived from observations and projections based on simulations that include uncertainties. Therefore, the visualized values should be regarded as approximations, and it is not appropriate to attempt to extract a single, precise value for a specific location. However, despite these limitations, we believe these maps can be helpful in visualizing the effects of climate change and better understanding its impacts.

We think that these map types should be part of a national atlas because of the topic's relevance and importance, with its broad implications on society, economy, biodiversity, and environment, from which no one is excluded.

With these maps, we are making a small contribution to achieving SDG 13, on climate action, and we encourage other cartographers to work on visualizing the effects of climate change in order to raise awareness and clearly illustrate its impact.

The map of the zero degree line was successfully employed outside the atlas in an Instagram post. This suggests that this type of visualization can be effective in social media and beyond the atlas. However, not all maps are equally suited to this purpose, and further investigation is needed.

In future work, it will be important to test through user surveys whether the visualizations in the AoS are truly effective and understandable, and to further improve our maps considering the guidelines that can be found in the literature. In addition, other types of visualization, such as storytelling approaches, may be considered to better connect the themes and provide more impactful information to the user.

Author Contributions: Conceptualization, Luca Gaia, Andreas Neumann and Lorenz Hurni; data curation, Luca Gaia; formal analysis, Luca Gaia; investigation, Luca Gaia; methodology, Luca Gaia and Andreas Neumann; resources, Luca Gaia; software, Luca Gaia; supervision, Luca Gaia, Andreas Neumann and Lorenz Hurni; validation, Luca Gaia; visualization, Luca Gaia; writing—original draft preparation, Luca Gaia; writing—review & editing, Luca Gaia, Andreas Neumann and Lorenz Hurni. All authors have read and agreed to the published version of the manuscript.

Funding: This research received no external funding.

Data Availability Statement: The raw data for the map about the glacial lakes can be accessed through the works of Mölg et al. (https://doi.pangaea.de/10.1594/PANGAEA.934190, accessed on 11 February 2025) and Steffen et al. (https://doi.org/10.3929/ethz-b-000554650). The raw data for the map about the plant phenology can be accessed at https://opendata.swiss/en/dataset/phanologische-beobachtungen (accessed on 11 February 2025). The DHM provided by Swisstopo can be accessed at https://www.swisstopo.admin.ch/de/hoehenmodell-dhm25 (accessed on 27 December 2024). The raw data for the zero degree line, as well as the code, will be made available by the authors on request. The final maps can be found in the AoS https://www.atlasderschweiz.ch/downloads/ (accessed on 11 February 2025).

Acknowledgments: We thank Simon Scherrer, Thomas Schlegel, Francesco Isotta, and Sven Kotlarski from MeteoSwiss for providing the data for the zero degree line and for their valuable inputs about the topic and during map creation. Special thanks go to Regula Gehrig from MeteoSwiss for supplying the data for the phenological map and offering her expertise on phenology. We also thank Yann Vitasse from WSL for his valuable insights into phenology. Our appreciation goes to Nico Mölg and Elias Hodel for providing the data on glacial lakes. Additionally, we are grateful to René Sieber and Christian Wohlert for their constructive feedback and thorough review of the manuscript. Finally, we express our sincere thanks to the entire AoS team for their technical support in producing the maps.

Conflicts of Interest: The authors declare no conflicts of interest.

References

1. IPCC. Summary for Policymakers. In *Climate Change 2023: Synthesis Report.Contribution of Working Groups I, II and III to the Sixth Assessment Report of the Intergovernmental Panel on Climate Change*; Lee, H., Romero, J., Eds.; IPCC: Geneva, Switzerland, 2023; pp. 1–34.
2. NCCS (Pub.). *CH2018—Climate Scenarios for Switzerland*; National Centre for Climate Services: Zurich, Switzerland, 2018; p. 24.
3. Moser, S.C. Communicating climate change: History, challenges, process and future directions. *WIREs Clim. Change* **2009**, *1*, 31–53. [CrossRef]
4. Sheppard, S.R.J. *Visualizing Climate Change: A Guide to Visual Communication of Climate Change and Developing Local Solutions*, 1st ed.; Routledge: London, UK, 2012.
5. Courtney, S.L.; McNeal, K.S. Seeing is believing: Climate change graph design and user judgments of credibility, usability, and risk. *Geosphere* **2023**, *19*, 1508–1527. [CrossRef]
6. Fish, C.S. Cartographic content analysis of compelling climate change communication. *Cartogr. Geogr. Inf. Sci.* **2020**, *47*, 492–507. [CrossRef]
7. McKendry, J.E.; Machlis, G.E. Cartographic design and the quality of climate change maps. *Clim. Change* **2008**, *95*, 219–230. [CrossRef]
8. Kraak, M.-J.; Roth, R.E.; Ricker, B.; Kagawa, A.; Sourd, G.L. *Mapping for a Sustainable World*; United Nations: New York, NY, USA, 2020.
9. Fish, C.S. Elements of Vivid Cartography. *Cartogr. J.* **2021**, *58*, 150–166. [CrossRef]
10. Fish, C.S.; Kreitzberg, K.Q. Mapping in an Echo Chamber: How Cartographic Silence Frames Conservative Media's Climate Change Denial. *Ann. Am. Assoc. Geogr.* **2023**, *113*, 2480–2496. [CrossRef]
11. Tsang, S.J. Communicating Climate Change: The Impact of Animated Data Visualizations on Perceptions of Journalistic Motive and Media Bias. *J. Broadcast. Electron. Media* **2023**, *67*, 161–182. [CrossRef]
12. Becsi, B.; Hohenwallner-Ries, D.; Grothmann, T.; Prutsch, A.; Huber, T.; Formayer, H. Towards better informed adaptation strategies: Co-designing climate change impact maps for Austrian regions. *Clim. Change* **2019**, *158*, 393–411. [CrossRef]
13. Kaye, N.R.; Hartley, A.; Hemming, D. Mapping the climate: Guidance on appropriate techniques to map climate variables and their uncertainty. *Geosci. Model Dev.* **2012**, *5*, 245–256. [CrossRef]

14. Liu, N.; Feng, Y.; Wang, S.; Meng, L. The Impact of Maps on Climate Information Communication: An Empirical User Study on Social Media Trustworthiness. *Abstr. ICA* **2024**, *7*, 1–2. [CrossRef]
15. Sheppard, S.R.J. Landscape visualisation and climate change: The potential for influencing perceptions and behaviour. *Environ. Sci. Policy* **2005**, *8*, 637–654. [CrossRef]
16. Schrot, O. From Information to Participation. Interactive Landscape Visualization as a Tool for Collaborative Planning. Ph.D. Thesis, ETH, Zürich, Switzerland, 2007. [CrossRef]
17. IPCC. IPCC WGI Interactive Atlas. Available online: https://interactive-atlas.ipcc.ch (accessed on 23 December 2024).
18. Prairie Climate Center. Clima Atlas of Canada. Available online: https://climateatlas.ca/ (accessed on 23 December 2024).
19. Gasser, M.; Hotea, M.D. *Landschaften des Wissens—50 Jahre Kartensammlung an der ETH-Bibliothek*; Michael Imhof Verlag: Petersberg, 2022.
20. Atlas of Switzerland. Atlas of Switzerland—Download. Available online: https://www.atlasderschweiz.ch/downloads/ (accessed on 23 December 2024).
21. Sieber, R.; Hurni, L. Future National Atlases—Strategies for Tearing Down the User's Firewall. *Abstr. Int. Cartogr. Assoc.* **2022**, *5*, 11. [CrossRef]
22. *Klimaatlas der Schweiz*; Schweizerische Meteorologische Anstalt: Wabern, Switzerland, 2000.
23. SCNAT. Klimaatlas der Schweiz. Available online: https://scnat.ch/de/uuid/i/e6fddc02-4ac3-5066-a264-8110c8262321-Klimaatlas_der_Schweiz (accessed on 23 December 2024).
24. Bihari, Z.; Babolcsai, G.; Bartholy, J.; Ferenczi, Z.; Gerhat-Kerenyi, J.; Haszpra, L.; Homoki-Ujvary, K.; Kovacs, T.; Lakatos, M.; Nemeth, A.; et al. *Climate—National Atlas of Hungary—Natural Environment*; Kocsis, K.E.-i.-C., Ed.; MTA CSFK Geographical Institute: Budapest, Hungary, 2018.
25. United Nations General Assembly. *Resolution adopted by the General Assembly on 25 September 2015: Transforming our world: The 2030 Agenda for Sustainable Development (A/RES/70/1)*; United Nations: New York, NY, USA, 2015.
26. Kent, A.J.; Vujakovic, P.; Eades, G.; Davis, M. Putting the UN SDGs on the Map: The Role of Cartography in Sustainability Education. *Cartogr. J.* **2020**, *57*, 93–96. [CrossRef]
27. Gaia, L.; Neumann, A.; Sieber, R.; Hurni, L. Cartographic visualization of the effects of climate change: A practical application for the Atlas of Switzerland. *Abstr. Int. Cartogr. Assoc.* **2024**, *7*, 42. [CrossRef]
28. Swisstopo. DHM25. Available online: https://www.swisstopo.admin.ch/de/hoehenmodell-dhm25 (accessed on 23 December 2024).
29. Scherrer, S.C.; Gubler, S.; Wehrli, K.; Fischer, A.M.; Kotlarski, S. The Swiss Alpine zero degree line: Methods, past evolution and sensitivities. *Int. J. Climatol.* **2021**, *41*, 6785–6804. [CrossRef]
30. CH2018. *CH2018—Climate Scenarios for Switzerland, Technical Report*; National Centre for Climate Services: Zurich, Hungary, 2018; p. 271.
31. Rouault, E.; Warmerdam, F.; Schwehr, K.; Kiselev, A.; Butler, H.; Łoskot, M.; Szekeres, T.; Tourigny, E.; Landa, M.; Miara, I.; et al. GDAL. Available online: https://zenodo.org/records/14046734 (accessed on 27 December 2024).
32. Bloch, M. Mapshaper. Available online: https://github.com/mbloch/mapshaper (accessed on 23 December 2024).
33. Visvalingam, M.; Whyatt, J.D. Line Generalization by Repeated Elimination of Points. *Cartogr. J.* **1993**, *30*, 46–51. [CrossRef]
34. Sieber, R.; Serebryakova, M.; Schnürrer, R.; Hurni, L. Atlas of Switzerland Goes Online and 3D—Concept, Architecture and Visualization Methods. In *Progress in Cartography: EuroCarto 2015*; Gartner, G., Jobst, M., Huang, H., Eds.; Springer International Publishing: Cham, Switzerland, 2016; pp. 171–184.
35. Cleland, E.E.; Chuine, I.; Menzel, A.; Mooney, H.A.; Schwartz, M.D. Shifting plant phenology in response to global change. *Trends Ecol. Evol.* **2007**, *22*, 357–365. [CrossRef] [PubMed]
36. Schwartz, M.D. *Phenology: An Integrative Environmental Science*, 2nd ed.; Springer: Dordrecht, The Netherlands, 2013.
37. MeteoSwiss. Plant Phenology. Available online: https://www.meteoswiss.admin.ch/weather/weather-and-climate-from-a-to-z/phenology.html (accessed on 23 December 2024).
38. Menzel, A.; Sparks, T.H.; Estrella, N.; Koch, E.; Aasa, A.; Ahas, R.; Alm-Kübler, K.; Bissolli, P.; Braslavská, O.G.; Briede, A.; et al. European phenological response to climate change matches the warming pattern. *Glob. Change Biol.* **2006**, *12*, 1969–1976. [CrossRef]
39. Meier, M.; Vitasse, Y.; Bugmann, H.; Bigler, C. Phenological shifts induced by climate change amplify drought for broad-leaved trees at low elevations in Switzerland. *Agric. For. Meteorol.* **2021**, *307*, 108485. [CrossRef]
40. MeteoSwiss. Phenological Observations. Available online: https://opendata.swiss/en/dataset/phanologische-beobachtungen (accessed on 23 December 2024).
41. Agarwal, V.; Van Wyk de Vries, M.; Haritashya, U.K.; Garg, S.; Kargel, J.S.; Chen, Y.J.; Shugar, D.H. Long-term analysis of glaciers and glacier lakes in the Central and Eastern Himalaya. *Sci. Total Environ.* **2023**, *898*, 165598. [CrossRef] [PubMed]
42. Zhang, T.; Wang, W.; An, B. Heterogeneous changes in global glacial lakes under coupled climate warming and glacier thinning. *Commun. Earth Environ.* **2024**, *5*, 374. [CrossRef]

43. Mölg, N.; Huggel, C.; Herold, T.; Storck, F.; Allen, S.; Haeberli, W.; Schaub, Y.; Odermatt, D. Inventory and evolution of glacial lakes since the Little Ice Age: Lessons from the case of Switzerland. *Earth Surf. Process. Landf.* **2021**, *46*, 2551–2564. [CrossRef]
44. Steffen, T.; Huss, M.; Estermann, R.; Hodel, E.; Farinotti, D. Volume, evolution, and sedimentation of future glacier lakes in Switzerland over the 21st century. *Earth Surf. Dyn.* **2022**, *10*, 723–741. [CrossRef]
45. Lorenzoni, I.; Nicholson-Cole, S.; Whitmarsh, L. Barriers perceived to engaging with climate change among the UK public and their policy implications. *Glob. Environ. Change* **2007**, *17*, 445–459. [CrossRef]
46. Johannsen, I.M.; Lassonde, K.A.; Wilkerson, F.; Schaab, G. Communicating Climate Change: Reinforcing Comprehension and Personal Ties to Climate Change Through Maps. *Cartogr. J.* **2017**, *55*, 85–100. [CrossRef]
47. Chess, C.; Johnson, B.B. Information is not enough. In *Creating a Climate for Change: Communicating Climate Change and Facilitating Social Change*; Moser, S.C., Dilling, L., Eds.; Cambridge University Press: Cambridge, UK, 2007; pp. 223–234.
48. Schroth, O.; Angel, J.; Sheppard, S.; Dulic, A. Visual Climate Change Communication: From Iconography to Locally Framed 3D Visualization. *Environ. Commun.* **2014**, *8*, 413–432. [CrossRef]
49. Shepherd, I.D.H. Travails in the Third Dimension: A Critical Evaluation of Three-Dimensional Geographical Visualization. In *Geographic Visualization*; John Wiley & Sons: Chichester, UK, 2008; pp. 199–222.
50. Brath, R. 3D InfoVis is Here to Stay: Deal with It. In Proceedings of the 2014 Ieee Vis International Workshop on 3dvis (3dvis), Paris, France, 9 November 2014; pp. 25–31. [CrossRef]
51. Niedomysl, T.; Elldér, E.; Larsson, A.; Thelin, M.; Jansund, B. Learning Benefits of Using 2D Versus 3D Maps: Evidence from a Randomized Controlled Experiment. *J. Geogr.* **2013**, *112*, 87–96. [CrossRef]
52. Lai, P.C.; Kwong, K.-H.; Mak, A.S.H. Assessing the Applicability and Effectiveness of 3D Visualisation in Environmental Impact Assessment. *Environ. Plan. B Plan. Des.* **2010**, *37*, 221–233. [CrossRef]
53. Häberling, C.; Bär, H.; Hurni, L. Proposed Cartographic Design Principles for 3D Maps: A Contribution to an Extended Cartographic Theory. *Cartogr. Int. J. Geogr. Inf. Geovisualiza.* **2008**, *43*, 175–188. [CrossRef]
54. Aitken, O.; Moore, A.; Diaz-Rainey, I.; Nguyen, Q.; Cox, S.; Bodeker, G. Geospatial Metaphors in the Communication of Climate Change-Related Flooding Uncertainty. *Abstr. Int. Cartogr. Assoc.* **2024**, *7*, 4. [CrossRef]
55. MeteoSchweiz. Nullgradgrenze im Atlas der Schweiz. Available online: https://www.meteoschweiz.admin.ch/ueber-uns/meteoschweiz-blog/de/2024/06/nullgradgrenze-im-atlas-der-schweiz.html (accessed on 23 December 2024).
56. ETH. Atlas of Switzerland. Climate Change in Switzerland: 1864–2099 in Fast Motion. Available online: https://www.instagram.com/p/C9heidQKddP/?utm_source=ig_web_button_share_sheet&igsh=M2M0Y2JmOTAyOA (accessed on 27 December 2024).
57. Lorenz, S.; Dessai, S.; Forster, P.M.; Paavola, J. Tailoring the visual communication of climate projections for local adaptation practitioners in Germany and the UK. *Philos. Trans A Math. Phys. Eng. Sci.* **2015**, *373*, 20140457. [CrossRef] [PubMed]

International Journal of
Geo-Information

MDPI

Article

How Hydrological Extremes Affect the Chlorophyll-a Concentration in Inland Water in Jiujiang City, China: Evidence from Satellite Remote Sensing

Wei Jiang [1], Xiaohui Ding [2,*], Fanping Kong [3], Gan Luo [1], Tengfei Long [4], Zhiguo Pang [1], Shiai Cui [4], Jie Liu [1] and Elhadi Adam [5]

1 State Key Laboratory of Simulation and Regulation of Water Cycle in River Basin, China Institute of Water Resources and Hydropower Research, Beijing 100038, China; jiangwei@iwhr.com (W.J.); luogan@jxnu.edu.cn (G.L.); pangzg@iwhr.com (Z.P.); liujie2000y09@163.com (J.L.)
2 School of Software and Internet of Things Engineering, Jiangxi University of Finance and Economics, Nanchang 330032, China
3 Beijing Capital Air Environmental Science & Technology Co., Ltd., Beijing 100176, China; kongfanping@aircapital.cn
4 Aerospace Information Research Institute, Chinese Academy of Sciences, Beijing 100094, China; longtf@aircas.ac.cn (T.L.); cuisa@aircas.ac.cn (S.C.)
5 School of Geography, Archaeology and Environmental Studies, University of the Witwatersrand, Jan Smut Ave, Johannesburg 2050, South Africa; elhadi.adam@wits.ac.za
* Correspondence: dingxiaohui@gdas.ac.cn; Tel.: +86-010-6878-5404

Abstract: From 2020 to 2022, hydrological extremes such as severe floods and droughts occurred successively in Jiujiang city, Poyang Lake Basin, posing a threat to regional water quality safety. The chlorophyll-a (Chl-a) concentration is a key indicator of river eutrophication. Until now, there has been a lack of empirical research exploring the Chl-a trend in inland water in Jiujiang in the context of hydrological extremes. In this study, Sentinel-2 satellite remote sensing data sourced from the Google Earth Engine (GEE) cloud platform, along with hourly water quality data collected from monitoring stations in Jiujiang city, Jiangxi Province, China, are utilized to develop a quantitative inversion model for the Chl-a concentration. The Chl-a concentrations for various inland water types were estimated for each quarter from 2020 to 2022, and the spatiotemporal distribution was analyzed. The main findings are as follows: (1) the quantitative inversion model for the Chl-a concentration was validated via in situ measurements, with a coefficient of determination of 0.563; (2) the spatial estimates of the Chl-a concentration revealed a slight increasing trend, increasing by 0.1193 μg/L from 2020 to 2022, closely aligning with the monitoring-station data; (3) an extreme drought in 2022 led to less water in inland water bodies, and consequently, the Chl-a concentration displayed a significant upward trend, especially in Poyang Lake, where the mean Chl-a concentration increased by approximately 1 μg/L from Q1 to Q2 in 2022. These findings revealed the seasonal changes in the Chl-a concentrations in inland waters in the context of extreme hydrological events, thus providing valuable information for the sustainable management of water quality in Jiujiang city.

Keywords: chlorophyll-a concentration; inland water; Sentinel-2; human activities; seasonal change; Google Earth Engine

1. Introduction

With the rapid advancement of urbanization and climate change, the health of water environments is increasingly threatened by the continuous discharge of wastewater [1,2].

Lakes and rivers, as essential freshwater resources, play a vital role in maintaining the overall health of surrounding ecosystems [3–6]. Currently, the Chl-a concentration is recognized as a key parameter for assessing water quality [7,8]. Under extreme hydrological conditions, monitoring Chl-a concentrations has become critical for the healthy management of rivers and lakes [9–11]. During floods, large amounts of nutrients are washed into water bodies, which can trigger a sudden and significant increase in Chl-a levels due to rapid algal blooms [10]. On the other hand, under drought conditions, a reduced water volume and increased nutrient concentration can also lead to abnormal changes in Chl-a concentrations [11]. By continuously monitoring Chl-a concentrations during these extreme events, an in-depth understanding of the response of aquatic ecosystems to extreme hydrological conditions can be achieved, and targeted strategies can be developed to safeguard the ecological functions of rivers and lakes.

A variety of methods have been developed to measure Chl-a concentrations in lakes and rivers, such as field sampling, laboratory analyses, and satellite remote sensing [12,13]. Despite their benefits, none of these methods alone can satisfy the need for large-scale, high-precision, real-time monitoring. Field sampling and laboratory analysis offer high accuracy but are labor intensive and time consuming, limiting their widespread application [14]. In recent years, advancements in and the expansion of satellite remote sensing have provided significant opportunities for large-scale seasonal water quality monitoring [15]. Satellite data, such as those from the Sentinel-2 satellite, offer global and continuous observations, making them an attractive tool for monitoring Chl-a concentrations [16–20]. Nevertheless, the accuracy of remote sensing-based inversion models for Chl-a concentrations is often constrained by a lack of synchronous ground-based data for model construction and validation [19].

Fortunately, the integration of the Internet of Things (IoT) and remote sensing technology has demonstrated promising potential for improving the accuracy of water quality monitoring [21]. IoT-based wireless sensor networks (WSNs) provide real-time ground data that can be used to enhance remote sensing models for water quality parameters such as Chl-a [22–26]. For example, remote sensing data and information collected with IoT sensors can be combined to accurately measure suspended sediment concentrations [27]. In China, numerous new water quality monitoring stations equipped with IoT and WSN technology have been established to monitor key water quality parameters [28]. These advancements make it possible to achieve accurate, cost-effective, and efficient water quality monitoring. Compared with traditional manual sampling, IoT and WSN technologies can realize the real-time collection of water quality data, which are obtained at the same time as data collected by satellites, thus avoiding modeling errors caused by differences in sampling time [29].

In recent years, studies have shown that global climate change and human activities have put significant pressure on regional river and lake environments, but the degree of this pressure varies across regions and times [30,31]. Jiujiang city is an important river and lake area in the Yangtze River Basin and is also home to an important channel through which Poyang Lake merges into the Yangtze River. Changes in water environment quality are significantly affected by climate change and human water diversion activities. In 2020, extreme floods occurred in Poyang Lake; the flood duration and intensity were close to the scale of the catastrophic flood in 1998, and the water level exceeded the historical high. In 2022, extreme drought events occurred in Poyang Lake, Jiujiang city, from June to August. On 23 September 2022, the water level of Poyang Lake decreased to 7.1 m, which is the new lowest record since records began in 1951 [32].

Previous studies have explored long-term changes in Chl-a concentrations in the Poyang Lake Basin [7,33,34], but research on the seasonal changes in Chl-a concentrations in inland waters in Jiujiang under extreme hydrological events is lacking. Therefore, the purpose of this study was to explore the spatiotemporal pattern of the Chl-a concentration in inland water in Jiujiang during extreme hydrological events on the basis of remote sensing and water quality monitoring data. To achieve this goal, a quantitative inversion model for the Chl-a concentration in inland water in Jiujiang was developed based on Sentinel-2 imagery and water quality monitoring-station data. Then, seasonal Chl-a concentration maps were generated over a three-year period (2020 to 2022) via the Google Earth Engine (GEE) cloud computing platform. Finally, the patterns of the Chl-a concentrations in inland water in Jiujiang city were explored in typical lake areas. This study provides vital support for the management of inland waters in Jiujiang city with respect to extreme hydrological changes under climate change impact.

2. Study Area and Materials

2.1. Study Area

Jiujiang city is located in the northern part of Jiangxi Province, on the southern bank of the middle and lower reaches of the Yangtze River, with a total area of 19,100 square kilometers (Figure 1). According to the statistical yearbook of Jiujiang in 2023, the permanent population of Jiujiang city was 4.536 million [35]. Jiujiang city has a dense river network and a developed water system, as an important river and lake area in the Yangtze River Basin. Poyang Lake, which is China's largest freshwater lake, is the core component of Jiujiang's river and lake water environment, and most of the lake water is located in Jiujiang [36]. The water quality of the Jiujiang Lake area of Poyang Lake is significantly affected by seasonal factors and surrounding human activities [37]. The water quality is relatively good during the flood season because of the large amount of water and frequent water exchange; during the dry season, the water quality tends to decline because of the reduction in water area and the relatively increased concentration of pollutants [34,38].

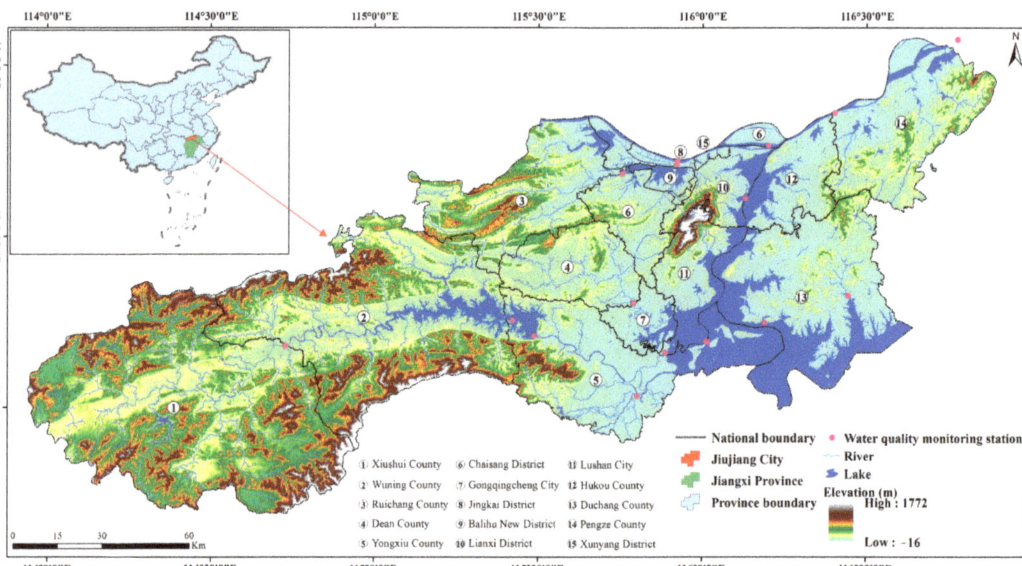

Figure 1. Map of the study area showing the distribution of water quality monitoring stations.

2.2. Data and Processing

To explore the seasonal variation in Chl-a concentration in rivers and lakes in Jiujiang city, research has focused on ground water quality monitoring-station data and Sentinel-2 satellite remote sensing data.

(1) Water quality monitoring-station data.

To ensure effective protection and continuous monitoring of the water environment, a network of 34 online water quality monitoring stations was established throughout the area; however, only 18 water quality monitoring stations (Table 1) are equipped to monitor water quality parameters, such as water temperature, pH, Chl-a, dissolved oxygen, conductivity, turbidity, ammonia nitrogen, total phosphorus, total nitrogen, algae density and water quality category. These stations serve as vital sources for collecting ground-based synchronous monitoring data, which are essential for tracking water quality over time.

Table 1. Water quality monitoring stations used in this study.

No.	Station Name	County	Water Area	Monitoring Time
1	Yaogang	Lianxi District	Yangtze River	1 July 2018 to 17 November 2022
2	Hukou (Right Bank)	Lianxi District	Yangtze River	1 July 2018 to 17 November 2022
3	Hamashitou	Lianxi District	Poyang Lake	1 September 2020 to 17 November 2022
4	Saicheng Lake	Chaisang District	Chang River	1 July 2018 to 17 November 2022
5	Mai Gou	Chaisang District	Chang River	1 September 2020 to 17 November 2022
6	Wuning Jiepai	Wuning County	Zhelin Lake	1 September 2020 to 17 November 2022
7	Sandu (Lake Reservoir)	Xiushui County	Zhelin Lake	1 July 2018 to 17 November 2022
8	the Dam of Zhelin Lake	Yongxiu County	Zhelin Lake	1 July 2018 to 17 November 2022
9	Jianjiao Village	Yongxiu County	Yangliu Jin River	1 September 2020 to 17 November 2022
10	Wu Cheng Xiu River	Yongxiu County	Xiu River	1 July 2018 to 17 November 2022
11	Wu Cheng Gan River	Yongxiu County	Gan River	1 July 2018 to 17 November 2022
12	Liao River	Yongxiu County	Liao River	1 July 2018 to 17 November 2022
13	Caojia	Duchang County	Tutang River	1 September 2020 to 17 November 2022
14	Duchang	Duchang County	Poyang Lake	1 July 2018 to 17 November 2022
15	Hongguang Village	Pengze County	Yangtze River	1 September 2020 to 17 November 2022
16	Zhaixia	Gongqing County	Boyang River	1 September 2020 to 17 November 2022
17	Maochang	Gongqing County	Boyang River	1 September 2020 to 17 November 2022
18	Xiangkou	Pengze County	Yangtze River	1 July 2018 to 17 November 2022

The water quality monitoring data obtained in this study are from 1 July 2018 to 17 November 2022. At each station, samples are obtained 5 times a day, at 00:00 a.m., 4:00 a.m., 12:00 p.m., 16:00 p.m. and 20:00 p.m. In fact, owing to the different constructions and operation times of the water quality monitoring stations in Table 1, from 1 July 2018 to 31 August 2020, only 10 water quality monitoring stations provided data regularly. Moreover, due to the effects of extreme climate events and instrument maintenance, some monitoring-station data were missing. To match the remote sensing transit time of the Sentinel-2 satellite (11:00–12:00 a.m.), the water quality data collected at 12:00 p.m. were selected as the model construction data. In fact, due to the missing Chl-a data in the station records, even if the satellite provides cloud-free imaging, those stations with missing monitoring data cannot be used to construct the sample dataset.

(2) Sentinel-2 satellite remote sensing imagery.

The Sentinel-2 satellite was launched by the European Space Agency in 2015 and then again in 2017. Sentinel-2 has wide-bandwidth, high-resolution and multispectral characteristics and can achieve global 5-day returns through satellite networking; thus, it is widely used in the quantitative monitoring of river and lake water-quality parameters [39,40]. The Sentinel-2 multispectral instrument (MSI) samples 13 spectral bands: visible and NIR at 10 m, red edge and SWIR at 20 m, and atmospheric bands at a 60 m spatial resolution [41]. To avoid the influence of the atmosphere on the remote sensing inversion of water quality parameters, the surface reflectance product of Sentinel-2 was selected. This product is obtained by running the Sen2Cor correction algorithm, which can eliminate the influence of the atmosphere on the quantitative inversion of water quality products [42]. This product has been integrated into the GEE. Owing to the frequent occurrence of clouds and rain in Jiujiang, a total of 248 effective Sentinel-2 images were selected from the GEE platform to establish a quantitative inversion model for the Chl-a concentration through a cloud mask algorithm based on information from the quality assurance (QA) band and effective monitoring data from ground water quality stations. Furthermore, the median filtering method was used to identify and select cloud-free imagery on the basis of the median reflectance values for each pixel across multiple images within a quarter, ensuring sufficient data accuracy. Then, combined with the constructed remote sensing quantitative inversion model for the Chl-a concentration, the seasonal spatial maps of the Chl-a concentration in Jiujiang city from 2020 to 2022 were generated using these images.

3. Method

The overall methodology flowchart is presented in Figure 2. The first step is to construct a Chl-a concentration sample library using Sentinel-2 satellite data and water quality station monitoring data. The second step involves constructing a remote sensing quantitative inversion model for the Chl-a concentration by analyzing the Chl-a spectral characteristic response band, applying an empirical regression method, and evaluating the accuracy of the model. In the third step, the time series of Sentinel-2 remote sensing data and a quantitative inversion model are combined to generate seasonal Chl-a maps, analyze the spatiotemporal variation characteristics of Chl-a concentrations in different inland water types, and reveal the impact of extreme hydrological events on the changes in Chl-a concentrations in inland water in Jiujiang.

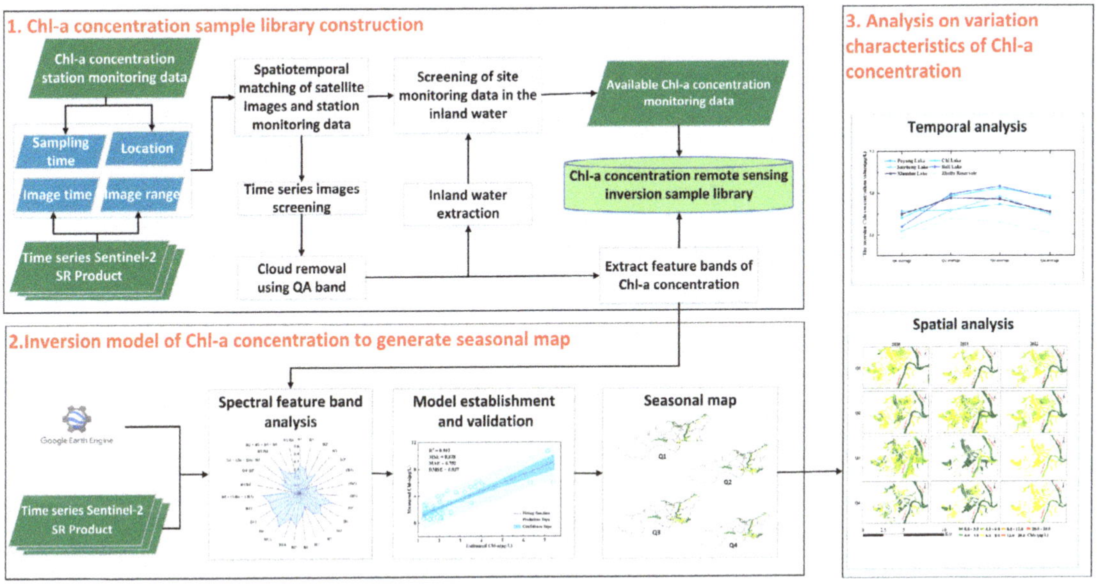

Figure 2. Overall flowchart of this study.

3.1. Chl-a Concentration Sample Library Construction

To ensure the accuracy and compatibility of the Chl-a remote sensing inversion model, it is very important to construct a set of Chl-a sample libraries that are consistent in time and space. In this study, a process for obtaining a Chl-a remote sensing inversion sample library on the basis of Sentinel-2 images and water quality station monitoring data is proposed, as shown in Figure 2.

Firstly, on the basis of the imaging time and the range of satellite remote sensing images and combined with the location and time information for the water quality monitoring station, the spatiotemporal matching method was used to screen the available satellite remote sensing data, and the Chl-a monitoring-station data that were consistent with the satellite transit time were selected.

Secondly, for satellite remote sensing images, the satellite image cloud removal method based on the QA band is used to remove the influence of cloud coverage in the station area. The inland water extraction method for Sentinel-2 data employed in a previous study was used to eliminate the errors in station monitoring caused by seasonal water fluctuations [41], and the available Chl-a monitoring data from station monitoring were screened.

Finally, according to the spectral response mechanism of Chl-a, the feature bands of the Chl-a concentration were extracted on the basis of Sentinel-2 images, including single bands and band combinations. Furthermore, combined with the available Chl-a monitoring data from station monitoring, a Chl-a remote sensing inversion sample library was constructed, and it contained 44 valid samples with a mean value of 2.15 μg/L and a standard deviation of 1.43 μg/L.

3.2. Chl-a Concentration Inversion Model

To construct the inversion model, the remote sensing reflectance values for each band of the pixels corresponding to the sampling points was calculated. According to previous studies, water bodies experiencing pronounced eutrophication often exhibit elevated Chl-a concentrations, resulting in remarkable absorption peaks near 440 and 670 nm and high reflectance in the near-infrared spectrum [43,44]. The B1 (443 nm) and B4 (665 nm) bands were first considered in the construction process of the inversion model. Owing to the various types of water bodies and complex hydrological environments, all of the bands, first-order differential bands, and some band combinations from the Sentinel-2 images were also included in the selection of the optimal bands/band combinations. As listed in [42,44], band combinations such as B4/B5, (1/B4–1/B5) × B6, B7 × (1/B4–1/B5), B5 × (B6 + B4)/B2, B3/B4, B2 + B3 + B4 + B5, and B3/B6 are good indicators of the Chl-a concentration in inland water. Selection was carried out by calculating the Pearson correlation coefficient between the information for each spectral band/band combination and the corresponding ground-based Chl-a measurements. The bands/band combination with the highest correlation coefficient was chosen as the independent variable in the model. Multiple fitting functions, such as polynomial, logarithmic, linear, and exponential functions, were applied to the dataset to identify the best relationship between the satellite-derived data and the Chl-a concentrations. Ultimately, the model with the highest coefficient of determination (R^2) was selected as the optimal inversion model, as it demonstrated the strongest ability to predict the Chl-a concentration in the study area. In the regression process, all of the valid samples were used to develop the inversion model via the regression method. The performance of the developed model was measured with several indicators, as described in Section 3.3.

As presented in Figure 3, the correlation between the band combination (B4/B5) and the Chl-a concentration was the strongest, with a correlation coefficient of −0.41. To ensure sufficient evaluation accuracy, this band combination was used to construct the inversion model using multiple linear regression. The Chl-a concentration ($Chla$) inversion model was constructed as follows:

$$Chla = -224.23 \times \left(\frac{B4}{B5}\right)^3 - 648.87 \times \left(\frac{B4}{B5}\right)^2 + 604.18 \times \frac{B4}{B5} - 176.44 \quad (1)$$

The accuracy of the proposed model was assessed by comparing the predicted values with the corresponding measured values. The fitting result is presented in Figure 4 in the form of a scatter plot. The results indicate that the Chl-a inversion model has notably high accuracy, with an R^2 value of 0.56. The MSE, MAE, and $RMSE$ are 0.878, 0.752, and 0.937, respectively. These indicators of model performance are already the best of all alternative models. Consequently, the coordinated use of Sentinel-2 images and synchronous ground monitoring data in the Chl-a concentration inversion model is considered feasible.

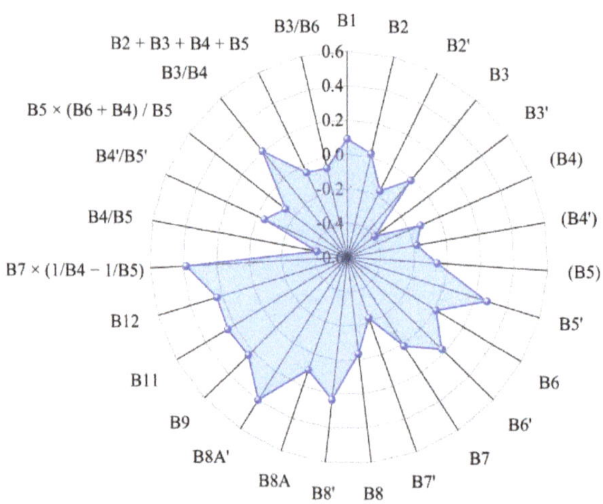

Figure 3. The values of the correlation coefficients between bands/band combinations and the measured Chl-a concentration.

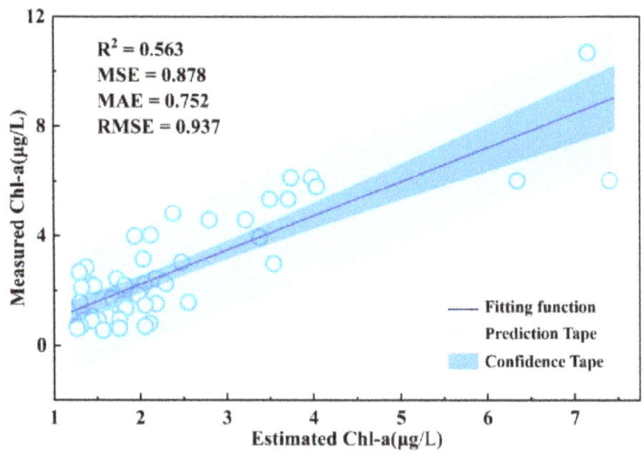

Figure 4. Scatter plot fitting results for the Chl-a concentration.

3.3. Inversion Model Accuracy Assessment Indices

To comprehensively evaluate the performance of the prediction model, four metrics are used: coefficient of determination (R^2), mean squared error (MSE), mean absolute error (MAE), and root mean squared error ($RMSE$) [45]. These metrics are applied to analyze model performance from different perspectives, such as goodness of fit, squared errors, absolute errors, and the magnitude of errors. The principles and formulas are as follows:

$$R^2 = 1 - \frac{\sum_{i=1}^{n}(y_i - \hat{y}_i)}{\sum_{i=1}^{n}(y_i - \overline{y})} \quad (2)$$

where y_i is the i-th observed value, $\hat{y}_i$ is the i-th predicted value, $\overline{y}$ is the mean of the observed values, the numerator represents the residual sum of squares, and the denominator represents the total sum of squares. The R^2 value ranges from 0 to 1. R^2 = 0 indicates poor model performance, whereas R^2 = 1 indicates perfect performance. The coefficient of determination reflects the model's ability to explain the variance in the dependent variable.

$$MSE = \frac{1}{n}\sum_{i=1}^{n}(y_i - \hat{y}_i)^2 \quad (3)$$

where n is the sample size. The MSE is the average squared difference between the predicted and actual values and reflects the overall error level of the model. Small MSE values indicate good model performance, but the metric is sensitive to outliers because of the squared term.

$$MAE = \frac{1}{n}\sum_{i=1}^{n}|y_i - \hat{y}_i| \quad (4)$$

The mean absolute error quantifies the average absolute difference between the predicted and actual values, providing a straightforward reflection of the magnitude of the error. Small MAE values indicate small prediction errors, and this metric is less sensitive to outliers than the MSE is.

$$RMSE = \sqrt{\frac{1}{n}\sum_{i=1}^{n}(y_i - \hat{y}_i)^2} \quad (5)$$

The $RMSE$ is the square root of the MSE, which represents the actual magnitude of the prediction error. The $RMSE$ has the same units as the original data, making it easier to interpret than other metrics, and small $RMSE$ values indicate high prediction accuracy.

By analyzing these metrics collectively, the model's prediction performance can be scientifically and comprehensively evaluated, providing robust quantitative support for its effectiveness.

4. Results

Because clouds and precipitation have significant impacts on the inversion of Chl-a concentrations, images from the same quarter and the corresponding ground monitoring data are integrated to produce quarterly Chl-a concentration products, from 2020 to 2022, with the GEE platform and the optimized inversion model. These quarterly products are obtained through the synthesis of monthly products. In each year over the three-year study period, there are four quarterly products, namely, the first-quarter (Q1) product, second-quarter (Q2) product, third-quarter (Q3) product, and fourth-quarter (Q4) product. These quarterly products of the Chl-a concentration are presented in Figure 5. The Chl-a concentration in the water bodies of Jiujiang city was distributed primarily from 0 to 12 μg/L. Specifically, the Chl-a concentrations in most water bodies are lower than 8 μg/L. From the perspective of temporal changes, the overall Chl-a concentration in the water bodies of Jiujiang city gradually increased, increasing by 0.1193 μg/L from 2020 to 2022. These increasing trends are most significant in areas where the Chl-a concentration exceeds 4 μg/L.

As the largest water body in Jiujiang city, the regions of Poyang Lake with Chl-a concentrations greater than 4 μg/L are distributed mainly in the northern, eastern and western nearshore areas (Figure 6). Most of these areas are near the river inlets of Poyang Lake. Specifically, the Chl-a concentrations are higher in southern Poyang Lake than in the northern areas and higher in the eastern lake areas than in the western areas in every quarter. The increase in the Chl-a concentration in the nearshore areas of Poyang Lake has resulted mainly from human activities, such as industrialization and accelerated urbanization. The increased water discharge accelerates phytoplankton growth due to the large amount of nutrients (e.g., nitrogen and phosphorus) contained in wastewater. In addition, the water located in nearshore areas often remains stagnant for long periods [46,47]. Eutrophic substances can accumulate gradually in these water bodies, resulting in the rapid growth of phytoplankton and relatively high Chl-a concentrations. Thus, these water bodies tend to be in a state of eutrophication or even organic pollution. In particular, the water level of Poyang Lake reached a historically low level in 2022 due to increased water consumption and climate change (Figure 6). This low water level significantly contributed to the abnormal increase in the Chl-a concentration.

To verify the reliability of the proposed model, the monthly observation data for the Chl-a concentration collected in five monitoring areas (Benhu, Wucheng, Xingzi, Laoyemiao, and the Poyang Lake outlet) of Poyang Lake from January 2020 to September 2022 were used to calculate the average and standard deviation Chl-a concentration in the water body of Poyang Lake in each quarter (Figure 7). The Chl-a concentration in the water body of Poyang Lake clearly increased over the three-year study period. This increasing trend resulted from increased wastewater discharge and decreased water levels, which were caused by the high intensity of human activities and climate change [48,49]. As presented in Figure 6, the water area of Poyang Lake changed the most during the transition periods from Q4 of 2020 to Q1 of 2021, from Q3 of 2021 to Q4 of 2021, and from Q2 of 2022 to Q3 of 2022. Thus, the standard deviation of the monitored Chl-a concentration were correspondingly largest in Q4 of 2020, Q3 of 2021, and Q3 of 2022, as illustrated in Figure 7. At a low water level, the increased wastewater discharge and the high depth of light penetration provide good nutritional and light conditions for the proliferation of algae, triggering an increase in the Chl-a concentration. In general, the measured results are consistent with the inversion results of the proposed model. Therefore, the credibility of the proposed model is verified.

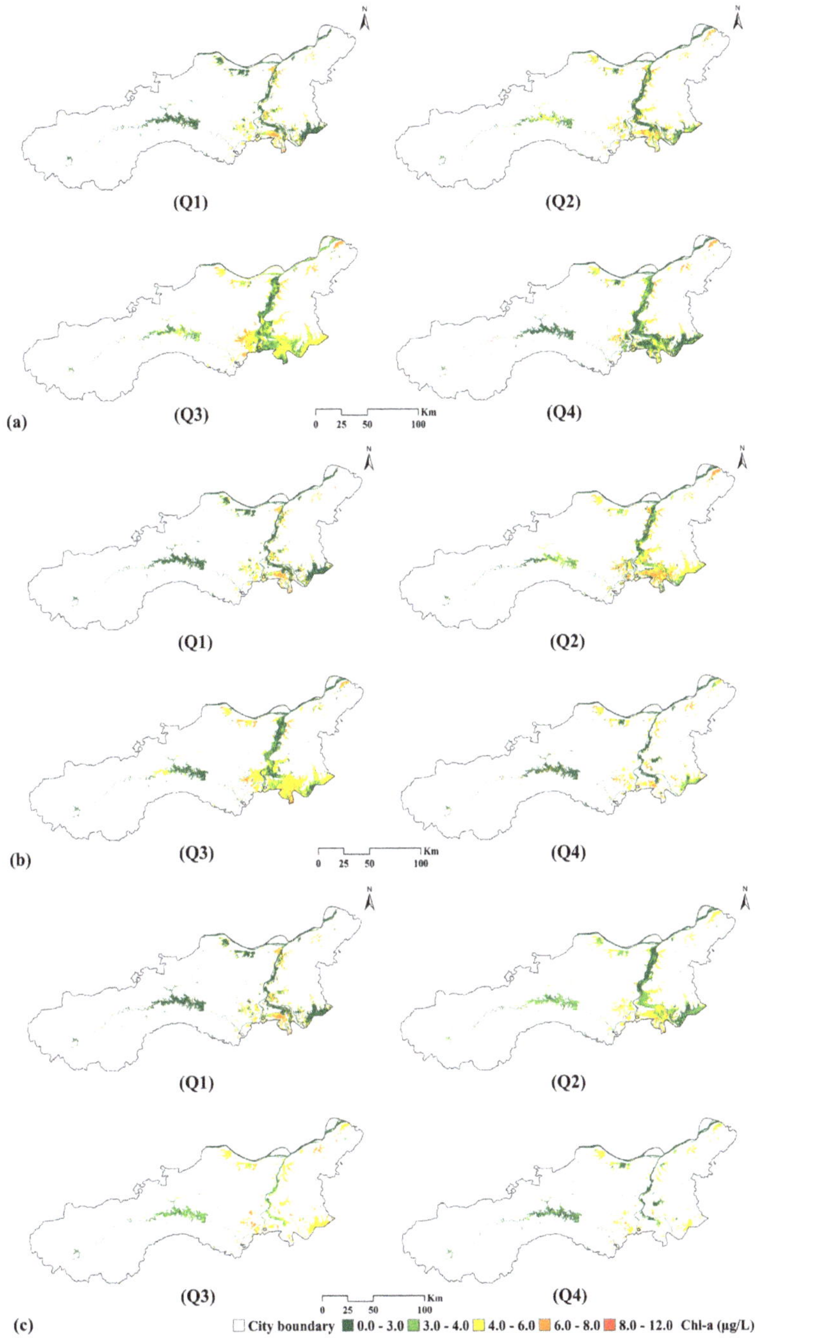

Figure 5. Spatial distribution maps of the Chl-a concentration. (**a**), (**b**), and (**c**) are the four quarters (Q1, Q2, Q3, and Q4) of the Chl-a concentration inversion results in 2020, 2021 and 2022, respectively.

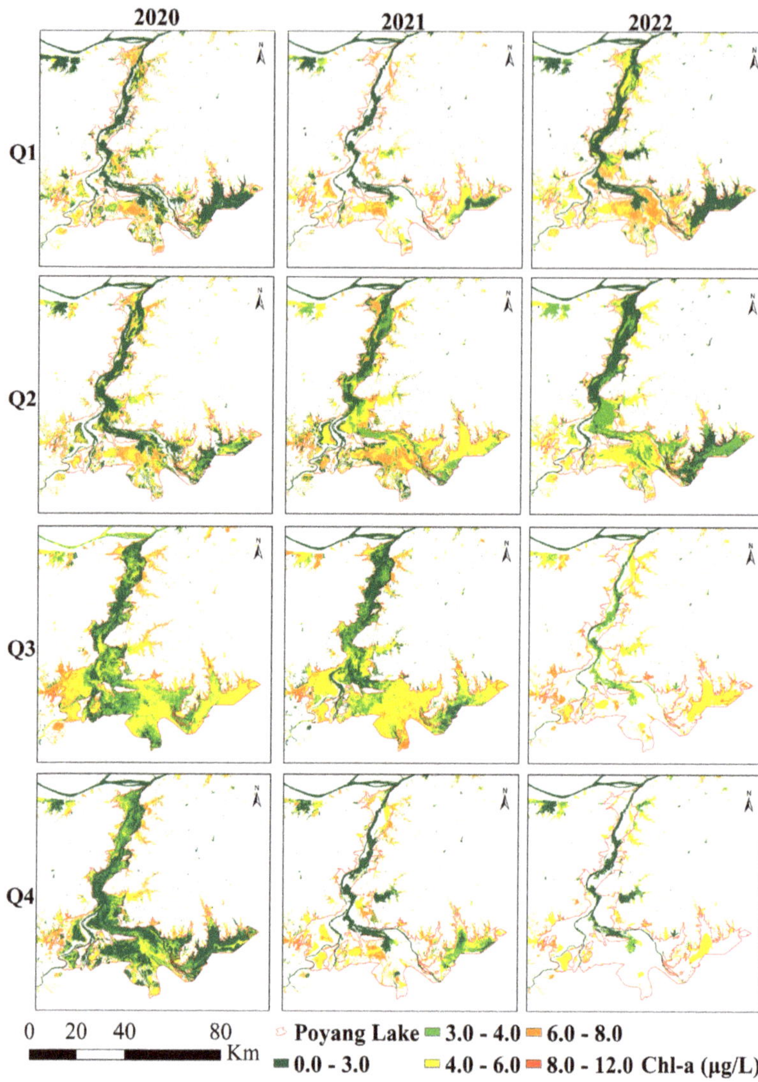

Figure 6. Spatial distribution of the Chl-a concentration in Poyang Lake in the four quarters (Q1, Q2, Q3, and Q4) of 2020, 2021, and 2022.

Furthermore, the variations in the Chl-a concentrations of six main water bodies (Poyang Lake, Xinmiao Lake, Zhelin Reservoir, Chihu Lake, Saicheng Lake and Bali Lake) in Jiujiang city are shown in Figure 8. Over the past three years, the Chl-a concentrations in these lakes have exhibited different change characteristics, as shown in Figure 8A. Specifically, the Chl-a concentrations in Bali Lake, Poyang Lake, Saicheng Lake and the Zhelin Reservoir slightly increased. However, a slight decreasing trend in the Chl-a concentration was detected in Xinmiao Lake. In general, the Chl-a concentration in Chihu Lake did not change over the three-year study period. The Chl-a concentrations in the Zhelin Reservoir were the lowest in most quarters because of the good ecological environment there. As shown in Figure 8B, the Chl-a concentrations of these lakes are generally higher in Q2 and Q3 than in Q1 and Q4. The maximum average Chl-a concentrations in these lakes usually occur in Q3, with decreases in Q4.

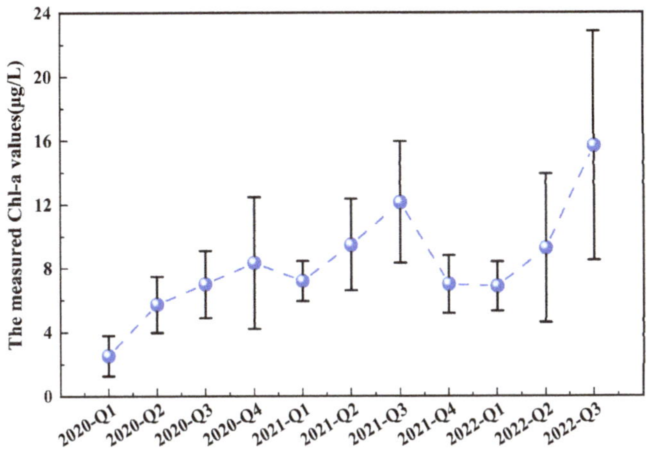

Figure 7. The measured mean Chl-a concentrations and error intervals at the monitoring stations in Poyang Lake.

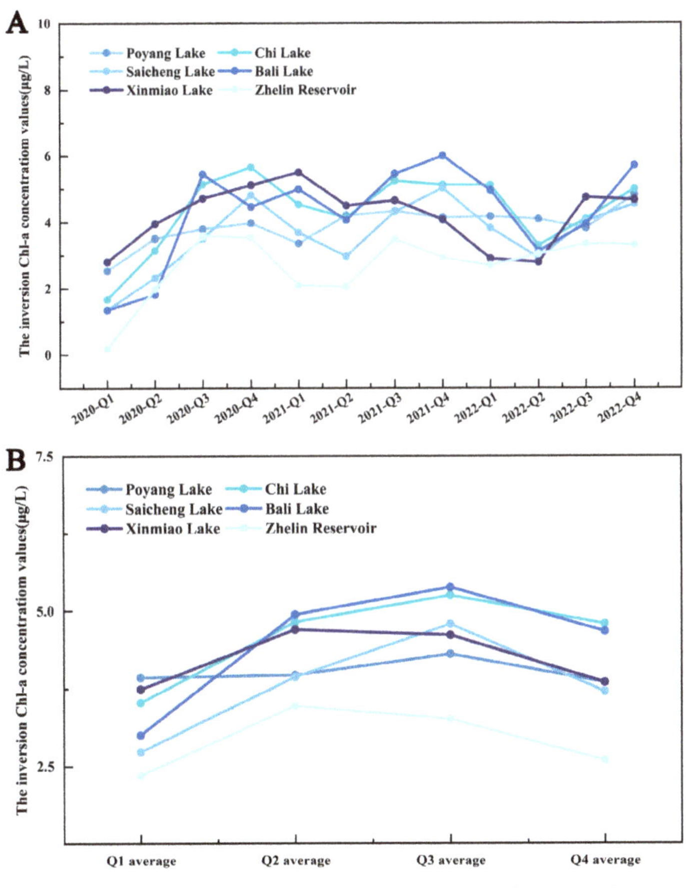

Figure 8. (**A**) Quarterly changes from 2020 to 2023, and (**B**) four quarterly average changes in the Chl-a concentration in each lake.

To further illustrate the effectiveness of the proposed method, the spatiotemporal distributions of the Chl-a concentration in three typical water body types (natural lakes, reservoirs, and aquaculture ponds) are presented in Figures 9–11. As shown in Figure 9, Saicheng Lake is a typical lake adjacent to the urban area of Jiujiang. The Chl-a concentration in the western part of this lake was generally higher than that in the eastern part in each quarter from 2020 to 2022. Agricultural nonpoint source pollutants, domestic sewage and aquaculture pollution are transported to Saicheng Lake via inflowing rivers, which are located mainly west of the lake. In the Zhelin Reservoir, the Chl-a concentrations were relatively high in Q2 and Q3 from 2020 to 2022 (Figure 10). Moreover, many aquaculture ponds greatly influence the health of the surrounding water bodies in Jiujiang. The Chl-a concentrations in aquaculture ponds can be effectively obtained via the proposed model. As presented in Figure 11, the Chl-a concentrations in these aquaculture ponds showed an overall increasing trend from 2020 to 2022. In addition, the Chl-a concentrations in Q1 and Q2 were relatively higher than those in Q3 and Q4. Owing to aquaculture pollution, the Chl-a concentrations in aquaculture ponds are generally higher than those in other types of water bodies.

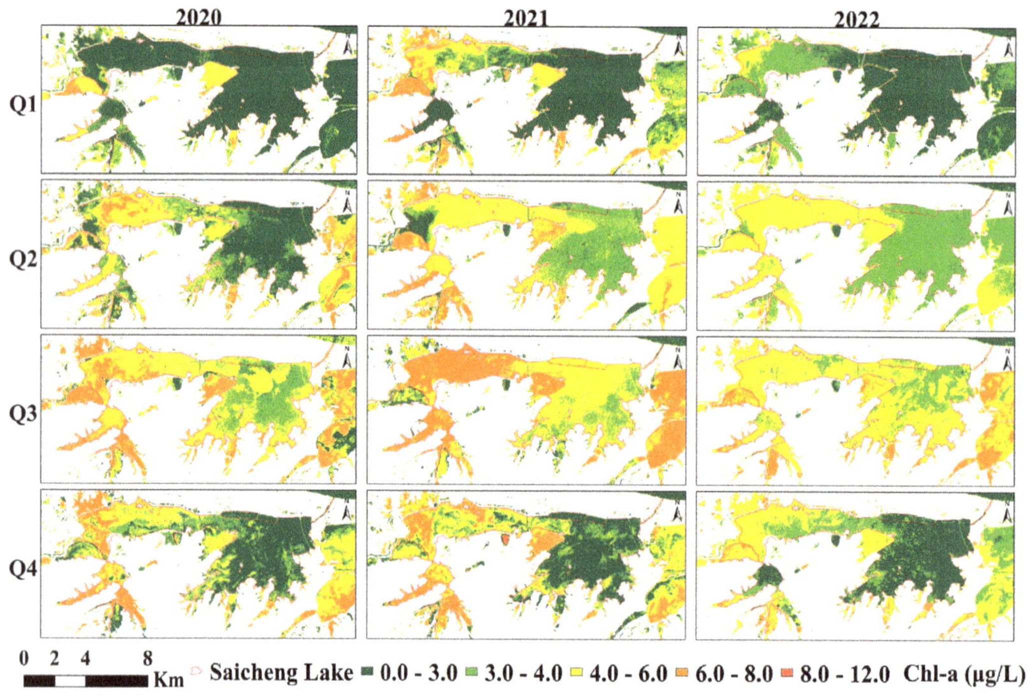

Figure 9. The spatial distribution of the Chl-a concentration in Saicheng Lake in four quarters (Q1, Q2, Q3, and Q4) from 2020 to 2022.

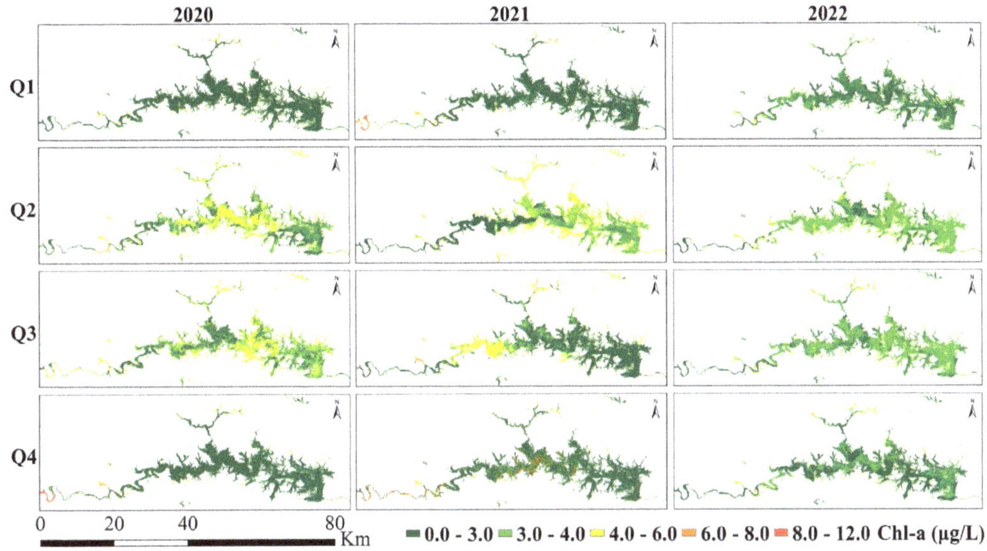

Figure 10. The quarterly spatial distribution of the Chl-a concentration in Zhelin Reservoir (four quarters: Q1, Q2, Q3, and Q4) from 2020 to 2022.

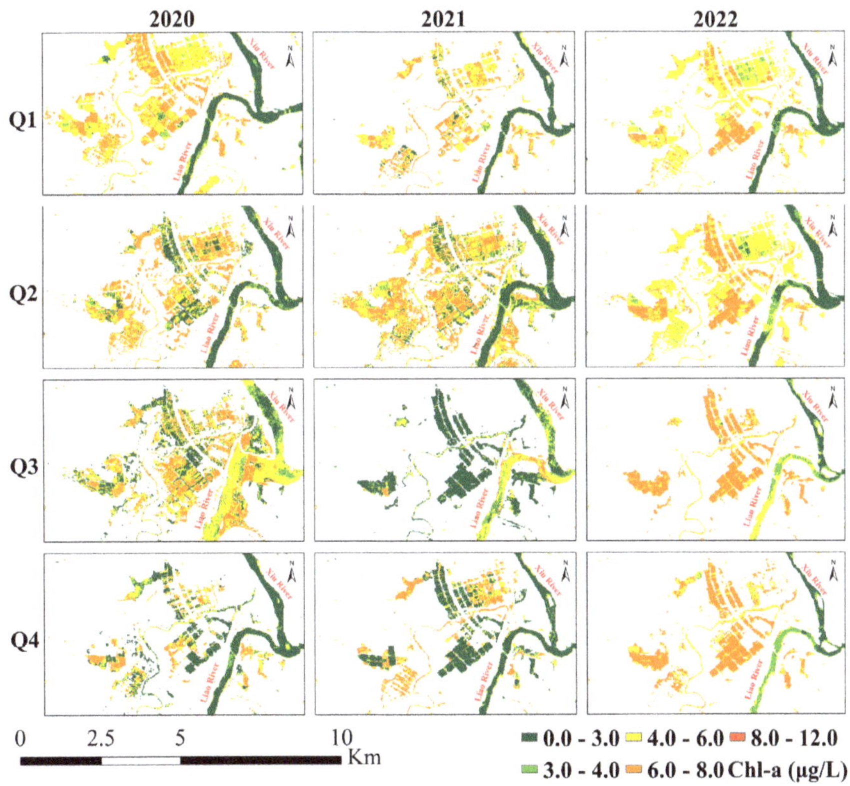

Figure 11. The spatial distribution changes of the Chl-a concentration in aquaculture ponds over four different quarters (Q1, Q2, Q3, and Q4) of 2020, 2021, and 2022.

5. Discussion

The impact of extreme hydrological events on the water quality of inland water bodies has been widely studied [11,50,51]. Since the Chl-a concentration in rivers and lakes is influenced by various factors, such as temperature, light, zooplankton, and human wastewater discharge, the response of Chl-a concentrations in inland water bodies to extreme hydrological events varies across regions [7,52]. In this study, satellite remote sensing and ground-based observation data were combined, a quantitative remote sensing inversion model for Chl-a concentrations was established and the seasonal variation characteristics of Chl-a concentrations in different types of inland water bodies in the Jiujiang region were analyzed in the context of extreme hydrological events from 2020 to 2022. The results confirmed that extreme hydrological events (floods and droughts) led to increased Chl-a concentrations in inland water bodies.

Moreover, the empirical model used in this study is stable and suitable for time series comparisons. As illustrated in Figure 12, the changes in the Chl-a concentration in the Xiu River over different quarters can be clearly observed. By constructing a spatiotemporally matched water quality sample library, uncertainties related to the station distribution are reduced. This approach not only compensates for the lack of observational data for small rivers and lakes but also provides decision support for Chl-a concentration monitoring and the development of water quality early-warning systems for the Jiujiang region.

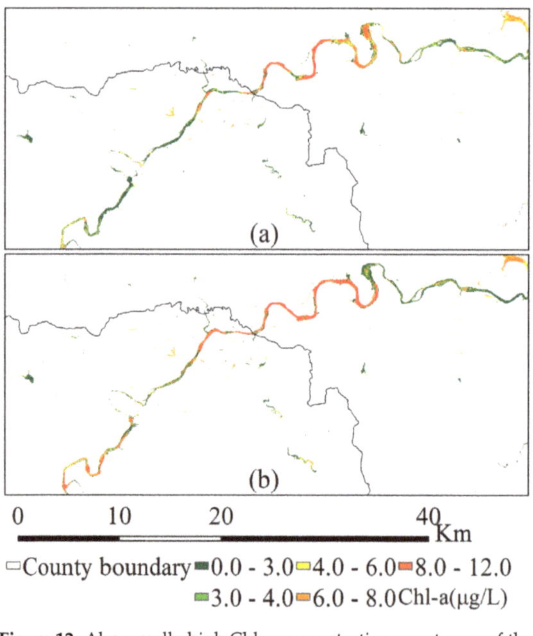

Figure 12. Abnormally high Chl-a concentrations upstream of the Xiu River in Q4 of 2020 (**a**) and Q1 of 2021 (**b**).

Generally, changes in the Chl-a concentration in inland water bodies are caused by natural and human activities [7]. An analysis of site monitoring data from 2008 to 2018 revealed that the Chl-a concentration in Poyang Lake has tended to increase, and the Chl-a concentration in winter (dry season) is significantly lower than that in summer (rainy season) and autumn (water withdrawal season) [8]. The results shown in Figure 8 of this study are consistent with this trend reported in a previous study [8], and this increasing trend in the Chl-a concentration continued from 2020 to 2022. Extreme hydrological events have the most significant impact on river and lake water levels. A previous study reported that water level is an important factor affecting the Chl-a concentration in Poyang Lake and revealed that higher water levels in summer and autumn correspond to higher Chl-a concentrations, whereas lower water levels in winter and spring correspond to lower Chl-a concentrations [53,54]. Figure 6 also reveals similar seasonal changes in the Chl-a concentration. In addition, through the satellite remote sensing inversion of Chl-a concentration changes in different types of inland water bodies, the seasonal variation in the Chl-a concentration (Figure 10) in closed lakes, such as Zhelin Lake, is smaller than that in open lakes (Poyang Lake), which also confirms, to a certain extent, that hydrological exchange contributes to the increase in the Chl-a concentration [55,56]. For aquaculture areas that are highly affected by human activities, the Chl-a concentration clearly tends to increase (Figure 11), and the impact of aquaculture activities on water quality safety cannot be ignored [57,58].

The extreme hydrological events in Poyang Lake seriously threaten drinking water safety, agricultural irrigation and ecological security [8]. In this study, Sentinel-2 satellite remote sensing data were used to generate seasonal Chl-a concentration maps for inland water bodies in the Jiujiang area from 2020 to 2022, confirming that extreme hydrological events have positive and lasting impacts on the Chl-a concentration in different inland waters, mainly increasing Chl-a levels. With the planning and construction of the Poyang Lake water conservancy project hub in the future [59,60], the ability of humans to regulate the extreme hydrology of Poyang Lake will improve. These engineering measures can be used to reduce the impact of extreme hydrological events on the Chl-a concentrations of rivers and lakes and further ensure water quality safety in Jiujiang city.

Limited by satellite remote sensing imaging data and ground water quality station observation data, the results of the Chl-a concentration remote sensing quantitative inversion model and inversion accuracy assessment in this study are still uncertain to some degree. The main influencing factors are as follows: (1) the number and distribution of Chl-a sample libraries are limited. Owing to the spatiotemporal matching of satellite remote sensing and water quality monitoring sites and the elimination of factors such as cloud cover and seasonal fluctuations in water areas, there are only 44 available samples, and the samples were obtained from 18 monitoring sites, which do not cover all inland water types, resulting in uncertainty in the model construction [61]. (2) Relatively low model inversion accuracy. An empirical remote sensing-based quantitative inversion model for Chl-a was constructed on the basis of the correlation between Chl-a and each characteristic band of the Sentinel-2 satellite imagery. The coefficient of determination (R^2) of this model is 0.563, which is lower than that reported in previous studies [62,63]. The main reason for this difference is that the samples cover various inland water types, and the spatiotemporal variation in the Chl-a concentration is significant, especially in closed-lake areas. The inverted Chl-a concentration (Figure 9) was generally higher than that for open lakes (Figure 6). (3) The influence of lake bottom reflectivity in shallow water areas. Since most of the lakes in the Jiujiang area are shallow lakes, the lake bottom reflectivity has an impact on the inversion of the Chl-a concentration and may lead to overestimates of the Chl-a concentration, especially in the dry season [64]. Moreover, in areas with water and land interfaces, the Chl-a concentration is higher than that in the central lake area, which may be why the concentration range shown on the inverted Chl-a concentration map is greater than the measured range.

In the future, as water quality observation data continue to accumulate and additional artificial water quality observation data are collected, the shortcomings of insufficient model sample quantities can be effectively overcome; moreover, when these data are combined with machine learning methods, the model inversion accuracy can be further improved [61].

In addition, the lake bottom reflectivity in shallow water areas can be corrected by applying actual water quality spectrum data for inland water bodies to improve the inversion accuracy of the Chl-a concentration.

6. Conclusions

In this study, Sentinel-2 satellite images were integrated with ground-based water quality monitoring data to develop a Chl-a concentration inversion model. The GEE platform was employed to quantitatively retrieve seasonal Chl-a concentrations for the period of 2020–2022. The key findings are as follows:

(1) A sample library for Chl-a concentration remote sensing inversion was constructed via the time–space matching method. On the basis of the characteristic bands of the Chl-a spectrum, a Chl-a remote sensing quantitative inversion model was constructed. The coefficient of determination (R^2) of the model was 0.563.

(2) Affected by extreme hydrological events, the seasonal concentration of Chl-a in inland waters in the Jiujiang area increased, which was consistent with the variation characteristics of the Chl-a concentration observed in the region. The maximum average Chl-a concentrations in the studied lakes usually occur in Q3.

(3) The Chl-a concentration in Poyang Lake, a lake connected to rivers, was higher than that in Zhelin Lake, a closed lake system. Additionally, the Chl-a concentrations in aquaculture ponds were generally higher than those in the other types of inland water bodies.

In this study, a novel approach for analyzing seasonal variations in Chl-a concentrations in inland water bodies at a high level of spatial detail is introduced. While the current inversion model contains inherent uncertainties, the ongoing accumulation of sample data can facilitate continuous optimization. Future efforts should focus on developing more universal Chl-a concentration inversion models by using machine learning methods to support sophisticated river and lake water quality management.

Author Contributions: Conceptualization, Wei Jiang and Xiaohui Ding; methodology, Wei Jiang; software, Xiaohui Ding; validation, Fanping Kong, Gan Luo, Tengfei Long, Zhiguo Pang and Shiai Cui; formal analysis, Wei Jiang and Xiaohui Ding; investigation, Wei Jiang, Xiaohui Ding, Jie Liu and Elhadi Adam; resources, Wei Jiang; data curation, Xiaohui Ding and Gan Luo; writing—original draft preparation, Wei Jiang and Xiaohui Ding; writing—review and editing, Wei Jiang; visualization, Xiaohui Ding and Gan Luo; supervision, Wei Jiang; project administration, Xiaohui Ding; funding acquisition, Wei Jiang. All authors have read and agreed to the published version of the manuscript.

Funding: This research was funded by the National Key R&D Program of China (2023YFE0110300), the National Natural Science Foundation of China (42301450 and 42201467), the China Scholarship Council (202303340010), the Key Research and Development Program Project of Jiangxi Province (20232BBE50021), Water Conservancy Technology Demonstration Project (SF-202405) and the Scientific and Technological Innovation Talent Project of the Disaster Reduction Center of IWHR (WH0166B012023).

Data Availability Statement: The data that support the findings of this study are available on request from the corresponding author.

Acknowledgments: The authors thank editors and three anonymous reviewers for their valuable comments to improve our manuscript.

Conflicts of Interest: Author Fanping Kong is employed by Beijing Capital Air Environmental Science & Technology Co., Ltd. The remaining authors declare that the research was conducted in the absence of any commercial or financial relationships that could be construed as a potential conflict of interest.

References

1. Wang, Q.; Yang, Z. Industrial water pollution, water environment treatment, and health risks in China. *Environ. Pollut.* **2016**, *218*, 358–365. [CrossRef] [PubMed]
2. Wang, Q.G.; Gu, G.; Higano, Y. Toward integrated environmental management for challenges in water environmental protection of Lake Taihu Basin in China. *Environ. Manag.* **2006**, *37*, 579–588. [CrossRef]
3. Qin, B.; Gao, G.; Zhu, G.; Zhang, Y.; Song, Y.; Tang, X.; Xu, H.; Deng, J. Lake eutrophication and its ecosystem response. *Chin. Sci. Bull.* **2013**, *58*, 961–970. [CrossRef]
4. Han, Y.; Zhang, K.; Lin, Q.; Huang, S.; Yang, X. Assessing lake ecosystem health from disturbed anthropogenic landscapes: Spatial patterns and driving mechanisms. *Ecol. Indic.* **2023**, *147*, 110007. [CrossRef]
5. Wang, J.; Xu, M.; Zhang, G.; Zhang, T.; Wang, D.; Zang, H.; Hou, X. Progress and future of China's aquatic ecological environment protection. *River* **2024**, *3*, 119–128. [CrossRef]
6. Xia, J.; Xue, Y.; Li, P.; Song, J.; Wang, G.; Luo, W. Opportunities and challenges of interdisciplinarity in river water environmental ethics and integrated river basin management. *River* **2024**, *3*, 1–7. [CrossRef]
7. Wang, J.; Zhang, Y.; Yang, F.; Cao, X.; Bai, Z.; Zhu, J.; Chen, E.; Li, Y.; Ran, Y. Spatial and temporal variations of chlorophyll-a concentration from 2009 to 2012 in Poyang Lake, China. *Environ. Earth Sci.* **2015**, *73*, 4063–4075. [CrossRef]
8. Li, B.; Yang, G.; Wan, R. Multidecadal water quality deterioration in the largest freshwater lake in China (Poyang Lake): Implications on eutrophication management. *Environ. Pollut.* **2020**, *260*, 114033. [CrossRef]
9. Hansen, C.H.; Burian, S.J.; Dennison, P.E.; Williams, G.P. Evaluating historical trends and influences of meteorological and seasonal climate conditions on lake chlorophyll a using remote sensing. *Lake Reserv. Manag.* **2020**, *36*, 45–63. [CrossRef]
10. Liao, A.; Han, D.; Song, X.; Yang, S. Impacts of storm events on chlorophyll-a variations and controlling factors for algal bloom in a river receiving reclaimed water. *J. Environ. Manag.* **2021**, *297*, 113376. [CrossRef] [PubMed]
11. Jang, M.T.G.; Alcântara, E.; Rodrigues, T.; Park, E.; Ogashawara, I.; Marengo, J.A. Increased chlorophyll-a concentration in Barra Bonita reservoir during extreme drought periods. *Sci. Total Environ.* **2022**, *843*, 157106. [CrossRef]
12. Meroni, M.; Rossini, M.; Guanter, L.; Alonso, L.; Rascher, U.; Colombo, R.; Moreno, J. Remote sensing of solar-induced chlorophyll fluorescence: Review of methods and applications. *Remote Sens. Environ.* **2009**, *113*, 2037–2051. [CrossRef]
13. Rotta, L.; Alcântara, E.; Park, E.; Bernardo, N.; Watanabe, F. A single semi-analytical algorithm to retrieve chlorophyll-a concentration in oligo-to-hypereutrophic waters of a tropical reservoir cascade. *Ecol. Indic.* **2021**, *120*, 106913. [CrossRef]
14. Palmer, S.C.; Kutser, T.; Hunter, P.D. Remote Sensing of Inland Waters: Challenges, Progress and Future Directions. *Remote Sens. Environ.* **2015**, *157*, 1–8. [CrossRef]
15. Guojin, H.E.; Guizhou, W.; Tengfei, L.; Yan, P.; Wei, J.; Ranyu, Y.; Weili, J.; Zhaoming, Z. Opening and sharing of big earth observation data: Challenges and countermeasures. *Bull. Chin. Acad. Sci. (Chin. Version)* **2018**, *33*, 783–790.
16. Tufillaro, N.; Grötsch, P.; Lalović, I.; Moitié, S.D.; Zeitlin, L.; Zurita, O. Monitoring water quality in the lower Kansas River using remote sensing. *River* **2024**, *3*, 284–303. [CrossRef]
17. Sagan, V.; Peterson, K.T.; Maimaitijiang, M.; Sidike, P.; Sloan, J.; Greeling, B.A.; Maalouf, S.; Adams, C. Monitoring inland water quality using remote sensing: Potential and limitations of spectral indices, bio-optical simulations, machine learning, and cloud computing. *Earth-Sci. Rev.* **2020**, *205*, 103187. [CrossRef]
18. Sun, X.; Zhang, Y.; Shi, K.; Zhang, Y.; Li, N.; Wang, W.; Huang, X.; Qin, B. Monitoring water quality using proximal remote sensing technology. *Sci. Total Environ.* **2022**, *803*, 149805. [CrossRef] [PubMed]
19. Topp, S.N.; Pavelsky, T.M.; Jensen, D.; Simard, M.; Ross, M.R. Research trends in the use of remote sensing for inland water quality science: Moving towards multidisciplinary applications. *Water* **2020**, *12*, 169. [CrossRef]
20. Chawla, I.; Karthikeyan, L.; Mishra, A.K. A review of remote sensing applications for water security: Quantity, quality, and extremes. *J. Hydrol.* **2020**, *585*, 124826. [CrossRef]
21. Prapti, D.R.; Mohamed Shariff, A.R.; Che Man, H.; Ramli, N.M.; Perumal, T.; Shariff, M. Internet of Things (IoT)-based aquaculture: An overview of IoT application on water quality monitoring. *Rev. Aquac.* **2022**, *14*, 979–992. [CrossRef]
22. Ullo, S.L.; Sinha, G.R. Advances in smart environment monitoring systems using IoT and sensors. *Sensors* **2020**, *20*, 3113. [CrossRef]
23. Abraham, S.; Beard, J.; Manijacob, R. Remote environmental monitoring using Internet of Things (IoT). In Proceedings of the 2017 IEEE Global Humanitarian Technology Conference (GHTC), San Jose, CA, USA, 19–22 October 2017; pp. 1–6.
24. Kamaruidzaman, N.S.; Rahmat, S.N. Water monitoring system embedded with internet of things (IoT) device: A review. *IOP Conf. Ser. Earth Environ. Sci.* **2020**, *498*, 12068. [CrossRef]
25. Jan, F.; Min-Allah, N.; Düştegör, D. Iot based smart water quality monitoring: Recent techniques, trends and challenges for domestic applications. *Water* **2021**, *13*, 1729. [CrossRef]
26. Gholizadeh, M.H.; Melesse, A.M.; Reddi, L. A comprehensive review on water quality parameters estimation using remote sensing techniques. *Sensors* **2016**, *16*, 1298. [CrossRef]

27. Zeng, M.; Peng, J.; Jiang, L.; Feng, J. Temporal and spatial distribution of suspended sediment concentration in lakes based on satellite remote sensing and internet of things. *IEEE Access* **2021**, *9*, 87849–87856. [CrossRef]
28. Jiang, J.; Tang, S.; Han, D.; Fu, G.; Solomatine, D.; Zheng, Y. A comprehensive review on the design and optimization of surface water quality monitoring networks. *Environ. Model. Softw.* **2020**, *132*, 104792. [CrossRef]
29. Park, J.; Kim, K.T.; Lee, W.H. Recent advances in information and communications technology (ICT) and sensor technology for monitoring water quality. *Water* **2020**, *12*, 510. [CrossRef]
30. Lan, Y.; Jin, M.; Yan, C.; Zou, Y. Schemes of groundwater exploitation for emergency water supply and their environmental impacts on Jiujiang City, China. *Environ. Earth Sci.* **2015**, *73*, 2365–2376. [CrossRef]
31. Free, G.; Bresciani, M.; Pinardi, M.; Ghirardi, N.; Luciani, G.; Caroni, R.; Giardino, C. Detecting climate driven changes in chlorophyll-a in deep subalpine lakes using long term satellite data. *Water* **2021**, *13*, 866. [CrossRef]
32. Ma, M.; Qu, Y.; Lyu, J.; Zhang, X.; Su, Z.; Gao, H.; Yang, X.; Chen, X.; Jiang, T.; Zhang, J. The 2022 extreme drought in the Yangtze River Basin: Characteristics, causes and response strategies. *River* **2022**, *1*, 162–171. [CrossRef]
33. Wu, Z.; He, H.; Cai, Y.; Zhang, L.; Chen, Y. Spatial distribution of chlorophyll a and its relationship with the environment during summer in Lake Poyang: A Yangtze-connected lake. *Hydrobiologia* **2014**, *732*, 61–70. [CrossRef]
34. Li, B.; Yang, G.; Wan, R.; Zhang, Y.; Dai, X.; Chen, Y. Spatiotemporal variability in the water quality of Poyang Lake and its associated responses to hydrological conditions. *Water* **2016**, *8*, 296. [CrossRef]
35. Jiujiang Municipal Bureau of Statistics; Jiujiang Investigation Team of the National Bureau of Statistics. *Jiujiang Statistical Yearbook 2023*; China Statistics Press: Beijing, China, 2023.
36. Wang, W.; Yang, P.; Xia, J.; Zhang, S.; Hu, S. Changes in the water environment and its major driving factors in Poyang Lake from 2016 to 2019, China. *Environ. Sci. Pollut. Res.* **2023**, *30*, 3182–3196. [CrossRef] [PubMed]
37. Li, J.; Zheng, Z.; Liu, G.; Chen, N.; Lei, S.; Du, C.; Xu, J.; Li, Y.; Zhang, R.; Huang, C. Estimating effects of natural and anthropogenic activities on trophic level of inland water: Analysis of Poyang Lake Basin, China, with Landsat-8 observations. *Remote Sens.* **2023**, *15*, 1618. [CrossRef]
38. Ma, M.; Jia, J.; Hu, Y.; Yang, J.; Lu, Y.; Shi, K.; Gao, Y. Changes in chlorophyll a and its response to nitrogen and phosphorus characteristics over the past three decades in Poyang Lake, China. *Ecohydrology* **2021**, *14*, e2270. [CrossRef]
39. Drusch, M.; Del Bello, U.; Carlier, S.; Colin, O.; Fernandez, V.; Gascon, F.; Hoersch, B.; Isola, C.; Laberinti, P.; Martimort, P. Sentinel-2: ESA's optical high-resolution mission for GMES operational services. *Remote Sens. Environ.* **2012**, *120*, 25–36. [CrossRef]
40. Virdis, S.G.; Xue, W.; Winijkul, E.; Nitivattananon, V.; Punpukdee, P. Remote sensing of tropical riverine water quality using sentinel-2 MSI and field observations. *Ecol. Indic.* **2022**, *144*, 109472. [CrossRef]
41. Jiang, W.; Ni, Y.; Pang, Z.; Li, X.; Ju, H.; He, G.; Lv, J.; Yang, K.; Fu, J.; Qin, X. An effective water body extraction method with new water index for sentinel-2 imagery. *Water* **2021**, *13*, 1647. [CrossRef]
42. Llodrà-Llabrés, J.; Martínez-López, J.; Postma, T.; Pérez-Martínez, C.; Alcaraz-Segura, D. Retrieving water chlorophyll-a concentration in inland waters from Sentinel-2 imagery: Review of operability, performance and ways forward. *Int. J. Appl. Earth Obs. Geoinf.* **2023**, *125*, 103605. [CrossRef]
43. Pahlevan, N.; Smith, B.; Schalles, J.; Binding, C.; Cao, Z.; Ma, R.; Alikas, K.; Kangro, K.; Gurlin, D.; Hà, N. Seamless retrievals of chlorophyll-a from Sentinel-2 (MSI) and Sentinel-3 (OLCI) in inland and coastal waters: A machine-learning approach. *Remote Sens. Environ.* **2020**, *240*, 111604. [CrossRef]
44. Ansper, A.; Alikas, K. Retrieval of chlorophyll a from Sentinel-2 MSI data for the European Union water framework directive reporting purposes. *Remote Sens.* **2018**, *11*, 64. [CrossRef]
45. Hodson, T.O. Root mean square error (RMSE) or mean absolute error (MAE): When to use them or not. *Geosci. Model Dev. Discuss.* **2022**, *15*, 5481–5487. [CrossRef]
46. Rakhimbekova, S.; O'Carroll, D.M.; Robinson, C.E. Factors controlling phosphorus mobility in nearshore aquifers adjacent to large lakes. *Sci. Total Environ.* **2021**, *799*, 149443. [CrossRef] [PubMed]
47. Noll, M.R.; Szatkowski, A.E.; Magee, E.A. Phosphorus fractionation in soil and sediments along a continuum from agricultural fields to nearshore lake sediments: Potential ecological impacts. *J. Great Lakes Res.* **2009**, *35*, 56–63. [CrossRef]
48. Duan, W.; He, B.; Nover, D.; Yang, G.; Chen, W.; Meng, H.; Zou, S.; Liu, C. Water quality assessment and pollution source identification of the eastern Poyang Lake Basin using multivariate statistical methods. *Sustainability* **2016**, *8*, 133. [CrossRef]
49. Xu, J.; Bai, Y.; You, H.; Wang, X.; Ma, Z.; Zhang, H. Water quality assessment and the influence of landscape metrics at multiple scales in Poyang Lake basin. *Ecol. Indic.* **2022**, *141*, 109096. [CrossRef]
50. Hrdinka, T.; Novický, O.; Hanslík, E.; Rieder, M. Possible impacts of floods and droughts on water quality. *J. Hydro-Environ. Res.* **2012**, *6*, 145–150. [CrossRef]
51. Mosley, L.M. Drought impacts on the water quality of freshwater systems; review and integration. *Earth-Sci. Rev.* **2015**, *140*, 203–214. [CrossRef]

52. Shi, X.; Yu, H.; Zhao, S.; Sun, B.; Liu, Y.; Huo, J.; Wang, S.; Wang, J.; Wu, Y.; Wang, Y. Impacts of environmental factors on Chlorophyll-a in lakes in cold and arid regions: A 10-year study of Wuliangsuhai Lake, China. *Ecol. Indic.* **2023**, *148*, 110133. [CrossRef]
53. Wang, S.; Gao, Y.; Jia, J.; Kun, S.; Lyu, S.; Li, Z.; Lu, Y.; Wen, X. Water level as the key controlling regulator associated with nutrient and gross primary productivity changes in a large floodplain-lake system (Lake Poyang), China. *J. Hydrol.* **2021**, *599*, 126414. [CrossRef]
54. Wu, Z.; Lai, X.; Zhang, L.; Cai, Y.; Chen, Y. Phytoplankton chlorophyll a in Lake Poyang and its tributaries during dry, mid-dry and wet seasons: A 4-year study. *Knowl. Manag. Aquat. Ecosyst.* **2014**, *412*, 06. [CrossRef]
55. Liu, L.; Huang, Q.; Wu, Y.; Arhonditsis, G.; Wang, T.; Cao, Y.; Fang, C. An ensemble modeling framework to elucidate the regulatory factors of chlorophyll-a concentrations in the Nanji wetland waters of Poyang Lake. *Ecol. Inform.* **2024**, *82*, 102729. [CrossRef]
56. Huang, H.; Zhang, J. Prediction of chlorophyll a and risk assessment of water blooms in Poyang Lake based on a machine learning method. *Environ. Pollut.* **2024**, *347*, 123501. [CrossRef]
57. Wang, L.; Pu, H.; Sun, D. Estimation of chlorophyll-a concentration of different seasons in outdoor ponds using hyperspectral imaging. *Talanta* **2016**, *147*, 422–429. [CrossRef]
58. Abd-Elrahman, A.; Croxton, M.; Pande-Chettri, R.; Toor, G.S.; Smith, S.; Hill, J. In situ estimation of water quality parameters in freshwater aquaculture ponds using hyperspectral imaging system. *ISPRS-J. Photogramm. Remote Sens.* **2011**, *66*, 463–472. [CrossRef]
59. Wu, Y.; Guo, L.; Xia, Z.; Jing, P.; Chunyu, X. Reviewing the Poyang lake hydraulic project based on humans' changing cognition of water conservancy projects. *Sustainability* **2019**, *11*, 2605. [CrossRef]
60. Yao, J.; Gao, J.; Yu, X.; Zhang, Q. Impacts of a proposed water control project on the inundation regime in China's largest freshwater lake (Poyang Lake): Quantification and ecological implications. *J. Hydrol. Reg. Stud.* **2022**, *40*, 101024. [CrossRef]
61. Song, K.; Li, L.; Li, S.; Tedesco, L.; Duan, H.; Li, Z.; Shi, K.; Du, J.; Zhao, Y.; Shao, T. Using partial least squares-artificial neural network for inversion of inland water chlorophyll-a. *IEEE Trans. Geosci. Remote Sens.* **2013**, *52*, 1502–1517. [CrossRef]
62. Pang, S.; Zhu, L.; Liu, C.; Ju, J. Causes and impacts of decreasing chlorophyll-a in Tibet plateau lakes during 1986–2021 based on Landsat image inversion. *Remote Sens.* **2023**, *15*, 1503. [CrossRef]
63. Harvey, E.T.; Kratzer, S.; Philipson, P. Satellite-based water quality monitoring for improved spatial and temporal retrieval of chlorophyll-a in coastal waters. *Remote Sens. Environ.* **2015**, *158*, 417–430. [CrossRef]
64. Cannizzaro, J.P.; Carder, K.L. Estimating chlorophyll a concentrations from remote-sensing reflectance in optically shallow waters. *Remote Sens. Environ.* **2006**, *101*, 13–24. [CrossRef]

MDPI AG
Grosspeteranlage 5
4052 Basel
Switzerland
Tel.: +41 61 683 77 34

ISPRS International Journal of Geo-Information Editorial Office
E-mail: ijgi@mdpi.com
www.mdpi.com/journal/ijgi

www.ingramcontent.com/pod-product-compliance
Lightning Source LLC
Chambersburg PA
CBHW061453220926
51850CB00004B/83
* 9 7 8 3 7 2 5 8 4 7 5 3 2 *